Structural Geology:
A Quantitative Introduction

Tackling structural geology problems today requires a quantitative understanding of the underlying physical principles, and the ability to apply mathematical models to deformation processes within the Earth.

Accessible, yet rigorous, this unique textbook demonstrates how to approach structural geology quantitatively using calculus and mechanics, and prepares students to interface with professional geophysicists and engineers who appreciate and utilize the same tools and computational methods to solve multidisciplinary problems. Clearly explained methods are used throughout the book to quantify field data, set up mathematical models for the formation of structures, and compare model results to field observations.

An extensive online package of coordinated laboratory exercises enables students to consolidate their learning and put it into practice by analyzing structural data and building insightful models. Designed for single-semester undergraduate courses, this pioneering text prepares students for graduates studies and careers as professional geoscientists.

David D. Pollard is a Professor Emeritus in Geology at Stanford University. He holds a Ph.D. in Geology from Stanford University and a Diploma of Imperial College (University of London). He has been on the faculty at Stanford since 1983, where he taught an undergraduate course in structural geology from which this textbook emerged. He co-authored *Fundamentals of Structural Geology*, published by Cambridge University Press in 2005, which won the Best Publication of the Year Award from the Structural Geology and Tectonics Division of the Geological Society of America in 2007. He is a Fellow of the Geological Society of America and the American Geophysical Union.

Stephen J. Martel is a Professor in Earth Sciences at the University of Hawai'i. He holds a Ph.D. in Applied Earth Sciences from Stanford University. Since joining the faculty in Hawai'i in 1992, he has taught both structural geology and engineering geology. He previously worked at the Bureau of Economic Geology at The University of Texas at Austin, and at Lawrence Berkeley National Laboratory. Particular research interests include landslides, nuclear waste disposal, neotectonics, fault mechanics, rock fracture, detailed geologic mapping, and the influence of topography on stresses in rock masses. He is a Fellow of the Geological Society of America.

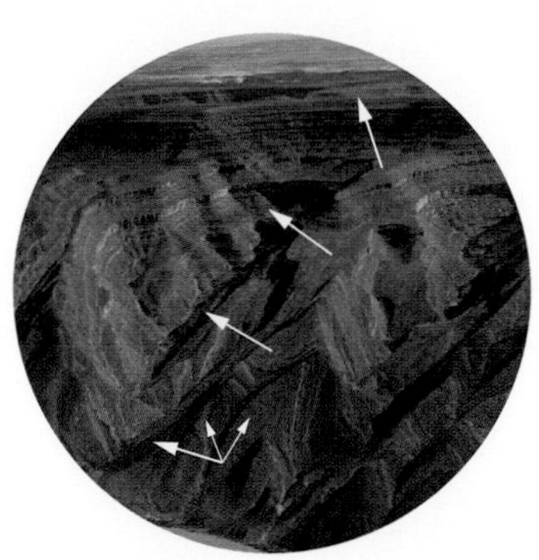

Structural Geology

A Quantitative Introduction

David D. Pollard
Stanford University, California

Stephen J. Martel
University of Hawai'i, Manoa

CAMBRIDGE
UNIVERSITY PRESS

University Printing House, Cambridge CB2 8BS, United Kingdom

One Liberty Plaza, 20th Floor, New York, NY 10006, USA

477 Williamstown Road, Port Melbourne, VIC 3207, Australia

314–321, 3rd Floor, Plot 3, Splendor Forum, Jasola District Centre, New Delhi – 110025, India

79 Anson Road, #06–04/06, Singapore 079906

Cambridge University Press is part of the University of Cambridge.

It furthers the University's mission by disseminating knowledge in the pursuit of education, learning, and research at the highest international levels of excellence.

www.cambridge.org
Information on this title: www.cambridge.org/9781107035065
DOI: 10.1017/9781139547222

© David D. Pollard and Stephen J. Martel 2020

This publication is in copyright. Subject to statutory exception and to the provisions of relevant collective licensing agreements, no reproduction of any part may take place without the written permission of Cambridge University Press.

First published 2020

Printed in Singapore by Markono Print Media Pte Ltd

A catalogue record for this publication is available from the British Library.

ISBN 978-1-107-03506-5 Hardback

Additional resources for this publication at www.cambridge.org/SGAQI.

Cambridge University Press has no responsibility for the persistence or accuracy of URLs for external or third-party internet websites referred to in this publication and does not guarantee that any content on such websites is, or will remain, accurate or appropriate.

CONTENTS

PART IV

BOXES

SYMBOLS

a	acceleration vector
α	azimuth angle (Greek alpha)
b	Burgers vector for a dislocation
B	bending modulus for a thin elastic layer
$\mathbf{c}(t)$	parametric representation of a curved line; a vector-valued function of the arbitrary parameter t
c	specific heat capacity
C_t	conventional triaxial compressive strength
C_u	uniaxial compressive strength
CPO	Crystallographic Preferred Orientation
D_c	differential strength in compression
$\mathrm{d}\mathbf{c}$	differential tangent vector to a curved line
$\mathrm{d}f$	total differential of a scalar function $f(x, y, z)$
$\frac{\mathrm{D}f}{\mathrm{D}t}$	material time derivative of a scalar function $f(x, y, z, t)$
$\vec{\nabla} f$	gradient of a scalar function $f(x, y, z)$; a vector
$\nabla \mathbf{v}$	gradient of a vector function $\mathbf{v}(x, y, z)$; a tensor
Δp, ΔP	driving pressure (Greek capital delta)
$\Delta\sigma$	differential stress
$\Delta\sigma_{\mathrm{I}}$, $\Delta\sigma_{\mathrm{II}}$,	driving stress for mode I, mode II fracture
$\Delta\mathbf{u}$	displacement discontinuity vector
Δx, Δy, Δz	lengths of the edges of a volume element
$\bar{D}$	sample mean
D	rate of deformation tensor; rate of stretch tensor
$\mathbf{D}'$	deviatoric rate of deformation tensor
DEM	Digital Elevation Model
EBSD	Electron Backscatter Diffraction
E	Young's modulus of elasticity
E, N, U	geographic coordinates: East, North, Up
E, F, G	coefficients of the first fundamental form for a curved surface
E	Lagrangian finite strain tensor
$\begin{matrix} E_{xx} & E_{xy} & E_{xz} \\ E_{yx} & E_{yy} & E_{yz} \\ E_{zx} & E_{zy} & E_{zz} \end{matrix}$	components of the Lagrangian finite strain tensor
$\underset{3\times 3}{[\mathrm{E}]}$	matrix form of the Lagrangian finite strain tensor
[=]	"has units"
{=}	"has dimensions"
$\boldsymbol{\varepsilon}$	small strain tensor (Greek epsilon)
$\begin{matrix} \varepsilon_{xx}, \varepsilon_{xy}, \varepsilon_{xz} \\ \varepsilon_{yx}, \varepsilon_{yy}, \varepsilon_{yz} \\ \varepsilon_{zx}, \varepsilon_{zy}, \varepsilon_{zz} \end{matrix}$	components of the small strain tensor
$\underset{3\times 3}{[\varepsilon]}$	matrix form of the small strain tensor
$\dot{\varepsilon}$	rate of axial strain
f_{H_2O}	fugacity of water
f, **F**	force vector

$\mathbf{F}_{\text{grav}}$	gravitational body force vector
$\mathbf{F}_{\text{buoy}}$	buoyant body force vector
$\mathbf{F}$	spatial gradient of deformation tensor
$\frac{\partial x}{\partial X}\ \frac{\partial x}{\partial Y}\ \frac{\partial x}{\partial Z}$ $\frac{\partial y}{\partial X}\ \frac{\partial y}{\partial Y}\ \frac{\partial y}{\partial Z}$ $\frac{\partial z}{\partial X}\ \frac{\partial z}{\partial Y}\ \frac{\partial z}{\partial Z}$	components of the gradient of deformation tensor
$\underset{3\times 3}{[\mathrm{F}]}$	matrix form of the gradient of deformation tensor
$\mathbf{g}$	acceleration of gravity vector
$\mathbf{g}^*$	representative acceleration of gravity vector for the lithosphere
g_x^*, g_y^*, g_z^*	components of the representative gravity acceleration vector
G	shear modulus of elasticity
$\mathbf{G}$	displacement gradient tensor
$\frac{\partial u_x}{\partial X}\ \frac{\partial u_x}{\partial Y}\ \frac{\partial u_x}{\partial Z}$ $\frac{\partial u_y}{\partial X}\ \frac{\partial u_y}{\partial Y}\ \frac{\partial u_y}{\partial Z}$ $\frac{\partial u_z}{\partial X}\ \frac{\partial u_z}{\partial Y}\ \frac{\partial u_z}{\partial Z}$	Cartesian components of the displacement gradient tensor
$\underset{3\times 3}{[\mathrm{G}]}$	matrix form of the displacement gradient tensor
GPS	Global Positioning System
η	Newtonian viscosity (Greek eta)
I	first fundamental form for a curved surface
II	second fundamental form for a curved surface
[I]	identity matrix
$\hat{\mathbf{i}}, \hat{\mathbf{j}}, \hat{\mathbf{k}}$	base vectors for the Cartesian coordinate system
k	thermal conductivity
$\mathbf{k}$	curvature vector for a curved line
kg	kilogram; base unit for mass in SI
K	kelvin; base unit for temperature in SI
$K_{\mathrm{I}}, K_{\mathrm{II}}, K_{\mathrm{III}}$	mode I, mode II, and mode III stress intensity
K_{IC}	mode I fracture toughness
κ	scalar curvature for a curved line (Greek kappa)
κ	thermal diffusivity
κ_{n}	normal curvature of a curved surface
κ_1, κ_2	principal values (eigenvalues) of normal curvature for a curved surface
κ_{g}	Gaussian curvature
κ_{m}	mean normal curvature
L	length dimension
$\mathbf{L}$	spatial gradient of velocity tensor
$\frac{\partial v_x}{\partial x}\ \frac{\partial v_x}{\partial y}\ \frac{\partial v_x}{\partial z}$ $\frac{\partial v_y}{\partial x}\ \frac{\partial v_y}{\partial y}\ \frac{\partial v_y}{\partial z}$ $\frac{\partial v_z}{\partial x}\ \frac{\partial v_z}{\partial y}\ \frac{\partial v_z}{\partial z}$	components of the spatial gradient of velocity tensor
$\underset{3\times 3}{[\mathrm{L}]}$	matrix form of the spatial gradient of velocity tensor
L, M, N	coefficients of the second fundamental form for a curved surface
LNB	Level of Neutral Buoyancy

λ, λ_o	current and initial wavelength of a buckled layer (Greek lambda)
λ	Lame's constant of elasticity (Greek lambda)
Λ	bulk viscosity (Greek capital lambda)
m	meter; base unit for length in SI
m	mass
M	mass dimension
M	bending moment on the cross section of a thin layer
M_0	seismic moment
M_w	seismic moment magnitude
μ_i	coefficient of internal friction (Greek mu)
μ_c	coefficient of friction
$\hat{\mathbf{n}}$	unit normal vector
$\hat{\mathbf{N}}$	unit normal vector to a surface
ν	Poisson's ratio of elasticity (Greek nu)
[O]	shape operator for a curved surface
p	pressure, thermodynamic pressure
$\bar{p}_o$	pressure in a static liquid
$\bar{p}$	mean normal pressure in a flowing liquid
P_c	confining pressure in a conventional triaxial test
P_p	pore fluid pressure
p, x, X	position vectors
$\vec{PQ}$	vector from point P to point Q
ϕ	inclination angle (Greek phi)
ϕ	angle of shearing
Φ	Airy stress function (Greek capital phi)
Ψ	Stokes stream function (Greek capital psi)
q	fold limb slope amplification factor
q	heat flux vector
Q	activation energy
Q	volume flow rate
r, θ, ϕ	spherical coordinates (Greek theta, phi)
R	universal gas constant
R	flexural rigidity of a bending elastic layer
Re	Reynolds number
R	pure rotation tensor
U	pure stretch tensor
ρ	mass density (Greek rho)
ρ	radius of curvature of a curved line
ρ_d	dislocation density
s	second; base unit for time in SI
s	estimated standard deviation
s^2	estimated variance
s	spherical variance
SI	International System of Units
S	stretch tensor
S_1, S_2, S_3	principal values of the stretch tensor
S	shear force on the cross section of a thin layer

S_o	inherent shear strength
S_f	frictional strength
$\mathbf{s}(u, v)$	parametric representation of a curved surface; vector-valued function of the arbitrary parameters u and v
SfM	Structure from Motion
$\boldsymbol{\sigma}$	stress tensor (Greek sigma)
$\sigma_{xx}, \sigma_{xy}, \sigma_{xz}$ $\sigma_{yx}, \sigma_{yy}, \sigma_{yz}$ $\sigma_{zx}, \sigma_{zy}, \sigma_{zz}$	Cartesian components of the stress tensor
$\underset{3\times3}{[\sigma]}$	matrix form of the stress tensor
$\sigma_1, \sigma_2, \sigma_3$	principal values (eigenvalues) of the stress tensor
$\hat{\mathbf{l}}, \hat{\mathbf{m}}, \hat{\mathbf{n}}$	principal vectors (eigenvectors) of the stress tensor
$\underset{3\times3}{[\sigma']}$	matrix form of the deviatoric stress tensor
$\underset{3\times3}{[\sigma^e]}$	matrix form of the effective stress tensor
σ_{ys}, k	yield strength
σ_s	maximum shear stress
t	time
t	arbitrary parameter for a curved line
$\mathbf{t}$	traction vector acting on a surface
$\mathbf{t}$	tangent vector for a dislocation line
$\hat{\mathbf{t}}$	unit tangent vector
T	temperature
T	time dimension
T_u	uniaxial tensile strength
$\mathbf{T}$	tangent vector to curve
$\theta_{\mathbf{v}\hat{\mathbf{i}}}, \theta_{\mathbf{v}\hat{\mathbf{j}}}, \theta_{\mathbf{v}\hat{\mathbf{k}}}$	direction angles between vector **v** and base vectors (Greek theta)
$\cos\theta_{\mathbf{v}\hat{\mathbf{i}}}, \cos\theta_{\mathbf{v}\hat{\mathbf{j}}}, \cos\theta_{\mathbf{v}\hat{\mathbf{k}}}$	direction cosines of direction angles
θ_c	Coulomb angle; orientation of potential shear fractures
Θ	temperature dimension (Greek capital theta)
u, v	arbitrary parameters for curved surface
$\mathbf{u}$	displacement vector
u_x, u_y, u_z	components of the displacement vector
$\underset{1\times3}{[\mathrm{u}]}$	matrix form of the displacement vector
UTM	Universal Transverse Mercator projection
UDF	undefined angle
$\mathbf{v}$	velocity vector
v_x, v_y, v_z	components of the velocity vector
$\underset{1\times3}{[\mathrm{v}]}$	matrix form of the velocity vector
$\mathbf{W}$	rate of spin tensor; vorticity tensor
x, y, z	Cartesian coordinates
$\omega_x, \omega_y, \omega_z$	small rotation angles (Greek omega)
$\boldsymbol{\Omega}$	small rotation tensor (Greek capital omega)

PREFACE

Structural geology is a core course in the curriculum for undergraduate students majoring in Geology at the college and university level. Usually, structural geology is a junior or senior level course, taken after students complete introductory and core courses in geology and the supporting courses in mathematics and physics that are appropriate for a major in the *science* part of a more broadly conceived Science, Technology, Engineering, and Mathematics (STEM) curriculum. This textbook is an *introduction* to structural geology for the undergraduate major that builds upon those formative geology courses, and makes extensive use of the relevant concepts and tools from the supporting courses in mathematics and physics.

This textbook also is appropriate for geology students whose first course in structural geology was primarily descriptive and qualitative. In addition, the quantitative approach used here has proven to be accessible and useful for students from other disciplines, such as geophysics, petroleum engineering, and civil engineering, who are likely to be working with structural geologists in their professional careers. Both authors have welcomed students from other disciplines in their structural geology courses, and both have found that these students enrich the experience for the geology students.

Although this textbook is a first course in structural geology, it takes a decidedly different approach to the subject matter than other "first course" textbooks, which focus on descriptions of structures and *qualitative* explanations for their formation. Our goal is to provide a balance between description and analysis of structures, so we offer *quantitative* explanations for their formation, based on the physics of deformation. Despite this difference in approach, the topics we cover are similar to those in other "first course" textbooks. For example, chapters are devoted to the basic categories of geologic structures including fractures, faults, folds, fabrics, and intrusions. However, the shift to a quantitative treatment of the formation of structures necessarily relies on more equations to build the student's knowledge base. We find that carefully labeled diagrams complement the equations substantially, so we include many diagrams in the textbook.

The mathematical pre-requisite for this book is a course in calculus that includes differential calculus and integral calculus of functions of one variable. Some calculus courses include analytic geometry and vector calculus, while others introduce aspects of linear algebra. Some of the elementary concepts from analytic geometry, vector calculus, and linear algebra are used in this textbook, but they are at a level that does not require a pre-requisite course. Instead, we introduce the necessary concepts and motivate readers to learn them by offering direct applications to structural geology.

The differential calculus of more than one variable is used throughout the book, but a course in multivariate calculus is not considered a pre-requisite. We introduce the few necessary extensions from differential calculus of one variable to multiple variables, including the partial derivative, the gradient vector, and the material time derivative. Finally, although differential equations appear throughout the book, a course in ordinary and partial differential equations is not a pre-requisite. Differential equations appear solely for displaying the underlying physical concepts and relationships. Solutions are provided where they illustrate applications to structural geology, but solution methods are left to more advanced textbooks and courses.

We recognize that some college and university students struggle with spatial thinking tasks encountered in their first structural geology class. They are challenged to learn to "think in 3D." The authors of this textbook have found that a modern graphical user interface and a computational engine like MATLAB provide many helpful tools and needed support for this learning process. Scripts with dynamic three-dimensional graphical output are run, modified, and rerun using MATLAB to obtain spatial feedback, to alter incorrect mental models, and to build intuition. These tools, along with an elementary understanding of vector calculus and differential geometry, open the door for thinking in 3D.

The goal of this textbook is to build confidence in students that they know not only what the common geologic structures are, and how to name, describe, and map them, but they also know how to apply a set of fundamental physical principles of deformation to explain the origins of these structures. To promote this goal, most of the analyses in this book follow a *step-by-step* procedure, starting with the most basic principles and leading to a result that can be compared to observations or data. This approach results in many equations, but each of them adds incrementally to the mathematical derivations, and to understanding the physics of the tectonic processes. Memorization of equations is not the authors' objective for this book. Instead, we advocate *reading* the equations as an integral part of the text to build confidence and understanding.

Commitment to the step-by-step procedure described in the previous paragraph presented the authors with a significant challenge. If we analyzed all of the structures described in other "first course" textbooks, this book would be too big for a typical first course. Instead, we selected a subset of those structures that admit an analysis at the introductory level. As a consequence, this textbook is *tutorial rather than encyclopedic*. For each of the five categories of structures we identify a "canonical problem" that illuminates the underlying physics and provides a template for the student to use in the analysis of other similar structures. The canonical problems also are the building blocks for developing a sound physical intuition that should help students analyze other structures in the future.

Each chapter of this textbook ends with a section on Further Reading. This is aimed at students who desire to expand their horizons and delve into related textbooks, monographs, and review papers. The Further Reading section also provides faculty with resources for enriching their lectures and conversations with students. The books and review papers listed in these sections would form a good working library for a practicing structural geologist in academia, industry, or a government laboratory.

This textbook contains abundant color photographs of outcrops, hand samples, and thin sections. These, and all the diagrams, graphs, and maps are freely available for instructors and students to download for teaching and learning purposes from the textbook website: www.cambridge.org/SGAQI. This material comes largely from the senior author's photographic collection, and from the Ph.D. theses and published papers of his students. The choice to use "in house" graphical material, data, maps, and analysis results was made because of accessibility and familiarity. We encourage instructors to provide their students with materials from their own collections, and to enrich their courses with results from their own research. Also available from this website are the .kmz files referred to in the captions of selected figures, so readers can take virtual field trips to these outcrops and map areas using Google Earth.

The book is supported by online student exercises, which are also available at the website given above. Students are encouraged to work through the online exercises after reading and addressing the chapter review questions. For many of the online exercises, students write MATLAB® scripts to solve quantitative problems and present graphical results. Other online exercises ask students to derive key mathematical relationships using paper and pencil. Solutions for selected online exercises and sample MATLAB® scripts are available to instructors for download.

This textbook was originally conceived as one of a pair of books by the authors; the other being a Lab Manual of practical and field-based instruction together with student exercises and activities. Writing of the Lab Manual is underway and we anticipate that it will be published within the next year or two. In the meantime, we intend to post some of the draft exercises and activities at www.cambridge.org/SGAQI so that instructors can start testing them out in their classes. These include introductory exercises for mapping, orthographic projections, stereonets and three-point problems, rotations and cross sections. Please continue to check back to the website regularly for new materials. We welcome any feedback on any of the online resources posted there.

This textbook has four parts. Part I (Chapter 1) summarizes the scope of structural geology. Part II (Chapters 2 and 3) reviews and summarizes the mathematical tools and physical principles used in this textbook. Part III (Chapters 4–6) covers the three major styles of deformation: brittle, ductile, and viscous. Part IV (Chapters 7–11) covers the five broad categories of geologic structures: fractures, faults, folds, fabrics, and intrusions. For each category we introduce the canonical model for that structure and derive the resulting stress, strain, displacement, or velocity fields.

This textbook contains more material than could reasonably be presented in a one-quarter or one-semester course. At Stanford University, the senior author developed the following schedule for a one-quarter (10 week, 20 lecture) course:

- Chapter 1 – lecture 1
- Chapter 2 – lectures 2 and 3
- Chapter 3 – lectures 4 and 5
- Chapter 4 – lectures 6, 7, and 8
- Chapter 6 – lectures 9, 10, and 11
- Chapter 7 – lectures 12, 13, and 14
- Chapter 8 – lectures 15, 16, and 17
- Chapter 11 – lectures 18, 19, and 20

This selection emphasizes brittle and viscous deformation and uses fractures, faults, and intrusions as the representative structures. An alternative selection substitutes Chapters 5 and 10 for Chapters 6 and 11, and thereby includes ductile deformation and rock fabrics instead of viscous deformation and intrusions. Another alternative is to be more selective within chapters and cover more deformation styles and structures, while omitting some of the analyses.

ACKNOWLEDGEMENTS

David Pollard thanks the students who enrolled in his course at Stanford University and provided valuable motivation and feedback for developing his lecture notes into a textbook. He also thanks the Ph.D. students and colleagues who carried out research with him and provided materials that appear here as graphical data, diagrams, photographs, and maps. These students and colleagues shared the pleasures of discovering new structures in the field, mapping structures with many different high-tech instruments, and solving boundary and initial value problems of continuum mechanics using both analytical and numerical models to reveal new insights about structures. He especially thanks his co-author, Steve Martel, for many enjoyable Skype conversations about the manuscript, and for countless exchanges of edited drafts that eventually grew into the published book.

Steve Martel is fortunate to have had the privilege to learn from, and work with, David Pollard and many of his graduate students at Stanford University. He also is grateful for the wonderful opportunity to work with many undergraduate and graduate students at the University of Hawai'i, as well as colleagues there and around the world. The ongoing collaborative experience of trying to understand the world better through a combination of new discoveries and the use of tools developed by our forerunners has been fascinating, humbling, and rewarding.

The authors thank Richard Stultz for helping to set up the tabletop experiments used to illustrate the mechanical behavior of materials in Chapters 4, 5, 6, and 9, and for his expertise in taking and editing the photographs of these experiments. The authors also thank Ryan McCarty for his expertise in digitizing, colorizing, and editing all the maps and many of the diagrams. His graphics skill and artistry have added a significant dimension to the textbook. David Pollard thanks Professors Bernard Hallet and Darrel Cowan of the Department of Earth and Space Sciences, University of Washington, for hosting his sabbatical in 2010, during which the outlines of this textbook emerged.

Finally, we thank those scientists and engineers who, in past generations, laid down the cornerstones of continuum mechanics, and those philosophers who established the self-correcting methodology of science advocated in this textbook that compensates for human fallibility, technological innovation, and random discovery. Using that methodology we may, in the words of Charles S. Peirce, confidently "pile the ground before the foot of the outworks of truth with the carcasses of this generation, and perhaps of others to come after it, until some future generation, by treading on them, can storm the citadel" (Peirce, 2011).

PART I

Part I (Chapter 1) summarizes the scope of structural geology by addressing the following questions:

- What forces cause deformation of rock in Earth's lithosphere and asthenosphere?
- What are the three major mechanical styles for this deformation?
- What five broad categories of geologic structures result from this deformation?
- What is the methodology advocated in this textbook for analyzing geologic structures?

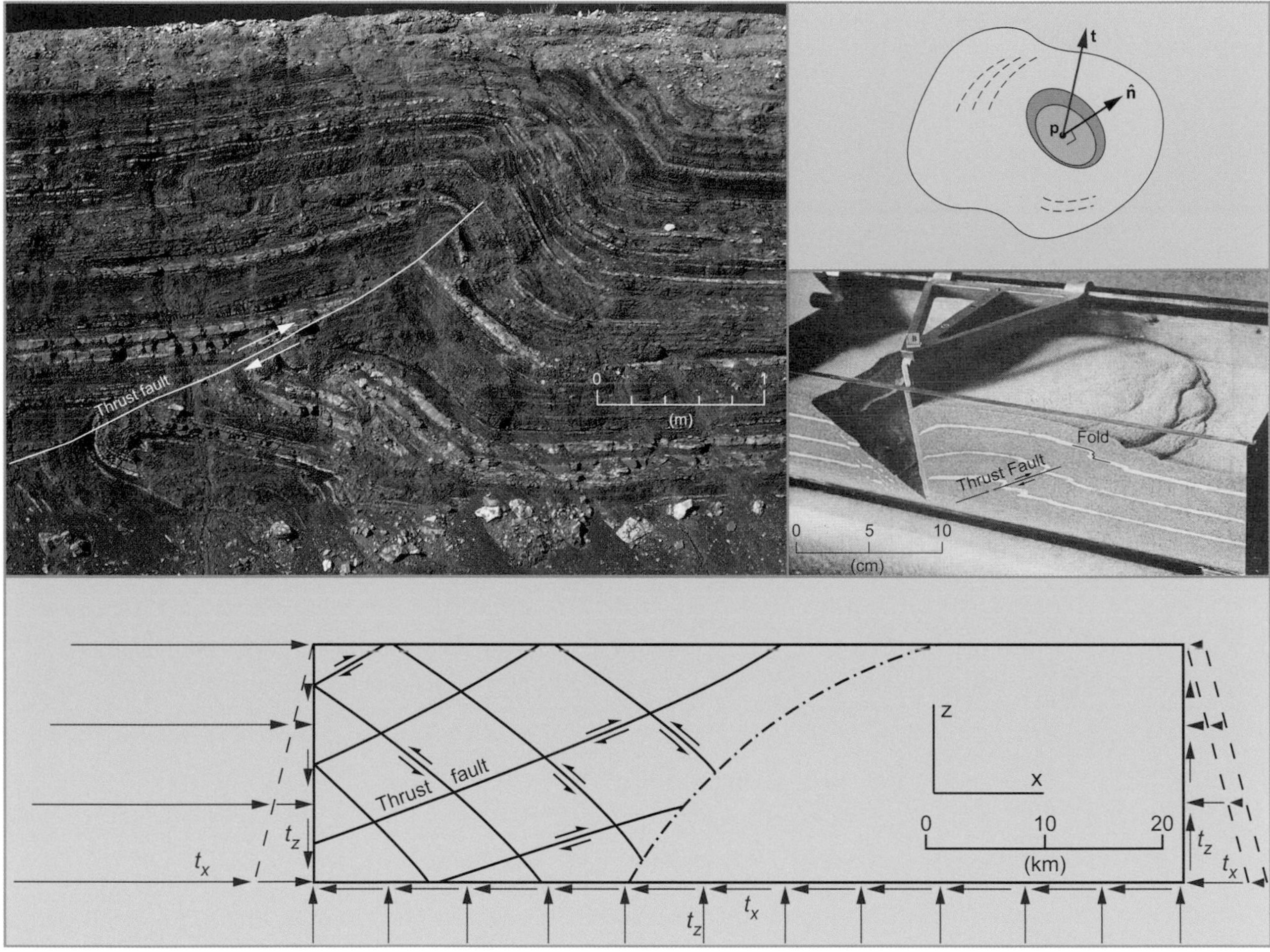

Chapter 1
Scope of Structural Geology

Introduction

Chapter 1 sets the stage for a quantitative introduction to structural geology. We begin by identifying forces that cause deformation in Earth's lithosphere and asthenosphere. Then, we describe three different styles of deformation, and five broad classes of geologic structures that result from this deformation. To lay out the methodology for studying geologic structures, we introduce what we mean by a complete mechanics and by canonical models of structural geology. Then we examine the roles of physics and mathematics in studying the origins of geologic structures. Finally, we describe applications of structural geology to problems facing our society and the careers that utilize structural geology to solve those problems.

Geologic structures develop when rock deforms from its original configuration into some different configuration. For example, sedimentary rocks typically form in horizontal and tabular layers, but later these layers may tilt into a geologic structure called a *fold* (**Figure 1.1**). To understand the process of folding one must be able to characterize both the initial and the final configurations, identify the material properties of the rock that resisted the deformation, and deduce the forces that caused the deformation. Characterizing the geometry of structures and their progenitors requires data obtained by geologic mapping, and a good understanding of geologic history. Tests carried out in a rock mechanics laboratory measure the material properties of rock. The underlying theory that relates forces to deformation comes from that part of physics called

Figure 1.1 This fold is located in the Rainbow Basin Natural Area, 13 km north of Barstow in the Mojave Desert of southern California. These sedimentary rocks formed in horizontal layers during the Miocene epoch. Today they tilt about 20° from horizontal. Because the two limbs of the fold are inclined toward each other, this is a syncline.

mechanics, and the language of mechanics is mathematics. Thus, structural geology is a blending of geological knowledge and field techniques, with the results from laboratory tests and the physical principles and mathematical methods of mechanics.

Structural geologists study geologic structures and seek to understand Earth history by placing structures in the proper sequence and dating the tectonic events that generated the structures. They use mechanical principles to understand how and why the structures formed. Structural geologists also help explore for and produce Earth's resources, mitigate geologic hazards, and manage a sustainable environment. As such, structural geologists work in academia, in industry, for governmental agencies, and as private consultants. They work in laboratories, analyze data using computers, and collect geologic data and make maps in the field. In the course of their work structural geologists interface with other earth scientists, with engineers, and with the general public. In these many ways structural geologists contribute to our society.

1.1 DEFORMATION OF EARTH'S LITHOSPHERE AND ASTHENOSPHERE

Geophysical data support a primary division of the solid Earth into an inner core, outer core, mantle, asthenosphere, and lithosphere (**Figure 1.2**). Structural geologists focus their attention on deformation of rock in the two outer shells, which are distinguished on the basis of their mechanical properties. The lithosphere is the stronger upper layer and the asthenosphere is the weaker lower layer. The boundary between the lithosphere and the asthenosphere is associated with a temperature of about 1,300 °C. Under the oceans the base of the lithosphere is at a depth of 50 to 150 km, and under the continents it is at a depth of 50 to 300 km. The base of the asthenosphere is not well defined, but may be as shallow as 200 km, or as deep as 700 km. In any case, the combined thickness of the lithosphere and asthenosphere is a very small fraction of Earth's equatorial radius, which is about 6,380 km.

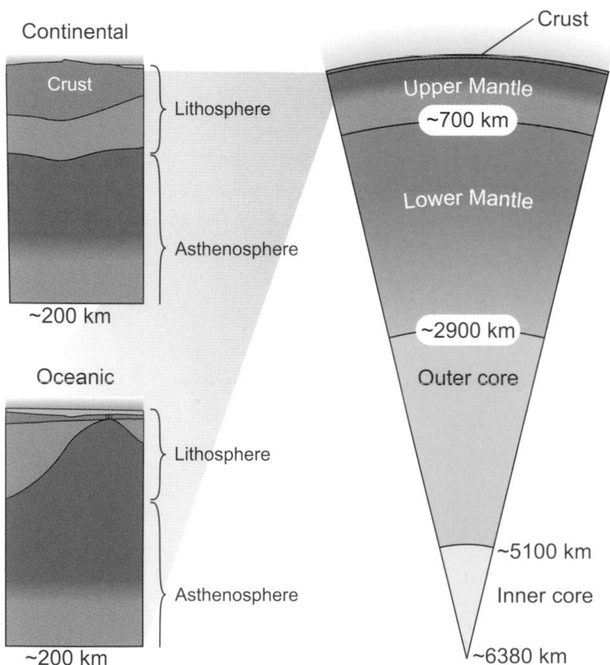

Figure 1.2 Schematic illustration of Earth's primary shells (on the right) including the inner and outer core, lower mantle, upper mantle, and crust. The mantle and crust are classified based on different *composition*. The two outer shells called the asthenosphere and lithosphere (on the left) are classified based on different *strength*. See Press et al. (2004) in Further Reading.

The lithosphere is characterized by *solid mechanical behavior* with localized deformation exemplified by brittle fracturing and faulting in the upper part, and with more distributed deformation exemplified by plastic flow in the lower part. The asthenosphere is characterized by *fluid mechanical behavior* with broadly distributed viscous flow. The change in mechanical behavior from solid-like to fluid-like depends upon the composition of the rock, and also on the temperature and pressure, both of which increase with depth. Increased pressure tends to increase the strength of rock, whereas increased temperature tends to decrease the strength. Within the lithosphere the effect of pressure dominates; hence the apparent strength and rigidity of this outer shell. Below the lithosphere–asthenosphere boundary the effect of temperature dominates; hence the apparent weakness and mobility of this underlying shell. To elucidate these different mechanical behaviors we devote Chapter 4 of this book to *elastic* solid behavior, Chapter 5 to *plastic* solid behavior, and Chapter 6 to *viscous* fluid behavior.

The strong lithosphere composes the plates of plate tectonics (**Figure 1.2**). The *idealized* concept of plate tectonics is that perfectly rigid plates move laterally, because they are carried along by large-scale convection of the flowing asthenosphere. New material is added to the lithosphere from the upwelling asthenosphere beneath oceanic ridges, and lithosphere is consumed back into the down going asthenosphere along subduction zones. The study of Earth's deformation at global length scales that are relevant to plate tectonics is called geodynamics and textbooks that cover that discipline (see Further Reading at the end of this chapter) overlap some of the topics of structural geology.

If tectonic plates really were perfectly rigid, structural geologists would have very little to study, because only those rocks below the lithosphere–asthenosphere boundary would be deformed. In fact, Earth's lithosphere does deform and this deformation produces a diverse suite of structures including *fractures*, *faults*, *folds*, *fabrics*, and *intrusions*. We devote Chapters 7 through 11 to these categories of structures. They form in many different geologic and tectonic settings, and under broad ranges of temperature, pressure, and rate of deformation. They develop over length scales from millimeters to kilometers and time scales from milliseconds to millions of years. It would be daunting to investigate geologic structures across these immense

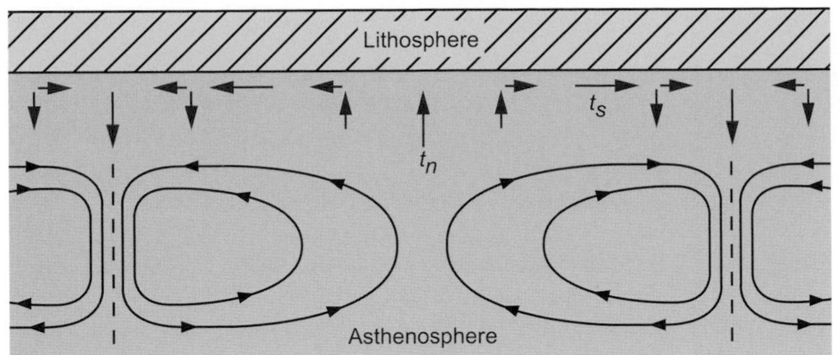

Figure 1.3 Schematic (not to scale) conceptual model of convecting asthenosphere and overlying lithosphere from one of the classic papers on crustal deformation. Closed curves parallel to velocity vectors (black) are flow lines for convection cells; discrete arrows (red) are normal and shear tractions (forces per unit area) acting on the base of lithosphere. Modified from Hafner (1951).

time and length scales, under such a wide range of conditions, and in such disparate rocks without a unifying theory. Continuum mechanics provides the unifying theory, and therefore is given special attention in this textbook.

Earth's two outer shells (**Figure 1.2**) also may be distinguished on the basis of their chemical composition and mineralogy. The study of Earth's composition in this broad sense is the subject of petrology and geochemistry. The upper shell, referred to as the crust, is made up of various *sedimentary*, *metamorphic*, and *igneous* rocks that are relatively rich in silica, and are composed primarily of the minerals feldspar and quartz. The lower shell, called the mantle, is primarily peridotite, an *igneous* rock that is relatively poor in silica and is composed mostly of the minerals olivine and pyroxene. Due in large part to the different densities of these common minerals, crustal rocks typically are less dense than mantle rocks. The oceanic crust is up to about 10 km thick and the continental crust is about 50 km thick, whereas the mantle is about 1,800 km thick. Thus, the lithosphere typically includes the crust and a portion of the upper mantle, whereas the asthenosphere is a portion of the mantle that is a few to several hundred kilometers thick, below the lithosphere–asthenosphere boundary.

Because the objective of structural geology is to characterize and understand the *deformation* of rock to form geologic structures, the classification of Earth's outer shells based on mechanical properties is more germane than the classification based on composition and mineralogy. However, the physical properties of rock do depend upon composition and mineralogy, so both classifications are useful for structural geologists. Because the lithosphere is more accessible, and it contains the majority of Earth's resources that can be produced, and it is the site of most geologic hazards, *we focus primarily on structures in the lithosphere*.

1.1.1 Tectonic Surface Forces: Tractions

Why do Earth's lithosphere and asthenosphere deform? In the context of Newtonian mechanics *forces* cause the deformation of rock in Earth's lithosphere and asthenosphere. In general, these forces are categorized as surface forces and body forces, and both play important roles. We focus on the surface forces here and take up body forces in the next section.

On the base of the lithosphere, surface forces per unit area, called tractions, cause the vertical and lateral motion of the

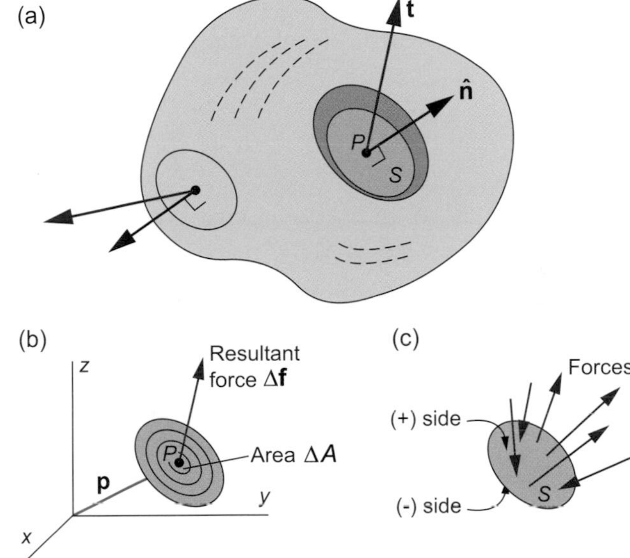

Figure 1.4 The traction vector illustrated. (a) The traction vector **t** acts at the point P on the surface S with outward unit normal $\hat{\mathbf{n}}$. (b) The point P is located by the position vector **p**. On successively smaller patches of S the surface area is ΔA and the resultant force is $\Delta\mathbf{f}$. (c) The mechanical action of the rock on the positive (+) side of the surface S acting on the negative (–) side of S is represented by distributed forces with resultant $\Delta\mathbf{f}$. The traction vector is defined in equation (1.1).

tectonic plates (**Figure 1.3**). Pressure and viscous drag of the convecting asthenosphere induces these tractions on the base of the lithosphere, and these tractions are, in part, responsible for the geologic structures that are the central topic of structural geology.

The traction vector, **t**, is defined as the *surface force per unit area* acting at a point P on a surface S with a given orientation, either within a rock mass or on its exterior (**Figure 1.4a**). The surface could be a physical surface, like bedding in sedimentary rock or a compositional boundary in metamorphic rock, or it could be an arbitrarily defined surface. The orientation of the surface is specified using the outward unit normal vector $\hat{\mathbf{n}}$. The point P is located by the position vector **p** (**Figure 1.4b**), using a

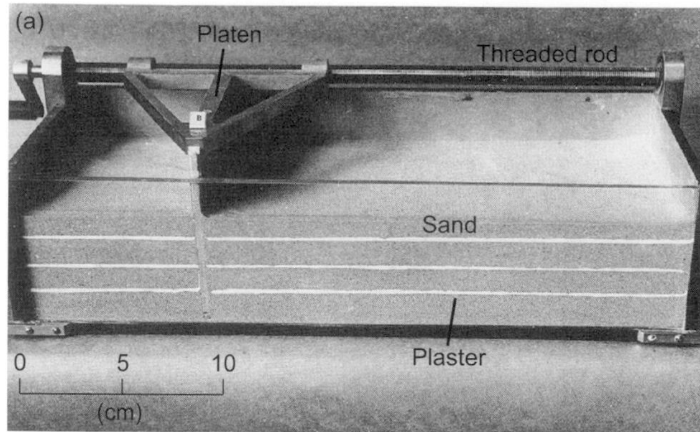

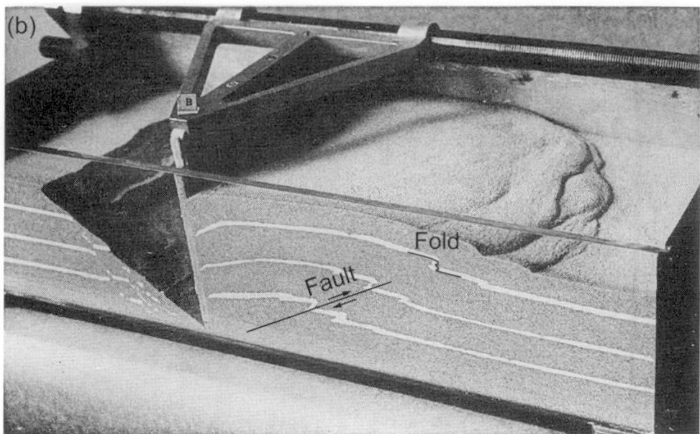

Figure 1.5 Fold and thrust mountain building in the laboratory. (a) Photograph of initial stage of sandbox experiment: layers of sand; interlayers of white plaster. (b) Photograph of final stage after platen moved to the right: asymmetric folds and thrust faults are highlighted by the white plaster. Modified from Hubbert (1951).

Cartesian coordinate system (x, y, z). Note that we use bold letters to represent vectors.

The traction **t** represents the mechanical action of the rock on the positive (+) side of the surface S, acting on the rock on the negative (–) side of S (**Figure 1.4c**). This mechanical action is quantified using the forces distributed on S, but these could be quite diverse in magnitude and direction if S is large. To focus on the point P, we consider some number, k, of successively smaller patches of S that all include the point P (**Figure 1.4b**). For each patch the vector sum (resultant) of the distributed forces is called $\Delta\mathbf{f}$ and the area is called ΔA. The traction is defined as the ratio of resultant force to surface area, in the limit as k goes to an infinite value:

$$\mathbf{t}(\mathbf{p}, \hat{\mathbf{n}}) = \lim_{k\to\infty} [\Delta\mathbf{f}/\Delta A] \tag{1.1}$$

In this limit we insist that the longest dimension of S goes to zero, so the patch goes to a point, not a line. Also, although both $\Delta\mathbf{f}$ and ΔA go toward zero as the patch goes toward a point, *one of the basic tenets of continuum mechanics is that the ratio, $\Delta\mathbf{f}/\Delta A$, approaches a well-defined limit.*

On the left side of (1.1) the quantities $(\mathbf{p}, \hat{\mathbf{n}})$ are a reminder that the traction vector depends upon the *location*, **p**, of the point and the *orientation*, $\hat{\mathbf{n}}$, of the surface. At the same point in a rock mass, the traction may differ on surfaces with different orientations passing through that point. Thus, the traction vectors in **Figure 1.3** are profoundly different from the velocity vectors in that figure, which have a unique magnitude and direction at a given point. The traction vector at a given point is unique only because we have specified the surface to be the bottom of the lithosphere. If the surface were taken as a vertical plane through that point, instead of the horizontal plane, the traction vector would be different. We explore the traction vector in greater detail in later chapters, because it is one of the most important physical quantities in structural geology.

The tractions acting on the base of the lithosphere (**Figure 1.3**) induce *stresses* within the lithosphere that contribute to the development of geologic structures. We identify the tractions on the boundary of this deforming rock mass and ask: what are the stresses within the lithosphere and how do these stresses *cause* the rock to deform? This is an example of a boundary-value problem, and much of this textbook is devoted to setting up and evaluating such problems, so there will be many opportunities to study the physical quantity we call stress in later chapters.

As an example of a boundary-value problem, consider the initial and final stages of a classic laboratory experiment at the centimeter scale (**Figure 1.5**), meant to simulate fold and thrust mountain building. The box is filled with loose, dry sand with thin layers of white plaster that act as markers. A rigid steel platen moves along a threaded rod from left to right as the crank is turned, and the platen pushes the sand layers laterally to form asymmetric folds and thrust faults. The tractions imposed by the platen, and those imposed by the base and sides of the sandbox, induce stresses within the sand that *cause* the folds and faults.

In **Figure 1.6** the solution to a boundary-value problem, meant to address faulting during fold and thrust mountain building, is illustrated using the calculated state of stress to define the orientation of *thrust faults*. Tractions on the left and right sides of the model, labeled t_x, represent the horizontal tectonic forces that *cause* the faulting. These tractions might be related to the tractions on the base of the lithosphere (**Figure 1.3**), or they might be due to local disturbances within the lithosphere, but they are greater on the left side than on the right side. On the base of the model the traction component t_z supports the weight of the overlying rock, and the component t_x balances the net horizontal traction applied to the model sides. Given this geometry, the material properties of rock, and the prescribed boundary tractions, the principles of continuum mechanics enable one to calculate the stresses within the rock mass, and deduce the orientation and location of faults. One set of thrust faults dips to the left and curves gently upward, similar to those in the sandbox experiment (**Figure 1.5**).

Natural examples of fold and thrust structures exist at the scale of entire mountain belts, but also at the outcrop scale (**Figure 1.7**), where they are easier to record in a photograph. In this example, thinly bedded sedimentary rocks are cut and offset by a small thrust fault, and the offset decreases to zero at

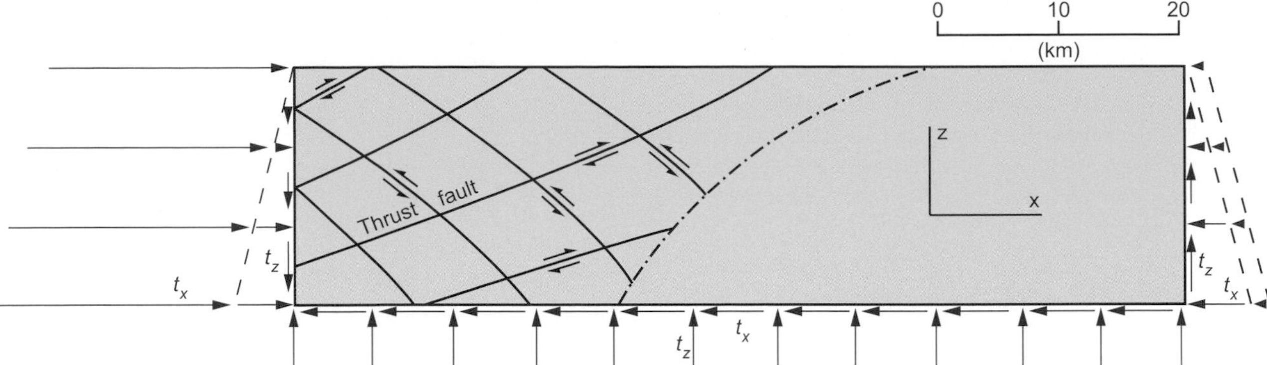

Figure 1.6 Illustration of a boundary-value problem with applied tractions t_x and t_z on a rectangular block of elastic material. Note 20 km length scale. According to this model stresses within the block cause the thrust faults. Modified from Hafner (1951).

Figure 1.7 Road cut on State Hwy 276 near Mt Ellsworth in the southern Henry Mountains, UT. Thrust fault offsets and folds the sedimentary strata.

the fault tip. To the right of the fault tip the beds are folded, but continuous. One could define a rectangular domain surrounding this rock mass in its undeformed state, prescribe traction boundary conditions on the sides of that domain, and begin to investigate the formation of these structures in the way illustrated in Figure 1.6. We encounter examples of this procedure later in the textbook.

1.1.2 Tectonic Body Forces: Gravity and Buoyancy

Why do Earth's lithosphere and asthenosphere deform? In the previous section we pointed to tectonic surface forces as one cause of this deformation. Here we consider body forces due to *gravity* and *buoyancy* that drive more dense rock downward, and less dense rock upward. Recall that Earth's crust is composed of rock that typically is less dense than the rock in Earth's mantle. Buoyancy, in the broadest sense, explains why crustal rocks sit on top of mantle rocks. For another example, we note that less dense salt moves upward relative to more dense clastic sedimentary rock as salt domes rise in sedimentary basins. Similarly, less dense magma rises to form igneous intrusions in denser host rock.

The gravitational body force, $\mathbf{F}_{\text{grav}}$, acting on a rock mass is:

$$\mathbf{F}_{\text{grav}} = m\mathbf{g} \tag{1.2}$$

In this vector equation, m is the mass of material making up the body, and the gravitational force vector is directed downward, the same direction as the gravitational acceleration, $\mathbf{g}$ (Figure 1.8).

To evaluate the gravitational force we take the gravitation acceleration as uniform and constant over the length and time scales considered, so $\mathbf{g} = \mathbf{g}^*$, and the magnitude of this vector is $g^* = 9.8\ \text{m s}^{-2}$, representative of values in Earth's lithosphere. The Cartesian coordinates are oriented with z positive upward, so the only non-zero component of gravitational acceleration is

$g_z = -g^*$. As an example consider a spherical body of radius a, with mass density, ρ_i. The mass *inside* the sphere is the volume of the sphere times the mass density: $m = \frac{4}{3}\pi a^3 \rho_i$.

The buoyant force is determined by Archimedes' principle. This famous relationship states that the buoyant force is equal to the weight of the material displaced by the body. Therefore, we write the buoyant force, $\mathbf{F}_{buoy}$, as:

$$\mathbf{F}_{buoy} = -m_d\mathbf{g} \tag{1.3}$$

In this vector equation m_d is the displaced mass and $\mathbf{g}$ is the acceleration of gravity. By definition the acceleration vector is directed *downward* and mass is non-negative, so the negative sign on the right side assures that the buoyant force vector is positive, and therefore directed *upward*: it buoys up the less dense body. For example, the mass displaced by a spherical body of radius a is calculated using the volume of the sphere and the mass density, ρ_o, of the material *outside* the sphere, such that $m_d = \frac{4}{3}\pi a^3 \rho_o$.

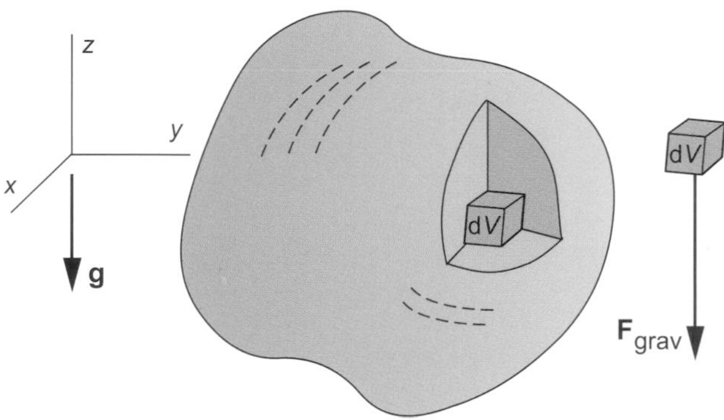

Figure 1.8 The gravitational body force per unit volume, $\mathbf{F}_{grav}$, is proportional to the mass, *m*, and the gravitational acceleration, **g**, which defines the direction *down*. The body force acts in the down direction on the small volume element, d*V*.

Adding the gravitational force (1.2) and buoyant force (1.3), and substituting for the respective masses, the resultant body force is:

$$\begin{aligned}\mathbf{F}_{grav} + \mathbf{F}_{buoy} &= (m - m_d)\mathbf{g} \\ &= \langle 0,\ 0,\ \tfrac{4}{3}\pi a^3(\rho_o - \rho_i)g^* \rangle, \text{ spherical body}\end{aligned} \tag{1.4}$$

In the second line we write the vector equation in terms of the components of the resultant body force. The angular brackets contain the three values, separated by commas, which are the *components* of force in the *x*, *y*, and *z* coordinate directions (**Figure 1.8**). If the inside density is less than the outside density, the resultant body force is positive (directed upward), the sphere is *buoyant*, and it would tend to rise. If the density inside and outside are equal, the body force is zero, so the sphere is *neutrally buoyant* and would tend to be *stationary*. If the density inside exceeds that outside, the body force is negative, and the sphere would tend to sink. We say "tend to" because the material surrounding the sphere must deform for the sphere to move.

Structures that form due to gravitational and buoyant body forces can be produced in laboratory experiments. Images from such an experiment (**Figure 1.9**) show the different forms of a buoyant liquid (black oil) rising in a more dense liquid (clear syrup). The two fluids initially fill the rigid box with the less dense fluid on top, and then the box is inverted (stage a). In stages b through d, the interface progressively becomes wavier. In stages e through g, the waves pinch off and rise with a thinning tail. These structures may be analogous to *salt diapirs*, bodies of salt that are buoyant and rise through the overlying sediments. In stages g and h, the tops of the model diapirs flatten against the top of the box and spread laterally. Ultimately, the oil forms a horizontal layer (not shown) at the top of the box.

Structures that form due to gravitational and buoyant body forces also can be studied using boundary-value problems from fluid mechanics. For example, **Figure 1.10** illustrates the velocity field for a spherical body of viscous liquid rising in a more dense viscous liquid because of the buoyant force. The velocity

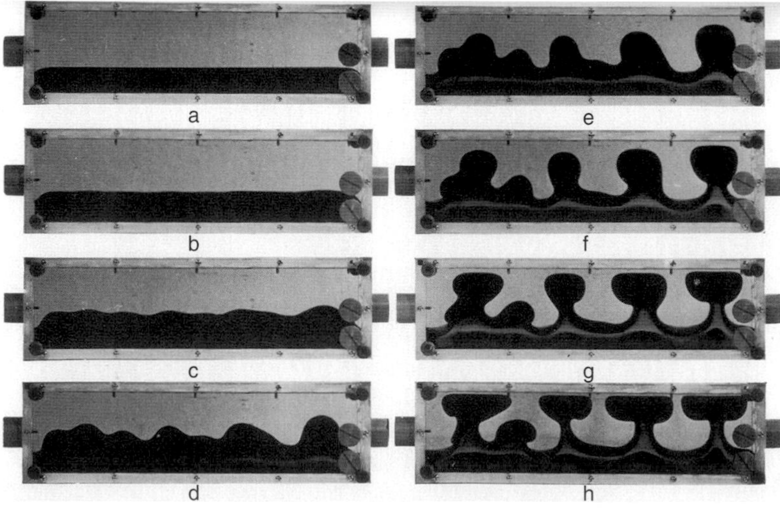

Figure 1.9 Rigid transparent box filled with two fluids. In stages a through h less dense oil (black) rises in more dense syrup (clear) forming diapirs with tails. Modified from Ramberg (1967).

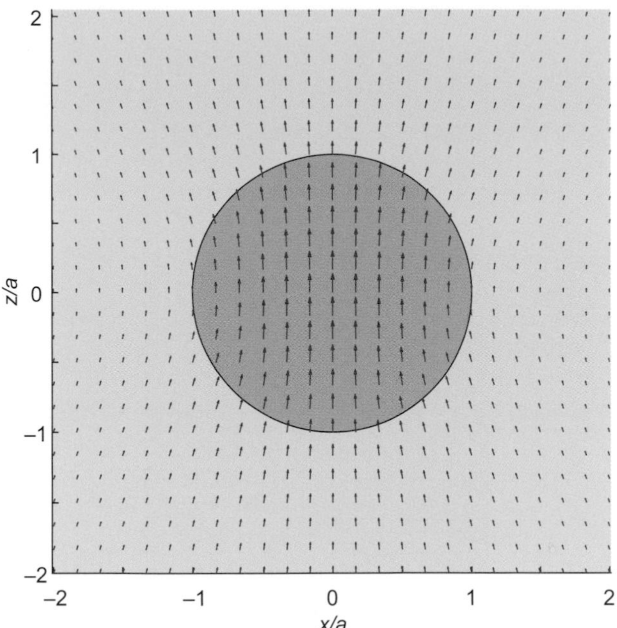

Figure 1.10 Cross section of a sphere of viscous fluid rising in a fluid of different viscosity and greater density. Velocity vectors (arrows) illustrate how the two fluids move as the sphere rises relative to a stationary observer a great distance from the sphere. The equations describing this velocity field are derived in Section 11.6.

field inside and outside the sphere match at the boundary. This solution to the viscous boundary-value problem provides a model for rising salt or magma bodies in a deformable host rock. We explore this solution for the viscous sphere in greater detail in Chapters 2 and 11. In this and many other ways, *gravitational* and *buoyant forces* contribute to deformation of the lithosphere and asthenosphere.

1.2 STYLES OF DEFORMATION

How do Earth's lithosphere and asthenosphere deform? Broadly speaking the answer includes three distinctive styles of deformation during which rock *fractures* like a brittle solid, *flows* like a ductile solid, or *flows* like a viscous liquid. These different styles of deformation, or combinations of them, produce the geologic structures that are the subject matter of structural geology. We devote one chapter to each of these styles of deformation, but here we introduce them using evocative pictures that suggest the profound differences between brittle and ductile solids, and between ductile and viscous flow.

1.2.1 Brittle

Fractures are the most obvious sign of brittle deformation, and fractures that *opened* as they formed are called veins or joints. Veins are sealed with minerals deposited from groundwater, whereas joints are barren. For example, the *sedimentary rock* shown in **Figure 1.11** includes a bed of gray limestone cut by veins that opened and were filled with white calcite. The limestone also is broken by a host of joints with various orientations, and none of these fractures offset the bedding surface, which locally is approximately planar. The joints gap open, and although some of this opening may be due to weathering, some of it records the fact that the joints opened as the limestone was pulled apart in the plane of the bed.

Faults are another sign of brittle deformation. They are distinguished from veins and joints because they *sheared* as they formed. A fault crops out across the center of **Figure 1.11** and dips steeply to the left. We identify this structure as a fault because the gray limestone to the left is juxtaposed against a black shale that crops out on the right side of the fault at the same level as the limestone. These strata were deposited at different times, and each was continuous across the plane of the fault before it slipped. Striations, called slickenlines, on a prominent surface of the fault suggest the shearing was directed along the dip of the fault, but the sense of relative motion is not obvious in this image. Did the gray limestone slide up the fault, so it is older than the shale, or did it slide down the fault, so it is younger? These and other questions could be addressed by detailed mapping that identifies the positions of these beds in the stratigraphy.

In Chapter 4 we review data from laboratory tests that help to define the conditions of temperature and pressure under which rocks fracture and fault as they deform. We also describe the strength of rock and its resistance to fracturing and faulting. Test results such as these help structural geologists interpret outcrops with rock broken by veins, joints, and faults.

1.2.2 Ductile

The most obvious sign of ductile deformation is pronounced distortion without fracture. For example, the outcrop photograph in **Figure 1.12** shows a *metamorphic rock*, called the Moine Schist, made up of highly contorted layers, alternately rich in quartz (white) and mica (gray). Originally this was a sedimentary rock, and these were alternating planar layers of sandstone and shale. Metamorphism has altered the quartz-rich layers, for example, from sand-sized grains of quartz with open pores providing perhaps 20% porosity, to larger interlocked crystals of quartz with near zero porosity. This alteration occurred by plastic deformation and re-crystallization, but the temperature of metamorphism was insufficient to melt the rock. On the other hand the temperature and pressure of metamorphism were sufficient to prevent fracturing as these layers were contorted into very ornate folds.

In Chapter 5 we review data from laboratory tests that help to define the conditions of temperature and pressure under which rocks flow as ductile solids as they deform. We also describe the strength of rock under these conditions, and the physical mechanisms that enable minerals such as quartz to deform without fracturing. These test results help structural geologists interpret outcrops, such as that in **Figure 1.12**, with folded and sheared rock that did not fracture or fault as it deformed.

Figure 1.11 Three types of brittle fractures in Lower Jurassic limestone beds at Lilstock Beach, Somerset, England. Vertical calcite filled veins (white) and open joints (black) cut the light gray limestone on the left. A fault with striations runs up the middle of the image juxtaposing the gray limestone with the black shale on the right. Scale is approximate in the foreground. See Engelder and Peacock (2001) for detailed maps, outcrop photos, and structural interpretations of this area. Google Earth file: Figure 1.11 Lilstock Beach England fractures.kmz. UTM: 30 U 485890.63 m E, 5672212.54 m N (see **Box 1.1** for an explanation of UTM).

1.2.3 Viscous

Fissure eruptions near the summit of Kilauea Volcano on the island of Hawaii produce ample evidence for viscous deformation: molten rock behaves like a viscous liquid. A nighttime image (**Figure 1.13**) shows a curtain of fire erupting from a fissure in the background and lava flowing over a cliff that surrounds a crater in the foreground. There can be no doubt that the mechanical behavior of lava is that of a liquid, but its slow progress down the slope of the volcano, compared to water flowing down a similar slope, suggests lava has a greater resistance to flow. Resistance to flow is measured by the material property called *viscosity*, and the viscosity of basaltic lava is as much as six orders of magnitude greater than that of water.

In Chapter 6 we review laboratory tests that measure viscosity and melting as a function of temperature and chemistry. Because rock is a multi-component system, usually consisting of several different minerals, it has a range of melting temperatures. At the solidus temperature all constituents are glass or crystals, and at the liquidus temperature all constituents have melted. For basaltic lava (**Figure 1.13**), the solidus is about 1,000 °C and the liquidus is about 1,200 °C.

1.3 GEOLOGIC STRUCTURES

What are the products of deformation in Earth's lithosphere? We organize the products of deformation into five different categories of geologic structures: fractures, faults, folds, fabrics, and intrusions. The tectonic surface forces (tractions) and body forces (gravity and buoyancy) *cause* brittle, ductile, and viscous deformation, and the ensuing relative motions produce these structures. We devote a chapter to each category, but introduce them here with a few images.

1.3.1 Fractures

Opening fractures are the most common structures that form by brittle deformation in Earth's lithosphere. They occur within individual mineral grains at scales of microns to millimeters, and in outcrops such as the Cedar Mesa Sandstone (**Figure 1.14**) at scales of meters to kilometers. In this example the fractures are vertical and are organized into two nearly orthogonal sets, labeled A and B, based on their orientation. They are regularly spaced, but appear to be confined to the mesa-forming sandstone unit.

Box 1.1 Universal Transverse Mercator (UTM) Projection and Google Maps

The acronym UTM in the caption for **Figure 1.11** stands for Universal Transverse Mercator. This is the name of a mathematical projection used to map the horizontal position of a point on Earth's nearly spheroidal surface to a flat piece of paper. This is the standard projection method for modern geological maps, and we use it for all the geological maps in this textbook. Given the UTM information from the caption (30 U 485890.63 m E, 5672212.54 m N), one can precisely locate the point, which is on top of a Royal Navy lookout tower perched on a cliff facing the Bristol Channel in Somerset, SW England. Low tides expose the outcrop pictured in **Figure 1.11** on Lilstock Beach at the base of that cliff.

Many of the outcrop images in this textbook include UTM coordinates in the caption. Readers can use this information to locate and visit those sites using navigation software such as Google Maps. However, the default coordinates for Google Maps are latitude and longitude. Using one of the many tools available on the web to convert from UTM coordinates to latitude and longitude, we find the lookout tower is located at (51°12′4.1″ N, 3°12′7.1″ W). Recall that the three numbers, each followed by a superscript sign, refer to degrees, minutes, and seconds for this geographic coordinate system. The tower is a bit more than 51 degrees north of the equator, and a bit more than 3 degrees west of the line of longitude that passes through the front door of the observatory at Greenwich, England. The awkwardness of using degrees, minutes, and seconds for the coordinates of a point on a map is mitigated somewhat by converting to decimal degrees (51.201138°, –3.201959°). Type this latitude and longitude in decimal degrees in the search box of Google Maps, click on the search icon, and select the satellite option to view the lookout tower near Lilstock Beach.

Note: The instructions given here have been tested using Google Maps on a Windows desktop PC in September 2018. Different software versions and operating systems may require somewhat different procedures.

The UTM projection system divides Earth's surface into 60 *zones* bounded by lines of longitude. Each zone is 6 degrees wide, and they are numbered sequentially to the east from the 180 degree line of longitude. The line of longitude through the middle of a zone is the *central meridian*. On the Earth image shown in **Figure B1.1** the numbers along the equator are the zone numbers. Each longitudinal zone is subdivided into 20 *bands* bounded by lines of latitude. Each band is 8 degrees wide, and the bands are lettered sequentially toward the North Pole with N being the first band north of the equator. The first number and first letter in the UTM designation for the lookout tower near Lilstock Beach (**Figure 1.11**) are the zone number, 30, and band letter, U.

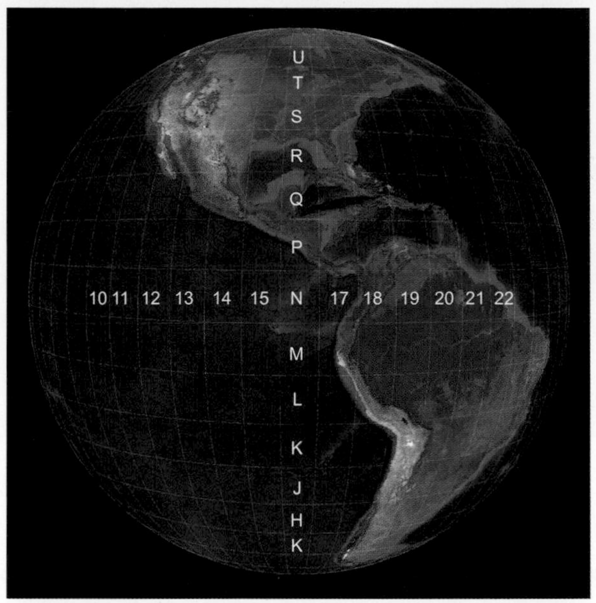

Figure B1.1 Image of Earth with zones bounded by lines of longitude and numbered along the equator, and bands bounded by lines of latitude and lettered from N at the equator.

The second and third numbers (485890.63 m E and 5672212.54 m N) in the UTM designation for the lookout tower are the *Easting* and *Northing*, both measured in meters. To understand these coordinates, one must know the location of the origin. The point at the intersection of the equator and the central meridian of a zone has coordinates (500,000 m E, 0 m N) for locations in the northern hemisphere. Thus, the origin is 500,000 m west of the central meridian and on the equator. To avoid negative coordinates for locations in the southern hemisphere, the point at the intersection of the equator and the central meridian is given the coordinates (500,000 m E, 10,000,000 m N). Thus, the origin is 500,000 m west of the central meridian and 10,000,000 m south of the equator for the southern hemisphere. Taking the coordinates (485890.63 m E, 5672212.54 m N) from **Figure 1.11** as an example, we understand that the lookout tower is about 486 km *east* and 5,672 km *north* of the origin for zone 30. Using UTM coordinates measured with a well-known distance scale (meters) has significant advantages over geographic coordinates measured with degrees.

With few exceptions, in this textbook the geographic coordinates East, North, and Up (E, N, U) are associated, respectively, with Cartesian axes (x, y, z). Both coordinate systems are right-handed, with E and x aligned with the thumb; N and y aligned with the index finger; U and z aligned with the middle finger.

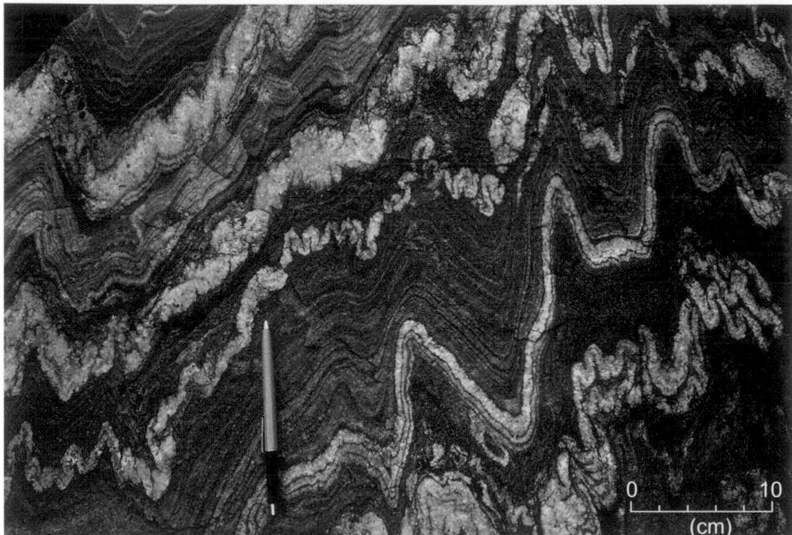

Figure 1.12 Folded quartz-rich layers (white) and micaceous layers (gray) in the Moine Schist, Scotland. See Hudleston and Treagus (2010) for photographs and interpretations of folds.

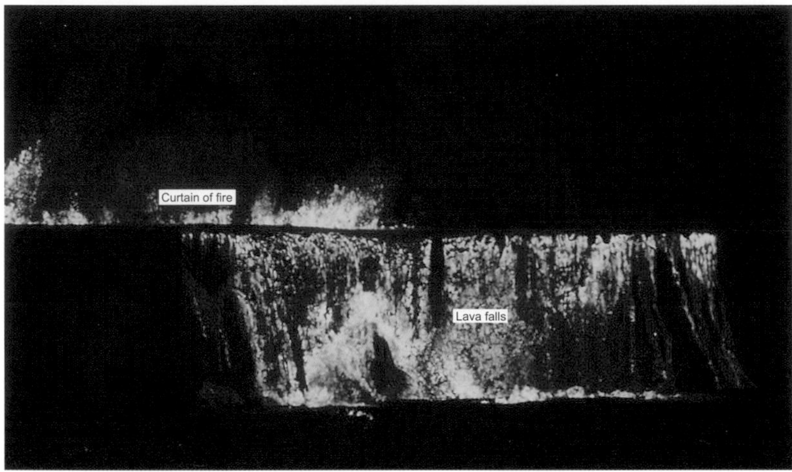

Figure 1.13 Lava erupting from a fissure, as a curtain of fire, and then flowing over a cliff, as at the lava falls, into a pit crater on Kilauea Volcano, HI. Photograph by R. T. Holcomb.

In Chapter 7 we describe and analyze opening fractures that are barren, and are called joints, like those at Cedar Mesa (**Figure 1.14**), and also those that are filled with hydrothermal minerals, and are called veins (**Figure 1.11**). Some opening fractures are filled with igneous rock, and these are called dikes. Dikes feed the fissure eruptions at Kilauea Volcano (**Figure 1.13**).

1.3.2 Faults

Fractures are distinguished from faults because the relative motion of the rock on either side of a fault is dominantly parallel to the plane of the fault. In other words, faulting is characterized by *shearing* and fracturing is characterized by *opening*. The outcrop pictured in **Figure 1.15** shows the Lake Edison Granodiorite (blue-gray), an igneous rock that has been cut by younger dikes (white) that opened and were filled with quartz and feldspar. The dikes serve as good markers for the still younger fractures (labeled B–B) that only opened, and for faults (labeled A–A) that offset the dikes by as much as a few decimeters. Faults are not as common as fractures, and they can be much more complicated structures, but they also are indicators of brittle deformation. We describe and interpret faults in Chapter 8.

1.3.3 Folds

Sedimentary rocks are deposited in nearly horizontal and planar layers. It is surprising, therefore, to come across an outcrop like the one pictured in **Figure 1.16** where sedimentary strata are steeply inclined and crumpled into ornate chevron-shaped folds. This package of rocks is composed of alternating chert and shale beds, with the cherts about a decimeter thick and the intervening shale a few to several centimeters thick. The chert beds are more resistant to erosion and weathering, so they project from the outcrop. The shale beds are less resistant to erosion, so they

Figure 1.14 Two orthogonal sets of vertical joints (e.g. A–A and B–B) cut the mesa-capping sandstone bed in the Cedar Mesa Sandstone overlooking Monument Valley, UT. Joint spacing varies from about 10 to 40 m. UTM: 12 S 592430.00 m E, 4119974.00 m N. Google Earth file: Figure 1.14 Cedar Mesa UT joints.kmz. (See **Box 1.2** for an explanation of KMZ and KML.)

retreat into the outcrop. A few faults disrupt the bedding. The limbs of the folds are nearly straight and the hinges of the folds form acute angles ranging from about 30° to 50°. Straight limbs and angular hinges are characteristic of chevron folds. This is one of many different fold geometries, some of which we describe further in Chapter 9.

1.3.4 Fabrics

When rock is deformed under metamorphic conditions of elevated temperature and pressure, the constituents change shape with little or no fracturing and faulting, and yet the minerals do not melt to form a liquid. This is neither brittle deformation, nor viscous deformation. Rather it is ductile deformation. The deformation of rock under these conditions can produce alignments of mineral grains, or other heterogeneities, to form linear fabrics and planar fabrics. In **Figure 1.17** objects that originally were pebbles, presumably with approximately equant shapes, now are shaped like pancakes with one very short dimension, and two rather long dimensions. The alignment of these pancake-shaped pebbles defines a *planar* fabric in this metamorphic conglomerate. Chapter 10 is devoted to the description of fabrics and the mechanisms of ductile deformation.

1.3.5 Intrusions

Molten rock (magma) forms deep in the lower lithosphere or asthenosphere and moves upward into the lithosphere due to buoyancy. In the Henry Mountains of southeastern Utah magma rose to within a few kilometers of Earth's surface and intruded the nearly flat-lying sedimentary rocks of the Colorado Plateau. There, the magma intruded between particular strata and spread out as thin horizontal sills. Some of these sills spread far enough laterally to bend the overlying strata upward, and thus thicken to form laccoliths. The distal edge of the Trachyte Mesa laccolith is shown in **Figure 1.18**. Here the Entrada Sandstone is bent and stretched over the edge of the laccolith. Intrusions are described and analyzed in Chapter 11.

1.4 METHODOLOGY OF STRUCTURAL GEOLOGY

How does one practice structural geology? A partial answer to this question comes from looking back to one of the founding events of the discipline. In 1788 John Playfair and Sir James Hall accompanied James Hutton on a geological field trip to Siccar Point on the east coast of Scotland. There they observed what

Box 1.2 Keyhole Markup Language (KML) and Google Earth

The acronym KMZ in the caption for **Figure 1.14** is the compressed (*zipped*) variant of KML, which stands for Keyhole Markup Language. Keyhole, Inc developed this file format to work with geographic information and produce annotated visualizations such as images and maps. Google acquired Keyhole in 2004, and KML has become the international standard used by geospatial software such as Google Earth. Some outcrop images and maps in this textbook include a .kmz file name in the caption, so the reader can use Google Earth for a virtual field trip to these sites by opening the file in available from the Cambridge website www.cambridge.org/SGAQI.

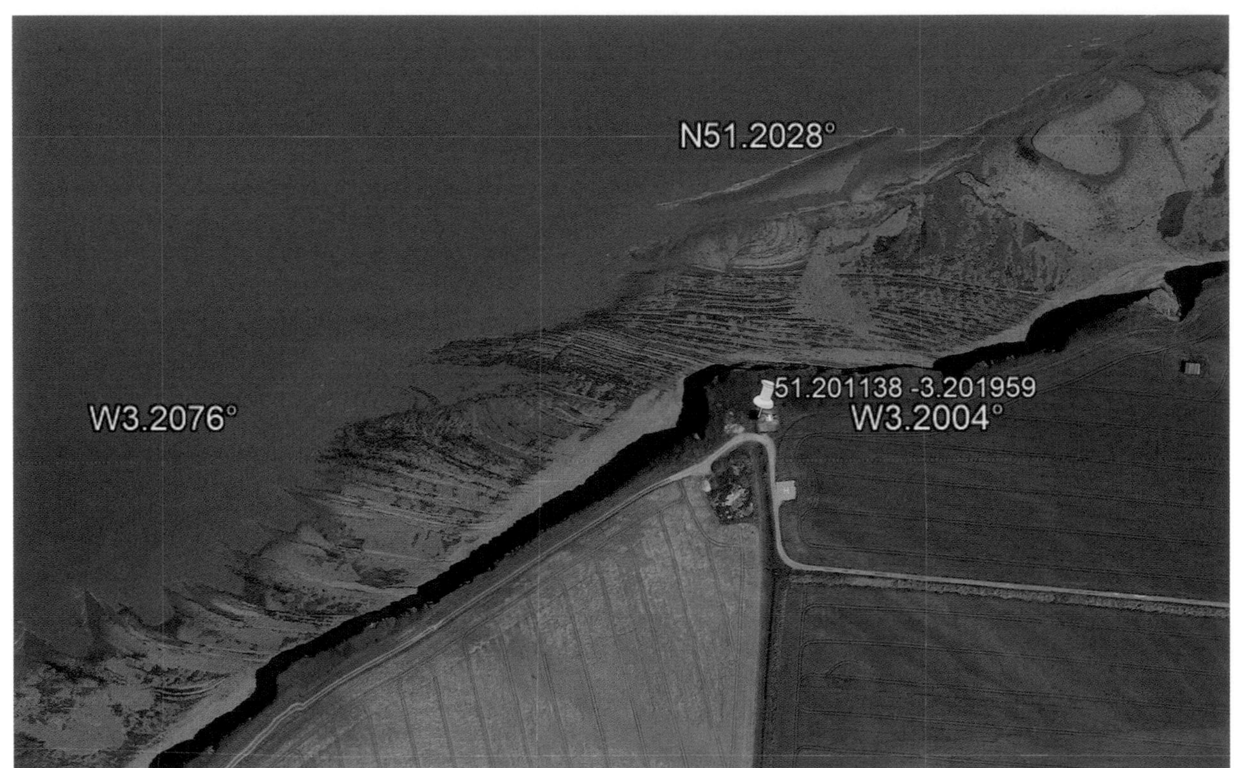

Figure B1.2 Google Earth image near (51.201138°, –3.201959°). The yellow pushpin (Placemark) is on top of the Royal Navy lookout tower facing the Bristol Channel in SW England.

Note: The instructions given here have been tested using Google Earth Pro version 7.3 on a Windows desktop PC in September 2018. Different software versions and operating systems may require somewhat different procedures.

The default coordinate system for Google Earth is latitude and longitude, expressed in decimal degrees. These geographic coordinates use a mathematical model for Earth's shape defined by the World Geodetic System of 1984 (WGS84). This model is an oblate ellipsoid, a three-dimensional surface formed by rotating an ellipse about its minor axis, which passes through the north and south poles. To go to a particular location with known geographic coordinates open Google Earth, type or copy and paste the decimal degrees in the Search box, for example 51.201138 N 3.201959 W, and click Search (see **Figure B1.2**). This location is for the Royal Navy lookout tower facing the Bristol Channel in SW England (see **Box 1.1**). The displayed grid system in Google Earth conforms to decimal degrees if you click on Tools in the Menu Bar and select Options; then select Decimal Degrees under Show Lat/Long; and click OK. To display the coordinate grid, click on View in the Menu Bar and select Grid.

To go to a particular location with known UTM coordinates open Google Earth. Click on Tools in the Menu Bar and then on Options; select Universal Tranverse Mercator under Show Lat/Long; select Meters/Kilometers under Units of Measurement; and then click OK. To display the coordinate grid, click on View in the Menu Bar and select Grid. Navigate and zoom until zone 12 and band S fill the window. Click on View in the Menu Bar, select Reset, and then select Tilt and Compass to reset the view with North up and no tilt. Click on Add in the Menu Bar and select Placemark. In the New Placemark dialogue box enter the Name "Cedar Mesa UT joints"; enter the

Zone, Easting, and Northing for this site from the caption for **Figure 1.14**; and click OK. The placemark should move to the new location defined by the UTM coordinates and the name should be displayed next to the pin. Navigate to center the pin and zoom in to a view similar to **Figure 1.14**. Right click on the pin and select Properties. In the Edit Placemark dialogue box select the View tab and click the Reset button. This makes the current view the one you return to when you double click on the pin, or when you double click on the name Cedar Mesa UT joints in the Places sidebar.

Google Earth provides tools for measuring distance and orientation on an image. To access these tools click Tools in the Menu Bar and select Ruler. In the Ruler dialogue box make sure the Line tab is selected. Move the cursor onto the image and click on a target location. Now move the cursor to another location and note that a yellow line extends from the first location to the location of the cursor. As you move the cursor the Length and Heading numbers in the dialogue box change. Note that the heading is the direction (azimuth) from the first location to the current location measured from 0 to 360 degrees counter-clockwise from North. Click on a second location and the Length and Heading appear in the dialogue box. We employ this tool to measure distance and direction on images associated with review questions at the end of this chapter.

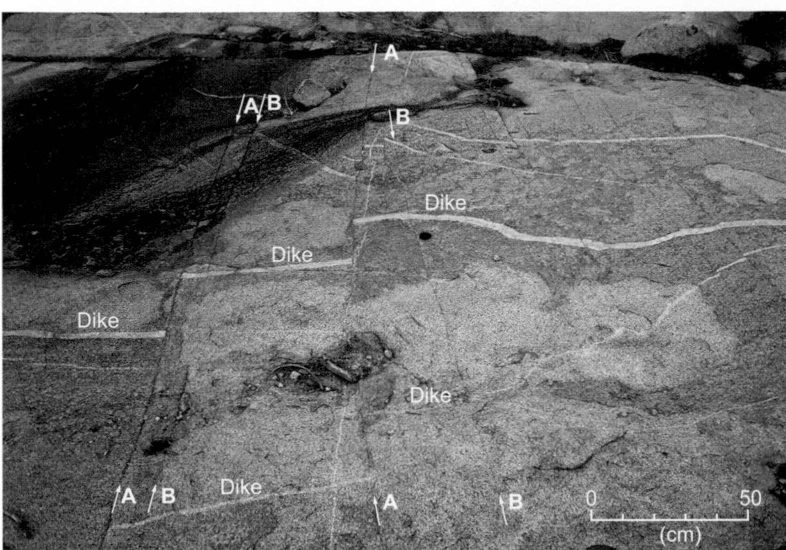

Figure 1.15 Small faults offset older dikes from a few centimeters to a few decimeters in a left-lateral sense. Outcrop in the Lake Edison Granodiorite of the central Sierra Nevada, CA. Scale is approximate in foreground. See Segall and Pollard (1983) for detailed maps and structural interpretations.

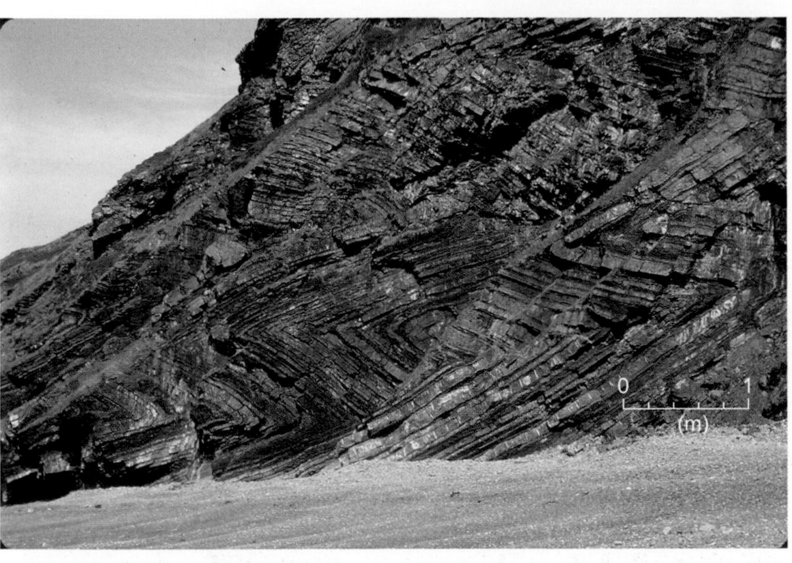

Figure 1.16 Sedimentary strata, composed of chert beds about a decimeter thick alternating with thinner shale beds and forming chevron folds. Outcrop at Millook Haven, south of Bude, Cornwall on the southwest coast of England. Scale is approximate. Google Earth file: Figure 1.16 Millook Haven England folds.kmz. UTM: 30 U 388858.40 m E, 5625668.10 m N.

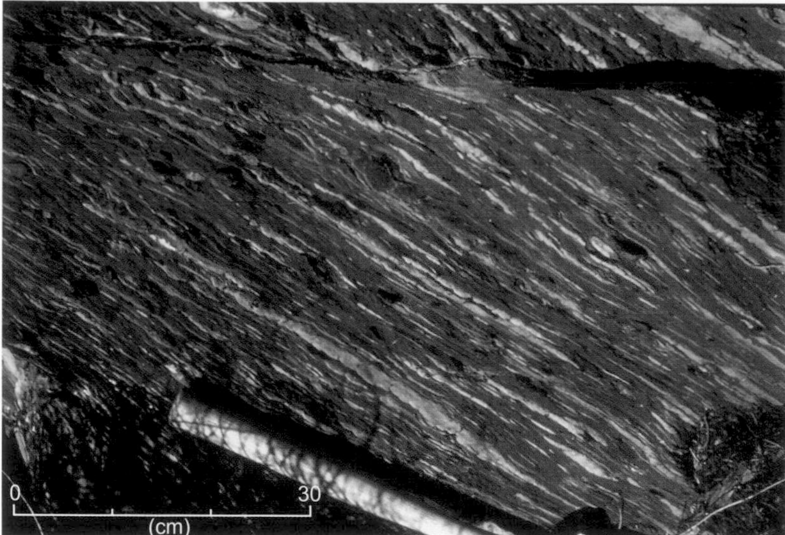

Figure 1.17 Metamorphic conglomerate from the Panamint Mts., CA, with flattened pebbles forming a planar fabric.

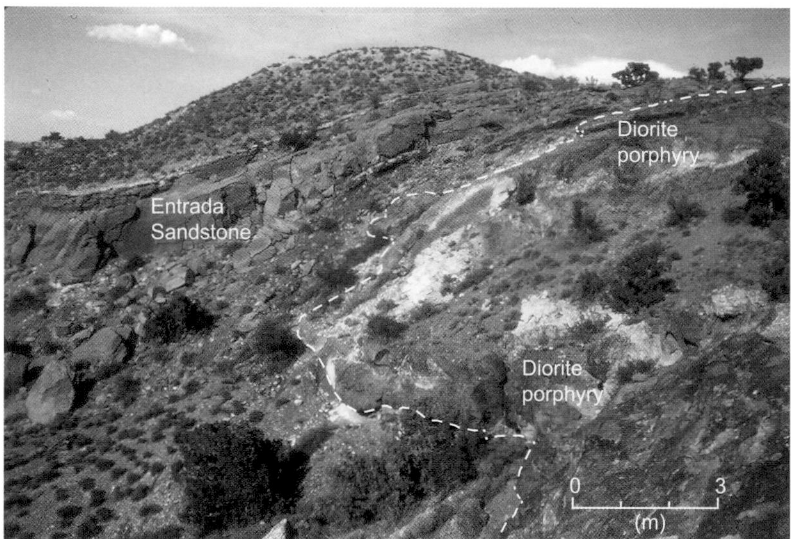

Figure 1.18 Trachyte Mesa laccolith, Henry Mts., UT. The gray igneous rock (diorite porphyry) with overlying reddish Entrada Sandstone pushed up, stretched, thinned, and bent by the intruding magma. Dashed white line marks outer contact of laccolithic intrusion. Scale is approximate in foreground. See Morgan et al. (2008) for detailed maps and fabric interpretation. Google Earth file: Figure 1.18 Trachyte Mesa UT laccolith.kmz. UTM: 12 S 535977.31 m E, 4199851.14 m N.

now is called Hutton's unconformity to honor the insight he brought to the fledgling study of geology by interpreting this outcrop (**Figure 1.19**). His interpretation went something like the following. The sedimentary strata that now are almost vertical, were deposited in horizontal layers. Later they were tilted to near vertical and then eroded. On this jagged erosional surface (the unconformity) a conglomerate was deposited that includes fragments of the older underlying rock. The red sandstone then was deposited as horizontal layers on the conglomerate. The sandstone now dips gently to the left in this photograph, so this whole package of rocks was tilted again, and then eroded to expose the outcrop seen today.

Hutton and his colleagues did not have the tools to quantify the timing of these events, but they appreciated that an immense span of time must have passed, and that enormous forces were required to deform these rocks. John Playfair wrote of their visit to Siccar Point:

> The mind seemed to grow giddy by looking so far into the abyss of time; and while we listened with earnestness and admiration to the philosopher [Hutton] who was now unfolding to us the order and series of these wonderful events, we became sensible how much further reason may sometimes go than imagination may venture to follow. (From McIntyre and McKirdy, 2012)

Their sense of wonder must have been palpable given that Church dogma of the day put the age of the Earth at 6,000 years, and described the only natural event of any consequence during that time as the biblical story of the great flood. Here at Siccar Point, Hutton saw the geologic evidence for a much longer and

Figure 1.19 Outcrop of Hutton's unconformity at Siccar Point, Scotland. Nearly vertical Silurian strata are overlain by a basal conglomerate and the gently inclined Devonian strata. Dashed white line marks Hutton's angular unconformity, a 65 Ma hiatus in time. Scale approximate in foreground. Photograph ©Lorne Gill/SNH. Google Earth file: Figure 1.19 Sicar Point Scotland unconformity.kmz. UTM: 30 U 543626.74 m E, 6198671.76 m N.

more interesting history. It now is understood that the vertical Silurian sedimentary rocks were deposited about 435 million years ago, and the Devonian red sandstone was deposited about 370 million years ago, so the gap in time represented by Hutton's unconformity is about 65 million years.

Four aspects of scientific methodology relevant to the practice of structural geology are highlighted in the visit to Siccar Point by Hutton and his colleagues. (1) They made *field observations* to motivate and constrain their thoughts, while standing on the outcrop. (2) They used basic geologic knowledge to deduce the *relative ages* of the observed strata: younger sediments were deposited on top of older sedimentary rocks. (3) They used knowledge of current processes like erosion and deposition to inform their interpretations of the rock record: the *principle of uniformitarianism.* (4) They used the geometry of the structures to deduce the motions of the tilted strata: a *kinematic description* of the geologic events. Their combination of field observations, relative ages, uniformitarianism, and kinematic reasoning produced a compelling story that was a key building block in the founding of the science of geology.

The four aspects of methodology introduced in the previous paragraph are as vital today as they were in 1788, but they do not provide answers to many of the questions we now ask about geologic structures. Therefore, we build upon the methodology followed by Hutton and adopt a framework of continuum mechanics to understand the physical processes that deform rocks. We defer the deposition of the strata to sedimentologists, the erosion to geomorphologists, and the timing of events to geochronologists, while focusing as structural geologists on the geologic structures and their origins. This mechanical framework enables us to answer the ultimate questions: how and why do geologic structures form? The answers require a quantification of the forces that cause deformation, the properties of rock that resist deformation, and the accelerations, velocities, and displacements that provide the kinematics of deformation. All of these are part of the framework of continuum mechanics.

The insights that led to the development of continuum mechanics came from Sir Isaac Newton and were published in his seminal work *The Principia* in 1687 (see Newton et al., 1999, in Further Reading). Thus, the principles of what we now call Newtonian mechanics were available in 1788, but Hutton and his colleagues did not exploit those principles. In fact, even intimate knowledge of Newton's contributions would not have enabled them to understand the deformation at Siccar Point, because the mechanics of Newton focuses on the motions of discrete bodies, such as the planets, that were treated as widely separated *points of mass* that never come in contact with one another. What was needed was a mechanics of *continuous bodies* applied to the rock making up Earth's crust, in which every particle is acted upon by its neighbors, and every particle can move relative to its neighbors, based on the physical properties of the continuum.

Many mathematicians, physicists, and engineers have participated in the development of continuum mechanics, and some of their insights will be exploited in this textbook to help us understand the development of geologic structures. In the decades just before Hutton's visit to Siccar Point in 1788, Leonhard Euler developed the spatial description of motion that is used today in fluid mechanics, and Joseph-Louis Lagrange developed the referential description of motion that is used today in solid mechanics. In the first few decades after Hutton's death in 1797, Claude-Louis Navier formalized the equations of motion for the elastic solid, and Augustine-Louis Cauchy published his general equations of motion for the continuum, and introduced the stress tensor. By the end of the nineteenth century, continuum mechanics had developed to the point where it was being used to address many practical problems of science and engineering. Today, structural geologists have the opportunity to employ continuum mechanics as an integral part of their scientific methodology, and this textbook adopts that approach (see **Figure 1.20**).

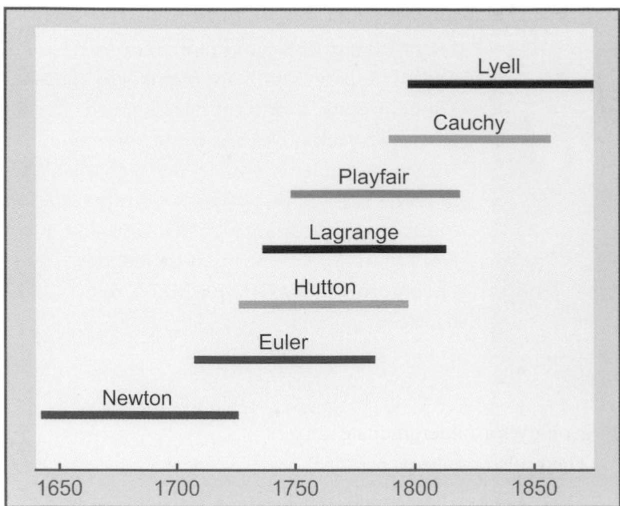

Figure 1.20 Timeline with key players during the founding of structural geology and continuum mechanics.

1.4.1 A Complete Mechanics

Hutton and his colleagues talked about the kinematics of the deformation that they inferred from the geometric arrangement of strata at Siccar Point. The older Silurian strata were deposited as horizontal layers, but later were tilted, and today remain nearly vertical. Kinematics is that branch of mechanics that is concerned with the motion of individual particles, and of rigid and deformable bodies made up of particles. Hutton articulated a story in words, describing the tilting of the strata, which was sufficient for understanding the geologic history of the strata at Siccar Point in 1788. This was a huge step forward relative to Church dogma and the biblical story.

However, it was Newton who recognized the importance of the mass of particles, and who provided the quantitative relationship between the causative force and the resulting kinematic quantity, the acceleration. Cauchy subsequently put Newton's laws in the context of the continuum, and made it possible to incorporate kinematics into a complete mechanics that includes the forces, the material properties, and the kinematics. In other words, Newton and Cauchy provided tools to do more than explore the sequence by which geologic structures form: their contributions enable us to address *how* and *why* geologic structures form.

A complete mechanics uses the physics of deformation to analyze and interpret field observations of geologic structures. In this way, a coherent and self-consistent, if idealized, description of the development of a geologic structure may be achieved. This methodology was articulated in a paper published as part of the 20th Anniversary Special Issue of the *Journal of Structural Geology* (Fletcher and Pollard, 1999). It also underlies the organization and intent of the advanced textbook, *Fundamentals of Structural Geology* (Pollard and Fletcher, 2005).

A complete mechanics does *not* imply that all of the details and complexities in the process leading to a particular structure or structural type are included. However, it does require an explicit choice of model geometry, material properties, boundary and initial conditions, and the fundamental laws of physics, to produce a set of equations from which all results follow, *including* the kinematics of deformation. When the model results do not conform in important ways with the available data from geological field observations, a more refined model is formulated using a somewhat different choice of geometry, material properties, and/or boundary conditions, but the model must include all of these aspects of the mechanics; it must be *complete*. Thus, a sequence of quantitative models are put forward that successively provide an improved understanding of the geologic structure. This understanding includes answers to questions, left unanswered by Hutton at Siccar Point, about *how* and *why* structures form.

In their 1999 paper Fletcher and Pollard proposed, as a goal for the twenty-first century, a significant revision of the educational objectives for courses in structural geology, so that students are prepared to use the full complement of mechanical tools necessary to formulate a plan for field data collection, to set up and solve relevant boundary-value problems, and to evaluate the correspondence between models and data. This textbook is designed to make that methodology available as a *first course* in structural geology.

1.4.2 The Canonical Models

Canonical models in structural geology are those that reduce the physical processes generating geologic structures to the simplest form possible, without too much loss of generality. From the etymology of canon (Greek, *kanōn*) we understand that this word means rod, rule, or *measuring stick*. In other words, we expect these problems to be used as measuring sticks from which we draw some understanding of the more complicated geologic structure. Because geologic structures develop when rocks physically deform, we anticipate that well-defined and idealized models from that part of physics called mechanics will serve as the measuring sticks.

In parlance drawn from another branch of geology, we intend these canonical problems to stand like the *type locality* of a named sedimentary unit. The type locality serves as the standard of reference by which that unit is identified at other localities. For this textbook it is not a locality or physical object, but a well-defined problem of mechanics that serves as a *standard of reference* for the physical process that led to the development of a geologic structure.

We acknowledge that geologic structures did not develop exactly as in the canonical problem, just as the stratigrapher understands that the sedimentary unit at another exposure will differ somewhat from that at the type locality. None-the-less, we expect to gain insight and build intuition from each canonical problem. Of equal importance, when more robust and detailed models emerge, the canonical problems provide the necessary foundation for evaluating them and understanding them.

1.4.3 The Role of Physics

The tectonic processes that structural geologists investigate, and the canonical models that they use to discuss these processes, are governed by equations of motion that come to us from Newton's laws. Newton points the way through his insightful discovery that associates force, **F**, and mass, m, and acceleration, **a**, in the vector equation:

$$\mathbf{F} = m\mathbf{a} \tag{1.5}$$

One of the most important underlying themes of this book is the elaboration of this relationship, to describe the brittle, ductile, and viscous deformation of rock and magma in Earth's lithosphere and asthenosphere.

The physics course for science and engineering majors at most universities and colleges includes classical mechanics based on Newton's laws. However, these laws usually are used to consider the motion of individual particles and rigid bodies, not the relative motion of different particles in a deformable body such as flowing lava (**Figure 1.13**), or folding strata (**Figure 1.16**). To address deformation of rock and the formation of geologic structures, Newton's laws must be *recast* and applied to a continuous body using the principles of continuum mechanics. Because that topic usually is not part of the introductory physics course, we introduce and explain the necessary elementary concepts of continuum mechanics in this textbook.

1.4.4 The Role of Mathematics

The equations of motion that describe folding schist (**Figure 1.12**) or faulting granite (**Figure 1.15**) utilize mathematical concepts from university- and college-level calculus for scientists and engineers. Therefore, throughout the textbook we use the concepts and tools from such a course, including: functions and limits, derivatives and integrals, series expansions, analytical geometry, vector algebra and vector calculus, partial derivatives and gradients. The interpretation of key equations in the textbook are facilitated by numerical and graphical results found using MATLAB, our choice for a computation and graphics engine. In addition, skills are developed in verifying that equations are correct, such as dimensional consistency and expected behavior in limiting cases.

We use elementary concepts from linear algebra and differential geometry, which may or may not be covered in the first calculus course for scientists and engineers. Linear algebra and differential geometry are *not* considered a pre-requisite course for this textbook, so the concepts (e.g. matrix algebra, eigenvalue and eigenvector problems, and parametric representations of curves and surfaces) are introduced and explained here using examples from structural geology.

We anticipate that students completing a course using this textbook will have a clear understanding of the place and importance of mathematics in the daily work of a structural geologist. For this reason, the book's sub-title is *A Quantitative Introduction*. This textbook includes about 500 numbered equations. The authors have not memorized these equations, and we do not expect students to commit them to memory. On the other hand, they are an important part of the story we are telling about deformation and geologic structures. Some of these equations will be repeated so many times that a student may find they can recite them from memory, but the role for these equations is to organize the fundamental principles and relevant data of deformation in a precise and reproducible manner.

1.5 APPLICATIONS AND CAREERS

The founders of geology, including James Hutton, John Playfair, and Sir James Hall, were interested in discovering how geologic structures informed their understanding of Earth's history. Because we live on a planet with an active lithosphere and asthenosphere, geologic structures are one of the key components used to understand Earth's billion-year history. In that context structural geology has become one of the pillars of a curriculum in geology for undergraduate students.

There also are very practical reasons for studying structural geology. Exploring for and producing Earth's resources employs many geologists with a strong background in structural geology. All of the basic energy sources that come from rock (hydrocarbon, nuclear, and geothermal), as well as the common and rare minerals and elements that play important roles in modern manufacturing, are found in association with geologic structures.

Career opportunities in the hydrocarbon industry put some structural geologists in the exploration work flow to help determine the location and nature of hydrocarbon sources and traps. Other structural geologists are on the production side, working with the geophysicists who gather and analyze seismic reflection data, and with the petroleum engineers who simulate hydrocarbon flow to wells that tap the reservoirs.

Structural geologists contribute to the mitigation of geologic hazards, including earthquakes, volcanic eruptions, and landslides. For example, earthquakes that shake our cities, doing tremendous damage to the infrastructure and killing people, originate on faults that slip dynamically and set off seismic waves that can travel through the Earth. By understanding the nature of faults and their mechanical behavior, structural geologists work with seismologists and engineering geologists to assess earthquake hazards. Also, by understanding the plumbing systems of volcanos, and the mechanical behavior of molten rock, structural geologist work with volcanologists and geophysicists to assess volcano hazards. Structural processes near the surface of the Earth also operate in landslides, and a structural geologist can help mitigate or avoid the damage done by landslides.

Structural geologists also work toward achieving a sustainable environment. Detecting and cleaning up groundwater contamination involves an understanding of the fracture systems in aquifers, because these fractures in many places are the super highways for groundwater flow. Also, the long-term containment of nuclear waste in granitic rock, or salt, or other buried repositories requires a clear understanding of the role of geologic structures in underground fluid flow, and the deformation of rock on a geologic time scale.

For those motivated to teach, career opportunities exist for structural geologists in academia. Structural geology is a key component in the standard curriculum for majors in geology. Thus, teaching positions for a structural geologist exist at most colleges and universities that offer a geology major.

Opportunities also exist at major universities for those interested in research and graduate education. The approach developed in this book should help prepare structural geologists in a broad array of areas as their careers unfold.

Recapitulation

Structural geologists describe, model, and explain the deformation of rock in Earth's lithosphere and asthenosphere. The primary objectives of structural geology are to *document* the deformed architecture of these outer shells of Earth, and to *understand* the tectonic processes that cause the structures to form. Because the mechanical behavior of rock changes with temperature and pressure, and because both of these physical quantities increase substantially with depth in the lithosphere and asthenosphere, structural geologists must consider rock deformation that varies from *brittle* to *ductile* to *viscous*: the whole range from solid to liquid material behavior. Given that range of behaviors, it should not be surprising that a wide variety of structures exist, including five at the broadest level of categorization: *fractures, faults, folds, fabrics,* and *intrusions.*

Structural geology is a blending of geological knowledge and field techniques, with the results from laboratory tests and the physical principles and mathematical methods of mechanics. This textbook demonstrates how to integrate field observations, the relative ages of observed structures; the principle of uniformitarianism; and a complete framework of continuum mechanics to understand the physical processes that deform rocks. *Canonical problems* reduce the physical processes generating geologic structures to the simplest form possible, without too much loss of generality. Geologic structures did not develop exactly as in these canonical problems, but they are insightful, and build intuition. Following this methodology, a sequence of quantitative models successively provide an improved understanding of geologic structures.

A career as a structural geologist has many desirable features, ranging from the enjoyment of working outdoors, to the pleasure of observing deformed minerals in thin sections under the microscope, to the challenge of solving complex mathematical and computational problems, to the satisfaction of making a substantial contribution toward discovering new mineral resources or mitigating geologic hazards. Opportunities exist for fulfilling careers in industry, academia, government laboratories, and private consulting.

REVIEW QUESTIONS

The following questions are designed to highlight the expected *learning outcomes* for this chapter. Each question is taken directly from the material in the chapter and, for the most part, in the same sequence that it appears in the chapter. If an answer is not forthcoming, students are advised to read the relevant section of the chapter and discover the answer.

1.1. Many petrologists and geochemists subdivide Earth's outer shells into *crust* and *mantle*, whereas many structural geologists focus on two outer shells called the *lithosphere* and *asthenosphere*. Describe the differences between these two classification schemes and give the estimated thicknesses for the shells.

1.2. The convecting asthenosphere imposes *tractions* on the base of the lithosphere that cause the motion of tectonic plates and the deformation of rock within these plates. The traction vector varies with *position* of the surface, **p**, and with the *orientation*, $\hat{\mathbf{n}}$, of the surface on which it acts. Describe these two vectors.

1.3. The traction vector is defined as the limiting value of a ratio of two physical quantities. What are those quantities, and how do they, and their ratio, vary in this limit?

1.4. M. King Hubbert's sandbox is used to describe a *boundary-value problem* in solid mechanics. What physical quantity is used to define the boundary conditions for this problem? What physical quantity is calculated in the interior of the body by solving this problem?

1.5. Body forces act on rock in Earth's lithosphere and asthenosphere, causing geologic structures to form. Describe and distinguish *gravitational* and *buoyant* body forces using words and two vector equations.

1.6. What does it mean for a body of rock, say salt in a salt intrusion, to be *neutrally buoyant*?

1.7. Three distinctive styles of deformation are *brittle*, *ductile*, and *viscous*. Give a one sentence description of each that would serve to identify the style from field observations.

1.8. In this chapter, and throughout the textbook, we organize the products of rock deformation into five broad categories of geologic structures: *fractures*, *faults*, *folds*, *fabrics*, and *intrusions*. Give a one sentence description of each category.

1.9. The following Google Earth image is taken from the file: Question 01 09 Cedar Mesa.kmz. Open the file; click on the Placemark for this question to open the popup box; copy and paste the questions into your answer document; and address the assigned questions. The Google Earth image data are from Landsat / Copernicus.

1.10. The founding of geology as a modern scientific discipline can be traced back to the field trip led by James Hutton to Siccar Point on the east coast of Scotland in 1788. Four key aspects of scientific methodology were used to interpret this outcrop. What were they, and how were they used to understand Hutton's angular unconformity?

1.11. In 1687, Sir Isaac Newton provided the fundamental mechanical insight regarding the motion caused by the resultant force acting on a particle of given mass, but this was not used by Hutton to explain the deformation at Siccar Point in 1788. What branch of Newtonian mechanics was developed in the eighteenth and nineteenth centuries that we use today to address how and why geologic structures form?

1.12. Why do physically based explanations of structural processes have an advantage over explanations that are not based on physics?

MATLAB EXERCISES FOR CHAPTER 1: A TUTORIAL

A quantitative introduction to structural geology includes the fundamental mechanical theory, data analysis, and mathematical modeling. We have chosen MATLAB as the computational and graphical engine to make the theoretical equations throughout the textbook come to life through numerical examples and three-dimensional dynamic figures. The online tutorial helps to make this platform a valuable partner for investigating and understanding tectonic processes and their structural products. The tutorial includes an introduction to the MATLAB user interfaces; the basic mathematical operations on constants, variables, and arrays; special functions for vectors and tensors; gridding and plotting; and programming and executing scripts. The exercises at the end of the tutorial serve as a review of the basic MATLAB functionality necessary to solve the exercises in subsequent chapters. We urge students to adopt this tool to understand the physical concepts, to analyze structural data, and to build models of tectonic processes. Becoming an accomplished MATLAB user develops crucial skills in reasoning, organization, mathematics, computer programming, and logic. Solidifying these skills puts students in an excellent position to collaborate with other scientists and to succeed as a structural geologist. www.cambridge/SGAQI

FURTHER READING

For citations in figure captions see the reference list at the end of the book.

Fletcher, R. C., and Pollard, D. D., 1999. Can we understand structural and tectonic processes and their products without appeal to a complete mechanics? *Journal of Structural Geology*, 21, 1071–1088.
This paper introduces the concept of a complete mechanics in the practice of structural geology.

McIntyre, D. B., and McKirdy, A., 2012. *James Hutton: The Founder of Modern Geology*. National Museums Scotland, Edinburgh, Scotland.
In this very readable and informative book, the authors make the case for James Hutton being the founder of modern geology.

Mukherjee, S., 2015. *Atlas of Structural Geology*. Elsevier, Amsterdam.
This atlas contains a multitude of color photographs, each with a brief description, location information, and cited references, for outcrop-scale folds, shear zones, faults, boudins, veins, and other structures.

Newton, Isaac, Cohen, I. B., and Whitman, A., 1999. *The Principia, Mathematical Principles of Natural Philosophy, a New Translation by I. Bernard Cohen and Anne Whitman*. University of California Press, Berkeley, CA.
This authoritative and comprehensive translation makes *The Principia* accessible to the modern reader.

Pollard, D. D., and Fletcher, R. C., 2005. *Fundamentals of Structural Geology*. Cambridge University Press, Cambridge.
A complete mechanics is the underlying theme for this advanced textbook in structural geology.

Press, F., Siever, R., Grotzinger, J., and Jordan, T., 2004. *Understanding Earth*. W.H. Freeman and Company, New York.
This beautifully illustrated textbook provides undergraduate students with a course in physical geology that is a pre-requisite for courses in structural geology.

Ragan, D. M., 2009. *Structural Geology: An Introduction to Geometrical Techniques*, 4th edition. Cambridge University Press, Cambridge.
This is a comprehensive and very readable treatment of the geometric concepts useful in structural geology.

Turcotte, D. L., and Schubert, G., 2002, *Geodynamics*, 2nd edition. Cambridge University Press, Cambridge.
Geodynamics approaches many of the same topics covered in structural geology, but typically uses geophysical data rather than geological data to constrain mechanical models.

PART II

Part II (Chapters 2 and 3) reviews and summarizes the mathematical tools and physical principles used in this textbook. However, these chapters do not present isolated concepts in mathematics and physics. Instead, they integrate the tools and principles with applications to geologic structures. For example:

- Chapter 2 introduces scalar, vector, and tensor quantities using the classic study by Ernst Cloos of deformed ooliths from the Paleozoic limestone of the South Mountain fold in Maryland and Pennsylvania; and
- Chapter 3 derives the equations for conservation of linear momentum using a fault that crops out in the Lake Edison Granodiorite of the Sierra Nevada, California.

Chapter 2
Mathematical Tools

Introduction

Chapter 2 highlights and reviews the mathematical tools used in this textbook and offers a first glimpse of some of the geological applications of the mathematics. These tools include using position vectors for mapping geologic structures, using stereographic projections to investigate the orientations of structures, and representing curved geological surfaces with vector functions. Like any learned activity, whether a foreign language, a musical instrument, or a sport, retention of expertise requires regular practice. The mathematical concepts reviewed in this chapter appear throughout the book to foster this practice. Including the mathematical analysis of geological data, and the formulation of mathematical models to explain how and why geological structures develop, greatly enrich the practice of structural geology.

The mathematical pre-requisite for this book is a course in calculus that includes differential calculus and integral calculus of functions of one variable. Some calculus courses include analytic geometry and vector calculus, while others introduce aspects of linear algebra. Some of the elementary concepts from analytic geometry, vector calculus, and linear algebra are used in this textbook, but they are at a level that does not require a pre-requisite course. Instead, we introduce the necessary concepts in this chapter and motivate readers to learn them by offering direct applications to structural geology. The differential calculus of more than one variable also is used throughout the book, but a course in multivariate calculus is not considered a pre-requisite. We introduce the few necessary extensions from differential calculus of one variable to multiple variables, including the partial derivative, the gradient vector, and the material time derivative in this chapter. Finally, although differential equations appear throughout the book, a course in ordinary and partial differential equations is not a pre-requisite. Differential equations appear solely for displaying the underlying physical concepts and relationships. Solutions are provided where they illustrate applications to structural geology, but solution methods are left to more advanced textbooks and courses.

This textbook contains nearly 500 numbered equations. The authors have not memorized these equations, and we do not expect students to commit them to memory. They are simply part of the story we are telling about deformation and geologic structures. Some of these equations will be repeated so many times that a student may find they can recite them from memory, but the important role for these equations is to organize the fundamental principles and relevant data of deformation in a precise and reproducible manner.

2.1 CHARACTERISTICS OF SCALAR, VECTOR, AND TENSOR QUANTITIES

All of the physical quantities used to describe tectonic processes and their structural products have an underlying mathematical representation. These mathematical representations fall into three distinct categories that we call scalars, vectors, and tensors. Mathematicians refer to all of these quantities as tensors and distinguish the categories by referring to scalars as tensors of order 0. They refer to vectors as tensors of order 1, and to tensors as tensors of order 2. We adopt the distinct names scalar, vector, and tensor because the science and engineering literature uses these names, and we do not need the deeper meaning of the mathematical terminology.

To introduce the characteristics of scalars, vectors and tensors we consider a classic example of structural analysis carried out by Ernst Cloos on the Cambrian and Ordovician limestones of South Mountain fold in Maryland and Pennsylvania (**Figure 2.1**). The major structure is the anticline labeled South Mountain on the geological map. This structure crops out north of the Potomac River for a few tens of kilometers along an azimuth just east of north.

On a cross section (A′–A) that is roughly perpendicular to the strike of bedding near Boonsboro, Maryland (**Figure 2.2**), the eastern limb of the anticline dips slightly, whereas to the west of Monument Knob the western limb dips steeply. Cloos identified smaller folds farther to the west, also with steep western limbs, and in some places, the sedimentary units are overturned. The sedimentary rocks originated as roughly planar and horizontal layers, so the folding is a clear indication of deformation. Small round growth structures called *oöids* by Cloos, and now called ooliths, provide a beautiful record of deformation in these sedimentary

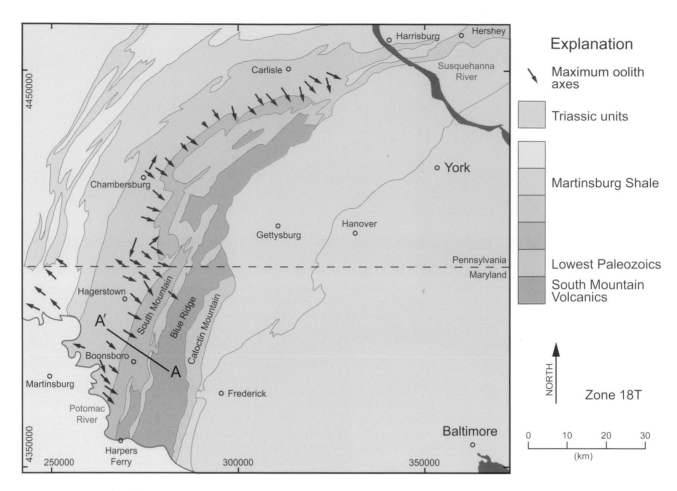

Figure 2.1 Map of geology and structures west of Baltimore, Maryland, and between the Potomac and Susquehanna Rivers. South Mountain anticline trends just east of north. Line A′–A marks cross section shown in Figure 2.2. Small arrows point in the direction of the long axes of deformed ooliths. Modified from Cloos (1947).

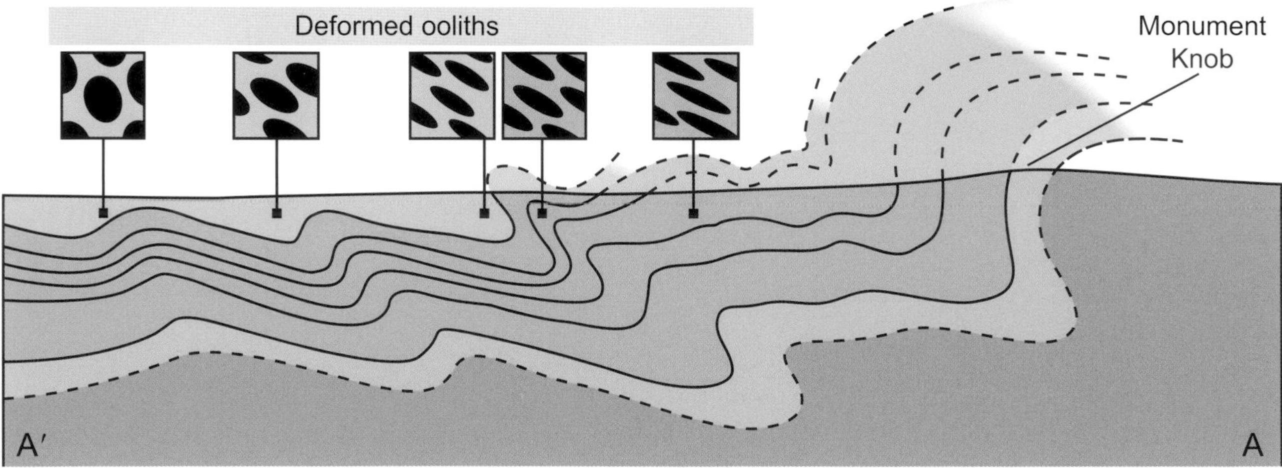

Figure 2.2 Geologic cross section of South Mountain fold. See line A′–A on Figure 2.1 for location. Black ellipses, greatly exaggerated in size, represent the average shapes of deformed ooliths at particular outcrops. Modified from Cloos (1947).

rocks. Black ellipses above the cross section schematically illustrate the average orientation and shape of ooliths at selected locations in the train of folds (**Figure 2.2**).

Where undeformed, ooliths are roughly *spherical* (**Figure 2.3a**), and range from 0.25 to 2 mm in diameter. Ooliths usually are composed of calcium carbonate in concentric layers around a shell fragment or sand grain, and they can have a radiating fibrous pattern. They form by chemical precipitation in shallow, wave-agitated water. Where the surrounding sedimentary rock is deformed, the ooliths also deform and become approximately

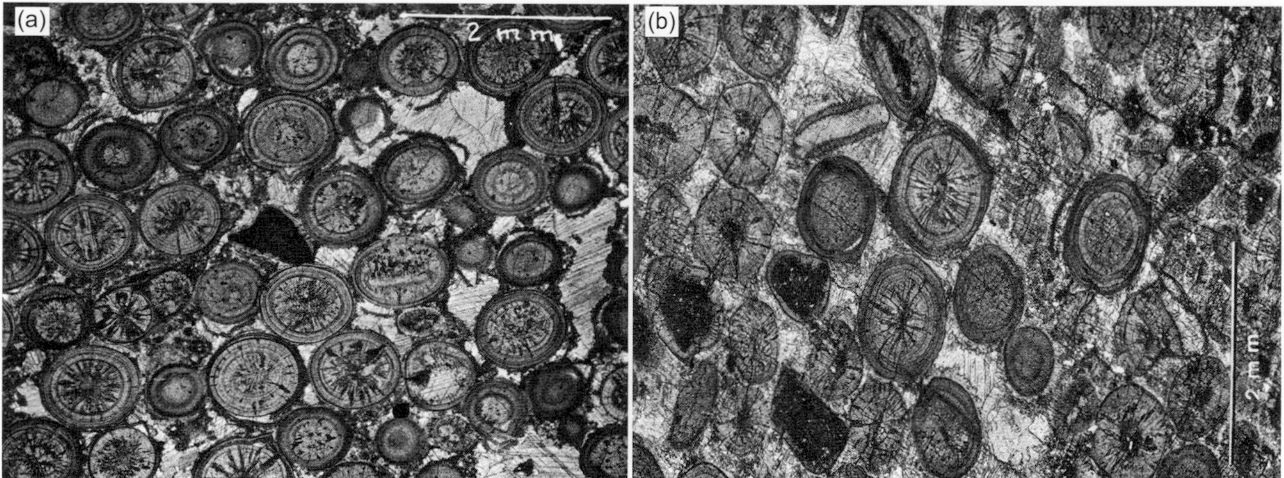

Figure 2.3 Thin section images of ooliths from South Mountain fold. 2 millimeter scale bar. (a) Nearly undeformed ooliths with mean axial ratio = 1.16. (b) Deformed ooliths with mean axial ratio = 1.56. Modified from Cloos (1971).

ellipsoidal in shape (**Figure 2.3b**). The small arrows drawn on the geologic map (**Figure 2.1**) give the average trend of the long axes of deformed ooliths at particular outcrops. Most of these arrows point to the southeast, approximately perpendicular to the axis of the South Mountain fold.

In the following sections we use the ooliths at South Mountain fold to describe the key characteristics of scalar, vector, and tensor quantities. We also illustrate some of the useful ways to visualize these quantities.

2.1.1 Scalars

The mass density, ρ, of an oolith in the South Mountain fold (**Figure 2.2**) is an example of a scalar quantity. Because ooliths are composed primarily of calcium carbonate, their approximate density is:

$$\rho = 2.7 \times 10^3 \,\mathrm{kg\,m^{-3}}.$$

The *magnitude*, given by a single number, serves to quantify a scalar. Most scalars also are associated with physical units, in this case kilograms per meter cubed, which come from the standardized SI system of units that we describe in Section 3.1. Because negative mass density has no physical significance, the number that quantifies mass density does not include a sign, but it is always non-negative. The scalar we call length also is restricted to non-negative quantities, so it carries no sign. On the other hand, elevation is a scalar that is defined with a *magnitude* and a *sign*, because elevation may be either positive or negative, indicating distance above or below a *datum*. The datum can be mean sea level, or the datum for the standard ellipsoidal model for Earth, or some other specified datum.

The spatial variations of a scalar quantity in two dimensions are visualized using a graph of a surface (**Figure 2.4a**) that represents those variations. In this example the scalar, z, is a function of the two variables, $z = f(x, y)$. For each choice of a coordinate pair (x, y), one value of z exists, so any line constructed perpendicular to the (x, y)-plane intersects the surface at just one point. The three-dimensional surface corresponding to the function can be quite complicated, but any plane, $z = c$ (constant), intersects the surface along a curve and the projection of that curve onto the plane $z = 0$ is called a level curve or contour. The value of the function is a constant at all points on a particular contour. The family of contours, usually separated by a fixed contour interval, makes up a contour map (**Figure 2.4b**). At the point ($x = 0.5$, $y = -1.3$) the three-dimensional surface has a value −4.77, so this plots on the contour map as the red dot between the −4 and −6 contours.

Geologists may use topographic maps with elevation contours to locate and map outcrops. In this case, elevation is a function of the two horizontal geographic coordinates, Easting and Northing. Structural geologists use contour maps for many different purposes. In Chapter 3 we use color-filled contour maps (e.g. **Figure 3.12**) to illustrate how displacement in each coordinate direction varies near a lattice defect in a crystal at the nanometer scale. These defects play an important role in the plastic deformation of crystals. Displacement is a vector quantity, but the vector components in each coordinate direction are scalars, which may be contoured. In another example (**Figure 3.14**), a component of displacement at Earth's surface due to slip on the Hector Mine fault during a big earthquake is contoured at the 10 km scale, and the contour patterns are compared to those from a theoretical model of the displacements. We use contour maps to visualize the variations of scalar quantities in two dimensions.

2.1.2 Vectors

A vector may be quantified *geometrically* (**Figure 2.5a**) as an arrow with a *tail* (the unadorned end) and a *head* (the pointed end). The distance from the tail to the head represents the *magnitude* of the vector, and the arrowhead points in the direction of the vector. We use boldface letters, for example **v**, to name vectors. Mathematical vectors are completely characterized by magnitude and direction and not, for example, by a scale, a location, or dimensions other than length. Vectors are mathematically equivalent if they have the same magnitude and direction (**Figure 2.5b**).

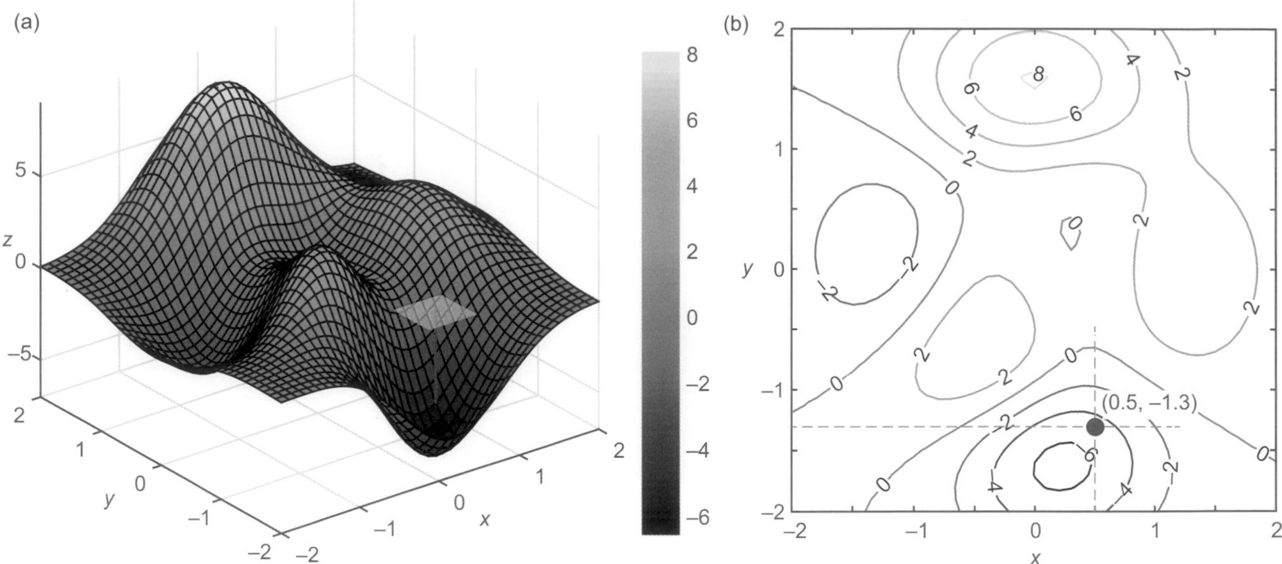

Figure 2.4 Scalar function of two variables, $z = f(x, y)$. (a) Three-dimensional image of a surface representing the function with a wire-frame mesh and color. Color bar indicates value of z. (b) Contour map of surface. Red dot indicates point identified by vertical red line in (a) where $f(0.5, -1.30) = -4.77$. See Ragan (2009) in Further Reading.

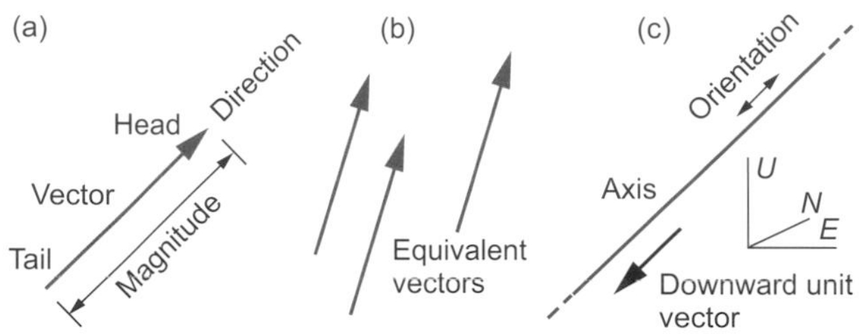

Figure 2.5 (a) Geometrical representation of a vector as a red arrow with head and tail. (b) Equivalent vectors: same magnitude (length) and same direction. (c) Geometric representation of an axis as a blue line with a particular orientation. Unit vector (black) adopts the geologic convention that represents the axis as a vector with a direction that is horizontal or downward. See Varberg et al. (2006) in Further Reading.

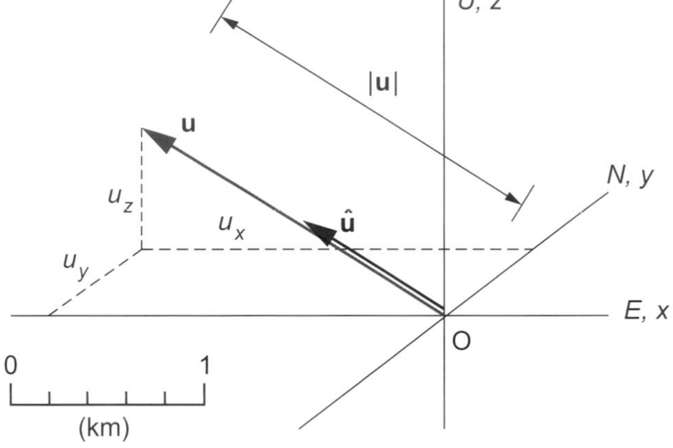

Figure 2.6 Schematic illustration of displacement vector **u** (blue arrow) extending from the initial location at the origin of coordinates to the current location of the oolith at the western edge of the cross section in Figure 2.2. Also shown are the vector magnitude |**u**| and a unit vector $\hat{\mathbf{u}}$ (black arrow, shifted up slightly) in the same direction as **u**.

A vector is different mathematically from an axis (**Figure 2.5c**), because the vector has both a *length* and a particular *direction*. The axis, in contrast, is a line with a particular *orientation*, but the two possible directions along that line are not distinguished. Lineations are geological structures, which, like an axis, have a measurable orientation, but no direction. For example, metamorphic and igneous rocks commonly contain alignments of tabular mineral grains that form a rock fabric, which is the subject of Chapter 10. Because one of two oppositely directed vectors of unit magnitude may define the orientation of a lineation, one must adopt a convention that makes the choice. We adopt the geologic convention (**Figure 2.5c**) that represents the orientation of a lineation as a downward (or horizontally) directed unit vector in the geographic coordinates named East, North, and Up (*E*, *N*, *U*).

The displacement, **u**, is a vector representing the motion of a particular oolith from its original location in the undeformed sedimentary basin to its current location in the folded strata of South Mountain. As an example, consider an oolith represented by the black ellipse at the western (left) edge of the cross section in **Figure 2.2**. We don't know precisely where it was before the folding, but we estimate it was a couple of kilometers to the southeast of the current location and perhaps half a kilometer deeper. To illustrate the estimated displacement vector we adopt a coordinate system in **Figure 2.6** and use a vector arrow to visualize the displacement. The tail of this arrow is at the origin, the presumed original location of the oolith.

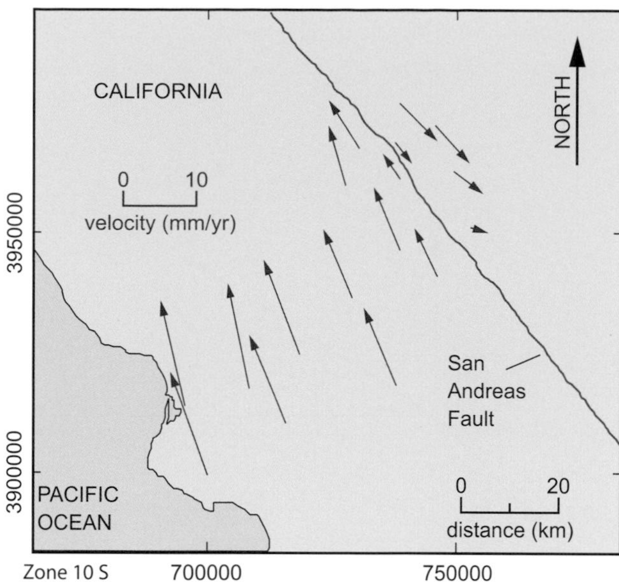

Figure 2.7 Velocity vector field (purple arrows) for plate motion along San Andreas Fault using geodetic data from 1966 to 1984. Note the different velocity vector scale and geographic distance scale. Modified from Harris and Segall (1987).

In **Figure 2.6** the Cartesian axes (x, y, z) are chosen to be a right-handed system, so you can visualize the axes by pointing your thumb, first finger, and middle finger of your right hand in orthogonal directions. Also, z is vertical and (x, y) lie in the horizontal plane. Note that the geographic coordinates East, North, and Up (E, N, U) are associated, respectively, with the Cartesian axes (x, y, z). We adopt these conventions for the textbook, with a few exceptions.

The vector magnitude and vector direction quantify a vector. For example, the displacement vector magnitude (length of the vector arrow) is $|\mathbf{u}|$, and the direction is the unit vector, $\hat{\mathbf{u}}$, in the direction of the displacement vector (**Figure 2.6**):

$$|\boldsymbol{u}| = 2.19 \times 10^3\,\mathrm{m}, \hat{\mathbf{u}} = -\langle 0.912, 0.342, 0.228\rangle$$

The vector magnitude is non-negative by definition, and it is associated with the appropriate physical units, in this case meters. The *angular brackets* indicate the three numbers within the brackets are the vector components. These components are the projections of the vector, $\hat{\mathbf{u}}$, onto the x-, y-, and z-axes, respectively. The signs of these components indicate whether the projected length lies along the positive or negative axis of the respective coordinate.

The geometric representation of vectors as arrows is useful for plots of physical vector fields (**Figure 2.7**). As distinguished from mathematical vectors (**Figure 2.5b**), physical vectors have dimensions, units, a location, and a physical interpretation. A scale is required to indicate the actual magnitude of physical vectors, either because the length of the arrow is not at true scale, or because the vector does not have dimensions of length. Most physical vectors used in structural geology have particular positions in space and/or particular moments in time. For example, the velocity vectors for tectonic plate motion in central California (**Figure 2.7**) are functions of their geographic positions and the period 1966 to 1984.

2.1.3 Tensors

The geometric shapes of the undeformed and deformed ooliths at South Mountain fold (**Figure 2.3**) provide data that is necessary to quantify the deformation using a tensor. To describe the relevant tensor quantity we first introduce the stretch of material lines that radiate from the center of an oolith (**Figure 2.3b**) to its perimeter. For the *undeformed* oolith, the length, L_u, of these lines is the same in all radial directions, because the shape approximates a sphere. For the *deformed* oolith, a given material line has a length, L_d, and the stretch of that line is the ratio of deformed to undeformed lengths: L_d/L_u. If the stretch is greater than one, the material line lengthened; if the stretch is one, the line did not change length; and if the stretch is less than one, and greater than zero, the line shortened. The stretch does not carry physical units, because it is a ratio of lengths.

Because the deformed oolith approximates an *ellipsoid*, the stretch varies with orientation of the material line and the deformation gradient tensor quantifies the stretch in all possible orientations for a given oolith. We define this tensor mathematically in Section 5.5.1, but here we describe the deformation in enough detail to introduce some basic concepts of a tensor. First, we adopt a Cartesian coordinate system with a vertical z-axis and a y-axis that is approximately parallel to the trend of the folds at South Mountain (**Figure 2.1**). For these coordinates the stretch in the y-direction is zero, so we view the deformed oolith in the (x, z)-plane (**Figure 2.8**). In this plane the shape of the undeformed oolith is a circle, and the shape of the deformed oolith is an ellipse. The long axis of the ellipse plunges downward to the right, which is toward the southeast as seen on the cross section of South Mountain anticline (**Figure 2.2**). We choose the radius of the undeformed oolith to be $L_u = 1$ mm, and assume the deformation is homogeneous on the scale of the oolith. In other words, the deformation is identical everywhere near the oolith.

An ellipsoid may have three unique axes that have the *greatest*, *intermediate*, and *least* axial lengths, and these axes are *orthogonal*. In this case, the distribution of stretch within the oolith has a greatest, intermediate, and least magnitude. These are the principal values of stretch, symbolized as S_1, S_2, and S_3. The principal directions of stretch are orthogonal and they are defined by three unit vectors: $\hat{\mathbf{u}}_1, \hat{\mathbf{u}}_2$, and $\hat{\mathbf{u}}_3$. These vectors are parallel to the ellipsoidal axes, but their directions along those axes are arbitrary. For the oolith depicted in **Figure 2.8** the principal stretches and their directions are:

$$S_1 = 1.3, \hat{\mathbf{u}}_1 = \langle 0.8944, 0 - 0.4472\rangle$$

$$S_2 = 1.0, \hat{\mathbf{u}}_2 = \langle 0, 1, 0\rangle$$

$$S_3 = 0.8, \hat{\mathbf{u}}_3 = \langle -0.44720, -0.8944\rangle$$

The *angular brackets* indicate the three numbers within the brackets are vector components. Note that S_2 has a magnitude of one, indicating no stretch. Also, the only non-zero component

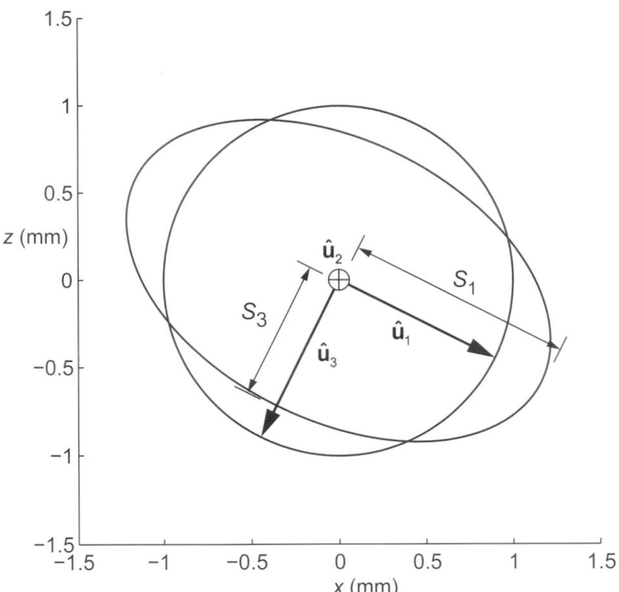

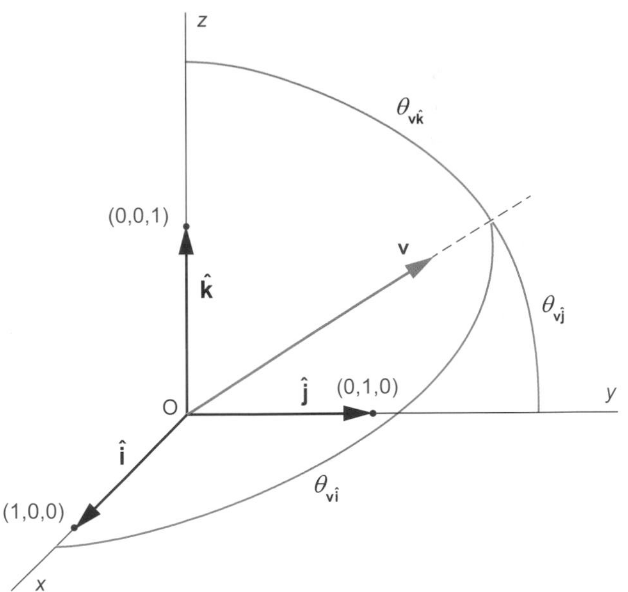

Figure 2.8 Cross section of undeformed oolith (blue circle) and deformed oolith (red ellipse) viewed in the (x, z)-plane. See Figure 2.3 for images of ooliths. Radius of the undeformed oolith is 1 mm. Magnitudes of principal stretches are S_1, $S_2 = 1$, and S_3; unit principal direction vectors are $\hat{\mathbf{u}}_1, \hat{\mathbf{u}}_2$, and $\hat{\mathbf{u}}_3$. We assume the deformation is homogeneous at the millimeter scale.

Figure 2.9 Right-handed Cartesian reference frame (x, y, z), unit base vectors $\left(\hat{\mathbf{i}}, \hat{\mathbf{j}}, \text{and } \hat{\mathbf{k}}\right)$, and direction angles $\left(\theta_{v\hat{i}}, \theta_{v\hat{j}}, \theta_{v\hat{k}}\right)$ between the arbitrary vector, **v**, and the base vectors. See Varberg et al. (2006) in Further Reading.

of $\hat{\mathbf{u}}_2$ is the second component, which has a magnitude of one, indicating this vector is directed along the positive *y*-axis. The second components of $\hat{\mathbf{u}}_1$ and $\hat{\mathbf{u}}_3$ both are zero, indicating these principal directions are in the (*x*, *z*)-plane (**Figure 2.8**). The principal stretch magnitudes are *independent* of the choice of coordinate system. The principal stretch directions also do not vary with the choice of coordinate system, but the components of the unit vectors do depend upon this choice.

Most scalars, vectors, and tensors used in structural geology are *field quantities*, meaning they can vary with *position* and with *time*. For example, the distribution of mass density defines a scalar field for each thin section shown in **Figure 2.3**. The distribution of velocity defines a vector field for the geodetic data near the San Andreas Fault (**Figure 2.7**). The spatial variation in the oolith shapes at different positions on the cross section of the South Mountain fold (**Figure 2.2**) defines a tensor field. The ooliths at these sites were approximately spherical before the folding and faulting, so the stretch also varied in time. Variations such as these motivate structural geologists to document and analyze the spatial and temporal changes of all the relevant scalar, vector, and tensor quantities used to understand tectonic processes and their structural products.

2.2 ALGEBRAIC REPRESENTATION OF VECTORS

Here we describe the properties, mathematical representations, and elementary mathematical manipulations of vectors that find applications in structural geology. The algebraic representation of a vector facilitates a host of useful operations (e.g. scalar and vector products) and computations; it admits vectors that can be moved and those that are fixed in a reference frame; and it admits vector functions, which are necessary for the quantitative description of geologic structures such as folds using differential geometry.

2.2.1 Vectors and their Components

A vector is thought of as a line segment with an *initial point* at the tail and a *terminal point* at the head. These points are located in right-handed Cartesian coordinates in which the coordinate axes (*x*, *y*, *z*) are mutually perpendicular and have an origin, *O*. Because we associate with each coordinate axis the range of numbers from $-\infty$ to $+\infty$, and we place zero at the origin, these axes have a sense of direction from negative toward positive. We also distinguish the positive half of a coordinate axis from the negative half.

Vectors are denoted by upright boldface letters: **v**, **p**, **f**, ... Vectors also are denoted by an ordered pair of italic capital letters with an over-arrow: $\overrightarrow{OP}$, $\overrightarrow{PQ}$, $\overrightarrow{AB}$, ... For example, the line segment from the initial point *P* to the terminal point *Q* is the vector $\overrightarrow{PQ}$. Vectors with a magnitude equal to one are denoted by a caret (little hat) over the boldface letter: $\hat{\mathbf{u}}, \hat{\mathbf{n}}, \hat{\mathbf{v}}$, ... These are unit vectors, and they provide a unique *direction*.

Three special unit vectors, $\hat{\mathbf{i}}, \hat{\mathbf{j}}$, and $\hat{\mathbf{k}}$, are defined for a Cartesian coordinate system (**Figure 2.9**), all with tails fixed at the origin, and each with its head at a unit distance along one of the three positive coordinate axes. Thus, the components of these vectors are (1, 0, 0), (0, 1, 0), and (0, 0, 1), respectively. These

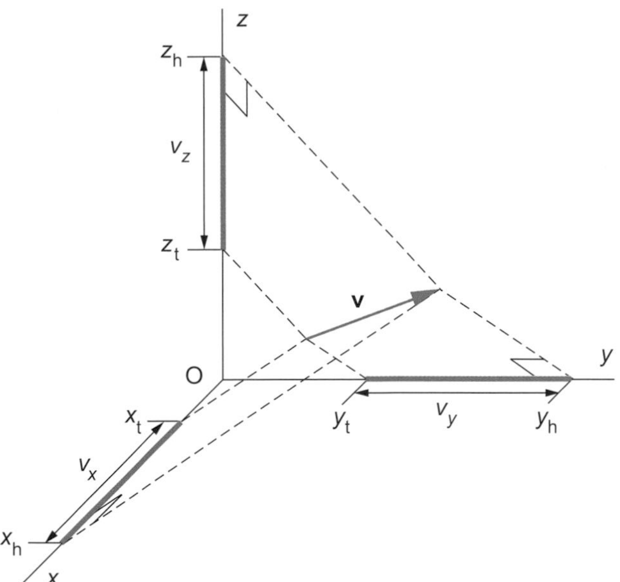

Figure 2.10 Projections of the arbitrary vector, **v**, onto the Cartesian coordinate axes are the rectangular components, (v_x, v_y, v_z). See Varberg et al. (2006) in Further Reading.

so-called base vectors are used to define any other vector in the Cartesian system. Base vectors are dimensionless and unitless quantities. The base vectors are fixed vectors because their tails always are at the origin.

The direction of an arbitrary vector, **v**, is determined by the direction angles, which are most easily visualized if the origin is translated to the tail of vector, so this tail and those of the unit base vectors share the same point (**Figure 2.9**). A direction angle is measured in the plane defined by the vector and a base vector, and it is the smaller of the two angles between the vector and that base vector. Thus, the range of a direction angle is $0° \leq \theta \leq 180°$. Direction angles in three dimensions are unsigned, in contrast to angles measured in two-dimensional coordinate planes that may be assigned a positive (e.g. counter-clockwise) or negative (e.g. clockwise) sign by convention. We use the Greek theta, θ, for direction angles with two subscripts that refer to the two vectors that define the angle. For example, $\theta_{\mathbf{v}\hat{\mathbf{j}}}$ is the direction angle between the vector **v** and the unit base vector $\hat{\mathbf{j}}$.

Suppose the tail of an arbitrary non-zero vector, **v**, is at a point with coordinates (x_t, y_t, z_t) and the head is at a point (x_h, y_h, z_h). All vectors, except the zero vector, have a direction from tail to head. The projected magnitudes and directions of the vector **v** onto the coordinate axes (**Figure 2.10**) are: $(x_h - x_t)$, $(y_h - y_t)$, $(z_h - z_t)$. If the quantity in parentheses is positive, the projected direction is in the respective positive coordinate direction. For example, in **Figure 2.9** the vector **v** is directed such that all three quantities are positive. The projections of the arbitrary vector, **v**, onto the coordinate axes are the vector components. They are symbolized using the non-bold letter chosen for the vector itself and a single subscript that refers to the respective coordinate axis:

$$v_x = (x_h - x_t),\ v_y = (y_h - y_t),\ v_z = (z_h - z_t) \quad (2.1)$$

The rectangular components are signed quantities, and the sign indicates whether the projected vector is directed in the positive or negative coordinate direction.

The vector, **v**, may be written algebraically in either of two forms:

$$\mathbf{v} = v_x\hat{\mathbf{i}} + v_y\hat{\mathbf{j}} + v_z\hat{\mathbf{k}} \text{ or } \mathbf{v} = \langle v_x, v_y, v_z \rangle \quad (2.2)$$

The first form is a vector equation utilizing the components to scale the base vectors. In the second form one reads $=\langle\rangle$ as: "has the components" and it is understood that these components would be associated with the respective base vectors in the vector equation. Note that the components carry the dimensions and units of physical vectors, because the base vectors are dimensionless and unitless quantities.

The vector magnitude of an arbitrary vector, **v**, is calculated in terms of its components as:

$$|\mathbf{v}| = \sqrt{v_x^2 + v_y^2 + v_z^2} \quad (2.3)$$

The direction cosine is the cosine of a direction angle, and it is defined as the ratio of the vector component to the vector magnitude:

$$\cos\theta_{\mathbf{v}\hat{\mathbf{i}}} = v_x/|\mathbf{v}|,\ \cos\theta_{\mathbf{v}\hat{\mathbf{j}}} = v_y/|\mathbf{v}|,\ \cos\theta_{\mathbf{v}\hat{\mathbf{k}}} = v_z/|\mathbf{v}| \quad (2.4)$$

The range of the direction cosines is $-1 \leq \cos\theta \leq +1$. Values outside this range would be erroneous. Given the vector magnitude and the direction in terms of the direction cosines, the vector components are:

$$v_x = |\mathbf{v}|\cos\theta_{\mathbf{v}\hat{\mathbf{i}}},\ v_y = |\mathbf{v}|\cos\theta_{\mathbf{v}\hat{\mathbf{j}}},\ v_z = |\mathbf{v}|\cos\theta_{\mathbf{v}\hat{\mathbf{k}}} \quad (2.5)$$

Given the range of the direction cosines, the rectangular components may be positive or negative and their absolute value is less than or equal to that of the vector magnitude. Again, values outside this range would be erroneous.

2.2.2 Scalar Product of Two Vectors

The multiplication of vectors may be accomplished in different ways and each of these has a distinct name and definition. The scalar product (or *dot product*) of two arbitrary vectors, **v** and **w**, is defined as the sum of the products of the respective components:

$$\mathbf{v} \cdot \mathbf{w} = v_x w_x + v_y w_y + v_z w_z \quad (2.6)$$

As the name implies, the result of this operation is a *scalar*, not a vector. Note that the order of multiplication does not change the result: **v**·**w** = **w**·**v**.

While (2.6) serves to calculate the scalar product, an equivalent definition provides more geometric insight (**Figure 2.11**):

$$\mathbf{v} \cdot \mathbf{w} = |\mathbf{v}||\mathbf{w}|\cos\theta_{\mathbf{vw}},\ 0 \leq \theta_{\mathbf{vw}} \leq \pi \quad (2.7)$$

Note that $\cos\theta_{\mathbf{vw}}$ is the direction cosine for the direction angle, which is the smaller angle between these vectors in the plane that contains them. You can slide **w** so its tail coincides with the tail of **v** to define this plane. The product $|\mathbf{v}|\cos\theta_{\mathbf{vw}}$ is the orthogonal projection of **v** onto the line parallel to **w**. In other words, $|\mathbf{v}|\cos\theta_{\mathbf{vw}}$ is the component of **v** in the direction of **w**. Therefore, the geometric interpretation of (2.7) is the component of **v** in the direction of **w** multiplied by the magnitude of **w**. The scalar product (2.7) also may be interpreted as the product of the component of **w** in the direction of **v** multiplied by the magnitude of **v**.

Suppose you want to know how much the sedimentary beds at Comb Ridge monocline (**Figure 2.12**) have rotated from their original horizontal orientation. You have survey data that gives the current orientation of a particular bedding surface as its unit normal vector $\hat{\mathbf{n}}$. The rotation is the direction angle, $\theta_{\hat{\mathbf{n}}\hat{\mathbf{k}}}$, between the unit base vector $\hat{\mathbf{k}}$ for the vertical axis and the unit normal vector. Because the bedding was originally horizontal, the normal vector at that time was vertical and parallel to $\hat{\mathbf{k}}$. From (2.7) the scalar product $\hat{\mathbf{n}}\cdot\hat{\mathbf{k}} = |\hat{\mathbf{n}}||\hat{\mathbf{k}}|\cos\theta_{\hat{\mathbf{n}}\hat{\mathbf{k}}} = \cos\theta_{\hat{\mathbf{n}}\hat{\mathbf{k}}}$. From (2.5) $\cos\theta_{\hat{\mathbf{n}}\hat{\mathbf{k}}} = n_z$, that is the z-component of the unit normal vector. Solving for the direction angle, which is the rotation angle, $\theta_{\hat{\mathbf{n}}\hat{\mathbf{k}}} = \cos^{-1} n_z$. For this bedding surface at Comb Ridge, $\theta_{\hat{\mathbf{n}}\hat{\mathbf{k}}} \approx 35^\circ$. This angle also is the *dip* of the bedding, as measured at the outcrop with an inclinometer.

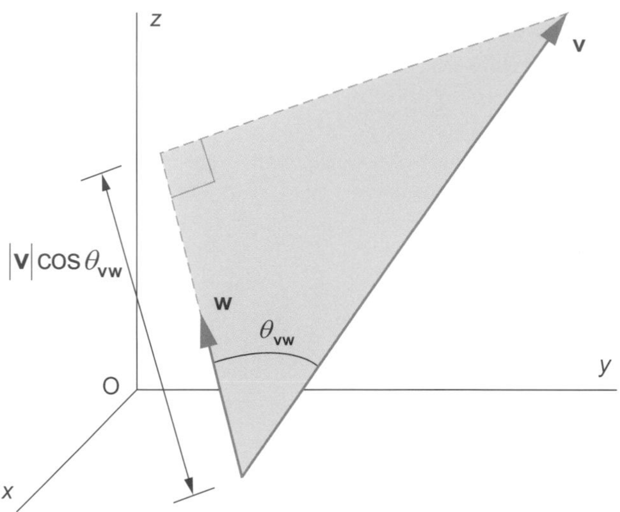

Figure 2.11 Geometric interpretation of the scalar product of the arbitrary vectors **v** and **w**. The direction angle between **v** and **w** is θ_{vw}. See Varberg et al. (2006) in Further Reading.

2.2.3 Vector Product of Two Vectors

The vector product (or *cross product*) of two vectors yields another vector. This product has many uses in structural geology, one of which is determining the orientation of layers in folded sedimentary strata. To characterize these orientations throughout a fold, structural geologists use the unit normal vector to the surface. An example of a large fold in sedimentary rock is shown in the aerial photograph of Raplee anticline (**Figure 2.13**). The white normal vectors on the sedimentary surfaces illustrate the change in orientation of these surfaces from the crest of the fold (upper right) down the limb of the anticline to the San Juan River. The two vectors, **v** and **w**, are parallel to the top of a particular sedimentary bed. We show in a later section how these two vectors are measured using data taken with a compass and inclinometer at the outcrop. After describing what a vector product is, we use those two vectors to calculate the unit normal vector, $\hat{\mathbf{n}}$.

Figure 2.12 Comb Ridge monocline near Bluff, Utah. The unit normal to bedding is $\hat{\mathbf{n}}$ and the base vector for the z-axis is $\hat{\mathbf{k}}$. These are not drawn to scale. The rotation angle for the sedimentary bedding, is found using $\hat{\mathbf{n}}\cdot\hat{\mathbf{k}}$ to be $\theta_{\hat{\mathbf{n}}\hat{\mathbf{k}}} \approx 35^\circ$ based on the scalar product (2.7). Google Earth file: Figure 2.12 Comb Ridge UT monocline.kmz. UTM 12 S 617668.32 m E, 4124689.42 m N.

Figure 2.13 Raplee anticline on the San Juan River near Mexican Hat, UT. Unit normal vectors **n̂** on various sedimentary surfaces change orientation over the fold. Vectors **v** and **w** are parallel to a particular bedding surface and are used to define the normal vector to that surface using the vector product (2.10). Photograph by I. Mynatt. Google Earth file: Figure 2.13 Raplee Ridge UT anticline. kmz. UTM: 12 S 604042.03 m E, 4113941.08 m N.

The vector product of two arbitrary vectors, **v** and **w**, is defined using the components:

$$\mathbf{v} \times \mathbf{w} = \left(v_y w_z - v_z w_y\right)\hat{\mathbf{i}} + \left(v_z w_x - v_x w_z\right)\hat{\mathbf{j}} + \left(v_x w_y - v_y w_x\right)\hat{\mathbf{k}} \quad (2.8)$$

The presence of the base vectors on the right-hand side of (2.8) indicates the result of this operation is another vector. The direction of the vector $\mathbf{v} \times \mathbf{w}$ is normal to the plane containing **v** and **w**, and points in the direction of the thumb on your right hand when your fingers curl *from* **v** *to* **w** (**Figure 2.14**). You can slide **w** so its tail coincides with the tail of **v** to employ this right-hand convention. Note that the product $\mathbf{w} \times \mathbf{v}$ produces a vector in the opposite direction, so $\mathbf{v} \times \mathbf{w} = -(\mathbf{w} \times \mathbf{v})$. In other words, the vector product depends on the order of the two vectors.

In **Figure 2.14** $\theta_{\mathbf{vw}}$ is the smaller angle between the vectors in the plane that contains them. The order of the subscripts uses the convention "from **v** to **w**" to emphasize the order of multiplication. To gain geometric insight, the magnitude of the vector product is:

$$|\mathbf{v} \times \mathbf{w}| = |\mathbf{v}||\mathbf{w}| \sin\theta_{\mathbf{vw}},\ 0 \le \theta_{\mathbf{vw}} \le \pi \quad (2.9)$$

Given the range of $\theta_{\mathbf{vw}}$ from 0° to 180°, the sine of the direction angle in (2.9) ranges from 0 to +1, so the magnitude of the vector product is non-negative and ranges from zero to the product of the magnitudes of the two vectors. If the vectors **v** and **w** are parallel, the vector product is zero; if they are perpendicular, the vector product takes on its greatest magnitude.

Now we have the tools to calculate the unit normal vector, **n̂**, to the sedimentary surface in **Figure 2.13**. The direction of **n̂** is the same as the direction of the vector product, $\mathbf{v} \times \mathbf{w}$, and we divide by the magnitude of this vector product to find the unit normal vector:

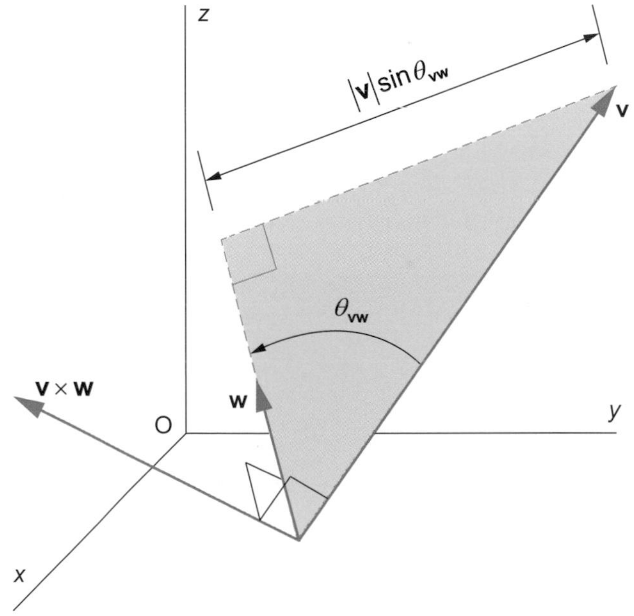

Figure 2.14 Geometric interpretation of the vector product of the arbitrary vectors **v** and **w**. The direction angle between **v** and **w** is $\theta_{\mathbf{vw}}$ and the direction of $\mathbf{v} \times \mathbf{w}$ is determined using the right-hand rule (see text). See Varberg et al. (2006) in Further Reading.

$$\hat{\mathbf{n}} = \frac{\mathbf{v} \times \mathbf{w}}{|\mathbf{v} \times \mathbf{w}|} \quad (2.10)$$

In the next section we introduce some of the many vector tools that aid the structural geologist when mapping in the field, and in analyzing field data.

2.3 MAPPING GEOLOGIC STRUCTURES WITH VECTORS

The position vector might just as well be called the mapping vector by structural geologists, because it is used to locate points with special significance on a geologic map. Once a point has been located with respect to the chosen coordinate system, a geologist might measure the orientation of structures that crop out there. Unit vectors provide the means to quantify these orientations. For example, the unit tangent vector defines the orientation of a lineation and the unit normal vector defines the orientation of a metamorphic foliation or sedimentary bedding surface. These and many other examples demonstrate the utility of vectors in structural mapping.

2.3.1 Position Vectors for Mapping

A position vector, **p**, is distinguished from an arbitrary vector because it is a fixed vector with a tail always at the origin (**Figure 2.15**). In other words, the tails of all position vectors have the coordinates: $(x_t, y_t, z_t) = (0, 0, 0)$. The *components* of the vector **p** are proportional to the vector magnitude $|\mathbf{p}|$ and to the respective direction cosine:

$$p_x = |\mathbf{p}|\cos\theta_{\mathbf{p}\hat{\mathbf{i}}}, p_y = |\mathbf{p}|\cos\theta_{\mathbf{p}\hat{\mathbf{j}}}, p_z = |\mathbf{p}|\cos\theta_{\mathbf{p}\hat{\mathbf{k}}} \quad (2.11)$$

The dimensions of the position vector are carried by the components, which always have dimensions of length. The position vector serves to define the location of a point, P, in the Cartesian reference frame because the head of the vector has coordinates (x_h, y_h, z_h), which are the three coordinates of that point, and the three components of the position vector.

Figure 2.15 Structural mapping at Sheep Mountain anticline, Wyoming using local coordinates (*x, y, z*) and position vectors. The position vector **p** is directed to an outcrop, *P*(*x, y, z*) in the Madison Limestone on the bank of the Big Horn River where a bed-parallel fault and slickenlines are exposed (**Figure 2.16**). Photograph by N. Bellahsen. Google Earth file: Figure 2.15 Sheep Mountain WY anticline.kmz. UTM: 12 T 728000.08 m E, 4944412.97 m N. See Bellahsen et al. (2006).

While adequate for detailed mapping and model construction, local coordinate systems do not place the geological structures in a global context. To achieve this global context we use the Global Positioning System (GPS) to determine locations on Earth's surface (see **Box 2.1**). Most GPS receivers have onboard computers that report locations using UTM, the Universal Transverse Mercator Projection that transforms points on Earth's surface to a plane (see **Box 1.1**). Most modern maps use UTM coordinates, and most cars, phones, and cameras now have inexpensive GPS receivers. For reporting structural data with field locations, the local coordinates (x, y, z) are transformed to UTM coordinates (Easting, Northing, and Elevation). We symbolize these geographic coordinates as (E, N, U), where U is up and directed opposite to the gravitational acceleration vector, **g**.

2.3.2 Unit Vectors for Orientation Data

When working in the field, geologists define the orientation of planar and linear structural elements relative to the geographic coordinate system (E, N, U). Planar elements approximate locally the orientation of structures such as bedding surfaces, foliations, faults (**Figure 2.16**), fractures, and dikes. Linear elements approximate locally the orientation of structures such as slickenlines (**Figure 2.16**), lineations, and the intersections of two planar elements. We say "approximate locally" because natural structures generally are rough and curved. However, at the point of measurement, the planar element is *tangent* to the curved surface and the linear element is *tangent* to the curve.

The orientation of a planar structural element is defined using a unit vector that is normal to the plane, such as the unit normal vector $\hat{\mathbf{n}}$ in **Figure 2.16**. The orientation of a linear structural element is defined using a unit vector that is parallel to the line, such as $\hat{\mathbf{u}}$ in **Figure 2.16**. The structural geologist measures the orientations of these elements at an outcrop using a compass and inclinometer, which provide angles that can be related to these unit vectors. We begin by describing these angles, and then in Section 2.3.4 show how to transform them to the components of the unit vectors.

Consider an inclined planar element that approximates the fault surface in **Figure 2.16**. Looking down on an imaginary horizontal plane, the azimuth (compass bearing) is the clockwise angle measured from North to the intersection of the horizontal and inclined planes. This angle is symbolized using the Greek alpha, α (**Figure 2.17a**), and it is measured in the field using a compass. The intersection of the horizontal plane and the planar element defines a horizontal line, referred to as the line of strike. The compass bearing of this line could be measured in either of two opposite directions, so we use the right-hand convention to specify the strike direction. If you look along the line of strike such that the planar element slopes down to your right, you are looking in the strike direction. We use a subscript on the Greek alpha to indicate the particular geologic direction, for example α_s for *strike* direction. The azimuth of strike is recorded with *three* digits, including leading zeroes as necessary, and usually omitting the degree symbol: turning clockwise from 000 (North) to

Box 2.1 Global Positioning System (GPS)

The Global Positioning System (GPS) provides locations on Earth's surface by measuring the travel time of radio signals from a set of satellites to a receiver on the ground. The travel time measurement requires very precise atomic clocks on the satellites and sophisticated electronic techniques for synchronizing the clock in the receiver to the satellite clocks. Using these synchronized clocks, the time a radio signal was sent from a satellite, and the time that signal arrived at the receiver are recorded. The difference in those times is the travel time for the signal from the satellite. The velocity of the radio signal depends upon conditions in the atmosphere, which are used to calculate a corrected velocity. Knowing the corrected velocity, v, and the travel time, t, the distance, d, from the receiver to that satellite is computed as $d = vt$.

Computing one location requires a minimum of three satellite distances, but most systems use four or more distances. The redundant data refine the precision of the location and help to reduce the error in the location. The United States government operators of GPS provide the positions of the satellites as a function of time. The computed distances from the receiver to each satellite provide the data necessary to calculate the location of the receiver on Earth's surface. To record an elevation, GPS must refer to a datum: elevation is positive above that datum and negative below it. The datum used for GPS is the World Geodetic System datum for 1984, called WGS-84, which is a mathematical model of Earth's surface. The model is an oblate ellipsoid: a three-dimensional surface formed by rotating an ellipse about its minor axis, which is parallel to Earth's rotation axis.

The classical methods of mapping usually relied upon a base that was a topographic map or aerial photograph, or upon painstaking surveying using optical transits and graduated rods. Determining a location on a topographic map depends upon interpretations of the shapes of elevation contours, and aerial photographs lack quantitative location data. Today, locations are determined to better than ± 1 m using portable GPS receivers (**Figure B2.1**). These are an excellent choice for mapping most geological structures. Receivers that are more precise are available for studies of active crustal deformation related to earthquake and volcanic hazards (Segall, 2010, Further Reading) and for geodetic surveys (Hofmann-Wellenhof et al., 2001, Further Reading).

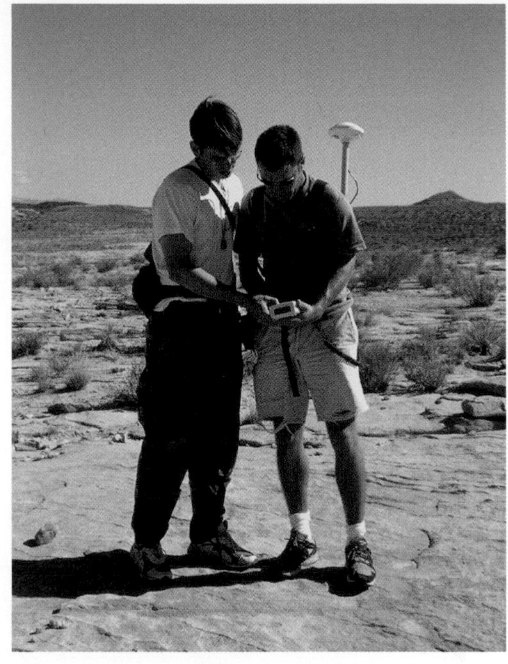

Figure B2.1 Geologists map structures at the Valley of Fire, Nevada, using GPS to determine UTM coordinates. The white GPS antenna is on a short yellow mast extending from the geologist's backpack. The yellow device in the hands of the geologists is the computer used to enter data and calculate the location of the receiver.

090 (East) to 180 (South) to 270 (West) to just less than 360 for slightly west of north.

Looking at an imaginary vertical plane that is perpendicular to the strike direction (**Figure 2.17a**), the inclination is the angle measured downward from horizontal and symbolized using the Greek phi, ϕ. The inclination is measured using an inclinometer in the field, and it is the second angle needed to uniquely specify the orientation of a planar element. This angle, known as the dip, is symbolized as ϕ_d. The dip direction is 90° clockwise from the strike direction in the horizontal plane. By convention *two* digits are used to specify the dip, including a leading zero as necessary, and usually omitting the degree symbol: a dip of 7 degrees would be written 07, and the dip ranges from 00 to 90.

Two special cases are noteworthy. For a horizontal planar element the dip is zero, so the strike is not uniquely defined, and we would write (UDF, 00) for the strike and dip. For a vertical planar element the dip is ninety degrees, and either one of two possible strike directions is suitable. For example, the strike and dip of a vertical plane with line of strike and dip recorded in the field as (136, 90), could equally well be recorded as (316, 90). To summarize, two angles are required to specify the orientation of a planar element. For example, the strike and dip of the planar element in **Figure 2.17a** are $(\alpha_s, \phi_d) = (040, 35)$. Some field instruments facilitate measuring the dip direction and dip (**Figure 2.17**), and these two angles also define the orientation of a planar element.

The angle measured from the horizontal plane to the planar element in a vertical plane that does not contain the dip direction is referred to as the apparent dip. In **Figure 2.17b**, for example, the vertical plane contains the direction E. The apparent dip is

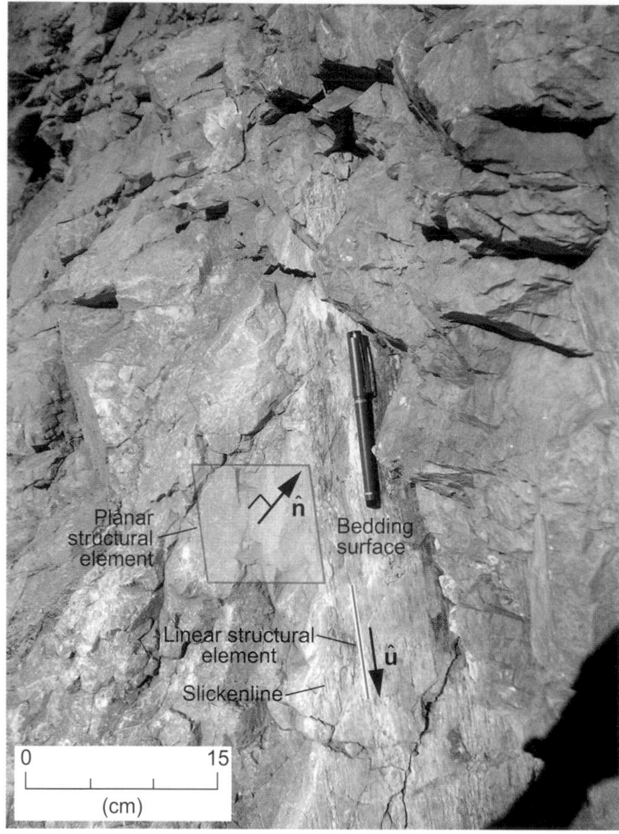

Figure 2.16 Bedding surface in Madison Limestone at Sheep Mountain, WY, that has slipped to become a fault with slickenlines (parallel to gray line). See **Figure 2.15** for location at $P(x, y, z)$. Pen lies on the fault and parallel to the slickenlines. Unit vector $\hat{\mathbf{n}}$ is normal to the planar structural element; unit vector $\hat{\mathbf{u}}$ is parallel to the linear structural element. Google Earth file: Figure 2.16 Sheep Mountain WY slickenlines.kmz. UTM: 12 T 727386.77 m E, 4943996.61 m N.

symbolized as ϕ_a, and is always less than the true dip, $\phi_a < \phi_d$. If the vertical plane contains the strike direction, the apparent dip is zero. Apparent dips are commonly observed in the field where exposures cut obliquely across geologic structures, and these must be corrected to record the true dip.

The orientations of *linear* elements are defined with respect to the geographic coordinate system (E, N, U) using two angles (**Figure 2.18**). Consider an imaginary vertical plane that contains the linear element. The trend is the azimuth (compass bearing) of this vertical plane in the direction of inclination of the linear element. The trend is symbolized as α_t, and is evaluated using three digits. The inclination of the linear element is measured in the vertical plane from horizontal down to the linear element. This angle is referred to as the plunge; it is symbolized as ϕ_p; and it is evaluated using two digits.

Two angles, the trend and plunge (α_t, ϕ_p), are necessary to define the orientation of a linear element, and these would be recorded, for example, as (356, 58). The trend and plunge of the linear element in **Figure 2.18** are (α_t, ϕ_p) = (120, 50). Two special cases are noteworthy. For a vertical linear element the plunge is ninety degrees, and the trend is undefined, so this would be recorded as (UDF, 90). For a horizontal element the plunge is zero degrees, so two possible trends exist and either one can be used. Thus, the same horizontal linear element could be recorded as (022, 00) or (202, 00).

The *ranges* of azimuths and inclinations are restricted by geologic convention:

$$0^\circ \le \alpha < 360^\circ, \ 0^\circ \le \phi \le 90^\circ \tag{2.12}$$

In other words, azimuth includes all compass directions without repetition, and negative values are avoided. The inclination is restricted to range from horizontal to vertical downward.

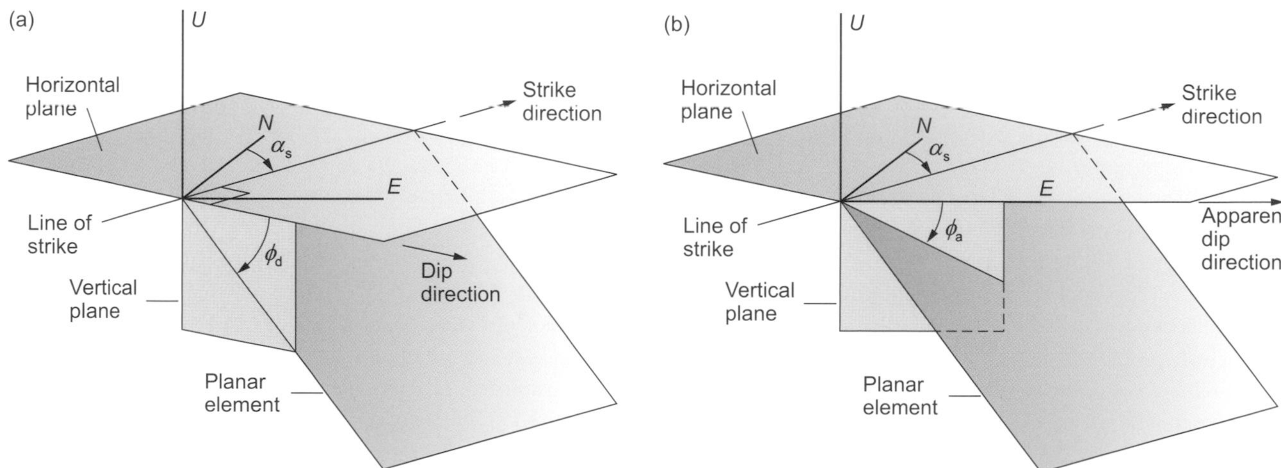

Figure 2.17 (a) Graphical explanation for the strike direction, dip direction, and dip of a planar element. Azimuth of strike is α_s and angle of dip is ϕ_d. (b) Graphical explanation for the *apparent* dip of a planar element. Angle of apparent dip is ϕ_a. See Ragan (2009) in Further Reading.

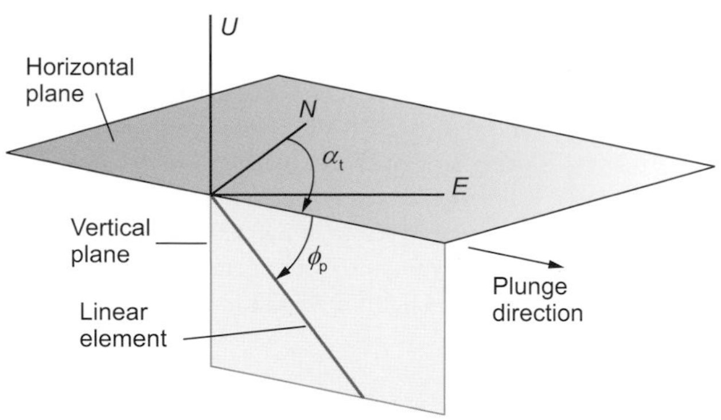

Figure 2.18 Graphical explanations for the trend and plunge of a linear element. Azimuth of trend is α_t and angle of plunge is ϕ_p. See Ragan (2009) in Further Reading.

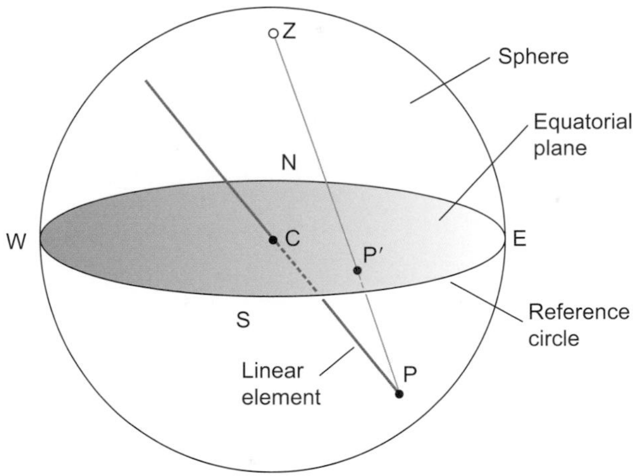

Figure 2.19 Visualization of the three-dimensional stereographic projection of a linear element through the center, C, of the sphere. Zenith, Z, of sphere is slightly rotated toward the reader. Viewed from the zenith of the sphere, the point P projects onto the equatorial plane at the point P′. See Ragan (2009) in Further Reading.

2.3.3 Stereographic Projection of Structural Elements

After gathering orientation data (azimuth and inclination) in the field, a structural geologist uses a graphical technique called stereographic projection to visualize, compare, and analyze these data. For example, one might measure the strike and dip of the top of the Madison Limestone at Sheep Mountain (**Figure 2.15**) at many locations on the fold, and then use these data to analyze the geometric form of the fold. As another example, one might measure the trend and plunge of slickenlines (**Figure 2.16**) at various exposures of bedding surface faults on this fold and use these data to analyze the direction of fault slip during folding.

Here we introduce the stereographic projection and explain how to plot linear elements. This technique also is sufficient to plot any planar structural element, because their orientation is determined by the normal line to the plane. Because each orientation measurement plotted on a stereographic projection is divorced from its geographic location, important spatial information is lost when employing this projection. None-the-less, used in combination with maps and other spatial analysis tools, the stereographic projection has an important role to play in structural geology.

The stereographic projection may be visualized using a *transparent sphere* (**Figure 2.19**). An arbitrarily oriented linear element passes through the center, C, of the sphere and intersects the lower hemisphere at point P. In this view, the sphere is tilted slightly toward the observer, so you can see the zenith at point Z. The *equatorial plane* (blue) similarly is tilted toward the observer and its perimeter, the *reference circle*, is labeled with the compass directions north (N), east (E), south (S) and west (W). Now imagine looking at P on a line of sight from the zenith. Along this line the point P projects to the point P′ (red dot) on the equatorial plane, so the trend is in the (S-E) quadrant. For horizontal linear elements the projection point would plot on the reference circle, and for vertical linear elements it would plot at the central point C. Because this particular P′ plots between the reference circle and the center we understand that the linear element is inclined. In general, the location of point P′ represents the orientation of the linear structural element: each point on the equatorial plane represents the orientation of a different linear element.

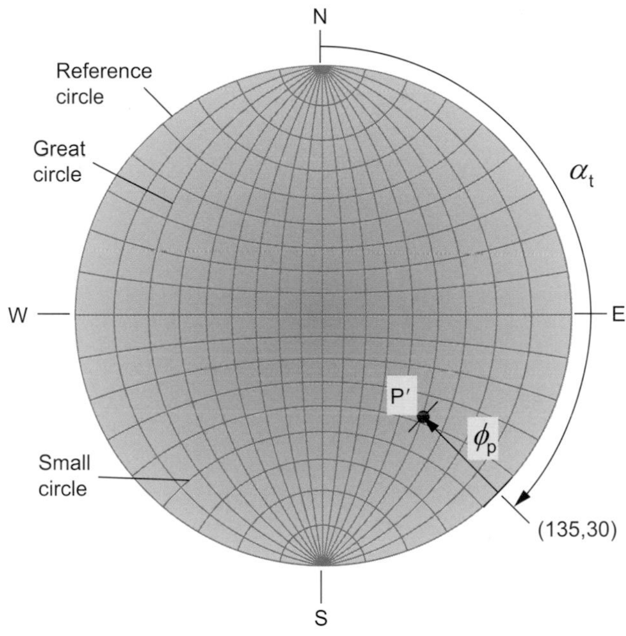

Figure 2.20 An orthogonal network of small circles and great circles composes the stereonet. The point P′ (see **Figure 2.19**, red dot) is plotted using the trend and plunge of the linear element: $(\alpha_t, \phi_p) = (135, 30)$. See Ragan (2009) in Further Reading.

To determine the exact orientation of a linear element, an orthogonal network of great circles and small circles, called the stereonet, is placed on the equatorial plane (**Figure 2.20**). The stereonet is used to plot points and measure their angular

Figure 2.21 Normal fault juxtaposes the older Navajo Sandstone and the younger and down-dropped Carmel Formation. The strike and dip of the fault surface were measured using a compass and inclinometer. The unit normal vector is $\hat{\mathbf{n}}$. Scale is approximate in foreground. Google Earth file: Figure 2.21 Chimney Rock UT normal fault.kmz. UTM: 12 S 542854.00 m E, 4342837.00 m N.

relations. Here the point P′ from **Figure 2.19** is plotted as a red dot with trend and plunge $(\alpha_t, \phi_p) = (135, 30)$. The azimuth (compass bearing) is plotted from N clockwise around the reference circle using the small circle intersections that are at intervals of ten degrees. The inclination is plotted on a straight line from the reference circle toward the center. This can be done by rotating the stereonet, so the straight line overlays the N–S line and using the small circle intersections with the N–S line to measure the inclination. Computer applications are available to do this plotting, but a structural geologist should understand how data are plotted, and how to interpret the distribution of data on a stereographic projection.

For an application of the stereographic projection to geologic data consider the Chimney Rock fault array that crops out on the northern San Rafael Swell in southeastern Utah. These faults are described in more detail in Section 8.2.1. Normal faults of this array juxtapose the Navajo Sandstone and the overlying Carmel Formation (**Figure 2.21**). In this outcrop the fault surface is exposed on the Navajo Sandstone and the down-dropped Carmel Formation is seen to the left of the fault. The strike and dip of this fault surface was measured using a compass and inclinometer, and that operation was repeated at more than 500 outcrops of faults in this array.

The strike and dip (α_s, ϕ_d) of any geological surface are converted to the trend and plunge (α_t, ϕ_p) of the normal to that surface using radians:

$$\alpha_t = \alpha_s - \frac{\pi}{2}, \phi_p = \frac{\pi}{2} - \phi_d \tag{2.13}$$

If the strike is in the range $0 \le \alpha_s < \pi/2$, then one uses $\alpha_t = \alpha_s + 3\pi/4$ to avoid negative values of trend following the geological convention (2.12).

The orientation data for the Blueberry fault (blue circles) and Glass fault (red circles) are plotted as normals to the local fault plane on the stereonet in **Figure 2.22** using (2.13). With a few exceptions, the Blueberry fault normals plot as a cluster of points in the S-E quadrant, and the Glass fault normals plot in a cluster in the N-E quadrant. The separation of the two clusters identifies these faults as members of two different fault sets. If either of these faults were perfectly planar all of their orientation data would plot at the same point on the stereonet. Some of the scatter in the points reflects measurement imprecision, but typically that is only a few degrees. Therefore, the majority of the scatter reflects actual changes in the orientation of the fault surface from

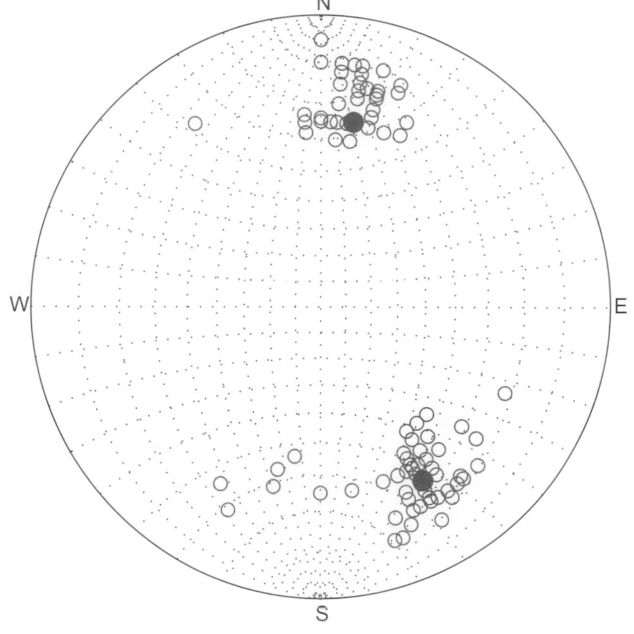

Figure 2.22 Stereographic projection of normals to the Blueberry fault (blue circles) and Glass fault (red circles) at Chimney Rock, UT. Filled circles are mean orientations for each fault. Modified from Maerten et al. (2001).

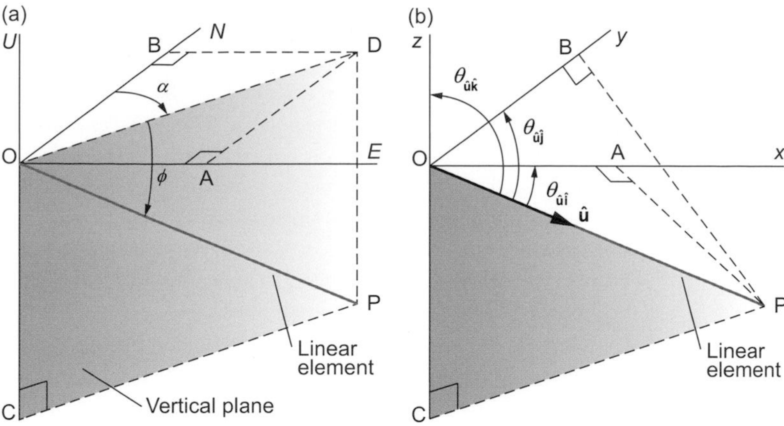

Figure 2.23 (a) The azimuth and inclination (α, ϕ) for a linear structural element are illustrated using the geographic coordinates (E, N, U). (b) Direction angles $\left(\theta_{\hat{u}\hat{i}}, \theta_{\hat{u}\hat{j}}, \theta_{\hat{u}\hat{k}}\right)$ are illustrated between the unit vector $\hat{\mathbf{u}}$ that is parallel to the linear element, and the base vectors for the coordinate axes (x, y, z).

one outcrop to another. The stereographic projection enables one to distinguish members of the two fault sets and to evaluate the non-planar geometry of the fault surfaces.

The filled blue circle and the filled red circle on the stereonet (**Figure 2.22**) represent the *mean orientation* of the Blueberry and Glass faults, respectively. To calculate the mean orientations one must convert the geographic angles (α_t, ϕ_p) to unit vectors, a task we take up in the next section.

2.3.4 Transformation of Field Orientations to Unit Vectors

A compass and inclinometer are used to measure azimuth and inclination of structural elements in the field. However, these angles are not the best choices for data analysis, or for comparison to mechanical models. These analyses and comparisons usually are done most efficiently and effectively using unit vectors and their associated direction angles and direction cosines. Here we show how to determine the components of the unit vector that defines the orientation of a structure given its azimuth and inclination.

Consider a linear structural element, or the normal to a planar structural element, defined by a radial line segment with length OP (**Figure 2.23**). The geographic coordinate system is (E, N, U) and the azimuth and inclination of the structural element are (α, ϕ). In what follows we use projections of OP onto the horizontal plane and onto the coordinate axes. OP projects onto the horizontal plane using $OD = OP \cos\phi$. Given the restriction, $0° \leq \phi \leq 90°$, on the range of the inclination, OD always is positive. OD projects onto the E and N axes, respectively, using $OA = OD \sin\alpha$ and $OB = OD \cos\alpha$. OA or OB may be zero, if OD coincides with the N or E axis. If not zero, OA and OB are positive or negative depending upon whether OD projects onto the positive or negative E and N axes. The line segment OP projects onto the negative U axis using $OC = -OP \sin\phi$. The line segment OC is equal to or less than zero, because the plunge is horizontal or downward.

The unit orientation vector $\hat{\mathbf{u}}$ that is parallel to the linear structural element is drawn with its tail at the origin and direction angles, $\left(\theta_{\hat{u}\hat{i}}, \theta_{\hat{u}\hat{j}}, \theta_{\hat{u}\hat{k}}\right)$ (**Figure 2.23**). Note that the geographic coordinates (E, N, U) and the Cartesian coordinates are *ordered* such that $(x, y, z) = (E, N, U)$. Because a direction angle is the *smaller* angle between two vectors or two intersecting lines, and because the geologic convention requires the inclination to be measured from the horizontal plane *downward*, the direction angles have the following restricted ranges:

$$0° \leq \theta_{\hat{u}\hat{i}} \leq 180°, 0° \leq \theta_{\hat{u}\hat{j}} \leq 180°, 90° \leq \theta_{\hat{u}\hat{k}} \leq 180° \quad (2.14)$$

The direction cosines for $\hat{\mathbf{u}}$ are defined from the projections of OP onto the coordinate axes, and these are written in terms of α and ϕ using relationships from the previous paragraph:

$$\cos\theta_{\hat{u}\hat{i}} = \frac{OA}{OP} = \frac{OA}{OD}\frac{OD}{OP} = \sin\alpha\cos\phi \quad (2.15)$$

$$\cos\theta_{\hat{u}\hat{j}} = \frac{OB}{OP} = \frac{OB}{OD}\frac{OD}{OP} = \cos\alpha\cos\phi \quad (2.16)$$

$$\cos\theta_{\hat{u}\hat{k}} = -\frac{OC}{OP} = -\sin\phi \quad (2.17)$$

Because $\hat{\mathbf{u}}$ is a unit vector, the direction cosines are equal to the respective components of this vector. Given the azimuth and inclination of any linear structural element, equations (2.15) to (2.17) define the unit orientation vector.

Given the direction cosines (components) of the unit orientation vector, $\hat{\mathbf{u}}$, the geographic angles (azimuth and inclination) of the linear structural element are recovered from (2.15) to (2.17) as:

$$\alpha = \tan^{-1}\left(\frac{OA}{OB}\right) = \tan^{-1}\left(\frac{\cos\theta_{\hat{u}\hat{i}}}{\cos\theta_{\hat{u}\hat{j}}}\right) \quad (2.18)$$

$$\phi = \sin^{-1}\left(\frac{OC}{OP}\right) = \sin^{-1}\left(-\cos\theta_{\hat{u}\hat{k}}\right) \quad (2.19)$$

These equations are used to put the unit orientation vector back into the geologic context, where azimuth and inclination are the expected angles.

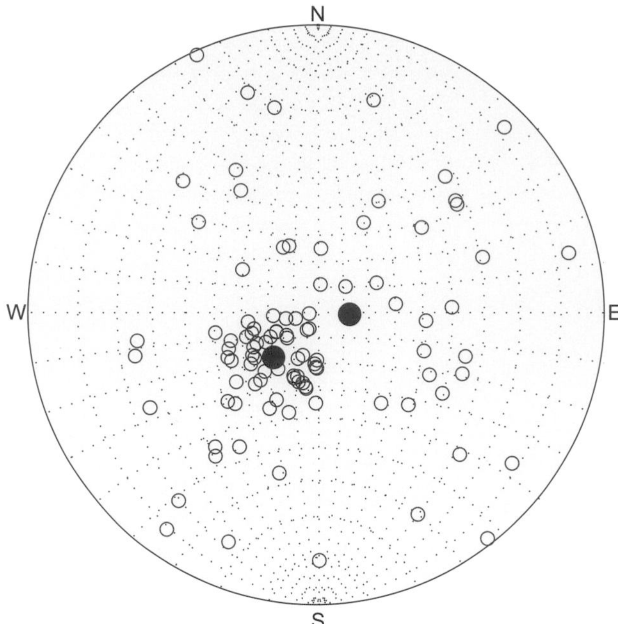

Figure 2.24 Stereonet showing scattered set of orientation data (blue circles) with spherical variance s = 0.48 and vector mean (filled blue circle) at α_t = 093, ϕ_p = 77. Clustered set of orientation data (red circles) has spherical variance s = 0.03 and vector mean (filled red circle) at α_t = 225, ϕ_p = 65. See Fisher et al. (1987) and Mardia and Jupp (2000) in Further Reading.

In Figure 2.22 normals to the Blueberry fault are plotted as open blue circles on the stereonet. The mean orientation of these normals is found using the components of the unit orientation vectors defined in (2.15) through (2.17). To compute the mean orientation, we define the ith unit vector for the Blueberry fault as:

$$\hat{\mathbf{u}}(i) = \left\langle \cos\theta_{\hat{\mathbf{u}}\hat{\mathbf{i}}}, \cos\theta_{\hat{\mathbf{u}}\hat{\mathbf{j}}}, \cos\theta_{\hat{\mathbf{u}}\hat{\mathbf{k}}} \right\rangle = \langle u_x(i), u_y(i), u_z(i) \rangle$$

The resultant (non-unit) vector, $\mathbf{U} = \langle U_x, U_y, U_z \rangle$, is found by adding the n unit vectors for the Blueberry fault. This is accomplished by adding each component of these vectors:

$$U_x = \sum_{i=1}^{n} u_x(i), U_y = \sum_{i=1}^{n} u_y(i), U_z = \sum_{i=1}^{n} u_z(i)$$

The magnitude of the resultant vector is $U = \left(U_x^2 + U_y^2 + U_z^2\right)^{1/2}$, and the direction cosines of the resultant vector are $\cos\theta_{\mathbf{U}\hat{\mathbf{i}}} = U_x/U$, $\cos\theta_{\mathbf{U}\hat{\mathbf{j}}} = U_y/U$, $\cos\theta_{\mathbf{U}\hat{\mathbf{k}}} = U_z/U$. These direction cosines are substituted into (2.18) and (2.19) to find the azimuth and inclination of the resultant vector, which are used to plot the solid blue circle on Figure 2.22. This resultant vector is directed in the mean orientation of the n unit vectors.

The spherical variance, s, is a good measure of the clustering of orientation data plotted as points on a stereonet. The spherical variance is defined as:

$$s = \frac{n - U}{n} \tag{2.20}$$

For tightly *clustered* points, the magnitude of the resultant vector, U, approaches the number of data points n, so $s \to 0$. For example, s = 0.03 for the data plotted as red circles on Figure 2.24. In comparison, s = 0.48 for the widely *scattered* blue circles in this figure. For uniformly scattered orientations the magnitude of the resultant vector approaches $n/2$, so $s \to$ ½. The spherical variance for the Blueberry fault data is 0.037, and that for the Glass fault is 0.198, so the Blueberry fault data are more tightly clustered by this measure. The spherical variance is an example of direction statistics, more of which can be found in textbooks on that subject (see Fisher et al., 1987, in Further Reading).

2.4 GEOLOGIC STRUCTURES REPRESENTED BY VECTOR FUNCTIONS

The *precise* and *reproducible description* of the geometry of geologic structures is one of the primary tasks of structural geologists. Broadly speaking geologic structures fall into two geometric categories, curved lines and curved surfaces. Locally, curves resemble straight lines and curved surfaces resemble planes, so we begin by focusing attention on straight lines, and then consider planes. The vector functions used to characterize lines and planes provide the data for geometric analysis of geologic structures using stereographic and statistical techniques. These vector functions also provide the input for mathematical modeling of the formation of geologic structures using mechanics.

2.4.1 Vector Representation of a Linear Structure

The alignment of tabular mineral grains in a metamorphic or igneous rock defines a linear fabric that can be related to deformation or flow. Fabrics are the subject of Chapter 10. As an example, consider the outcrop near a contact between igneous rock and sedimentary rock in the Henry Mountains of southern Utah (Figure 2.25) where grains of the mineral plagioclase are deformed into tabular streaks. These streaks on the outcrop surface locally define a line in three-dimensional space that one can characterize in the field by measuring its orientation and relating this geometrically to the orientation of the contact. In this case the line, or lineation, is parallel to the contact. This lineation is related to shearing deformation as the flow of magma into this intrusion was resisted by viscous drag against the stationary host rock. Viscous flow of magma is described in Chapter 6 and intrusions are the subject of Chapter 11. The orientation of the white streaks provides an estimate of the local direction of shearing and flow of the magma.

While the orientation of a lineation is a useful measure for interpreting the local kinematics of intrusion or faulting, a quantitative description of the line itself in three-dimensional space provides additional information to relate these structures to the mechanics of their formation. In this section we describe how a

Figure 2.25 (a) Contact between diorite porphyry (gray) and Entrada Sandstone (red) from the Henry Mountains, UT. (b) Lineations composed of stretched plagioclase grains (white streaks) within the diorite porphyry at the contact. Google Earth file: Figure 2.25 Trachyte Mesa UT contact.kmz. UTM 12 S 536014.00 m E, 4199841.00 m N.

lineation can be characterized geometrically using vector functions. This characterization requires field data on the locations of points along a particular lineation, which would be provided by a set of position vectors relative to a local frame of reference. Such vectors are obtained, for example, using total station surveying equipment, ground-based LIDAR, or tape and compass.

Using Cartesian coordinates, an arbitrarily oriented straight line of infinite length may be defined by any two points on that line, say P and Q, given by their position vectors $\mathbf{p}$ and $\mathbf{q}$ (Figure 2.26). A vector parallel to the straight line and extending from $\mathbf{p}$ to $\mathbf{q}$ is defined as the difference between $\mathbf{q}$ and $\mathbf{p}$:

$$\overrightarrow{PQ} = \mathbf{q} - \mathbf{p} = (q_x - p_x)\hat{\mathbf{i}} + (q_y - p_y)\hat{\mathbf{j}} + (q_z - p_z)\hat{\mathbf{k}}$$

The magnitude of this vector, $\left|\overrightarrow{PQ}\right|$, is found using (2.3). Because the two points are at arbitrary locations along the line, we define the direction from P toward Q with the unit vector:

$$\hat{\mathbf{u}} = \frac{\mathbf{q} - \mathbf{p}}{|\mathbf{q} - \mathbf{p}|} \tag{2.21}$$

The unit vector is placed with its tail at P, so it extends from P toward Q (Figure 2.26). The *magnitude*, $\left|\overrightarrow{PQ}\right|$, and the *direction*, $\hat{\mathbf{u}}$, quantify the length and orientation of the line segment between P and Q.

The vector equation for the infinitely long straight line containing P and Q (Figure 2.26) is constructed using the position vector for the point P and the unit vector (2.21):

$$\mathbf{c}(t) = \mathbf{p} + t\hat{\mathbf{u}}, \ -\infty \le t \le +\infty \tag{2.22}$$

The form $\mathbf{c}(t)$ signals that this is a vector-valued function of t, not simply a vector. As t varies over the infinite range indicated in (2.22) the heads of the position vectors $\mathbf{c}(t)$ trace out the infinite

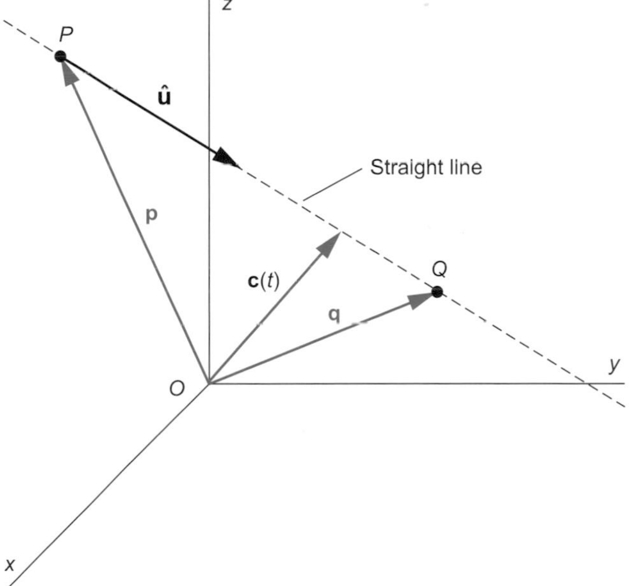

Figure 2.26 A straight line of arbitrary orientation defined by two points, P and Q, with corresponding position vectors $\mathbf{p}$ and $\mathbf{q}$. Direction from P to Q is given by the unit vector $\hat{\mathbf{u}}$ defined in (2.21). See Varberg et al. (2006) in Further Reading.

line. For a line segment of *finite length*, for example with end points at P and Q, a restricted range would be used: $0 \le t \le |\mathbf{q} - \mathbf{p}|$.

All vectors, except the zero vector, have a direction: geometrically from tail to head, and algebraically from initial point to terminal point. A line (2.22), or line segment, is referred to as an axis because it has an orientation, but not a direction. To

Figure 2.27 Aerial photograph of Sheep Mountain anticline, taken looking southeast. Major lithologic formations labeled in white. Note that the tops of more resistant beds are exposed as large roughly planar bedding surfaces inclined away from Sheep Mountain. From Lovely et al. (2010). Google Earth file: Figure 2.27 Sheep Mountain WY anticline nose.kmz. UTM: 12 T 721397.00 m E, 4948519.00 m N.

prescribe the orientation of the line we used the vector $\overrightarrow{PQ}$ to define the unit vector $\hat{\mathbf{u}}$, but we could have used the vector $\overrightarrow{QP}$ to define the unit vector $-\hat{\mathbf{u}}$. From a mathematical point of view $\hat{\mathbf{u}}$ and $-\hat{\mathbf{u}}$ are parallel to one another and work equally well for prescribing the orientation of the line. Therefore, although the vector-valued function $\mathbf{c}(t)$ is composed of vectors, the line or line segment defined by (2.22) does not distinguish the two possible directions.

Structural geologists follow the convention that the angle of inclination of a lineation, the plunge, is measured from the horizontal plane *downward*. Here we take the z-axis as vertical and positive upward as in **Figure 2.26**. To follow the convention for angle of inclination one computes the unit vector $\mathbf{u}$ using (2.21) and then applies the following test and resolution: if $\hat{\mathbf{u}}_z > 0$, then use $-\hat{\mathbf{u}}$, else use $\hat{\mathbf{u}}$. In other words, if the unit vector is directed upward, replace it using the downward directed unit vector with the same orientation. Then, for example, one would use the components of the unit vector, which are the direction cosines, to compute the trend and plunge of the lineation using (2.18) and (2.19).

2.4.2 Vector Representation of a Planar Structure

The sedimentary formations at Sheep Mountain are folded into a spectacular anticline (**Figure 2.27**) that crops out north of the town of Greybull in northcentral Wyoming on the eastern flank of the Bighorn Basin. This fold is described and analyzed in more detail in Chapter 9. The aerial photograph looks to the southeast along the anticline. Beds dip steeply, at nearly 90°, on the left (northeastern) limb, and they dip gently, usually between 10° and 40°, on the right (southwestern) limb. On this aerial photograph about 1 kilometer of stratigraphy is exposed, from the Pennsylvanian Amsden at the crest of the fold, to the Cretaceous Frontier Formation on the distal limb. Curved bedding surfaces in these formations locally resemble planes, and here we focus attention on those planes.

Figure 2.28 Gypsum Springs bedding outcrop on the southwest flank of Sheep Mt. anticline, WY. The bedding surface strikes away from the viewer and dips to the right. *P*, *Q*, and *R* are three non-collinear points associated with position vectors. Scale approximate in foreground. Modified from Lovely et al. (2010). Google Earth file: Figure 2.28 Sheep Mountain WY fold limb.kmz. Approximate location UTM: 12 T 721525.65 m E, 4948119.44 m N.

We also use the Sheep Mountain anticline to show how structural geologists use a modern data acquisition system, known as Airborne Laser Swath Mapping (ALSM) to quantify orientations of sedimentary units across the fold. An outcrop in the Gypsum Springs formation on the southwest limb is shown in **Figure 2.28**. The ALSM laser reflected off of bedding surfaces such as this one and provided a digital elevation map with 1 m resolution. We use ALSM data for three points on this outcrop to describe the vector representation of this surface, and to introduce the mathematical technique for determining its orientation from the three points. This is referred to in the literature of structural geology as a three-point problem.

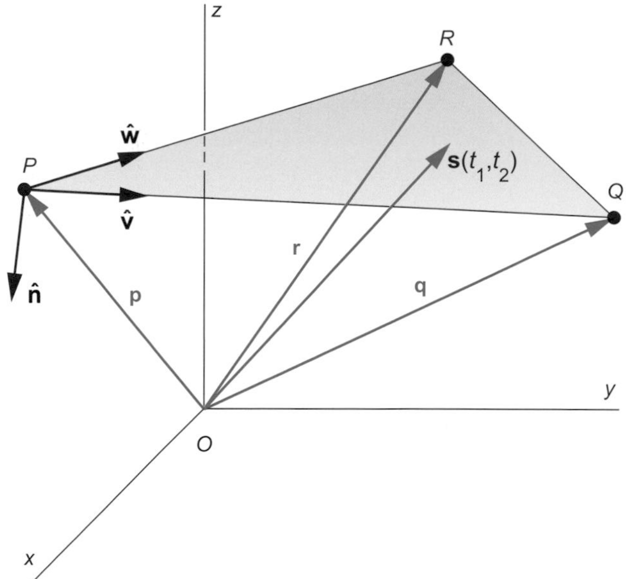

Figure 2.29 Arbitrary plane defined by three points, *P*, *Q*, and *R* and corresponding position vectors **p**, **q**, and **r**. Colored area is underside of small triangular patch of the plane. Directions from *P* to *Q* and from *P* to *R* are given respectively by the unit vectors $\hat{\mathbf{v}}$ and $\hat{\mathbf{w}}$. Unit normal vector to the plane, $\hat{\mathbf{n}}$, is directed downward.

A plane can be defined by any three *non-collinear* points on that plane (**Figure 2.29**), say *P*, *Q*, and *R*, given by the three position vectors **p**, **q**, and **r**. Following the procedure used to derive (2.21), unit vectors directed from *P* toward *Q* and from *P* toward *R* are:

$$\hat{\mathbf{v}} = \frac{\mathbf{q}-\mathbf{p}}{|\mathbf{q}-\mathbf{p}|}, \hat{\mathbf{w}} = \frac{\mathbf{r}-\mathbf{p}}{|\mathbf{r}-\mathbf{p}|} \tag{2.23}$$

Both of these unit vectors lie in the plane defined by the points *P*, *Q*, and *R*.

A vector equation for the plane can be constructed using the position vector for the fixed point, *P*, and the two unit vectors defined in (2.23) as follows:

$$\mathbf{s}(t_1, t_2) = \mathbf{p} + t_1\hat{\mathbf{v}} + t_2\hat{\mathbf{w}}, \ -\infty \le t_1, t_2 \le +\infty \tag{2.24}$$

The form $\mathbf{s}(t_1, t_2)$ signals that this is a vector-valued function of t_1 and t_2, not simply a vector. As t_1 and t_2 vary over the range indicated, the heads of the position vectors $\mathbf{s}(t_1, t_2)$ trace out all of the points on the infinite plane. Geologic structures are approximated locally by a plane, but are not infinite in extent. One could redefine the ranges of the parameters, t_1 and t_2, to cover that portion of the plane that approximates the geologic structure. Alternatively, one could use the polygon of ALSM points to define the portion of the plane at the exposed outcrop, or focus on the orientation of the plane at the point of interest by calculating the unit normal vector.

The orientation of the plane is given by its normal vector, which is found using the vector product (2.8) of the unit vectors, $\hat{\mathbf{v}}$ and $\hat{\mathbf{w}}$. The vector product is not necessarily a unit vector, but it may be normalized to define the unit normal vector to the plane (**Figure 2.29**):

$$\hat{\mathbf{n}} = \frac{\hat{\mathbf{v}} \times \hat{\mathbf{w}}}{|\hat{\mathbf{v}} \times \hat{\mathbf{w}}|} \tag{2.25}$$

Using (2.9) the denominator is equal to $\sin\theta_{\hat{\mathbf{v}}\hat{\mathbf{w}}}$, and the non-collinear condition insures that the denominator is not zero.

The unit normal vector for a given plane has two possible directions, $\hat{\mathbf{n}}$ and $-\hat{\mathbf{n}}$. For vector products, $\hat{\mathbf{v}} \times \hat{\mathbf{w}} = -(\hat{\mathbf{w}} \times \hat{\mathbf{v}})$, so the direction of the normal is reversed simply by changing the order of the two vectors in the product. Structural geologists have adopted the convention that the normal vector should be directed either horizontally or downward. To follow that convention compute the unit vector $\hat{\mathbf{u}}$ using (2.25), and then apply the test and resolution introduced in the last section: if $\hat{u}_z > 0$, then use $-\hat{\mathbf{u}}$, else use $\hat{\mathbf{u}}$. The components of this unit vector, which are the direction cosines, are used to compute the trend and plunge of the normal using (2.18) and (2.19).

2.5 VECTOR QUANTITIES

The description of geologic structures begins at the outcrop where data are collected to quantify orientations of *lines* and *planes* that locally approximate the structure. Taking a broader view, structures that are approximated locally as straight lines actually are *curves* in three dimensions, and structures that are approximated locally as planes actually are *curved surfaces* in three dimensions. Vectors are very useful for describing curves and curved surfaces, but to take advantage of this utility we need to augment the mathematics reviewed in Section 2.4 by including vector calculus.

2.5.1 Vector Representation of a Curved Lineation

A curve may be thought of as a set of points, arranged *side-by-side* in an orderly and continuous distribution. We already know how to define a point using a position vector, so this concept is extended to curves by defining a set of position vectors that represent all of the points along the curve. The spatial continuity of this set of points is assured by defining the set of position vectors using a continuous vector-valued function:

$$\mathbf{c}(t) = c_x(t)\hat{\mathbf{i}} + c_y(t)\hat{\mathbf{j}} + c_z(t)\hat{\mathbf{k}} \tag{2.26}$$

Here t is a scalar variable that indicates position along the curve: as t increases the heads of successive position vectors locate successive points along the curve. The three scalar functions, $c_x(t)$, $c_y(t)$, and $c_z(t)$, are the components of the position vectors that define the curve. As t varies these components vary and define the successive position vectors for the points along the curve. The variable t is called the arbitrary parameter of the curve and (2.26) is referred to as the parametric representation of the curve. The word *arbitrary* that modifies parameter is used to distinguish that parameter from the *natural* parameter of the curve. The natural parameter identifies position on the curve, but also measures the arc length along the curve.

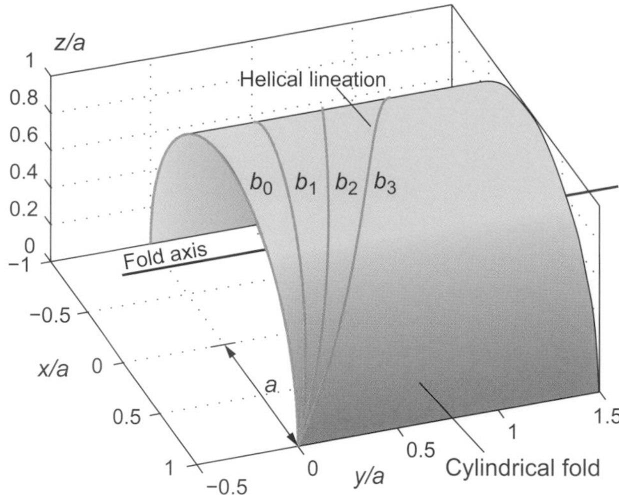

Figure 2.30 Cylindrical folded surface of radius *a*. Helical lineations lie on the surface and have pitches b_0 to b_3. The fold axis, when moved in a half-circle, generates the folded surface.

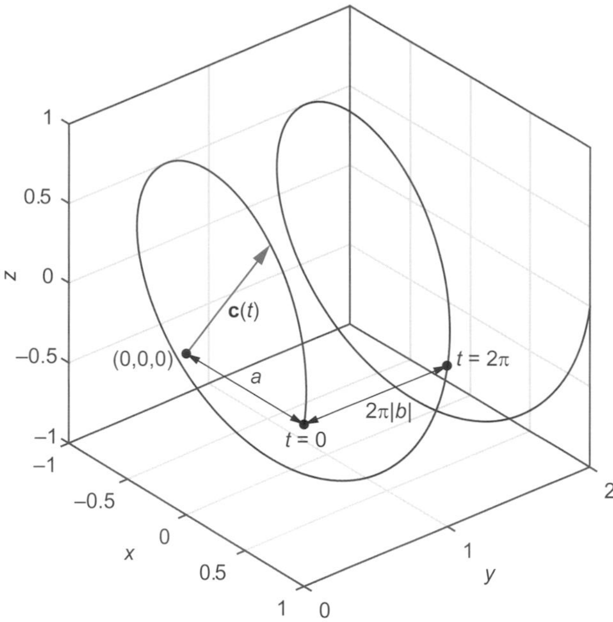

Figure 2.31 Circular helix defined in (2.27) by the vector function, **c**(*t*), with arbitrary parameter *t*, radius *a* = 1, and pitch *b* = 1/2π. Helix wraps around an invisible cylinder with an axis along the *y*-axis and radius *a*. Refer to **Figure 2.30** for illustration of helix on a cylindrical fold.

As an example of a three-dimensional geological curve recall the slickenlines on the Madison Limestone bedding surface at Sheep Mountain anticline (**Figure 2.16**). Although this exposure only reveals a small portion of the bedding surface, we know from other exposures (**Figure 2.15**) that this surface is folded over the anticline in what roughly approximates a cylindrical shape. If slickenlines are developed on this surface they would form a continuous curve that could be quantified using a vector function (2.26). In **Figure 2.30** we have idealized the bedding surface as a half cylinder and drawn several possible curves to approximate the slickenlines. To understand the shapes of these curves, we develop the concept of a parametric representation further by introducing a particular curve called the *circular helix*.

The parametric representation for the circular helix in **Figure 2.31** is defined as the following vector function:

$$\mathbf{c}(t) = a(\cos t)\hat{\mathbf{i}} + bt\hat{\mathbf{j}} + a(\sin t)\hat{\mathbf{k}} \qquad (2.27)$$

The components of this vector function are $c_x(t) = a(\cos t)$, $c_y(t) = bt$, and $c_z(t) = a(\sin t)$, and the ranges of the constants, a and b, are $a > 0$ and $-\infty < b < +\infty$. The points on this curve lie on a right cylinder of radius a with the cylindrical axis coincident with the y-axis. As the arbitrary parameter increases from $t = 0$ to $t = 2\pi$ the position vectors for points on the curve advance in the y-direction a distance $2\pi|b|$, and the x and z components return to their original values. As t continues to increase the points continue to advance in the y-direction while encircling the y-axis. Lineations in nature, such as the slickenlines on the Madison Limestone bed at Sheep Mountain (**Figure 2.16**), may be adequately approximated by a *segment* of a circular helix. They do not continue over the crest of the anticline (**Figure 2.15**) because the amount of shearing on bedding planes goes to zero at the crest.

Recall that the local orientation of a curvilinear geologic structure is measured as the orientation of the line element that is *tangent* to the lineation. We make use of the unit tangent vector, $\hat{\mathbf{t}}(t)$, along a curve to define the local orientation (**Figure 2.32**). Some care is needed to distinguish the symbol for the arbitrary parameter of a curve, t, from that for the unit tangent vector, $\hat{\mathbf{t}}$. The relationship between the unit tangent vector and the curve at any point $\mathbf{c}(t)$ can be thought of intuitively in terms of a straight line passing through that point. For example, consider the *secant line* that passes through $\mathbf{c}(t)$ and through a point a short distance along the curve $\mathbf{c}(t + \Delta t)$. In the limit as $\Delta t \to 0$ the secant line becomes parallel to the curve and

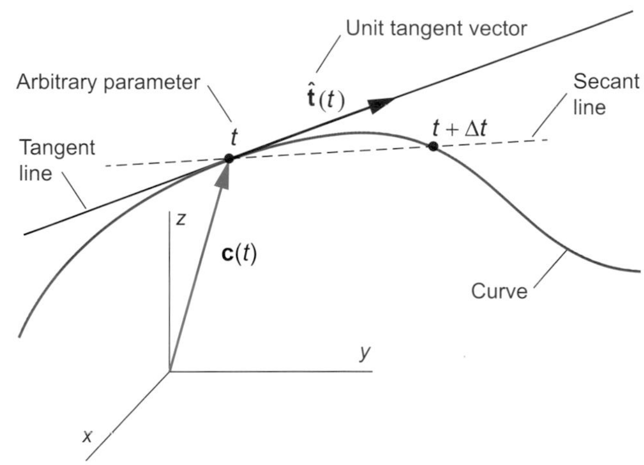

Figure 2.32 Diagram to define the unit tangent vector, $\hat{\mathbf{t}}(t)$, to a curve using the position vector function **c**(*t*). See Varberg et al. (2006) in Further Reading.

coincident with the tangent line. The tangent line is the one straight line, of an infinite number of differently oriented straight lines through the point, that has the closest contact with the curve, and best quantifies the orientation of the curve at that point. One can say that the tangent line is the *best fitting* straight line to the curve at that point.

The unit tangent vector is defined using the derivative of the vector function $\mathbf{c}(t)$:

$$\hat{\mathbf{t}}(t) = \frac{d\mathbf{c}}{dt} \Big/ \left|\frac{d\mathbf{c}}{dt}\right| \tag{2.28}$$

The derivative is divided by the magnitude of the derivative to produce the unit tangent vector. One takes the derivative of a vector function of one variable by taking the derivative of each component with respect to that variable, and then using these as the components of the new vector:

$$\frac{d\mathbf{c}(t)}{dt} = \frac{dc_x(t)}{dt}\hat{\mathbf{i}} + \frac{dc_y(t)}{dt}\hat{\mathbf{j}} + \frac{dc_z(t)}{dt}\hat{\mathbf{k}} \tag{2.29}$$

The function $\mathbf{c}(t)$ is differentiable at some particular value of the variable, say $t = t_o$, if each component is differentiable at that point. Note that the derivative of a vector function is a vector function: higher order derivatives may be calculated following the same procedure, and the standard formulae for derivatives apply.

For the circular helix (**Figure 2.31**) the derivative and absolute value of the derivative of the vector function are found using (2.27):

$$\frac{d\mathbf{c}}{dt} = -a(\sin t)\hat{\mathbf{i}} + b\hat{\mathbf{j}} + a(\cos t)\hat{\mathbf{k}}, \quad \left|\frac{d\mathbf{c}}{dt}\right| = \left(a^2 + b^2\right)^{1/2}$$

Substituting the derivatives into (2.28) we find the unit tangent vector for the circular helix:

$$\hat{\mathbf{t}}(t) = \left(a^2 + b^2\right)^{-1/2}\left[-a(\sin t)\hat{\mathbf{i}} + b\hat{\mathbf{j}} + a(\cos t)\hat{\mathbf{k}}\right]$$

The orientation of a lineation determined in the field using geographic angles can be related to the unit tangent vector, $\hat{\mathbf{t}}$, of that lineation. For example, the slickenlines in **Figure 2.16** may be approximated locally with a linear element whose orientation is measured using the trend, α_t, and plunge, ϕ_p. This linear element is parallel to the unit tangent vector of the three-dimensional curve representing the slickenline that passes through the point of measurement. Thus, the linear element and the unit tangent vector have the same direction angles at a given point. Furthermore, because $\hat{\mathbf{t}}$ is a unit vector, the scalar components are equivalent to the direction cosines. Therefore, employing (2.15) through (2.17), we have:

$$\begin{aligned} t_x &= \cos\theta_{\hat{t}\hat{i}} = \sin\alpha_t \cos\phi_p \\ t_y &= \cos\theta_{\hat{t}\hat{j}} = \cos\alpha_t \cos\phi_p \\ t_z &= \cos\theta_{\hat{t}\hat{k}} = -\sin\phi_p \end{aligned} \tag{2.30}$$

In this way orientation data, (α_t, ϕ_p), taken at different points along a geological lineation can be related to the components of the unit tangent vectors at points on a three-dimensional curve.

2.5.2 Vector Representation of a Curved Surface

The interplay of sunlight and shadow on the fault surface in **Figure 2.33** indicate that the surface is curved. How would you describe such a curved surface quantitatively? In general, the analytical description of curved surfaces is approached from the intuitive notion of a set of points arranged in some continuous fashion in three-dimensional space. Sufficiently close to any particular point on the surface, the neighboring points are distributed such that they resemble a *plane*. This leads to the definition of a curved surface as a continuous vector function of two scalar variables called the *parameters* of the surface. As the two parameters vary, the heads of the successive position vectors sweep out the curved surface in three-dimensional space. Using this concept, one can describe geological surfaces in a *precise* and *reproducible* manner.

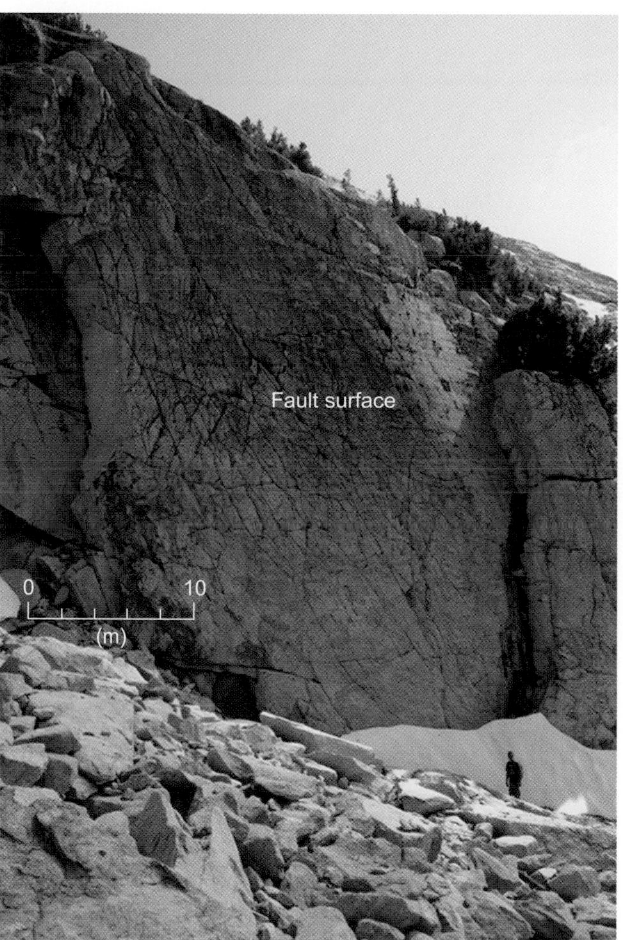

Figure 2.33 Fault surface exposed in granodiorite of the Sierra Nevada, CA. The curvature of this surface is highlighted by the sunlight and shadow, and the scale is indicated by the person in front of the snow bank. Photograph by W. A. Griffith. Google Earth file: Figure 2.33 Seven Gables CA fault surface. kmz. Approximate location UTM: 11 S 335937.29 m E, 4132478.97 m N.

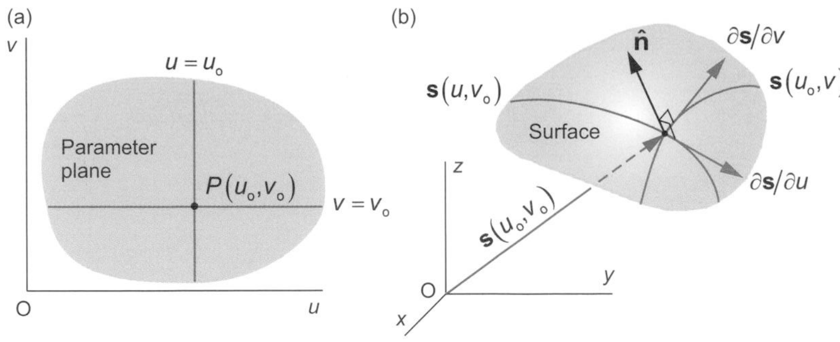

Figure 2.34 Parametric representation of a curved surface. (a) Two-dimensional parameter plane with parameters *u* and *v*. (b) Three-dimensional surface defined by vector function of two parameters, **s**(*u*, *v*).

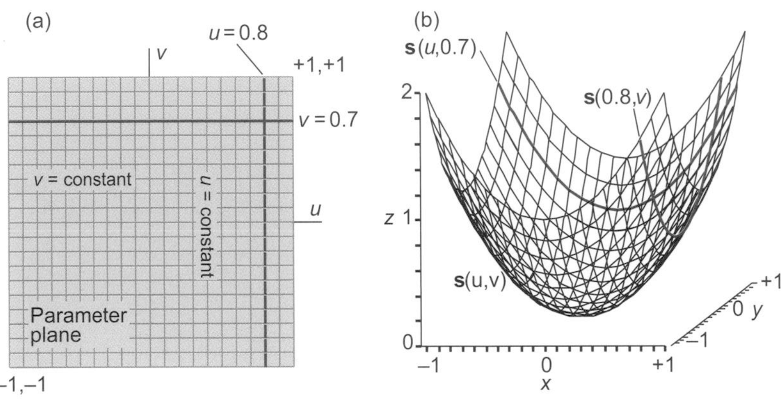

Figure 2.35 Differential geometry of an elliptic paraboloid. (a) Parameter plane with *u*- and *v*-coordinate lines *u* = 0.8 and *v* = 0.7 highlighted in blue. (b) Wire-frame diagram with *u*- and *v*-parameter curves **s**(u, 0.7) and **s**(0.8, v) highlighted in blue.

The concept described in the previous paragraph is implemented by letting the two axes (u, v) define the parameter plane (**Figure 2.34a**) on which the two coordinates are the parameters u and v. The three Cartesian axes (x, y, z) and the associated base vectors $\left(\hat{\mathbf{i}}, \hat{\mathbf{j}}, \hat{\mathbf{k}}\right)$ comprise the coordinate system for the curved surface (**Figure 2.34b**). The position vectors for the curved surface are written as a function of the two parameters:

$$\mathbf{s}(u,v) = s_x(u,v)\hat{\mathbf{i}} + s_y(u,v)\hat{\mathbf{j}} + s_z(u,v)\hat{\mathbf{k}} \quad (2.31)$$

The three scalar functions $s_x(u, v)$, $s_y(u, v)$, and $s_z(u, v)$ are the components of the vector function, $\mathbf{s}(u, v)$. These functions, along with the base vectors, determine the position vectors for all points on the curved surface. The vector equation (2.31) is called a parametric representation of the surface. An individual point (u_o, v_o) in the parameter plane (**Figure 2.34a**) maps onto the surface using the position vector $\mathbf{s}(u_o, v_o)$. Similarly, the coordinate lines $u = u_o$ and $v = v_o$ in the parameter plane map onto the curves $\mathbf{s}(u_o, v)$ and $\mathbf{s}(u, v_o)$ on the surface (**Figure 2.34b**).

As a specific example, consider the parameter plane (**Figure 2.35a**) with a rectangular grid of lines, u = constant and v = constant, parallel to the coordinate axes, and use the following parametric representation of a surface resembling a *geologic basin*:

$$\mathbf{s}(u,v) = u\hat{\mathbf{i}} + v\hat{\mathbf{j}} + \left(u^2 + v^2\right)\hat{\mathbf{k}} \quad (2.32)$$

This surface is called an elliptic paraboloid by mathematicians. Two sets of curved lines are used to represent the elliptic paraboloid in the wire-frame diagram (**Figure 2.35b**). The u- and v-parameter curves are the intersections of the surface with planes parallel to the (x, z)-plane and to the (y, z)-plane, respectively. The intersections of the curved surface with planes parallel to the (x, z)-plane form a set of parabolas, as do the intersections of the curved surface with planes parallel to the (y, z)-plane. A particular u-parameter curve, defined by the vector function $\mathbf{s}(u, 0.7)$, and a particular v-parameter curve, defined by $\mathbf{s}(0.8, v)$, are highlighted in blue. The point (0.8, 0.7) on the parameter plane maps to the point $\mathbf{s}(0.8, 0.7)$ on the curved surface at the intersection of the two highlighted curves. In this way every coordinate line and every point in the two-dimensional parameter plane have a corresponding curve and a corresponding point on the curved surface in three-dimensional space defined by (2.32).

For a variety of applications structural geologists need to quantify the slope of geological surfaces, such as the fault in **Figure 2.33**, or the tops of sedimentary strata in a basin. This requires the use of partial derivatives, which we introduce here using the elliptic paraboloid in **Figure 2.35b** as an example. Rather than using the parametric (vector) representation of the surface given above, we write z as a function of two variables:

$$z = f(x, y) = x^2 + y^2$$

The blue curve lying in the surface and in a (x, z)-plane is characterized by a variable x and a constant $y = 0.7$, so the function describing this curve is $z = x^2 + 0.49$. Because z is only a function of one variable for this curve, the slope of the tangent line to this curve at a given value of x is the first derivative, $dz/dx = 2x$. Similarly, the blue curve lying in the surface and in a (y, z)-plane is characterized by a variable y and a constant $x = 0.8$, so $z = 0.64 + y^2$, and the slope of the tangent line to this curve at a given value of y is the first derivative, $dz/dy = 2y$.

Using the partial derivative, one can calculate the slopes of the tangent lines to each blue curve in **Figure 2.35b** directly from the equation for the surface:

$$\frac{\partial z}{\partial x} = \frac{\partial f(x,y)}{\partial x} = 2x, \quad \frac{\partial z}{\partial y} = \frac{\partial f(x,y)}{\partial y} = 2y$$

The symbol ∂ is the partial derivative sign. The operation $\partial z/\partial x$ takes the derivative of the function $f(x, y)$ with respect to x, while holding y constant; the operation $\partial z/\partial y$ takes the derivative with respect to y, while holding x constant. At the intersection point (0.8, 0.7) of the two curves, the slopes are 1.6 and 1.4 respectively. Because both blue curves in **Figure 2.35b** lie in the surface, these slopes are the slopes of the surface at the intersection point *in* the respective directions of the curves. In other directions at the intersection point, the slope of the surface will differ.

We now return to the general case of a parametric representation of a surface as given in the vector equation (2.31). The tangent vectors to a surface in the directions of the u- and v-parameter curves are calculated using the partial derivatives of the vector function $\mathbf{s}(u, v)$. To calculate the partial derivative $\partial \mathbf{s}/\partial u$, for example, one takes the derivative with respect to u of each component of the vector function, while holding v constant. Then one uses these partial derivatives as the components of a new vector function. Thus, the partial derivatives of $\mathbf{s}(u, v)$ with respect to the two parameters are:

$$\begin{aligned}\frac{\partial \mathbf{s}(u,v)}{\partial u} &= \frac{\partial s_x}{\partial u}\hat{\mathbf{i}} + \frac{\partial s_y}{\partial u}\hat{\mathbf{j}} + \frac{\partial s_z}{\partial u}\hat{\mathbf{k}} \\ \frac{\partial \mathbf{s}(u,v)}{\partial v} &= \frac{\partial s_x}{\partial v}\hat{\mathbf{i}} + \frac{\partial s_y}{\partial v}\hat{\mathbf{j}} + \frac{\partial s_z}{\partial v}\hat{\mathbf{k}}\end{aligned} \tag{2.33}$$

For the elliptic paraboloid (2.32) the partial derivative with respect to u is $\partial \mathbf{s}/\partial u = 1\hat{\mathbf{i}} + 2u\hat{\mathbf{k}}$. This vector is tangent to a u-parameter curve and points in the direction of increasing u. Because it is independent of $\hat{\mathbf{j}}$, this vector lies in a (x, z)-plane, just like the u-parameter curve. Similarly, the partial derivative with respect to v is $\partial \mathbf{s}/\partial v = 1\hat{\mathbf{j}} + 2v\hat{\mathbf{k}}$. This vector is tangent to a v-parameter curve; points in the direction of increasing v; and lies in a (y, z)-plane like the v-parameter curve. These tangent vectors are not necessarily unit vectors.

The orientation of a surface at a particular point is uniquely determined by a vector that is directed perpendicular to the surface. Recall from Section 2.2.3 that the vector product of two non-parallel vectors defines a new vector that is perpendicular to the plane containing the two vectors. Thus, we use the vector product of the two tangent vectors (2.33) at any point on a surface to define the unit normal vector, at that point (**Figure 2.34b**):

$$\hat{\mathbf{n}} = \left(\frac{\partial \mathbf{s}}{\partial u} \times \frac{\partial \mathbf{s}}{\partial v}\right) \Big/ \left|\frac{\partial \mathbf{s}}{\partial u} \times \frac{\partial \mathbf{s}}{\partial v}\right| \tag{2.34}$$

The vector product of the two tangent vectors is divided by the magnitude of that product to produce a unit normal vector $\hat{\mathbf{n}}$. As calculated using (2.34) the unit normal vector may be directed upward. Because structural geologists have adopted the convention that the normal vector should be directed horizontally or downward, one should employ the following test (assuming z is positive upward): if $n_z > 0$, then use $-\hat{\mathbf{n}}$ for the unit normal vector, which also is directed perpendicular to the surface.

Orientation data, such as strike and dip, taken at scattered outcrops on geological surfaces like the tops of the folded strata at Raplee anticline (**Figure 2.13**), can be related to the components of the unit normal vectors at those outcrops. First, recall that the trend and plunge of the normal to a geological surface are found from the strike and dip of that surface using (2.13). Then, because $\hat{\mathbf{n}}$ is a unit vector, the scalar components are equivalent to the direction cosines. Using this fact, and employing (2.15) through (2.17), we use the trend and plunge angles to find the unit vector components:

$$\begin{aligned} n_x &= \cos\theta_{\hat{\mathbf{n}}\hat{\mathbf{i}}} = \sin\alpha_t \cos\phi_p \\ n_y &= \cos\theta_{\hat{\mathbf{n}}\hat{\mathbf{j}}} = \cos\alpha_t \cos\phi_p \\ n_z &= \cos\theta_{\hat{\mathbf{n}}\hat{\mathbf{k}}} = -\sin\phi_p \end{aligned} \tag{2.35}$$

2.5.3 Differentials and Total Differentials

Given the value of a scalar function or a vector function at one point in a scalar or vector field, what are the values at neighboring points? For example, suppose we know the value of the velocity in a deforming rock mass at one point, what are the values at nearby points? We asked questions like this again and again as important physical quantities related to deformation are defined in later chapters. To address this question we use the mathematical concepts of the *differential* and the *total differential*, so we review those concepts here. We begin with a function of one variable because the geometric interpretation is easily visualized, and then we generalize to a function of two variables.

The function of one variable is written $z = f(x)$, and the graph of this function is illustrated as a plane parabolic curve in **Figure 2.36a**. Now consider a particular point, P_0, on this graph where the value of the function, $z = f(x_0)$, is known, and ask how one estimates the change in value, Δz, from P_0 to the nearby point P, where $x = x_0 + \Delta x$. If the function is differentiable at P_0, then we compute the first derivative there, $df(x_0)/dx$, which is the slope of the tangent line to the curve at P_0. The tangent line is the best fitting line to the curve, and thus has the same slope as the curve. Because the tangent line is straight, we *estimate* the increment, dz, in the value of the function from P_0, for an increment of the independent variable, $dx = \Delta x$, as:

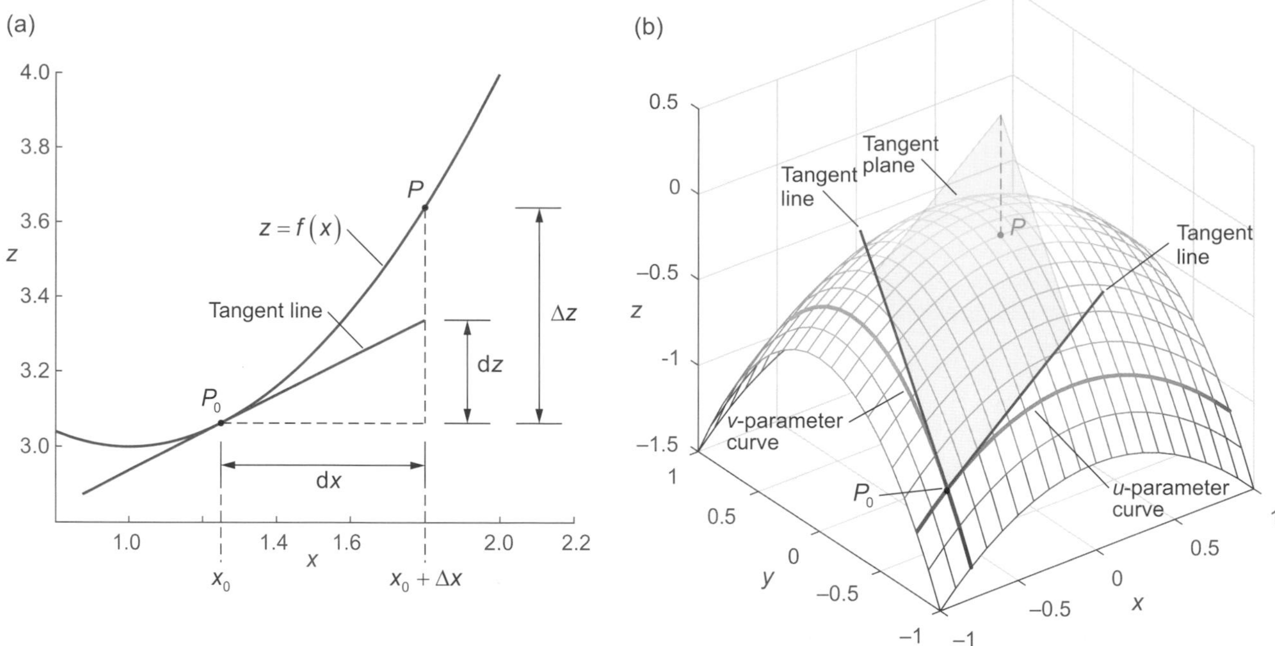

Figure 2.36 (a) Graph (blue curve) of a function of one variable, $z = f(x)$, with tangent line (red) at the point P_0 used to define the differential dz. (b) Graph of a function of two variables, $z = f(x, y)$, with tangent lines (red) to the u-parameter and v-parameter curves, and the tangent plane at the point P_0 used to define the total differential dz.

$$dz(x_0) \approx \left[\frac{df(x_0)}{dx}\right]dx$$

As drawn in **Figure 2.36a**, $dz(x_0)$ is not a very accurate estimate of the actual increment, Δz, because the tangent line diverges significantly from the curve over the distance Δx. However, in the limit as Δx goes to zero, the difference between Δz and $dz(x_0)$ goes to zero.

If the function of one variable, $z = f(x)$, is differentiable, then we generalize what we have established for the point P_0 in the preceding paragraph to all points on the curve illustrated in **Figure 2.36a** and define the differential dz as:

$$dz = \frac{dz}{dx}dx \tag{2.36}$$

The notation makes it appear that (2.36) has only two quantities, dx and dz. However, dz on the left side and dx by itself on the right side are *differential* quantities, whereas d/dx is the *derivative operator* and z is the function upon which it operates to produce the *derivative*. The differential dz is an accurate measure of how the function $z = f(x)$ varies in the neighborhood of any given point where the derivative of the function is defined.

For a function of two variables we write $z = f(x, y)$, and plot the graph of this function as the three-dimensional surface shown in **Figure 2.36b**. The example used here is an elliptic paraboloid, which is represented by the u- and v-parameter curves for that surface. Now consider a particular point, $P_0(x_0, y_0)$, on this surface where the value of the function is known, and ask how one estimates the change in the value from P_0 to a neighboring point. Suppose, for example, that the neighboring point is on the u-parameter curve that lies in the (x, z)-plane through the point P_0. Then, we follow a similar procedure for this curve that we used for the curve representing a function of one variable in **Figure 2.36a**. If the function is differentiable with respect to x at P_0, the first *partial* derivative, $\partial f(x_0, y_0)/\partial x$, is the slope of the tangent line to the u-parameter curve at that point. Because the tangent line is straight, we *estimate* the increment, dz, in the value of the function from P_0, for an increment of the independent variable, $dx = \Delta x$, as:

$$dz(x_0, y_0) \approx \left[\frac{\partial f(x_0, y_0)}{\partial x}\right]dx$$

As drawn in **Figure 2.36b**, this is not a very accurate estimate of the actual increment, because the tangent line diverges significantly from the curve over the finite distance, Δx. However, in the limit as Δx goes to zero, the difference between this estimate and the value of the function goes to zero. A similar estimate is made if the neighboring point is on the v-parameter curve that lies in the (y, z)-plane through the point P_0. The first *partial* derivative of the function with respect to y, $\partial f(x_0, y_0)/\partial y$, is the slope of the tangent line to the v-parameter curve at P_0.

If the function $z = f(x, y)$ is differentiable with respect to x and y, then the estimates describe in the previous paragraph for the arbitrary point P_0 can be extended to all points on a given u- or v-parameter curve. In each case the estimate pertains to the change in the value of the function for a change either in x or in y. On the other hand, what if one wanted to estimate the change in the value of the function from P_0 to a neighboring point P in an *arbitrary* direction (**Figure 2.36b**)? Note that the tangent lines for the

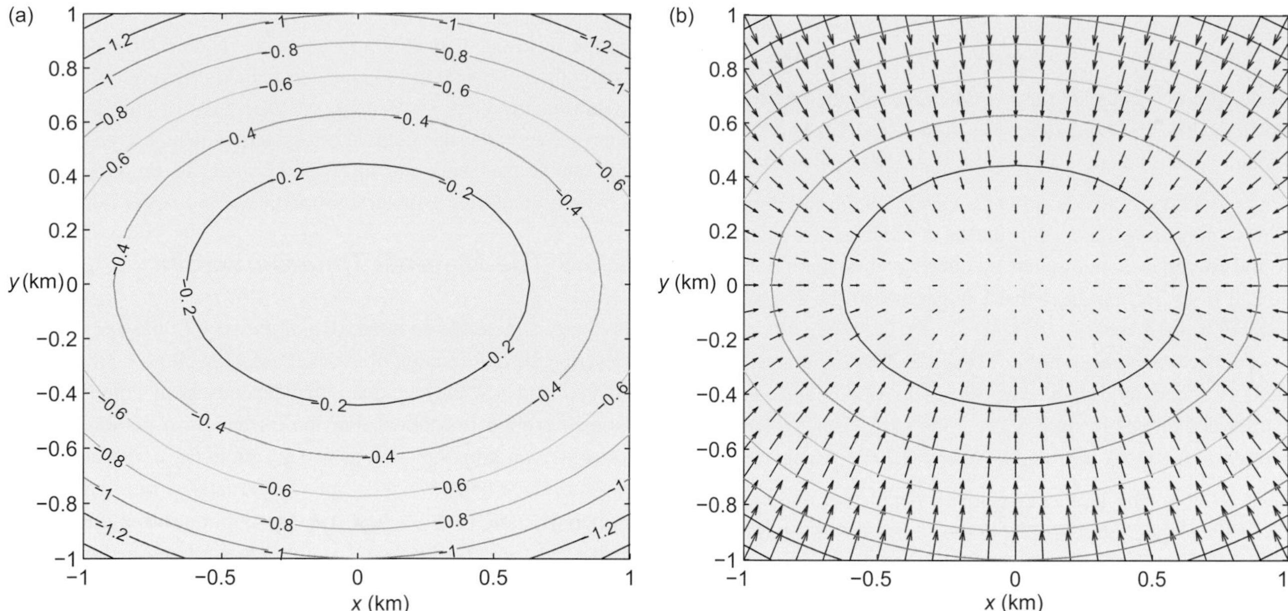

Figure 2.37 (a) The scalar function of two variables, $z = f(x, y)$, defined in (2.38) is used to construct a structure contour map of a subsurface dome. Contour interval is 0.2 km. (b) The gradient vector field (arrows) for the dome. This is underlain by the structure contours from (a). See Varberg et al. (2006) in Further Reading.

u- and v-parameter curves passing through P_0 define the tangent plane to the surface at that point, and we specify that the estimated value of the function at P lies on the tangent plane. The estimated value is composed of two components. The first component is the change in value of the function with respect to x, times the distance separating P_0 and P in the x-direction. The second component is the change in value of the function with respect to y times the distance separating P_0 and P in the y-direction. Adding the two components we have:

$$dz = \frac{\partial z}{\partial x}dx + \frac{\partial z}{\partial y}dy \tag{2.37}$$

This is called the total differential of the function $z = f(x, y)$. The total differential is an accurate measure of how the function varies in the neighborhood of any given point where the partial derivatives of the function are defined. The function $z = f(x, y)$ in (2.37) could be a physical scalar quantity such as mass density, that varies in two dimensions, for example representing the Easting and Northing coordinates on a map.

We also utilize vector quantities, such as the velocity vector, $\mathbf{v}$, that can vary in three dimensions. In this case each component of the vector is a function of the three independent variables, (x, y, z). Each vector component has a total differential and these are defined as:

$$\begin{aligned} dv_x &= \frac{\partial v_x}{\partial x}dx + \frac{\partial v_x}{\partial y}dy + \frac{\partial v_x}{\partial z}dz \\ dv_y &= \frac{\partial v_y}{\partial x}dx + \frac{\partial v_y}{\partial y}dy + \frac{\partial v_y}{\partial z}dz \\ dv_z &= \frac{\partial v_z}{\partial x}dx + \frac{\partial v_z}{\partial y}dy + \frac{\partial v_z}{\partial z}dz \end{aligned}$$

The total differential dv_y is the change in the vector component v_y in an arbitrary direction from a given point in the velocity vector field. That direction is determined by the relative values of dx, dy, and dz. The total differentials, dv_x and dv_z, have similar interpretations.

2.5.4 The Gradient of a Scalar Field is a Vector Field

In this section we introduce the gradient of a scalar field to quantify how that scalar field changes in magnitude with a change in position. As the title of this section states, the gradient of a scalar field is a vector field, but what are those vectors, and how do they quantify the spatial change in magnitude of the scalar field? As an example of a scalar field, consider a subsurface structural dome in a sequence of sedimentary rocks. Suppose the upper surface of one sedimentary bed in that dome is represented by the following function of two variables:

$$z = f(x, y) = -\tfrac{1}{2}(x^2 + 2y^2) \tag{2.38}$$

That surface is depicted on the contour map in **Figure 2.37a**. The values that are contoured are the scalar *depths*, z, from Earth's surface to the top of the sedimentary bed, which is a function of the two horizontal coordinates, x and y. Because z is positive upward with its origin at Earth's surface, the depths are negative numbers. Smaller negative numbers on the contours closer to the center of the map indicate shallower depths, so the sedimentary unit is folded into a dome-like shape. The map in **Figure 2.37a** is called a structure contour map, because it represents the three-dimensional shape of the geologic structure using contours (level lines) on a flat piece of paper.

The gradient of a scalar function of two variables, $z = z(x, y)$, is defined as:

$$\vec{\nabla} z = \frac{\partial z}{\partial x}\hat{\mathbf{i}} + \frac{\partial z}{\partial y}\hat{\mathbf{j}} \tag{2.39}$$

The symbol ∇ is called the del operator, and the arrow over the del indicates that the gradient of a scalar is a vector. Like the scalar, z, the gradient is a function of the two spatial coordinates, $\vec{\nabla} z = \vec{\nabla} z(x, y)$, so (2.39) defines a vector field in two dimensions. The vectors are composed of partial derivatives of z with respect to the coordinates, multiplied by the respective unit base vectors. Recall from Section 2.2.1 that $\hat{\mathbf{i}}$ is parallel to the x-axis and $\hat{\mathbf{j}}$ is parallel to the y-axis.

For the structural dome in **Figure 2.37a**, the gradient vector field is found by substituting (2.38) in (2.39) and taking the partial derivatives to find: $\vec{\nabla} z = -x\hat{\mathbf{i}} - 2y\hat{\mathbf{j}}$. This vector field is plotted as a set of arrows in **Figure 2.37b**. Along the x-axis, $\vec{\nabla} z = -x\hat{\mathbf{i}}$, so the gradient vector is parallel to the x-axis, and directed toward the crest of the dome. Along the y-axis, $\vec{\nabla} z = -2y\hat{\mathbf{j}}$, so the gradient vector is parallel to the y-axis, and also directed toward the crest of the dome. Elsewhere, the gradient vector arrows are oblique to the x- and y-axes and directed toward the crest of the dome. The magnitude of the gradient vector at any position is given by $\left|\vec{\nabla} z\right| = \left[(-x)^2 + (-2y)^2\right]^{1/2}$. These magnitudes are indicated by the lengths of the vector arrows in **Figure 2.37b**, which go to zero at the origin and increase away from the origin, reflecting the steepening slope of the dome with distance from its crest.

Notice in **Figure 2.37b** that the gradient vectors are everywhere *perpendicular* to the level lines (contours). This is generally true for the gradient of a two-dimensional scalar field. The gradient vectors defined using (2.39) lie in the (x, y)-plane, because they do not have a z component. However, if they are projected parallel to the z-axis onto the three-dimensional curved surface defined in (2.38), they would be directed *up* the steepest slope of the surface of the dome. In other words, the scalar function $z = f(x, y)$ at any particular point on the dome increases most rapidly in the direction indicated by the gradient vector. Also, the magnitude of this vector at any point is equal to the steepest slope of the surface at that point. In these ways the gradient vectors quantify geometric properties of the surface of the dome.

A common application of structure contour maps in the hydrocarbon industry utilizes depth data taken from reflection seismic surveys to map out deformed sedimentary horizons. Structural domes, such as that depicted in **Figure 2.37a**, are identified on the structure contour maps as possible traps for hydrocarbons. The gradient vectors (**Figure 2.37b**) on the structure contour map indicate possible migration routes for hydrocarbons, which are driven up the steepest slope of the sedimentary strata in the dome by buoyancy.

The concept of a gradient of a scalar may be extended to functions in three dimensions, for example $f(x, y, z)$, again using the del operator:

$$\vec{\nabla} f = \frac{\partial f}{\partial x}\hat{\mathbf{i}} + \frac{\partial f}{\partial y}\hat{\mathbf{j}} + \frac{\partial f}{\partial z}\hat{\mathbf{k}} \tag{2.40}$$

The arrow over the del indicates that this gradient is a vector. The partial derivatives on the right side of (2.40) are the rates of change of the scalar field in the respective coordinate directions. Scalar functions with continuous first partial derivatives, operated on by the del operator, produce vector fields. At any point in this field the gradient vector has the direction of the maximum increase in the scalar field, and it is perpendicular to the surface on which the scalar quantity is constant. In these ways the gradient quantifies geometric properties of the scalar field.

2.5.5 The Material Time Derivative

Velocity and acceleration are two of the vector quantities used to describe the deformation of a rock mass. How those quantities are related, and how they are quantified depends upon which description of motion is adopted. For the formation of geologic structures we can limit attention to two descriptions of motion, which we introduce here. We then define the material time derivative, which is used to link these two descriptions of motion. The material time derivative plays an important role in the fundamental laws of mass and momentum conservation that underlie much of the mechanics used to understand the formation of geologic structures. These laws will be introduced in Sections 3.5 and 3.6, but the material time derivative is a mathematical construct, so we introduce it here.

In a colloquial sense the two descriptions of motion can be distinguished by considering the viewpoints of two geologists observing a fissure eruption and lava flow on Kilauea Volcano (**Figure 1.13**). Geologist one focuses their attention on the rock cut by the fissure, and records a movie of rock particles on either side of the fissure as it opens. The initial positions of the particles before the eruption are known, and their velocities are calculated as a function of these initial positions. This is a referential description of motion because the kinematic quantity is referred to the initial positions. This also is a *material* description of motion, because the kinematic quantity is associated with given particles of the material (rock). The other geologist focuses their attention on the flowing lava and records a movie of congealed chunks of lava flowing by a given position. The initial positions of these chunks are not known, but the velocities are calculated as a function of time as different chunks move through the given position. This is a spatial description of motion, because the kinematic quantity is identified with a particular position in space.

The geological example we use to quantify these descriptions of motion, and to derive the material time derivative, is the buoyant rise and overturn of salt in an intrusion called a diapir. The Hänigsen dome is a noteworthy example of a salt diapir, located in northwestern Germany (**Figure 2.38a**). An interpretation of an early stage in the development of this diapir at ~85 Ma is portrayed here. In this cross section the stratigraphy of the intruded rock has been simplified to three salt layers. The older sedimentary strata are truncated and deformed by the intrusion, which is overlain by younger sedimentary strata. One side of the diapir displays a normal stratigraphic sequence with the oldest salt overlain by the middle salt and cap rock (the youngest salt is missing). The other side of the diapir is considerably more complex, because the sequence of salt layers is overturned, so the youngest is overlain successively

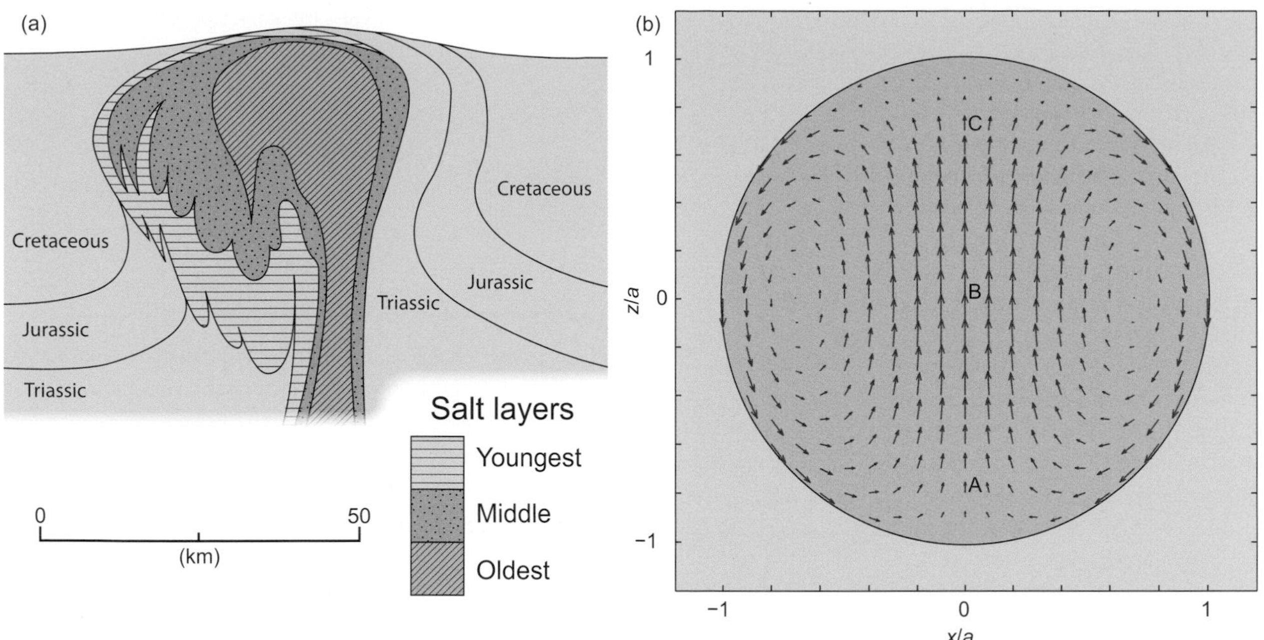

Figure 2.38 (a) Simplified cross section of the Hänigsen salt dome. Modified from Talbot and Jackson (1987). (b) Velocity vectors for the interior of a sphere of viscous liquid rising in a surrounding liquid of different viscosity and greater density. This is the velocity field as viewed by an observer moving with the same upward velocity as the sphere. Positions A, B, and C are referred to in the text. The derivation of this velocity field is in Section 11.6.

by the middle and oldest, and the diapir truncates the surrounding sedimentary rock.

The rise and overturn of salt, inferred from the cross section (**Figure 2.38a**), can be understood in a general way by studying the velocity field in a sphere of viscous liquid (**Figure 2.38b**) that is rising due to buoyancy within a liquid of different viscosity and greater density. The origin of the two-dimensional Cartesian coordinates is at the sphere center, with the x-axis horizontal and the z-axis vertical. The vector arrows illustrate the local velocity field that would be viewed by an observer stationed at the center of the sphere and moving with the same upward velocity, V_s, as the entire sphere. This is a spatial description of motion because the velocity is specified at each position in space relative to that origin, $v = v(x, z)$.

The viscous drag between the liquid inside and the liquid outside the sphere creates an interior velocity field (**Figure 2.38b**) in which the vectors near the centerline are directed upward; those near the top are directed toward the sides; those along the sides are directed downward; and those near the bottom are directed toward the centerline. The vector magnitudes are greatest near the center and along the sides, and least near the top, near the bottom, and near two points midway between the center and the sides. Because the velocity at any position in this field does not change with time, this is called a steady flow. The rendering of the velocity field in **Figure 2.38b** suffices for all time.

Given the steady circulation pattern within the viscous sphere (**Figure 2.38b**), one might suppose that *particles* of liquid neither accelerate nor decelerate. This supposition is not correct. To demonstrate this qualitatively, consider a particle of liquid moving upward near the centerline of the sphere. As the particle flows from location A to B, it must *accelerate* because the velocity at A is less than the velocity at B. From location B to C the particle must *decelerate* because the velocity at B is greater than the velocity at C. Also, for example, as a particle circulates within the right side of the sphere, it changes direction from upward, to horizontal, to downward, to horizontal, and back to upward. Clearly, the velocity vectors of particles change in both magnitude and direction as the particles circulate within the viscous sphere. In this paragraph we are discussing the flow in the viscous sphere using a material description of motion, one that considers the velocity of a *particle* rather than the velocity at a particular *position* in space. This brings into focus the two different descriptions of motion for the same flow regime.

The material and spatial descriptions of motion are reconciled quantitatively using the material time derivative of velocity. Because velocity is a vector quantity, each velocity component has a material time derivative, and each of these derivatives depends upon the gradient in that component. To emphasize the physical concepts behind this derivative we begin by considering only the *vertical velocity component*, which in general is a function of all three spatial coordinates and time, $v_z = v_z(x, y, z, t)$. The material time derivative is defined as:

$$\frac{\mathrm{D}v_z}{\mathrm{D}t} = \frac{\partial v_z}{\partial t} + v_x \frac{\partial v_z}{\partial x} + v_y \frac{\partial v_z}{\partial y} + v_z \frac{\partial v_z}{\partial z} \tag{2.41}$$

The notation $\mathrm{D}/\mathrm{D}t$ is used to emphasize that this is a special form of derivative that is different from $\mathrm{d}/\mathrm{d}t$ or $\partial/\partial t$. The first term on the right side of (2.41) is the partial derivative of v_z with respect to time, taken at a *fixed position* in space. This is the time derivative of the local velocity, which is equal to the particle acceleration *only if* the last three terms are zero. Those last three terms describe variations in velocity with time (accelerations) of a *particle* that momentarily is at that fixed position in space, but is subject to a gradient in the component v_z in each coordinate direction. Together, the four terms on the right side of (2.41) account for the acceleration of the particle at that fixed position in space. In other words, Dv_z/Dt is the particle acceleration in the z-direction at a given position.

To provide a specific example of the material time derivative defined in (2.41), we focus attention on the centerline of the rising viscous sphere (**Figure 2.38b**) where $x=0, \ -a \leq z \leq a$. From the vector arrows along the centerline, velocity components v_x and v_y are zero. Also, because the flow is steady, the vertical velocity, v_z, is not a function of time. Therefore, the material time derivative (2.41) reduces to:

$$\frac{Dv_z}{Dt} = v_z \frac{\partial v_z}{\partial z}$$

Using the velocity distribution within the sphere that is derived in Section 11.6, the local velocity along the centerline is:

$$v_z = \tfrac{1}{2}\left(\frac{\zeta}{1+\zeta}\right) V_s \left[1 - \left(\frac{z}{a}\right)^2\right] \quad \text{on } x=0, \ -a \leq z \leq a$$

Here $\zeta = \eta_o/\eta_i$, where η_o and η_i are the viscosity *outside* and *inside* the sphere, V_s is the rise velocity of the sphere, and a is the sphere radius. The local velocity is proportional to the rise velocity and varies along the centerline from zero at the lower edge of the sphere, to a maximum at the center, to zero at the upper edge (**Figure 2.39**).

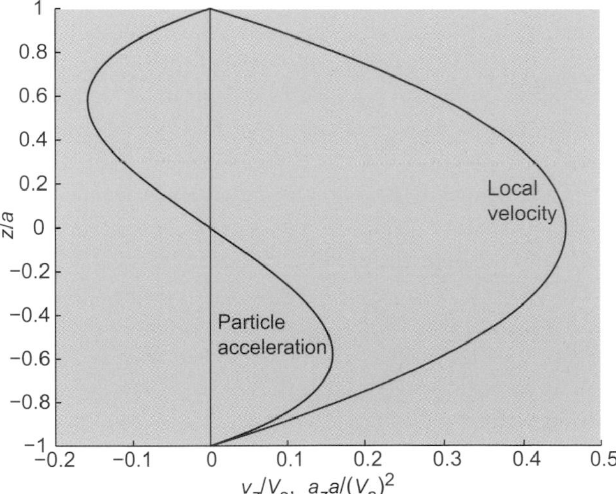

Figure 2.39 Local velocity and particle acceleration along centerline of viscous sphere rising in viscous surroundings. Refer to **Figure 2.38b** for the velocity vectors on a cross section of the sphere.

The change in local velocity in the vertical direction along the centerline is:

$$\frac{\partial v_z}{\partial z} = -\frac{1}{a^2}\left(\frac{\zeta}{1+\zeta}\right) V_s z \qquad x=0, \ -a \leq z \leq a$$

Multiplying the local velocity and its derivative, the material time derivative is:

$$\frac{Dv_z}{Dt} = \frac{V_s^2}{2a}\left(\frac{\zeta}{1+\zeta}\right)^2 \left[\left(\frac{z}{a}\right)^3 - \frac{z}{a}\right] \qquad x=0, \ -a \leq z \leq a$$

This is the particle acceleration at any fixed position along the centerline. It is proportional to the rise velocity squared and inversely proportional to the radius of the sphere. The distribution of the particle acceleration along the centerline is shown in **Figure 2.39**. Consistent with what can be inferred from the local velocity vector field (**Figure 2.38b**), the particle acceleration is positive in the lower half of the sphere and negative in the upper half, and it goes to zero at the bottom, middle, and top of the sphere.

Generalizing the material time derivative of velocity to consider all three components of the velocity vector, we have:

$$\begin{aligned}
\frac{Dv_x}{Dt} &= \frac{\partial v_x}{\partial t} + v_x\frac{\partial v_x}{\partial x} + v_y\frac{\partial v_x}{\partial y} + v_z\frac{\partial v_x}{\partial z}\\
\frac{Dv_y}{Dt} &= \frac{\partial v_y}{\partial t} + v_x\frac{\partial v_y}{\partial x} + v_y\frac{\partial v_y}{\partial y} + v_z\frac{\partial v_y}{\partial z}\\
\frac{Dv_z}{Dt} &= \frac{\partial v_z}{\partial t} + v_x\frac{\partial v_z}{\partial x} + v_y\frac{\partial v_z}{\partial y} + v_z\frac{\partial v_z}{\partial z}
\end{aligned} \tag{2.42}$$

The left sides of these equations are the three components of the particle acceleration, $\mathbf{a} = \langle a_x, a_y, a_z \rangle$. The first term on each right side is the time rate of change of the local velocity component. The second through fourth terms on each right side account for the variations in velocity with time (accelerations) of a *particle* that momentarily is at that fixed position in space, but has a velocity that varies spatially. That spatial gradient in velocity is carried through the fixed position, resulting in a local change in velocity with time. In Section 3.6 we find that the material time derivative of velocity (2.42) plays a prominent role in the equations of motion as constrained by conservation of linear momentum.

The material time derivative can be applied to a *scalar* quantity. Using **mass density**, ρ, as an example, the material time derivative is:

$$\frac{D\rho}{Dt} = \frac{\partial \rho}{\partial t} + v_x\frac{\partial \rho}{\partial x} + v_y\frac{\partial \rho}{\partial y} + v_z\frac{\partial \rho}{\partial z} \tag{2.43}$$

The first term on the right side is the time rate of change of density at a fixed point in space. The second term accounts for the time rate of change of density due to material with a gradient in density in the x-direction being carried through that point in space by the velocity in the x-direction. The third and fourth terms have similar interpretations for the y- and z-directions, respectively. We describe the terms in (2.43) from a physical point of view in Section 3.5 when introducing conservation of mass and deriving the equation of continuity.

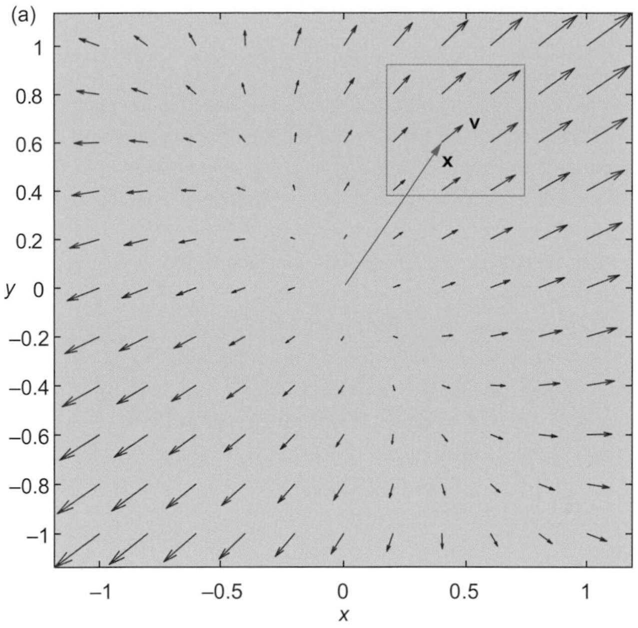

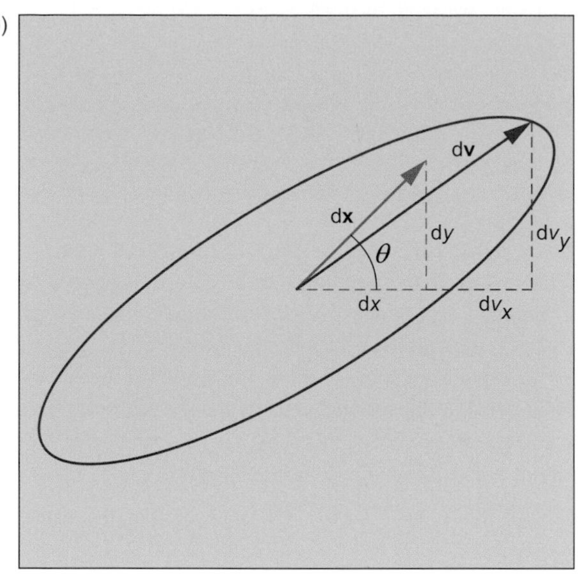

Figure 2.40 (a) Selected local velocity vectors for the field equations (2.44). Velocity **v** at position **x** is described in the text along with other vectors in the gray box. (b) Differential space centered on position **x** in (a) with infinitesimal material line overlain by d**x** that locates a neighboring point with velocity **v** + d**v**. The head of the relative velocity vector, d**v**, lies on an ellipse that graphically represents the spatial gradient in velocity tensor.

2.6 TENSOR QUANTITIES

In Section 2.1, we pointed out that originally spherical ooliths were deformed into ellipsoidal shapes as they were caught up in the deformation that formed South Mountain fold (**Figure 2.2** and **Figure 2.3**). The ellipsoidal shape enables one to visualize rock deformation in terms of the stretch of material line segments oriented in all possible directions from the center of the oolith. The mathematical construct we call the deformation gradient tensor quantifies the information embodied in the shape of the ellipsoid. Because an ellipsoid has three semi-axes that may be of different lengths, and because these axes are oriented in three orthogonal directions, this tensor must carry enough information to calculate these directions and the corresponding stretches. In this section, we describe the properties of this and other tensors that find application in structural geology.

2.6.1 The Gradient of a Vector Field is a Tensor Field

Recall from Section 2.5.3 that the gradient of a scalar field is a vector field. At any point in a deforming rock mass the gradient vector is directed toward the maximum increase of the scalar field, and the gradient vector magnitude is the spatial rate of increase of the scalar function in that direction. Thus, the gradient $\vec{\nabla} f$ defined in (2.40) provides quantitative information about spatial variations in the *scalar* field $f(x, y, z)$. In this section, we seek quantitative information about the spatial variations of a *vector* field, and show that the gradient of a vector field defines a tensor field that carries that information.

To illustrate the spatial variation of a vector field, we consider a two-dimensional velocity field in which both components are linear functions of the coordinates x and y:

$$\begin{aligned} v_x &= 1.3x + 0.5y \\ v_y &= 0.5x + 0.8y \end{aligned} \tag{2.44}$$

Some of the velocity vectors for this field are shown as arrows in **Figure 2.40a**. Focusing on one of the vectors, $\mathbf{v} = \langle 0.82, 0.68 \rangle$, located at the position vector $\mathbf{x} = \langle 0.4, 0.6 \rangle$, notice that every nearby vector in the gray box is different from **v** in both magnitude and direction. At some nearby positions (upper right corner of gray box) the velocity increases, at other positions (lower left) it decreases. At some nearby positions (upper left) the velocity vector rotates counter-clockwise; at other positions (lower right) it rotates clockwise. Now let the gray box shrink down about the position **x**, so the local change in velocity from **x** to any neighboring position, **x** + d**x**, an infinitesimal distance away, is given by the differential quantity d**v**. The velocity **v** and the velocity **v** + d**v** are not restricted in their magnitudes, but the relative velocity, d**v**, is infinitesimal.

The total differentials of each velocity component (see Section 2.5.3) are the components of the relative velocity, d**v**:

$$\begin{aligned} \mathrm{d}v_x &= \frac{\partial v_x}{\partial x}\mathrm{d}x + \frac{\partial v_x}{\partial y}\mathrm{d}y \\ \mathrm{d}v_y &= \frac{\partial v_y}{\partial x}\mathrm{d}x + \frac{\partial v_y}{\partial y}\mathrm{d}y \end{aligned} \tag{2.45}$$

The coefficients on the right sides of these equations are partial derivatives of the velocity components with respect to the spatial coordinates. The velocity components in (2.44) yield the following partial derivatives: $\partial v_x/\partial x = 1.3$; $\partial v_x/\partial y = 0.5$; $\partial v_y/\partial x = 0.5$; $\partial v_y/\partial y = 0.8$. Thus, taking **dx** as a unit direction vector and the angle between the x-axis and **dx** as θ (**Figure 2.40b**), the components of **dv** for all neighboring positions about **x** are:

$$\begin{aligned} dv_x &= 1.3\cos\theta + 0.5\sin\theta \\ dv_y &= 0.5\cos\theta + 0.8\sin\theta \end{aligned}$$

As the angle varies over the range $0 \le \theta \le 2\pi$, the heads of these relative velocity vectors trace out an ellipse (**Figure 2.40b**) that graphically represents how the two-dimensional velocity field changes in all directions about the position **x**.

To capture the spatial variation of the velocity vector **v** at any point in a two-dimensional velocity field, we use the concept of a gradient, and symbolize this using the del operator, $\nabla\mathbf{v}$. The four partial derivatives in (2.45) form an array of numbers that are the components of a tensor quantity:

$$\nabla\mathbf{v} = \begin{Bmatrix} \dfrac{\partial v_x}{\partial x} & \dfrac{\partial v_x}{\partial y} \\ \dfrac{\partial v_y}{\partial x} & \dfrac{\partial v_y}{\partial y} \end{Bmatrix} \quad (2.46)$$

Note that the del operator is not written with an overlying arrow, because the gradient of a vector is not a vector. However, each component of the velocity vector, **v**, is a scalar quantity, and in Section 2.5.3 we described how the del operator with an overlying arrow, $\vec{\nabla}$, is used to define the gradient of a scalar, which is a vector. Taking the scalar velocity component, v_y, as an example, the gradient vector is:

$$\vec{\nabla}v_y = \frac{\partial v_y}{\partial x}\hat{\mathbf{i}} + \frac{\partial v_y}{\partial y}\hat{\mathbf{j}}$$

The components of this gradient vector are the same partial derivatives that appear in the second row on the right side of (2.46). The first row on the right side of (2.46) is composed of the two partial derivatives that are the components of the gradient vector $\vec{\nabla}v_x$.

The gradient of the two-dimensional velocity vector, $\nabla\mathbf{v}$, is an array with two rows and two columns (2.46), consisting of the components of the gradient vectors for the two components of **v**. Therefore, the tensor with components defined in (2.46) is called the velocity gradient tensor, and the del operator is used symbolically on the left side of that equation. We employ this tensor in Section 6.6.2 to characterize the kinematics of viscous flow, and in Section 10.3.2 to quantify the deformation in a ductile shear zone.

The particular example of a vector field written in (2.44) and illustrated in **Figure 2.40a** is special because the velocity components are *linear* functions of the spatial coordinates. This means that the components of the spatial gradient of velocity tensor are constants, so this tensor field is uniform in space. In other words, the spatial gradient of velocity tensor has a *homogeneous* distribution in the (x, y)-plane. The one ellipse illustrated in **Figure 2.40b** represents the spatial variation in velocity at every position in the (x, y)-plane. Note, however, that a homogeneous gradient of velocity tensor is associated with a velocity vector field that looks quite complicated in terms of changing velocity magnitudes and directions with position.

In general, the velocity components may be *non-linear* functions of the spatial coordinates, which means that the components of the gradient of velocity tensor (2.46) are functions of the spatial coordinates. For these cases, at every position in the (x, y)-plane, an ellipse with a different shape and orientation would graphically represent how the local velocity field changes. In other words, the spatial gradient of velocity tensor has a *heterogeneous* distribution in the (x, y)-plane.

The velocity vector field in (2.44) also is special because $\partial v_x/\partial y = \partial v_y/\partial x$. In this case, the terms in the upper right and lower left of the array in (2.46) are equal, and we refer to this gradient of velocity as a *symmetric* tensor. In Section 2.6.5 we explore the special mathematical properties of symmetric tensors. In Section 6.6.2 we describe the mathematical properties of non-symmetric tensors and introduce the rate of deformation tensor and rate of spin tensor. All of these tensors play important roles in quantifying deformation as geologic structures develop.

The gradient of a vector field in two dimensions extends to three dimensions without introducing new concepts. For example, taking the velocity vector field as a function of the three Cartesian coordinates, $\mathbf{v} = \mathbf{v}(x, y, z)$, the components of the velocity gradient tensor are organized in an array as follows:

$$\nabla\mathbf{v} = \begin{Bmatrix} \dfrac{\partial v_x}{\partial x} & \dfrac{\partial v_x}{\partial y} & \dfrac{\partial v_x}{\partial z} \\ \dfrac{\partial v_y}{\partial x} & \dfrac{\partial v_y}{\partial y} & \dfrac{\partial v_y}{\partial z} \\ \dfrac{\partial v_z}{\partial x} & \dfrac{\partial v_z}{\partial y} & \dfrac{\partial v_z}{\partial z} \end{Bmatrix} \quad (2.47)$$

Each row of this array contains the components of the gradient vector for one of the scalar components of velocity. For example, the second row is composed of the three components of $\vec{\nabla}v_y$. Each of the nine components of the spatial gradient of velocity tensor (2.47) may be a function of the three spatial coordinates.

2.6.2 Matrix Representation of Tensors and Vectors

From (2.47) we understand that a tensor is a set of nine numbers, which we organize in an array with three rows and three columns. This organization, and the mathematical properties of tensors, enable us to use linear algebra to work with tensor quantities as matrices (see Ferguson, 1994; Strang, 2016). Because the components of vectors form an array with one row and three columns, or three rows and one column, they too are amenable to computations using linear algebra. We introduce elementary properties of matrices as applied to tensors and vectors in this section.

In general, a matrix is a rectangular array of numbers, and each number is an element of the matrix. We use square brackets and non-bold letters to indicate a matrix quantity such as [M], [v], or [T]. We number the rows and columns of an array sequentially from the upper left element. The row number and column number are used as subscripts to identify each element, and in general we place no restrictions on the number of rows or columns. If the matrix [M] has m rows and n columns, we say it is of *order m* by n, and write that in symbolic form as $m \times n$, where m and n are whole numbers. When it is helpful to clarify matrix order, we write the corresponding $m \times n$ under the square brackets. Thus, the arbitrary matrix [M] is:

$$\underset{m\times n}{[\mathrm{M}]} = \begin{bmatrix} M_{11} & M_{12} & \cdots & M_{1n} \\ M_{21} & M_{22} & \cdots & M_{2n} \\ \cdots & \cdots & \cdots & \cdots \\ M_{m1} & M_{m2} & \cdots & M_{mn} \end{bmatrix} \tag{2.48}$$

If $m = n$, the matrix is called a square matrix. In this case, the elements $M_{11}, M_{22}, \ldots, M_{\mathrm{nn}}$ constitute the principal diagonal of the square matrix [M]. To quantify tensors in the physical context, we restrict attention to square matrices that are of order 2 by 2 or 3 by 3.

If m or n equal 1, the matrix [M] in expanded form is a row matrix or a column matrix:

$$\underset{1\times n}{[\mathrm{M}]} = [M_1\ M_2\ \ldots\ M_n], \text{or } \underset{m\times 1}{[\mathrm{M}]} = \begin{bmatrix} M_1 \\ M_2 \\ \vdots \\ M_m \end{bmatrix} \tag{2.49}$$

Note that the row matrix has n columns and the column matrix had m rows. In both cases, only one subscript is required. In some literature all row and column matrices are referred to as "vectors," even if they have more than three elements, but we restrict usage of the term *vector* to the physical context, with only two or three elements.

We define vectors and tensors using right-handed Cartesian coordinates (**Figure 2.9**) with three orthogonal coordinates (x, y, z). Symbolically we use lower-case bold upright letters for vectors (e.g. **u**, **v**, **w**). For some tensors we use upper-case bold upright letters (e.g. **R**, **S**, **T**); for others we use bold Greek letters (e.g. **σ**, **ε**, **ω**). The components of the arbitrary vector, **v**, are the elements of a *row matrix* or a *column matrix*:

$$\underset{1\times 3}{[\mathrm{v}]} = [v_x\quad v_y\quad v_z], \text{or } \underset{3\times 1}{[\mathrm{v}]} = \begin{bmatrix} v_x \\ v_y \\ v_z \end{bmatrix} \tag{2.50}$$

The *elements* of these matrices have symbolic names using the same letter as the vector with one subscript in the sequence (x, y, z) from left to right, or top to bottom. The components of the arbitrary tensor, **T**, are the elements of a *square matrix*:

$$\underset{3\times 3}{[\mathrm{T}]} = \begin{bmatrix} T_{xx} & T_{xy} & T_{xz} \\ T_{yx} & T_{yy} & T_{yz} \\ T_{zx} & T_{zy} & T_{zz} \end{bmatrix} \tag{2.51}$$

The *elements* of this matrix have symbolic names using the same letter as the tensor with two subscripts indicating, respectively, the row and column occupied by that element. For example, the component T_{yz} occupies the 2nd row (associated with the y coordinate) and 3rd column (associated with the z coordinate).

Subscripts on elements of a square matrix representing a tensor quantity usually carry information about the physical nature of that quantity. For example, the stress tensor, **σ**, is written in matrix form as:

$$\underset{3\times 3}{[\sigma]} = \begin{bmatrix} \sigma_{xx} & \sigma_{xy} & \sigma_{xz} \\ \sigma_{yx} & \sigma_{yy} & \sigma_{yz} \\ \sigma_{zx} & \sigma_{zy} & \sigma_{zz} \end{bmatrix} \tag{2.52}$$

The component σ_{yz} refers to a force per unit area (stress) that acts *on* a plane perpendicular to the y-axis and *in* the direction of the x-axis (**Figure 2.41**). Similarly, the component σ_{xx} refers to a force per unit area (stress) that acts *on* a plane perpendicular to the x-axis and *in* the direction of the x-axis. In this way, the subscripts provide important insight about the planes *on* which the stress acts and the coordinate direction *in* which the stress acts. This is an on-in convention. We describe this tensor in more detail in Section 4.6.

The stress tensor, **σ**, is an example of another important mathematical property of some of the tensors: the elements form a symmetric matrix. This means that elements in symmetric

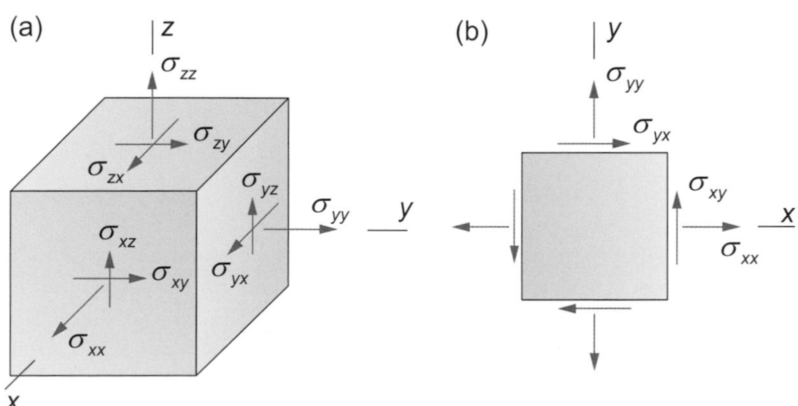

Figure 2.41 (a) Components of the stress tensor (force per unit area) acting on a cubic element. For each arrow shown, an arrow of equal length acting in the opposite direction exists on the parallel hidden face, so the net force on the element is zero. (b) Components of the stress tensor illustrated in 2D acting on faces of the cubic element that are perpendicular to the (x, y)-plane.

Box 2.2 Matrix Laboratory (MATLAB)

The authors of this textbook have found that MATLAB is an invaluable tool for teaching structural geology. It is particularly helpful for students exposed for the first time to the quantitative approach to structural geology taken in this textbook. In fact, MATLAB produced many of the figures in this book, and MATLAB helped the authors understand and test most of the mathematical concepts described in the book. Therefore, we strongly advocate adopting this tool and writing MATLAB scripts to analyze structural data and build models of tectonic processes. We have found that a computational engine like MATLAB provides helpful tools and support for this learning process (see Allmendinger et al., 2012, in Further Reading). Scripts with dynamic three-dimensional graphical output (e.g. **Figure B2.2**) offer spatial feedback to alter incorrect mental models and build intuition. These tools, along with an elementary understanding of vector calculus and differential geometry open the door for thinking in three dimensions (see Davis and Titus, 2017).

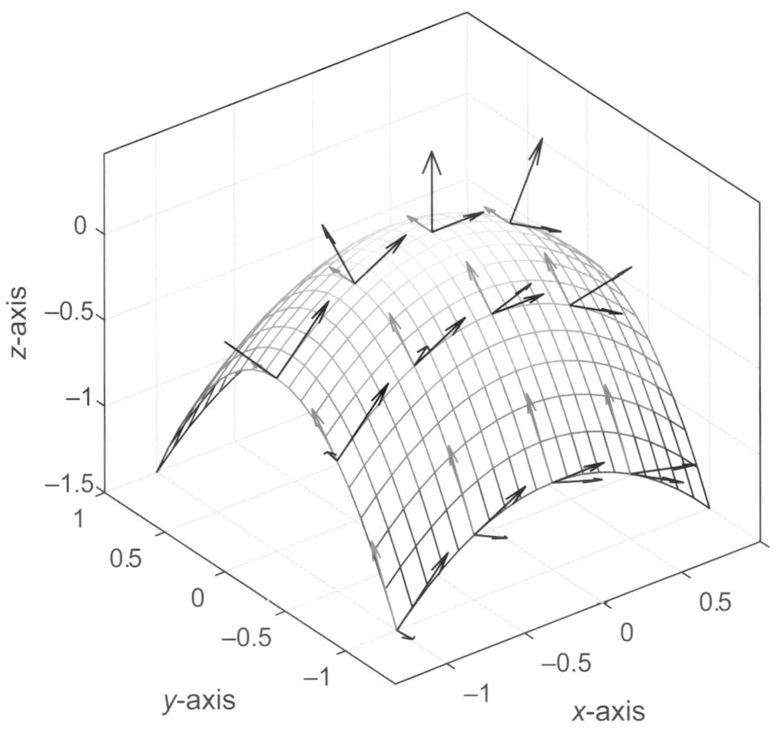

Figure B2.2 MATLAB model of folded strata approximated as an elliptic paraboloid (multi-colored wire mesh) with tangent vectors (blue and green) parallel to *u*- and *v*-parameter curves respectively, and unit normal vectors (red).

Becoming an accomplished user of MATLAB puts students in a good position to succeed in careers in structural geology. In part, that follows from the enhanced ability to communicate with professionals from related disciplines such as geophysics, rock mechanics, civil and petroleum engineering, who also use quantitative methods, and many of whom use MATLAB. In part, that follows from the help that MATLAB offers to gain confidence in otherwise abstract mathematical concepts (e.g. eigenvalue problems or boundary-value problems) by evaluating functions and graphing results. In part, that also follows from the intuition building that accompanies three-dimensional graphical output (**Figure B2.2**) and watching it rotate in real time on your computer monitor. The close connection between visualization and understanding is emphasized and enhanced using MATLAB's graphical user interface.

MATLAB provides an interactive environment for numerical computations along with a wide variety of data analysis, graphical, and visualization tools. Cleve Moler, then a professor at the University of New Mexico, conceived MATLAB in the 1970s to make FORTRAN matrix libraries more accessible to his students. In 1983 Moler joined forces with John Little to write a professional version with graphics functionality. In 1984 Moler (Chief Mathematician) and Little (President) founded MathWorks Inc. to develop MATLAB as a commercial computational and graphical engine. MATLAB has been evolving and improving rapidly since then, with more than 3 million users worldwide, primarily in science and engineering fields, and more than 5,000 colleges and universities using MATLAB for teaching and research in 2018.

The user interacts with MATLAB through its graphical user interface (GUI) by typing individual commands into the MATLAB command window, or by writing *scripts* and *functions* that consists of a set of commands performed in sequence. Both scripts and functions are written by the user, and may be saved for future use. MATLAB scripts enable one to change the value of one or more variables and to reevaluate a number of commands in order to explore the outcomes, usually with graphical output. MATLAB scripts act upon variables defined in the script, or defined in the MATLAB workspace. The variables calculated in a script remain in the MATLAB workspace until cleared. Scripts are learning tools, designed to address classic "what if" questions.

MATLAB functions have two key attributes that scripts lack: (1) an optional list of specific input variables (or arguments); and (2) an optional list of specified output variables. MATLAB contains many built-in functions such as trigonometric functions. Importantly, only the variables explicitly listed in the output list of a function are included in the workspace. Thus, functions can be used like "black boxes" that accept input and return output, but where the inner computational workings are concealed. Most functions are canned routines employing well-known mathematics, because how it operates is not in dispute.

The bulk of the MATLAB files we provide with this textbook are scripts, because we want the reader to investigate the inner workings of the MATLAB codes and not accept them like "black boxes." We believe that working with scripts promotes a better teaching and learning environment for students of structural geology, and prepares students for more advanced courses in this subject (see Pollard and Fletcher, 2005, in Further Reading).

positions across the principal diagonal are equal: $\sigma_{xy} = \sigma_{yx}$, $\sigma_{yz} = \sigma_{zy}$, $\sigma_{zx} = \sigma_{xz}$. In Section 2.6.6 we describe some of the important mathematical consequences of matrix symmetry. In Section 3.7 we describe one of the underlying physical principles of deformation, conservation of angular momentum, and point out that this leads to the symmetry of the stress tensor.

2.6.3 Elementary Matrix Operations

Addition and subtraction are possible only if the two matrices are of the same order, because each element in the first matrix adds to, or subtracts from, the corresponding element of the second matrix:

$$\underset{3\times3}{[\mathrm{S}]} + \underset{3\times3}{[\mathrm{T}]} = \begin{bmatrix} S_{xx}+T_{xx} & S_{xy}+T_{xy} & S_{xz}+T_{xz} \\ S_{yx}+T_{yx} & S_{yy}+T_{yy} & S_{yz}+T_{yz} \\ S_{zx}+T_{zx} & S_{zy}+T_{zy} & S_{zz}+T_{zz} \end{bmatrix} \quad (2.53)$$

To multiply a matrix by the scalar c one multiplies each element by that scalar:

$$c\underset{3\times3}{[\mathrm{T}]} = \begin{bmatrix} cT_{xx} & cT_{xy} & cT_{xz} \\ cT_{yx} & cT_{yy} & cT_{yz} \\ cT_{zx} & cT_{zy} & cT_{zz} \end{bmatrix} \quad (2.54)$$

Multiplication of matrices of different order is possible under certain restricted conditions. For example, one can multiply the 3 by 3 matrix [T] by the 3 by 1 matrix [w] as follows:

$$\underset{3\times3}{[\mathrm{T}]}\underset{3\times1}{[\mathrm{w}]} = \begin{bmatrix} T_{xx} & T_{xy} & T_{xz} \\ T_{yx} & T_{yy} & T_{yz} \\ T_{zx} & T_{zy} & T_{zz} \end{bmatrix} \begin{bmatrix} w_x \\ w_y \\ w_z \end{bmatrix} = \begin{bmatrix} T_{xx}w_x + T_{xy}w_y + T_{xz}w_z \\ T_{yx}w_x + T_{yy}w_y + T_{yz}w_z \\ T_{zx}w_x + T_{zy}w_y + T_{zz}w_z \end{bmatrix} \quad (2.55)$$

This is analogous to taking the scalar product of two vectors as in (2.6), except here the matrix [T] is thought of as composed of three row vectors, and the scalar product of each of those is taken with the column vector [w]. The result is a 3 by 1 column vector.

A 3 by 3 matrix results from the multiplication of two 3 by 3 matrices. Each element of the resulting matrix is the scalar product of a row from the first matrix and a column from the second matrix, as if they were vectors. The subscripts of the resulting matrix elements reveal which row and column form the scalar product. For example, consider the matrix multiplication [S][T] = [U]:

$$\begin{bmatrix} S_{xx} & S_{xy} & S_{xz} \\ S_{yx} & S_{yy} & S_{yz} \\ S_{zx} & S_{zy} & S_{zz} \end{bmatrix} \begin{bmatrix} T_{xx} & T_{xy} & T_{xz} \\ T_{yx} & T_{yy} & T_{yz} \\ T_{zx} & T_{zy} & T_{zz} \end{bmatrix} = \begin{bmatrix} U_{xx} & U_{xy} & U_{xz} \\ U_{yx} & U_{yy} & U_{yz} \\ U_{zx} & U_{zy} & U_{zz} \end{bmatrix} \quad (2.56)$$

The element in the second row and third column of [U] is U_{yz} and its subscripts indicate this element is composed of the scalar product of the second row of [S] and the third column of [T]: $U_{yz} = S_{yx}T_{xz} + S_{yy}T_{yz} + S_{yz}T_{zz}$.

Guided by (2.55) and (2.56) some generalizations follow about matrix multiplication. When multiplying two matrices, the number of columns of the first matrix must equal the number of rows of the second matrix. The number of rows of the first matrix and the number of columns of the second determine the order of the resulting matrix. Written schematically in terms of matrix order, the condition for two matrices to be conformable for multiplication is:

$$[m \times n][n \times o] = [m \times o] \quad (2.57)$$

A check of the number of rows and columns in each matrix reveals whether or not they are conformable for multiplication, and the order of the resulting matrix. The matrix multiplication [T][w] in (2.55) is consistent with the condition for conformability (2.57), because [T] has three columns and [w] has three rows. The resulting matrix has three rows and one column; it is a column vector.

Using the matrices [T] and [w] from (2.55), we ask: is it possible to reverse the sequence of multiplication and calculate [w][T]? Because [w] has one column and [T] has three rows, the condition of conformability (2.57) is not satisfied, so this multiplication is not possible. Therefore, we conclude that matrix multiplication is not necessarily commutative. The multiplication of two scalar quantities, a and b, is commutative: $ab = ba$. Unlike this example, one cannot necessarily change the sequence of matrix multiplication. For this reason, a special terminology specifies the sequence of multiplication. For the product [T][w], the matrix [T] is the *premultiplier* and the vector [w] is the *postmultiplier*. One says [T] premultiplies [w], or [w] postmultiplies [T]. These words help to ensure that one communicates the correct sequence.

The example of matrix multiplication in (2.56) suggests the possibility that [S][T] = [T][S], because both [S] and [T] are square matrices of the same order. Consider this possibility by letting [T] premultiply [S] and call the product [V]:

$$\begin{bmatrix} T_{xx} & T_{xy} & T_{xz} \\ T_{yx} & T_{yy} & T_{yz} \\ T_{zx} & T_{zy} & T_{zz} \end{bmatrix}\begin{bmatrix} S_{xx} & S_{xy} & S_{xz} \\ S_{yx} & S_{yy} & S_{yz} \\ S_{zx} & S_{zy} & S_{zz} \end{bmatrix} = \begin{bmatrix} V_{xx} & V_{xy} & V_{xz} \\ V_{yx} & V_{yy} & V_{yz} \\ V_{zx} & V_{zy} & V_{zz} \end{bmatrix}$$

The element in the second row and third column of [V] is V_{yz} and its subscripts indicate this element is composed of the scalar product of the second row of [T] and the third column of [S]: $V_{yz} = T_{yx}S_{xz} + T_{yy}S_{yz} + T_{yz}S_{zz}$. The same element of the matrix [U] from (2.56) is $U_{yz} = S_{yx}T_{xz} + S_{yy}T_{yz} + S_{yz}T_{zz}$. The two multiplications, [S][T] and [T][S] produce the same result only if all of the corresponding elements of [U] and [V] are equal, for example $U_{yz} = V_{yz}$. This very stringent requirement indicates that the multiplication of conformable matrices is not necessarily commutative.

2.6.4 Transpose, Orthogonal, Identity, and Inverse Matrix

Several different tensors play prominent roles in the mechanics and kinematics of deformation. These tensors, written as square matrices, are manipulated using important mathematical operations from linear algebra that we introduce in this section. The transpose, orthogonal, identity, and inverse of a square matrix arise in these operations.

We use the apostrophe to indicate the transpose of a matrix, so [T′] is the transpose of the arbitrary matrix [T] defined in (2.51). To find the transposed matrix, one interchanges each row with the corresponding column of the original matrix. Thus, row one becomes column one; row 2 becomes column two; and so forth. Taking the transpose does not change the elements on the principal diagonal of a square matrix. For example, using the square matrix [T] introduced in (2.51) as the original matrix, the transposed matrix is:

$$\underset{3\times 3}{[\mathrm{T}']} = \begin{bmatrix} T_{xx} & T_{yx} & T_{zx} \\ T_{xy} & T_{yy} & T_{zy} \\ T_{xz} & T_{yz} & T_{zz} \end{bmatrix} \tag{2.58}$$

For a symmetric matrix, like the stress tensor defined in (2.52), the transpose is equal to the original matrix, $[\sigma'] = [\sigma]$, because elements in symmetric positions across the principal diagonal are equal. The transpose of the 3 by 1 row matrix representing the vector **v** in the first of (2.50) is the corresponding 1 by 3 column matrix in the second of (2.50).

The addition of the transpose of a square matrix and the matrix itself is a *symmetric* matrix. For example, using an arbitrary 2 by 2 matrix [W] we find:

$$\underset{2\times 2}{[\mathrm{W}']} + \underset{2\times 2}{[\mathrm{W}]} = \begin{bmatrix} W_{xx} & W_{yx} \\ W_{xy} & W_{yy} \end{bmatrix} + \begin{bmatrix} W_{xx} & W_{xy} \\ W_{yx} & W_{yy} \end{bmatrix} = \begin{bmatrix} 2W_{xx} & W_{yx}+W_{xy} \\ W_{xy}+W_{yx} & 2W_{yy} \end{bmatrix} \tag{2.59}$$

The elements of the secondary diagonal in the resulting summation are equal.

The product of the transpose of a square matrix and the matrix itself is a *symmetric* matrix:

$$\underset{2\times 2}{[\mathrm{W}']}\,\underset{2\times 2}{[\mathrm{W}]} = \begin{bmatrix} W_{xx} & W_{yx} \\ W_{xy} & W_{yy} \end{bmatrix}\begin{bmatrix} W_{xx} & W_{xy} \\ W_{yx} & W_{yy} \end{bmatrix} = \begin{bmatrix} W_{xx}^2 + W_{yx}^2 & W_{xx}W_{xy} + W_{yx}W_{yy} \\ W_{xy}W_{xx} + W_{yy}W_{yx} & W_{xy}^2 + W_{yy}^2 \end{bmatrix} \tag{2.60}$$

The elements of the secondary diagonal in the resulting product are equal. The product [W][W′] also is symmetric, but matrix multiplication is not necessarily commutative, so this product is not necessarily the same as (2.60).

The transpose of the sum of two matrices is equal to the sum of the transpose of those matrices:

$$([\mathrm{S}] + [\mathrm{T}])' = [\mathrm{S}'] + [\mathrm{T}'] \tag{2.61}$$

In addition, the transpose of the product of two matrices is equal to the product of the transpose of those matrices *in the reverse order*:

$$([\mathrm{S}][\mathrm{T}])' = [\mathrm{T}'][\mathrm{S}'] \tag{2.62}$$

In other words, on the left side [S] premultiplies [T], and then one takes the transpose of this product. On the right side the transposed matrix [T′] premultiplies the transposed matrix [S′].

Deformation at any point in a deforming rock mass may include a pure rotation, which results in no change in length of material lines in any orientation at that point, and no change in the angular relationships of these lines. In Section 5.5 we introduce and describe in detail the role of rotation in deformation, so here we focus only on the special properties of rotation matrices. For example, the following matrix provides a pure rotation about the z-axis through an angle ω_z for material lines lying in the (x, y)-plane:

$$[\mathrm{R}] = \begin{bmatrix} \cos\omega_z & -\sin\omega_z & 0 \\ \sin\omega_z & \cos\omega_z & 0 \\ 0 & 0 & 1 \end{bmatrix} \tag{2.63}$$

To visualize this rotation consider three unit vectors that overlie material lines extending from the origin along the coordinate axes before rotation (**Figure 2.42**):

$$[\mathrm{u}] = \begin{bmatrix} 1 \\ 0 \\ 0 \end{bmatrix},\quad [\mathrm{v}] = \begin{bmatrix} 0 \\ 1 \\ 0 \end{bmatrix},\quad [\mathrm{w}] = \begin{bmatrix} 0 \\ 0 \\ 1 \end{bmatrix}$$

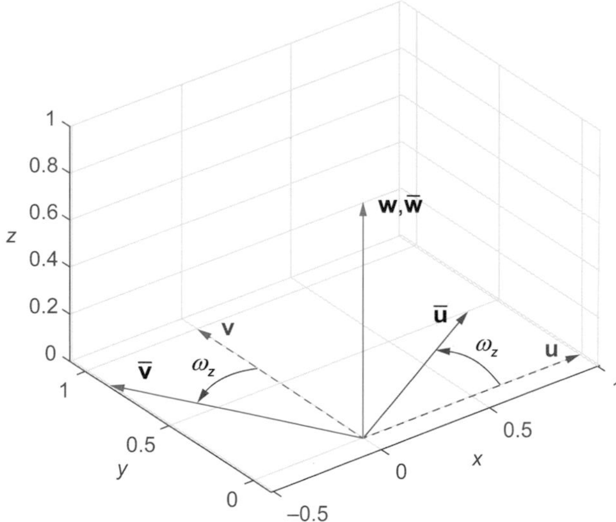

Figure 2.42 Pure rotation of material lines about the z-axis through angle ω_z as tracked by vectors **u**, **v**, and **w**, originally coincident with the coordinate axes. After rotation the new vectors are $\bar{\mathbf{u}}, \bar{\mathbf{v}}$, and $\bar{\mathbf{w}}$. The sense of rotation follows a right-hand rule: it is in the direction of the fingers of your right hand with your thumb pointing in the direction of the positive z-axis.

These are unit vectors and they are orthogonal. Using [R] from (2.63) as the pre-multiplier for each vector, we find the rotated vectors:

$$[\bar{\mathrm{u}}] = \begin{bmatrix} \cos\omega_z \\ \sin\omega_z \\ 0 \end{bmatrix}, [\bar{\mathrm{v}}] = \begin{bmatrix} -\sin\omega_z \\ \cos\omega_z \\ 0 \end{bmatrix}, [\bar{\mathrm{w}}] = \begin{bmatrix} 0 \\ 0 \\ 1 \end{bmatrix} \quad (2.64)$$

The rotated vectors, $\bar{\mathbf{u}}$ and $\bar{\mathbf{v}}$, are unit vectors, because their magnitudes from (2.3) are $(\pm\sin\omega_Z)^2 + (\cos\omega_Z)^2 = 1$. Furthermore, $\bar{\mathbf{w}}$ is unchanged by the rotation, so it remains a unit vector. Thus, the material lines associated with these vectors do not stretch. In addition, $\bar{\mathbf{u}}$ and $\bar{\mathbf{v}}$ remain orthogonal, because their scalar product from (2.7) is $\bar{\mathbf{u}}\cdot\bar{\mathbf{v}} = -\cos\omega_z\sin\omega_z + \sin\omega_z\cos\omega_z = 0$, and they remain in the (x, y)-plane because their z-component is zero. Also, $\bar{\mathbf{w}}$ remains perpendicular to the (x, y)-plane, because its x- and y-components are zero, so $\bar{\mathbf{w}}\cdot\bar{\mathbf{u}} = 0$ and $\bar{\mathbf{w}}\cdot\bar{\mathbf{v}} = 0$. Thus, the material lines do not change their angular relationships during the pure rotation.

Each rotation matrix is an orthogonal matrix, and orthogonal matrices have special properties. We showed above that the elements in each column of [R] are the components of orthogonal unit vectors. The elements in the rows of [R] also are the components of orthogonal unit vectors. Furthermore, either pre- or postmultiplying a rotation matrix by the transpose of that matrix produces a matrix with ones on the diagonal and zeros off the diagonal:

$$[\mathrm{R}'][\mathrm{R}] = [\mathrm{R}][\mathrm{R}'] = \begin{bmatrix} 1 & 0 & 0 \\ 0 & 1 & 0 \\ 0 & 0 & 1 \end{bmatrix} = [\mathrm{I}] \quad (2.65)$$

[I] is the identity matrix. The identity matrix, as the name implies, transforms any vector into itself. For example, premultiplying the arbitrary vector [v] using the identity matrix from (2.65), we find:

$$\underset{3\times3}{[\mathrm{I}]}\,\underset{3\times1}{[\mathrm{v}]} = \begin{bmatrix} 1 & 0 & 0 \\ 0 & 1 & 0 \\ 0 & 0 & 1 \end{bmatrix}\begin{bmatrix} v_x \\ v_y \\ v_z \end{bmatrix} = \begin{bmatrix} v_x \\ v_y \\ v_z \end{bmatrix} = \underset{3\times1}{[\mathrm{v}]}$$

Postmultiplying [v] by [I] does not satisfy conformability for multiplication (2.57), so that operation is not possible. On the other hand, pre- or postmultiplying the matrix representation of a tensor by the identity matrix returns the same matrix. Using the elements of the 3 by 3 square matrix [T] defined in (2.51), we have:

$$[\mathrm{I}][\mathrm{T}] = [\mathrm{T}][\mathrm{I}] = [\mathrm{T}]$$

This is one of the rare examples of commutative matrix multiplication.

In (2.64) and **Figure 2.42** we showed that when the rotation matrix [R] premultiplies the orthogonal unit vectors, [u] and [v], two rotated orthogonal unit vectors are produced. The elements of the rotation matrix are the components of the tensor **R**. This illustrates a property of vectors and tensors that is more general, because it applies to vectors that are not necessarily unit vectors, and to tensors that are not necessarily orthogonal tensors. Simply stated, a tensor, **T**, is a linear transformation that transforms any vector, **v**, into another vector, **w**. In matrix form this general relationship is:

$$\underset{3\times3}{[\mathrm{T}]}\,\underset{3\times1}{[\mathrm{v}]} = \underset{3\times1}{[\mathrm{w}]} \quad (2.66)$$

If the relationship in (2.66) were rewritten for scalar quantities as $ab = c$, one would solve for b using the familiar reciprocal relationship: $a^{-1}ab = a^{-1}c$ or $b = a^{-1}c$. A similar reciprocal relationship exists for some matrices, in which case we have:

$$\underset{3\times3}{[\mathrm{T}^{-1}]}\,\underset{3\times3}{[\mathrm{T}]}\,\underset{3\times1}{[\mathrm{v}]} = \underset{3\times3}{[\mathrm{T}^{-1}]}\,\underset{3\times1}{[\mathrm{w}]} \text{ or } \underset{3\times1}{[\mathrm{v}]} = \underset{3\times3}{[\mathrm{T}^{-1}]}\,\underset{3\times1}{[\mathrm{w}]} \quad (2.67)$$

In this case, one refers to $[\mathrm{T}^{-1}]$ as the inverse of [T]. The defining property of an orthogonal matrix is the fact that the inverse is equal to the transpose:

$$[\mathrm{R}^{-1}] = [\mathrm{R}'] \quad (2.68)$$

This explains why the product of the transpose of an orthogonal matrix and the matrix itself, in either order, is the identity matrix (2.65).

Although we use the reciprocal relationship for scalars with impunity, the reciprocal relationship for vectors and tensors written as matrices may not hold, because the inverse of a square matrix [T] may not exist. In this case, one refers to the square matrix [T] as singular. For example, [T] is singular if the elements of one row, or of one column, are zero. [T] is singular if the corresponding elements of two rows, or two columns, are equal. [T] is singular if the elements of a row (or a column) are scalar multiples of another row (or column). In addition, [T] is singular if the elements of a

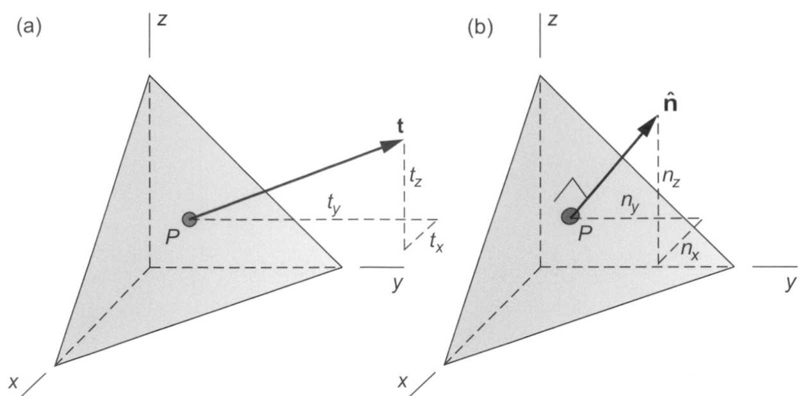

Figure 2.43 (a) Components of the traction vector **t** (force per unit area vector) acting on the face of a tetrahedron. Traction components on the hidden faces balance the force due to the traction. (b) Components of the unit normal vector $\hat{\mathbf{n}}$ on the face of the tetrahedron. See Pollard and Fletcher (2005) in Further Reading.

row (or a column) are linear combinations of the corresponding elements of the other two rows (or two columns).

If [T] is not singular, it is invertible, and the product of the square matrix and its inverse is the identity matrix:

$$[\mathrm{T}^{+1}][\mathrm{T}^{-1}] = [\mathrm{T}^{-1}][\mathrm{T}^{+1}] = [\mathrm{T}^{0}] = [\mathrm{I}]$$

This is one of those rare examples of commutative matrix multiplication. Also, we see that raising an invertible matrix [T] to the zeroth power produces the identity matrix.

2.6.5 Tensors as Linear Transformations: Cauchy's Formula

In Section 2.6.1 we described tensors as mathematical entities that quantify how a vector field changes in all directions from a given point (**Figure 2.40**). In Section 2.6.2 we used the components of tensors to populate the elements of square 2 by 2 or 3 by 3 matrices, and reviewed how to manipulate tensors using matrix algebra. In this section the *linear* properties of a tensor are described, and we emphasize that a tensor transforms one vector into another vector. We saw this mathematical relationship using the rotation tensor [R] in (2.64) which transformed orthogonal unit vectors into other orthogonal unit vectors (**Figure 2.42**). This elucidated the kinematics of deformation by showing that under a pure rotation, material lines do not stretch or shear.

The elements of the 3 by 3 square matrix [T] are the components of a tensor, **T**, if [T] satisfies two relationships when it premultiplies arbitrary vectors, represented here by the 3 by 1 column matrices [v] and [w]. The first relationship is:

$$\underset{3\times3}{[\mathrm{T}]}\left(\underset{3\times1}{[\mathrm{v}]} + \underset{3\times1}{[\mathrm{w}]}\right) = \underset{3\times3}{[\mathrm{T}]}\,\underset{3\times1}{[\mathrm{v}]} + \underset{3\times3}{[\mathrm{T}]}\,\underset{3\times1}{[\mathrm{w}]} \tag{2.69}$$

In other words, premultiplying the sum of two vectors by a tensor equals the sum of the premultiplication of each vector by the tensor. This is reminiscent of the familiar algebraic relationship for scalars: $a(b + c) = ab + ac$.

The second relationship involves the arbitrary scalar quantity d:

$$\underset{3\times3}{[\mathrm{T}]}\left(d\,\underset{3\times1}{[\mathrm{v}]}\right) = d\,\underset{3\times3}{[\mathrm{T}]}\,\underset{3\times1}{[\mathrm{v}]} \tag{2.70}$$

In other words, premultiplying the product of the scalar and the vector by a tensor equals the product of the scalar and the premultiplication of the vector by the tensor. Again this is reminiscent of the familiar algebraic relationship for scalars: $a(db) = d(ab)$. Satisfying (2.69) and (2.70) means that [T] is a linear transformation, and that it is a tensor.

To apply the properties of a tensor to a physical quantity that is the causative agent for rock deformation, we turn to the stress tensor. If the transpose of the stress tensor defined in (2.52) premultiplies a unit normal vector with components $\langle n_x, n_y, n_z \rangle$, a traction vector with components $\langle t_x, t_y, t_z \rangle$ is found that acts on the plane defined by the unit normal vector (**Figure 2.43**). The transposed stress tensor transforms the unit normal vector into the traction vector.

The matrix operation that defines the traction vector is the matrix multiplication: $[\mathrm{t}] = [\sigma'][\mathrm{n}]$. Recall that $[\sigma']$ is the transpose of the stress tensor, so we have:

$$\begin{bmatrix} t_x \\ t_y \\ t_z \end{bmatrix} = \begin{bmatrix} \sigma_{xx} & \sigma_{yx} & \sigma_{zx} \\ \sigma_{xy} & \sigma_{yy} & \sigma_{zy} \\ \sigma_{xz} & \sigma_{yz} & \sigma_{zz} \end{bmatrix} \begin{bmatrix} n_x \\ n_y \\ n_z \end{bmatrix} \tag{2.71}$$

Each element of the traction vector is a *linear* combination of the elements in the corresponding row of the stress tensor and the elements of the normal vector:

$$\begin{aligned} t_x &= \sigma_{xx}n_x + \sigma_{yx}n_y + \sigma_{zx}n_z \\ t_y &= \sigma_{xy}n_x + \sigma_{yy}n_y + \sigma_{zy}n_z \\ t_z &= \sigma_{xz}n_x + \sigma_{yz}n_y + \sigma_{zz}n_z \end{aligned} \tag{2.72}$$

Note that each traction vector component acts in the same direction as the stress tensor components that contribute to it (**Figure 2.44**). For example, study the subscripts in the second line of (2.72) to see that the stress components acting on x in y and on y in y, and on z in y contribute to the traction component acting in y. The set of equations (2.72) relating traction and stress components is Cauchy's formula.

Because the stress matrix is symmetric, the transpose is equal to the original matrix. Thus, we could have written (2.71) as $[\mathrm{t}] = [\sigma][\mathrm{n}]$, without taking the transpose of the stress tensor. However, that obscures the physics of this relationship because the traction vector component would not act in the same direction as the stress

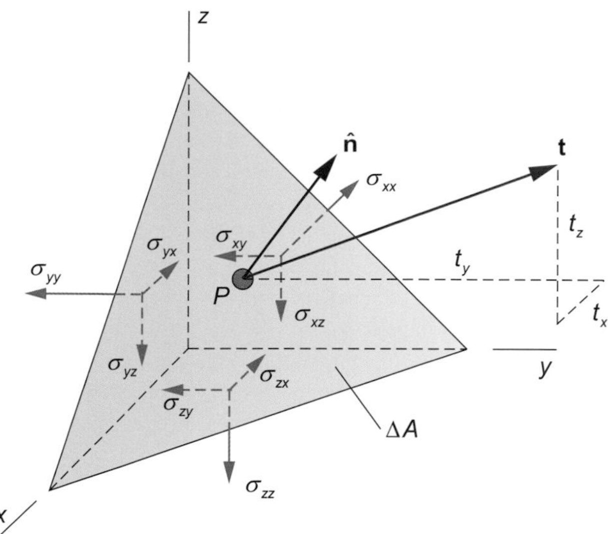

Figure 2.44 Three orthogonal faces of the Cauchy tetrahedron lie in the coordinate planes. The oblique face has area ΔA, with outward unit normal, $\hat{\mathbf{n}}$, and is acted upon by the traction, $\mathbf{t} = \langle t_x, t_y, t_z \rangle$. The stress components are σ with subscripts following the "on-in" convention. See Pollard and Fletcher (2005) in Further Reading.

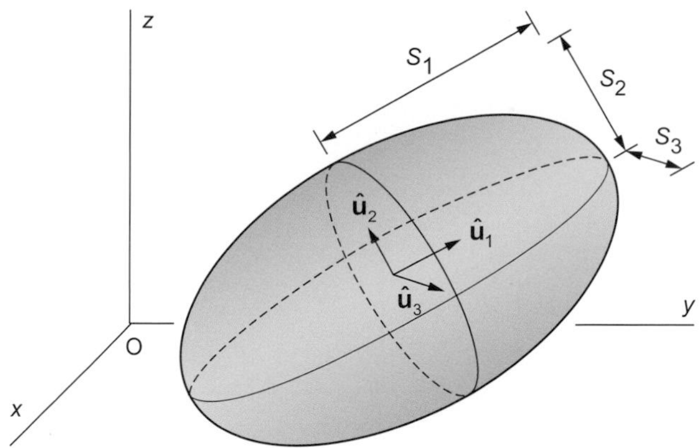

Figure 2.45 Schematic illustration of principal stretch magnitudes, S_1, S_2, and S_3, and directions, $\hat{\mathbf{u}}_1, \hat{\mathbf{u}}_2$, and $\hat{\mathbf{u}}_3$ for a deformed oolith (Figure 2.3).

tensor components that contribute to it (Figure 2.41). For example, the second of (2.72) would be $t_y = \sigma_{yx}n_x + \sigma_{yy}n_y + \sigma_{yz}n_z$. Recalling the on-in convention for subscripts, this equation suggests that some stress components acting in the x- and z-directions contribute to the traction component acting in the y-direction.

2.6.6 The Principal Values and Principal Vectors of a Symmetric Tensor

The ooliths described in Section 2.1 were nearly spherical in their undeformed state and nearly ellipsoidal in their deformed state (Figure 2.3). The three unit vectors parallel to the axes of a model deformed oolith (Figure 2.45) are $\hat{\mathbf{u}}_1, \hat{\mathbf{u}}_2$, and $\hat{\mathbf{u}}_3$. These three special unit vectors are the principal vectors or eigenvectors for the symmetric tensor that describes the deformation. The lengths of these axes relative to their original lengths are the principal values or eigenvalues of that tensor. For the deformed ooliths, these are the principal stretches: S_1, S_2, and S_3. In this section we describe the special properties of symmetric tensors that lead to principal values and directions.

Discussions of principal values and vectors commonly focus on equations of the following form:

$$\underset{3\times 3}{[\mathrm{T}]}\,\underset{3\times 1}{[\mathrm{u}]} = \lambda \underset{3\times 1}{[\mathrm{u}]} \qquad (2.73)$$

Here [T] is a *square symmetric matrix* (a tensor), [u] is a *unit vector* of arbitrary direction, and λ is a principal value (a scalar). Because λ is a scalar, λ[u] and [u] point in the same direction. Multiplication of the vector [u] by the scalar λ only changes the magnitude of the vector (unless $\lambda = 1$). For example, if $\lambda = 2$, the operation on the right side of (2.73) doubles the vector length, but its direction remains the same. What perhaps is surprising about (2.73) is that multiplication of certain unit vectors [u] by the tensor [T] changes the magnitude of that vector, but not the direction. Here we ask the question: for what unit vectors is this true?

As an example, consider how a particular tensor operates on unit vectors in two dimensions. The two-dimensional form of (2.73) is:

$$\underset{2\times 2}{[\mathrm{T}]}\,\underset{2\times 1}{[\mathrm{u}]} = \underset{2\times 1}{[\mathrm{v}]} = \lambda \underset{2\times 1}{[\mathrm{u}]} \qquad (2.74)$$

The chosen tensor is written in the form of a 2 by 2 symmetric matrix:

$$[\mathrm{T}] = \begin{bmatrix} 1.5 & 1 \\ 1 & 1.5 \end{bmatrix}$$

The vectors are shown with black and blue arrows in Figure 2.46a.

The first unit vector, $\hat{\mathbf{u}}^{(1)} = \langle 1,\ 0 \rangle$, is parallel to the x-axis, and we arrange these components in a 2 by 1 column matrix. Premultiplying this vector by the tensor [T], as indicated on the left side of (2.74), we find:

$$[\mathrm{T}]\left[\mathrm{u}^{(1)}\right] = \begin{bmatrix} 1.5 & 1 \\ 1 & 1.5 \end{bmatrix}\begin{bmatrix} 1 \\ 0 \end{bmatrix} = \begin{bmatrix} 1.5 \\ 1 \end{bmatrix} = \left[\mathrm{v}^{(1)}\right]$$

The resulting vector, $\mathbf{v}^{(1)}$, is not a simple scalar multiple of the original vector, because it is directed into the first quadrant, not along the x-axis. In other words, (2.74) does not hold. The second unit vector, $\hat{\mathbf{u}}^{(2)} = \langle 0,\ 1 \rangle$, is parallel to the y-axis before transformation and is directed into the first quadrant after the transformation. This vector also does not conform to (2.74).

Now consider two unit vectors that have special orientations relative to the chosen tensor. The unit vector $\hat{\mathbf{u}}^{(3)} = \left\langle \sqrt{2}/2, \sqrt{2}/2 \right\rangle$ is directed into the first quadrant at 45° to the x-axis (Figure 2.46b). The product of the chosen tensor and this vector is:

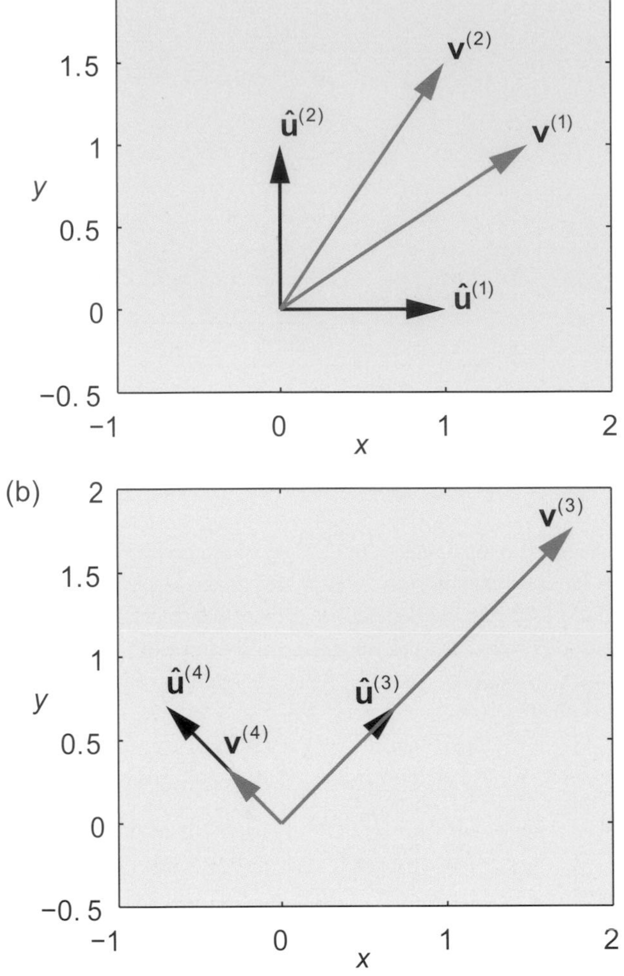

Figure 2.46 (a) Two unit vectors (black) directed along the coordinate axes and their transformed counterparts (blue). (b) Two special unit vectors (black) and their transformed counterparts (blue). See text for discussion of these vectors and the eigenvalue problem. See Strang (2016) in Further Reading.

$$[\mathrm{T}]\left[\mathrm{u}^{(3)}\right]=\begin{bmatrix}1.5 & 1\\ 1 & 1.5\end{bmatrix}\begin{bmatrix}\sqrt{2}/2\\ \sqrt{2}/2\end{bmatrix}=2.5\begin{bmatrix}\sqrt{2}/2\\ \sqrt{2}/2\end{bmatrix}$$

Here the resulting vector is a simple scalar multiple of the original vector.

$$[\mathrm{T}][\mathrm{u}^{(3)}]=[\mathrm{v}^{(3)}]=\lambda[\mathrm{u}^{(3)}]$$

The scalar so determined is one of the principal values, $\lambda_1 = 2.5$, associated with this tensor. The unit vector used in this calculation is parallel to one of the principal directions. The other principal direction is orthogonal to this one, so it is directed into the second quadrant at 135°. That happens to be the orientation of the fourth unit vector, $\hat{\mathbf{u}}^{(4)}$, so it also satisfies (2.74). The principal value in that direction is $\lambda_2 = 0.5$.

Generalizing these concepts to three dimensions, unit vectors in three particular directions maintain their directions when operated on by a tensor, although their magnitudes can change. These are the principal directions for that tensor. The signs of the principal values of a tensor are associated with a mathematical property that also has important physical implications. If all three principal stresses are positive, the tensor is called positive definite. If all three principal stresses are negative, the stress tensor is called negative definite. An ellipsoid can represent positive definite tensors graphically (**Figure 2.45**). The stretch tensor always is positive definite, because stretch is the ratio of final to initial length of a material line, and lengths are inherently positive. In contrast, principal stress components may be positive (tension), or negative (compression), so the stress tensor may not be positive definite, and cannot necessarily be represented graphically by an ellipsoid.

2.7 COORDINATE ROTATION FOR POINTS, VECTORS, AND TENSORS

Throughout this textbook we adopt right-handed Cartesian coordinates, (x, y, z). Usually, we direct the positive z-axis upward (**Figure 2.9**). However, the orientation of the coordinate axes is an arbitrary choice, and in many circumstances an alternative orientation of the coordinate axes is helpful. Accordingly, many applications in structural geology require a rotation of coordinates about a common origin.

The positions of points do not change with coordinate rotation, but the coordinates of the points may change. In addition, the magnitude and direction of vectors do not change with coordinate rotation, but the vector components may change. Similarly, the principal values and principal orientations of tensors do not change with coordinate rotation, but the tensor components may change. Here we use matrix methods to account for an arbitrary coordinate rotation in three dimensions, and show how to compute the new coordinates of points, and the new components of vectors and tensors. Allmendinger et al. (2012), in Further Reading at the end of this chapter, provides other valuable tools that utilize matrix methods.

To develop the matrix methods for coordinate rotation, we define a direction angle between each of the three axes of the *old* coordinates (x, y, z), and each of the three axes of the *new* coordinates $(\bar{x}, \bar{y}, \bar{z})$. The direction angle is the smaller angle between the two positive axes in the plane that they define. Nine such angles exist, and **Figure 2.47** illustrates three of these. Each direction angle has two subscripts: the first designates the new axis and the second designates the old axis. Thus, $\theta_{\bar{x}y}$ is the direction angle *to* the new $\bar{x}$-axis *from* the old y-axis. This is a to–from convention. Some textbooks follow this convention (e.g. Fung, 1969; Jaeger et al., 2007, in Further Reading), while other textbooks adopt a from–to convention (e.g. Malvern, 1969; Lai et al., 2010, in Further Reading) in which the first subscript designates the old axis and the second subscript designates the new axis. Although the outcome does not depend upon the convention, the matrix equations for coordinate rotation are different, as we point out below.

Each direction angle (**Figure 2.47**) is associated with a direction cosine, which has two subscripts that also follow the to–from

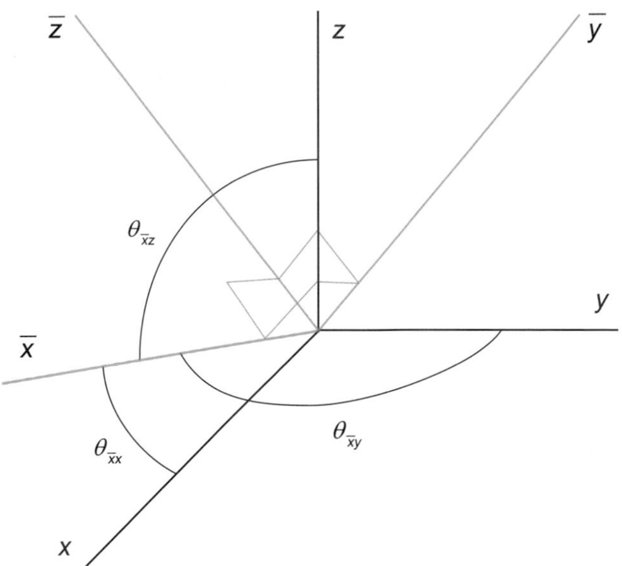

Figure 2.47 Old coordinate axes (x, y, z) and rotated new coordinate axes $(\bar{x}, \bar{y}, \bar{z})$. Direction angles, for example $\theta_{\bar{x}x}, \theta_{\bar{x}y}$, and $\theta_{\bar{x}z}$, quantify the angular relationships of new and old axes.

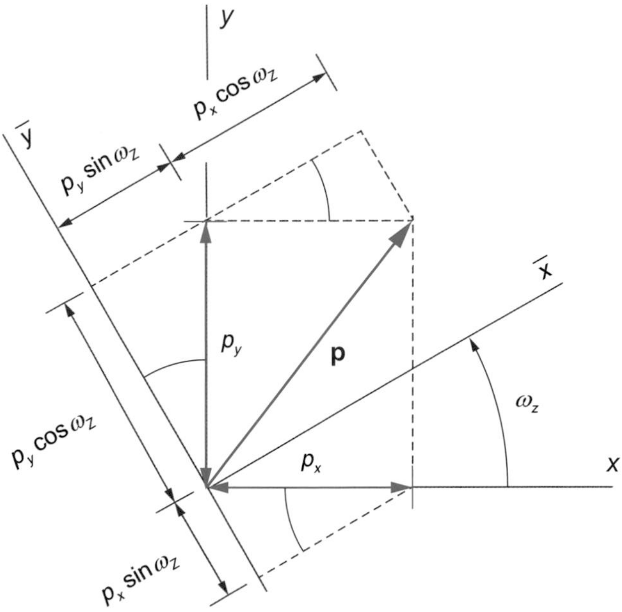

Figure 2.48 Rotation of Cartesian coordinates in two dimensions for vector quantities. The old coordinates are (x, y); the new coordinates are $(\bar{x}, \bar{y})$; the rotation angle is ω_z (marked by single arcs); and the position vector is $\mathbf{p} = \langle p_x, p_y \rangle$.

convention. For example, $C_{\bar{x}y} = \cos\theta_{\bar{x}y}$ is the direction cosine for the angle *to* the new $\bar{x}$-axis *from* the old y-axis. Nine direction cosines relate the three new axes and the three old axes.

Each new coordinate is a linear combination of the old coordinates and the direction cosines that have that new coordinate as their first subscript and the multiplicative old coordinate as the second subscript:

$$\begin{aligned}\bar{x} &= C_{\bar{x}x}x + C_{\bar{x}y}y + C_{\bar{x}z}z \\ \bar{y} &= C_{\bar{y}x}x + C_{\bar{y}y}y + C_{\bar{y}z}z \\ \bar{z} &= C_{\bar{z}x}x + C_{\bar{z}y}y + C_{\bar{z}z}z\end{aligned} \tag{2.75}$$

The first (left) subscript on a direction cosine matches the corresponding coordinate on the left side of (2.75), and the second (right) subscript matches the corresponding coordinate on the right side of (2.75). In matrix form, the equation for rotation of coordinates is:

$$\begin{bmatrix}\bar{x}\\ \bar{y}\\ \bar{z}\end{bmatrix} = \begin{bmatrix}C_{\bar{x}x} & C_{\bar{x}y} & C_{\bar{x}z}\\ C_{\bar{y}x} & C_{\bar{y}y} & C_{\bar{y}z}\\ C_{\bar{z}x} & C_{\bar{z}y} & C_{\bar{z}z}\end{bmatrix}\begin{bmatrix}x\\ y\\ z\end{bmatrix} \tag{2.76}$$

Subscripts on the direction cosines in (2.76) obey the standard convention for matrix subscripts with the first being common to the *row* and the second being common to the *column* of the square matrix.

From an algebraic perspective, the direction cosines in the square matrix of (2.76) are dimensionless weighting factors. For example, $C_{\bar{z}y}$ is a weighting factor that describes how much the old y-coordinate of a point contributes to the new $\bar{z}$-coordinate of that point. If the $\bar{z}$-axis and y-axis are coincident, $C_{\bar{z}y} = \cos 0^\circ = 1$ and the full weight of the old coordinate contributes to the new coordinate. If the two axes are orthogonal, $C_{\bar{z}y} = \cos 90^\circ = 0$, and the old coordinate contributes nothing to the new coordinate. From a geometric perspective, each direction cosine quantifies the projection of an old coordinate onto a new coordinate axis.

To understand how the direction cosines in (2.76) account for the change in value of the components of a vector, we consider a two-dimensional example in the (x, y)-plane for a rotation ω_z of coordinates about the z-axis (**Figure 2.48**). The position vector **p** has components p_x and p_y, and this figure illustrates the projection of these components onto the rotated coordinate axes $(\bar{x}, \bar{y})$. Projecting p_x onto the $\bar{x}$-axis yields $p_x\cos\omega_z$; projecting p_y onto the $\bar{x}$-axis yields $p_y\sin\omega_z$; and the sum of these contributions yields $p_{\bar{x}}$, thus accounting for the rotation. Projecting p_x onto the (negative) $\bar{y}$-axis yields $-p_x\sin\omega_z$; projecting p_y onto the (positive) $\bar{y}$-axis yields $p_y\cos\omega_z$; and the sum of these contributions yields $p_{\bar{y}}$, thus accounting for the rotation. The components of **p** with respect to the new $(\bar{x}, \bar{y})$coordinates are:

$$\begin{aligned}p_{\bar{x}} &= p_x\cos\omega_z + p_y\sin\omega_z \\ p_{\bar{y}} &= -p_x\sin\omega_z + p_y\cos\omega_z\end{aligned} \tag{2.77}$$

From the trigonometry of **Figure 2.48** we note that $\cos\theta_{\bar{x}x} = \cos\omega_z$; $\cos\theta_{\bar{x}y} = \cos(90 - \omega_z) = \sin\omega_z$; $\cos\theta_{\bar{y}x} = \cos(90 + \omega_z) = -\sin\omega_z$; and $\cos\theta_{\bar{y}y} = \cos\omega_z$. Substituting these trigonometric identities into (2.77), the new components of the position vector **p** are:

$$\begin{aligned}p_{\bar{x}} &= p_x\cos\theta_{\bar{x}x} + p_y\cos\theta_{\bar{x}y} \\ p_{\bar{y}} &= p_x\cos\theta_{\bar{y}x} + p_y\cos\theta_{\bar{y}y}\end{aligned} \tag{2.78}$$

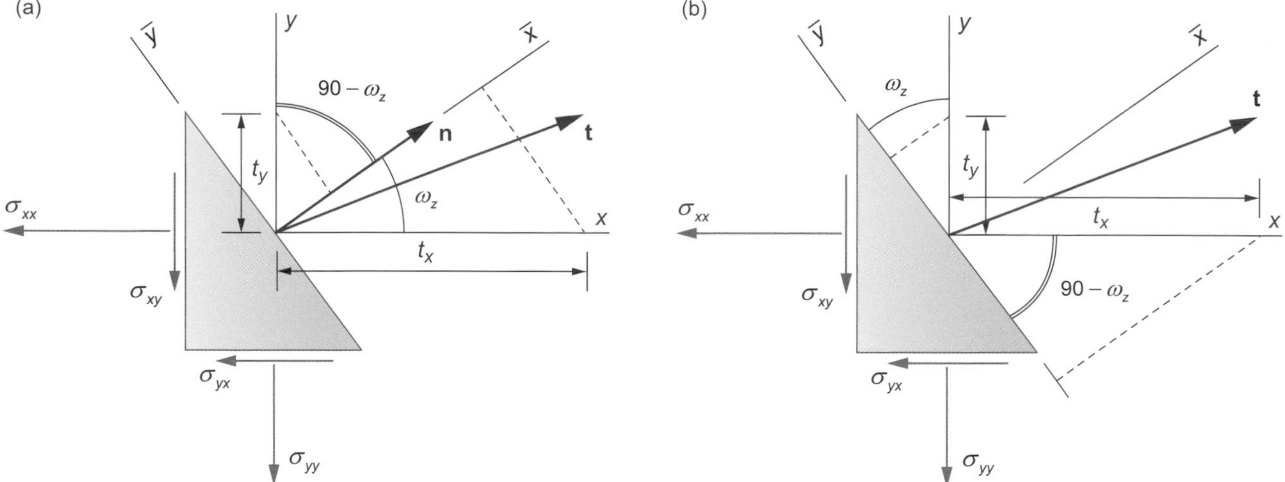

Figure 2.49 Resolution of the traction vector components of **t**, referred to the (x, y)-coordinates, onto the rotated $(\bar{x}, \bar{y})$-coordinates. (a) Resolution of t_x and t_y onto the $\bar{x}$-axis. (b) Resolution of t_x and t_y onto the $\bar{y}$-axis.

In matrix form, (2.78) is:

$$\begin{bmatrix} p_{\bar{x}} \\ p_{\bar{y}} \end{bmatrix} = \begin{bmatrix} C_{\bar{x}x} & C_{\bar{x}y} \\ C_{\bar{y}x} & C_{\bar{y}y} \end{bmatrix} \begin{bmatrix} p_x \\ p_y \end{bmatrix}$$

Note that this 2 by 2 matrix of direction cosines is taken from the upper left corner of the 3 by 3 matrix of direction cosines in (2.76).

The results of the previous paragraph motivate the general three-dimensional matrix equation for the change in components of a vector quantity due to rotation of coordinates:

$$\underset{3\times1}{[\bar{p}]} = \underset{3\times3}{[C]}\,\underset{3\times1}{[p]} \tag{2.79}$$

The elements of $[\bar{p}]$ are the components of the vector referred to the new coordinates; [C] is the matrix of direction cosines (2.76); the elements of [p] are the components of the vector referred to the old coordinates. The matrix [C] must have the special properties of an *orthogonal* matrix, introduced in Section 2.6.4. If the matrix of direction cosines is defined using the from–to convention, and we call that matrix [D], the equation analogous to (2.79) for the change in components of a vector quantity due to rotation of coordinates is $[\bar{p}] = [D'][p]$.

Following the discussion leading from (2.66) to (2.67), we solve (2.79) for the old vector by pre-multiplying both sides of this equation by the inverse of the direction cosine matrix:

$$\underset{3\times3}{[C^{-1}]}\,\underset{3\times1}{[\bar{p}]} = \underset{3\times3}{[C^{-1}]}\,\underset{3\times3}{[C]}\,\underset{3\times1}{[p]} = \underset{3\times3}{[I]}\,\underset{3\times1}{[p]} = \underset{3\times1}{[p]}$$

Recall that the inverse of an orthogonal matrix equals the transpose of that matrix (2.68). Hence, the old vector components are calculated from the new vector components using the transpose of the direction cosine matrix:

$$\underset{3\times1}{[p]} = \underset{3\times3}{[C']}\,\underset{3\times1}{[\bar{p}]} \tag{2.80}$$

If the matrix of direction cosines is defined using the from–to convention, and we call that matrix [D], the equation analogous to (2.80) is $[p] = [D][\bar{p}]$.

The coordinates of a point, $P(x, y, z)$, equal the components of the position vector for that point, $\mathbf{p} = \langle p_x, p_y, p_z \rangle$. Thus, changes in coordinates (2.76), and changes in vector components (2.79), follow the same rule for a rotation of coordinate axes. Although we use the components of the position vector to derive (2.79), the components of all the physical vectors in this textbook adhere to these equations for rotation of axes. For mathematicians, the defining characteristic of vectors is that their components vary with rotation of coordinate axes according to (2.79).

It is important to distinguish the matrix [R], defined in (2.63), that rotates *vectors* (proxies for material lines) with *fixed* coordinate axes, from the matrix of direction cosines [C] that rotates *coordinate axes* and finds the new components of a *fixed* vector. For a rotation of axes about the z-axis through angle ω_z the matrix of direction cosines is:

$$[C] = \begin{bmatrix} \cos\omega_z & \sin\omega_z & 0 \\ -\sin\omega_z & \cos\omega_z & 0 \\ 0 & 0 & 1 \end{bmatrix} \tag{2.81}$$

Comparing (2.63) and (2.81), we note the only differences are the signs of the $\sin\omega_z$ elements. Inspection of the rows and columns of both matrices reveals that [R] is the transpose of [C].

To understand how the direction cosines in (2.76) account for changes in the components of a tensor quantity due to rotation of the coordinate axes, we employ the stress tensor and focus on a two-dimensional rotation about the z-axis through an angle ω_z. The Cauchy tetrahedron, introduced in **Figure 2.44**, reduces to a triangular element in the (x, y)-plane (**Figure 2.49**). Four stress components (σ_{xx}, $\sigma_{xy} = \sigma_{yx}$, σ_{yy}) act on the coordinate planes; the traction vector, **t**, acts on the oblique plane; and the two traction components are t_x and t_y. Determining the new stress components $(\sigma_{\bar{x}\bar{x}}, \sigma_{\bar{x}\bar{y}} = \sigma_{\bar{y}\bar{x}}, \sigma_{\bar{y}\bar{y}})$ involves two steps. The first step follows the derivation of (2.77) to resolve the traction vector components, t_x and t_y, onto the new axes (**Figure 2.49a,b**):

$$\begin{aligned} t_{\bar{x}} &= t_x \cos\omega_z + t_y \sin\omega_z \\ t_{\bar{y}} &= -t_x \sin\omega_z + t_y \cos\omega_z \end{aligned} \tag{2.82}$$

The second step introduces the stress components using Cauchy's formula (2.72), reduced to two dimensions:

$$t_x = \sigma_{xx} n_x + \sigma_{yx} n_y = \sigma_{xx} \cos\omega_z + \sigma_{yx} \sin\omega_z$$
$$t_y = \sigma_{xy} n_x + \sigma_{yy} n_y = \sigma_{xy} \cos\omega_z + \sigma_{yy} \sin\omega_z \tag{2.83}$$

The right sides of (2.83) follow by recalling that n_x and n_y are components of the unit normal vector, **n**, on the oblique plane (**Figure 2.49**), and these are related to the rotation angle as $n_x = \cos\omega_z$ and $n_y = \cos(90 - \omega_z) = \sin\omega_z$.

To find the stress components in the new coordinates we use (2.83) to substitute for the traction vector components on the right side of (2.82) and recognize that $t_{\bar{x}} = \sigma_{\overline{xx}}$ and $t_{\bar{y}} = \sigma_{\overline{xy}}$:

$$t_{\bar{x}} = \sigma_{xx} \cos^2\omega_z + 2\sigma_{xy} \sin\omega_z \cos\omega_z + \sigma_{yy} \sin^2\omega_z = \sigma_{\overline{xx}}$$
$$t_{\bar{y}} = (\sigma_{yy} - \sigma_{xx}) \sin\omega_z \cos\omega_z + \sigma_{xy}(\cos^2\omega_z - \sin^2\omega_z) = \sigma_{\overline{xy}} \tag{2.84}$$

To find the other two stress components in the rotated coordinates we substitute $\omega_z + 90$ for ω_z in (2.84) and recognize that $t_{\bar{x}} = \sigma_{\overline{yy}}$ and $t_{\bar{y}} = -\sigma_{\overline{yx}}$:

$$t_{\bar{x}} = \sigma_{xx} \sin^2\omega_z - 2\sigma_{xy} \cos\omega_z \sin\omega_z + \sigma_{yy} \cos^2\omega_z = \sigma_{\overline{yy}}$$
$$t_{\bar{y}} = -(\sigma_{yy} - \sigma_{xx}) \cos\omega_z \sin\omega_z + \sigma_{xy}(\sin^2\omega_z - \cos^2\omega_z) = -\sigma_{\overline{yx}} \tag{2.85}$$

The equations in (2.84) and (2.85) are the classic equations for changes in the stress tensor components in the (x, y)-plane due to a rotation of coordinates about the z-axis (Timoshenko and Goodier, 1970, in Further Reading). Comparing (2.77) and (2.84), we note that changes in vector components due to coordinate rotation depend on $\sin\omega_z$ and $\cos\omega_z$ to the first power, whereas changes in tensor components depend on squares and products of these trigonometric functions.

To write the general matrix equation for the change in components of the stress tensor due to rotation of coordinates, we employ the direction cosine matrix [C] from (2.76) and its transpose [C′]:

$$\underset{3\times3}{[\bar{\sigma}]} = \underset{3\times3}{[\mathrm{C}]}\,\underset{3\times3}{[\sigma]}\,\underset{3\times3}{[\mathrm{C}']} \tag{2.86}$$

Pre-multiplying the stress matrix [σ] by [C], and post-multiplying that matrix product by [C′] accounts for the two steps described in the preceding paragraphs: (1) resolution of the traction vector components, t_x and t_y, onto the new axes (**Figure 2.49**); and (2) converting the traction components to the stress components using Cauchy's formula (2.72). If the matrix of direction cosines is defined using the from–to convention, and we call that matrix [D], the equation analogous to (2.86) is $[\bar{\sigma}] = [\mathrm{D}'][\sigma][\mathrm{D}]$.

Expanding (2.86) for the two-dimensional case illustrated in **Figure 2.49** while using the direction cosine matrix defined in (2.81) for the rotation angle ω_z, reproduces (2.84) and (2.85). Although we use the stress tensor here as an example, all the physical tensors in this textbook follow this prescription of change of components due to rotation of coordinates. For mathematicians, the defining characteristic of tensors is that tensor components vary with rotation of coordinate axes according to (2.86).

Recapitulation

The ooliths at South Mountain anticline provided an introduction to the three categories of physical quantities that we use throughout this textbook: *scalars, vectors,* and *tensors.* A single number serves to quantify a scalar, whereas three numbers (e.g. three components) are necessary to quantify a vector, and six numbers (e.g. six components) are necessary to quantify a symmetric tensor. Non-symmetric tensors require nine numbers (nine components). The scalars, vectors, and tensors that find application in structural geology are *field* quantities: they can vary in three-dimensional space and time. One visualizes the spatial variation of a scalar quantity in two dimensions using a contour map. A two-dimensional distribution of arrows, scaled to the vector magnitude, provides for the visualization of a vector quantity. Contour maps of the principal values help one to visualize the spatial variation of a tensor in two dimensions.

The *position vector* and *unit vector* find many applications in structural geology, both in the field for locating outcrops, and in the laboratory for manipulating orientation data. Linear and planar structural elements are oriented using geographic angles (azimuth and inclination) in the field, and these transform to direction angles and direction cosines for analysis and modeling. The *stereographic projection* provides a good visualization tool for orientation data and the unit direction vector and spherical variance enable one to quantify the clustering of such data about a mean orientation.

While field data quantify the orientations of geologic structures that locally approximate a line or plane, these structures are curves and curved surfaces in three dimensions. *Vector calculus* provides the tools to quantify the geometry of curves and curved surfaces. For example, the components of the unit tangent vector to a curve are the direction cosines, which transform to the azimuth and inclination at any position on the curve. Vector-valued functions of one and two parameters help characterize the geometry of linear and planar fabrics.

The spatial variation of structures and of the physical quantities related to their formation in two and three dimensions necessitates the use of *partial derivatives*, as well as *differentials*. For example, partial derivatives enable one to analyze a topographic map or structure contour map using the *gradient* vector field. Given the vector representation of a curved surface, be it a fault or a folded sedimentary formation, the partial derivative is used to calculate the tangent vectors at any position on the surface, and the vector product of the tangent vectors provides the unit normal that specifies the local orientation of the fault or strata. The *material time derivative* plays an important role in the fundamental laws of mass and momentum conservation that underlie much of the mechanics used to understand the formation of geologic structures.

Because the scalars, vectors, and tensors used in structural geology vary in three-dimension space, the mathematical concept of a spatial gradient provides a quantitative link among these physical quantities. In general, *the gradient of a scalar field is a vector field* that quantifies how that scalar field changes in magnitude with a change in position. At any point in the scalar field, the gradient vector has the direction of the maximum increase in the scalar field, and it is perpendicular to the surface on which the scalar quantity is constant. In general, *the gradient of a vector field is a tensor field* that quantities how that vector field changes in magnitude and direction with a change in position. At any point in the vector field, the corresponding tensor carries that information for all possible directions from that point.

The elementary concepts of *matrix algebra* provide valuable tools to manipulate vectors and tensors. Arranging the components of a vector in a 3 by 1 column matrix, and the components of a tensor in a 3 by 3 square matrix opens the door for matrix multiplication. For example, pre-multiplying the vector by the tensor is a linear transformation that transforms the vector into another vector. This operation leads to Cauchy's formula, one of the most useful relationships in structural geology. Several different special matrices, including the transpose, orthogonal, identity, and inverse matrix arise in relevant matrix operations. The tools of linear algebra enable one to find the principal *values* and principal *directions* of a symmetric tensor, and these quantities serve to characterize the magnitudes and orientations of the tensor quantities. Changes in the components of vector and tensor quantities, due to the rotation of coordinate axes, obey equations that utilize the matrix of direction cosines and its transpose.

The authors hope that readers of this chapter have come to appreciate how the mathematical analysis of structural data gathered in the field, and the mathematical manipulation of physical quantities used for modeling tectonic processes enrich the practice of structural geology. In the last century the task of analyzing structural data largely was accomplished using techniques that employed descriptive geometry, and required precise and laborious drawing on paper using pencils and drafting instruments. Today these same tasks are accomplished using matrix algebra and vector calculus on computers with applications like MATLAB that provide beautiful color renditions of the geometry, as well as contour maps and vector fields at the touch of a button.

REVIEW QUESTIONS

The following questions are designed to highlight the expected *learning outcomes* for this chapter. Each question is taken directly from the material in the chapter and, for the most part, in the same sequence that it appears in the chapter. If an answer is not forthcoming, students are advised to read the relevant section of the chapter and discover the answer.

2.1. This textbook contains nearly 500 numbered equations. How many of those do the authors expect a student to memorize?

2.2. All physical quantities that are used to understand tectonic processes have one of three mathematical representations: *scalar*, *vector*, or *tensor*. Name one example of each representation, and give the appropriate *SI units* for that physical quantity.

2.3. Any vector quantity may be written $\mathbf{v} = v_x\hat{\mathbf{i}} + v_y\hat{\mathbf{j}} + v_z\hat{\mathbf{k}}$. Use words and a drawing to describe the quantities on the right side of this vector equation and the coordinate system.

2.4. The *position* vector is called the *mapping* vector in this textbook. Explain how it is used and why this justifies the name.

2.5. A particular unit vector is used to define the orientation of a linear structural element. What is this unit vector, and how is it quantified using data gathered in the field, for example, on the *trend* and *plunge* of a lineation?

2.6. A particular unit vector is used to define the orientation of a planar structural element. What is this unit vector, and how is it quantified using data gathered in the field, for example, on the *strike* and *dip* of a fault surface?

2.7. *Linear structural elements* may be characterized geometrically using a vector-valued function such as $\mathbf{c}(t) = \mathbf{p} + t\hat{\mathbf{u}}$. Describe all the quantities in this equation and explain using words and a drawing how the function characterizes a lineation.

2.8. *Planar structural elements* may be characterized geometrically using a vector-valued function such as $\mathbf{s}(t_1, t_2) = \mathbf{p} + t_1\hat{\mathbf{v}} + t_2\hat{\mathbf{w}}$. Describe all the quantities in this equation and explain using words and a drawing how the function characterizes a bedding surface.

2.9. A curved lineation (e.g. slickenline on a fault) is represented using the continuous vector function $\mathbf{c}(t) = c_x(t)\hat{\mathbf{i}} + c_y(t)\hat{\mathbf{j}} + c_z(t)\hat{\mathbf{k}}$, where t is the arbitrary parameter of the curve. Show how to calculate from field data (*trend* and *plunge* of the lineation), the unit vector that best quantifies the orientation of the lineation.

2.10. A curved surface (e.g. top of a folded sedimentary bed) is represented using the continuous vector function $\mathbf{s}(u,v) = s_x(u,v)\hat{\mathbf{i}} + s_y(u,v)\hat{\mathbf{j}} + s_z(u,v)\hat{\mathbf{k}}$, where u and v are the two parameters. Show how to calculate from field data (*strike* and *dip* of the bed), the unit vector that best quantifies the orientation of the bedding surface.

2.11. Use the model salt dome (**Figure 2.38b**) to describe a *spatial description of motion* and explain why, despite being an example of steady flow, particles accelerate and decelerate as they circulate in the spherical body. In addressing this question, compare and contrast the spatial description with a *material description of motion*.

2.12. *Cauchy's formula* is used, for example, to calculate the traction vector acting on a fault by multiplying the transpose of the stress tensor and the unit normal vector to the fault plane using matrix multiplication: $[\mathrm{t}] = [\sigma'][\mathrm{n}]$. Write out the components of the vectors and tensor in matrix form. Identify the SI units of these quantities.

2.13. Draw a carefully labeled two-dimensional sketch of the traction and stress components acting on a right triangular element with hypotenuse parallel to a fault plane and two sides parallel to the coordinate planes.

2.14. Explain why the stretch tensor always can be represented graphically by an ellipsoidal (recall the deformed oolith from **Figure 2.45**), but only some stress tensors can be represented by an ellipsoid.

2.15. Explain the differences between the rotation matrix [R], defined in (2.63), and the rotation matrix [C], defined in (2.81), in terms of what is being rotated and what is fixed.

2.16. The general matrix equation (2.79) for the change in components of a *vector* quantity due to rotation of coordinates utilizes the transpose of the matrix of direction cosines [C′], defined in (2.75). However, the general matrix equation (2.86) for the change in components of the tensor quantity due to rotation of coordinates employs both [C′] and [C]. Use **Figure 2.49** to explain why the tensor components require the second matrix multiplication.

MATLAB EXERCISES FOR CHAPTER 2: MATHEMATICAL TOOLS

Field data gathered to document the size and shape of geologic structures are quantified using mathematical constructs called scalars, vectors, and tensors. For example, the classic three-point problem to determine the orientation of a fault is solved using vectors, and the transformation of ooliths at South Mountain fold from spherical to ellipsoidal is quantified using the deformation gradient tensor. MATLAB provides the computational tools for solving many practical problems in structural geology. These exercises build an intuitive and a visual understanding of position vectors for mapping structures in the field and plotting orientation data on stereographic projections. To describe geologic structures in three dimensions, we use elementary concepts from differential geometry, including curvature vectors for curved lineations, and the gradient vector for folds. We explore the relationship between vector and tensor fields using the spatial gradient of velocity tensor. This builds intuition about the solid state flow of rock under high pressures and temperatures. MATLAB is particularly well suited for calculations with matrices, for example, the evaluation of traction vectors acting on faults given the local stress tensor. These exercises build a student's confidence in using computation to address problems in structural geology. www.cambridge/SGAQI

FURTHER READING

For citations in figure captions see the reference list at the end of the book.

Agterberg, F., 2014. *Geomathematics: Theoretical Foundations, Applications and Future Developments*. Springer, New York.
This textbook is largely devoted to statistical methods of data analysis in geology with many examples from the mining industry.

Allmendinger, R. W., Cardozo, N., and Fisher, D. M., 2012. *Structural Geology Algorithms: Vectors and Tensors*. Cambridge University Press, Cambridge.

This is a very approachable coverage of vectors and tensors used in structural geology. Problems are solved using linear algebra, and MATLAB® is used for computation and graphics.

Davis, J. C., 2002. *Statistics and Data Analysis in Geology*, 3rd edition. John Wiley & Sons, New York.
This third edition of a widely used book for geologists has many worked examples and exercises, and the analyzed data are available online.

Davis, J. R., and Titus, S. J., 2017. Modern methods of analysis for three-dimensional orientational data. *Journal of Structural Geology* 96, 65–89.
This review article explains how to apply statistics to complete three-dimensional orientation data gathered by geologists on structures such as foliation–lineation pairs, folds, and ellipsoids.

Ferguson, J., 1994. *Introduction to Linear Algebra in Geology*. Chapman & Hall, London.
Written for geologists, this book focuses on applications of linear algebra to solve problems in structural geology, petrology, and petroleum geology.

Fisher, N. I., Lewis, T., and Embleton, B. J. J., 1987. *Statistical Analysis of Spherical Data*. Cambridge University Press, Cambridge.
With a broad coverage of statistical concepts relevant to the earth sciences, this book also provides many applications with data and computational methods.

Fung, Y. C., 1969. *A First Course in Continuum Mechanics*. Prentice-Hall, Inc., Englewood Cliffs, NJ.
This textbook provides students of science and engineering with a first course on the fundamentals of solid and fluid mechanics.

Hofmann-Wellenhof, B., Lichtenegger, H., and Collins, J., 2001. *Global Positioning System: Theory and Practice*, 4th edition. Springer-Verlag, Wien, Germany.
Designed as a text for senior or graduate students, this book covers the basic concepts and techniques of GPS, including the reference systems, satellite orbits and signals, and data processing, along with methods of surveying with GPS.

Jaeger, J. C., Cook, N. G. W., and Zimmerman, R. W., 2007. *Fundamentals of Rock Mechanics*, 4th edition. Blackwell Publishing, Oxford.
This fourth edition of a classic book on rock mechanics deserves to be on the shelf of every structural geologist who utilizes rock mechanics data and concepts to address geological problems.

Lai, W. M., Rubin, D. H., and Krempl, E., 2010. *Introduction to Continuum Mechanics*. Elsevier, New York.
Presented as a first course in continuum mechanics for undergraduate students of engineering, this textbook provides many worked examples and exercises.

Malvern, L. E., 1969. *Introduction to the Mechanics of a Continuous Medium*. Prentice-Hall, Inc., Englewood Cliffs, NJ.
This authoritative book on continuum mechanics covers many of the facets of that discipline that have found application in structural geology, with key concepts and equations presented in vector, indicial, and matrix notation.

Mardia, K. V., and Jupp, P. E. 2000. *Directional Statistics*. John Wiley & Sons, New York.
Both statistics on the circle and those on the sphere are included in the comprehensive coverage of this book, along with practical applications to the earth sciences.

Olver, F. W. J., Olde Daalhuis, A. B., Lozier, D. W., Schneider, B. I., Boisvert, R. F., Clark, C. W., Miller, B. R., and Saunders, B. V. (Eds.) *Release 1.0.17 of 2017-12-22, [DLMF] NIST Digital Library of Mathematical Functions*. National Institute of Standards and Technology, US Department of Commerce.
See the online version of this monumental mathematical resource at: http://dlmf.nist.gov/

Pollard, D. D., and Fletcher, R. C., 2005. *Fundamentals of Structural Geology*. Cambridge University Press, Cambridge.

This advanced textbook covers differential geometry to characterize geologic structures using vector calculus, and continuum mechanics to model tectonic processes leading to the development of geologic structures.

Ragan, D. M., 2009. *Structural Geology: An Introduction to Geometrical Techniques*, 4th edition. Cambridge University Press, Cambridge.
This textbook covers much more than geometrical techniques, for example, with chapters on stress, strain, deformation, and flow that include mathematical manipulations of vectors and tensors as well as abundant line drawings, stereographic projections, and Mohr circles.

Segall, P., 2010. *Earthquake and Volcano Deformation*. Princeton University Press, Princeton, NJ.
This advanced textbook goes beyond the elementary mechanics and structural geology covered here, with a focus on geophysical data related to active deformation associated with volcanic and earthquake hazards.

Strang, G., 2016. *Introduction to Linear Algebra*, 5th edition. Wellesley-Cambridge Press, Wellesley, MA.
This fifth edition of a very popular textbook covers subjects with practical applications in many disciplines of engineering and science.

Timoshenko, S. P., and Goodier, J. N., 1970. *Theory of Elasticity*, 3rd edition. McGraw-Hill Book Company, New York.
This third edition of a classic book on elasticity covers most aspects of two- and three-dimensional theory with applications to a host of engineering problems, some of which are relevant to rock deformation.

Varberg, D. E., Purcell, E. J., and Rigdon, S. E., 2006. *Calculus*, 9th edition. Pearson/Prentice Hall, New York.
Calculus is covered in this ninth edition of a widely used textbook at the level expected as a pre-requisite for a course in quantitative structural geology.

Varberg, D. E., Purcell, E. J., and Rigdon, S. E., 2007. *Calculus with Differential Equations*. Pearson/Prentice Hall, New York.
With a similar coverage of basic calculus as the previous book, this includes introductions to derivatives of functions of two or more variables, multiple integration, vector calculus, and a chapter on elementary differential equations.

Chapter 3
Physical Concepts

Introduction

Chapter 3 reviews fundamental physical concepts that contribute to understanding the development of geologic structures. We begin by defining the units and dimensions of physical quantities encountered in structural geology. We point out that equations composed of these quantities must have consistent units and dimensions to be part of a valid explanation of a tectonic process. Next, we introduce the concept of a material continuum, and describe displacement and stress fields that demonstrate the continuum is an effective way to idealize rock at length scales from nanometers to tens of kilometers. Then, we consider the conservation laws for mass, momentum, and energy. We use them to derive the fundamental equations of continuity, motion, and heat transport in a material continuum. These equations underlie the three different styles of rock deformation and the canonical models for the five categories of geologic structures.

The tectonic processes that shape the architecture of Earth's lithosphere and the geologic structures produced by these processes are the central subjects of a course in structural geology. Broadly speaking, tectonic processes are both physical and chemical: for example, sedimentary strata may be contorted physically into spectacular folds, and also altered chemically, as they are transformed from sedimentary to metamorphic rock. For the most part, the structural geologist focuses on the physics of this process, the geochemist on the chemical alterations, and the petrologist on the transformation of one rock type to another. Many interesting and challenging problems require integration of the physical, chemical, and petrological concepts and data at the research level. For a first course in structural geology, however, this textbook focuses on the physical domain.

3.1 UNITS OF MEASURE

The mechanical relationships used to understand rock deformation depend on four base quantities: length, mass, time, and temperature. Each of these is associated with a particular base unit of measurement and a corresponding unit symbol from the International System of Units (Système internationale d'unités), which is abbreviated **SI** (**Box 3.1**):

$$\begin{array}{l} \text{length } [=] \text{ meter (m)} \\ \text{mass } [=] \text{ kilogram (kg)} \\ \text{time } [=] \text{ second (s)} \\ \text{thermodynamic temperature } [=] \text{ kelvin (K)} \end{array} \tag{3.1}$$

The equal sign in square brackets [=] is read "has the unit(s)," and the unit symbol is given in parentheses.

The physical quantities listed in (3.1) are called *base quantities*, because they are used to make up the other quantities in **Table 3.1**. These other quantities are the so-called derived quantities and they also have particular derived units that are listed in **Table 3.1**, first in terms of the base units, and second with their own unique unit, if that is accepted in SI. The table also lists a few common quantities and common units of measure that are used in structural geology, but are not part of SI.

The last column of **Table 3.1** gives the symbols used to specify the dimensions of the physical quantities. The dimensions of the mechanical base quantities are:

$$\begin{array}{l} \text{length } \{=\} \text{ L} \\ \text{mass } \{=\} \text{ M} \\ \text{time } \{=\} \text{ T} \\ \text{thermodynamic temperature } \{=\} \ \Theta \end{array} \tag{3.2}$$

The equal sign in curly brackets {=} is read "has the dimension(s)." The dimensions of the derived SI quantities are products of powers of the dimensions of the base quantities. Unlike units, the dimensions do not depend upon the choice of a measurement system like SI, but are uniquely defined for each physical quantity. For example, the International Prototype Metre has a length, by definition, of 1 m; this is approximately equal to 3.937×10^1 inches, 3.281×10^0 feet, and 6.214×10^{-4} miles. Thus, one could debate how to express this distance in terms of units, but this distance always has the dimension L. Similarly, the pressure due to the weight of the atmosphere at Earth's surface is called 1 atm, which equals 101 325 Pa, 14.696 psi, and 29.92 inHg. Again, one could debate the choice of units, perhaps even concluding that the non-standard unit atm is preferable to the SI unit Pa, but this pressure always has the dimensions $\mathrm{M\,L^{-1}T^{-2}}$.

Box 3.1 International System of Units (SI)

A treaty called The Convention of the Metre was signed in 1875 by seventeen countries to establish the International Bureau of Weights and Measures (Bureau international des poids et mesures, abbreviated BIPM) with headquarters in Sèvres, France: www.bipm.org/en/home/.

As of August 2018, 60 Member States and 42 Associate States and Economies compose the BIPM. The Bureau is responsible for providing the standards for units of measurement. For example, the International Prototype Meter from 1889 to 1960 was the distance between two lines on a bar of platinum-iridium alloy, preserved under carefully controlled conditions at Sèvres, France. Since 1983 the meter has been defined as the distance traveled by light in a vacuum for a time of (1/299 792 458) seconds. The number in the denominator is the value of the speed of light in meters per second. Although the original motivations for the Bureau were commercial and political, it has had significant positive impact on the practice of science and engineering. The underlying concepts were: (1) recognition of certain fundamental quantities (e.g. length, mass, and time); (2) adoption of particular units of measurement for these quantities (meter, kilogram, and second); and (3) implementation of these units using a base 10 (decimal) numbering system.

The treaty of 1875 also established an international meeting called the General Conference of Weights and Measures (Conférence générale des poids et mesures, abbreviated CGPM) that meets every four years and is charged with the propagation and improvement of the standards for units of measurement. In 1960 at the 11th CGPM the International System of Units (Système internationale d'unités, abbreviated **SI**) was established. The seven dimensionally independent base units for SI are the meter, kilogram, second, kelvin, ampere, mole, and candela (**Figure B3.1**). In 2014 the CGPM adopted a resolution to work on revising the definitions of the base units using physical constants related to the motion and energy of electrons, atoms, and photons, rather than material objects. The second was defined in this way in 1967, and the meter in 1983, but the kilogram remained the only base unit tied to a material object, a cylinder of platinum-iridium alloy preserved at Sèvres, France. Then, on November 16, 2018 the kilogram was defined in terms of fundamental physical constants. Today, SI is the metric system used by scientists and engineers around the world.

Figure B3.1 Logo for SI with base unit abbreviations for the kilogram, meter, second, ampere, kelvin, mole, and candela.

A brochure is available for downloading from BIPM that introduces SI, defines the base and derived units, and describes the conventions for writing unit symbols. The US National Institute of Standards and Technology also publishes a comprehensive guide to SI (see Thompson and Taylor, 2008, in Further Reading) including definitions of the base and derived units, rules and style conventions for units, and an extensive table of conversion factors from archaic units to SI: www.nist.gov/physical-measurement-laboratory/special-publication-811. This document should reside on the computer or mobile device of all scientists and engineers.

While some of the fundamental units date back over 200 years, practitioners of metrology, the science of measurement, continue to revise and upgrade the standards and procedures with the goal of achieving stable and reproducible units based upon atomic or quantum phenomena. For everyday usage by structural geologists, common calibrated objects and devices (e.g. the tape measure, laboratory balance, digital watch, and thermometer) adequately approximate the standards.

Proper names used for SI units (kelvin, newton, pascal, joule, and watt) on **Table 3.1** are not capitalized, although the corresponding symbols (K, N, Pa, J, and W) are. Also, when some physical quantities are derived, the units cancel exactly. For example, the kinematic quantities called stretch and strain are ratios of lengths, so they are unitless and dimensionless. In determining the dimensions of stretch, one would calculate $L L^{-1} = L^0 = 1$, so this is indicated in **Table 3.1** by the number 1 in the fourth column. The unit used for Celsius temperature is the degree Celsius, °C, and this is equal in magnitude to the SI base unit, the kelvin, K. However, Celsius temperature shifts the SI temperature scale, so 0 °C and 100 °C are approximately the freezing and boiling points of water at atmospheric pressure.

Table 3.1 Physical quantities, units, symbols, and dimensions

Base quantity	Base unit	Unit symbol(s)	Dimensions
length	meter	m	L
mass	kilogram	kg	M
time	second	s	T
thermodynamic temperature	kelvin	K	Θ
Derived quantity	**Derived unit**	**Unit symbol(s)**	**Dimensions**
plane angle	radian	$\mathrm{m\,m^{-1}} = \mathrm{rad}$	1 (dimensionless)
area	square meter	$\mathrm{m^2}$	L^2
volume	cubic meter	$\mathrm{m^3}$	L^3
displacement	meter	m	L
velocity	meter per second	$\mathrm{m\,s^{-1}}$	LT^{-1}
acceleration	meter per second squared	$\mathrm{m\,s^{-2}}$	LT^{-2}
stretch, strain	meter per meter	$\mathrm{m\,m^{-1}}$ (unitless)	1 (dimensionless)
mass density	kilogram per cubic meter	$\mathrm{kg\,m^{-3}}$	ML^{-3}
force, weight	newton	$\mathrm{kg\,m\,s^{-2}} = \mathrm{N}$	MLT^{-2}
traction, pressure, stress	pascal	$(\mathrm{kg\,m\,s^{-2}})\,\mathrm{m^{-2}} = \mathrm{N\,m^{-2}} = \mathrm{Pa}$	$ML^{-1}T^{-2}$
elastic modulus	pascal	$(\mathrm{kg\,m\,s^{-2}})\,\mathrm{m^{-2}} = \mathrm{N\,m^{-2}} = \mathrm{Pa}$	$ML^{-1}T^{-2}$
Poisson's ratio	strain/strain	$\mathrm{m\,m^{-1}/m\,m^{-1}}$ (unitless)	1 (dimensionless)
dynamic viscosity	pascal second	$(\mathrm{kg\,m\,s^{-2}})\,\mathrm{m^{-2}\,s} = \mathrm{Pa\,s}$	$ML^{-1}T^{-1}$
temperature	kelvin	K	Θ
heat, work	joule	$(\mathrm{kg\,m\,s^{-2}})\,\mathrm{m} = \mathrm{N\,m} = \mathrm{J}$	ML^2T^{-2}
power	watt	$(\mathrm{kg\,m\,s^{-2}})\,\mathrm{m\,s^{-1}} = \mathrm{J\,s^{-1}} = \mathrm{W}$	$ML^2\,T^{-3}$
thermal conductivity	watt per meter kelvin	$(\mathrm{kg\,m\,s^{-2}})\,\mathrm{m\,s^{-1}\,m^{-1}\,K^{-1}}$ $= \mathrm{J\,s^{-1}\,m^{-1}\,K^{-1}} = \mathrm{W\,m^{-1}\,K^{-1}}$	$MLT^{-3}\Theta^{-1}$
specific heat capacity	joule per kilogram kelvin	$(\mathrm{kg\,m\,s^{-2}})\,\mathrm{m\,kg^{-1}\,K^{-1}}$ $= \mathrm{J\,kg^{-1}\,K^{-1}}$	$L^2T^{-2}\Theta^{-1}$
thermal expansion	inverse kelvin	$\mathrm{K^{-1}}$	Θ^{-1}
Celsius temperature	degree Celsius	°C, 1 °C = K – 273.15	Θ
Common quantity	**Common unit**	**Unit symbol(s)**	**Dimensions**
volume	liter	L, $1\,\mathrm{L} = 0.001\,\mathrm{m^3} = 1\,\mathrm{dm^3}$	L^3
time	annum	a, $1\,\mathrm{a} = 3.15576 \times 10^7\,\mathrm{s}$	T
	day	d, 1 d = 86 400 s	T
	hour	h, 1 h = 3 600 s	T
	minute	min, 1 min = 60 s	T
plane angle	degree of arc	°, 1° = (π/180) rad	1 (dimensionless)
	minute of arc	′, 1′ = (π/10 800) rad	1 (dimensionless)
	second of arc	″, 1″ = (π/648 000) rad	1 (dimensionless)

Table 3.2 SI prefixes, symbols, and factors

Prefix	Symbol	Powers
nano-	n	10^{-9}
micro-	µ	10^{-6}
milli-	m	10^{-3}
centi-	c	10^{-2}
deci-	d	10^{-1}
deka-	da	10^{1}
hecto-	h	10^{2}
kilo-	k	10^{3}
mega-	M	10^{6}
giga-	G	10^{9}
tera-	T	10^{12}

A few non-standard units are mentioned in **Table 3.1** because of their common usage, and the fact that they are accepted for use with SI. Note that these are related to the base or derived units, but only by including a numerical factor other than 1. The liter is exactly 1/1000 of the standard SI measure of volume, 1 m^3, and has the appeal that 1 L of water has a mass of approximately 1 kg. The degree, minute, and second of arc are used for the first two quantities in the geographic coordinate system (latitude, longitude, elevation). The annum is used by geologists for number of the years before present when referring to the age of rocks, minerals, and fossils. The plane angle is related to the circular arc length, s, and radial length, r, through the geometric relationship $\theta = s/r$, so strictly speaking this quantity is a ratio of lengths and thus without units. However, the circular arc is arbitrarily divided into 360 degrees, which is equivalent to 2π radians, and both are considered dimensionless.

One of the beauties of SI is the base 10 or decimal numbering. Thus, large multiples and small subdivisions of the units are easily expressed as powers of ten. A prefix is added to the unit of measurement to express the corresponding order of magnitude of the value. **Table 3.2** lists the prefixes in common usage in structural geology.

3.2 ACCURACY, PRECISION, AND SIGNIFICANT FIGURES

All measurements are subject to error, and numbers used in calculations that are based on measurements should reflect the accuracy and/or precision of the measurements. Accuracy refers to how well a number matches the true value of the quantity being measured. In contrast, precision refers to the reproducibility of a measurement. To understand these concepts we calculate the sample mean, $\bar{D}$, and the estimate of variance, s^2, for a set of hypothetical measurements of the distance between the two lines on the bar of platinum-iridium alloy stored at Sèvres, France, which was the International Prototype Meter from 1889 to 1960.

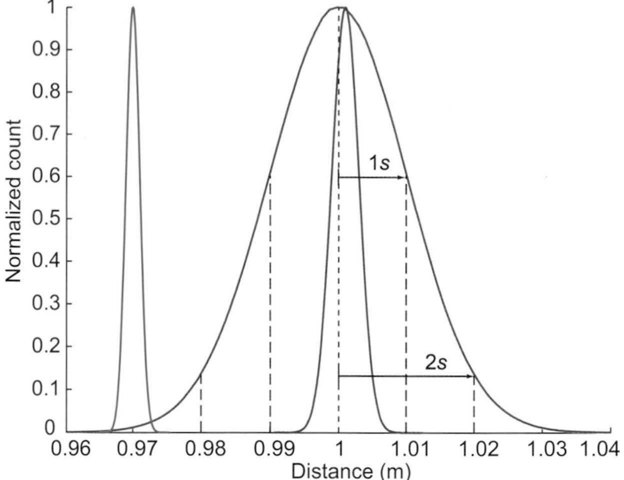

Figure 3.1 Histograms for different samples representing normally distributed measurements of the distance between the two lines on the one meter standard bar with sample mean, $\bar{D}$, and estimated standard deviation, s. Red $\bar{D} = 1, s = 0.01$; blue $\bar{D} = 1.001, s = 0.002$; green $\bar{D} = 0.97, s = 0.001$.

By definition the distance measured on the Prototype Meter is 1 m, and by assumption we insist that the data, D_i, are distributed about the mean in a special way, called a normal distribution (e.g. see the bell-shaped curves of **Figure 3.1**). Repeating the measurement n times, the sample mean is:

$$\bar{D} = \frac{1}{n}\sum_{i=1}^{n} D_i \tag{3.3}$$

In other words, this is the average value of the n distance measurements. Statisticians (see Davis, 2002, in Further Reading) distinguish the sample mean, $\bar{D}$, from the population mean, which would be the average value of all possible measurements of the distance between the two lines.

Given the sample mean (3.3), the estimated variance is:

$$s^2 = \frac{1}{n-1}\sum_{i=1}^{n} (D_i - \bar{D})^2 \tag{3.4}$$

This looks like the average of the squared differences between each measurement and the sample mean, except the fraction in front of the summation is $1/(n-1)$ instead of $1/(n)$. Because (3.4) uses the sample mean rather than the population mean, it underestimates the sum of the squared differences and using $1/(n-1)$ helps to correct this. The spread of the data about the sample mean is characterized using the estimated standard deviation, s, which is the square root of the estimated variance.

Suppose you measure the distance between the two lines on the bar many times using a tape measure that is awkward to use, and you obtain a sample mean and estimated standard deviation of 1 m and 0.01 m, respectively. These measurements are *accurate* because the sample mean is exactly the known value of the distance. However, the measurements are *not precise*, because the standard deviation is 1 cm. Most geologists would expect to measure meter-long structures in outcrop to better than 1 cm using a steel tape with a millimeter scale. Plotting the data as a histogram

(**Figure 3.1**, red curve), you have a broad bell-shaped curve that peaks at the mean value of 1 m. We have assumed that the data are normally distributed and this is what a normal distribution looks like, a symmetric bell-shaped curve centered on the sample mean. A characteristic of all normal distributions is that 68% of the measurements fall within one standard deviation of the mean. For this case that would be between 0.99 m and 1.01 m (between the two inner dashed black lines). In general 95% of the measurements fall within two standard deviations of the mean, and for this case that is between 0.98 m and 1.02 m (between the two outer black dashed lines).

Alternatively, suppose the measurements are made with an instrument that is not as awkward to use. You might obtain a mean length of 1.001 m and a standard deviation of 0.002 m (**Figure 3.1**, blue curve). The value of the mean is within a tenth of one percent of the actual value. The standard deviation has improved from 1 cm to 2 mm. Thus, 95% of the measurements are within 4 mm of the sample mean. This data set would be judged to be both *accurate* and *precise*. Although good instruments and careful scientists should yield measurements that are both accurate and precise, this is not always the case. Measurements can be highly reproducible and inaccurate, particularly if the measurement device is not correctly calibrated, or if it is read in a consistent but incorrect manner. A sample mean of 0.97 m for the distance between the two lines on the prototype meter with a standard deviation of 0.001 m would be an example of a *precise*, but *inaccurate* measurement (**Figure 3.1**, green curve).

Measurements in the field should be checked, if at all possible, to assure that the accuracy and precision are sufficient for the problem at hand. This usually entails checking the calibration on an instrument before using it and taking multiple measurements in the field. One example of a field check is a back sight with a compass between two surveying points A and B: the azimuth measured from A to B should be 180° from the measurement B to A. Another field check might be to survey three points that are aligned (the human eye is usually an excellent instrument for checking such alignments). If the position vectors for these points fail to line up when plotted on a map, then the measurements, the plotting, or both must be errant. Correctly identifying a measurement error does not preclude a plotting error, and vice-versa.

Significant figures are used to indicate the precision of a measurement. The standard procedure is to record the number of digits that are known accurately, and then record one more whose value is in doubt. All these digits are considered significant. *The number of digits to the right of the leftmost non-zero digit, inclusive, is the number of significant figures.* Any zeros that appear to the left of the leftmost non-zero digit are not significant. Zeros appearing to the right of the rightmost non-zero digit are significant. For example, 0.002 has one significant figure, 0.0021 has two significant figures, and 0.00200 has three significant figures. One should exercise care not to include extra zeros that are not merited. Using scientific notation, these values would be written: 2×10^{-3}, 2.1×10^{-3}, 2.00×10^{-3}.

Commonly, physical quantities are not measured directly, but instead are calculated from indirect measurements. The results of calculations should not be reported with more significant figures than the least precise datum used in the calculations. For example, suppose a series of measurements yields lengths of 0.1 km, 0.217 km, 0.20 km, and 0.36 km. The mean length should be reported as 0.2 km, not 0.21925 km since the original measurements had one significant figure, not two. All calculations performed should use all the significant figures in each measurement, and any rounding off of results should occur after the last calculation, not before. For cases where a number is to be rounded to one digit less, the rules are as follows:

(1) If the last digit is less than 5, then the preceding digit is left unchanged. For example, 8.42 is rounded down to 8.4 if two significant figures are to be reported.
(2) If the last digit is greater than 5, the preceding digit is increased by 1. For example, 3.6 is rounded up to 4.
(3) If the last significant digit equals 5, then the preceding digit is always rounded up, so 8.65 would round up to 8.7.

When calculations are complete and two or more figures are to the right of the last significant digit, rounding focuses on the next digit to the right. For example, if rounding to two significant figures, 6.7503 is viewed as 6.7(503) and is rounded up to 6.8, whereas 6.7488 is viewed as 6.7(488) and is rounded down to 6.7. If rounding to three significant figures, 6.7503 is viewed as 6.75(03) and is rounded down to 6.75, whereas 6.7488 is viewed as 6.74(88) and is rounded up to 6.75.

3.3 DIMENSIONAL ANALYSIS

We use the unique dimensional definitions in the last column of **Table 3.1** when carrying out a dimensional analysis of equations that are meant to describe physical processes. The test for dimensional homogeneity is the first step in such an analysis. Discovering and evaluating dimensionless products contained in these equations provides a better understanding of the physical process. The graphical presentation of data, or results from modeling, usually is most effective if the quantities plotted against one another have been converted into dimensionless quantities. We illustrate these aspects of dimensional analysis using examples drawn from the literature of structural geology.

3.3.1 Dimensional Homogeneity

When faced with an unfamiliar equation that represents some physical process or mechanism, the first test is to demonstrate dimensional homogeneity. This concept was first enunciated by Joseph Fourier in 1822, along with his law of heat conduction, which we describe in Section 3.8.1. For dimensional homogeneity *every term in a physically meaningful equation must have the same dimensions.* By *term* we mean those products or quotients that are separated by plus or minus signs when the equation is expanded algebraically. For example, the equation $A = B[CD - (E/F)] + G$ has four terms: A, BCD, BE/F, and G. If the dimensions of A on the left side are force, that is A {=} MLT^{-2}, then the three terms on the right side also must have dimensions of force. Dimensional homogeneity does not guarantee that an equation is an accurate description of the physical process. Perhaps some important mechanical aspects of the process were omitted, either knowingly to simplify the mathematical challenges, or

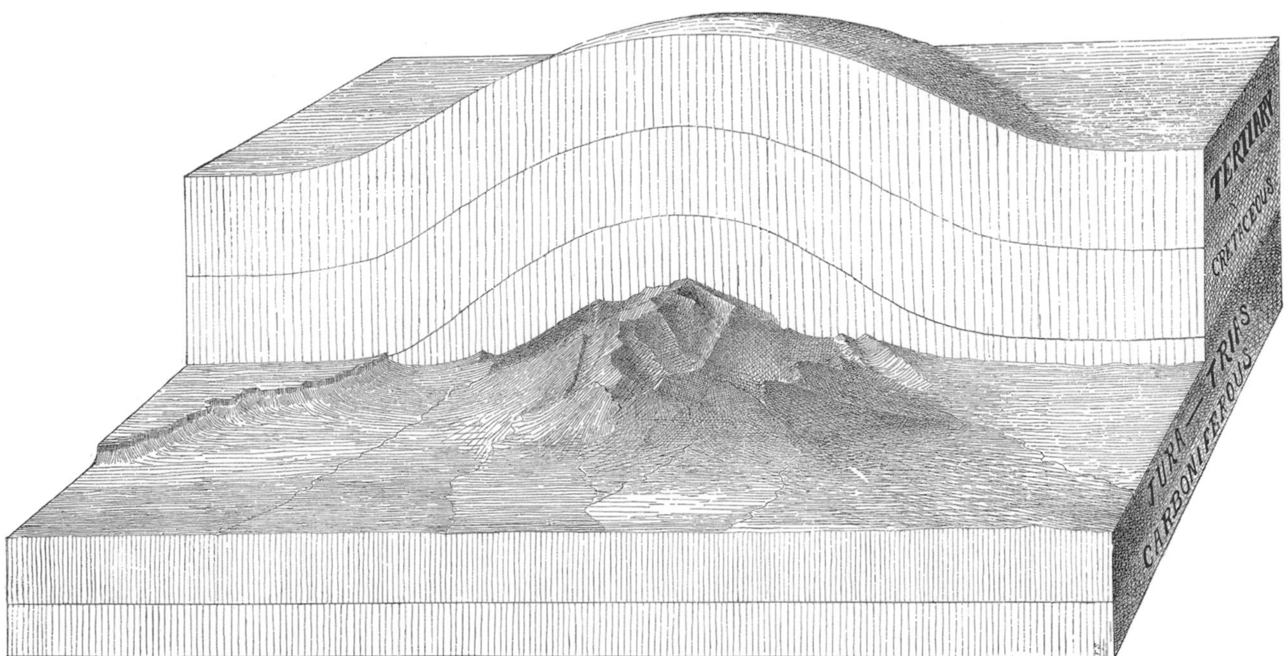

Figure 3.2 Frontispiece from *Report on the Geology of the Henry Mountains* used to illustrate "the form of the displacement and the progress of erosion" at Mt. Ellsworth. Modified from Gilbert (1877).

Figure 3.3 Photograph looking northeast toward Mt. Ellsworth, southern Henry Mountains, UT. Massive gray rock near the summit is igneous; layered tan and reddish rock on the flanks are inclined sedimentary formations; horizontal reddish sedimentary strata are seen in right foreground. See Jackson and Pollard (1988). UTM 12 S 529096.00 m E, 4172539.00 m N.

unknowingly out of ignorance of those aspects of the process. Other common errors that do not affect dimensional homogeneity are the inclusion of an incorrect dimensionless constant, or an incorrect plus or minus sign.

In the next two sections we test the dimensional homogeneity for equations that describes a process of elastic deformation and viscous deformation. These processes are considered in more detail in Chapters 4 and 6, respectively.

3.3.2 Bending Strata Over a Laccolith

Igneous intrusions are responsible for the formation of many mountain ranges in the western United States by bending and uplifting the overlying sedimentary rocks (**Figure 3.2**). The molten rock flowed upward through Earth's crust from a deeper magma reservoir and then insinuated itself between two sedimentary units, spreading laterally as a thin sill until it gained enough leverage on the overlying strata to bend them upward into a dome. This physical process was conceived and described by G. K. Gilbert in 1875 during his geological investigation of the Henry Mountains in southeastern Utah. The remnants of one of the sedimentary domes in the Henry Mountains may be seen at Mt. Ellsworth, which is viewed photographically from the southwest in **Figure 3.3**.

Here we examine a pair of equations that describe the vertical displacement, u_z, of the sedimentary layer immediately over the

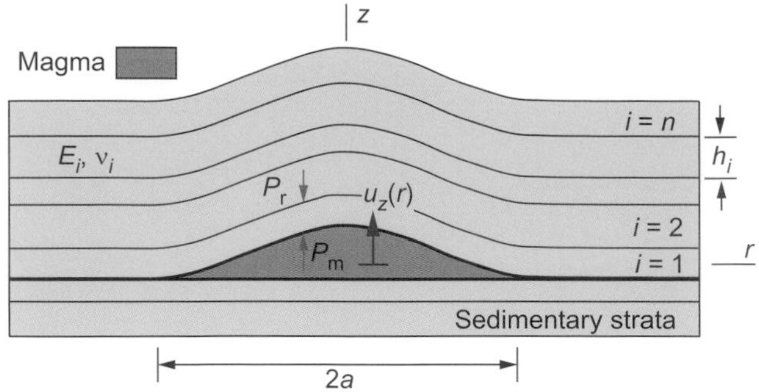

Figure 3.4 Schematic cross section of a model laccolith with diameter 2a under a stack of n sedimentary layers. The ith layer has elastic properties E_i and ν_i, and thickness h_i. P_m is magma pressure and P_r is pressure due to the weight of rock. See Pollard and Johnson (1973).

model laccolith (**Figure 3.4**) and ask: are these equations dimensionally homogeneous? The equations are:

$$u_z = \frac{\Delta P}{64R}\left(a^4 - 2a^2r^2 + r^4\right) \quad (3.5)$$

$$R \equiv \sum_{i=1}^{n} \frac{E_i h_i^3}{12(1 - \nu_i^2)} \quad (3.6)$$

These equations look complicated because of the many different symbols representing different physical quantities. The *first step* in dimensional analysis is to identify the number of terms. In (3.5) the left-hand side contains one term and the right-hand side, if expanded by multiplying through by the leading quotient, contains three terms. Equation (3.6) defines the resistance to bending, called the flexural rigidity, R. This equation has one term on the left-hand side and n terms on the right-hand side, because the summation is carried out from $i = 1$ to $i = n$. Here n is the number of layers over the laccolith. Fortunately, these n terms are identical with respect to dimensions, so we only need to consider one typical term.

The *second step* in dimensional analysis is to identify the constants, and the independent and dependent variables, and to establish their dimensions. For the constants in (3.5) and (3.6) the dimensions are taken from **Table 3.1**:

$$\Delta P = P_m - P_r = \text{driving pressure } \{=\}\ ML^{-1}T^{-2}$$
$$a = \text{laccolith radius } \{=\}\ L$$
$$h_i = \text{layer thickness } \{=\}\ L$$
$$E_i = \text{elastic modulus } \{=\}\ ML^{-1}T^{-2}$$
$$\nu_i = \text{Poisson's ratio } \{=\}\ 1$$

P_m is the pressure on the bottom layer due to the magma and P_r is the pressure on the bottom layer due to the overburden weight. The positive difference between these pressures is the driving pressure that bends the stack of layers upward into a bell-shaped arc, consequently creating room for the magma. The driving pressure is resisted by the flexural rigidity of the sedimentary strata over the laccolith. The laccolith is circular in plan with radius a. The sedimentary layers over the laccolith have thicknesses, h_i, elastic moduli, E_i, and Poisson's ratios, ν_i. The elastic moduli and Poisson's ratios are measures of stiffness and compressibility that we define and discuss in more detail in Sections 4.4.1 and 4.10.

The independent variable, r, and dependent variable, u_z, in (3.5) and their dimensions are:

$$r = \text{radial coordinate } \{=\}\ L$$
$$u_z = \text{vertical displacement } \{=\}\ L$$

The radial coordinate ranges from 0 at the laccolith center to a at the laccolith perimeter. The vertical displacement is that along the middle surface of the layer immediately over the magma. The distribution of displacement, $u_z = u_z(r)$, is the unknown that may be calculated given all the constants and the independent variable.

The *third step* in dimensional analysis is to combine the dimensions of the individual constants and variables, with appropriate powers, to find the dimensions of each term. Beginning with (3.6) we determine the dimensions of R using one typical term in the summation.

$$R\ \{=\}\ \frac{ML^{-1}T^{-2}L^3}{1\left(1 - 1^2\right)}\ \{=\}\ ML^2T^{-2}$$

The denominator from (3.6) contains the numbers 12 and 1, but all numbers are dimensionless, so these are replaced in the dimensional expression with the number 1. Poisson's ratio also is dimensionless, so it is replaced with the number 1. If the resulting denominator were evaluated using the standard rules of algebra we would be faced with dividing by zero. However, when comparing dimensions in this dimensional expression we recognize that the denominator contains only dimensionless terms and therefore is ignored, as if dividing by 1.

The dimensions of each term in (3.5), including R from the previous paragraph, are:

$$L\{=\}\frac{ML^{-1}T^{-2}}{ML^2T^{-2}}\left(L^4 + L^2L^2 + L^4\right)\{=\}L^{-3}\left(L^4 + L^4 + L^4\right)\{=\}L$$

One does not actually add the three terms in parentheses, but merely compares their dimensions. We conclude that (3.5) is homogeneous with respect to dimensions, and that each term has dimensions of length. One could question if these equations correctly describe the physics of layer bending over a laccolith, but these equations do pass the dimensional homogeneity test.

3.3.3 Flow of Magma in a Dike

The second example of dimensional analysis is taken from fluid dynamics (see Kundu et al., 2015, in Further Reading), the theory used to model the flow of molten rock, one of the subjects of Chapter 6. Suppose you are told that the following equation describes the flow of magma in a dike to Earth's surface where it erupts from a volcanic fissure (**Figure 3.5**):

$$\rho\frac{\partial v_z}{\partial t} = -\frac{\partial p}{\partial z} + \eta\frac{\partial^2 v_z}{\partial x^2} + \rho g \quad (3.7)$$

The symbols in (3.7), represent different physical quantities, and also partial derivatives with respect to time, $\partial/\partial t$, and partial

Figure 3.5 Photograph of a nighttime fissure eruption on Kilauea Volcano, Hawaii. Echelon "curtains of fire" rise tens of meters into the air and feed lava flows descending the slope of the volcano. Photograph by R. Holcomb.

derivatives with respect to the two spatial coordinates, $\partial/\partial z$ and $\partial^2/\partial x^2$. This equation contains four terms, one on the left-hand side and three on the right-hand side. The combinations of symbols within each term are different, so how could these terms have the same dimensions? The answer is not obvious, until we take a closer look and do a careful dimensional analysis.

The constants in (3.7) and their respective dimensions taken from **Table 3.1** are:

$$\rho = \text{mass density } \{=\} \, \text{ML}^{-3}, \; \eta = \text{viscosity } \{=\} \, \text{ML}^{-1}\text{T}^{-1}, \\ g = \text{acceleration of gravity } \{=\} \, \text{LT}^{-2}.$$

The dependent variables in (3.7), both of which refer to unknown quantities in the magma, are:

$$v_z = \text{velocity in } z \, \{=\} \, \text{LT}^{-1}, \; p = \text{pressure } \{=\} \, \text{ML}^{-1}\text{T}^{-2}$$

The z coordinate is vertical, so v_z is the upward velocity of the magma, toward Earth's surface (**Figure 3.6**). This is the only non-zero component of velocity, and it varies only in the x coordinate, across the dike. The magma is pushed upward by the pressure gradient, $(p_2 - p_1)/L$, if the pressure gradient exceeds the static gradient due to the weight per unit volume of magma, ρg. The pressure gradient is negative because $p_1 > p_2$.

The independent variables in (3.7) are time, t, and spatial coordinates, z and x. These are included in partial differential operators, so we consider their dimensions in this form:

$$\frac{\partial}{\partial t} \{=\} \, \text{T}^{-1}, \; \frac{\partial}{\partial z} \{=\} \, \text{L}^{-1}, \; \frac{\partial^2}{\partial x^2} \{=\} \, \text{L}^{-2}$$

Finally, grouping the dimensions appropriate for each term in (3.7) we have:

$$\text{ML}^{-3}\text{T}^{-1}\text{LT}^{-1} \{=\} \\ \text{L}^{-1}\text{ML}^{-1}\text{T}^{-2} + \text{ML}^{-1}\text{T}^{-1}\text{L}^{-2}\text{LT}^{-1} + \text{LT}^{-2}\text{ML}^{-3}$$

Adding the powers of like dimensions in each term, we reduce this dimensional expression to:

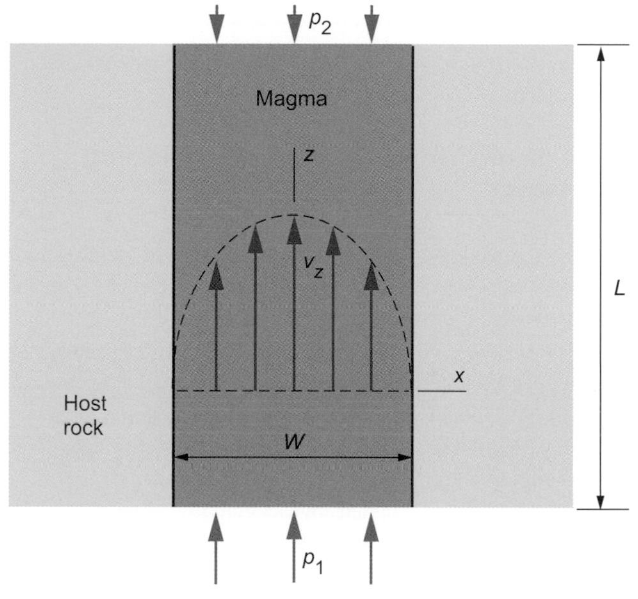

Figure 3.6 Schematic drawing of viscous magma flowing upward in a dike of constant width, W, subject to a pressure drop from p_1 to p_2 over a length L. The velocity in the z-coordinate direction, v_z, is shown as vectors that decrease to zero at the contact between the magma and host rock.

$$\text{ML}^{-2}\text{T}^{-2} \{=\} \, \text{ML}^{-2}\text{T}^{-2} + \text{ML}^{-2}\text{T}^{-2} + \text{ML}^{-2}\text{T}^{-2}$$

We conclude that the partial differential equation (3.7) is dimensionally homogeneous and all of the terms have dimensions of mass times acceleration divided by volume, that is $\text{MLT}^{-2}\text{L}^{-3}$.

Considering Newton's second law, $\mathbf{F} = m\mathbf{a}$, we understand the terms of (3.7) have dimensions that are force per unit volume. Reviewing the physical quantities in these four terms we characterize them as force per unit volume due to: (1) acceleration of the fluid; (2) the pressure gradient along the conduit; (3) the viscous

Figure 3.7 Offset vehicle tracks across right-lateral rupture from the 1999 Hector Mine Earthquake in southeastern California. Modified from Treiman (2009).

drag on the walls of the conduit; and (4) gravitational attraction. If these were all the forces acting on the magma, (3.7) would be a complete equation of motion based on a balance of forces. We consider applications and solutions for this equation in Chapter 6 with the assurance that the dimensional homogeneity test has been passed.

3.3.4 Dimensionless Graphs and Scaling Relations

In the presentation of field data as well as the results from laboratory experiments and theoretical analyses, structural geologists use many different kinds of graphs. Perhaps the most elementary is the plot of one dependent variable on the ordinate versus one independent variable on the abscissa. Here we describe how and why one should plot variables as dimensionless ratios, and we show how this leads to scaling relations that reveal some of the underlying physics of geologic structures.

For an example of a dimensionless graph we use data for slip on a large strike-slip fault during the Hector Mine earthquake. On October 16, 1999 this earthquake (M_w 7.1) rumbled across the Mojave Desert, rupturing the ground surface along the Lavic Lake, Bullion, and Mesquite Lake faults with an aggregate trace length of almost 50 km. This event produced numerous horizontal and vertical offsets of geomorphic and cultural features that cross the rupture trace (**Figure 3.7**), allowing geologists to measure fault slip at more than 400 points, and to map the traces of the rupture on aerial photographs at a scale of 1:10,000. This earthquake likely would have caused widespread destruction and loss of life in a more developed and populated location. In this desert terrain it provided a wonderful opportunity to investigate the characteristics of earthquake ruptures that involve multiple faults and a variety of deformation features at Earth's surface.

From the data base of measured offsets we plot the horizontal component of slip, Δu, as red asterisks for the main rupture from 86 stations along the Lavic Lake fault (**Figure 3.8a**). The estimated uncertainty associated with the measurements is plotted as error bars. The Lavic Lake fault trace is about 35 km long and the greatest reported horizontal slip is 5.25 m with a right-lateral sense. In the data base the offsets are associated with distance along the rupture trace from north to south as projected onto a simplified fault geometry consisting of five straight segments that approximate the broadly sinusoidal trace of the rupture.

The zero value of projected distance along the rupture trace was set arbitrarily about one kilometer north of the first data point. The measured offsets on closely spaced multiple fault segments were added, but in many cases the recorded values represent a minimum estimate of the total offset, owing to lack of exposure and absence of suitable markers. Furthermore, some of the offsets were measured on secondary structures that branch off of, or are a considerable distance from, the main rupture trace. Therefore, certain low slip values, well away from the ends of the rupture trace, may not be representative of the total slip on the underlying fault.

On **Figure 3.8a** the horizontal slip also is plotted as blue circles for 50 stations along the Johnson Valley fault that ruptured during the Landers earthquake on June 28, 1992. This large earthquake (M_w 7.3) ruptured the Johnson Valley fault, and also the Homestead Valley, Emerson, and Camp Rock faults. The Johnson Valley fault trace is about 20 km long, and the greatest reported horizontal slip is 3.10 m with a right-lateral sense. While the combined plots for the two faults is instructive, it is difficult to compare these data because the faults have different lengths and different maximum slips.

Comparison of the slip distributions for the Johnson Valley and Lavic Lake faults are greatly facilitated by dividing the distance and slip for each fault by a characteristic length, and preparing a *dimensionless graph* (**Figure 3.8b**). The first step is to transform the projected distances, so the first data point is at zero. The next step is to acknowledge the apparent symmetry of the slip about the mid-point of the rupture trace and transform the projected distances, so $x = 0$ is at the mid-point of each fault trace. The final step is to normalize both the distance along the rupture traces, and the horizontal slips, by a

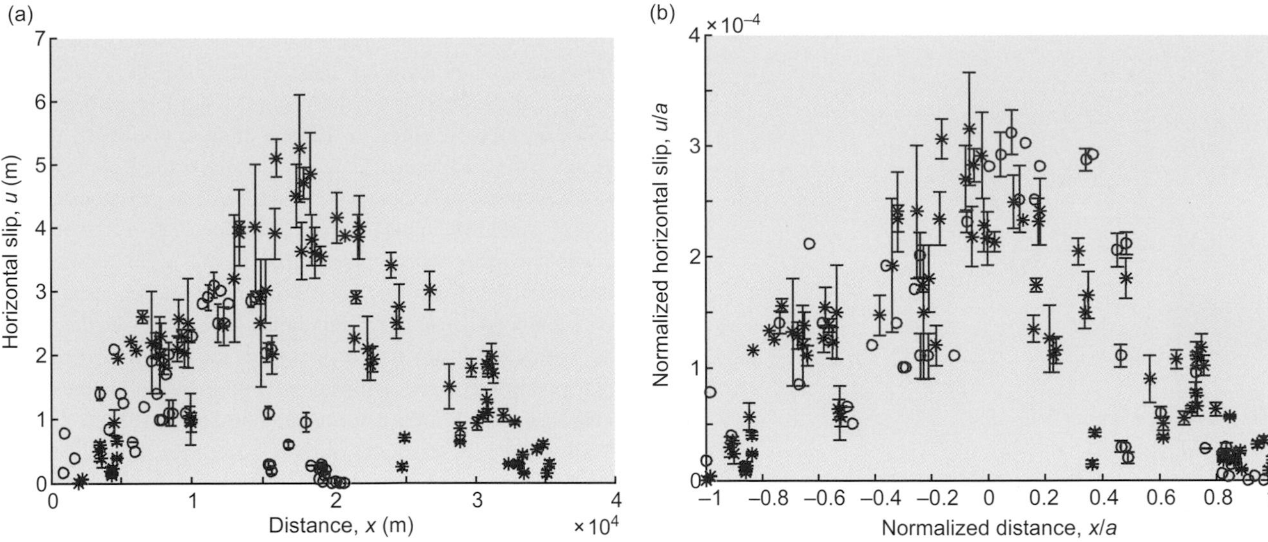

Figure 3.8 (a) Plot of horizontal slip, Δu, versus distance, x, along the Johnson Valley (blue circles) and Lavic Lake (red asterisk) faults determined from lateral offset of geomorphic and cultural features along the main ruptures from the Landers and Hector Mines earthquakes, respectively. (b) Plot of normalized horizontal slip, $\Delta u/a$, versus normalized distance, x/a, along the Johnson Valley and Lavic Lake faults. Data from Treiman et al. (2002).

characteristic length. In this case the characteristic length is the half-length, a, of each fault trace, so we plot x/a on the abscissa and $\Delta u/a$ on the ordinate in **Figure 3.8b**. The quantities plotted on the abscissa and ordinate are dimensionless and therefore have no units.

The normalized slip distributions on **Figure 3.8b** have significant scatter, but most of the data from the two faults fall along a curve from zero at $x/a = -1$, to maximum values of about $\Delta u/a = 3 \times 10^{-4}$ at $x/a = 0$, and then back to zero at $x/a = +1$. The fact that data from both faults follow the same distribution suggests that a single *scaling relation* might satisfy both slip events. Clearly the slip increases with fault length, and the similarity of the normalized slip distributions suggests the slip scales *linearly* with fault half-length. Indeed, in Section 8.4 we describe a model for faults using elasticity theory that relates slip to half-length as:

$$\Delta u = 2\Delta\sigma(1 - \nu)(a^2 - x^2)^{1/2}/G$$

At the center of the model fault, the maximum slip scales linearly with the half-length as $\max(\Delta u) = 2\Delta\sigma a(1-\nu)/G$. The *scale factor*, $2\Delta\sigma a(1-\nu)/G$, is composed of the measurable quantities including the stress drop, $\Delta\sigma$, Poisson's ratio, ν, and the elastic shear modulus, G. We learn more about these quantities in Section 4.10.

The dimensionless plot (**Figure 3.8b**) helps us to appreciate one aspect of the physics of faulting (fault slip scales *linearly* with fault length) without constructing any models. Once the model is in hand, the dimensionless graph provides a good way to display data that are, or are not, consistent with the model results. The dimensionless graph also enables the direct comparison of slip data from two faults that have different lengths and maximum slips.

3.4 MATERIAL CONTINUUM

The equations of motion that describe the deformation of rock, whether as a brittle solid, a ductile solid, or a viscous fluid, are formulated using the concept of a material continuum. The real numbers provide a familiar example of a continuum. Given any two different real numbers, say 25.687 22 and 25.687 27, one can come up with a third number that falls between them, for example 25.687 24. This is true for two different real numbers with as many significant figures as you care to write down. Therefore, the set of real numbers are continuous, and we say that the real numbers form a mathematical continuum. The real number line illustrates this concept graphically. Again, the notion is that you can pick any two distinct points on the line and always find a third point between these, no matter how much you zoom in on the number line.

The concept of the mathematical continuum also is apparent when one evaluates or plots a continuous function, say $y = \sin(x)$. Using the same two real numbers, 25.687 22 and 25.687 27, as angles measured in degrees of arc for the independent variable x, one evaluates the dependent variable as $y = 0.433\,458\,1$ and $y = 0.433\,458\,9$, respectively. The number 25.687 24, corresponds to the intermediate value, $y = 0.433\,458\,4$. One can plot this function with y on the ordinate and x on the abscissa and, similar to the number line, zoom in *ad infinitum* by changing the scale of the abscissa, to observe finer and finer pieces of the continuous sinusoidal curve.

While the mathematical continuum described in the preceding paragraph is a familiar and readily accepted concept, the material continuum requires some explanation and justification (Box 3.2). To understand the material continuum, recall that real numbers represent both space and time, and thus are taken as continuous. For example, when we adopt the Cartesian coordinate system to

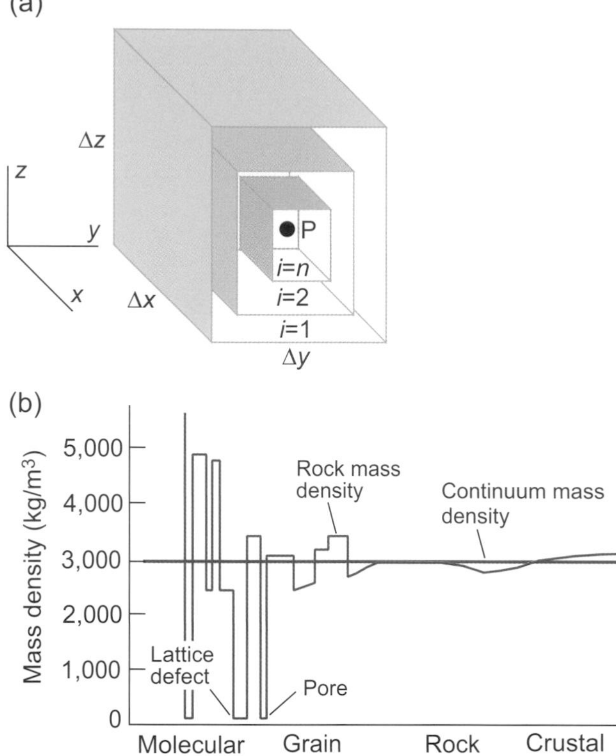

Figure 3.9 (a) Rectangular blocks of successively smaller volume as *i* goes from 1 to *n*, used to define mass density at the point P according to (3.8). (b) Schematic plot of mass density in Earth's crust as a function of volume.

demarcate space, the values of the coordinates along each axis are continuous. The analogue watch with a sweeping second hand provides an understandable demonstration of the fact that time is not discontinuous.

As a concrete example of the material continuum, consider the definition and continuity of mass density. To define the mass density at an arbitrary point P in the material continuum (**Figure 3.9**), consider a rectangular block with sides parallel to the coordinate axes and lengths $\Delta x \leq \Delta y \leq \Delta z$. This block contains the point P and has a volume ΔV_1 and mass Δm_1. Now consider n blocks, all containing the point P, but with successively lesser volume, and name these $i = 1$ to $i = n$. In particular, the nth block has a volume ΔV_n and a mass Δm_n, so the mass density of that block is the ratio $\Delta m_n / \Delta V_n$. In the spirit of calculus, we now consider the limiting value of the mass to volume ratio as n becomes very large and ΔV_n becomes very small. Both the mass and the volume would necessarily become very small, but we postulate that the ratio of these quantities converges toward the value called the *mass density of the material continuum at the point P*:

$$\rho(\mathrm{P}) = \lim_{n \to \infty} \frac{\Delta m_n}{\Delta V_n} \tag{3.8}$$

In taking this limit we require the longest side of the block to approach a length of zero, that is $\Delta z \to 0$, so the block converges toward a small volume and not toward a plane or a line.

The density of a rock mass varies with the volume of rock, as shown schematically in **Figure 3.9b**. From crustal to hand specimen scales, minor variations exist about a value close to 3×10^3 kg m^{-3}. At the grain scale, the volume could include pores, and inside one of those pores, the density drops to that of air, about 1 kg m^{-3}. At the molecular scale, the volume could include mostly space between molecules, or it could include mostly atoms, so the density would be highly variable. Despite these variations, the continuum mass density is 3×10^3 kg m^{-3} right down to an infinitesimal volume, and right down to a mathematical point. By adopting the material continuum, we do not negate the granular or molecular structure of rock, but we do acknowledge the remarkably successful application of the material continuum to investigations of rock deformation. In Sections 3.4.2 through 3.4.4 we describe examples of those successful applications.

The definition of mass density (3.8) does not preclude spatial variation. In that case the material is described as *non-uniform* or *heterogeneous* with respect to mass density, which is a function of the coordinates, $\rho = \rho(x, y, z)$. The definition of mass density (3.8) also does not preclude variation in time, in which case the material is said to have a *non-constant* mass density, so $\rho = \rho(t)$.

3.4.1 Discontinuities in the Material Continuum

The concept of a material continuum does not preclude a discontinuity in any one of the relevant physical quantities across a *finite* number of surfaces, which can include closed surfaces that bound holes. Such discontinuities are abundant in nature. For example, an abrupt change in mass density can occur across bedding surfaces in sedimentary rock (**Figure 3.10**), across compositional boundaries in metamorphic rock, and across the contacts between intruded igneous rock and the host rock. The definition of mass density (3.8) can be applied, but the location of discontinuities and the jump in density across them must be explicitly defined.

For example, suppose the origin of coordinates is at a bedding interface in **Figure 3.10**, where limestone overlies sandstone, and the z-axis is perpendicular to this interface and positive upward. There, we would have one measure of mass density employing (3.8) for points just above the (x, y)-plane, and a different measure just below that plane:

$$\rho(x, y, z > 0) = \rho(\text{limestone})$$
$$\rho(x, y, z < 0) = \rho(\text{sandstone})$$

By a similar argument the displacement discontinuity across the fractures in the limestone bed seen in **Figure 3.11** could be explicitly defined and incorporated into the continuum model. As the fracture opened, rock particles on the right side of the fracture displaced to the right, whereas originally neighboring particles on the left side displaced to the left (**Figure 3.11**, inset). The component of displacement in the x-direction, u_x, has opposite signs on opposite sides of the fracture. Placing the origin of coordinates at the fracture with the (y, z)-plane parallel to the fracture plane, the opening displacement discontinuity, Δu_x, is:

$$\Delta u_x = u_x(x = 0^+, y, z) - u_x(x = 0^-, y, z)$$

Box 3.2 Continuum Mechanics

Given that the space-time framework for the description of a physical process is based on real numbers, and therefore continuous, it is natural to view the physical quantities attached to such a framework as continuous. For example, as a material deforms, one could expect to evaluate the velocity within that material at *any point* and at *any time*. Given this velocity distribution, and treating mass density as continuous, one could expect to evaluate the momentum at any point and time. These are very appealing concepts, because they admit the use of the powerful methods of calculus to construct mathematical models based on mechanical principles of deformation (see Malvern, 1969; Lai et al., 2010, in Further Reading). These models explicitly address the spatial and temporal variations in velocity and momentum throughout a rock mass as geologic structures form and develop. Continuum mechanics is the application of fundamental physical principles, including conservation of mass and momentum and energy, to the material continuum.

As described in Section 3.4, the material continuum is a *theoretical construct* for which we define mass density, momentum, and energy at every point and for all time. It follows that continuum mechanics also is a *theoretical construct*. Therefore, the solutions for problems posed in continuum mechanics by structural geologists are models of the behavior of idealized materials, chosen because they approximate the behavior of rock as measured in the laboratory, or as inferred from field observations of geologic structures. Mapped geologic structures usually provide the model geometries, and the solutions are constrained by boundary or initial conditions, again inferred from field or laboratory observations of geologic structures. The interdependence of models and observations is key to the successful application of continuum mechanics in structural geology (**Figure B3.2**).

Physical quantities that describe the kinematics of rock deformation include displacement, velocity, acceleration, strain, and rate of deformation. Physical quantities that describe the causative agents of deformation include force, traction, and stress. Continuous functions of space and time represent all of these physical quantities in the material continuum, except across a *finite* number of surfaces where certain quantities are *discontinuous*. These discontinuities are particularly important for applications to structural geology, because most geologic structures (e.g. fractures, faults, folds, and intrusions) involve discontinuities. For example, the displacement field is discontinuous across an opening fracture, or a slipping fault, or a bedding plane in a flexural slip fold, or the contacts of a dike. Thus, we adopt continuum mechanics, while recognizing that discontinuities can play a critical role in the development of geologic structures.

Although the founders of continuum mechanics mostly were physicists and applied mathematicians, this discipline is associated with schools of engineering today, and the authors of most modern textbooks on continuum mechanics are engineers (e.g. Malvern, 1969; Lai et al., 2010, see Further Reading). This transition occurred because continuum mechanics proved to be a very useful tool to study practical problems in engineering, and because physicists moved on to consider the atomistic nature of materials, and phenomena where special relativity is important. Today, earth scientists borrow continuum mechanics from the engineers and apply it to a host of problems including geodynamics (Turcotte and Schubert, 2002), crustal deformation (Segall, 2010), and reservoir geomechanics (Zoback, 2010) (see Further Reading). In this textbook we take continuum mechanics as the theoretical basis for studies in structural geology.

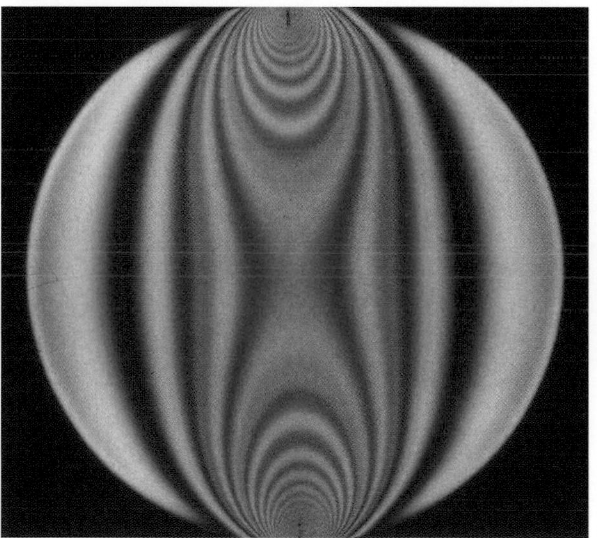

Figure B3.2 Image from a laboratory experiment of a thin disk of photoelastic material loaded by point forces at the top and bottom. Compare this pattern (constant color = constant maximum shear stress) to patterns in **Figure 3.13**.

Here $x = 0^+$ refers to the initial position of particles slightly on the positive side of the (y, z)-plane, and $x = 0^-$ refers to initial positions slightly on the negative side. The difference in displacement components at these two positions defines the discontinuity, and the magnitude of the opening of the two fracture surfaces.

Adoption of the material continuum for rock with discontinuities in density or displacements is straightforward, but it may be limited by practical considerations such as bookkeeping to record all the relevant discontinuities, or by computer memory to accommodate the data, or by computer processing time to solve the equations describing the deformation with all the discontinuities. For example,

Figure 3.10 A prominent limestone bed (gray) overlies a sandstone bed (red) on the western limb of Raplee anticline, UT. The bedding interface is a discontinuity in mass density that could be explicitly defined when building a structural model at this scale. Approximate location UTM 12 S 603875.00 m E, 4115539.00 m N.

Figure 3.11 A fractured limestone bed on the western limb of Raplee anticline, UT. The opening fracture is a discontinuity in displacement that is explicitly defined when building a model at this scale. The opposing arrows on the inset illustrate the displacement discontinuity.

at the scale of the entire Raplee anticline the individual beds of limestone and sandstone are too numerous to model the changes in density and numbers of fractures explicitly. At this scale one could choose to use the average density as representative of the more complex distribution known to exist in nature. Similarly, at the scale of the anticline one could choose to model the fractured strata as a continuum with uniform anisotropic properties to approximate the mechanical response of the fractured strata.

3.4.2 Continuum at the Molecular Scale: Edge Dislocations

The material continuum construct ignores the molecular structure of solids and fluids. This begs the question: at what length scale does the continuum postulate break down because of the molecular structure of matter? An answer comes from a remarkable experiment to investigate the displacement field near a defect in silicon dioxide (silica) using a combination of high-resolution electron microscopy and image analysis (**Figure 3.12**). The defect in this case is called an edge dislocation and the lattice spacing is 0.271 nm in x and 0.192 nm in y. The images are color contour maps of the displacement component magnitudes: u_x in the x-direction, and u_y in the y-direction (**Figure 3.12a,c**). The dislocation is at the center of each image and the color range for u_x is 0 nm to 0.192 nm, and for u_y is –0.271 nm to 0 nm.

Color contour maps for the same range of displacements from a model of an edge dislocation (**Figure 3.12b,d**) were constructed using *elasticity theory*, a branch of continuum mechanics that we introduce in Chapter 4. Although the theoretical displacement fields have the crispness that comes from a mathematical

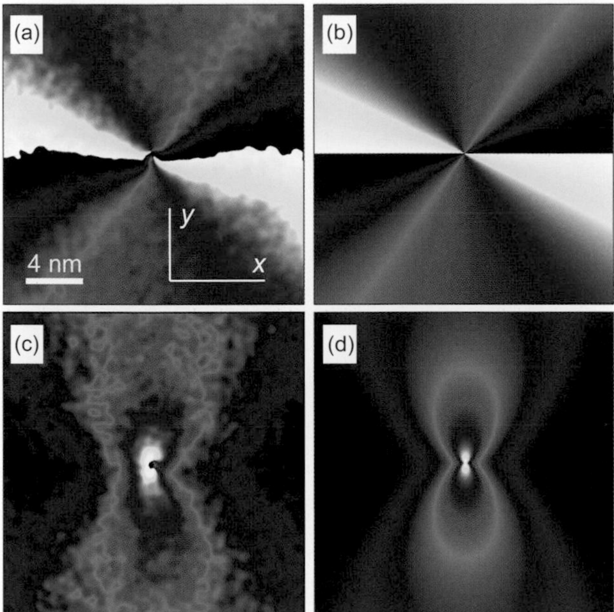

Figure 3.12 Contour maps of displacement magnitudes near an edge dislocation in silica. (a) Image of displacement component u_x in silica. (b) Displacement component u_x from continuum theory. (c) Image of displacement component u_y in silica. (d) Displacement component u_y from continuum theory. Modified from Hytch et al. (2003).

calculation, and the experimental fields have a roughness related to the measurement techniques, the correlations in terms of the color patterns are astounding and insightful. The deviation of the experimental displacements from the continuum theory displacements was 0.003 nm at a radial distance of 7.5 nm from the dislocation. This error is just a few percent of the nominal displacement magnitude of 0.1 nm in the vicinity of the edge dislocation, so the agreement between the experimental data and the theory is excellent. Furthermore, from the similarity of the patterns of contours one may conclude that the displacements from the continuum theory match the measured displacements down to radial distances of a few nanometers from the edge dislocation.

To directly answer the question posed at the beginning of this section: the continuum postulate breaks down, because of the molecular structure of matter, below length scales of a few nanometers. From the nanometer scale up to the scale of single mineral grains, the material continuum is an appropriate idealization for analysis of deformation. However, the material continuum may be problematic at scales of several grains, because many rocks, especially clastic sedimentary rocks, are porous, and because rocks typically are made up of several different kinds of minerals. We address issues related to grain scale heterogeneity in the next section.

3.4.3 Continuum at the Grain Scale: Porous Sandstone

The challenges to the continuum concept introduced at the grain scale are exemplified using a medium sandstone, which ranges in grain size from 0.25 mm to 0.50 mm, has a porosity typically in the range from 10% to 20%, and may include detrital minerals other than the predominant quartz. Considering spatial variations in mass density, for example, the sample volume may fall within a pore, within different minerals, or some combination thereof. Continuum mechanics can be applied at this scale if one maps out every pore and grain with appropriate discontinuities in physical quantities at the grain boundaries.

An instructive example of continuum theory at the grain scale uses disks of a transparent model material called CR-39 shaped to represent individual sand grains (**Figure 3.13a**). When the collection of model grains is subjected to a load, they deform and the deformation is recorded optically using a technique called photoelasticity. The black and white swaths within the model grains are contours of the greatest shear stress. The black areas between grains are the pores, where the shear stress is zero. The patterns of contours within the model grains look very complicated, but systematic patterns can be identified by studying the contours in a circular disk loaded with point forces along the vertical diameter (**Figure 3.13b**). In this figure the upper half shows contours calculated using *elasticity theory*, the branch of continuum mechanics we introduce in Chapter 4. The lower half of the circular disk shows the photoelastic pattern and the two sets of contours are a nearly perfect match.

The model sandstone grains in **Figure 3.13a** that have only two neighboring and opposed grains in direct contact, have a pattern of shear stress contours that looks very similar to that in the circular disk (**Figure 3.13b**). Good examples are identified by A, B, and C on the photograph of the laboratory experiment. The similarity of the distributions of greatest shear stress from the laboratory photoelastic experiment and the theoretical calculation is another demonstration of the efficacy of the continuum theory. This theory can be employed at the grain scale, but each grain must be accounted for in terms of its geometry and physical properties. This would be a daunting task if more than a few hundred grains were considered, so in that application one would resort to treating the sandstone as a continuum, without grains and pores, but using average properties.

3.4.4 Continuum at the Crustal Scale: Hector Mine Earthquake

Is there an upper limit to the length scale for applying continuum theory to understand geological structures? If the length scale exceeds the problematic range of grain scale heterogeneity, rock can be adequately described as a continuum with the proviso that there may be a finite number of surfaces across which some physical quantities are discontinuous.

As an example at the 10 to 100 km scale, consider the M_w 7.1 Hector Mine earthquake of 16 October 1999 in southeastern California (**Figure 3.14**). The data contoured in this figure are from a satellite ranging system using Synthetic Aperture Radar Interferometry (InSAR). This technology measures the distance from the satellite to the ground using the travel time of a reflected radar signal. In this case the satellite was ERS-2 and the images were taken from the same position of the satellite on

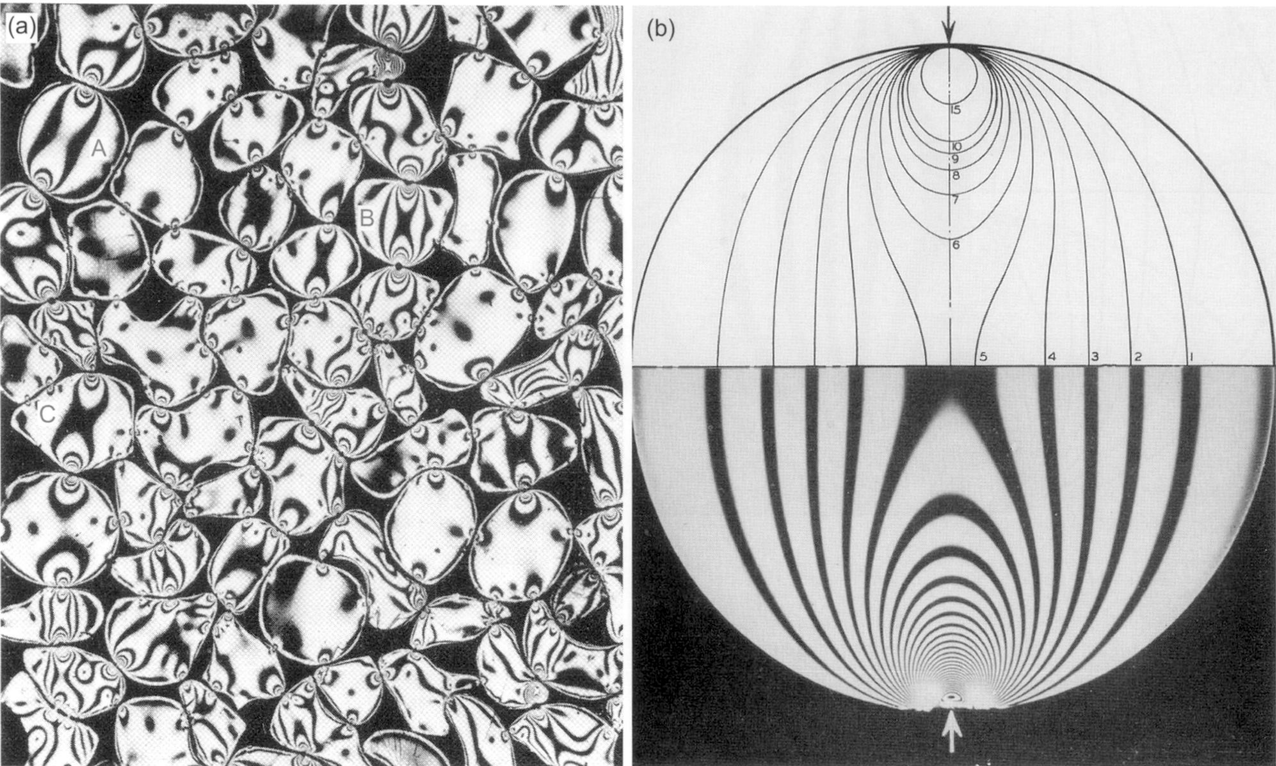

Figure 3.13 (a) Photoelastic laboratory experiment of the shear stress patterns in irregular disks designed to represent sand grains in sandstone. Grains labeled A, B, and C are discussed in text. Modified from Gallagher et al. (1974). (b) Comparison of theoretical (upper) and experimental (lower) shear stress pattern in a circular disk with diametrically opposed point forces. Modified from Frocht (1948).

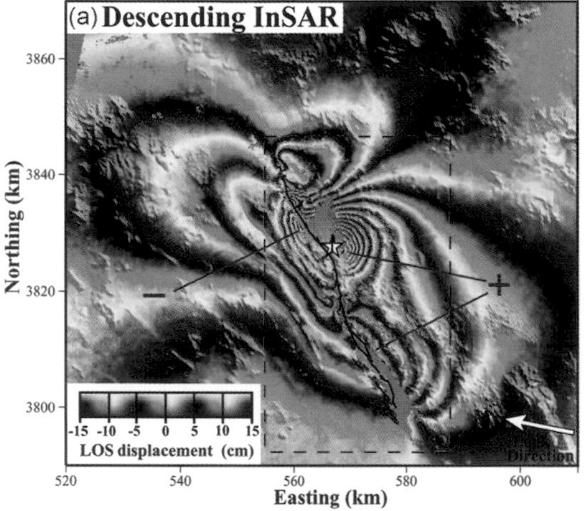

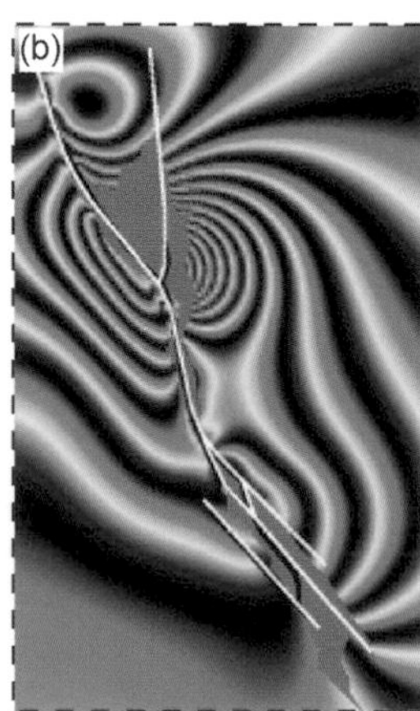

Figure 3.14 Contour maps of the displacement of Earth's surface during the M_w 7.1 Hector Mine, California, earthquake. (a) Displacements in the look direction of the satellite using synthetic aperture radar interferometry. From Jónsson et al. (2002). (b) The same displacement component within the dashed box on (a) calculated using continuum theory. From Maerten et al. (2005).

15 September 1999 and 20 October 1999, before and after the earthquake. By taking the difference of each distance measurement to the same location on Earth's surface, the displacement of those locations in the interval of time may be calculated. These displacements are attributed to slip on the Hector Mine fault.

The interferogram (**Figure 3.14a**) displays color contours of the displacement at Earth's surface in the look direction of the satellite (see white arrow in the lower right corner) for the time period between the two satellite passes. The displacement vector scale is a series of color cycles such that each complete cycle from red to yellow to blue represents a change of displacement of 10 cm. In this remarkable image one sees the displacement field associated with fault slip during the earthquake (and perhaps with additional slip during the month-long interval between the radar images) over a region of about 7,200 km^2. The contours of displacement make continuous curves that truncate on the fault segments that slipped during the earthquake. These faults are examples of the finite number of surfaces within a material continuum across which discontinuities exist in the displacement vector.

The continuum theory used to construct a model of the Hector Mine displacement field (**Figure 3.14b**) is *linear* elasticity, which underlies the deformation we discuss in more detail in Section 4.10. The model includes six discrete fault segments based on observations of the ruptures at Earth's surface mapped by geologists. The fault segments dip 83° to the east and are composed of a mesh of 612 triangular elements that approximate the curvature of the fault segments. Across each element the displacement components are discontinuous in the strike and dip directions. This model produces a contour map of displacement that is remarkably similar to that taken from the InSAR data.

In summary, the three examples we have described, and a multitude of other examples recorded in the literature of structural geology, demonstrate the efficacy of the material continuum over a range of length scales from 10^{-9} m to 10^{5} m, fourteen orders of magnitude. Only near the scale of individual mineral grains in porous rock, or within a rock mass that contains more surfaces of discontinuity than can reasonably be accounted for with modern computers and software, need one consider other methods to describe and model deformation. Despite the heterogeneity of rock masses, one usually can include the most significant surfaces of discontinuity when constructing models, and thereby continue to exploit the power of the material continuum. We accept the continuum postulate and, in what follows, we develop the necessary refinements to cover the topics of mechanics that are relevant to structural geology.

3.5 CONSERVATION OF MASS

One of the foundational principles of classical mechanics is that mass is not created or destroyed during physical deformation or chemical reaction. Today we understand that the mass of an object does change at velocities near the speed of light, and that mass and energy are interconvertible in nuclear reactions. However, relativistic velocities are not achieved during tectonic processes, and nuclear reactions are rare enough in Earth's lithosphere, so the classical principle of mass conservation remains a good starting point for investigations in structural geology.

The concept of mass conservation may be understood qualitatively by considering the folded Moine Schists observed in outcrop near Loch Monar, Scotland (**Figure 3.15**). The interfaces between the gray schist and white quartz/feldspar layers are interpreted to have been planar before this rock was deformed and the layers crumpled into contorted shapes. In the following thought experiment we imagine the process of folding at depth in Earth's lithosphere and provide a frame of reference using a two-dimensional Cartesian coordinate system (x, y). The origin is placed adjacent to the photograph to indicate it does not move as the rock mass deforms.

The *made-up* image in **Figure 3.15a** represents an earlier stage during the deformation when the folds had a lesser amplitude and greater wavelength than the actual image in **Figure 3.15b**. The made-up image was created by reversing the deformation: stretching by a factor 1.25 in x, shortening by a factor 0.80 in y, and cropping both images. This deformation is illustrated by the different shapes of the coin in the lower left corner of both images. The center of the small square element is fixed in space at the point (x_o, y_o) and each side is 1 cm long, so the volume of this element is 10^{-6} m^3.

At time t_1 (**Figure 3.15a**) the volume element contains both schist and a portion of a quartz/feldspar layer, while at time t_2 (**Figure 3.15b**) it contains only schist. As the folds developed, rock particles with different densities moved in and out of the element. Quartz has a mass density of about 2.6×10^3 kg m^{-3}, and the densities of quartz/mica schists range from about 2.7×10^3 to 3.0×10^3 kg m^{-3} (Clark, 1966). Therefore, the mass within this element at a given time could range from about 2.6×10^{-3} kg to 3.0×10^{-3} kg, depending upon which constituents of the Moine Schist occupied the volume.

The velocity at the center of the element (**Figure 3.15**) also may change from time t_1 to t_2 in both magnitude and direction. To keep track of the rate of change of mass, we subtract the rate of mass going out of the element from the rate of mass coming in:

$$\begin{pmatrix} \text{rate of} \\ \text{mass} \\ \text{increase} \end{pmatrix} = \begin{pmatrix} \text{rate of} \\ \text{mass} \\ \text{in} \end{pmatrix} - \begin{pmatrix} \text{rate of} \\ \text{mass} \\ \text{out} \end{pmatrix} \tag{3.9}$$

If the rate of mass going out is less than that coming in, the difference is positive and this measures the rate of mass *increase* in the element. A negative difference signals a rate of mass *decrease*. This strict bookkeeping does not admit the spontaneous creation or destruction of mass within the element, so (3.9) is a statement that defines mass conservation in words with reference to the volume element at (x_o, y_o). We quantify this qualitative statement in a later section, but first we consider more closely the description of motion visualized in **Figure 3.15**.

3.5.1 Spatial Description of Motion

In Chapter 1 we began by stating that geologic structures develop when rock is deformed from its original configuration into some different configuration. That deformation involves motion and relative motion of rock particles, so to quantitatively describe rock deformation we need a mathematical description of motion. Leonhard Euler (1707–1783) and Joseph-Louis Lagrange (1736–1813), each developed a mathematical description of motion, and both are used today. In Chapters 4 and 5 we employ the referential description of motion that Lagrange developed, and use that to understand the kinematics of elastic and plastic deformation of rock in a solid state. Here we introduce the spatial description of motion that Euler developed, and use that to understand how mass is transported and conserved as geologic structures develop. We also use Euler's description in Chapter 6 to understand the kinematics of flow of magma and salt.

Consider the center of a rectangular volume element at an arbitrary point designated by the position vector, $\mathbf{x} = \langle x, y, z \rangle$ in

(a)

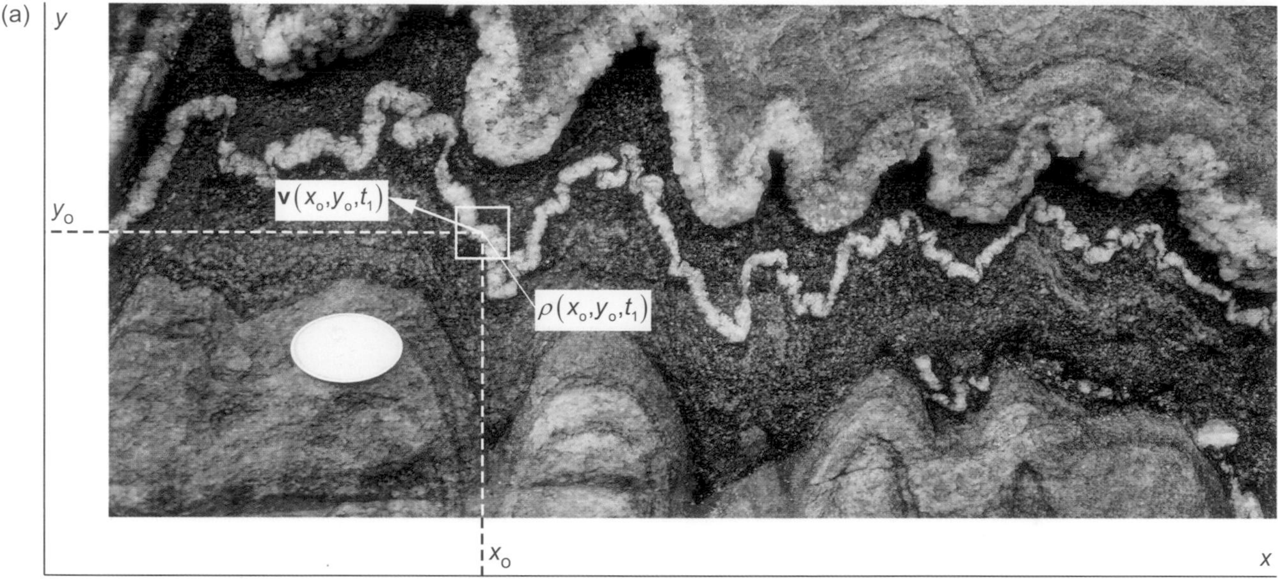

(b)

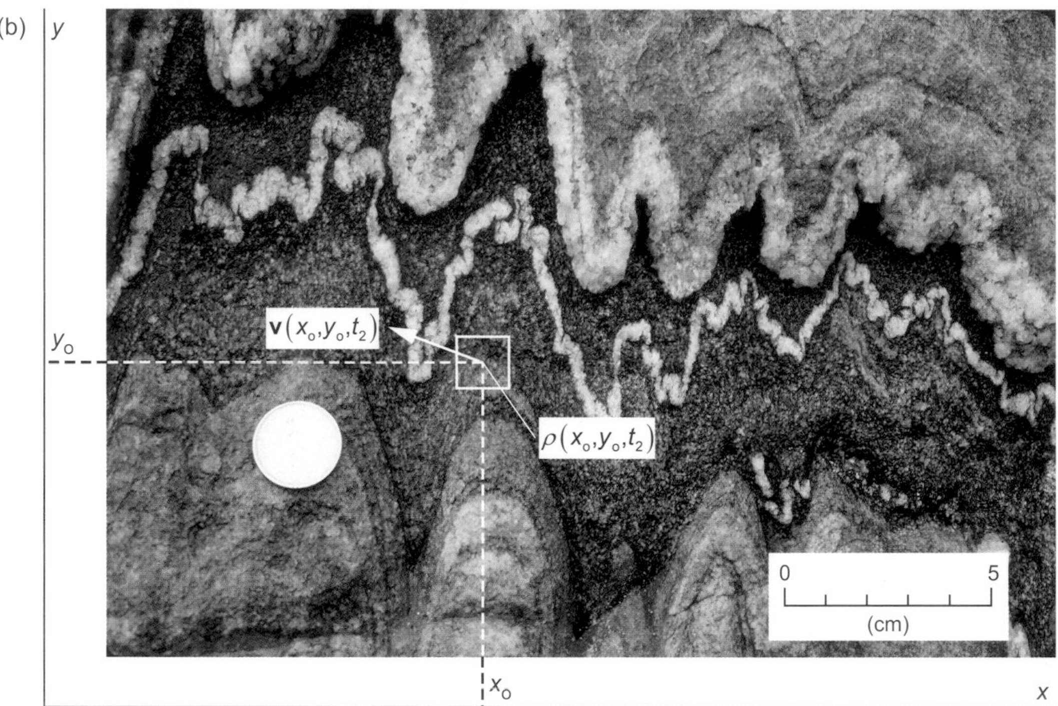

Figure 3.15 Multilayer folds in Moine Schists from an outcrop near Loch Monar, Scotland; coin 2.3 cm. Small white square is fixed in space at the central point (x_o, y_o) as folds develop. (a) Velocity vector and density at (x_o, y_o) at earlier time, t_1. (b) Velocity vector and density at (x_o, y_o) at later time, t_2.

three-dimensional space (**Figure 3.16**). Recall that we introduced the right-handed Cartesian axes (x, y, z) in **Figure 2.6** and pointed out that z usually is taken as vertical, so x and y lie in the horizontal plane. The normals to the sides of the element are parallel to the coordinate axes, and the lengths of the sides are the small quantities Δx, Δy, and Δz. The position vector, $\mathbf{x}$, and the corresponding volume element are fixed in space, while particles of the deforming rock move through the element. The velocity, $\mathbf{v}$, of a particle at the center of the volume element is a function of that position and time, and so $\mathbf{v} = \mathbf{v}(\mathbf{x}, t)$. At a given time, every position in the velocity field is associated with a particular velocity. As time advances from some initial time to some later time, different particles are associated with the position, $\mathbf{x}$, and these particles may have different velocities. This description of motion is called *spatial*, because the velocity field is defined with respect to given *positions* in three-dimensional space, not with respect to a given particle as it travels through that space.

Using the spatial description of motion, $\mathbf{v} = \mathbf{v}(\mathbf{x}, t)$ is referred to as the local velocity. Choice of the word *local* follows from the fact that different particles pass through the volume element at $\mathbf{x}$,

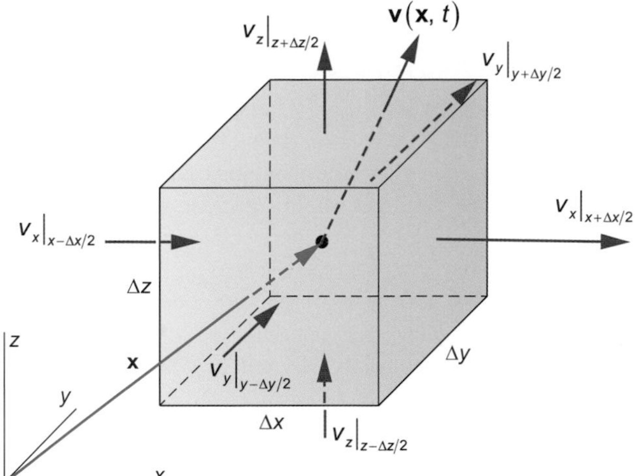

Figure 3.16 Spatial description of motion. The local velocity vector is **v**; the fixed volume element has edges of length Δx, Δy, and Δz; **x** is the position vector for the element center. See Malvern (1969) in Further Reading.

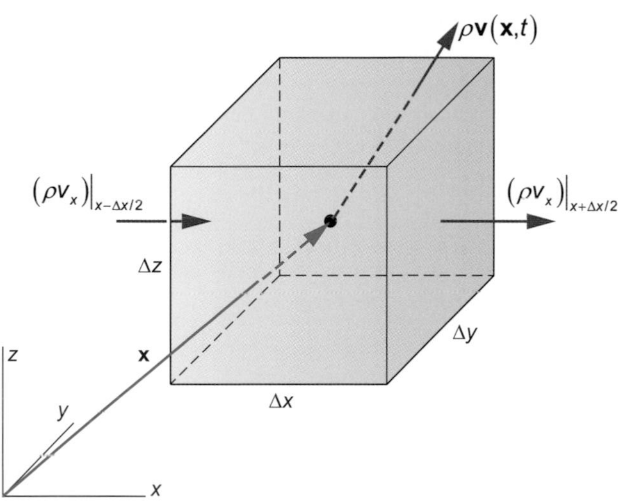

Figure 3.17 Diagram used to consider temporal variations in mass density and establish the equation of continuity. The mass rate per unit volume vector is $\rho\mathbf{v}$. The fixed volume element has edges of length Δx, Δy, and Δz; ρ is mass density; **x** is the position vector for the element center; **v** is the local velocity. See Malvern (1969) in Further Reading.

so $\mathbf{v} = \mathbf{v}(\mathbf{x}, t)$ is not the velocity of a particular particle, unless that particle happens to be exactly at **x** at the given time. In contrast, for the referential description of motion introduced in Section 4.9.1, a velocity is associated with each particle, so that is called the particle velocity.

3.5.2 Equation of Continuity

Consider the center of a rectangular volume element at an arbitrary point designated by the position vector, $\mathbf{x} = \langle x, y, z\rangle$ in three-dimensional space (**Figure 3.17**). The normals to the sides of the element are parallel to the coordinate axes, and the lengths of the sides are the small quantities Δx, Δy, and Δz. The volume element is fixed in space, so **x** does not vary with time, but particles of deforming rock move through the element. The mass density, ρ, and the local velocity, **v**, both are functions of position and time, and so is the rate of mass per unit volume, $\rho\mathbf{v} = \rho\mathbf{v}(\mathbf{x}, t)$. The components of this vector in the x-direction and evaluated at the left and right faces of the element are illustrated in **Figure 3.17** by arrows at the center of each face, but we understand mass is moving through the entire face. To gain conceptual understanding at this point we consider only these x-components of the rate of mass, but later we generalize the derivation to three dimensions.

The rate of mass *in* through the left face of the volume element (**Figure 3.17**) is proportional to density, the component of velocity acting perpendicular to that face, and the surface area of that face: $\rho v_x \Delta y \Delta z$, where both ρ and v_x are evaluated at $x - \Delta x/2$. The rate of mass *out* through the right face is given by the same expression with ρ and v_x evaluated at $x + \Delta x/2$. Because the other components of the mass rate per unit volume, ρv_y and ρv_z, are directed parallel to the left and right faces of the volume element, they cannot carry mass in or out through these faces. The rate of accumulation of mass within the element is given by the temporal derivative of density at the element center multiplied by the volume of the element: $(\partial\rho/\partial t)\Delta x\Delta y\Delta z$. The partial derivative is used because density is a function of position and time, $\rho = \rho(\mathbf{x}, t)$.

Equating the rate of accumulation of mass to the difference between the rates of mass in and out we have:

$$[(\partial\rho/\partial t)|_x]\Delta x\Delta y\Delta z = [(\rho v_x)|_{x-\Delta x/2} - (\rho v_x)|_{x+\Delta x/2}]\Delta y\Delta z$$

This equation is a quantitative form of the word equation (3.9) considering only the x-component of the mass rate. Now divide both sides by the volume and consider a set of n successively smaller elements such that in the limit as $n \to \infty$ the largest dimension of the element approaches zero and the element converges to the central point at **x**.

$$\frac{\partial\rho}{\partial t} = \lim_{n\to\infty}\left[\frac{(\rho v_x)|_{x-\Delta x/2} - (\rho v_x)|_{x+\Delta x/2}}{\Delta x}\right] = -\frac{\partial}{\partial x}(\rho v_x)$$

In this limit the quantity in square brackets becomes the negative partial derivative of the x-component of mass rate per unit volume. Both the temporal change of density and the spatial change of the component of mass rate are the respective values at the point **x** in this one-dimensional description of mass conservation.

Conservation of mass due to motion in the x-coordinate direction reveals interesting concepts that underlie this physical law. Because both density and velocity may be functions of x, the partial derivative of their product is:

$$-\frac{\partial}{\partial x}(\rho v_x) = -\left(\rho\frac{\partial v_x}{\partial x} + v_x\frac{\partial\rho}{\partial x}\right) \tag{3.10}$$

The first term in parentheses on the right side of (3.10) describes the rate of change of mass per unit volume if the density does not

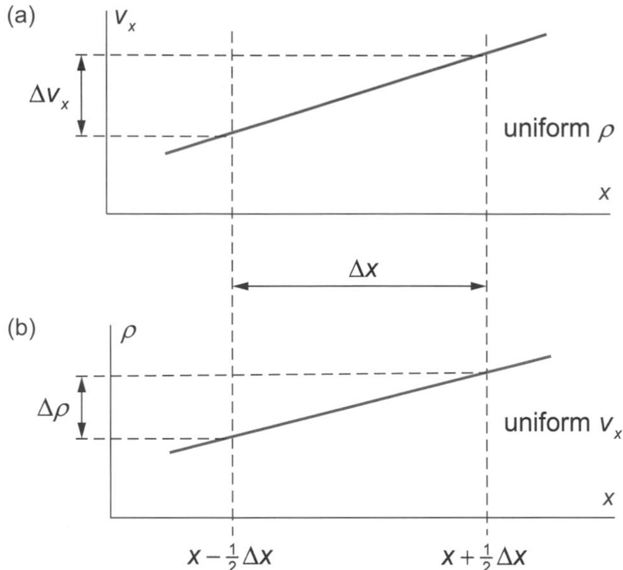

Figure 3.18 Graphs illustrating the two mechanisms for mass change in a fixed volume element. (a) A spatial increase of velocity with uniform density stretches the material. (b) A spatial increase of density with uniform velocity carries less mass in than out of the element.

vary with x, but the velocity does vary with x (**Figure 3.18a**). For example, if the velocity increases with x, the material in the vicinity of **x** is stretched, and rate of mass out exceeds the rate of mass in, so the mass per unit volume decreases with time (note the leading minus sign) in proportion to $\partial v_x/\partial x$.

The second term in parentheses on the right side of (3.10) describes the rate of change of mass per unit volume if the velocity does not vary with x, but the density does vary with x (**Figure 3.18b**). For example, if the density increases with x, motion in the x-direction carries less mass in than out of the fixed volume element, so the mass per unit volume decreases with time (again note the leading minus sign) in proportion to $\partial\rho/\partial x$.

Density at a fixed point in a material continuum can change by either one (or both) of the two independent mechanisms described in the preceding two paragraphs, while mass is conserved. In summary, the one-dimensional time rate of change of mass density in the x-direction is:

$$\frac{\partial\rho}{\partial t} = -\left(\rho\frac{\partial v_x}{\partial x} + v_x\frac{\partial\rho}{\partial x}\right)$$

The one-dimensional relationship may be generalized for mass rates through all six faces of the volume element (**Figure 3.17**):

$$\frac{\partial\rho}{\partial t} = -\rho\left(\frac{\partial v_x}{\partial x} + \frac{\partial v_y}{\partial y} + \frac{\partial v_z}{\partial z}\right) - \left(v_x\frac{\partial\rho}{\partial x} + v_y\frac{\partial\rho}{\partial y} + v_z\frac{\partial\rho}{\partial z}\right) \tag{3.11}$$

Both mechanisms of mass change illustrated in **Figure 3.18** operate in the three coordinate directions. This scalar equation is a *spatial* description of mass conservation because it describes changes at a fixed point in space and both the density and the velocity components are expressed as functions of the spatial coordinates. The relationship in (3.11) is called the Equation of Continuity. It specifies that mass is *conserved*: it is neither created nor destroyed at each and every point in the continuum. We use this famous equation later in this chapter to derive the general equations of motion for all solids and liquids. They, in turn, are used to derive the more specialized equations of motion for the *elastic* solid in Section 4.11, and for the *viscous* liquid in Section 6.8.

Moving the last three terms on the right side of (3.11) to the left side, and recalling the application of the material time derivative to a scalar quantity (Section 2.5.5), the equation of continuity is written:

$$\frac{1}{\rho}\frac{\mathrm{D}\rho}{\mathrm{D}t} = -\left(\frac{\partial v_x}{\partial x} + \frac{\partial v_y}{\partial y} + \frac{\partial v_z}{\partial z}\right) \tag{3.12}$$

Here $\mathrm{D}\rho/\mathrm{D}t$ is the material time derivative of the mass density. The terms in parentheses on the right side of (3.12) measure the rate of flow of material *away* from any particle. The left side is the change in density with respect to time in the neighborhood of that particle, per unit density. Where the velocity field is *divergent* (the sum of terms in parentheses is positive), the density decreases; where the velocity field in *convergent*, the density increases with time.

Testing the equation of continuity for dimensional homogeneity, on the left side of (3.12) we have:

$$\frac{1}{\rho}\frac{\mathrm{D}\rho}{\mathrm{D}t}\{=\}\mathrm{M}^{-1}\mathrm{L}^{3}\mathrm{M}\mathrm{L}^{-3}\mathrm{T}^{-1} = \mathrm{T}^{-1}$$

On the right side the gradients in velocity components all have the same dimensions, so for example:

$$\frac{\partial v_x}{\partial x}\{=\}\mathrm{L}\mathrm{T}^{-1}\mathrm{L}^{-1} = \mathrm{T}^{-1}$$

The equation of continuity is dimensionally homogeneous. As geologic structures develop and mass is conserved, models of this development must honor (3.12), or what is written in an expanded form as (3.11), at every point.

3.5.3 Incompressible Materials

One of the most common postulates employed in setting up models for the development of geologic structures is that the model rock or model magma is incompressible. This requires that near any particle in the material continuum and for all relevant times the velocity gradients are constrained such that a stretch in one coordinate direction is compensated exactly by contractions in the other coordinate directions, resulting in no net volume change.

For a material to be perfectly *incompressible*, the velocity gradients on the right side of (3.12) sum to zero:

$$\frac{\partial v_x}{\partial x} + \frac{\partial v_y}{\partial y} + \frac{\partial v_z}{\partial z} = 0, \text{ so } \frac{\mathrm{D}\rho}{\mathrm{D}t} = 0 \tag{3.13}$$

Thus, the mass density in the neighborhood of each and every particle is *constant*. This condition is widely employed, and usually tacitly accepted, for models of tectonic processes where the dominant behavior is ductile deformation (Chapter 5), and for models of magma flow where the dominant behavior is viscous deformation (Chapter 6). The corresponding continuum theories are plasticity and viscous fluid dynamics.

The condition specified in (3.13) assures conservation of mass for the *incompressible*, but *deformable* material continuum that is heterogeneous with respect to density. This is best appreciated by returning to the expanded form of the continuity equation (3.11), now written as:

$$\frac{\partial \rho}{\partial t} + \left(v_x \frac{\partial \rho}{\partial x} + v_y \frac{\partial \rho}{\partial y} + v_z \frac{\partial \rho}{\partial z} \right) = 0$$

The density in any fixed volume element (**Figure 3.18**) can change with time, for example as material with different density is carried into that element. However, if the material continuum is *homogeneous* with respect to density and *incompressible*, but *deformable*, we have:

$$\frac{\partial \rho}{\partial t} = 0 \tag{3.14}$$

This is the most constrained condition for the material continuum subject to conservation of mass: the mass density is *uniform in space* and *constant in time*.

3.6 CONSERVATION OF LINEAR MOMENTUM

The conservation of linear and angular momentum lead to Cauchy's laws of motion, which describe the deformation of geologic structures, whether the rock deformed as a brittle elastic solid, a ductile plastic solid, or a viscous fluid. As such, these conservation laws are of fundamental importance to our understanding to all tectonic processes. As with the equation of continuity, the equations derived here apply strictly to a rock mass that is isothermal and isochemical, but we show by example that they lead to considerable insight into many tectonic processes that only approximate these conditions.

Recall from the elementary particle dynamics of Newton that linear momentum, **p**, is a vector quantity defined as the product of particle mass and velocity: $\mathbf{p} = m\mathbf{v}$ (**Figure 3.19**). The rate of change with respect to time of linear momentum is the resultant force, **f**, acting on the particle:

$$\frac{\mathrm{d}(m\mathbf{v})}{\mathrm{d}t} = \mathbf{f} \tag{3.15}$$

Momentum is conserved in the sense that the rate of change with respect to time is directly and only related to the resultant force: it is not spontaneously created or destroyed by any other mechanism. Note that all of the vector quantities we have mentioned here have the same direction. In this sense the motion is *linear*: it proceeds along a line parallel to the applied force.

The relationship between momentum and force (3.15) is appropriate for an isolated particle, with mass concentrated at a point. For the material continuum, which is the basis for modeling geologic structures, we replace $m\mathbf{v}$ with the linear momentum per unit volume, $\rho\mathbf{v}$, and consider how this quantity changes with time at any point in the deforming rock mass. Those changes in momentum are directly related to the resultant surface forces that we introduced in Section 1.1.1, and they are associated with the traction vector and the stress tensor that we described in Section 2.6.3. The time rate of change in linear momentum also is related to the resultant body forces due to gravity and buoyancy that we introduced in Section 1.1.2. In this way we generalize (3.15) to apply to the material continuum.

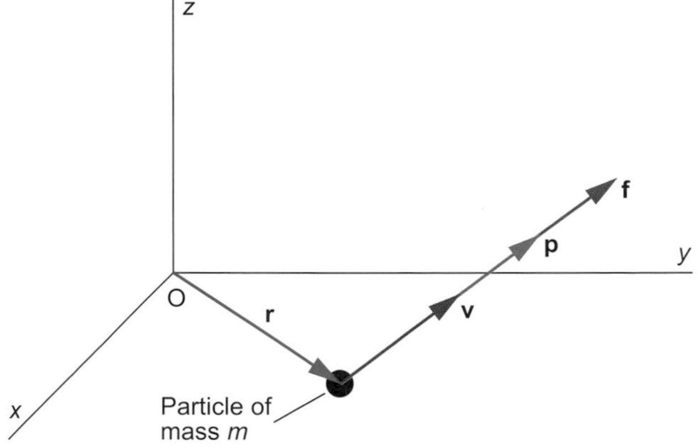

Figure 3.19 Particle of mass m at position **r** acted upon by resultant force **f** and moving with velocity **v** has linear momentum **p**.

The concept of momentum conservation for the material continuum may be understood qualitatively by considering a fault observed in an outcrop of the Lake Edison Granodiorite. **Figure 3.20a** is a *made-up* image of that outcrop, created by sliding the image on one side of the fault laterally along the fault to its original position. This represents a time t_1 as the fault was just beginning to slip, so the white aplite dike and black xenolith appear to be continuous across the fault. The actual outcrop image, shown in **Figure 3.20b**, is representative of the configuration at time t_2 as the fault slip was about to stop. The two pieces of the dike and of the xenolith are offset across the fault. In the following thought experiment we imagine the process of faulting at depth in Earth's crust as slip accelerates and then decelerates, so granodiorite passes through the fixed white square (**Figure 3.20**), which is a two-dimensional representation of a small volume element. We provide a frame of reference using a Cartesian coordinate system (x, y), with origin placed adjacent to the photograph to indicate it does not move as the rock mass deforms.

Linear momentum acting at the center of the fixed volume element (**Figure 3.20**) is the vector quantity $\rho\mathbf{v}$, and this may change from time t_1 to t_2 in both magnitude and direction. The rate of momentum increase is equal to the resultant surface and body forces according to (3.15), but in the deforming rock mass momentum also can be carried *in* and *out* of the fixed volume element. Thus, to the resultant forces we must add the momentum carried in and subtract the momentum carried out, so the total rate of momentum increase is:

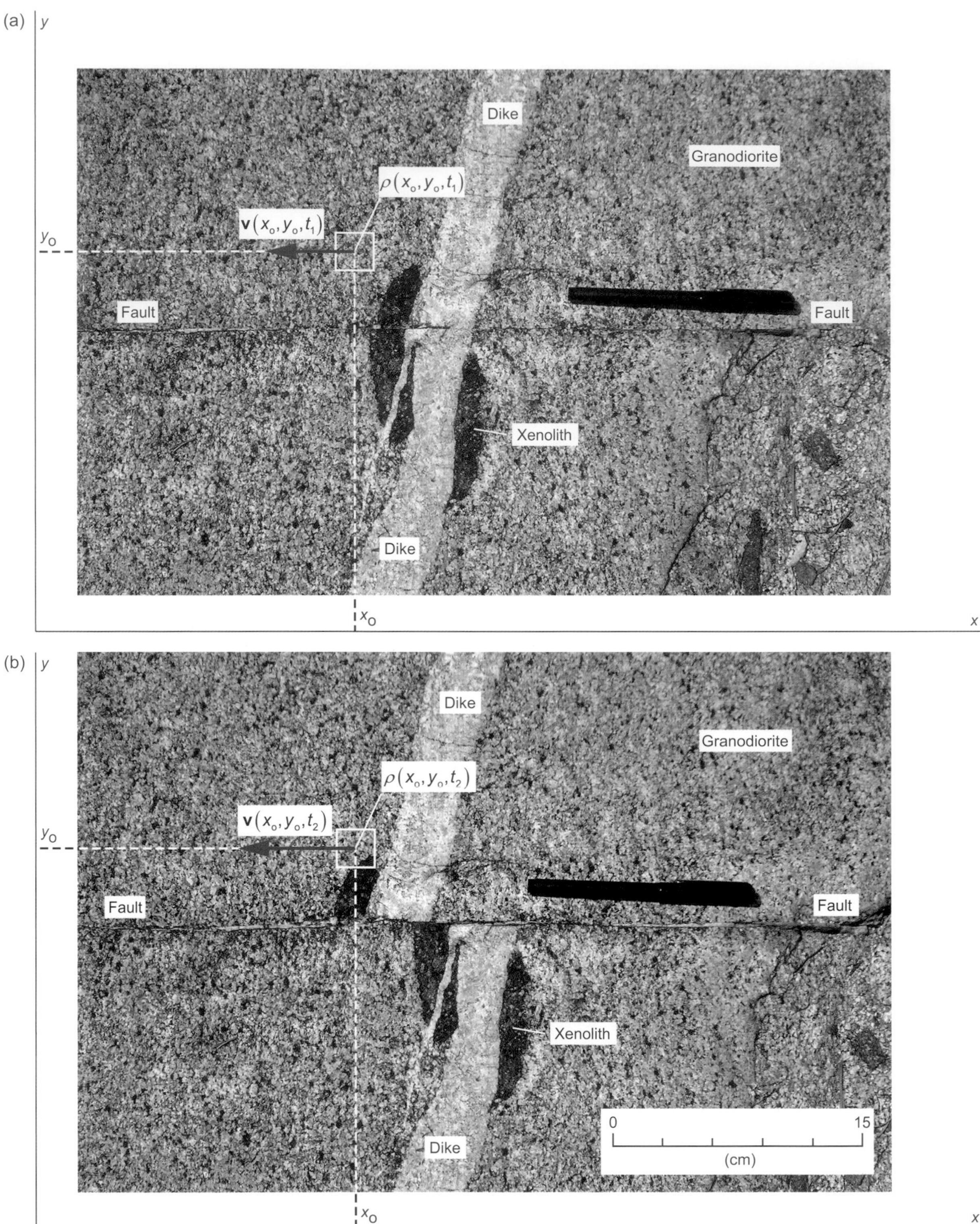

Figure 3.20 Fault in Lake Edison Granodiorite, Sierra Nevada, CA; pen 14 cm. Small white square is fixed in space at the point (x_o, y_o). Rock with density, ρ, moves through this square with velocity, $\mathbf{v}$, as slip develops on the fault. (a) Velocity vector and density at an earlier time; (b) Velocity vector and density at a later time. Image (a) created from outcrop image (b) by cutting and sliding along the fault.

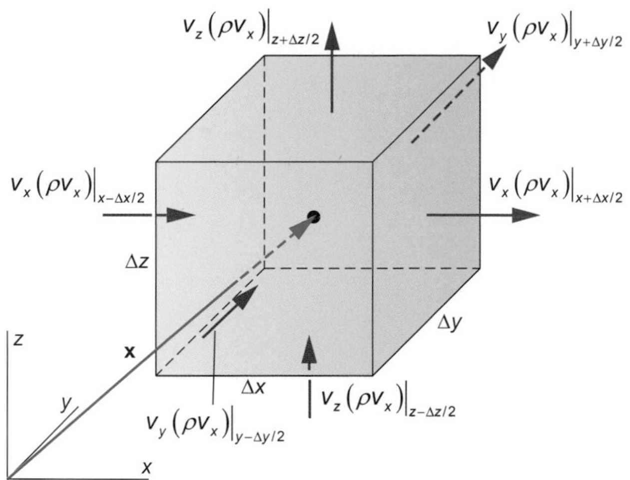

Figure 3.21 Schematic diagram used to consider temporal variations in linear momentum and establish the equations of motion. The linear momentum per unit volume is $\rho\mathbf{v}$. The fixed volume element has edges of length Δx, Δy, and Δz. The x-component of momentum, ρv_x, is carried through the faces of the element by the component of velocity acting normal to that face. See Malvern (1969) in Further Reading.

$$\begin{pmatrix}\text{rate of}\\ \text{momentum}\\ \text{increase}\end{pmatrix} = \begin{pmatrix}\text{rate of}\\ \text{momentum}\\ \text{in}\end{pmatrix} - \begin{pmatrix}\text{rate of}\\ \text{momentum}\\ \text{out}\end{pmatrix} + \begin{pmatrix}\text{resultant}\\ \text{surface and}\\ \text{body forces}\end{pmatrix} \tag{3.16}$$

If the rate of momentum going out is less than that coming in, the difference is positive and this contributes to the rate of momentum increase in the volume element: a negative difference corresponds to a rate of momentum decrease. This strict bookkeeping does not admit the spontaneous creation or destruction of momentum, so (3.16) is a statement that defines momentum conservation in words with reference to a fixed volume element at the point (x_o, y_o).

3.6.1 Cauchy's First Equations of Motion

The qualitative concept expressed by (3.16) can be quantified using the principles of calculus and generalized for all rock masses. To do this we accumulate the quantities represented in the terms of (3.16), take the limit as the volume element shrinks to a point, and define the conditions for conservation of momentum at that point. Then, we exploit the concept of the material continuum to assert that the resulting equations apply at each and every point of the rock mass under investigation.

Consider the center of a volume element at an arbitrary point designated by the position vector, $\mathbf{x} = \langle x, y, z\rangle$ in three-dimensional space (**Figure 3.21**). The normals to the faces of the element are parallel to the coordinate axes, and the lengths of the edges are the small quantities Δx, Δy, and Δz. The volume element is *fixed* in space, so $\mathbf{x}$ does not vary with time, but deforming rock or flowing magma moves through the element. The mass density, ρ, and the velocity, $\mathbf{v}$, both are functions of position and time, and so is the momentum per unit volume, $\rho\mathbf{v} = \rho\mathbf{v}(\mathbf{x}, t)$. This vector has components in the three coordinate directions, but for clarity we focus attention on the x-component, ρv_x, and later generalize to include all components. The rate of increase of the x-component of momentum within the element is quantified as:

$$\left(\frac{\partial}{\partial t}\rho v_x\right)\Delta x\Delta y\Delta z \tag{3.17}$$

This is one component of the first term on the left side of (3.16).

To evaluate the first term on the right side of (3.16) the *rate* of the x-component of momentum entering the element through the left face is taken as the product of the velocity component acting perpendicular to that face, the momentum per unit volume, and the area of that face: $v_x(\rho v_x)|_{x-\Delta x/2}\Delta y\Delta z$. Here the density and velocity components are evaluated at the center of the left face, $x - \Delta x/2$. The x-component of momentum also can enter through the front face, carried by the y-component of velocity, $v_y(\rho v_x)|_{y-\Delta y/2}\Delta z\Delta x$, and through the bottom face, carried by the z-component of velocity, $v_z(\rho v_x)|_{z-\Delta z/2}\Delta x\Delta y$. The second term on the right side of (3.16) is evaluated by considering the rate of the x-component of momentum leaving the element through the right, back, and top faces. Subtracting the rate of momentum out from in, we have:

$$\left(v_x\rho v_x|_{x-\Delta x/2} - v_x\rho v_x|_{x+\Delta x/2}\right)\Delta y\Delta z + \left(v_y\rho v_x|_{y-\Delta y/2} - v_y\rho v_x|_{y+\Delta y/2}\right)\Delta z\Delta x + \left(v_z\rho v_x|_{z-\Delta z/2} - v_z\rho v_x|_{z+\Delta z/2}\right)\Delta x\Delta y$$

After dividing through by the volume of the element, consider a set of n successively smaller elements such that in the limit as $n \to \infty$ the largest dimension of the element approaches zero and the element converges to the central point at $\mathbf{x}$. The finite differences become partial derivatives, so we are left with:

$$-\left(\frac{\partial}{\partial x}v_x\rho v_x + \frac{\partial}{\partial y}v_y\rho v_x + \frac{\partial}{\partial z}v_z\rho v_x\right) \tag{3.18}$$

This quantity accounts for the convective transport of the x-component of momentum, ρv_x. It represents mathematically what is described in the first two terms on the right side of (3.16). In each term we see the x-component of momentum multiplied by one of the velocity components. The change in convective transport across the element is accounted for by the partial derivatives.

The third term on the right side of (3.16) is evaluated by considering the resultant of the surface forces, which are related to the components of the stress tensor acting on the volume element, and the body force, which is force acting on the volume element due to gravity. The resultant force is a vector with components in the three coordinate directions, but for clarity we focus attention on the x-component, f_x, and later generalize to include all components.

The stress components (**Figure 3.22**) are the surface forces per unit area that act in the x-coordinate direction. We use an on-in convention so, for example, σ_{yx} is the force per unit area acting *on* the face with normal parallel to the y-coordinate axis and *in* the

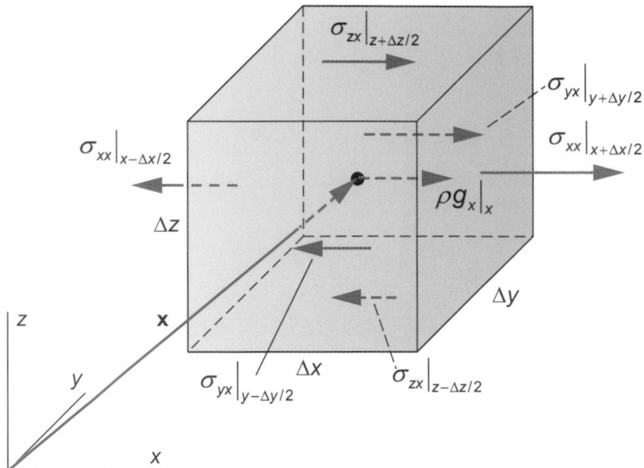

Figure 3.22 Schematic diagram used to consider temporal variations in linear momentum due to surface and body forces acting on a stationary volume element (blue) with edges of length Δx, Δy, and Δz. The position vector for the element center is **x**; ρ is mass density; g_x is the x-component of gravitational acceleration; and σ is a stress component identified by the two subscripts. See Malvern (1969) in Further Reading.

x-direction. A stress component that acts normal to a face is called a normal stress, and one that acts tangential to a face is called a shear stress. On opposite faces the stress components act in opposite directions. The body force per unit volume due to gravity is the product of the density and the gravitational acceleration, $\rho\mathbf{g}$, and the x-component is ρg_x. The resultant force in the x-direction is the sum of the stress components, multiplied by the area of the respective faces, and the body force multiplied by the volume of the element:

$$\left[\sigma_{xx}\big|_{x+\mathrm{d}x/2} - \sigma_{xx}\big|_{x-\mathrm{d}x/2}\right]\Delta y\Delta z + \left[\sigma_{yx}\big|_{y+\mathrm{d}y/2} - \sigma_{yx}\big|_{y-\mathrm{d}y/2}\right]\Delta z\Delta x + \left[\sigma_{zx}\big|_{z+\mathrm{d}z/2} - \sigma_{zx}\big|_{z-\mathrm{d}z/2}\right]\Delta x\Delta y + \rho g_x\Delta x\Delta y\Delta z$$

Again, dividing through by the volume of the element, we consider a set of n successively smaller elements such that in the limit as $n \to \infty$ the largest dimension of the element approaches zero and the element converges to the central point at **x**. The finite differences become partial derivatives leaving:

$$\frac{\partial\sigma_{xx}}{\partial x} + \frac{\partial\sigma_{yx}}{\partial y} + \frac{\partial\sigma_{zx}}{\partial z} + \rho g_x \tag{3.19}$$

These quantities account for the resultant force acting on the element.

To complete the description of momentum conservation we equate (3.17) to the sum of (3.18) and (3.19). This gives the quantitative form of the word equation (3.16) considering only the x-component of the momentum rate:

$$\frac{\partial}{\partial t}\rho v_x = -\left(\frac{\partial}{\partial x}v_x\rho v_x + \frac{\partial}{\partial y}v_y\rho v_x + \frac{\partial}{\partial z}v_z\rho v_x\right) + \left(\frac{\partial}{\partial x}\sigma_{xx} + \frac{\partial}{\partial y}\sigma_{yx} + \frac{\partial}{\partial z}\sigma_{zx}\right) + \rho g_x \tag{3.20}$$

Recalling that the density and the velocity components are functions of the spatial variables and time, the left-hand side and the first term in parentheses on the right side can be expanded using the standard rule for the derivative of a product. Neglecting the last two terms in (3.20) for the moment, and taking the derivatives of the products of density and velocity we have:

$$\rho\frac{\partial v_x}{\partial t} + v_x\frac{\partial\rho}{\partial t} = -\rho\left(v_x\frac{\partial v_x}{\partial x} + v_y\frac{\partial v_x}{\partial y} + v_z\frac{\partial v_x}{\partial z}\right) + v_x\left[-\rho\left(\frac{\partial v_x}{\partial x} + \frac{\partial v_y}{\partial y} + \frac{\partial v_z}{\partial z}\right) - \left(v_x\frac{\partial\rho}{\partial x} + v_y\frac{\partial\rho}{\partial y} + v_z\frac{\partial\rho}{\partial z}\right)\right] + \cdots$$

Referring to the equation of continuity (3.11), we see that the second term on the left side is exactly equal to the second term on the right side, so they cancel. Moving the first term on the right to the left side and now including the last two terms on the right side of (3.20), we have:

$$\rho\left[\frac{\partial v_x}{\partial t} + \left(v_x\frac{\partial v_x}{\partial x} + v_y\frac{\partial v_x}{\partial y} + v_z\frac{\partial v_x}{\partial z}\right)\right] = \left(\frac{\partial}{\partial x}\sigma_{xx} + \frac{\partial}{\partial y}\sigma_{yx} + \frac{\partial}{\partial z}\sigma_{zx}\right) + \rho g_x \tag{3.21}$$

This is the quantification of the word equation (3.16) for the x-component of the momentum rate.

The last step in the derivation of Cauchy's equations of motion is the recognition that the terms in square brackets on the left side of (3.21) are the material time derivative of the velocity component v_x. Recall from Section 2.5.5 that this derivative is written $\mathrm{D}v_x/\mathrm{D}t$, and it is defined as:

$$\frac{\mathrm{D}v_x}{\mathrm{D}t} = \frac{\partial v_x}{\partial t} + v_x\frac{\partial v_x}{\partial x} + v_y\frac{\partial v_x}{\partial y} + v_z\frac{\partial v_x}{\partial z}$$

The material time derivative represents the x-component of particle acceleration, written in terms of the *spatial description of motion*. The y- and z-components of the momentum rate are found using a similar derivation to what produced (3.21). Writing all three components of the momentum rate using the material time derivative, we have:

$$\begin{aligned}\rho\frac{\mathrm{D}v_x}{\mathrm{D}t} &= \left(\frac{\partial\sigma_{xx}}{\partial x} + \frac{\partial\sigma_{yx}}{\partial y} + \frac{\partial\sigma_{zx}}{\partial z}\right) + \rho g_x\\ \rho\frac{\mathrm{D}v_y}{\mathrm{D}t} &= \left(\frac{\partial\sigma_{xy}}{\partial x} + \frac{\partial\sigma_{yy}}{\partial y} + \frac{\partial\sigma_{zy}}{\partial z}\right) + \rho g_y\\ \rho\frac{\mathrm{D}v_z}{\mathrm{D}t} &= \left(\frac{\partial\sigma_{xz}}{\partial x} + \frac{\partial\sigma_{yz}}{\partial y} + \frac{\partial\sigma_{zz}}{\partial z}\right) + \rho g_z\end{aligned} \tag{3.22}$$

Augustine-Louis Cauchy derived these equations of motion in the middle of the nineteenth century. Today we call them Cauchy's first law or Cauchy's momentum equations. They guarantee momentum conservation with reference to a stationary volume element using a spatial description of motion. They incorporate the continuity equation (3.11), so conservation of mass is an integral part of these equations of motion. As mentioned at the beginning of this section, these conservation laws are of fundamental importance for understanding all tectonic processes.

3.6.2 Assessing Cauchy's First Law

The equations of motion (3.22) appear daunting at first glance, so we begin by testing for dimensional homogeneity (see Section 3.3). Focusing on the first of (3.22), on the left side we have mass density times the material time derivative of the x-component of velocity, so the dimensions are:

$$\rho \frac{\mathrm{D}v_x}{\mathrm{D}t} \{=\} \mathrm{ML}^{-3}\mathrm{LT}^{-1}\mathrm{T}^{-1} = \mathrm{ML}^{-2}\mathrm{T}^{-2}$$

On the right side, spatial gradients in the stress components have the same dimensions as (force/area)/length. Writing the dimensions of force as the dimensions of mass × acceleration, we have:

$$\frac{\partial \sigma_{xx}}{\partial x} \{=\} \mathrm{MLT}^{-2}\mathrm{L}^{-2}\mathrm{L}^{-1} = \mathrm{ML}^{-2}\mathrm{T}^{-2}$$

For the last term on the right side, we have mass density times acceleration, so the dimensions are:

$$\rho g_z \{=\} \mathrm{ML}^{-3}\mathrm{LT}^{-2} = \mathrm{ML}^{-2}\mathrm{T}^{-2}$$

The equations of motion (3.22) are dimensionally homogeneous.

Taking a more physical point of view, note that the dimensions on the right sides of each equation in the previous paragraph are $\mathrm{ML}^{-2}\mathrm{T}^{-2} = \mathrm{MLT}^{-2}\mathrm{L}^{-3}$. In other words, the dimensions are those of mass times acceleration divided by volume, or force divided by volume. Using these words to reconstruct the equations of motion in physical terms, we have:

$$\frac{\text{mass} \times \text{acceleration}}{\text{volume}} = \frac{\text{surface force}}{\text{volume}} + \frac{\text{body force}}{\text{volume}} \tag{3.23}$$

With this interpretation, the equations of motion reduce to Newton's second law, $m\mathbf{a} = \mathbf{F}$, (per unit volume) as applied to the *material continuum*. As geologic structures develop, the equations of motion (3.22) hold at every point, so deformation of the material continuum that we use to model these structures follows these relationships.

We emphasize that conservation of linear momentum (3.16), and Cauchy's equations of motion (3.22) that follow from this fundamental physical principle, are developed without regard for particular material properties other than mass density. Thus, they apply quite generally to the entire spectrum of deformation, from rocks that deform and fracture as a brittle solid (Chapter 4), to those that deform plastically as a ductile solid (Chapter 5), to those that flow as a viscous liquid (Chapter 6). This is perhaps the most compelling attribute of these equations: they describe the deformation of all rocks and magmas in Earth's lithosphere and asthenosphere.

Cauchy's equations of motion apply right down to the nanometer length scale of 10^{-9}m where the structures are lattice defects in individual mineral grains (Section 3.4.2). They also apply to motion of tectonic plates that span length scales of a thousand kilometers, 10^6 m, or more (Section 3.4.4). The range of applicable time scales is similarly impressive, from less than microseconds, 10^{-6} s, to the age of the Earth, 10^{15} s. However, special relativity, proposed by Albert Einstein in 1905 (see Resnick and Halliday, 1977, in Further Reading), provides limitations for application of Newton's laws and Cauchy's equations of motion to physical systems. For example, suppose observer S is at rest in the inertial reference frame (x, y, z, t), and observer S' is moving with another inertial reference frame (x', y', z', t') at a uniform velocity v along the common x and x' axes. Suppose that observer S' measures the length of a rod, $\Delta x'$, the mass of a particle, m', and a clock interval, $\Delta t'$. The rod, particle, and clock are traveling with velocity v, so they appear to be at rest relative to observer S'. However, observer S would record the following *relativistic* values for the moving rod, particle, and clock:

$$\Delta x = \Delta x' \sqrt{1-(v/c)^2},\ m = \frac{m'}{\sqrt{1-(v/c)^2}},\ \Delta t = \frac{\Delta t'}{\sqrt{1-(v/c)^2}}$$

As the velocity, v, approaches the speed of light, c, the length of the rod contracts; the mass of the particle increases; and the clock interval increases for observer S.

Velocities associated with rock deformation in Earth's lithosphere, even during great earthquakes (see Section 8.6.2), typically do not exceed a few kilometers per second, say 3×10^3 m s^{-1}. On the other hand, the velocity of light is 299,792,458 m s^{-1}, or about 3×10^8 m s^{-1}. Thus, the ratio $(v/c)^2$ is about 1×10^{-10}, which is negligible compared to one. Therefore, we are justified in using measures of length, mass, and time throughout this textbook that ignore relativistic effects.

Because Cauchy's equations of motion are derived by postulating isothermal and isochemical conditions, applications must be predicated on a careful assessment of the possible effects of temperature variations and chemical reactions, which can play significant roles in the development of geologic structures. We take up non-isothermal conditions in Section 3.8, but choose to ignore the effects of chemical reactions as beyond the scope of a first course in structural geology. We recommend, however, courses in geochemistry and petrology that usually are part of an undergraduate curriculum in geology, where students learn about chemical reactions and mineralogical changes that may be relevant to the development of geologic structures.

3.7 CONSERVATION OF ANGULAR MOMENTUM

The definition of angular momentum for particles uses the concept of linear momentum, described in Section 3.6, as the product of particle mass and velocity: $\mathbf{p} = m\mathbf{v}$ (**Figure 3.19**). Recall from the physics of particle dynamics that angular momentum, $\boldsymbol{\Phi}$, is the vector product of the position vector, $\mathbf{r}$, for a point mass (**Figure 3.23**) and the linear momentum, $\mathbf{p}$, of that point mass such that: $\boldsymbol{\Phi} = \mathbf{r} \times \mathbf{p}$. The rate of change with respect to time of angular momentum is equal to the resultant torque, $\boldsymbol{\tau}$, acting on the point mass:

$$\frac{\mathrm{d}\boldsymbol{\Phi}}{\mathrm{d}t} = \boldsymbol{\tau} \tag{3.24}$$

Angular momentum is conserved in the sense that the rate of change with respect to time is directly and only related to the

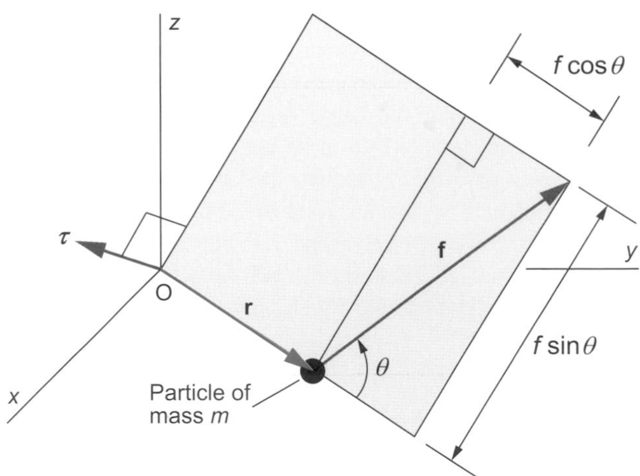

Figure 3.23 Schematic diagram to define the torque, **τ**, with respect to the origin, O, of the Cartesian coordinate system. The position vector, **r**, and net force, **f**, act on the particle of mass *m*.

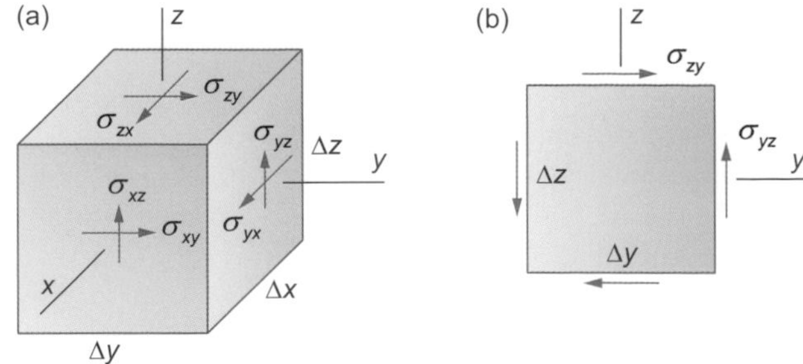

Figure 3.24 Schematic diagrams used to consider relations among the shear stress components acting on a volume element with side lengths Δx, Δy, and Δz centered at the origin. (a) Three-dimensional shear stress components. (b) Two-dimensional shear stress components. See Malvern (1969) in Further Reading.

resultant torque: it is neither spontaneously created nor destroyed by any other mechanism.

For the material continuum one might suppose that conservation of angular momentum would result in a set of equations at least as daunting as those for conservation of linear momentum (3.22), but remarkably it does not. In the next section we show how angular momentum is adapted for the material continuum. This leads to Cauchy's second equation of motion, which is expressed very simply in terms of the shear stresses acting on a volume element at any point in the continuum.

3.7.1 Cauchy's Second Law

To understand the physics behind the conservation of angular momentum for the material continuum we consider the relationship between torque and the forces that are related to the stress components acting at a point in the continuum. Recall from elementary particle dynamics that the torque due to a given force is the vector product of that force and the position vector locating the point of action of the force (**Figure 3.23**): $\boldsymbol{\tau} = \mathbf{r} \times \mathbf{f}$. The magnitude of the torque is:

$$|\boldsymbol{\tau}| = |\mathbf{r}||\mathbf{f}| \sin\theta,\ 0 \le \theta \le \pi \qquad (3.25)$$

Here θ is the smaller angle between the position vector and force vector (**Figure 3.23**). We consider the special case where the time rate of change of angular momentum (3.24) is zero, so the resultant torque is zero: $\boldsymbol{\tau} = 0$. We also consider the special case where the stress state does not vary spatially over the volume element we are considering. The restrictions applied for these special cases lead to a simpler derivation, but the same result would be found for a spatially varying state of stress, and for a temporally varying angular momentum.

To apply (3.25) to the material continuum consider a volume element centered at the coordinate origin with only shear stress applied to the faces (**Figure 3.24a**). The side lengths of this element are Δx, Δy, and Δz, and it is small enough that the stresses are uniform throughout. Thus, the shear stresses shown on the front, right, and top faces would have equal and opposite counterparts on the back, left, and bottom faces that are not shown. Also not shown are any normal stress components, because they are associated with forces directed along the coordinate axes, for which $\theta = 0$, so $\sin\theta = 0$, and by (3.25) they produce no torque.

To calculate the torque about the x-axis, consider the shear stresses acting in the (y, z)-plane (**Figure 3.24b**). The magnitude of the force associated with the shear stress σ_{yz} acting on the right face of the two-dimensional element is $\sigma_{yz}\Delta x\Delta z$, and the magnitude of the position vector for that force, which acts at the center of the face, is $\Delta y/2$. The smaller angle between these two vectors is $\theta = \pi/2$, so $\sin\theta = 1$ and the magnitude of the torque is $\sigma_{yz}\ \Delta x\Delta y\Delta z/2$. Using the right-hand rule, this torque vector is directed along the positive x-axis. By a similar argument the magnitude of the torque associated with the shear stress σ_{zy} acting on the top face is $\sigma_{zy}\ \Delta x\Delta y\Delta z/2$. The two shear stresses, σ_{yz} and σ_{zy}, are referred to as conjugate shear stresses because they act in the same plane on orthogonal faces of the volume element. Again, using the right-hand rule, this torque vector is directed along the negative x-axis, so the net torque along the x-axis is:

$$\sigma_{yz}\Delta x\Delta z(\Delta y/2)\mathbf{i} - \sigma_{zy}\Delta x\Delta y(\Delta z/2)\mathbf{i} = 0$$

This requires the two shear stresses to be equal: $\sigma_{yz} = \sigma_{zy}$.

Similar results are found for the shear stresses on adjacent faces that produce torques about the y- and z-axes. For the spatially varying stress field we consider the limit as the volume element shrinks about the origin and conclude that at every point in the material continuum the conjugate shear stresses must be equal:

$$\sigma_{xy} = \sigma_{yx},\ \sigma_{yz} = \sigma_{zy},\ \sigma_{zx} = \sigma_{xz} \qquad (3.26)$$

These are Cauchy's second law.

3.7.2 Assessing Cauchy's Second Law

The derivation of Cauchy's second law in Section 3.7.1 considers a special case in which body forces are neglected and the stress components are taken as homogeneous over the length scale of the volume element (**Figure 3.24**). Both of these restrictions may be relaxed and Cauchy's equations still applies, so it has very broad application to the mechanics of continua.

The relationships among the shear stress components expressed in (3.26) mean that stress is a symmetric tensor. That is, when viewed as a 3 by 3 matrix, the corresponding off-diagonal elements are equal:

$$\begin{bmatrix} \sigma_{xx} & \sigma_{xy} & \sigma_{xz} \\ \sigma_{yx} & \sigma_{yy} & \sigma_{yz} \\ \sigma_{zx} & \sigma_{zy} & \sigma_{zz} \end{bmatrix}$$

For example, the component in the second row and third column, σ_{yz}, is equal to the component in the third row and second column, σ_{zy}. The symmetry of the stress tensor applies to the full range of rock deformation that we consider in this textbook, including the deformation of elastic solids, plastic solids, and viscous liquids.

3.8 CONSERVATION OF ENERGY

Igneous intrusions, such as the dike shown in **Figure 3.25**, put magma at roughly 1000 °C against host rock at roughly 100 °C, and this huge temperature difference drives the flow of heat from the magma into the host rock. The adjacent host rock may be metamorphosed by the increase in temperature, which changed the color of the sandstone at this outcrop, and made it more resistant to erosion. Also, cooling of igneous intrusions can generate thermal stresses that induce fracturing, and thereby change important physical properties such as strength and permeability. Heating of the host rock in some places is associated with the deposition of valuable ore deposits. We return to some of these processes in Chapter 11 and here refer the reader to Bird et al., (2007) and Carslaw and Jaeger (1986) in Further Reading. In this section we use the effects of the magmatic intrusion seen in **Figure 3.25** to introduce some of the fundamental physical concepts related to energy conservation and heat flow.

In general, heat flows from hotter to colder regions, but different mechanisms are responsible for this heat transfer: (1) electromagnetic radiation (e.g. when your skin is warmed by sun light); (2) convection (e.g. when your coffee is cooled by stirring in cold cream); and (3) conduction (e.g. when your hand is warmed by picking up a hot pan). Of these, convection and conduction have many applications in structural geology. With reference to **Figure 3.25**, heat was transferred upward by convection (flow) of magma in the dike, and heat was transferred laterally by conduction through the host rock. Convection usually is associated with flow of a liquid, like magma, that carries the heat. Conduction usually is associated with heat transfer within a solid, like rock. However, in a solid and permeable rock, heat also may be transferred by convection (flow) of groundwater or hydrocarbons through the rock pores and through fractures. Also, in a static magma the dominant mechanism for heat transfer could be conduction. To introduce conservation of energy, we focus on *conductive heat transfer*, but the underlying principle of energy conservation also applies to convection.

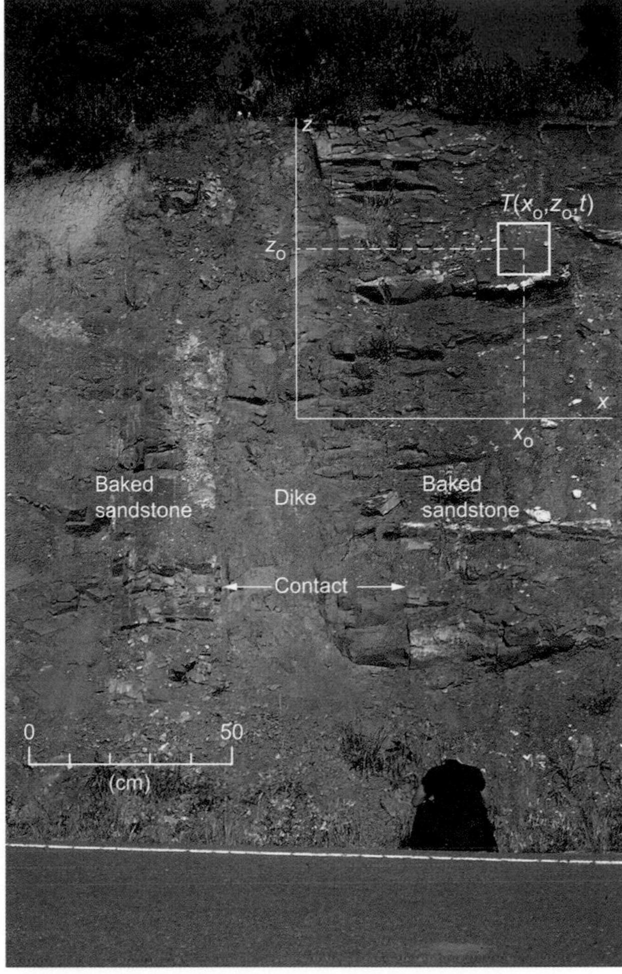

Figure 3.25 Vertical igneous dike cutting sandstone near Walsenburg, CO. Sandstone within ~1 m of both sides of the dike is darker in color, harder, and more resistant to erosion, all indications of baking by heat flow from the dike. Element at fixed position, (x_o, z_o), during heat flow from the dike into the sandstone. The dependent variable is the temperature, which is a function of position and time, $T(x_o, z_o, t)$.

3.8.1 Fourier's Law

Heat is transferred by conduction from hotter to colder regions of a solid, and this is expressed formally by Fourier's law for heat conduction, which was published by Joseph Fourier in 1822:

$$\mathbf{q} = -k\left(\frac{\partial T}{\partial x}\mathbf{i} + \frac{\partial T}{\partial y}\mathbf{j} + \frac{\partial T}{\partial z}\mathbf{k}\right) \qquad (3.27)$$

In this vector equation $\mathbf{q}$ is the heat flux vector, k is the thermal conductivity, and the partial derivatives of temperature, T, with

Table 3.3 Laboratory values of thermal conductivity

Thermal conductivity ($J\,m^{-1}\,s^{-1}\,K^{-1}$)		
Rock type	From	To
Quartzite	3.1	8.0
Gneiss	1.7	4.8
Basalt	2.6	3.6
Granite	2.6	3.8
Limestone	1.8	3.3
Sandstone	1.5	4.3
Shale	1.2	2.3
Coal	0.2	4.6

From Clark (1966), Table 21-1

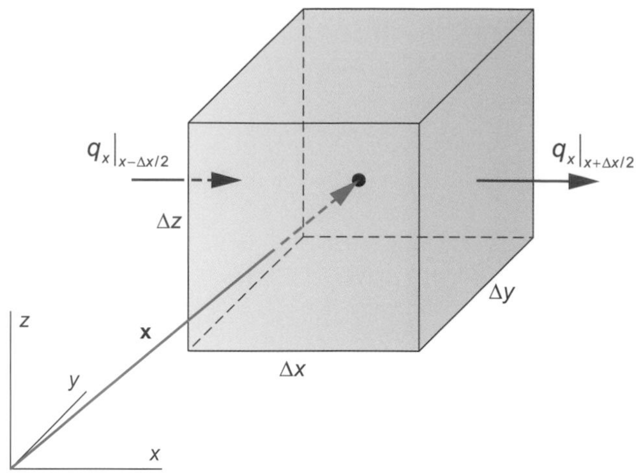

Figure 3.26 Fixed volume element used to constrain flow of heat such that energy is conserved during conduction in one dimension, parallel to the x-axis.

respect to the three coordinates are the components of the temperature gradient vector. Note that k is a physical property of the material and **k** is the unit vector in the direction of the z-axis. Fourier's law states that the heat flux vector is proportional to the temperature gradient vector, and the proportionality constant is the thermal conductivity. Because of the negative sign on the right side, the vectors on either side of (3.27) are oppositely directed. The scalar temperature increases most rapidly in the direction of the temperature gradient vector. The heat flux vector is oppositely directed: down the temperature gradient.

Fourier's law contains dimensions and units that must be understood to use this relationship correctly. Recalling that heat energy and work are dimensionally equivalent (see **Table 3.1**), and that work is force times displacement, the dimensions of heat are those of force times length, $MLT^{-2}L$. The SI units of heat are ($kg\,m\,s^{-2}$) m = N m = J (joule). The joule is the SI unit for work, energy, and quantity of heat. Heat flux is a measure of the heat transferred across a surface of given area in a given time. Therefore, the units of heat flux on the left side of (3.27) are $\mathbf{q}$ [=] $J\,m^{-2}\,s^{-1}$. The units of temperature gradient on the right side of (3.27) are $\partial T/\partial x$ [=] $K\,m^{-1}$. For homogeneity of units (and dimensions) the units of thermal conductivity must be k [=] $J\,m^{-1}\,s^{-1}\,K^{-1}$. **Table 3.1** provides the corresponding dimensions.

Laboratory values of thermal conductivity for various rock types are listed in **Table 3.3**. The range for any particular rock type is not large, and the range over the entire set of rock types, excluding coal, is less than one order of magnitude. The thermal conductivities of familiar materials such as steel, glass, and wood are about 46, 0.7, and 0.05 $J\,m^{-1}\,s^{-1}\,K^{-1}$, respectively. In other words, steel is a better conductor of heat than rock, glass is about the same as rock, and wood is a poorer conductor of heat than rock.

3.8.2 Equation of Heat Conduction

In Fourier's law, as written in (3.27), the conductivity is postulated to be the same in all directions, so the material is said to be *isotropic* with respect to thermal conductivity. This may not always be the case for rock, but it is a good starting point for this discussion of energy conservation. Fourier's law provides the relationship between heat flux and temperature gradient, but does not directly address the question: what is the temperature distribution as a function of position and time in a rock mass with a spatially variable temperature?

To address this question consider heat flow through the white square near the dike in **Figure 3.25**. At the time of dike emplacement heat was convected upward as magma flowed into the dike, and the host rock deformed as the dike opened. We suppose this happened rapidly enough, so very little heat was lost into the host rock. Now consider the time after the magma came to rest, and heat was transferred only by conduction perpendicular to the dike contact. Our objective is to determine the temperature at every location as a function of time, $T(x_o, z_o, t)$.

The white square in **Figure 3.25** is a two-dimensional representation of a volume element used to consider energy conservation in a continuum (**Figure 3.26**). To keep track of the rate of heat energy increase in the element, we subtract the heat flux going out from that coming into the element:

$$\begin{pmatrix}\text{rate of}\\ \text{energy}\\ \text{increase}\end{pmatrix} - \begin{pmatrix}\text{heat flux}\\ \text{in}\end{pmatrix} - \begin{pmatrix}\text{heat flux}\\ \text{out}\end{pmatrix} \tag{3.28}$$

If the rate of heat going out is less than that coming in, the difference is positive and this contributes to the rate of energy *increase* in the volume element: a negative difference corresponds to a rate of energy *decrease*. This strict bookkeeping does not admit the spontaneous creation or destruction of energy, so (3.28) is a statement that defines energy conservation in words, with reference to the stationary volume element. The frame of reference is the Cartesian coordinate system (x, y, z); the volume element is fixed at location **x** for all time; and heat flows only by conduction in the x-direction.

To quantify the rate of energy increase at the position **x**, the center of the fixed volume element (**Figure 3.26**), we use the following:

$$\rho C \frac{\partial T}{\partial t} \tag{3.29}$$

Here C is a material property called specific heat capacity, which measures the change in energy per kilogram per kelvin. Therefore, the units of heat capacity are $\mathrm{J\,kg^{-1}\,K^{-1}}$. The units and dimensions of the other quantities (mass density, temperature, and time) have been discussed and are recorded in **Table 3.1**. Considering all of these physical quantities, the units of rate of energy increase per unit volume are $\mathrm{J\,m^{-3}\,s^{-1}}$.

To evaluate the first term on the right side of (3.28) the x-component of heat flux entering the element through the left face is multiplied by the area of that face: $q_x|_{x-\mathrm{d}x/2}\Delta y\Delta z$. The second term on the right side of (3.28) is evaluated as $q_x|_{x+\mathrm{d}x/2}\Delta y\Delta z$. The units of heat flux in the x-direction are $\mathrm{J\,m^{-2}\,s^{-1}}$. Subtracting the heat flux *out* from the heat flux *in*, and dividing through by the volume of the element, $\Delta x\Delta y\Delta z$, we have $(q_x|_{x-\mathrm{d}x/2} - q_x|_{x+\mathrm{d}x/2})/\Delta x$. Now consider a set of n successively smaller volume elements such that in the limit as $n \rightarrow \infty$ the largest dimension of the element approaches zero, so the element converges to the central point at $\mathbf{x}$:

$$\lim_{n\to\infty}\left[\frac{q_x|_{x-\mathrm{d}x/2} - q_x|_{x+\mathrm{d}x/2}}{\Delta x}\right] = -\frac{\partial q_x}{\partial x}$$

In this limit the quantity in square brackets becomes the negative partial derivative of the heat flux component in x with respect to x. Equating this change in heat flux and the rate of energy increase from (3.29) we have:

$$\rho C\frac{\partial T}{\partial t} = -\frac{\partial q_x}{\partial x} \tag{3.30}$$

For the conduction of heat in one dimension, as illustrated in **Figure 3.26**, the time rate of change of temperature is proportional to the spatial rate of change of heat flux in the x-direction.

Recall that we are looking for the temperature distribution in space and time, but the relationship we have just derived (3.30) has two dependent variables, temperature and heat flux. The next step is to use Fourier's law to replace the heat flux. Taking the first component of the temperature gradient from (3.27), $q_x = -k\,(\partial T/\partial x)$, and substituting into (3.30), we find:

$$\rho C\frac{\partial T}{\partial t} = k\frac{\partial^2 T}{\partial x^2} \tag{3.31}$$

This is the one-dimensional conductive heat flow equation, which applies if $T = T(x, t)$ only. The rate of increase of energy at any point in the continuum depends upon the second derivative of the temperature with respect to position in the direction of heat flow, in this case the x-direction.

The derivative on the right side of (3.31) is a measure of the *curvature* of the graph of temperature versus position. If temperature varies linearly with position, the temperature will not vary with time and the flow of heat is just sufficient to maintain a temperature distribution that does not vary with time. If the temperature versus time graph is non-linear, the implication is that temperature will vary with time. We provide a solution to this equation with application to heat transfer from a dike in Chapter 11.

The one-dimensional heat flow equation given in (3.31) is generalized to the three-dimensional conductive heat transport equation following the same steps as outlined above:

$$\rho C\frac{\partial T}{\partial t} = k\left(\frac{\partial^2 T}{\partial x^2} + \frac{\partial^2 T}{\partial y^2} + \frac{\partial^2 T}{\partial z^2}\right) \tag{3.32}$$

Testing for dimensional homogeneity by examining each distinct term using **Table 3.1**, we find:

$$\rho C\frac{\partial T}{\partial t}\{=\}\mathrm{ML^{-3}{\cdot}L^2T^{-2}\Theta^{-1}{\cdot}\Theta T^{-1}} = \mathrm{ML^{-1}T^{-3}}$$
$$k\frac{\partial^2 T}{\partial x^2}\{=\}\mathrm{MLT^{-3}\Theta^{-1}{\cdot}\Theta L^{-2}} = \mathrm{ML^{-1}T^{-3}}$$

The equation is dimensionally homogeneous. The rate of increase of energy at any point depends upon the second derivatives of the temperature field with respect to the three spatial coordinates.

Strictly speaking equation (3.32) applies to a body that is homogeneous and isotropic with respect to conductivity, and homogeneous with respect to mass density and specific heat capacity. These material properties are considered to be independent of temperature and time. This famous equation has attracted the attention of many applied mathematicians and engineers over the years, and many solutions have been compiled in textbooks (see e.g. Bird et al., 2007; Carslaw and Jaeger, 1986, in Further Reading) that find application in structural geology.

Recapitulation

Tectonic processes in Earth's lithosphere and asthenosphere are, for the most part, physical processes, so in this introductory textbook we focus on the *physics* of rock deformation as it informs the development of geologic structures. Structural geologists should be aware that geochemical alterations and petrological changes during deformation can affect the mechanical properties of rock, and they should use this information to refine their understanding of the tectonic processes that lead to the development of geologic structures.

To get started we identify the physical quantities, their SI units and dimensions. The SI base units of mass, length, time and temperature are used to derive units for all other quantities of interest to structural geologists. While some may choose to use a system of units different than SI, particular physical quantities have only one set of dimensions. Thus, *dimensional analysis* transcends the choice of units and enables one to evaluate whether an equation meets the first test for being physically meaningful: it must be *homogeneous* with respect to dimensions. Also, dimensionless graphs provide a good way to compare field data from geologic structures with different length and time scales.

The concept of a *material continuum* underlies most theoretical constructs used to analyze geologic structures. The displacement field at the nanometer scale around a dislocation in silica, and the stress field within model sand grains subject to point forces at grain contacts, and the displacement field at Earth's surface during a major earthquake, vary systematically in patterns that match those derived using continuum mechanics theory. Models based on this construct do not have to be continuous everywhere. For example, they can have discontinuities in displacement across faults or fractures. In this way most, if not all, geologic structures can be modeled successfully using the continuum construct.

Continuum mechanics is based on fundamental laws of physics known as the *conservation laws*. We derive those laws in order to understand the basis for theoretical models used to investigate rock deformation and the formation of geologic structures. Because the same conservation laws apply, regardless of whether rock deforms as a brittle elastic or plastic solid, or as a viscous fluid, or as a material with any conceivable combination of these behaviors, these laws are generally applicable to all tectonic processes. The conservation laws for mass, momentum, and energy are the theoretical heart of structural geology.

Starting with the conservation law for mass we derive the equation of continuity and explore how this general relationship is specialized for incompressible rock, and a rock mass that is homogeneous with respect to density. Starting with the conservation laws of linear and angular momentum, and incorporating continuity, we derive the equations of motion credited to Cauchy. Also, we combine Fourier's law for heat conduction with a statement of energy conservation to find the general equations for heat conduction. Taken together, the conservation equations for mass, momentum, and energy form the basic building blocks for all of the theoretical analyses that are carried out in this textbook to investigate geologic structures.

REVIEW QUESTIONS

The following questions are designed to highlight the expected *learning outcomes* for this chapter. Each question is taken directly from the material in the chapter and, for the most part, in the same sequence that it appears in the chapter. If an answer is not forthcoming, students are advised to read the relevant section of the chapter and discover the answer.

3.1. Write down the units for traction using SI notation, starting with the *base* units only. Then, simplify the units using other derived units, and end with a definition of the *pascal*.

3.2. Explain the difference between *precision* and *accuracy* using your own example of a geological measurement in the laboratory or in the field.

3.3. Suppose you read in the literature of structural geology that the following equation describes the flow of magma in a sill:

$$\rho \frac{\partial v_x}{\partial t} = -\frac{\partial p}{\partial x} + \eta \frac{\partial^2 v_x}{\partial z^2}$$

Carry out an analysis to show that this equation is *dimensionally homogeneous* (or not), and explain the importance of dimensional homogeneity for physical equations.

3.4. The following Google Earth image is taken from the file: Question 03 04 Mount Ellsworth laccolith.kmz. Open the file; click on the Placemark for this question to open the popup box; copy and paste the questions into your answer document; and address the assigned questions. The Google Earth image data are from Landsat / Copernicus.

3.5. Describe the *material continuum*. This construct does not preclude a finite number of surfaces with a discontinuity in a relevant physical quantity. Use the displacement as that quantity and define the *displacement discontinuity* associated with a fault.

3.6. Density at a fixed point in a material continuum can change by two *independent* mechanisms while mass is conserved. Use the following equation for one-dimensional flow in the y-direction to describe those two mechanisms:

$$\frac{\partial \rho}{\partial t} = -\left(\rho \frac{\partial v_y}{\partial y} + v_y \frac{\partial \rho}{\partial y}\right)$$

3.7. Can the mass density at a fixed point in space change with time as an *incompressible* rock mass deforms? Use the definition of incompressibility and the equation of continuity to justify your answer.

3.8. *Cauchy's first equations of motion* underlie all physically meaningful descriptions of rock deformation, and therefore are of central importance to structural geology. Consider the following example from these equations:

$$\rho \frac{\mathrm{D}v_z}{\mathrm{D}t} = \left(\frac{\partial \sigma_{xz}}{\partial x} + \frac{\partial \sigma_{yz}}{\partial y} + \frac{\partial \sigma_{zz}}{\partial z}\right) + \rho g_z$$

Explain how each term in this equation relates to *Newton's second law*, $\mathbf{F} = m\mathbf{a}$, applied to the material continuum.

3.9. Draw a cubic element with faces perpendicular to the Cartesian coordinate axes. Draw all of the stress components acting on this element that appear in the equation written in Question 3.8. Label the coordinates and stress components.

3.10. *Cauchy's second equations of motion* have important implications for the stress tensor. Write out this tensor in matrix form and describe those implications for the elements of the matrix.

3.11. Conduction is one mechanism for transport of heat in the lithosphere. *Fourier's law* for heat conduction is written as the following vector equation:

$$\mathbf{q} = -k\left(\frac{\partial T}{\partial x}\mathbf{i} + \frac{\partial T}{\partial y}\mathbf{j} + \frac{\partial T}{\partial z}\mathbf{k}\right)$$

Describe each quantity in this equation and write down its SI units. Explain the negative sign on the right side with reference to the direction of heat conduction relative to gradients in temperature.

MATLAB EXERCISES FOR CHAPTER 3: PHYSICAL CONCEPTS

The basic physical concepts that govern the deformation of solids (e.g. rock) and fluids (e.g. magma) are the conservation laws of mass, momentum, and energy. To address questions about the origin and development of geologic structures, and to build models of tectonic processes, structural geologists build upon the concepts that embody these physical laws. To this end, we begin with exercises designed to familiarize students with the International System of Units (SI), and the conversion of archaic units (e.g. foot, pound, pound per square inch) to SI units (e.g. meter, kilogram, newton per square meter). We explore how to quantify the precision of measurements and test equations for dimensional homogeneity, a requirement for physically meaningful relationships. The formation of multilayer folds in the Moine schist from Scotland provides an outcrop example for understanding the concept of mass conservation in a deforming continuum. Faulting in the Lake Edison Granodiorite from the Sierra Nevada of California provides an outcrop example for considering momentum conservation and Cauchy's laws of motion. These exercises enable students to comprehend the fundamental physical laws of rock deformation, so they can apply these laws with confidence while working in the field or analyzing structural data in the laboratory. www.cambridge/SGAQI

FURTHER READING

For citations in figure captions see the reference list at the end of the book.

Bird, R. B., Stewart, W. E., and Lightfoot, E. N., 2007. *Transport Phenomena*. John Wiley & Sons, Inc., New York.

This engineering textbook provides a thorough coverage of momentum and heat transfer in fluids and solids, including many examples of solutions to worked problems, some of which have applications to structural geology.

Carslaw, H. S., and Jaeger, J. C., 1986. *Conduction of Heat in Solids*. Clarendon Press, Oxford.
This classic textbook on conduction of heat in solids includes many examples of solved problems with a thorough exposition of the mathematical methods.

Clark, S. P., (Ed.), 1966. *Handbook of Physical Constants*. Memoir 97. The Geological Society of America, Inc., New York.
Although published half a century ago, this remains a valuable compilation of many physical properties for rocks and minerals.

Davis, J. C., 2002. *Statistics and Data Analysis in Geology*, 2nd edition. John Wiley & Sons, New York.
This second edition of the formative book on geostatistics progresses from elementary concepts to matrix algebra, followed by data and map analysis, and concludes with the analysis of multivariate data.

Jaeger, J. C., Cook, N. G. W., and Zimmerman, R. W., 2007. *Fundamentals of Rock Mechanics*, 4th edition. Blackwell Publishing, Oxford.
This fourth edition of a classic book on rock mechanics deserves to be on the shelf of every structural geologist who utilizes rock mechanics data and concepts to address geological problems.

Kundu, P. K., Cohen, I. M., and Dowling, D. R., 2015. *Fluid Mechanics*. Academic Press, London.
This update of a classic textbook in fluid mechanics is a comprehensive coverage of the fundamentals for advanced undergraduate or graduate students in relevant science and engineering disciplines, and includes applications to geophysical fluid mechanics.

Lai, W.M., Rubin, D.H., and Krempl, E., 2010. *Introduction to Continuum Mechanics*. Elsevier, New York.
Presented as a first course in continuum mechanics for undergraduate students of engineering, this textbook provides many worked examples and exercises.

Malvern, L. E., 1969. *Introduction to the Mechanics of a Continuous Medium*. Prentice-Hall, Inc., Englewood Cliffs, NJ.
This authoritative book on continuum mechanics covers many of the facets of that discipline that have found application in structural geology, with key concepts and equations presented in vector, indicial, and matrix notation.

Pollard, D. D., and Fletcher, R. C., 2005. *Fundamentals of Structural Geology*. Cambridge University Press, Cambridge.
This textbook for structural geologists lays out the connection between the conservation laws and the governing equations of elasticity and viscous fluid mechanics with applications to rock deformation.

Ramsay, J. G., and Lisle, R., 2000. *The Techniques of Modern Structural Geology. Volume 3: Applications of Continuum Mechanics in Structural Geology*. Academic Press, New York, pp. 701–1061.
This third volume in a series of books on structural geology includes an introduction to finite difference and finite element numerical methods for calculating heterogeneous stress and strain fields in models of deforming structures, as well as chapters on fault slip analysis and heterogeneous finite strain fields related to shear zones and folds.

Ranalli, G., 1987. *Rheology of the Earth: Deformation and Flow Processes in Geophysics and Geodynamics*. Allen & Unwin, London.
This textbook considers the constitutive properties of Earth materials from the viewpoints of continuum mechanics and solid-state physics, thereby covering the mechanical behavior at both large and small scales.

Resnick, R., and Halliday, D. 1977. *Physics*, Part 1, 3rd edition, John Wiley & Sons, New York.
This is one of the classic textbooks on university physics for students of science and engineering. Readers are advised to consult modern textbooks, but the fundamental concepts have not changed.

Segall, P., 2010. *Earthquake and Volcano Deformation*. Princeton University Press, Princeton, NJ.
This advanced textbook builds upon the elementary mechanics and structural geology covered here, with a focus on active deformation related to faulting.

Thompson, A., and Taylor, B. N., 2008, *Guide for the Use of the International System of Units (SI)*. National Institute of Standards and Technology, Special Publication 811, US Department of Commerce.
The standard reference for all things related to units and dimensions: a great resource found online at www.nist.gov/physical-measurement-laboratory/special-publication-811.

Turcotte, D. L., and Schubert, G., 2002, *Geodynamics*. 2nd edition, Cambridge University Press, Cambridge.
Geodynamics approaches many of the same topics covered here, but typically uses geophysical data to constrain mechanical models, rather than geological data.

Zoback, M. D., 2010. *Reservoir Geomechanics*. Cambridge University Press, Cambridge.
Insights from geophysics, rock mechanics, structural geology, and petroleum geology are integrated in this textbook with abundant data and useful applications to the hydrocarbon industry.

PART III

Part III (Chapters 4–6) covers the three major styles of rock deformation: brittle, ductile, and viscous. For each style we describe structures varying in scale from thin section to crustal, document laboratory experiments providing material property values, and review the constitutive and motion equations relevant to that style. For example:

- Chapter 4 illustrates the brittle behavior of limestone using the fractures and faults at Lilstock Beach, England;
- Chapter 5 investigates the ductile flow of a bed of salt toward an ascending salt dome in a sedimentary basin; and
- Chapter 6 applies the concept of laminar viscous flow to magma intruding the sills at Shonkin Sag, Montana.

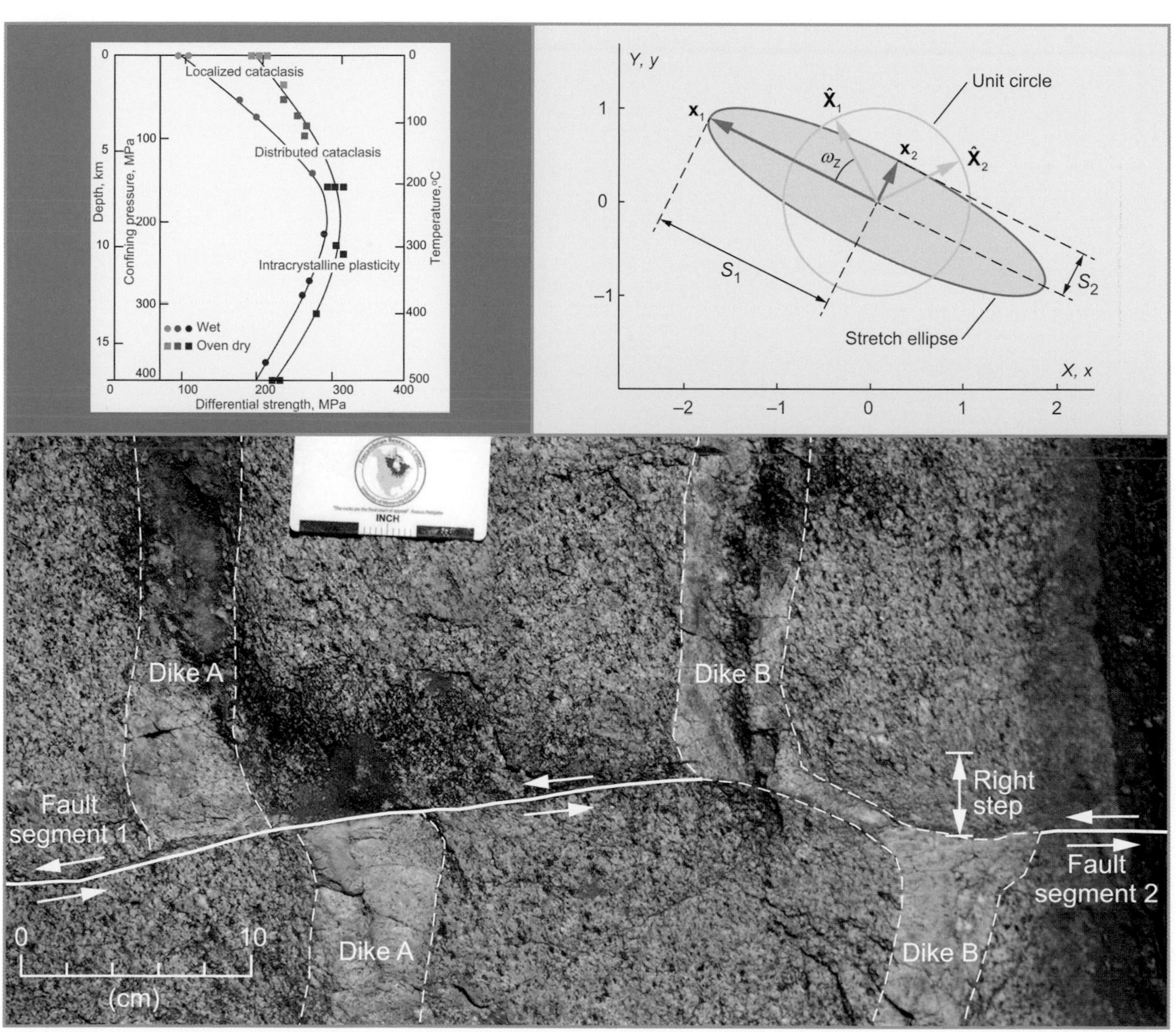

Chapter 4
Elastic–Brittle Deformation

Introduction

Chapter 4 introduces elastic–brittle deformation of rock using field observations of geologic structures, laboratory tests of mechanical behaviors, and theoretical constructs that relate stress to strain. We use solid Earth tides to demonstrate that linear elastic behavior is characteristic of rock deformation in much of Earth's lithosphere. Then, we use field and laboratory observations to show how elastic deformation of rock culminates in brittle fracture, which sets a limit on rock strength. Brittle deformation occurs as opening fractures and as shearing fractures. Laboratory tests measure the elastic stiffness and strength of rock, and examples show how strength depends on rock type, confining pressure, and pore fluid pressure. To quantify elastic deformation we introduce the stress tensor and small strain tensor. We link the components of these tensors using Hooke's law for linear elasticity. Then, we show how Newton's second law, embodied in Cauchy's equations of motion, is used to model elastic deformation and the development of fractures and faults.

Brittle elastic deformation of rock dramatically impacts society during earthquakes, volcanic eruptions, and slope failures. Such deformation also plays a major role in the recovery of hydrocarbons from oil and gas reservoirs, and in the management of groundwater aquifers, because both natural fractures and fractures induced for oil and gas extraction procedures provide underground conduits for the flow of hydrocarbons and water. Ore deposits commonly are found in fractures and faults, so understanding brittle deformation can be crucial for exploration and production of mineral resources. These important applications serve to motivate a better understanding of elastic–brittle deformation in Earth's lithosphere.

4.1 HOOKEAN ELASTIC SOLID

A rubber band (**Figure 4.1a**) displays a familiar behavior when pulled by a hanging weight (**Figure 4.1b**): it stretches along its length and thins across its width. Circles drawn on the unstretched band change into ellipses with long axes parallel to the band length and short axes parallel to the band width. When the weight is removed, the band returns to its initial length and width, and the ellipses return to circles, which are exactly the same size they were before the weight was hung. *Distributed* and *recoverable* deformation such as this is called elastic.

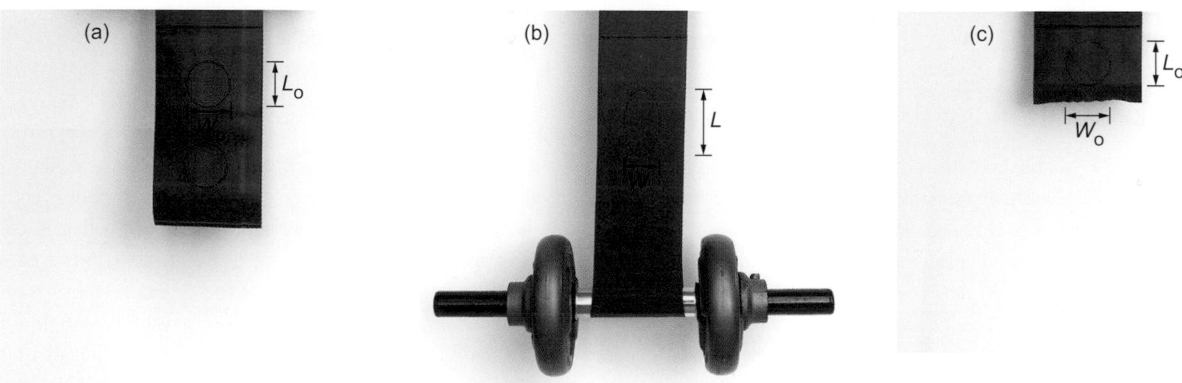

Figure 4.1 Three stages in the brittle elastic deformation of a rubber band. (a) Unloaded band with circles and reference line. (b) Band loaded with a weight stretches such that length increases and width decreases: circles become ellipses. (c) Upper half of fractured band. Both halves return to unloaded configuration and retain elasticity. Photography by Richard Stultz.

When the pull of the weight equals the strength of the rubber band, it breaks along a single fracture; the weight falls to the floor; the two halves of the band snap back to their original length and width; and the ellipses change back into circles (**Figure 4.1c**). If the broken ends of the band are brought into contact and pressed together, the fracture does not heal. *Localized* and *irrecoverable* deformation such as this is called brittle.

It is noteworthy that the two fractured pieces of the rubber band (**Figure 4.1c**) continue to exhibit distributed and recoverable elastic behavior. In other words, the process of fracturing damages the band along the fracture, but the vast majority of the band is not damaged, and retains its elasticity. Because elastic deformation precedes and apparently leads to brittle deformation, and yet the band remains elastic despite fracturing, these two phenomena are inextricably linked. Similar behavior occurs for fractures and faults in Earth's lithosphere: while damage may accumulate near these structures and change the mechanical behavior of the rock there, the majority of the intervening rock remains elastic. In the upper parts of Earth's lithosphere typical rock behavior is elastic and brittle, hence the chapter title: elastic–brittle deformation.

The English natural philosopher Robert Hooke (1635–1703) studied the relationship between the force acting on a solid object and the extension (change in length) of that object. He was particularly interested in watch mechanisms that, in those days, relied on mechanical devices including springs and wheels. According to Gordon (1976, p. 36), in 1676 Hooke published a document entitled *A decimate of the centesme of the inventions I intend to publish*. This strange title apparently reflected Hooke's concern about scientific priority: he wanted to establish a claim to certain ideas and "discoveries" before he had developed them sufficiently for a more formal publication. One section of this document has the immodest subtitle: *The true theory of elasticity of springiness*. The only entry under that heading is:

ceiiinosssttuu

This is an *anagram* of a Latin phrase that revealed, if unscrambled, Hooke's concept of the elastic solid.

In 1679 Hooke published the formal account of his investigations into elasticity as *De Potentia Restitutiva*, which includes the solution to the Latin anagram:

ut tension sic uis

Gordon (1976) translates this phrase:

as the extension, so the force

In this way Hooke established the proportionality between the force applied and the resulting extension: for example, if the force is doubled, the extension is doubled. In the context of a spring attached to a fixed wall, the displacement, u, of the free end is related to the applied force, F, as $F = ku$ where k is the spring constant of proportionality. Through countless experiments over the next three centuries engineers and scientists have demonstrated that Hooke's hypothesis adequately describes the behavior of many different solid materials over a broad range of conditions.

A modern understanding of elastic behavior for the continuum relies on the concepts of stress and strain, which were not developed when Hooke formulated his hypothesis. To introduce these concepts consider again the stretched band in **Figure 4.1**. The stress acting along the length of the band equals the applied force divided by the cross-sectional area, A:

$$\sigma = F/A \tag{4.1}$$

Stress has dimensions of force divided by area, $\sigma\ \{=\}\ \mathrm{MLT^{-2}L^{-2}} = \mathrm{ML^{-1}T^{-2}}$, and the SI units of stress are newtons per meter squared, which is defined as the pascal, $\sigma\ [=]\ \mathrm{N\,m^{-2}} = \mathrm{Pa}$. The weight hung on the band in **Figure 4.1b** is 19.2 N and the cross-sectional area of the band is $6.86 \times 10^{-5}\ \mathrm{m^2}$, so the stress is 2.8×10^5 Pa = 0.28 MPa. Usually, stress in laboratory tests is reported in the unit megapascal, MPa, because the vertical stress in Earth's lithosphere and asthenosphere due to the weight of the overlying rock increases with depth at about 25 MPa/km, and the intention is to use stresses representative of a few, to a few tens, of kilometers depth. Recall from Section 2.6.4 that stress is a tensor quantity with normal and shear components acting on the faces of a cubic element at any point in the band. The quantity in (4.1) that looks like a scalar, refers only to the normal stress component acting along the length of the band.

The strain (also called the unit extension) acting along the length of the band equals the change in length per unit initial length:

$$\varepsilon = (L - L_o)/L_o \tag{4.2}$$

The dimensions of strain are length divided by length, $\varepsilon\ \{=\}\ \mathrm{L/L} = 1$, so strain is dimensionless and carries no units. To measure the strain acting along the length of the band, we take the circle in **Figure 4.1a** for reference and measure its vertical diameter, L_o = 41.0 mm. When the weight is applied this circle transforms to an ellipse with a vertical length L = 61.5 mm. Therefore, the strain is $\varepsilon = 0.5$. The strain also is a tensor quantity, with longitudinal and shear components acting in different orientations at any point in the band, so (4.2) only refers to the longitudinal component acting along the length of the band.

Following Hooke's hypothesis, we assert that the stress is proportional to the strain:

$$\sigma = E\varepsilon \tag{4.3}$$

The proportionality constant is Young's modulus of elasticity, named after Thomas Young (1773–1829) who first described this material constant. The dimensions of Young's modulus are the same as those of stress, because the strain is dimensionless, so $E\ \{=\}\ \mathrm{ML^{-1}T^{-2}}$, and the units of Young's modulus are the pascal, $E\ [=]\ \mathrm{N\,m^{-2}} = \mathrm{Pa}$. For the elastic band in **Figure 4.1**, Young's modulus is $E = \sigma/\varepsilon = 0.56$ MPa.

To address the elasticity of rock, consider a cylindrical sample loaded along its axis by a force, F, applied to the two ends (**Figure 4.2**). We choose a cylindrical shape because that is the conventional shape used in laboratory tests. In those tests forces may be applied that pull on the ends of the cylinder, or that push on the ends. The normal stress acting along the axis of the sample is a tension (positive) if the force pulls, and a compression (negative) if the force pushes. The resulting axial strain is either an extension (positive) or a contraction (negative). Typical values of strain in laboratory tests of elastic rock, before brittle failure, range from about 10^{-6} to 10^{-2}. Young's modulus usually is reported in gigapascal

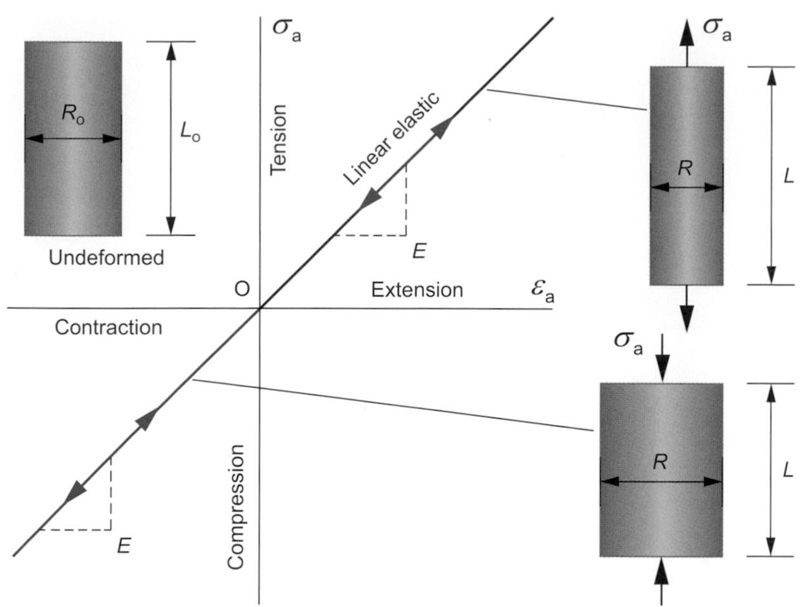

Figure 4.2 Axial stress, σ_a, is plotted versus axial strain, ε_a, for a cylinder of linear elastic material. The slope is Young's modulus, E. Tension tests (red) plot in the first quadrant and compression tests (blue) plot in the third quadrant. Extension and contraction of the cylindrical specimen is greatly exaggerated in the images on the right. See Jaeger et al. (2007) in Further Reading.

(GPa) because typical values for rock range from 1 to 100 GPa. Recall that the elastic band pictured in **Figure 4.1** has a Young's modulus of about 1 MPa, so rock is one thousand to one hundred thousand times stiffer than the band. Comparing their elastic stiffness, we would say the band is very *soft* and rock is very *stiff*.

The words *tension* and *compression* describe stress, as in (4.1), and the words *extension* and *contraction* describe strain, as in (4.2). These equations show that stress and strain have different dimensions, carry different units, and represent different physical concepts. Although in casual speech stress and strain may be used synonymously, in matters of physics and structural geology they are distinguished by consistently using the definitions and corresponding words described here. These words are used to label the axes in **Figure 4.2** to emphasize which word is used with which quantity and which sign of that quantity. This graph also serves to emphasize that the ranges of stress and strain in laboratory tests, and in Earth's lithosphere and asthenosphere, require one to use both quadrant one and quadrant three.

4.2 ELASTIC DEFORMATION OF EARTH'S LITHOSPHERE

Elasticity is easy to demonstrate with a rubber band because you can stretch it with your fingers. Rock, on the other hand, is so stiff that the deformation caused by your fingers is imperceptible. We do, however, have direct evidence that the rock making up Earth's lithosphere is elastic. Perhaps the most compelling demonstration of this is the tidal deformation caused by the gravitational attraction of the Moon. For anyone who spends a day at the beach, the ocean tides are visible evidence of Moon's attraction. Less well known, but also well documented, are the solid Earth tides. Earth's surface slowly moves up and down in response to Moon's changing gravitational attraction as Earth revolves and Moon orbits.

Near Leonard, Oklahoma a seismometer, located in the Arbuckle dolomite 840 m below the surface, recorded the velocity changes due to this motion (**Figure 4.3**). Although the seismometer changes location by about 30 cm for each revolution of Earth, the velocities are so small it takes very sensitive instruments to detect this motion. These recordings are ample proof that the deformation is periodic and recoverable. Furthermore, very successful mathematical models have been applied to study Earth tides and these are based on the postulate that the mechanical behavior is elastic. Just as a mark on the rubber band periodically moves back and forth as your pull increases and decreases, the Arbuckle dolomite moves up and down as the Moon's gravitational attraction increases and decreases.

When small changes in tectonic or gravitational forces in Earth's lithosphere are reversed on time scales from hours to months, the rock mass returns to its former configuration: it is elastic. In the next section we use an example from an outcrop of limestone on the coast of the Bristol Channel, England, to illustrate what happens when changes in loading are not small, and brittle fracture takes over from elastic deformation.

4.3 BRITTLE DEFORMATION AT THE OUTCROP AND GRAIN SCALES

When stress in the brittle part of Earth's lithosphere is increased to the rock strength, subsequent deformation is *localized* along discrete fractures. The growth of these brittle structures is *irrecoverable*: the two surfaces of a fracture do not heal when pressed together. These structures remain in the rock mass and may be brought to Earth's surface by uplift and erosion, where they can be mapped and studied by structural geologists, perhaps millions of years later. The elastic deformation that accompanied the fracturing usually does not leave a visible record, so one must

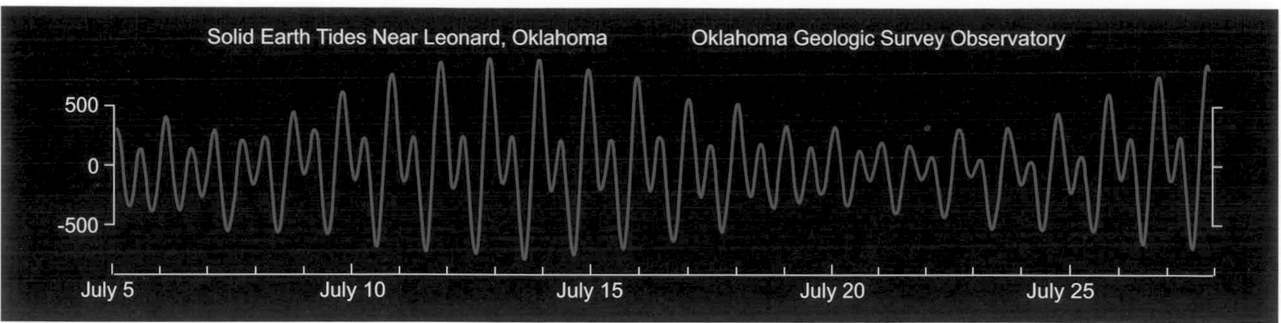

Figure 4.3 Graph of vertical velocity versus time for a borehole seismometer at 840 m depth in the Arbuckle dolomite near Leonard, OK (courtesy of the Oklahoma Geological Survey, www.okgeosurvey1.gov/tide.html). The complete seismogram was sampled every 1000 s to show Earth tides. Each pair of adjacent peaks corresponds to one revolution of Earth. The broad variation in amplitude corresponds to the 29.5 day lunar cycle around Earth. The greatest vertical change in position is about 30 cm.

Figure 4.4 Outcrop of limestone bedding surface at Lilstock Beach, England, with opening fractures (joints). The step in the bedding surface next to the geologist's boots marks a small shear fracture (fault), with the far side offset about 25 cm upward, relative to the near side. See Figure 1.11 for location. Scale is approximate in the foreground. See Engelder and Peacock (2001) for detailed maps, outcrop photos, and structural interpretations of this area.

infer its role in the fracturing process from laboratory experiments (Sections 4.4 and 4.7) and theoretical models (Section 4.11).

As an example of brittle deformation at the outcrop scale, consider the Limestone beds exposed at Lilstock Beach (Figure 4.4) on the southern coast of the Bristol Channel, England. These strata are broken by numerous fractures that provide compelling visual evidence of brittle deformation. The exposed bedding surface in this photograph steps upward about 25 cm just beyond the geologist's boots, along a nearly straight trace that extends across the field of view from left to right. From the offset bedding surface we infer that the two surfaces of this fracture sheared past one another, so this is a *shearing* fracture or fault. We infer that the stress reached the shear strength of the limestone bed as faulting initiated.

The traces of many vertical fractures that do not offset the bedding also are exposed at Lilstock Beach (**Figure 4.4**). Most of these are approximately straight, although a few are gently curved. We infer that the two surfaces of each fracture moved apart, so these are *opening* fractures. Geologists refer to an *opening* fracture as a joint, and infer that the stress reached the tensile strength of the limestone bed when each joint formed. Several sub-parallel opening fractures are termed a joint set. How many joint sets do you see in this photograph? Geologists use abutting relationships to determine the relative ages of two joint sets. The traces of abutting joints stop or start at joints of the second set and therefore are inferred to be *younger*.

As an example of brittle deformation at the grain scale, consider the electron backscatter image (**Figure 4.5**) from a sample of Aztec Sandstone, a Jurassic deposit of wind-blown sand grains, now exposed at the Valley of Fire State Park, NV. Focusing on the quartz grains (gray), note that the multitude of micro-fractures have irregular traces and many different orientations, so they cannot be grouped into a few sets like the joints at Lilstock Beach (**Figure 4.4**). However, many of the fractures extend from points where two grains are in contact. Also, where fractures intersect the edge of a grain, little evidence exists for offset, so most fractures appear to have opened without shearing. We infer that the stress reached the tensile strength of quartz when these micro-fractures formed. The geologic history of the Aztec Sandstone suggests that the sample shown in **Figure 4.5** was buried by no more than 3 km of younger sediments. Apparently the weight of the overlying sediments was sufficient to fracture the quartz grains.

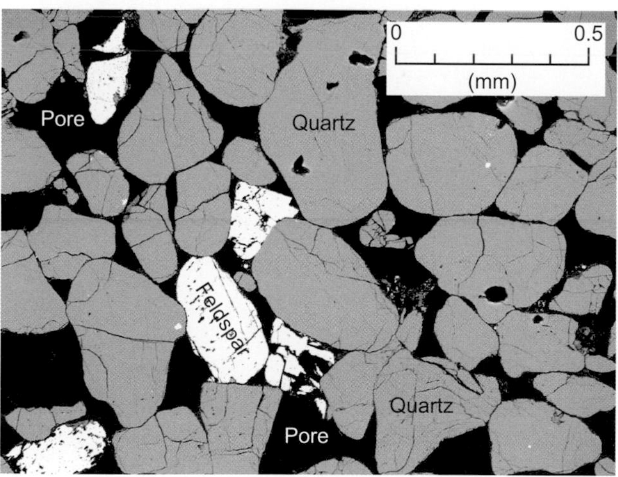

Figure 4.5 Electron backscatter image of Aztec Sandstone from the Valley of Fire State Park, NV. Sample was taken more than 1 m from nearest outcrop-scale structures (deformation bands). Black is pore space; medium gray is quartz; white is feldspar. Quartz grains are cut by open fractures that have irregular traces and a variety of orientations. Modified from Sternlof et al. (2005).

The abundance of joints and faults exposed at Earth's surface, and the multitude of fractures at the grain scale revealed by microscopic techniques, beg the question: how does rock behave mechanically under loading that leads to brittle fracture? Why, for example, did the limestone at Lilstock Beach (**Figure 4.4**) break by *shearing* to create a fault in one instance, but break by *opening* to create joints in many other instances? We address these questions in Sections 4.4 and 4.7 using laboratory experiments.

4.4 ELASTIC–BRITTLE DEFORMATION IN THE LABORATORY

Measurements of the physical properties of rock in the laboratory traditionally fall within the engineering discipline called rock mechanics, although geophysicists and structural geologists, as well as mining and civil engineers, contribute to this activity (**Box 4.1**). Structural geologists use these measurements to build conceptual models of rock deformation and to constrain mechanical models for the development of geologic structures.

The elastic response of rock to a changing state of stress is measured in the laboratory using a universal testing machine (**Figure 4.6**). These machines consist of lower and upper steel cross heads held in place with steel tie bars that together provide a nearly rigid frame. Within this frame, a hydraulically driven piston applies a force through a steel platen onto the circular ends of a cylindrical rock specimen. Because force is applied only parallel to the axis of the specimen, this is called a uniaxial test. If the fluid pressure, P, in the top of the hydraulic cylinder increases, the piston moves down and shortens the rock specimen by applying a *compressive* stress. If the fluid pressure in the bottom of the hydraulic cylinder increases, the piston moves up and stretches the rock specimen by applying a *tensile* stress. Tension tests require special grips that hold the specimen ends, whereas the platen simple presses against the specimen end in a compression test. The testing machine is called *universal* because it can apply either tension or compression. Modern testing machines use sophisticated electronics to control the motion of the piston. They can impose a prescribed force, or displacement, or displacement rate as a function of time.

A force transducer placed between the piston and the specimen converts mechanical deformation into an electronic signal that is calibrated to the total vertical force, F. A uniaxial test is designed to achieve a uniform axial stress, $\sigma_a = F/A$, within the rock specimen and this stress equals the applied force divided by the cross-sectional area, A, of the specimen. A displacement transducer attached to the platens converts vertical displacement into an electronic signal that is calibrated to that displacement. Suppose the original specimen length is L_o, the final length is L, and the change in length after a given loading is $L - L_o$. Then, the axial strain is $\varepsilon_a = (L - L_o)/L_o$. The strain is unlikely to be uniform at the mineral grain scale, so ε_a represents an average value at the specimen scale. Rock mechanics data usually are displayed on plots of axial stress versus axial strain (**Figure 4.7**).

Box 4.1 Rock Mechanics

Rock mechanics is the study of the mechanical behavior of rock and rock masses under conditions relevant to the design, construction, and sustainability of engineered structures such as buildings, bridges, dams, excavations, well bores, and tunnels (Jaeger et al., 2007; Cosgrove and Hudson, 2016; see Further Reading). Because these structures primarily exist at Earth's surface, or in the upper few kilometers of Earth's lithosphere, the relevant conditions typically are limited to stresses of a few hundred MPa, temperatures of a few hundred °C, and strain rates where the mechanical behavior of rock is elastic and brittle. Important exceptions do exist. For example, in geothermal and volcanic regions greater temperatures may be associated with inelastic and ductile behavior of rock involved in engineering projects. Also, well bores encounter conditions at depths approaching 10 km that require consideration of inelastic deformation.

Although engineered structures can be very large on a human length scale (**Figure B4.1**), they rarely approach the length scales of geologic structures involved, for example, in mountain building or plate tectonic movements. Also, the time scale of interest for most engineered structures is less than a few hundred years, whereas the structural geologist may be required to consider time scales that range to hundreds of millions of years. Thus, although common ground does exist in the elastic–brittle realm (Paterson and Wong, 2005; see Further Reading), the ductile and viscous behavior of rock at the elevated temperatures and pressures below the brittle to ductile transition (Wong and Baud, 2012; see Further Reading), is primarily of interest to structural geologists.

Rock mechanics investigations include *theoretical* analysis (e.g. Crouch and Starfield, 1983) using the same continuum mechanical principles and tools described in Chapter 3. The results of theoretical analyses in rock mechanics find applications in structural geology, solid earth geophysics, and petroleum engineering. Rock mechanics also involves *laboratory* studies of the mechanical properties of rock (e.g. Hoek and Martin, 2014; Kwasniewski et al., 2013; see Further Reading), similar to those described in Sections 4.4 and 4.7 of this chapter, and in Section 5.4 of the next chapter. Laboratory studies with applications to earthquake hazards (e.g.

Figure B4.1 This ten meter diameter tunnel boring machine completed a 723 m long tunnel in Turkey in one month. Photograph courtesy of The Robbins Company.

Lockner and Beeler, 2003; see Further Reading) share many techniques with those in engineering rock mechanics. Unlike the objectives of structural geology, rock mechanics also is concerned with the *design* of engineered structures (e.g. Bieniawski, 1984; see Further Reading). In this area of rock mechanics, the work overlaps with that of civil engineers and mining engineers.

Rock mechanics, as we know it today, began in the middle of the twentieth century. The first United States Symposium on Rock Mechanics was convened in 1956, and the International Society for Rock Mechanics started in 1962. In 1969 the senior author of this textbook was an MSc student at Imperial College, London, where he attended a seminar given by Professor J. C. Jaeger of Australian National University. Professor Jaeger was well-known at that time as the co-author of a renowned book, *Conduction of Heat in Solids*, but his seminar was on *rock mechanics*. During his seminar, Professor Jaeger mentioned a new book, co-authored with N. G. W. Cook, called *Fundamentals of Rock Mechanics*. At the first opportunity the senior author of this textbook went down to Foyles Bookshop on Charing Cross Road and purchased *Fundamentals of Rock Mechanics*. Over the ensuing years he has turned to this book time and again for its comprehensive and authoritative coverage of rock mechanics theory, applications, and laboratory results. In 2007 he was delighted to purchase the revised and updated fourth edition, prepared by Professor R. W. Zimmerman of Imperial College (Jaeger et al., 2007).

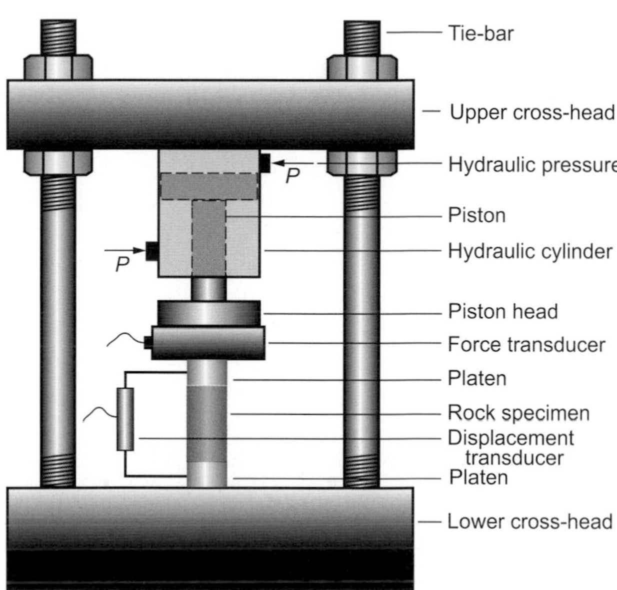

Figure 4.6 Schematic drawing of a universal hydraulic testing machine used to measure stiffness and strength in uniaxial tension and compression. Modified from Jaeger et al. (2007).

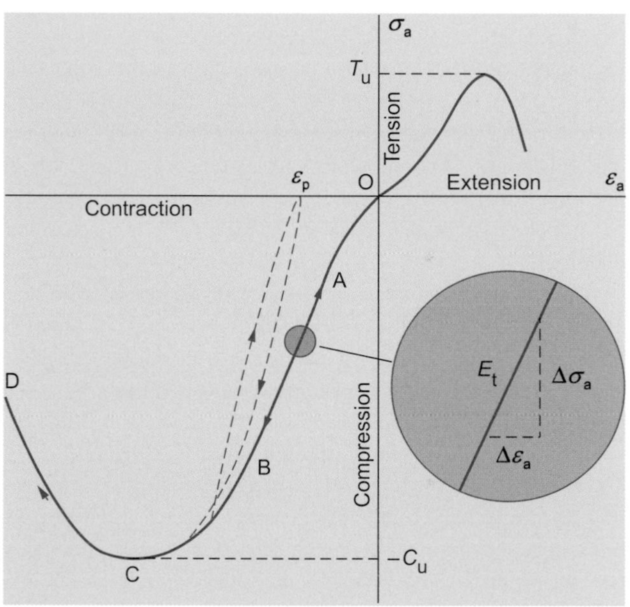

Figure 4.7 Generic plot of axial stress, σ_a, versus axial strain, ε_a, for a uniaxial tension test (quadrant 1, red) and uniaxial compression test (quadrant 3, blue). T_u is the uniaxial tensile strength and C_u is the uniaxial compressive strength. The point ε_p is a permanent strain. Points A, B, C, and D are described in the text. The blue inset shows a change in axial stress, $\Delta\sigma_a$, and in axial strain, $\Delta\varepsilon_a$, defining the tangent elastic modulus, E_t, as the local slope of the line. Modified from Jaeger et al. (2007).

The output from the force transducer and the displacement transducer (**Figure 4.6**) provide a record of axial stress and axial strain during tension and compression tests. A *generic* uniaxial curve of stress versus strain (**Figure 4.7**) illustrates several important concepts and facilitates the definition of key terms. For specimens loaded in compression, the concave downward segment OA represents non-linear deformation usually attributed to closing of microcracks within grains and along grain boundaries. This is elastic deformation if it is recoverable: the microcracks open upon unloading and the curve is retraced to the origin. Along the segment AB, deformation is recoverable: during cycles of loading and unloading in this range, the stress–strain graph retraces this line. Here the rock specimen responds mechanically to loading as a linear elastic solid.

4.4.1 Concepts of Stiffness, Compressibility, and Strength

The slope of the stress–strain curve is a measure of stiffness of the rock specimen: for a given change in axial stress, $\Delta\sigma_a$, a *stiffer* response is associated with a steeper slope and lesser change in axial strain, $\Delta\varepsilon_a$. A *softer* response is associated with a less steep slope and a greater change in axial strain. For the concave downward segment OA the specimen stiffens with increased strain, as the microcracks close. If the axial stress is a continuous and differentiable function of axial strain, $\sigma_a = f(\varepsilon_a)$, the tangent modulus is a measure of stiffness at a given strain, defined as:

$$E_t(\varepsilon_a) = \lim_{\Delta\varepsilon_a \to 0} \frac{\Delta\sigma_a}{\Delta\varepsilon_a} = \frac{d\sigma_a}{d\varepsilon_a} \quad (4.4)$$

In other words, the tangent modulus is the first derivative of axial stress with respect to axial strain. Particular values of the tangent modulus are reported with the associated axial strain, ε_a. For example $E_t(1.5 \times 10^{-4}) = 35$ GPa indicates a tangent modulus of 35 GPa at an axial strain of $\varepsilon_a = 1.5 \times 10^{-4}$.

For an idealized elastic–brittle solid the axial stress versus axial strain plot is a straight line for all strains (**Figure 4.2**). The measure of stiffness is the slope of this line and is called Young's modulus:

$$E = \frac{\Delta\sigma_a}{\Delta\varepsilon_a} = \frac{\sigma_a}{\varepsilon_a} = \text{uniform} \quad (4.5)$$

Because axial extension, ε_a, is dimensionless, the tangent modulus and Young's modulus have the same dimensions as the axial stress, σ_a. That is, E {=} $ML^{-1}T^{-2}$ and the SI units are E [=] N m^{-2} = Pa. Usually, the values of these moduli are reported with the unit gigapascal, GPa, and typical values for rock stiffness are in the range 1 to 100 GPa. Recall that the elastic band in **Figure 4.1** has a Young's modulus of about 0.5 MPa, so it is much softer than rock.

The specimen radius usually decreases when loaded in uniaxial tension and increases when loaded in uniaxial compression (**Figure 4.2**). This is a familiar response of elastic materials: stretch a rubber band and it becomes thinner (**Figure 4.1**). To quantify this phenomenon suppose the axial stress is compression, sufficient to shorten the specimen from L_o to L, so $L_o > L$. The axial strain, $\varepsilon_a = (L - L_o)/L_o$, is negative, which is a contraction. Typically, axial compression causes an increase in the radius of the specimen from R_o to R, with $R_o < R$. The radial strain, $\varepsilon_r = (R - R_o)/R_o$, is positive, which is an extension. To quantify this behavior a second elastic property is defined, called Poisson's ratio:

$$\nu = -(\varepsilon_r/\varepsilon_a) \quad (4.6)$$

Here the loading is a uniaxial stress. Because the strains in the numerator and denominator have opposite signs, their ratio is negative. The negative sign in front of the ratio is used in the definition (4.6), so Poisson's ratio is a positive number. As a ratio of dimensionless quantities, Poisson's ratio is dimensionless, and carries no SI units.

For the elastic band in **Figure 4.1** the longitudinal strain is $\varepsilon_L = (L - L_o)/L_o$ and the lateral strain is $\varepsilon_W = (W - W_o)/W_o$. The band is stretched by the weight acting along its length, so the longitudinal strain is positive, $\varepsilon_L > 0$, but the band is shortened along its width, so the lateral strain is negative, $\varepsilon_W < 0$. Poisson's ratio is the negative of the ratio of these two orthogonal strains, $\nu = -(\varepsilon_W/\varepsilon_L)$, so it is a positive, dimensionless number. In this case we found $\nu = 0.36$.

The value of Poisson's ratio quantifies interesting aspects of how a material deforms, namely its *compressibility*. If the volume of a rock specimen does not change during a uniaxial test, it is called an incompressible solid, and Poisson's ratio has the value $\nu = \frac{1}{2}$. The specimen responds to axial contraction by extending in the radial direction just enough to maintain a constant volume. If the radial strain is zero, Poisson's ratio has the value $\nu = 0$. This is a perfectly compressible solid. Familiar examples that approach these two extremes include cork, which is very compressible, and rubber, which is nearly incompressible. Experimental measurements of Poisson's ratio for rock typically range from 0.05 to 0.4, so some rocks are almost perfectly compressible.

The point B on the generic stress–strain curve (**Figure 4.7**), called the yield point, marks the transition from *linear* and *recoverable* deformation to *non-linear* and *irrecoverable* deformation. Segment BC of this curve is concave upward because inelastic deformation mechanisms within the sample act to progressively decrease the stiffness. These mechanisms usually are irrecoverable: unloading proceeds, for example, on the dashed path and results in a permanent axial strain, ε_p. Because the slope of the stress–strain curve is positive on the segment BC, and permanent strain is accumulating, the rock is said to be in a ductile state. The slope of the curve goes to zero at point C where the stress is a minimum (the greatest compression). Because the axial stress is a continuous and differentiable function of axial strain, $\sigma_a = f(\varepsilon_a)$, the first derivative is zero at the minimum. The magnitude of the stress at this point is the uniaxial compressive strength, C_u. Thereafter, as the specimen continues to shorten along the segment CD, the stress becomes less compressive, so the slope of the stress–strain curve is negative, and the rock is said to be in a brittle state. The test is terminated at point D, usually because of limitations of the testing equipment.

For specimens loaded in tension (**Figure 4.7**), the stress–strain curve plots in the first quadrant because axial stress is a tension (positive) and axial strain is an extension (positive). The generic stress–strain curve for the tension test may include segments analogous to O–A–B–C–D for the compression test. The tensile stress is limited by the uniaxial tensile strength, T_u, which usually is much less than the uniaxial compressive strength. After the axial stress passes through the maximum, the slope turns negative. *The negative stress–strain slope after the stress equals the strength in both compression and tension is a defining characteristic of brittle deformation.*

Definitions of uniaxial tensile strength, T_u, and uniaxial compressive strength, C_u, are:

$$T_u \equiv \max(\sigma_a), \quad \sigma_a > 0, \quad \sigma_r = 0 \quad (4.7)$$

$$C_u \equiv |\min(\sigma_a)|, \quad \sigma_a < 0, \quad \sigma_r = 0 \quad (4.8)$$

In all uniaxial tests the radial stress, σ_r, is zero. Strength is a scalar quantity and does not carry a sign, so compressive strength is defined by taking the absolute value of min(σ_a), because σ_a is negative for the compression test. The uniaxial tensile and compressive strengths have the same dimensions as stress, for example T_u {=} $ML^{-1}T^{-2}$, and strengths carry the same units as stress, for example C_u [=] N m^{-2} = Pa. Usually, strengths are reported with the units megapascal, MPa.

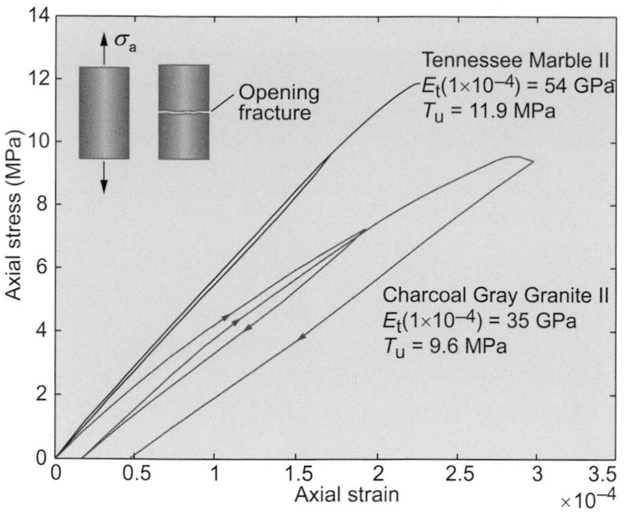

Figure 4.8 Uniaxial tension test data for Tennessee marble II and Charcoal Gray granite II. The tangent modulus, E_t, and the uniaxial tensile strength, T_u, indicate that the marble is stiffer and stronger than the granite. Modified from Wawersik (1968).

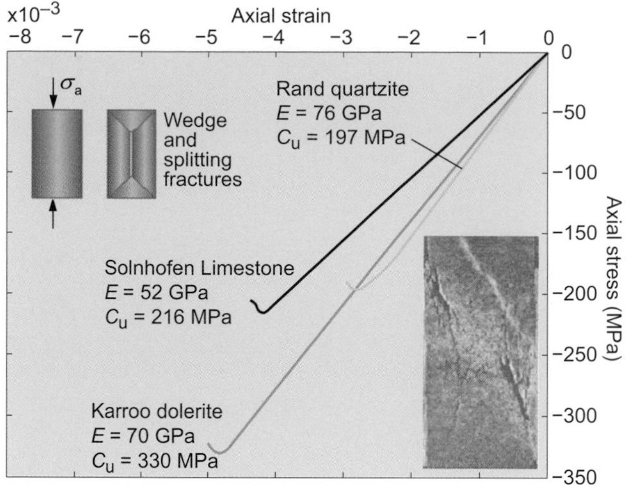

Figure 4.9 Uniaxial compression test data for three different rock types. The apparent Young's modulus, E, is given for a straight line approximation to these curves. The uniaxial compressive strength is C_u. Modified from Jaeger and Cook (1969). Tennessee Marble I sample on right after uniaxial compression test shows axial splitting fractures and incipient wedge fractures. Modified from Wawersik (1968).

The tangent modulus (4.4), Young's modulus (4.5), and the two uniaxial strengths, (4.7) and (4.8), have the same dimensions and units. This can lead to confusion about the important differences among these physical quantities. The elastic moduli correspond to the slope of the stress–strain curve (**Figure 4.7**); the strengths correspond to the extreme values of this curve. We compare the elastic moduli of different rocks with the qualitative terms *soft* and *stiff*, but we compare strengths using the qualitative terms *weak* and *strong*. In the next section we use stress–strain curves from rock mechanics tests to illustrate how the stiffness and strength of rock are measured and then provide tables of values for many different types of rock.

4.4.2 Rock Stiffness and Strength in Uniaxial Tension and Compression

Data from uniaxial tension tests on two different rocks are shown as plots of axial tension versus axial extension in **Figure 4.8**. These curves are typical of the ten specimens of each rock type that were tested. Tennessee marble II approximates a linear elastic solid as loads increase to ~10 MPa. The nearly straight unloading curve from ~10 MPa to 0 MPa almost coincides with the loading curve, and the axial strain goes to zero as the axial stress returns to zero. The tangent modulus at an axial strain of 1×10^{-4} is 54 GPa and this is nearly uniform up to an axial stress of 10 MPa. Loading to axial stress greater than 10 MPa causes a slight decrease in the tangent modulus, and close to the tensile strength the curve is concave downward. The uniaxial tensile strength for this specimen is 11.9 MPa, and brittle failure occurred almost immediately after the axial stress reached this value. The specimen broke into two parts along a macroscopic opening fracture oriented approximately perpendicular to the specimen axis.

The mechanical behavior of Charcoal Gray granite II (**Figure 4.8**) is somewhat non-linear and inelastic. Unloading from about 7 MPa proceeds along a stress–strain curve that is clearly separated from the loading curve, and results in a permanent axial strain. The tangent modulus at an axial strain of 1×10^{-4} is 35 GPa, softer than Tennessee marble II. Upon re-loading above 7 MPa the curve for the granite continues with a concave downward shape and passes through a maximum at a tensile strength of 9.6 MPa. Unloading just beyond the maximum and before macroscopic failure results in a permanent axial strain of nearly 0.5×10^{-4}. When the granite was tested to failure, the behavior was brittle (negative post-peak slope), and an opening fracture formed approximately perpendicular to the specimen axis (see **Figure 4.8** inset).

Plots of axial compression versus axial contraction for uniaxial compression tests on three different rocks are shown in **Figure 4.9**. Karroo dolerite and Solnhofen Limestone display almost perfectly linear and apparently recoverable behaviors up to axial contractions between -4×10^{-3} and -5×10^{-3}. The non-linear behavior characterized by segment OA of **Figure 4.7**, is absent. The data were not provided for unloading these samples, but the behavior was dominantly elastic and brittle. With very little additional contraction, both curves pass through a minimum and terminate, indicating abrupt loss of load carrying capacity. The Rand quartzite displays a nearly linear and apparently recoverable behavior until an axial contraction of about -2×10^{-3}, and then the slope decreases gradually to a minimum, followed closely by failure. In contrast to the limestone and dolerite, the quartzite displays more significant non-linear deformation before reaching the uniaxial compressive strength. Brittle failure in

Table 4.1 Laboratory values of elastic constants

	Young's modulus (GPa)		Poisson's ratio	
Rock Type	From	To	From	To
Quartzite	70	105	0.11	0.25
Gneiss	16	103	0.10	0.40
Basalt	16	101	0.13	0.38
Granite	10	74	0.10	0.39
Limestone	1	92	0.08	0.39
Sandstone	10	46	0.10	0.40
Shale	10	44	0.10	0.19
Coal	1.5	3.7	0.33	0.37

Data from Bieniawski (1984)

Table 4.2 Laboratory values of uniaxial strengths (MPa)

	Tensile		Compressive	
Rock type	From	To	From	To
Quartzite	17	28	200	304
Gneiss	3	21	73	340
Basalt	2	28	42	355
Granite	3	39	30	324
Limestone	2	40	48	210
Sandstone	3	7	40	179
Shale	2	5	36	172
Coal	1.9	3.2	14	30

Data from Bieniawski (1984)

uniaxial compression usually is a result of wedge and splitting fractures (see **Figure 4.9** inset).

Table 4.1 provides representative values of *apparent* Young's modulus for selected metamorphic, igneous, and sedimentary rocks. We say "apparent" because the complete stress–strain data are not presented and the modulus is likely to change with axial strain as illustrated in **Figure 4.7**. A few generalizations can be made from these laboratory results. Metamorphic rocks (quartzite and gneiss) and igneous rocks (basalt and granite) typically are stiffer than clastic sedimentary rocks (sandstone and shale), although weathered granite can be less stiff than well-indurated sandstone. Qualitatively, we describe rocks with Young's moduli around 100 GPa as very stiff, whereas rocks with values around 1 GPa are very soft. Rocks in the upper part of this range would ring when hit by a geologist's hammer, whereas those in the lower part would respond with a dull thud. From **Table 4.1** we draw the following conclusions: *laboratory specimens of rock have Young's moduli that range from about 1 GPa to 100 GPa with a "typical" value of about 50 GPa.*

Representative values of Poisson's ratio for rock are given in **Table 4.1**, and for all of the rock types listed these ratios vary only from 0.08 to 0.40. These values demonstrate that rock is a somewhat compressible solid. We draw the following conclusions from **Table 4.1**: *laboratory specimens of rock have Poisson's ratios that range from about 0.1 to 0.4, with a "typical" value of about 0.25.*

Table 4.2 provides selected uniaxial tensile and compressive strengths. Uniaxial tensile strengths range over about one order of magnitude among all the rock types tabulated. Different samples of the same lithology have a similar range. The tensile strength of crystalline rock can be as small as that for clastic sedimentary rock, but typically it is somewhat greater. From these data we conclude: *laboratory specimens of rock have uniaxial tensile strengths that range from about 2 to 40 MPa with a typical value of 10 MPa.*

Uniaxial compressive strengths shown in **Table 4.2** also range over about one order of magnitude. Crystalline rocks tend to have greater compressive strengths than clastic sedimentary rocks, but some granite is weaker than some shale. We draw the following conclusions from these data: *laboratory specimens have uniaxial compressive strengths that range from about 30 to 355 MPa with a typical value of 150 MPa.*

Table 4.2 reveals a crucial difference between uniaxial tensile and compressive strengths for any particular rock type. In general, the ranges do not overlap and typical values differ by an order of magnitude. Sandstone, for example, has a typical tensile strength of about 5 MPa and a typical compressive strength of about 110 MPa. Granite has a typical tensile strength of about 20 MPa and a typical compressive strength of about 180 MPa. *In the context of uniaxial strength, rock is about ten times stronger in compression than in tension.*

While uniaxial tests in the laboratory reveal important properties of rock, the state of stress in Earth's lithosphere is multi-axial, and the relevant length scale of deformation can be meters to kilometers. In the next section we use data from the field at the kilometer scale and with multi-axial stress states to evaluate rock stiffness.

4.5 FIELD ESTIMATES OF ROCK STIFFNESS AT THE KILOMETER SCALE

The igneous dike extending southward from the volcanic neck at Ship Rock, NM (**Figure 4.10**) is a natural testing machine for elastic deformation at the ten kilometer scale. Approximately 30 million years ago magma forced open a fracture, compressing the surrounding Mancos Shale to gain space for the magma. The current level of erosion is about 1 km below Earth's surface at the time of volcanic activity. The length and thickness of the dike, and the depth of intrusion, are key data for a preliminary analysis of the elastic stiffness of the sedimentary rock in this region.

The south dike is conceptualized in **Figure 4.11** as a crack in an elastic solid. The horizontal (x, y)-plane is cut by a crack of length L, lying along the x-axis and centered on the y-axis. The

Figure 4.10 Southern dike at Ship Rock, NM is roughly 9 km long and attains a thickness of about 10 m. For scale note the sport utility vehicle (SUV) on the dirt road to left and just below the center of this image. UTM: 12 S 693217.94 m E, 4062057.75 m N.

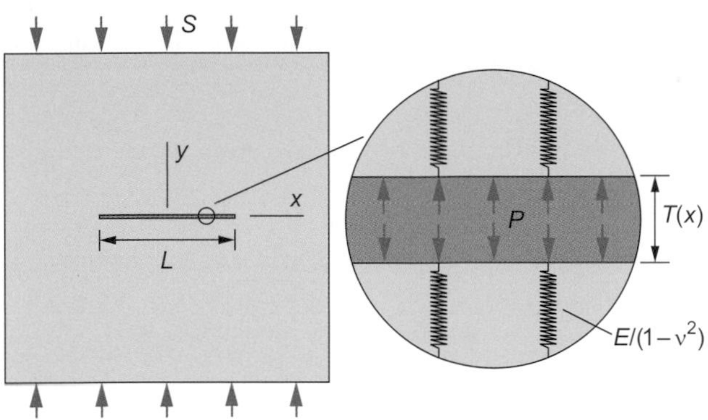

Figure 4.11 Conceptual model of a crack of length, *L*, and opening *T*(*x*) in an elastic material. Fluid pressure, *P*, acts inside the crack (see inset), while a remotely applied compressive stress of magnitude *S* acts across the plane of the crack. Elastic stiffness is $E/(1 - \nu^2)$, represented schematically by springs.

crack is opened by a pressure, P, which represents the magma pressure, and this opening is resisted by a compressive stress of magnitude S acting perpendicular to the crack, which represents the horizontal stress due to the weight of the overlying rock. The crack opening is a function of the x-coordinate, $T(x)$, varying from a maximum, T_m, at the center, $x = 0$, to zero at the crack tips, $x = \pm L/2$. The Mancos Shale may be thought of as a bed of elastic springs that are compressed to make room for the invading magma. The elastic stiffness of the shale resists the opening of the dike and this is characterized by both Young's modulus and Poisson's ratio in the form $E/(1 - \nu^2)$.

The maximum opening, T_m, of the pressurized crack is given by an equation derived in Section 7.3, so we only report the resulting equation here:

$$T(x = 0) = T_m = \frac{2L(P - S)}{E/(1 - \nu^2)} \quad (4.9)$$

The maximum opening is proportional to the crack length and to the difference between the pressure and stress. The opening is inversely proportional to the elastic stiffness. Because Poisson's ratio is about ¼, we take $(1 - \nu^2) \approx 1$. Using this approximation, and solving (4.9) for Young's modulus we have:

$$E \approx \frac{2(P - S)}{T_m/L} \quad (4.10)$$

Notice the similarity between this equation and equation (4.5) that defines Young's modulus for a laboratory test: $E = \sigma_a/\varepsilon_a$. Twice the difference between the pressure and the stress in (4.10) plays the same role as the axial stress, σ_a, in the laboratory test. The ratio of maximum thickness to length in (4.10) plays the same role as the axial strain, ε_a, in the laboratory test.

Both of the geometric quantities in (4.10) can be measured in the field, and for the south dike at Ship Rock we have $L \approx 9$ km and $T_m \approx 10$ m. Young's modulus in (4.10) is proportional to the difference between the pressure, P, and the compressive stress, S. To estimate these quantities we assume the dike breached Earth's surface 1 km above the outcrop. The magma pressure is estimated as that in a static column of magma 1 km tall: $P = \gamma_m D$ where $\gamma_m \approx 2.6 \times 10^4$ N m^{-3} is the unit weight of the magma and $D = 1$ km. Similarly, the least compressive stress is estimated as $S = \gamma_r D$ where $\gamma_r \approx 2.4 \times 10^4$ N m^{-3} is the unit weight of Mancos Shale. Using these estimates, the driving pressure for the south dike is $P - S \approx (\gamma_m - \gamma_r)D \approx 2$ MPa. Using these values in (4.10), the estimated Young's modulus of the sedimentary rock at Ship Rock is $E \approx 4$ GPa.

The model estimate of Young's modulus, ~4 GPa, is somewhat below the range for shale (10 to 44 GPa) as measured at the decimeter scale in the laboratory (**Table 4.1**). This is expected because heterogeneities at the kilometer scale (fractures, faults, bedding surfaces, etc.) are likely to make the larger rock mass *softer*. We conclude that using laboratory data for rock stiffness to model geologic structures is a good first approximation, even if the length scale of the structure is four to five orders of magnitude greater. However, one should be aware that laboratory values may exaggerate the stiffness by up to one order of magnitude.

4.6 THREE-DIMENSIONAL STRESS STATES

In the Introduction for Chapter 2, we pointed out that the ellipsoidal shape of deformed ooliths can be used to visualize the stretch of material line segments oriented in all possible directions at that point. The originally spherical ooliths caught up in the deformation that formed South Mountain fold (**Figures 2.2** and **2.3**) embody in their ellipsoidal shape the information that is carried quantitatively by a *tensor*. The ellipsoid has three semi-axes that may be of different lengths, and these axes are oriented in three orthogonal directions. The tensor carries enough information to calculate those three principal values and three principal axes. Here we focus on a different tensor, the stress tensor, and introduce the principal stresses, which enable us to describe stress states in their simplest form and to relate these stresses to the initiation and orientation of brittle fractures and faults.

The stress tensor $\boldsymbol{\sigma}$ is defined at a point by considering the nine stress components acting on the faces of a small cubic element centered at that point (**Figure 4.12a**). For illustration purposes, the stress components are represented by arrows positioned at the centers of these faces. However, each stress component represents the resultant of all forces acting on that face in the respective direction, divided by the area of the face. Because we want to define stress components at a point, we take the limit as the area goes toward zero and the cubic element shrinks down to that point. For example, the front face has an area ΔA_x and is called an x-face because the normal to it is parallel to the x-axis. If the resultant force on this face in the y-direction is ΔF_y, the stress component on the x-face in the y-direction is defined as:

$$\sigma_{xy} = \lim_{\Delta A_x \to 0} \left(\frac{\Delta F_y}{\Delta A_x} \right) \quad \textbf{(4.11)}$$

The eight other stress components are defined similarly.

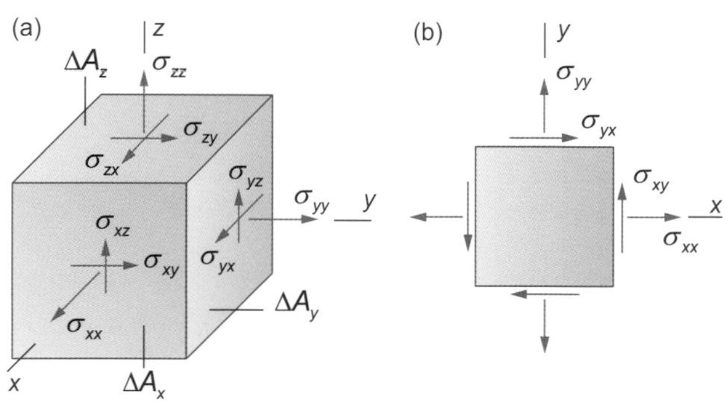

Figure 4.12 (a) Three-dimensional view of nine stress components acting on a cubic element at a point in a deforming solid or flowing liquid. Area of the x-face is ΔA_x. (b) Two-dimensional view of element cross section with four stress components. See Timoshenko and Goodier (1970) in Further Reading.

In **Figure 4.12a** only one arrow is visible for each of the 9 stress components, but in every case a corresponding arrow points in the opposite direction on the parallel, but hidden face of the cube. Each stress component is represented by the *pair* of arrows, so a total of 18 arrows represent the 9 stress components. Four of the pairs of arrows are shown in the two-dimensional drawing of **Figure 4.12b**. For each stress component the two subscripts refer, respectively, to the coordinate direction of the normal to the surface *on* which the stress acts, and the coordinate direction *in* which the stress acts. This is called the "on-in" convention for subscripts. For example, σ_{yy} acts *on* the y-faces *in* the y-direction, and σ_{xz} acts *on* the x-faces *in* the z-direction.

4.6.1 Normal and Shear Stresses

Two kinds of stress components are distinguished: a normal stress acts *perpendicular* to a face of the cubic element (**Figure 4.12**); a shear stress acts *parallel* to a face. The two subscripts are the same for a normal stress (e.g. σ_{yy}), and different for a shear stress (e.g. σ_{xz}). If the two arrows of a normal stress pull on the element, then the normal stress is called a tension; if they push, the normal stress is called a compression. By convention, tension is positive and compression is negative (**Figure 4.13**). The three normal stress components acting at any point (**Figure 4.12**) are independent of one another, and each may be tensile (positive), zero, or compressive (negative).

The six shear stresses acting at a point are not independent of one another. To conserve angular momentum the so-called conjugate shear stresses must be equal in magnitude:

$$\sigma_{xy} = \sigma_{yx},\ \sigma_{yz} = \sigma_{zy},\ \sigma_{zx} = \sigma_{xz} \quad \textbf{(4.12)}$$

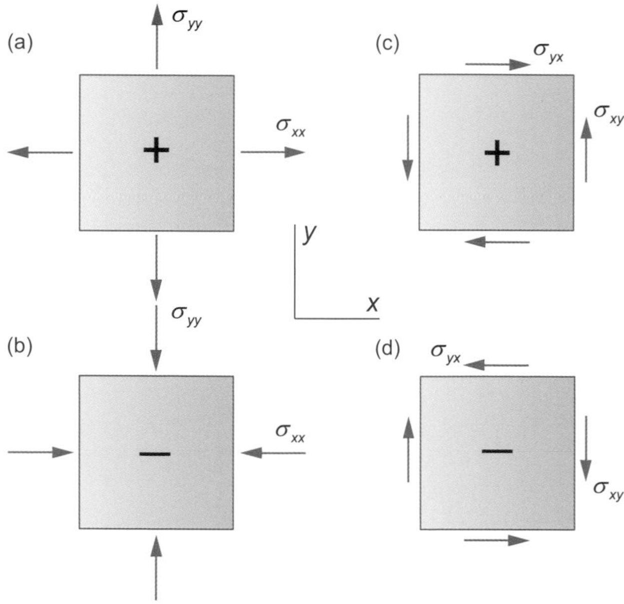

Figure 4.13 Sign conventions for stress components. Tensile normal stress is positive.

This is a restatement of Cauchy's second law that was derived in Section 3.8. In **Figure 4.13c**, σ_{yx} acting alone would induce a clockwise angular acceleration of the element, and σ_{xy} acting alone would induce a counter-clockwise acceleration. To satisfy the first equation in (4.12) these two shear stresses always act in concert with equal magnitudes, so the angular acceleration is zero. The relationships among the shear stress components expressed in (4.12) hold at all points in any deforming solid or flowing liquid. Because conjugate shear stresses are equal, they must have the same sign. By convention, the shear stresses as drawn in **Figure 4.13c** are positive and those in **Figure 4.13d** are negative.

Based on Cauchy's second equations of motion (4.12), the number of independent stress components is reduced from 9 to 6 for the general three-dimensional stress state (**Figure 4.12a**), and from 4 to 3 independent components for the general two-dimensional stress state (**Figure 4.12b**). The stress tensor components may be organized as the elements of a square matrix:

$$\underset{3\times 3}{[\sigma]} = \begin{bmatrix} \sigma_{xx} & \sigma_{xy} & \sigma_{xz} \\ \sigma_{yx} & \sigma_{yy} & \sigma_{yz} \\ \sigma_{zx} & \sigma_{zy} & \sigma_{zz} \end{bmatrix} \tag{4.13}$$

By (4.12), the matrix of stress components is a symmetric matrix: elements in symmetric positions across the principal diagonal are equal.

4.6.2 Principal Stresses

At a point in a deforming solid or flowing liquid, nine stress components act on a cubic element with edges that are parallel to the chosen Cartesian coordinate system (**Figure 4.12a**). These nine components are arranged in a symmetric matrix (4.13). If one chooses another orientation for the coordinate axes at that point, and considers a cubic element with edges parallel to those axes, the values of the stress components acting on the *reoriented* cubic element at the same point usually will change, but they are uniquely determined by the tensor transformation equations presented in Section 2.7. In other words, given the stress components for one orientation of the coordinate axes, we always can calculate the components for any other orientation. In this sense the stress tensor has an *existence* that is independent of the coordinate system, even though its components do depend on the coordinate system. This is analogous to the *existence* of a vector that is independent of the coordinate system: the vector components vary with the orientation of the coordinate system at a given point, but the vector itself does not change.

In a thought experiment, suppose you calculate the nine stress components at a given point for all possible orientations of the cubic element (**Figure 4.12a**). One special orientation would emerge (**Figure 4.14a**). In that orientation all shear stress components would be exactly zero, and two of the normal stress components would be extreme values. The normal stresses acting on this cubic element are the principal stresses, and they are named and ordered algebraically as:

$$\sigma_1 \geq \sigma_2 \geq \sigma_3 \tag{4.14}$$

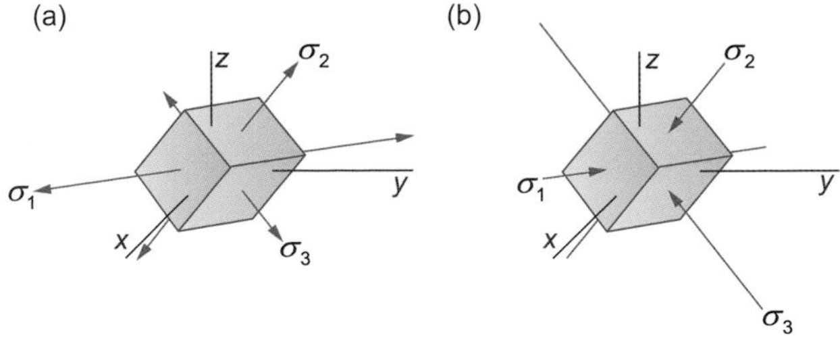

Figure 4.14 Cubic elements oriented to display the principal normal stresses: σ_1, σ_2, and σ_3. No shear stresses act on the faces of these elements. (a) All principal stresses are tensile: $[\sigma]$ is positive definite. (b) All principal stresses are compressive: $[\sigma]$ is negative definite.

Individually these are called the *greatest*, *intermediate*, and *least principal stress*. In general, all normal stresses acting on cubic elements of any orientation at a given point are bounded by the two extreme principal stresses, σ_1 and σ_3, and one or more principal stresses may be exactly zero.

The state of isotropic stress is the one exceptional case for the thought experiment described in the previous paragraph. For this case the normal stresses for *all* orientations of the cubic element are equal, $\sigma_{xx} = \sigma_{yy} = \sigma_{zz} = \sigma_i$, and the shear stresses are zero. For all orientations of the coordinate system at that point the stress matrix is:

$$\underset{3\times 3}{[\sigma]} = \begin{bmatrix} \sigma_i & 0 & 0 \\ 0 & \sigma_i & 0 \\ 0 & 0 & \sigma_i \end{bmatrix} \tag{4.15}$$

For this isotropic state of stress, the principal stress orientations are not defined, and the principal stress values are not distinguished from one another.

Two broad categories of stress states are recognized based on the *signs* of the principal stresses. Given the ordering of the principal stresses prescribed in (4.14), and the established sign convention (tension positive), it follows that if σ_3 is *tensile* (positive), then all three principal stresses are tensile at that point (**Figure 4.14a**). In this case the stress matrix (4.13) is referred to as positive definite. Furthermore, all normal stresses acting on cubic elements of any orientation at that point are tensile. It also follows that if σ_1 is *compressive* (negative), then all three principal stresses are compressive (**Figure 4.14b**). In this case the stress matrix (4.13) is negative definite, and all normal stresses acting on cubic elements of any orientation at that point are compressive. The state of stress may be neither positive definite nor negative definite. For these states of stress the signs of the extreme principal stresses are different, so σ_1 is *tensile* (positive) and σ_3 is *compressive* (negative). For these cases the intermediate principal stress, σ_2, may be tensile, zero, or compressive.

The three principal stresses always are oriented along three orthogonal axes (**Figure 4.14**), and each of these is referred to as a principal axis. We use the word *axis* rather than *direction*, because the principal stresses are not vectors, and the pair of arrows that represent a particular principal stress point in opposite directions. One direction is not preferred over the other, and the

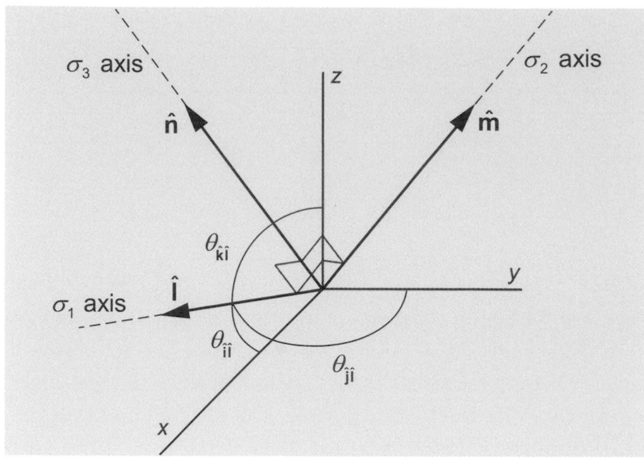

Figure 4.15 Unit vectors $\left(\hat{\mathbf{l}}, \hat{\mathbf{m}}, \hat{\mathbf{n}}\right)$ are parallel to the principal axes for the three principal stresses (σ_1, σ_2, σ_3). The direction angles for $\hat{\mathbf{l}}$, for example, are $\theta_{\hat{i}\hat{i}}, \theta_{\hat{j}\hat{i}}, \theta_{\hat{k}\hat{i}}$.

axis provides the *orientation* without having to specify a direction. On the other hand, when we calculate these orientations, we find three orthogonal unit vectors, $\left(\hat{\mathbf{l}}, \hat{\mathbf{m}}, \hat{\mathbf{n}}\right)$ that are parallel to the principal stress axes for (σ_1, σ_2, σ_3) at a given point:

$$\hat{\mathbf{l}} = \langle l_x, l_y, l_z \rangle, \hat{\mathbf{m}} = \langle m_x, m_y, m_z \rangle, \hat{\mathbf{n}} = \langle n_x, n_y, n_z \rangle \quad \textbf{(4.16)}$$

Because these unit vectors do have a direction, it is not uncommon for principal axes to be referred to as principal directions. Because the vectors in (4.16) are unit vectors, their components are the direction cosines. For example, l_y is the cosine of the angle $\theta_{\hat{j}\hat{i}}$, which is the smaller angle between the base vector $\hat{\mathbf{j}}$ for positive y-axis and the direction of the unit vector $\hat{\mathbf{l}}$, which is parallel to the principal axis of σ_1 (**Figure 4.15**).

The principal stress values and unit vectors directed parallel to the orientations of the principal stresses (**Figure 4.15**) are arranged in matrices that facilitate their computation and provide some insights about the state of stress. The matrix [V] of principal stress *values* is a *diagonal matrix*, which means all the off-diagonal elements are zero:

$$[\mathrm{V}] = \begin{bmatrix} \sigma_3 & 0 & 0 \\ 0 & \sigma_2 & 0 \\ 0 & 0 & \sigma_1 \end{bmatrix} \quad \textbf{(4.17)}$$

Note the order of the principal stresses in [V], varying from least to greatest from upper left to lower right along the principal diagonal. The matrix [O] of principal stress *orientations* is a full matrix:

$$[\mathrm{O}] = \begin{bmatrix} n_x & m_x & l_x \\ n_y & m_y & l_y \\ n_z & m_z & l_z \end{bmatrix} \quad \textbf{(4.18)}$$

The elements in each column of the matrix [O] are the direction cosines for the unit vector that is parallel to a particular principal axis (**Figure 4.15**). For example, the direction cosines in the third column of [O] are the components of the unit vector that define the orientation of σ_1. The square root of the sum of the squares of the direction cosines in each column of [O], which is the magnitude of the respective vector, must equal unity.

For the general triaxial stress state, the solution to the principal value problem is tedious to compute by hand, but MATLAB provides the function eig that calculates the principal values and principal axes. For example, suppose the stress matrix is [S]. The full matrix [O] of directions cosines (4.18), and the diagonal matrix [V] of principal values (4.17) are found using the following MATLAB statement:

$$[\mathrm{O}, \mathrm{V}] = \mathrm{eig}[\mathrm{S}] \quad \textbf{(4.19)}$$

This function works equally well for two-dimensional stress states. Calculation of the principal stress magnitudes and orientations, given the stress components, is a classic problem of mathematics called an eigenvalue problem. For these problems the matrices defined in (4.19) are related as: [S][O] = [O][V]. As a consequence of the stress matrix being *square*, with 3 rows and 3 columns, and *symmetric*, the principal values always are real numbers and the principal axes always are orthogonal. In summary, the state of stress at any point in a rock mass can be quantified by specifying the three principal stresses, along with the three unit vectors giving the principal stress orientations.

4.7 ELASTIC–BRITTLE DEFORMATION UNDER AXISYMMETRIC LOADING

In this section we describe laboratory tests in which stress is applied along the specimen axis and on its cylindrical surface to approximate conditions closer to those in Earth's lithosphere than the uniaxial tests described in Section 4.4.2. The most common apparatus used to apply stress in more than one direction on a rock specimen is shown schematically in **Figure 4.16**. This so-called triaxial pressure vessel is placed between the two platens of a universal testing machine (**Figure 4.6**), which provides the applied axial force, F, and induces the axial stress, $\sigma_a = -(F/A)$ in the cylindrical specimen with cross-sectional area, A. The fluid pressure in the region between the inner wall of the pressure vessel and the cylindrical jacket surrounding the rock specimen is called the confining pressure, P_c. The jacket, often rubber or malleable metal like copper, is impermeable to fluids and much more easily deformed than the specimen itself. The confining pressure provides the radial stress, $\sigma_r = -P_c$, which is equal in all radial directions.

For compression tests in a triaxial pressure vessel the principal stresses are $\sigma_1 = \sigma_2 = \sigma_r$ and $\sigma_3 = \sigma_a$. In other words, the axial stress is the least principal stress (greatest compression) and the specimen *shortens* in the axial direction. For extension tests the principal stresses are $\sigma_1 = \sigma_a$ and $\sigma_2 = \sigma_3 = \sigma_r$. For these tests the axial stress is the greatest principal stress (least compression) and the specimen *extends* in the axial direction. All normal stresses are equal that act in radial directions on the cylindrical specimens. In this regard the test should properly be called an axisymmetric test. However, the name *triaxial* is firmly embedded in the literature of rock mechanics, so we refer to these tests as conventional triaxial tests. To distinguish the important differences, a test with three independent and different principal stresses, $\sigma_1 > \sigma_2 > \sigma_3$, is called a

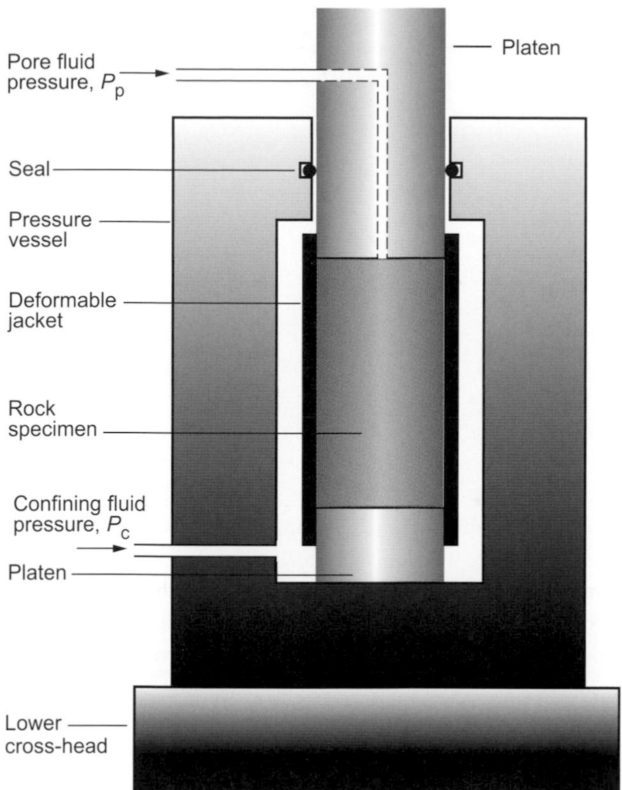

Figure 4.16 Schematic drawing of a pressure vessel used for conventional triaxial testing of cylindrical rock specimens. The pressure vessel is inserted between the platen and the lower cross-head of a universal testing machine (Figure 4.6). Modified from Paterson and Wong (2005).

true triaxial test. The true triaxial test provides the most general loading conditions and could, in principle, simulate any state of stress in Earth's lithosphere. However, the true triaxial test is technically very difficult to achieve in the laboratory. Most laboratory data that have been reported in the literature come from conventional triaxial compression tests, so we confine the discussion in the next section to those tests.

Figure 4.17 summarizes behaviors in axisymmetric (conventional triaxial) compression tests and serves to define important terms. The initial portions of all three stress–strain curves are linear and the strain is completely recovered upon unloading, so the rock is in an elastic state. The limit of purely elastic behavior is indicated by point A on the curve OABC. Thereafter, the slope decreases and unloading would reveal a permanent strain as illustrated by ε_p in **Figure 4.7**. From point A to point B on the curve OABC in **Figure 4.17** the rock is in a ductile state because the slope of the stress–strain curve is positive and permanent strain is accumulating. The extreme value (minimum) of axial stress is labeled as point B on this curve and defines the conventional triaxial compressive strength:

$$C_t \equiv |\min(\sigma_a)| \tag{4.20}$$

After point B the slope of the stress–strain curve is negative, so the specimen is losing its load carrying capacity, and is said to be in a brittle state.

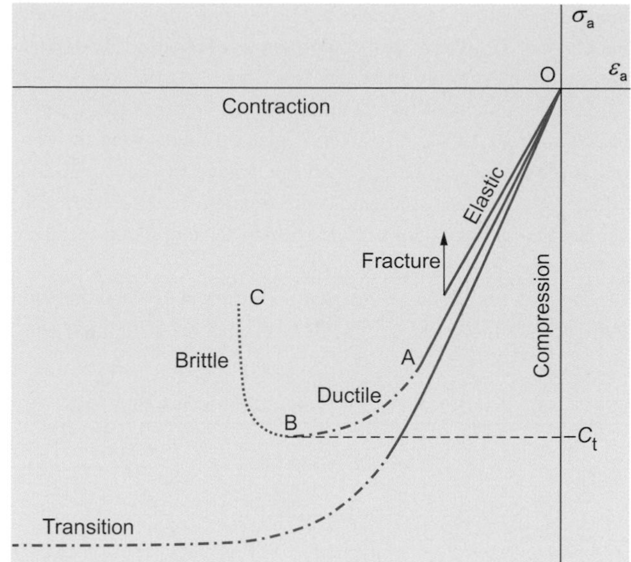

Figure 4.17 Schematic plot of axial stress, σ_a, versus axial strain, ε_a, for axisymmetric (conventional triaxial) compression tests. C_t is the compressive strength. For the curve O–C the state of the rock varies from elastic OA (solid line) to ductile AB (dash–dot line) to brittle BC (dotted line). See Jaeger et al. (2007) in Further Reading.

On **Figure 4.17** the curve with elastic deformation followed immediately by fracture is one end member of the generic behavior typified by the curve OABC. At this extreme the rock does not enter the ductile state and the brittle state is witnessed only by the dramatic loss of load carrying capacity associate with fracture. It is very difficult to capture the complete stress–strain curve for such brittle behavior, so the arrow figuratively indicates the stress going to zero and the strain being undefined. The other end member of the generic behavior OABC is illustrated by the curve marked *transition* which shows no degradation of stress, the slope is positive or zero throughout, and the test is terminated at a particular strain, because of the constraints of the testing machine and pressure vessel.

The variety of different relationships between stress and strain illustrated on **Figure 4.17** is collectively referred to as elastic–brittle deformation, the title of this chapter. Beyond the transition, for specimens that remain in the ductile state, tests do not end with fracture and the large permanent strain is associated with distributed deformation throughout the specimen, characteristic of elastic–ductile deformation, the subject of Chapter 5.

4.7.1 Stiffness and Strength in Conventional Triaxial Compression Tests

For Tennessee marble II under confining pressures that vary from 0 to 48.3 MPa (**Figure 4.18**), curves of axial stress versus axial strain change systematically in a manner that is typical for many rock types. The curves are slightly concave downward at very low stress, with the gradually increasing stiffness presumably reflecting closure of grain boundaries and small cracks. Curves then straighten out and maintain a nearly *uniform* tangent modulus of about 77 GPa to axial compressions of about –120 MPa. As the confining pressure increases from zero (the uniaxial test) to 48.3 MPa (equivalent to ~2 km depth), the compressive strength increases from 132 to 276 MPa. Apparently increasing the confining pressure has little effect on the tangent modulus, but does affect the compressive strength, presumably by retarding mechanisms that lead to failure such as micro-cracking and localized deformation at grain boundaries.

The tests on Tennessee marble II (**Figure 4.18**) approach the transition from brittle to ductile deformation with increasing confining pressure. In all cases the deformation is nearly linearly elastic (recoverable) up to axial stress of about –120 MPa. For the unconfined test, P_c = 0 MPa, the post-elastic deformation is dominantly brittle. From an axial stress of about –120 MPa to the compressive strength at –132 MPa, a very short concave upward segment is followed by a lone curve with a steep negative slope. In contrast, for a confining pressure of 48.3 MPa the post-elastic deformation approaches the *transition* to ductile behavior illustrated in **Figure 4.17**. From an axial stress of about –120 MPa to the minimum value at –276 MPa, a very broad concave upward segment with positive slope is followed by a nearly zero slope.

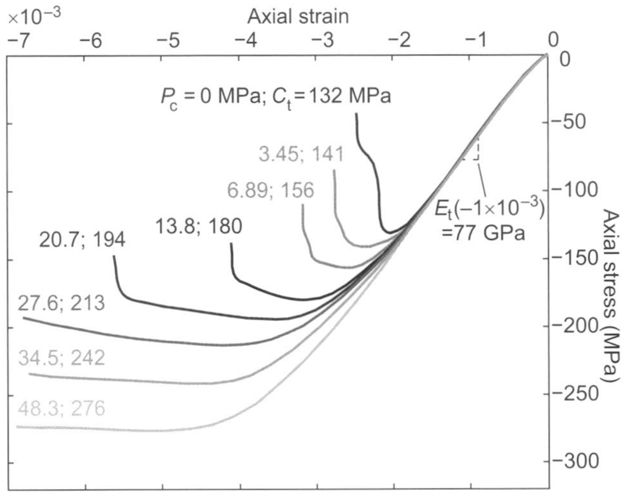

Figure 4.18 Conventional triaxial compression test data on Tennessee marble II plotted as axial stress versus axial strain. For each curve the confining pressures, P_c, and conventional triaxial compressive strengths, C_t, are given. The labeled quantities increase as the colors transition from violet to blue. E_t is the tangent modulus. Modified from Wawersik and Fairhurst (1970).

For confining pressures less than 13.8 MPa, failure of Tennessee marble II is predominantly by axial splitting fractures and wedge fractures (**Figure 4.19a**). For greater confining pressures the style of failure changes to shearing on two sets of inclined fractures that spread throughout the specimens (**Figure 4.19b, c**). This failure geometry is referred to as conjugate shear fractures. Clearly the style of brittle deformation changes and the compressive stress carrying capacity of the samples increases (**Figure 4.18**), but does the increase in compressive strength vary systematically with confining pressure? This would suggest that compressive strength increases systematically with depth in Earth's lithosphere. We address this question in the next section.

4.7.2 Strength Variations with Confining Pressure

The effect of confining pressure on strength is illustrated using conventional triaxial test data for a suite of rocks from the Tertiary basins of Japan, consisting primarily of claystone, siltstone, shale, and sandstone along with a few volcanic rocks (**Figure 4.20**). These results are recorded graphically on a plot of differential stress, $\Delta\sigma$, versus axial strain, ε_a. The differential stress, commonly used in rock mechanics tests, is a *scalar* quantity equal to the difference between the axial and radial stress:

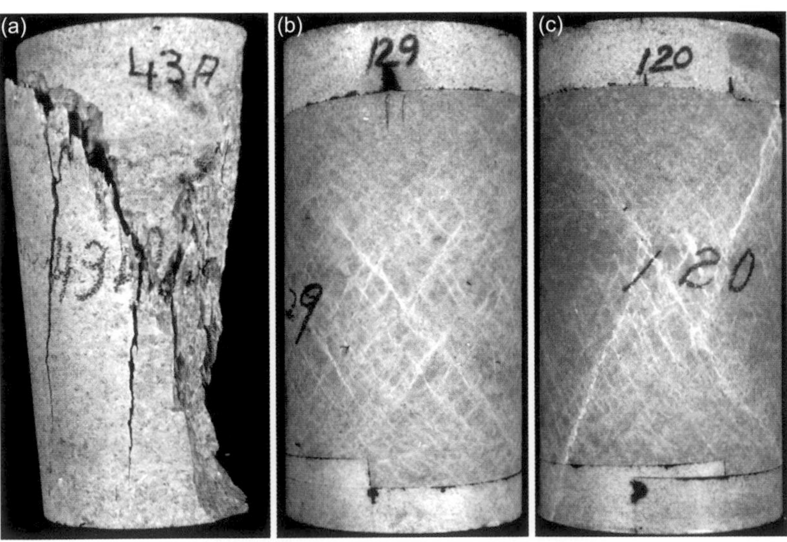

Figure 4.19 Photographs of failure styles in Tennessee marble II cylinders with two inch diameters. (a) uniaxial tests with axial splitting and wedge fractures. (b) Conventional triaxial test at P_c = 13.8 Pa with two sets of shear fractures. (c) Conventional triaxial test at P_c = 34.5 MPa with two sets of shear fractures. From Wawersik (1968).

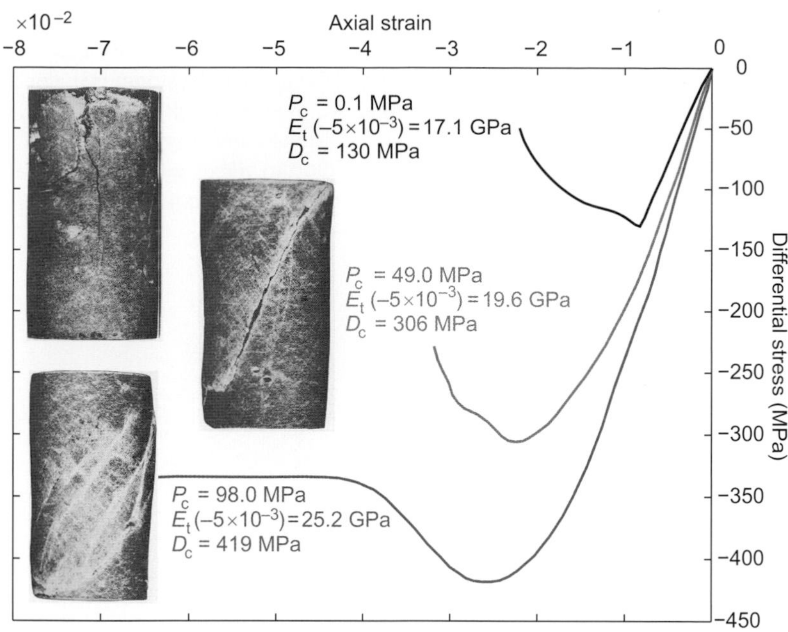

Figure 4.20 Differential stress versus axial strain for Ohtawa basalt at three different confining pressures, P_c. Images of specimens after tests display three different styles of brittle deformation described in the text. The tangent modulus, E_t, and differential strength in compression, D_c, are indicated for each confining pressure. Modified from Hoshino et al. (1972).

$$\Delta\sigma = \sigma_a - \sigma_r \tag{4.21}$$

The extreme value of the differential stress in a compression test is called the differential strength:

$$D_c \equiv |\min(\Delta\sigma)| \tag{4.22}$$

The differential strengths were evaluated at room temperature on dry samples and the confining pressures ranged from 0.1 MPa (atmospheric pressure) to 245 MPa (equivalent to about 10 km depth).

Figure 4.20 shows differential stress versus axial strain curves for Ohtawa basalt with confining pressures corresponding to depths of approximately 0 km, 2 km, and 4 km. The differential strengths increase from 130 to 419 MPa over this range of confining pressures. The behavior is nearly linear and elastic to axial strains of -0.5×10^{-2}. The tangent moduli at this strain increase from 17.1 to 25.2 GPa with the increase in confining pressure. The curves for Ohtawa basalt near the differential strength minimum broaden considerably with increasing confining pressure, and the negative slopes after passing the minimum are indicative of *brittle* behavior in all cases.

Three different styles of brittle failure for the Ohtawa basalt are shown in the photographic insets on **Figure 4.20**. Wedge fractures and axial splitting fractures develop at P_c = 0.1 MPa. A prominent inclined shear fracture with a background of shear deformation bands occurred at P_c = 49 MPa. A broadly distributed network of inclined shear deformation bands developed at P_c = 98 MPa, as well as a through-going shear fracture. The shear deformation bands are in two orientations, inclined at nearly equal but opposite angles to the specimen axis. The three different styles of brittle failure illustrated here are representative of those observed throughout the study of 100 different sedimentary and volcanic rocks. Although the transitions from one style to another occurred at different confining pressures for different rocks, there was a systematic progression from wedge fractures and axial splitting, to inclined shear fracture, to networks of conjugate deformation bands.

The effects of confining pressure on compressive strength are summarized in **Figure 4.21**, which plots data for six sedimentary rocks from the Japanese study on a graph of confining pressure, $\sigma_1 = \sigma_2 = -P_c$, versus axial compression *at failure*, $\sigma_3 = -C_t$. The uniaxial and conventional triaxial compressive strengths for two shales, claystone, fine-grained sandstone, and two medium-grained sandstones are included for confining pressures as great as 245 MPa, representative of depths as great as 10 km. The uniaxial compressive strengths plot along the right edge of the graph and range from 46 to 174 MPa. The strengths of two medium sandstones (XB and XC), both taken from the Maze formation at the same locality, differ by more than a factor of two, from 46 to 113 MPa. Presumably this difference reflects subtle differences in the constituents or weathering of these sandstones that are *not* reflected in their lithologic and formation names.

In general, the conventional triaxial compressive strengths increase with confining pressure (**Figure 4.21**), and the data for a particular rock (not including the uniaxial compressive strength) is reasonably approximated with a *straight* line. This is consistent with the Coulomb criterion, which is introduced in Section 4.8.2. However, straight lines that fit the conventional triaxial data do not intersect the ordinate at the uniaxial compressive strength. Presumably this reflects the change in failure mechanism from axial splitting and wedge fracture for the unconfined tests, to inclined shear fractures and networks of shear deformation bands for the confined tests (**Figure 4.22**).

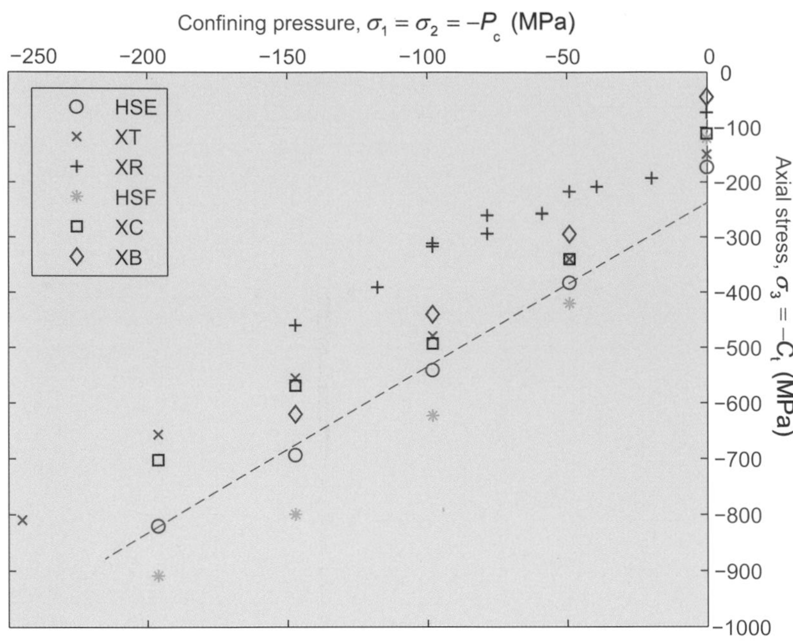

Figure 4.21 Conventional triaxial compressive strength for sedimentary rocks from Japan. HSE shale; XT shale; XR claystone; HSF fine sandstone; XC medium sandstone; XB medium sandstone. Least principal stress (ordinate) is plotted versus greatest principal stress (abscissa) in the third quadrant. Data from Hoshino et al. (1972).

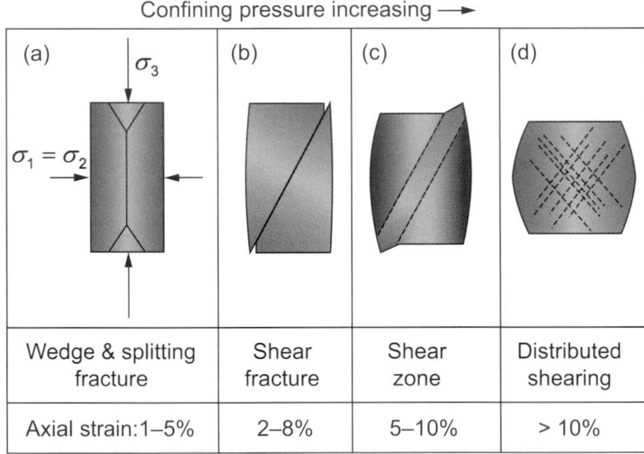

Wedge & splitting fracture	Shear fracture	Shear zone	Distributed shearing
Axial strain:1–5%	2–8%	5–10%	> 10%

Figure 4.22 Schematic drawings of failure styles in conventional triaxial compression tests. Confining pressure and axial strain (%) increase from left to right. Frames (a) and (b) depict brittle behavior; frame (c) depicts a transition to ductile behavior; and frame (d) is representative of ductile behavior. Modified from Griggs and Handin (1960).

In summary, the magnitude of the confining pressure in laboratory tests plays an important role in determining the *mechanisms* that govern compressive strength and the styles of brittle failure. Strength increases systematically with confining pressure and therefore with depth in Earth's lithosphere. Some use the term *fault* to describe the inclined shear fractures produced in laboratory tests, because of their apparent correspondence to geological faulting. We suggest using the term shear fracture in the laboratory context, and the term fault in the geological context, to clearly distinguish what are likely to be two different phenomena. This cautionary suggestion stems from the fact that the length and time scales may be quite different from laboratory to Earth's crust. Also, laboratory specimens usually are chosen to be free of the compositional and structural heterogeneities that often play roles in the development of faults in nature.

In the next section we explore how strength also varies with the magnitude of the fluid pressure in the rock pores, a phenomenon that can mitigate to some extent the increase in strength with depth. This helps us to understand how brittle fracture and faulting can occur at great depth in Earth's lithosphere.

4.7.3 Strength Variations with Pore Fluid Pressure

The effect of fluid pressure on the strength of soils that are saturated with water was elucidated by the soil scientist Karl Terzaghi in the middle of the twentieth century. He discovered that failure of soil samples in the laboratory depended upon a state of stress that was modified by the pore fluid pressure. His concept, called *effective* stress, describes how some of the load in the solid "skeleton" of a saturated soil is borne by the pore fluid. Laboratory tests showed that the effective stress concept applies when the pores are homogeneously and pervasively distributed at a fine scale, and interconnected such that local changes in fluid pressure during deformation return rapidly to a hydrostatic state by flow of the fluid within the pores. Porous and permeable rocks, such as sandstone, satisfy these requirements. In this section we define effective stress and demonstrate Terzaghi's concept using rock mechanics data (see Terzaghi, 1943, in Further Reading).

A port on the conventional triaxial pressure vessel (**Figure 4.16**) supplies a fluid under pressure, called the pore fluid pressures, P_p, directly to the end of the specimen, and thence to the interconnected pores of the rock. The pore pressure is less than the confining pressure, $P_p < P_c$, so the pore pressure does not inflate the jacket and stays within the rock pores. The effective stress $[\sigma^e]$ is defined by adding the pore fluid pressure, P_p, to the normal stress components and leaving the shear stress components unchanged:

$$[\sigma^e] = \begin{bmatrix} \sigma^e_{xx} & \sigma^e_{xy} & \sigma^e_{xz} \\ \sigma^e_{yx} & \sigma^e_{yy} & \sigma^e_{yz} \\ \sigma^e_{zx} & \sigma^e_{zy} & \sigma^e_{zz} \end{bmatrix} = \begin{bmatrix} \sigma_{xx} + P_p & \sigma_{xy} & \sigma_{xz} \\ \sigma_{yx} & \sigma_{yy} + P_p & \sigma_{yz} \\ \sigma_{zx} & \sigma_{zy} & \sigma_{zz} + P_p \end{bmatrix} \quad (4.23)$$

Recall that the sign convention established for stress components treats compressive stress as *negative*, and pressure is inherently *positive*. Thus, for example, at 1 km depth the vertical normal stress would be σ_{zz} = –25 MPa, the hydrostatic pore pressure would be P_p = 10 MPa, and the effective stress would be $\sigma^e_{zz} = \sigma_{zz} + P_p = -15$ MPa. In this way a *greater* pore pressure produces an effective stress that is *less* compressive.

Terzaghi's hypothesis is that volume change and strength depend upon the effective stress state. Two experimental observations, carried out using conventional triaxial procedures, have been used to test Terzaghi's hypothesis. The first observation comes from tests in which the pore pressure is zero, P_p = 0 MPa, and both the axial and radial stress are compressions, equal to the negative of the confining pressure: $\sigma_a = \sigma_r = -P_c$. For tests in the elastic range, the volume of the rock specimen changes in proportion to changes in the confining pressure. On the other hand, if the pore pressure is changed at the same rate as the confining pressure, the specimen volume does not change appreciably. These tests demonstrate that changes in the volume of rock specimens is dependent upon the effective confining pressure: $P^e_c = P_c - P_p$.

The second experimental observation is that the strength of rock specimens does not increase appreciably if the pore pressure is increased in concert with the confining pressure. We use conventional triaxial strength data for Berea Sandstone to illustrate this phenomenon (**Figure 4.23a**). Confining pressures were varied from 0 MPa to 200 MPa, representative of depths as great as 8 km. The tests were conducted at room temperature and pore pressures from 0 to 175 MPa. Each data point represents the principal stresses imposed on the specimen at failure. For zero pore pressure the strength increased from about 70 MPa to 630 MPa with increases in confining pressure from 0 MPa to 200 MPa. By increasing the pore pressure to 175 MPa at a confining pressure of 200 MPa, the strength decreased to about 340 MPa. Thus, increased pore pressure tends to mitigate the increase in strength due to increased confining pressure.

If Terzaghi's hypothesis has merit, the strength of Berea Sandstone should be the same for all specimens tested at the same *principal* effective stresses (4.23). To test this we re-plot the data using effective principal stresses. In **Figure 4.23b** the effective greatest principal stress, $\sigma^e_1 = \sigma_1 + P_c$ is plotted on the abscissa: this is equivalent to the negative of the effective confining pressure, $\sigma^e_1 = -P^e_c$. The effective least principal stress, $\sigma^e_3 = \sigma_3 + P_p$, is plotted on the ordinate: this is equivalent to the sum of the axial stress and pore pressure. The scattered data points from **Figure 4.23a** collapse onto a single straight line when plotted using effective principal stresses. Furthermore, at a given effective confining pressure, the strengths are the same, regardless of the actual confining pressure and pore pressure.

For example, the cyan asterisk in **Figure 4.23a** represents a test where P_c = 200 MPa and P_p = 150 MPa, yielding an effective confining pressure P^e_c = 50 MPa. To plot this point using effective principal stresses, the pore pressure is added to each principal stress, so the cyan asterisk moves 150 MPa to the right and 150 MPa up (see dashed arrows on **Figure 4.23a**). The two green x's in **Figure 4.23a** represent tests where P_c = 100 MPa and P_p = 50 MPa, also yielding an effective confining pressure P^e_c = 50 MPa. When plotted using effective principal stresses (**Figure 4.23b**), the cyan asterisk and the two green x's all move to a small cluster of points with a common effective confining pressure. All data points at a given effective confining pressure have essentially the same strength, consistent with Terzaghi's hypothesis.

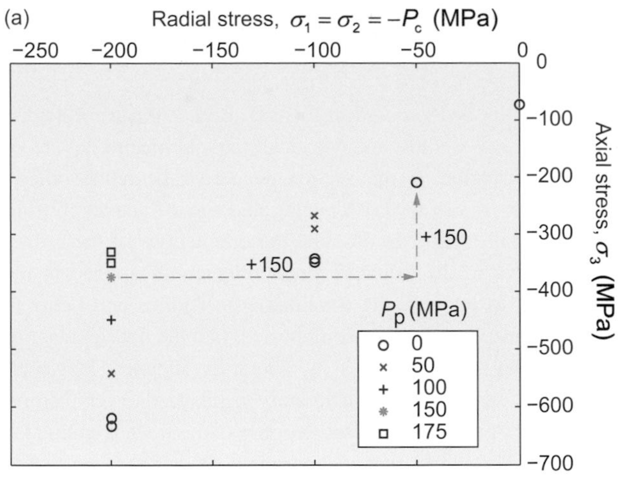

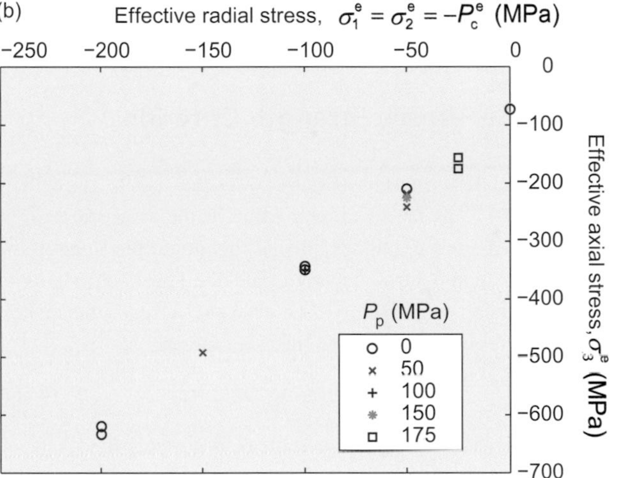

Figure 4.23 (a) Plot of principal stresses at failure during conventional triaxial tests on Berea Sandstone at a variety of pore pressures. (b) Plot of principal *effective* stresses at failure using same data as (a). See text for explanations. Data from Handin et al. (1963).

4.8 THE STATE OF STRESS DURING OPENING AND SHEAR FRACTURE

The limestone bed exposed at Lilstock Beach (**Figure 4.4**) displays a myriad of joints and one small fault, all manifestations of brittle deformation. In this section we derive theoretical criteria for the state of stress when opening and shear fractures form in the laboratory. Also, we describe the orientations of these fractures relative to the principal stress orientations applied in the laboratory. In general, the development of joints and faults in Earth's lithosphere is more complicated than the development of opening and shear fractures in laboratory samples. However, theoretical criteria based on laboratory results lead to very helpful insights, and provide important stepping stones toward criteria for jointing and faulting that we introduce in later chapters. For example, we offer more complete and nuanced descriptions and theoretical criteria for joints in Chapter 7 and for faults in Chapter 8.

4.8.1 The Tensile Strength Criterion

For uniaxial tension tests (**Figure 4.8**) the tensile strength is defined in (4.7) as the maximum value of the axial stress: $T_u = \max(\sigma_a)$. If the test continues beyond this point, the slope of the stress–strain curve turns negative, and the specimen begins to lose its load carrying capacity. The axial stress equals the greatest principal stress, $\sigma_a = \sigma_1$, so the fracture criterion is:

$$\sigma_1 = T_u \text{ (opening fracture)} \quad \textbf{(4.24)}$$

Typically, cylindrical specimens break into two parts along an opening fracture that is nominally perpendicular to the specimen axis (**Figure 4.8** inset). Therefore, this criterion stipulates that opening fractures form perpendicular to σ_1 (**Figure 4.24a**). In a multi-axial stress state the opening fracture would be parallel to the other two principal stresses, however, the magnitudes of σ_2 and σ_3 play no role in this criterion. That is not to say these principal stresses play no role in the initiation and propagation of a joint or vein in Earth's lithosphere. We return to that possibility in Chapter 7.

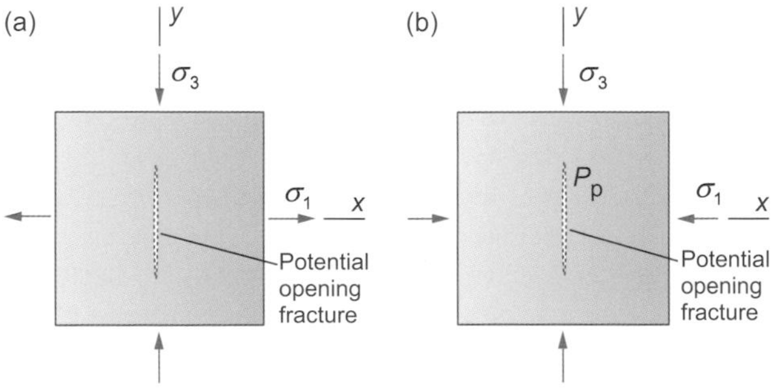

Figure 4.24 (a) Orientation of opening fracture relative to the principal stresses according to (4.24). (b) Orientation of opening fracture relative to the principal stresses according to (4.25); P_p is pore fluid pressure. The potential opening fractures are perpendicular to the greatest principal stress, σ_1.

If groundwater or hydrocarbons saturate the rock pores, and permeability is sufficient, Terzaghi's concept (4.23) may be used to re-write (4.24) in terms of the effective principal stress:

$$\sigma_1 + P_p = \sigma_1^e = T_u \text{ (opening fracture)} \quad \textbf{(4.25)}$$

This introduces the important possibility that opening fractures may form even when the greatest principal stress is compressive (negative). The pore fluid pressure must be great enough to overcome that compression *and* the tensile strength of the rock. The criterion (4.25) also places an upper bound on the magnitude of the pore fluid pressure, $P_p < T_u - \sigma_1$, because the fluid pressure would be relieved by expansion of the fluid into the opening joints when the criterion is satisfied.

The stress state in cylindrical laboratory specimens is designed to be *homogeneous*, in which case the principal stresses don't vary in magnitude or orientation. In contrast, the stress state in Earth's lithosphere may be heterogeneous. However, if the orientation of σ_1 varies smoothly and continuously with position, the tensile strength criteria, (4.24) and (4.25), are consistent with opening fractures forming along curved paths that everywhere are perpendicular to the local orientation of σ_1. Many examples of smoothly turning traces of joints are seen on the limestone bedding surface at Lilstock Beach (**Figure 4.4**). From these curved traces we infer how the local principal stress orientations varied as the joints formed.

A curved line that depicts the spatially varying direction of a principal stress is called a principal stress trajectory. In a two-dimensional stress state a set of trajectories represents the orientations of σ_1, and a second set represents the orientations of σ_3. An example of a set of σ_3 trajectories is shown in **Figure 4.25**. This illustration is taken from a laboratory experiment using

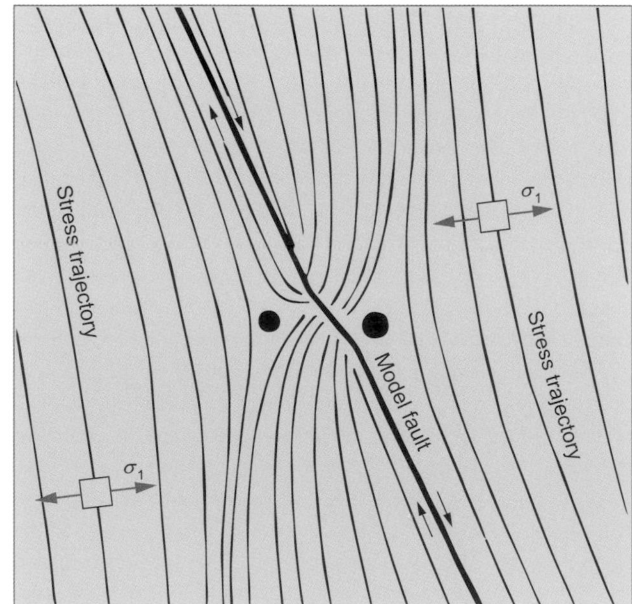

Figure 4.25 Result of laboratory photoelastic experiment that shows principal stress trajectories near a model fault with a geometric kink. From Petit et al. (2000).

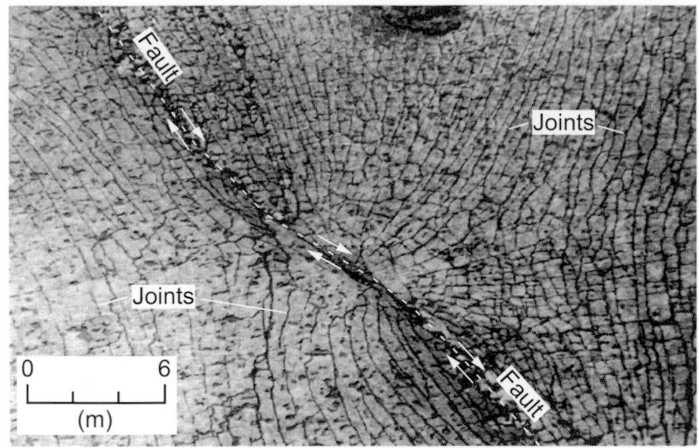

Figure 4.26 Aerial photograph of right-lateral fault with a geometric kink and nearby joint pattern at Nash Point, UK. Approximate location UTM: 30 U 461582.90 m E, 5694549.72 m N. From Petit et al. (2000). See also Rawnsley et al. (1998).

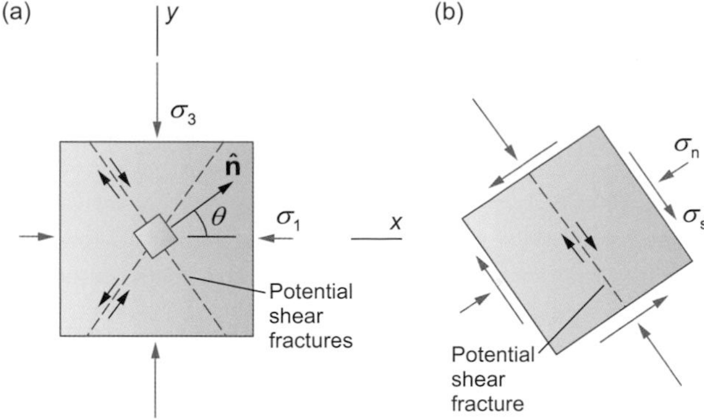

Figure 4.27 Schematic illustration used to derive the Coulomb criterion for brittle shear fracture. (a) Homogeneous principal stresses, σ_1 and σ_3, with potential shear fracture planes (dashed lines). (b) Normal and shear stresses, σ_n and σ_s, acting on potential shear fracture plane. Direction angle between greatest principal stress axis and unit normal vector $\hat{\mathbf{n}}$ is θ. Modified from Jaeger et al. (2007).

photoelastic techniques, which reveal the trajectories of σ_3 near a kink in a model fault. Recall that σ_1 is orthogonal to σ_3 at any point, so these trajectories should approximate the pattern of joint traces near such a kink in a fault, if (4.24) is a valid fracture criterion.

The prominent set of joints on the limestone bedding surface exposed at Nash Point, UK, exhibits a curving pattern adjacent to a small fault (**Figure 4.26**). The fault is not straight, but has a kink that apparently inhibited slip and distorted the pattern of local principal stresses. The pattern of joints is similar to the stress trajectories (**Figure 4.25**) from the photoelastic laboratory experiment. The correspondence between the stress trajectories and the joint pattern supports the interpretation that the joints formed in a stress field perturbed by slip on the kinked fault such that the trajectories are curved.

4.8.2 The Coulomb Shear Strength Criterion

In conventional triaxial compression tests at modest confining pressures, rock fails by the development of a shear fracture that is inclined to the specimen axis (e.g. **Figure 4.19** and **Figure 4.20**). Because these fractures are not parallel to a principal stress plane, both normal and shear stresses are resolved on the *would-be* shear fracture plane before failure. Therefore, a criterion for shear fracture must take both shear and normal stress into account. Such a criterion is attributed to Charles-Augustin de Coulomb, who studied the failure characteristics of engineering materials in the latter half of the eighteenth century. The basic hypothesis is that shear fracture is driven by the applied shear stress and resisted by a combination of the shear strength and the normal compressive stress acting across the potential fracture plane.

To understand Coulomb's criterion, consider a rock mass subject to a homogeneous state of stress (**Figure 4.27**). The intermediate principal stress, σ_2, plays no role in this criterion, so the figure shows the plane containing the greatest and least principal stresses, σ_1 and σ_3. The orientation of a potential shear fracture (dashed lines) is specified by a unit normal vector $\hat{\mathbf{n}}$ that makes an angle θ with the axis of the greatest principal stress. The symmetry of the stress tensor admits two orientations of shear fractures that are symmetrically inclined to σ_1.

The Coulomb criterion is a linear relationship between the magnitude of the shear stress, $|\sigma_s|$, and the normal stress, σ_n, acting on potential shear fracture planes (**Figure 4.27b**):

$$|\sigma_s| = S_o - \mu_i\sigma_n \quad \text{(shear fracture)} \tag{4.26}$$

Thus, the absolute value of the shear stress that could be sustained increases linearly as the normal stress becomes more compressive (more negative). The material constant S_o is the inherent shear strength, a measure of the resistance to shear fracture when the normal stress is zero. The absolute value of the shear stress is used in (4.26) because the development of shear fractures should not depend on the sense (sign) of shearing. The effect of the normal stress is modified by a constant, μ_i, called the coefficient of internal friction. The adjective "internal" is used because surfaces in frictional contact do not exist prior to fracture initiation. This criterion is applied across many disciplines including rock mechanics (**Box 4.1**), engineering geology (**Box 4.2**), solid earth geophysics, and structural geology.

If fluids such as groundwater or hydrocarbons saturate the rock pores, we use Terzaghi's concept (4.23) to re-write Coulomb's criterion in terms of the effective stress:

$$|\sigma_s| = S_o - \mu_i(\sigma_n + P_p) = S_o - \mu_i(\sigma_n^e) \quad \text{(shear fracture)} \tag{4.27}$$

In the stress analysis that follows most relationships are derived without including pore fluid pressure, but it should be understood that adding P_p to all normal stresses puts these relationships in the context of effective stresses.

Practical applications of Coulomb's criterion (4.26) are made by writing the normal and shear stresses, σ_n and σ_s, acting on the potential fracture planes (**Figure 4.27b**) in terms of the two

Box 4.2 Engineering Geology

Engineering geology is the study of rocks and geologic processes as they relate to engineering, especially civil and environmental engineering. Engineering geology focuses on the application of geology to the planning, design, construction, and monitoring of engineering projects constructed on or near Earth's surface. Engineering geology addresses rocks of all types and many geologic processes, but focuses on processes operating in the shallow lithosphere. Engineering geologists characteristically are generalists, but typically are required to apply skills in structural geology, geomorphology, petrology, stratigraphy, soil mechanics, and hydrogeology. They are involved in topics as diverse as the siting and construction of roads, dams, and nuclear power plants; the stability of slopes (**Figure B4.2**); the recognition and characterization of active faults; hazardous waste disposal; and underground construction. The practice of engineering geology is intertwined with the disciplines of environmental geology, geological engineering, and geotechnical engineering; all four disciplines benefit from collaboration and cooperation among geologist and engineers (Cosgrove and Hudson, 2016; see Further Reading).

Engineering geology emerged as a discipline in the late nineteenth century. In the first book titled *Engineering Geology*, published in 1880, the British geologist William Penning, described general uses of geology in civil engineering; this book emphasized geologic surveying and the collection of structural data in the field. In the first half of the twentieth century, Charles Berkey established the value of engineering geology in the construction of aqueducts, tunnels, bridges, and dams in the United States. Karl Terzaghi, a native Austrian widely regarded as the father of soil mechanics, promoted the value of engineering geology in the mid-twentieth century around the world. George Kiersch and Richard Jahns, two prominent American academic engineering geologists active in the mid-to-late twentieth century, contributed substantially to investigations of topics as diverse as dam failure, faulting hazards, slope stability, mineral supplies, as well as to geologic

Figure B4.2 Photograph of the 2018 landslide near Cusco, Peru courtesy of Galeria del Ministereio de Defensa del Perú (annotation added, www.flickr.com/photos/ministeriodedefensaperu/39935939755/in/dateposted/). The landslide reportedly spanned an area of about 300,000 m^2 and destroyed more than 100 structures. The slide mass is displaced down and to the right of the head scarp. The fissures to the right of the head scarp reflect deformation within the slide mass, and the toe marks the distal extent of the slide.

education, the editing of landmark books on geology, and professional service. The junior author of this textbook was fortunate to have been a student of Richard Jahns.

The mid-twentieth century was a boom period for the construction of major projects such as nuclear power plants, interstate highways, and large dams; several big dams also failed then. During this period of accelerating activity in the practice of engineering geology, publications addressing engineering geology began to flourish. For example, in 1962, the Geological Society of America began to publish a series of reviews of diverse topics in engineering geology; these reviews include many illuminating case histories. Among the prominent journals founded in the 1960s are the *Association of Engineering Geologists Bulletin* (now *Environmental & Engineering Geology Journal*), established in 1964; the *International Journal of Rock Mechanics and Mining Sciences & Geomechanics Abstracts* (now the *International Journal of Rock Mechanics and Mining Sciences*), established in 1964; and *Engineering Geology*, established in 1965. Contributions to rock mechanics and applied aspects of structural geology are prominent in these publications.

Engineering geology intersects structural geology when it addresses how rocks serve as a building material or a foundation. Rock mass strength on a variety of scales thus is a central issue for engineering geology, along with how fluid pressures and fractures affect rock mass strength. Specific topics in this book of particular relevance to engineering geologists include: (a) effects of pore pressure and confining pressure on rock strength (Sections 4.4 and 4.7); (b) laboratory strength measurements (Section 5.4); (c) characterization of slope stability by measurement of strain (Section 5.5); (d) fracture characterization (Section 7.1); (e) fracture mechanics (Sections 7.3 and 7.4); and (f) the kinematics of faulting (Section 8.5). Moreover, the general quantitative approach taken in this text is broadly relevant to engineering geologists.

extreme principal stresses, σ_1 and σ_3. This is accomplished using Cauchy's formula, which was introduced in Section 2.6.5. Here, we use the two-dimensional version of Cauchy's formula, which is derived in Section 7.2.2, but is given here without derivation:

$$\begin{aligned}\sigma_n &= \tfrac{1}{2}(\sigma_1+\sigma_3)+\tfrac{1}{2}(\sigma_1-\sigma_3)\cos 2\theta\\ \sigma_s &= -\tfrac{1}{2}(\sigma_1-\sigma_3)\sin 2\theta\end{aligned} \tag{4.28}$$

The normal and shear components of stress acting on potential shear fractures are related to the sum and difference of the extreme principal stresses, and to the orientation of the normal to fracture plane, specified by θ.

The Coulomb criterion (4.26) is based on the hypothesis that shear stress and normal compressive stress, respectively, work *for* and *against* shear fracture initiation. To evaluate the competition between these stresses, we rearrange the criterion such that $S_o = |\sigma_s| + \mu_i\sigma_n$, and then find the orientation of planes that require the greatest inherent shear strength, S_o, to resist shear fracture initiation. These planes are parallel to the potential shear fractures. Using (4.28) in (4.26) we find:

$$S_o = \pm\tfrac{1}{2}(\sigma_1-\sigma_3)\sin 2\theta + \mu_i\left[\tfrac{1}{2}(\sigma_1+\sigma_3)+\tfrac{1}{2}(\sigma_1-\sigma_3)\cos 2\theta\right] \tag{4.29}$$

The positive sign is used for θ in the first quadrant, and the negative sign is used for θ in the second quadrant. The greatest inherent shear strength is found using $dS_o/d\theta = 0$. This gives $\pm\cos 2\theta - \mu_i \sin 2\theta = 0$. Solving for the angle and calling it the Coulomb angle, θ_c, we find:

$$\theta_c = \tfrac{1}{2}\tan^{-1}\left(\frac{1}{\pm\mu_i}\right) \tag{4.30}$$

The orientations of potential shear fractures depend only on the internal friction, μ_i.

In evaluating the Coulomb angle using (4.30), the positive sign is associated with $0 < \theta_c \leq 45°$, and the negative sign is associated with $135° \leq \theta_c < 180°$ (**Figure 4.27a**). These two ranges account for the two sets of conjugate shear fractures. Said another way, often quoted in the literature of structural geology, the two sets of potential shear fractures form an acute angle, bisected by the axis of most compressive stress, σ_3. These fractures intersect in the axis of the intermediate principal stress, σ_2. In this way the orientations of all three principal stresses may be estimated from field observations of conjugate shear faults. For example, in **Figure 4.28** the greatest compression, σ_3, was nearly vertical; the least compression, σ_1, was approximately horizontal and parallel to the plane of the outcrop; and the intermediate principal stress, σ_2, was parallel to the line of intersection of the conjugate faults.

Although comparing the conjugate geometry of the faults in the A-1 sandstone (**Figure 4.28**) to that of the potential shear fractures according to the Coulomb criterion (**Figure 4.27a**) appears to be compelling, we have encountered very few field examples where such a comparison stood up to detailed scrutiny. In some field cases both conjugate fracture sets have the characteristics of opening fractures (joints), with no evidence of shearing (Section 7.1.1). In other cases shearing apparently occurred after the fractures initiated as joints, so the initiation process was not associated with shearing (Section 8.2.1). In still other cases members of one "conjugate" set propagated as opening fractures from members of the older set as the older set sheared (Section 7.1.4). These examples are not presented as an argument for abandoning the Coulomb criterion, but rather as a caution that it should be applied after detailed scrutiny of the field evidence.

Laboratory values of the coefficient of internal friction for seven different rocks are given in **Table 4.3** along with the corresponding Coulomb angle, θ_c, in the first quadrant. Values

Table 4.3 Laboratory values of coefficient of internal friction and angle of potential shear fracture (from Brace, 1964; Jaeger and Hoskins, 1966).

Rock type	μ_i	θ_c
Frederick diabase	1.7	15°
Westerly granite	1.4	18°
Witwatersrand quartzite	1.0	23°
Bowral trachyte	1.0	23°
Cheshire quartzite	0.9	24°
Carrara marble	0.7	28°
Gosford sandstone	0.5	32°

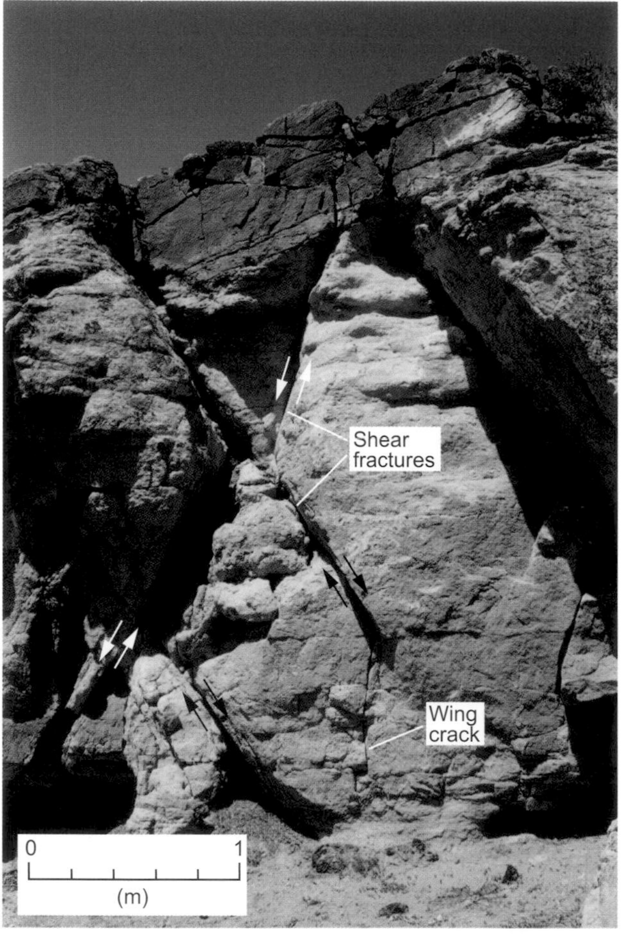

Figure 4.28 Conjugate shear fractures cutting the A-1 sandstone of the Frontier Formation on Emigrant Gap anticline, near Casper, WY. See Bergbauer and Pollard (2004). UTM: 13 T 362450.24 m E, 4757280.68 m N.

of μ_i range from 1.7 to 0.5, and these correspond to Coulomb angles from 15° to 32°. The lesser angles are associated with crystalline rocks and the greater angles are associated with sedimentary rocks.

4.9 DISPLACEMENT AND STRAIN FIELDS DURING BRITTLE DEFORMATION

The joints and faults exposed at Lilstock Beach (**Figure 4.29**) provide direct evidence relating to the kinematics of brittle deformation. By kinematics we mean the motion (*displacement*) and relative motion (*strain*) of rock particles. The objective of this section is to define the relationship between displacement and strain, and to establish the equations relating these kinematic quantities in a form that is relevant to elastic–brittle deformation.

The outcrop in **Figure 4.29** displays two types of structures indicative of brittle fracture: joints and faults. These structures are distinguished by the displacement directions of originally neighboring particles on the two sides of the structure. For a joint, the particles displace in opposite directions, *perpendicular* to the joint plane. Originally neighboring particles on the two sides of a fault also displace in opposite directions, but the directions are *parallel* to the fault plane. The kinematics of both joints and faults are characterized by a displacement discontinuity. That is, the otherwise continuous displacement field is interrupted by two nearly contiguous surfaces between which the direction of displacement vectors abruptly changes. For joints the displacement discontinuity is associated with *opening*; for faults the displacement discontinuity is associated with *shearing*. That is, the two surfaces move apart for a joint and they move parallel to one another for a fault. These discontinuities in the displacement field may be directly observed in outcrop by noting the direction of relative motion of the two surfaces, thus offering the structural geologists the opportunity to gather valuable kinematic data.

We now broaden our perspective and consider how the relative displacements across brittle fractures are related to the entire *field* of displacements near such structures. As an example we consider the opening of the dike exposed just south of the volcanic edifice called Ship Rock located in northwestern New Mexico (**Figure 4.10**). The dike outcrop is about 10 km long and the dike has a maximum thickness of about 10 m. When this dike formed the part we see in this outcrop was at a depth of about 1 km. We infer that the displacement discontinuity was dominantly opening, to accommodate the intrusion of magma. Following the relationships illustrated in **Figure 4.24**, we deduce that the least compressive stress, σ_1, in this region was approximately perpendicular to the vertical plane of the dike, and that the dike was driven open by a magma pressure acting on the surfaces of the dike that *exceeded* this least compressive stress.

As the dike south of Ship Rock opened, a displacement field developed in the surrounding Mancos Shale. We turn to a model of dike opening in an elastic solid to visualize that field (**Figure 4.30**). This model is derived from Cauchy's equations of motion in Section 7.2, so here we only use the resulting displacement field to illustrate the kinematics. The magnitudes of the displacement vectors are exaggerated in this figure, so they can be discerned. Based on the ratio of dike thickness to length, 10 m/10 km, the actual displacement magnitudes are less than 1/1000 of the dike length. The displacement vectors are directed away from the sides of the model dike and toward the dike tips.

Figure 4.29 Limestone and shale beds at Lilstock Beach, England, are cut by joints and faults. See Figure 1.11 for location. White arrows indicate directions of relative displacement of originally neighboring particles on the two surfaces of a joint and on either side of a fault. Displacement magnitudes are not represented by the lengths of these arrows. See Engelder and Peacock (2001).

Elastic–brittle deformation is characterized by small strains, and these strains are directly related to spatial changes in the displacements, such as those illustrated in Figure 4.30. For example, consider the displacement field along the y-axis, perpendicular to the model dike. The particle at A′ displaces away from the dike more than the particle at A″, so the material line between these particles shortens as the dike opens. This corresponds to a contractional strain. As a second example, consider the displacement field along the x-axis, parallel to the model dike. The particle at B′ displaces toward from the dike more than the particle at B″, so the material line between these particles extends as the dike opens. This corresponds to an extensional strain. As a third example, consider the displacement field near the x-axis, along the side of the model dike. The particle at C′ displaces away from the dike more than the particle at C″, and this relative motion corresponds to a shear strain. In general, the deformation associated with dike opening is characterized by a heterogeneous strain field that varies with position and includes contraction and extension, as well as shearing.

To fully appreciate the displacement and strain fields for structures such as dikes, faults, and joints, and to understand the kinematics of brittle deformation in general, we describe how relative motion is quantified in the next section. Then, we define the specific relationships among the components of displacement and components of strain. These relationships are the *kinematic equations* for small strains.

4.9.1 Referential Description of Motion

To study elastic–brittle deformation we consider the motion of particles in a rock mass with reference to their positions at a particular earlier time. This is called a referential description of motion because one refers back to the configuration of the particles at the earlier time and uses this information to determine where the particles are at later times. Recall that we introduced a very different description of motion in Section 3.5.1 where attention is focused on a particular *location* rather than a particular *particle*. That spatial description of motion was used to derive the continuity equations and Cauchy's equations of motion in Chapter 3. Both referential and spatial descriptions of motion have important roles to play in continuum mechanics, and they are not contradictory. For example, one can convert Cauchy's equations derived from a spatial viewpoint to a material viewpoint using the material time derivative of velocity introduced in Section 2.5.4.

Taking the material viewpoint, and using a referential description of motion, the position, $\mathbf{x}$, of a given particle is a function of the initial position, $\mathbf{X}$, and the elapsed time, t:

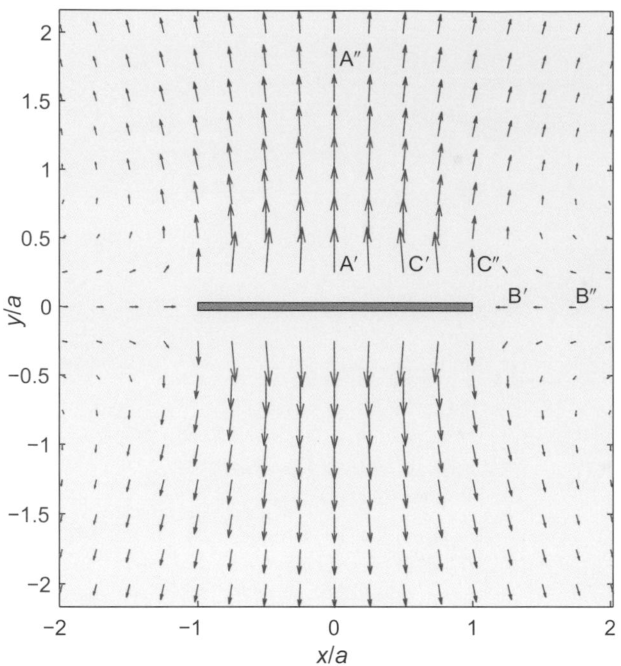

Figure 4.30 Displacement field for an opening dike. Vector displacements are greatly exaggerated in magnitude. Capital letters are described in the text with reference to the type of strain associated with spatial changes in displacement. Modified from Pollard and Segall (1987).

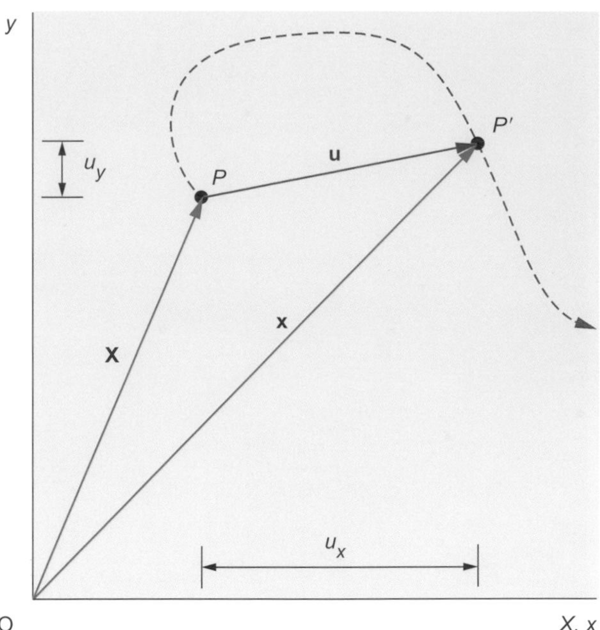

Figure 4.31 The displacement vector **u**; the position vector **X** for a particle in the initial (reference) position; and the position vector **x** for that particle at a later time. Dashed line depicts a *possible* path of this particle from the reference point *P* to the later point *P*′ and beyond. Modified from Malvern (1969).

$$\mathbf{x} = \mathbf{X} + \mathbf{u}(\mathbf{X}, t) \tag{4.31}$$

Here $\mathbf{u}(\mathbf{X}, t)$ is the displacement vector extending from the head of the position vector $\mathbf{X}$ to the head of the position vector $\mathbf{x}$ (**Figure 4.31**). In general the displacement is a function of all three coordinates and time, so as time progresses the particle moves along a path (dashed red curve) defined by this function. For applications of dynamic elastic theory, for example to particle motion during passage of a seismic wave, the continuous particle path is investigated. However, for the applications of quasi-static elastic theory considered in this textbook, we focus attention on an initial state and a final state. Stresses are applied that cause the configuration of particles to change from one state to the other, but the paths that particles take are not investigated.

Figure 4.31 illustrates a two-dimensional example of quasi-static deformation with a particle at point *P* at the initial time. That particle is at point *P*′ at the final time. Using the components of the initial position, (X, Y), and the components of the displacement vector, (u_x, u_y), the components of the final position are:

$$x = X + u_x(X, Y),\ y = Y + u_y(X, Y) \tag{4.32}$$

Note that the displacements are not explicit functions of time for quasi-static deformation, as they were for dynamic deformation in (4.31). Instead, the displacements are related to the strains in this section, constitutive laws are introduced in Section 4.10 to relate strains to stresses, and then in Section 4.11 the equations of motion are solved to determine all of these physical quantities.

4.9.2 Displacement Derivatives and Small Strains

Strain is a measure of the relative displacements between adjacent particles and it is composed of two fundamental kinematic quantities, stretching and shearing of material line segments. The rock mechanics experiments described in Sections 4.4 and 4.7 demonstrate that the strains before brittle failure are quite small (of order 10^{-3} to 10^{-2}). Under these circumstances the relationships connecting strains and displacements may be greatly simplified and we are justified in using these *small strain* approximations of the actual strain. In Section 5.5 we show that quantification of ductile deformation usually requires the actual strain with no approximations. Throughout this discussion of small strains the illustrations exaggerate the relative displacements and strains for the sake of clarity. The stretching and shearing of material lines in a deforming brittle rock often are too small to be seen, but they are easily measured using laboratory and field instruments.

Consider a particle at point *P* at the earlier time that has a position vector $\overrightarrow{OP} = X, Y$ (**Figure 4.32**). This particle moves to point *P*′ at the later time. The components of the position vector for *P*′ are related to the displacement components using (4.32):

$$\overrightarrow{OP'} = \overrightarrow{OP} + \overrightarrow{PP'} = \langle X + u_x(X, Y),\ \ Y + u_y(X, Y) \rangle \tag{4.33}$$

The vector $\overrightarrow{PQ}$ represents the material line extending from point *P* to point *Q*, it parallels the *X*-axis, and its length is $\left|\overrightarrow{PQ}\right| = \Delta X$. At the later time, this material line has *translated*, *rotated*, and *stretched*, so it connects points *P*′ and *Q*′, and is represented by the vector $\overrightarrow{P'Q'}$.

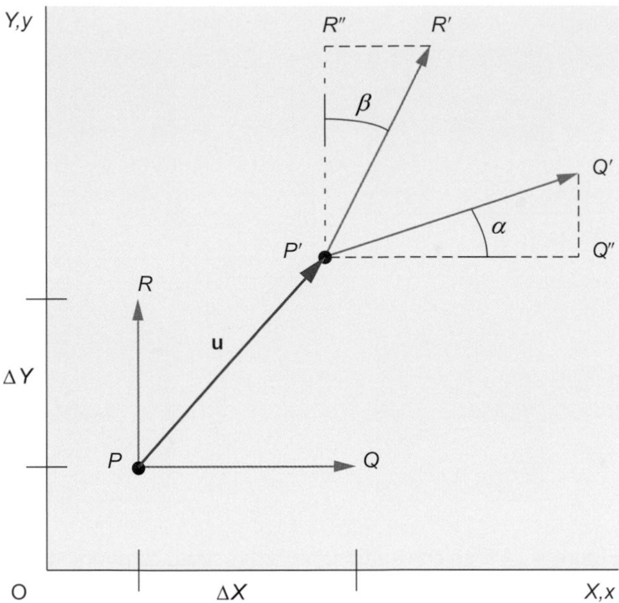

Figure 4.32 Sketch to define the kinematic relationships among the displacement and small strain components in two dimensions (exaggerated for the sake of illustration). Infinitesimal vectors extend from *P* to *Q* and from *P* to *R* in the initial state; from *P*′ to *Q*′ and from *P*′ to *R*′ in the final state.

To define the normal strain in the X-direction at P, the initial length of the material line and the vector representing it are diminished toward zero, that is $\Delta X \to 0$ (**Figure 4.32**). Furthermore, the actual length of the vector $\left|\overrightarrow{P'Q'}\right|$ is approximated as its *projected length* onto the X-axis, that is $\left|\overrightarrow{P'Q''}\right|$. The components of the position vector for point Q'' are:

$$\overrightarrow{OQ''} = \overrightarrow{OP} + \overrightarrow{PQ} + \overrightarrow{QQ''} = < X + \Delta X + u_x(X+\Delta X, Y),\ Y + u_y(X, Y) > \quad (4.34)$$

Subtracting (4.33) from (4.34), the vector representing the projected length is:

$$\overrightarrow{P'Q''} = \overrightarrow{OQ''} - \overrightarrow{OP'} = < \Delta X + u_x(X+\Delta X, Y) - u_x(X, Y) >, 0 \quad (4.35)$$

This vector has a zero y-component, so its length is that of the x-component.

The small normal strain at point P for a material line initially in the X-direction is defined as the fractional change in the *projected length* in the limit as the original length diminishes toward zero:

$$\varepsilon_{xx} \equiv \lim_{\Delta X \to 0} \frac{\left|\overrightarrow{P'Q''}\right| - \Delta X}{\Delta X} = \lim_{\Delta X \to 0} \frac{u_x(X+\Delta X,\ Y) - u_x(X,\ Y)}{\Delta X} = \frac{\partial u_x}{\partial X} \quad (4.36)$$

This limit is the definition of the partial derivative of u_x with respect to X. Thus, (4.36) defines the small normal strain for an infinitesimal material line that is parallel to the X-axis in the initial state.

The material line extending from P to R is parallel to the Y-axis and is represented by the vector $\overrightarrow{PR}$ with a length $\left|\overrightarrow{PR}\right| = \Delta Y$ (**Figure 4.32**). Following a procedure similar to that leading to (4.36), the small normal strain at the point P for the Y-direction is:

$$\varepsilon_{yy} = \frac{\partial u_y}{\partial Y} \quad (4.37)$$

Equation (4.37) defines the small normal strain for an infinitesimal material line that is parallel to the Y-axis in the initial state. The small normal strains defined in (4.36) and (4.37) differ from the actual normal strains because the projected lengths of the material lines are used to define the changes in length.

The shear strain is defined as one half the change in the angle between two material lines that are orthogonal in the initial state. In **Figure 4.32** the initial angle between the material lines is $\angle RPQ$ and the angle at the later time is $\angle R'P'Q'$, so one half the angular change is:

$$\tfrac{1}{2}(\angle RPQ - \angle R'P'Q') = \tfrac{1}{2}\left[\frac{\pi}{2} - \left(\frac{\pi}{2} - \alpha - \beta\right)\right] = \tfrac{1}{2}(\alpha + \beta) \quad (4.38)$$

The angles α and β are defined by first identifying their tangents and relating these to partial derivatives of the displacement components. For example, to calculate $\tan\alpha$ we use the lengths $\left|\overrightarrow{Q''Q'}\right|$ and $\left|\overrightarrow{P'Q''}\right|$. The components of the position vector for point Q' are:

$$\overrightarrow{OQ'} = \overrightarrow{OQ} + \overrightarrow{QQ'} = < X + \Delta X + u_x(X+\Delta X, Y),\ \ Y + u_y(X+\Delta X, Y) > \quad (4.39)$$

Subtracting (4.34) from (4.39) we have:

$$\overrightarrow{Q''Q'} = \overrightarrow{OQ'} - \overrightarrow{OQ''} = < 0,\ \ u_y(X+\Delta X, Y) - u_y(X, Y) > \quad (4.40)$$

This vector has an x component of zero, so its length is that of the y component. Using (4.40) and (4.35) the tangent of the angle α is defined in the limit as the length of the material line initially lying along the X-axis diminishes toward zero:

$$\tan\alpha = \lim_{\Delta X \to 0} \frac{\left|\overrightarrow{Q''Q'}\right|}{\left|\overrightarrow{P'Q''}\right|} = \lim_{\Delta X \to 0} \frac{[u_y(X+\Delta X, Y) - u_y(X, Y)]/\Delta X}{[\Delta X + u_x(X+\Delta X, Y) - u_x(X, Y)]/\Delta X} = \frac{(\partial u_y/\partial X)}{1 + (\partial u_x/\partial X)} \approx \frac{\partial u_y}{\partial X} \quad (4.41)$$

In the last step of (4.41), the partial derivative in the denominator is very small compared to one, so it is ignored.

By a similar procedure the tangent of the angle β is approximated as (**Figure 4.32**):

$$\tan\beta = \lim_{\delta Y \to 0} \frac{\left|\overrightarrow{R''R'}\right|}{\left|\overrightarrow{P'R''}\right|} = \frac{\dfrac{\partial u_x}{\partial Y}}{1 + \left(\dfrac{\partial u_y}{\partial Y}\right)} \approx \frac{\partial u_x}{\partial Y} \quad (4.42)$$

Because the angles α and β are very small, their tangents may be approximated by the angles themselves, based on the series

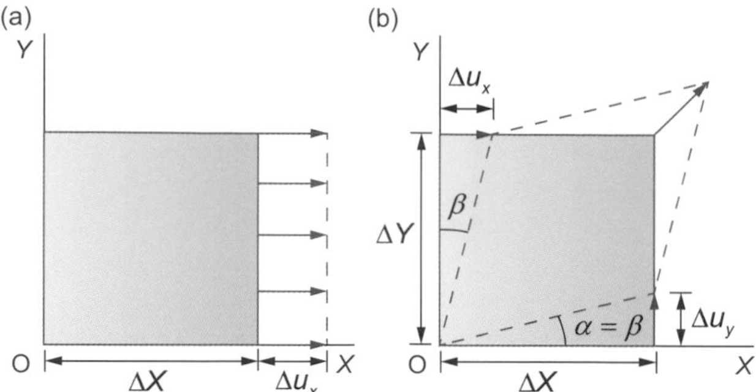

Figure 4.33 Examples of relative displacements associated with strains (exaggerated for the sake of illustration). (a) Relative displacements associated with positive normal strain in the x-direction. (b) Relative displacements associated with positive shear strain with respect to the x- and y-axes.

expansion for the tangent function: $\tan\alpha = \alpha + \frac{1}{3}\alpha^3 + \frac{2}{15}\alpha^5 + \frac{17}{315}\alpha^7 \cdots \approx \alpha$ for $\alpha << 1$ radian. With these approximations the two shear strains at point P, associated with material lines parallel to the X- and Y-axes at the initial time, are written using (4.41) and (4.42):

$$\varepsilon_{xy} = \frac{1}{2}\left(\frac{\partial u_x}{\partial Y} + \frac{\partial u_y}{\partial X}\right) = \varepsilon_{yx} \tag{4.43}$$

The small shear strain defined in (4.43) differ from the actual shear strains because of the approximations that ignore the partial derivative of displacement compared to one, and take only the first term in the series expansion for the tangent.

To gain an intuitive understanding of small strains, suppose u_x is the only non-zero displacement component and it varies only in the X-direction from a magnitude of zero along X = 0 to a magnitude Δu_x along $X = \Delta X$ (**Figure 4.33a**). Particles along $X = \Delta X$ move relative to particles along $X = 0$ as indicated by the displacement vector arrows. The square element with side length $\Delta X = \Delta Y$ at the initial time stretches in the X-direction, so the side lengths at the later time are $\Delta X + \Delta u_x$ and ΔY. In the limit as $\Delta X \to 0$ the small normal strain ε_{xx} is given by the partial derivative of displacement $\partial u_x/\partial X$ evaluated at $X = Y = 0$.

As a second example, suppose the square element with side length $\Delta X = \Delta Y$ at the initial time shears equally in both the X- and Y-directions, so the shape at the later time is a parallelogram with $\alpha = \beta$ (**Figure 4.33b**). This implies $\Delta u_x/\Delta Y = \Delta u_y/\Delta X$, and in the limit as $\Delta X \to 0$ and $\Delta Y \to 0$, the corresponding partial derivatives of displacement are equal: $\partial u_x/\partial Y = \partial u_y/\partial X$. The shear strains are given by one half the sum of these partial derivatives evaluated at $X = Y = 0$.

Positive normal strains are associated with stretching of infinitesimal material lines, whereas negative normal strains are associated with shortening (**Figure 4.34a, b**). Positive shear strains are associated with decreases in original right angles between infinitesimal material lines, whereas negative shear strains are associated with an angular increase (**Figure 4.34c, d**).

The kinematic equations that relate small strain components to partial derivatives of displacement components in three

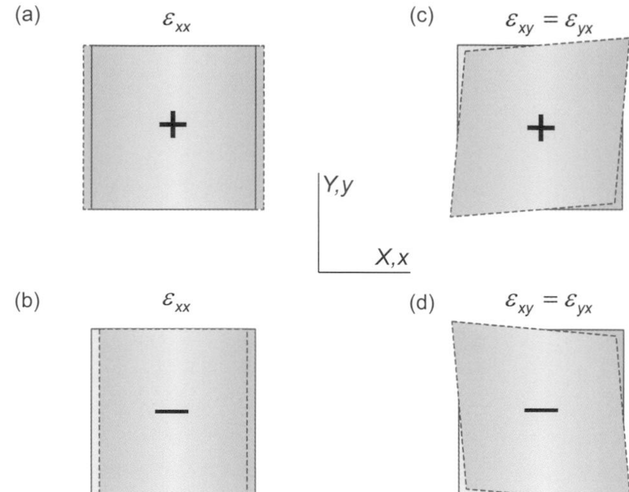

Figure 4.34 Sign conventions for small strain components. Square with solid border represents the initial shape; dashed border represents final state. (a) Stretching is positive; (b) shortening is negative; (c) decrease in a right angle is positive; (d) increase in a right angle is negative.

dimensions are found by the same procedures used above, so no new concepts are required. The small normal strains are:

$$\varepsilon_{xx} = \frac{\partial u_x}{\partial X}, \varepsilon_{yy} = \frac{\partial u_y}{\partial Y}, \varepsilon_{zz} = \frac{\partial u_z}{\partial Z} \tag{4.44}$$

The small shear strains are:

$$\begin{aligned}\varepsilon_{xy} &= \frac{1}{2}\left(\frac{\partial u_x}{\partial Y} + \frac{\partial u_y}{\partial X}\right), \varepsilon_{yz} = \frac{1}{2}\left(\frac{\partial u_y}{\partial Z} + \frac{\partial u_z}{\partial Y}\right), \varepsilon_{zx} = \frac{1}{2}\left(\frac{\partial u_z}{\partial X} + \frac{\partial u_x}{\partial Z}\right)\\ \varepsilon_{yx} &= \frac{1}{2}\left(\frac{\partial u_y}{\partial X} + \frac{\partial u_x}{\partial Y}\right), \varepsilon_{zy} = \frac{1}{2}\left(\frac{\partial u_z}{\partial Y} + \frac{\partial u_y}{\partial Z}\right), \varepsilon_{xz} = \frac{1}{2}\left(\frac{\partial u_x}{\partial Z} + \frac{\partial u_z}{\partial X}\right)\end{aligned} \tag{4.45}$$

Taken together (4.44) and (4.45) are referred to as kinematic equations for small strains.

The small strains are components of the tensor $\boldsymbol{\varepsilon}$, and they may be organized into a square matrix with three rows and three columns:

$$[\varepsilon] = \begin{bmatrix} \varepsilon_{xx} & \varepsilon_{xy} & \varepsilon_{xz} \\ \varepsilon_{yx} & \varepsilon_{yy} & \varepsilon_{yz} \\ \varepsilon_{zx} & \varepsilon_{zy} & \varepsilon_{zz} \end{bmatrix} \tag{4.46}$$

The normal strain components are found along the principal diagonal and the shear strain components in symmetric positions across the principal diagonal are equal: $\varepsilon_{xy} = \varepsilon_{yx}$, $\varepsilon_{yz} = \varepsilon_{zy}$, $\varepsilon_{zx} = \varepsilon_{xz}$. Thus, the small strain components form a symmetric matrix. As such the small strain has principal values and principal axes that are found using the same procedure that was introduced for the stress components in (4.17) and (4.18). The principal strains are ε_1, ε_2, and ε_3, and these act along three mutually orthogonal axes. As described for the principal stresses in Section 4.6.2, the orientations and values of the principal strains are calculated in MATLAB as an eigenvalue problem using [O, V] = eig[ε]. If the principal strains all are extensions (positive), the small strain matrix [ε] is positive definite; if they all are contractions (negative), this matrix is negative definite. The principal strains may have mixed signs, and one or more may be zero.

For two-dimensional problems, two of the principal strains are in the plane of interest, for example the (x, y)-plane, and they are calculated using:

$$\frac{1}{2}\left(\varepsilon_{xx} + \varepsilon_{yy}\right) \pm \left[\frac{1}{4}\left(\varepsilon_{xx} - \varepsilon_{yy}\right)^2 + \varepsilon_{xy}^2\right]^{1/2} \quad (4.47)$$

The $\pm$ signs provide the greater and the lesser principal strain in the (x, y)-plane, respectively. The third principal strain is oriented parallel to the z-axis. The principal strain axes in the (x, y)-plane are calculated using:

$$\frac{1}{2}\tan^{-1}\left[2\varepsilon_{xy}/\left(\varepsilon_{xx} - \varepsilon_{yy}\right)\right] \quad (4.48)$$

The two angles identified using (4.48) are orthogonal, and they are oriented with respect to the x-axis.

Applications of the kinematic equations for small strains require small partial derivatives of displacement, a condition generally met for rock that deforms in an elastic brittle manner. It is important to note that the displacements are not required to be small; it is the partial derivatives of displacement that are required to be small. In the structural geology literature the small strains commonly are referred to as infinitesimal strains. For some, this implies that these kinematic quantities are inconsequential and can be ignored. Yet, for example, the majority of the deformation outside the immediate fault zone for most major earthquakes is quite adequately described using small strains. Clearly the deformation associated with major earthquakes is not inconsequential, and indeed an understanding of small strains is a *pre-requisite* for investigating the development of most geologic structures, even those that require one to use the actual strains. Given equations for the actual strains, one can quantify exactly what is meant by small partial derivatives of displacement and evaluate the error introduced by the approximations leading to (4.44) and (4.45). We return to that task in Section 5.5.5.

The spatial derivatives of the displacement components that appear on the right sides of (4.45) can lead to rotations of material lines at a point, as illustrated in **Figure 4.32**. Suppose the two orthogonal material lines represented by the vectors $\overrightarrow{PQ}$ and $\overrightarrow{PR}$ in that figure both rotate counter-clockwise through the same small angle. The corresponding shear strain is zero, because the right angle between the two material lines in the initial state is a right angle in the final state. This constitutes a small pure rotation at the point P, because it involves no stretching or shearing. This rotation can be ignored when we formulate the constitutive equations for the elastic solid in the next section. However, small rotations do complete the kinematic story, and they have interesting applications in understanding the kinematics of geological structures such as faults. Therefore, we return to define, describe, and apply small rotations in Section 8.5.

4.9.3 Displacement Gradient Tensor and Small Strain Tensor

In general, small strains result from the geometric changes in length and orientation of infinitesimal material lines that radiate from a given particle, for example one initially at position $\mathbf{X}$. The particle moves through a displacement $\mathbf{u}(\mathbf{X}, t)$ to position $\mathbf{x}$ at time t (**Figure 4.31**). The displacement is not necessarily small, but we focus on infinitesimal material lines in order to define the strain at the arbitrary point P, the initial position of the particle in the material continuum. Infinitesimal vectors represent the material lines as they stretch and shear (**Figure 4.32**). The vector equation (4.31) describing the motion is repeated here for reference:

$$\mathbf{x} = \mathbf{X} + \mathbf{u}(\mathbf{X}, t)$$

Now consider the motion of a particle, initially at the point Q, which is in the neighborhood of P. The position vector $\mathbf{X} + \mathrm{d}\mathbf{X}$ locates this second particle, an infinitesimal distance from the particle at P. The second particle moves through a displacement $\mathbf{u}(\mathbf{X} + \mathrm{d}\mathbf{X}, t)$ to position $\mathbf{x} + \mathrm{d}\mathbf{x}$ at time t, so the vector equation describing the motion of this particle is:

$$\mathbf{x} + \mathrm{d}\mathbf{x} = \mathbf{X} + \mathrm{d}\mathbf{X} + \mathbf{u}(\mathbf{X} + \mathrm{d}\mathbf{X}, t)$$

Note that the second particle is located in an arbitrary direction from the first particle, so these equations apply to all particles in the neighborhood of the particle at P.

The displacement of the particle at $\mathbf{X} + \mathrm{d}\mathbf{X}$, *relative* to the particle at $\mathbf{X}$, is found by subtracting the first equation from the second equation displayed in the previous paragraph:

$$\mathrm{d}\mathbf{x} = \mathrm{d}\mathbf{X} + \mathbf{u}(\mathbf{X} + \mathrm{d}\mathbf{X}, t) - \mathbf{u}(\mathbf{X}, t) = \mathrm{d}\mathbf{X} + (\nabla\mathbf{u})\mathrm{d}\mathbf{X} \quad (4.49)$$

The right side follows from the fact that the displacement difference between the two particles, divided by the initial distance between those particles, is the *spatial gradient in displacement*:

$$\nabla\mathbf{u} = \frac{\mathbf{u}(\mathbf{X} + \mathrm{d}\mathbf{X}, t) - \mathbf{u}(\mathbf{X}, t)}{\mathrm{d}\mathbf{X}} \quad (4.50)$$

If the displacement difference in the numerator of (4.50) is zero, the displacement gradient is zero, and the infinitesimal material line between the two neighboring particles does not stretch. Because the vector equations (4.49) and (4.50) apply to all particles in the neighborhood, none of the radiating material lines stretch, and none change their orientation, so no shearing. This suggests a close relationship between the strain and the spatial gradient in displacement.

Recall from Section 2.6.1 that the gradient of a vector field is a tensor field, so the components of displacement gradients in (4.50) are the components of the displacement gradient tensor arranged in the following array:

$$\mathbf{G} = \begin{Bmatrix} \dfrac{\partial u_x}{\partial X} & \dfrac{\partial u_x}{\partial Y} & \dfrac{\partial u_x}{\partial Z} \\[2ex] \dfrac{\partial u_y}{\partial X} & \dfrac{\partial u_y}{\partial Y} & \dfrac{\partial u_y}{\partial Z} \\[2ex] \dfrac{\partial u_z}{\partial X} & \dfrac{\partial u_z}{\partial Y} & \dfrac{\partial u_z}{\partial Z} \end{Bmatrix} \quad (4.51)$$

Each row of this array contains the components of the gradient vector for one of the scalar components of displacement. For example, the second row is composed of the three components of the gradient vector for the component u_y:

$$\vec{\nabla} u_y = \frac{\partial u_y}{\partial X}\hat{\mathbf{i}} + \frac{\partial u_y}{\partial Y}\hat{\mathbf{j}} + \frac{\partial u_y}{\partial Z}\hat{\mathbf{k}}$$

Each of the nine components of the displacement gradient tensor (4.51) may be a function of the three spatial coordinates.

The components of the displacement gradient tensor (4.51) are the displacement derivatives that appear in (4.44) and (4.45), the kinematic equations for small strains. In fact, the small strain tensor may be written in terms of the displacement gradient tensor in matrix form:

$$[\varepsilon] = \tfrac{1}{2}\{[G] + [G']\} \tag{4.52}$$

In other words, the elements of the small strain matrix are equal to the elements of the matrix composed of one half the sum of the displacement gradient tensor and its transpose.

4.10 CONSTITUTIVE EQUATIONS FOR THE LINEAR ELASTIC SOLID

Neither the strains alone, nor the stresses alone, are sufficient to characterize elastic–brittle deformation. Here we tie the stresses and strains together using the elastic properties of solids to define a linear elastic constitutive law. These relationships describe the behavior of a rock mass that results from the internal *constitution* of the material. They are important for understanding elastic deformation because one may know the stress and want to calculate the strain, or vice versa. The constitutive equations also play a vital role in eliminating some of the dependent variables from Cauchy's first equations of motion that were introduced in Section 3.6. This is necessary to put those equations of motion in a form that can be solved and applied to the formation of geologic structures.

In (4.13) we arranged the stress components in a symmetric matrix, $[\sigma]$, and in (4.46) we arranged the small strain components in a symmetric matrix, $[\varepsilon]$. In the context of Newtonian mechanics the forces (per unit area) cause the deformations: *stress causes strain*. One might suppose that each element of $[\sigma]$ causes only the corresponding element of $[\varepsilon]$. For example, one might suppose that the normal stress σ_{yy} causes only the normal strain ε_{yy}. However, in (4.6) we learned that normal stress in one direction usually is accompanied by normal strains in the orthogonal directions, unless Poisson's ratio is zero. In this section we explicitly define the relationships among the stress and strain components taking Poisson's ratio into account.

Suppose a cubic element of elastic material is loaded first by an applied normal stress, only in the x-direction, and then only in y, and finally only in z (**Figure 4.35**). The applied stress σ_{xx} causes the element to lengthen in the x-direction by a displacement u_x (**Figure 4.35a**), so the small normal strain in x is ε_{xx}. According to (4.5) this strain is proportional to the applied stress, and the proportionality constant is the reciprocal of Young's modulus: $\varepsilon_{xx} = \sigma_{xx}/E$.

The applied stress σ_{yy} (**Figure 4.35b**) causes the element to lengthen in the y-direction, but also to shorten in the x-direction. The negative ratio of the normal strains in these two orthogonal

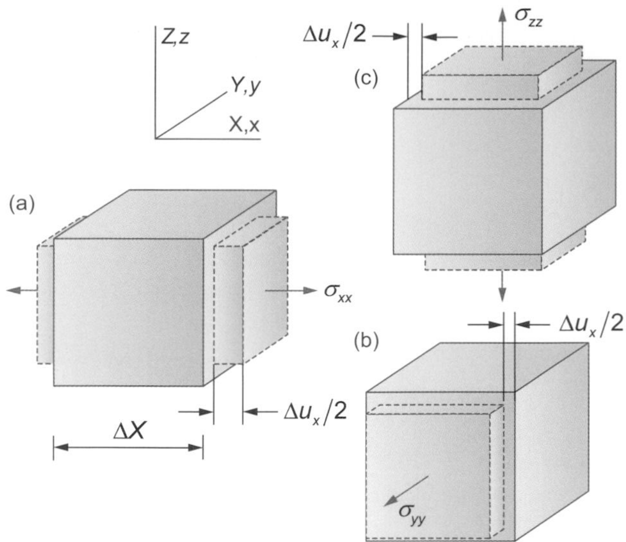

Figure 4.35 Cubic element with applied normal stresses in x, y, and z. The elements deform with normal strain in the x-direction due to the applied normal stress in x, y, and z.

directions is equal to Poisson's ratio from (4.6), that is $\nu = -\varepsilon_{xx}/\varepsilon_{yy}$. Thus, the strain in the x-direction due to the applied stress σ_{yy} is $\varepsilon_{xx} = -\nu\varepsilon_{yy}$. Again using (4.5), the small normal strain in y is proportional to the applied stress and inversely proportional to Young's modulus: $\varepsilon_{yy} = \sigma_{yy}/E$. Substituting for this normal strain, we have the small normal strain in x due to the applied stress, σ_{yy}, as $\varepsilon_{xx} = -\nu\sigma_{yy}/E$. A similar derivation leads to the small normal strain in x (**Figure 4.35c**) due to the applied normal stress, σ_{zz}, as $\varepsilon_{xx} = -\nu\sigma_{zz}/E$. Adding the contributions to the normal strain in the x-direction due to the applied normal stresses in the x-, y-, and z-directions we have: $\varepsilon_{xx} = [\sigma_{xx} - \nu\,(\sigma_{yy} + \sigma_{zz})]/E$. Relationships such as this make it clear that two elastic constants are involved in the strain–stress equations, and that an individual element of the strain matrix $[\varepsilon]$ is not necessarily related only to the corresponding element of the stress matrix $[\sigma]$.

Successively applying a normal stress in each coordinate direction (**Figure 4.35**), and accounting for all of the possible normal strains, we have:

$$\begin{aligned}
\varepsilon_{xx} &= \frac{1}{E}\left[\sigma_{xx} - \nu\left(\sigma_{yy} + \sigma_{zz}\right)\right] \\
\varepsilon_{yy} &= \frac{1}{E}\left[\sigma_{yy} - \nu\left(\sigma_{zz} + \sigma_{xx}\right)\right] \\
\varepsilon_{zz} &= \frac{1}{E}\left[\sigma_{zz} - \nu\left(\sigma_{xx} + \sigma_{yy}\right)\right]
\end{aligned} \tag{4.53}$$

These are linear relations that give the small normal strains in terms of the normal stresses and the two elastic constants, Young's modulus and Poisson's ratio. Each small shear strain is proportional to the respective shear stress:

$$\varepsilon_{xy} = \frac{1+\nu}{E}\sigma_{xy},\ \varepsilon_{yz} = \frac{1+\nu}{E}\sigma_{yz},\ \varepsilon_{zx} = \frac{1+\nu}{E}\sigma_{zx} \tag{4.54}$$

Equations (4.53) and (4.54) are a statement of Hooke's law for the elastic solid and provide the equations linking all the small strain components to the stress components. Recall from Section 4.1 that Robert Hooke solved his own anagram that described the linear relation between applied force and the resulting extension in 1679, so his name is associated with these linear relations between stress and small strain.

For the derivation of (4.53) and (4.54) we used the same elastic constants, Young's modulus and Poisson's ratio, regardless of direction. In other words, the orientation of the Cartesian coordinate system was completely arbitrary for this idealized elastic material. This implies that the solid is isotropic with respect to elastic constants. On the other hand, if Young's modulus (or Poisson's ratio) is measured for a sedimentary rock in a uniaxial test (**Figure 4.6**) using cylindrical specimens cored perpendicular and parallel to bedding, and different values are obtained for these two directions, that rock is anisotropic with respect to Young's modulus (or Poisson's ratio). A cubic specimen of rock that is anisotropic with respect to Young's modulus, when subjected to a uniform pressure, would contract more in the *softer* direction, which is the direction with the lesser Young's modulus. Because anisotropy with respect to elastic moduli is not significant for many rock types, much can be learned about geologic structures using *isotropic elasticity*, so that is where we focus attention.

Although Young's modulus and Poisson's ratio are most easily measured in laboratory experiments, other constants have been conceived that play useful roles in elasticity theory. The elastic shear modulus, G, is used to relate shear stress to shear strain. For example, noting the first such relationship in (4.54), the shear modulus is related to Young's modulus and Poisson's ratio as:

$$G = \frac{E}{2(1+\nu)} \tag{4.55}$$

For *perfectly compressible* material, $\nu = 0$, so $G = E/2$, and for *incompressible* material, $\nu = 1/2$, so $G = E/3$. Using (4.55), for example, the first of (4.54) would be rewritten using the shear modulus as $\varepsilon_{xy} = \sigma_{xy}/2G$. The shear strain is proportional to the corresponding shear stress and the proportionality constant is one over twice the shear modulus.

Lame's constant is helpful in calculating the stress components given the strain components. This constant is written in terms of Young's modulus and Poisson's ratio as:

$$\lambda = \frac{E\nu}{(1+\nu)(1-2\nu)} \tag{4.56}$$

For *perfectly compressible* material, $\nu = 0$ and $\lambda = 0$. For *incompressible* material, $\nu = ½$ and $\lambda = \infty$. The shear modulus and Lame's constant have the same dimensions as stress: for example, λ {=} $ML^{-1}T^{-2}$. Therefore, these constants carry the same units as stress: for example, G [=] $N\,m^{-2}$ = Pa. The values of these constants usually are reported with the unit gigapascal, GPa, and typical values for rock are in the range 1 to 100 GPa.

Hooke's law written as (4.53) and (4.54) may be rearranged to write the stress components as functions of the strain components using λ and G:

$$\begin{aligned}\sigma_{xx} &= 2G\varepsilon_{xx} + \lambda(\varepsilon_{xx}+\varepsilon_{yy}+\varepsilon_{zz})\\ \sigma_{yy} &= 2G\varepsilon_{yy} + \lambda(\varepsilon_{xx}+\varepsilon_{yy}+\varepsilon_{zz})\\ \sigma_{zz} &= 2G\varepsilon_{zz} + \lambda(\varepsilon_{xx}+\varepsilon_{yy}+\varepsilon_{zz})\end{aligned} \tag{4.57}$$

$$\sigma_{xy} = 2G\varepsilon_{xy},\ \sigma_{yz} = 2G\varepsilon_{yz},\ \sigma_{zx} = 2G\varepsilon_{zx} \tag{4.58}$$

Using (4.55) and (4.56) these stress–strain relations also may be written in terms of Young's modulus and Poisson's ratio. This demonstrates the important concept that *only two elastic constants are required for the constitutive equations of the isotropic elastic solid.*

If the strain components are known, Hooke's law in the form of (4.57) and (4.58) may be used to calculate the stress components. This begs the question: how might you *know* the strain components? The answers are: by solving the appropriate equations of motion, or by measurement in the laboratory or in the field. In Section 4.11 we present the relevant equations of motion and provide an example solution that considers slip on a fault.

4.11 EQUATIONS OF MOTION FOR THE ELASTIC SOLID

The general equations of motion for the material continuum were derived in Section 3.6 as Cauchy's first law and in Section 3.7 as Cauchy's second law. There we showed how they follow from the basic physical laws of conservation for linear and angular momentum. Restating Cauchy's equations:

$$\begin{aligned}\rho\frac{Dv_x}{Dt} &= \frac{\partial\sigma_{xx}}{\partial x} + \frac{\partial\sigma_{yx}}{\partial y} + \frac{\partial\sigma_{zx}}{\partial z} + \rho g_x^*\\ \rho\frac{Dv_y}{Dt} &= \frac{\partial\sigma_{xy}}{\partial x} + \frac{\partial\sigma_{yy}}{\partial y} + \frac{\partial\sigma_{zy}}{\partial z} + \rho g_y^*\\ \rho\frac{Dv_z}{Dt} &= \frac{\partial\sigma_{xz}}{\partial x} + \frac{\partial\sigma_{yz}}{\partial y} + \frac{\partial\sigma_{zz}}{\partial z} + \rho g_z^*\end{aligned} \tag{4.59}$$

$$\sigma_{xy} = \sigma_{yx},\ \sigma_{yz} = \sigma_{zy},\ \sigma_{zx} = \sigma_{xz} \tag{4.60}$$

The left sides of (4.59) are the product of mass per unit volume and the material time derivative of velocity (in other words, the particle acceleration) in each coordinate direction. The first three terms on the right sides of (4.59) are the spatial derivatives of stress components, which are surface forces per unit volume. The last term on each right side is the product of mass density and acceleration of gravity. These are the body forces per unit volume due to gravity in each coordinate direction. In other words, the equations in (4.59) are a statement of Newton's second law, $m\mathbf{a} = \mathbf{F}$, for a material continuum. Equations (4.60) show that the conjugate shear stress components are equal, so the stress tensor is symmetric.

Cauchy's equations, (4.59) and (4.60), have too many dependent variables relative to the number of equations. These six equations contain three velocity components and six stress components. To get to equations of motion that can be solved, we begin by focusing on the particle accelerations in (4.59). For example, the left side of the first of these equations is expanded using the definition of the material time derivative, and then it is approximated as follows:

$$\frac{Dv_x}{Dt} = \frac{\partial v_x}{\partial t} + v_x \frac{\partial v_x}{\partial x} + v_y \frac{\partial v_x}{\partial y} + v_z \frac{\partial v_x}{\partial z} \approx \frac{\partial v_x}{\partial t} = \frac{\partial^2 u_x}{\partial t^2} \quad \textbf{(4.61)}$$

If the products of velocity components and the spatial derivatives of those components are small compared to the time rate of change of velocity, they can be ignored. Next, the velocity is recognized as the time rate of change of displacement, $v_x = \partial u_x/\partial t$. Thus, $\partial^2 u_x/\partial t^2$, replaces Dv_x/Dt on the left side of the first equation in (4.59). In doing so, the displacement component becomes the dependent variable instead of the velocity component. Similar approximations are employed for the left sides of the second and third of (4.59).

Next, we reduce the number of dependent variables on the right sides of (4.59) by replacing stress components with displacement components using Hooke's law. For small strains, the derivatives of displacement with respect to the initial coordinates, (X, Y, Z), are approximately the same as the derivatives with respect to the final coordinates, (x, y, z), so we follow the customary practice and replace the upper case coordinates in (4.44) and (4.45) with lower case coordinates. Then, for example, the constitutive equation for the normal stress in the x-direction is written:

$$\sigma_{xx} = 2G\frac{\partial u_x}{\partial x} + \lambda\left(\frac{\partial u_x}{\partial x} + \frac{\partial u_y}{\partial y} + \frac{\partial u_z}{\partial z}\right)$$

Similarly, the shear stresses in the x-direction are written:

$$\sigma_{yx} = G\left(\frac{\partial u_y}{\partial x} + \frac{\partial u_x}{\partial y}\right), \sigma_{zx} = G\left(\frac{\partial u_z}{\partial x} + \frac{\partial u_x}{\partial z}\right)$$

Equations for the other stress components follow by similar steps.

Substituting the equations derived in the previous paragraph into (4.59), we find:

$$\begin{aligned}
\rho\frac{\partial^2 u_x}{\partial t^2} &= G\left(\frac{\partial^2 u_x}{\partial x^2} + \frac{\partial^2 u_x}{\partial y^2} + \frac{\partial^2 u_x}{\partial z^2}\right) + (G+\lambda)\left(\frac{\partial^2 u_x}{\partial x^2} + \frac{\partial^2 u_y}{\partial x \partial y} + \frac{\partial^2 u_z}{\partial x \partial z}\right) + \rho g_x^* \\
\rho\frac{\partial^2 u_y}{\partial t^2} &= G\left(\frac{\partial^2 u_y}{\partial x^2} + \frac{\partial^2 u_y}{\partial y^2} + \frac{\partial^2 u_y}{\partial z^2}\right) + (G+\lambda)\left(\frac{\partial^2 u_x}{\partial x \partial y} + \frac{\partial^2 u_y}{\partial y^2} + \frac{\partial^2 u_z}{\partial y \partial z}\right) + \rho g_y^* \\
\rho\frac{\partial^2 u_z}{\partial t^2} &= G\left(\frac{\partial^2 u_z}{\partial x^2} + \frac{\partial^2 u_z}{\partial y^2} + \frac{\partial^2 u_z}{\partial z^2}\right) + (G+\lambda)\left(\frac{\partial^2 u_x}{\partial x \partial z} + \frac{\partial^2 u_y}{\partial y \partial z} + \frac{\partial^2 u_z}{\partial z^2}\right) + \rho g_z^*
\end{aligned} \quad \textbf{(4.62)}$$

These are Navier's equations of motion for the linear elastic solid with isotropic elastic constants. They also are called Navier's *displacement* equations of motion, because the dependent variables are the displacement components.

Each distinct term in equations (4.62) must have the same dimensions to satisfy the condition of dimensional homogeneity (refer to **Table 3.1**). On the left sides of each equation we have mass density times the second derivative of displacement with respect to time, so the dimensions are $ML^{-3}LT^{-2} = (MLT^{-2})(L^{-3})$, which are dimensions of force per unit volume. On the right sides of (4.62) we have an elastic modulus times the second derivative of a displacement with respect to a coordinate, so the dimensions are $(ML^{-1}T^{-2})(LL^{-2}) = (MLT^{-2})(L^{-3})$. These terms also have dimensions of force per unit volume. The last term in each equation is mass density times acceleration of gravity, which has dimensions of force per unit volume. Thus, Navier's equations are dimensionally consistent and can be interpreted as a statement of Newton's second law, $m\mathbf{a} = \mathbf{F}$, for the *elastic* material continuum with isotropic elastic constants.

The independent variables in Navier's equations (4.62) are three spatial coordinates and time (x, y, z, t). The dependent variables are the three displacement components (u_x, u_y, u_z), so the three equations have three unknowns. These equations are solved for an elastic body of prescribed geometry subject to boundary conditions defined at every point on the exterior and interior boundaries in terms of the three displacement components as functions of time. The displacement components in the interior as functions of the spatial coordinates and time are the solution to this classic boundary-value problem:

$$u_x = u_x(x, y, z, t),\ u_y = u_y(x, y, z, t),\ u_z = u_z(x, y, z, t) \quad \textbf{(4.63)}$$

The six small strain components are found by taking the spatial derivatives of the displacement components using the kinematic equations, (4.44) and (4.45). Then, the constitutive equations, (4.57) and (4.58) are used to calculate the six stress components. In this way all of the relevant physical quantities are accounted for as functions of the coordinates and time.

Navier's equations of motion (4.62) are somewhat daunting in their complexity, but two simplifications, reducing the geometry to two dimensions, and eliminating the accelerations on the left sides, are taken up in the next section and lead to important solutions with applications to structural geology.

4.11.1 Two-dimensional, quasi-static deformation

The development of structures in Earth's brittle lithosphere may occur under conditions in which the left sides of Navier's equations (4.62) are very small compared to the right sides. In other words, the product of mass and acceleration are small compared to the surface and body forces per unit volume. In these cases it is appropriate to set the left sides to zero, which puts the problem in the realm of quasi-static equilibrium. The prefix "quasi" is used because this is not a problem of a perfectly *static* and *rigid* body. Instead, it is a problem that admits relative displacements and strains, but those develop such that the accelerations are small.

All geologic structures in Earth's lithosphere are three-dimensional, but for some structures a two-dimensional approximation is appropriate, especially in preliminary analyses, and this greatly reduces the mathematical complexity of the boundary-value problem governed by Navier's equations (4.62). For structures that are very long in one dimension relative to the other two dimensions, and if the geometry and loading do not change significantly in that long dimension, the structure may be treated as two-dimensional. Common examples would be the surface of a sedimentary layer, folded into a cylindrical fold, and blade-like joints that are confined to individual sedimentary strata (**Figure 4.36a**).

The special case of deformation that applies to two-dimensional structures is called plane strain. The (x, z)-plane is taken as the plane of interest in **Figure 4.36b**, and the out-of-plane

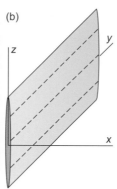

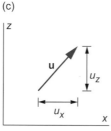

Figure 4.36 Geological structures approximating plane strain conditions, where displacement in the *z*-direction is zero. (a) Vertical joints near Raplee anticline, UT. See **Figure 2.13**. (b) Blade-shaped fracture. (c) Non-zero displacement components for plane strain. See Barber (2010) in Further Reading.

displacement component, u_y, is postulated to be zero. The two in-plane displacement components are functions of x and z only (**Figure 4.36c**):

$$u_x = u_x(x, y),\ u_y = 0,\ u_z = u_z(x, y) \tag{4.64}$$

Based on these restrictions on the displacement components, and ignoring body forces, Navier's displacement equations of motion (4.62) reduce to:

$$\begin{aligned}(2G+\lambda)\frac{\partial^2 u_x}{\partial x^2} + G\frac{\partial^2 u_x}{\partial z^2} + (G+\lambda)\frac{\partial^2 u_z}{\partial x \partial z} &= 0\\ G\frac{\partial^2 u_z}{\partial x^2} + (2G+\lambda)\frac{\partial^2 u_z}{\partial z^2} + (G+\lambda)\frac{\partial^2 u_x}{\partial x \partial z} &= 0\end{aligned} \tag{4.65}$$

These are the two-dimensional quasi-static equations of equilibrium for the isotropic and linear elastic solid.

All of the terms in (4.65) multiply an elastic modulus times a second derivative of displacement with respect to the coordinates, so the dimensions are $(ML^{-1}T^{-2})(LL^{-2}) = (MLT^{-2})(L^{-3})$, which are the dimensions of force per unit volume. Thus, Navier's equilibrium equations can be interpreted as the quasi-static expression of Newton's second law, $\mathbf{F} = 0$. In other words, the resultant force is taken as zero, because the mass times acceleration is negligible compared to the resultant force. The dependent variables in (4.65) are the two displacement components, u_x and u_z, and the independent variables are the two spatial coordinates, x and z. Presuming one has adequate laboratory data to determine the two elastic moduli, we have two equations and two unknowns.

The two equations (4.65) are solved for the two displacement components, u_x and u_z, subject to displacement boundary conditions on all external and internal boundaries. Then, the strain components in the (x, z)-plane are calculated from the kinematic equations, (4.44) and (4.45), reduced to plane strain conditions:

$$\varepsilon_{xx} = \frac{\partial u_x}{\partial x},\ \varepsilon_{zz} = \frac{\partial u_z}{\partial z},\ \varepsilon_{xz} = \frac{1}{2}\left(\frac{\partial u_x}{\partial z} + \frac{\partial u_z}{\partial x}\right) = \varepsilon_{zx} \tag{4.66}$$

All other strain components are zero. Finally, the stress components in the (x, z)-plane are calculated from Hooke's law, (4.57) and (4.58), reduced to plane strain conditions:

$$\sigma_{xx} = 2G\varepsilon_{xx} + \lambda(\varepsilon_{xx} + \varepsilon_{zz}),\ \sigma_{zz} = 2G\varepsilon_{zz} + \lambda(\varepsilon_{xx} + \varepsilon_{zz}),\quad \sigma_{xz} = 2G\varepsilon_{xz} = \sigma_{zx} \tag{4.67}$$

In order to prevent displacements in the z-direction, there must be a normal stress acting in that direction. This stress is computed using the last of (4.53) with $\varepsilon_{yy} = 0$ to find $\sigma_{yy} = \nu(\sigma_{xx} + \sigma_{zz})$. An example solution for the quasi-static, plane strain problem governed by Navier's displacement equations of motion is provided in the next section.

4.11.2 Example Solution: A Plane of Displacement Discontinuity

The two-dimensional plane of displacement discontinuity is a solution for the plane strain problem with many applications in structural geology. In the four schematic drawings of **Figure 4.37** the displacements of originally neighboring particles on opposing surfaces of a cut in the elastic material are indicated by vector

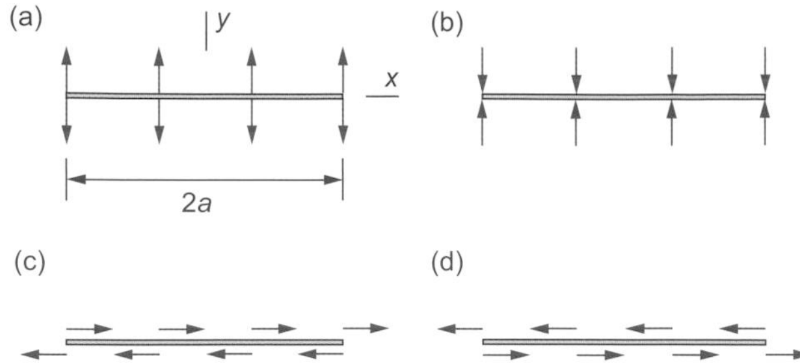

Figure 4.37 Models for geologic structures based on planar displacement discontinuities. (a) Opening fracture such as a dike, joint, or vein. (b) Closing structure such as a compaction band or solution surface. (c) Sliding fracture such as a right-lateral fault. (d) Sliding fracture such as a left-lateral fault. See Crouch and Starfield (1983) in Further Reading.

arrows. The lengths of these arrows are greatly exaggerated so they can be discerned. This cut is called a plane of displacement discontinuity because the displacements of originally neighboring particles on adjacent surfaces are in opposite directions. Note that we have taken the plane of interest as the (x, y)-plane, so the z-component of displacement is zero.

In **Figure 4.37a** the particles move apart, so a gap opens up along the plane. This is analogous to an opening fracture such as an igneous dike, a hydrothermal vein, or a joint. In **Figure 4.37b** the particles move toward the plane, analogous to the inward motion toward a pressure solution seam as mineral grains dissolve, or toward a compaction band as pores collapse. In **Figure 4.37c,d** the particles move parallel to the plane in dextral and sinistral relative motion, respectively, analogous to a right-lateral fault and a left-lateral fault.

In all cases illustrated in **Figure 4.37** the displacement discontinuity is uniform in magnitude along the plane. In nature the opening of a dike or vein is non-uniform, as is the slip on a fault. For these structures the displacement discontinuity typically is greatest near the center and always goes to zero at the terminations. A model with uniform displacement discontinuity captures the sense of relative displacement, but not the variable magnitude. In Section 7.2.5 we consider the opening of a fracture driven by a uniform internal pressure or remote tension and find a displacement discontinuity with an elliptical distribution. In Section 8.3.4 we consider the shearing of a fault driven by a uniform shear stress drop and again find a displacement discontinuity with an elliptical distribution. We expect the models illustrated in **Figure 4.37** to be good approximations for those with more realistic displacement discontinuities, except near the tips of the structures.

The displacement fields for the problems illustrated in **Figure 4.37** are given in the textbook by Crouch and Starfield (1983) (see Further Reading):

$$\begin{aligned} u_x &= \Delta u_x\left[2(1-\nu)\frac{\partial f}{\partial y} - y\frac{\partial^2 f}{\partial x^2}\right] + \Delta u_y\left[-(1-2\nu)\frac{\partial f}{\partial x} - y\frac{\partial^2 f}{\partial x \partial y}\right] \\ u_y &= \Delta u_x\left[(1-2\nu)\frac{\partial f}{\partial x} - y\frac{\partial^2 f}{\partial x \partial y}\right] + \Delta u_y\left[2(1-\nu)\frac{\partial f}{\partial y} - y\frac{\partial^2 f}{\partial y^2}\right] \end{aligned} \quad (4.68)$$

Here $f = f(x, y)$ is a displacement function, Δu_x and Δu_y are the displacement discontinuities in x and y across the plane of the structure, and ν is Poisson's ratio. The boundary conditions are given in terms of a uniform displacement discontinuity across the planar cut:

$$\begin{aligned} \Delta u_x &= u_x(|x| \le a, y = 0^-) - u_x(|x| \le a, y = 0^+) \\ \Delta u_y &= u_y(|x| \le a, y = 0^-) - u_y(|x| \le a, y = 0^+) \end{aligned} \quad (4.69)$$

A closing displacement discontinuity is positive; opening is negative; left lateral is positive; and right lateral is negative. The displacements induced by these boundary conditions decrease away from the surface of discontinuity (e.g. see **Figure 4.30**) and approach zero at distances that are very large compared to the length, $2a$.

The displacement function $f(x, y)$ in (4.68) is:

$$f(x,y) = \frac{-1}{4\pi(1-\nu)}\left[\begin{array}{l} y\left\{\tan^{-1}\left(\dfrac{y}{x-a}\right) - \tan^{-1}\left(\dfrac{y}{x+a}\right)\right\} \\ -(x-a)\ln\left\{\sqrt{(x-a)^2 + y^2}/C\right\} \\ +(x+a)\ln\left\{\sqrt{(x+a)^2 + y^2}/C\right\} \end{array}\right] \quad (4.70)$$

The constant C has the dimension length, and is used only to cancel the length dimension of the square root term, so one can take the natural logarithm of the quotient. The term $-\ln(C)$ goes to zero when derivatives are taken using (4.68), so it does not affect the displacement components.

4.11.3 Elastic Deformation Associated with Slip on the Tanna Fault

The displacement field associated with slip on a fault provides a good application of the elastic model for a plane of uniform displacement discontinuity (**Figure 4.37c** or **d**). We use displacement data derived from comparisons of the geographic positions of 71 geodetic stations after the great Kwantô earthquake of 1923 with those after the magnitude 7.3 Idu earthquake on November 26, 1930 (**Figure 4.38a**). The earthquake of 1930 occurred on the Idu Peninsula, which juts southward into the Pacific Ocean from the island of Honshu, Japan. At the time of the earthquake offsets as great as 3.8 m occurred on the Tanna Fault, which strikes roughly north–south near the middle of the peninsula, but is broadly curved over a distance of about 24 km.

The hash marks on the mapped fault (**Figure 4.38a**) indicate that the east side moved down relative to the west side near the south end of the rupture, and the west side moved down relative to the east side near the north end. However, the magnitude of the vertical displacements did not exceed 30 cm. In contrast, horizontal displacements near the fault were up to about 2 m in magnitude. The horizontal displacements near the fault are directed to the south on the west side, and to the north on the east side. Thus, the relative motion across the Tanna fault was predominantly left-lateral strike slip. Over distances of about 20 km on either side of the fault the horizontal displacement vectors form somewhat

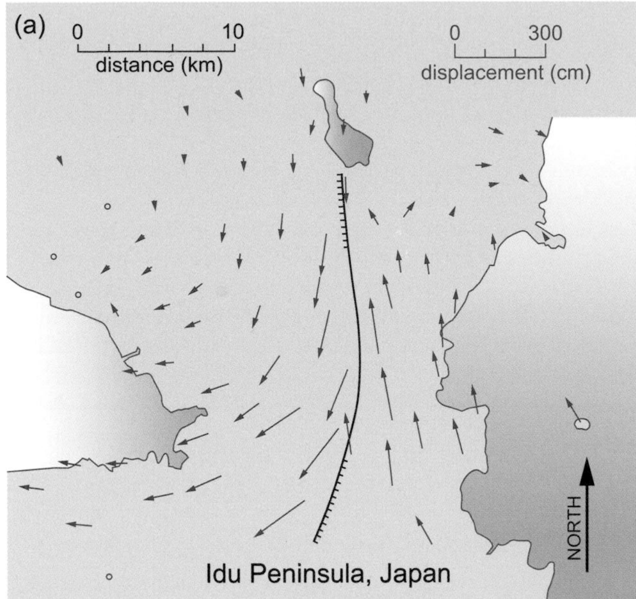

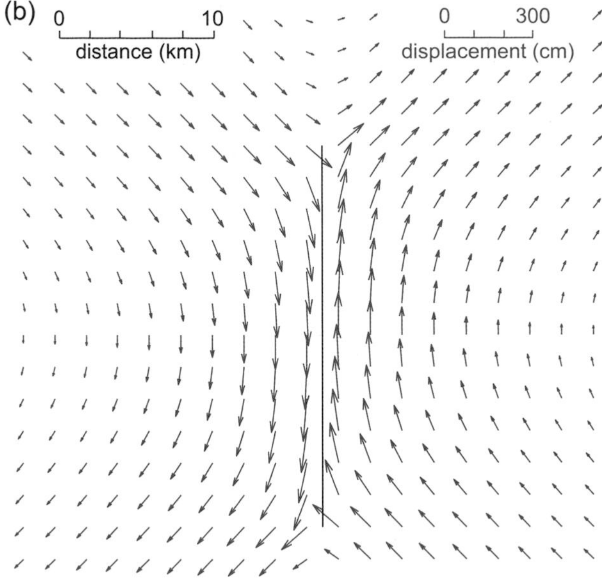

Figure 4.38 (a) Horizontal displacements from 71 geodetic stations associated with slip on the Tanna fault during the North Idu earthquake. From Chinnery (1961) after Yamaguti (1937). (b) Model fault displacement field for an elastic rock mass with a plane of uniform displacement discontinuity. Model details described in text. Modified from Pollard and Segall (1987).

ragged clockwise patterns. The displacement magnitudes decrease from about 2 m near the fault to less than 0.5 m at distances of ten to fifteen kilometers east and west of the fault.

The model displacement field (**Figure 4.38b**) was generated from a solution to the elastic boundary-value problem provided by substituting (4.70) into (4.68). Northing and Easting on the map are taken as the coordinates x and $-y$ for the model. For boundary conditions on the fault we take $\Delta u_x = 3.8$ m and $\Delta u_y = 0$. Because this is a two-dimensional solution, the vertical displacements are zero. The model fault is perfectly straight and scaled to the approximate length of the rupture on the Tanna fault, 24 km. The horizontal displacement vectors were computed at points on a square grid with spacing in x and y equal to 2 km.

Note the similarities between the displacement field for the Tanna fault (**Figure 4.38a**) and the model fault (**Figure 4.38b**). Displacements just to the west of the model fault are directed to the south; those just to the east are directed to the north. The model displacement magnitudes are approximately equal to those on the map from the Idu earthquake. On either side of the model fault, and over distances comparable to the fault length, the horizontal displacement vectors form clockwise patterns. The displacement magnitudes decrease from about 2 m near the model fault to less than 0.5 m at distances of ten to fifteen kilometers east and west of the fault. On the other hand the displacement field from the Idu earthquake lacks the regularity of the model displacement field. The differences between these two fields provide important clues about how one might improve the model.

Recapitulation

Mechanical testing of samples in the laboratory and measurements of solid Earth tides demonstrate that rock responds to loading approximately as a *linear elastic solid* over length scales ranging from centimeters to kilometers. That is, the strain increases linearly with stress, and this deformation is recoverable. The slope of the stress–strain graph is a measure of elastic stiffness, which varies from about 1 to 100 GPa. The load carrying capacity of rock is limited in uniaxial tension and compression by the tensile and compressive strengths, which are about 10 MPa and 100 MPa, respectively. Rock that breaks apart shortly after the stress reaches the strength is called *brittle*, and brittle deformation is manifest in outcrop as fractures and faults that have broken the rock along surfaces of displacement discontinuity. In conventional triaxial tests rock strength increases with confining pressure and decreases with pore fluid pressure. The Coulomb criterion provides a theoretical basis for strength under compression and shear stress.

Stress is a tensor quantity that measures the force per unit area acting on faces of an infinitesimal cubic element at a point in the material continuum. The shear stresses on conjugate faces of the cubic element are equal, resulting in six independent stress components. These are arranged in a symmetric 3 by 3 matrix, from which one can calculate the three principal stress magnitudes and orientations, and use these to characterize the state of stress at a point.

Small strain is a tensor quantity that measures changes in the length and orientation of infinitesimal material lines at a point in the material continuum. It also is made up of six independent components that may be arranged in a symmetric 3 by 3 matrix. The small strain components are dimensionless, because they are ratios of lengths, and they are related to partial derivatives of the displacements with respect to the coordinates through six kinematic equations. The gradient of the displacement vector field is the displacement gradient tensor, and the elements of the small strain matrix are equal to the elements of the matrix composed of one half the sum of the displacement gradient matrix and its transpose.

Stress and small strain for linear elastic materials are related by six constitutive equations, called *Hooke's law*. These equations include two elastic constants, Young's modulus and Poisson's ratio. Other elastic constants are the shear modulus and Lame's constant, but only two constants are required for an isotropic elastic solid.

Cauchy's general equations of motion are reduced to *Navier's equations* of motion for the linear elastic solid and these are used to model the development of geologic structures in the brittle portions of Earth's lithosphere. The displacement, strain, and stress fields vary systematically with position and time as brittle structures evolve. The two-dimensional problem of a plane of displacement discontinuity may be used to model: (1) opening fractures such as igneous dikes, hydrothermal veins, and joints; (2) closing structures such as pressure solution seams and compaction bands; and (3) shear fractures and faults. In this way a complete mechanical understanding of elastic–brittle deformation is obtained and used to interpret geologic data.

REVIEW QUESTIONS

The following questions are designed to highlight the expected *learning outcomes* for this chapter. Each question is taken directly from the material in the chapter and, for the most part, in the same sequence that it appears in the chapter. If an answer is not forthcoming, students are advised to read the relevant section of the chapter and discover the answer.

4.1. Consider a cylindrical bar of rock suitable for a uniaxial test in a universal testing machine. Define the axial stress and axial strain for this bar and give the SI units for each. Use these two quantities to write *Hooke's law* in its simplest, one-dimensional form.

4.2. The proportionality constant in Hooke's law for the uniaxial tension test is a measure of the *stiffness* of the rock sample. If sufficient tension is applied in this test, the sample will fracture and the peak stress is called the tensile *strength*. Compare and contrast stiffness and strength, including the SI units and representation on a plot of axial stress versus axial strain.

4.3. Describe how *Poisson's ratio* is determined using a uniaxial compression test. In doing so explain why a negative sign is needed in the definition. Also explain why a value of ½ for Poisson's ratio is associated with an incompressible material.

4.4. The following Google Earth image is taken from the file:
Question 04 04 Ship Rock dike.kmz. Open the file and answer the questions in the property box for this placemark.

4.5. Describe how the words *extension*, *tension*, *contraction*, and *compression* are associated with the physical quantities stress and strain. In doing so, indicate how the signs (positive or negative) of these quantities are assigned.

4.6. What is a joint set? Explain how *abutting* relationships are used to distinguish the relative ages of two joint sets in the same outcrop.

4.7. Draw a generic graph of axial stress versus axial strain for uniaxial tests of rock samples. Why do you need to use both quadrant one and quadrant three on this graph? What characterizes linear elastic deformation on this graph?

4.8. Write down the typical range of values for *Young's modulus* of elasticity for rock including the SI units. Name rock types that usually are at the *stiff* end of this range, and the rock types that usually are at the *soft* end of this range.

4.9. Uniaxial tests reveal a profound difference in rock strength in tension and compression. Cite typical values of *strength* in tension and compression for granite and sandstone. What are the implications of these data for structures such as joints and faults in Earth's brittle lithosphere?

4.10. Explain how the two *subscripts* used to identify a particular stress component relate to the coordinate system and the faces of a cubic element that are aligned with that system. Draw the cubic element in two dimensions for the (y, z)-plane including coordinates and negative representations of the stress components using arrows.

4.11. Use the *principal normal stresses* to describe what is meant by positive definite and negative definite states of stress. What are the values of the shear stresses acting on the cubic element upon which these principal stresses act?

4.12. What are the inferred principal stresses and their orientations within a cylindrical rock specimen in a conventional triaxial compression test? Why is this called "*conventional*" and what would the inferred stress state be in a true triaxial test? Why have we used the word "*inferred*" in these questions?

4.13. On a plot of axial stress versus axial strain, plot a curve in quadrant three that represents *elastic*, *ductile*, and *brittle* deformation during a single conventional triaxial compression test. Describe these behaviors in terms of the slopes of the curve.

4.14. Describe how the conventional triaxial compressive strength changes with increasing *confining pressure*. Sketch three cylindrical specimens that display: (a) wedge and splitting fractures; (b) shear fractures; and (c) distributed shearing. Indicate how the confining pressure changes to produce these different failure styles.

4.15. The *effective stress* tensor is defined by adding the pore fluid pressure to the normal stress components and leaving the shear stress components unchanged. Explain why a greater pore pressure produces an effective stress that is less compressive. Also, explain why rock strengths are the same for a given *effective confining pressure*, regardless of the actual confining pressure and pore pressure.

4.16. A fracture criterion for *opening* fractures is written $\sigma_1 = T_u$. Name and describe the physical quantities in this equation along with their SI units. Use this criterion to explain why a pattern of joints on a bedding surface might correspond to the pattern of principal stress trajectories for σ_3 as the joints formed.

4.17. A fracture criterion for *shear* fractures is written $|\sigma_s| = S_o - \mu_i\sigma_n$. Name and describe the physical quantities in this equation along with their SI units. Use this criterion to explain how a pattern of conjugate shear fractures might correspond to the orientations of the principal stresses during brittle deformation.

4.18. **Figure 4.30** illustrates the displacement field around an opening fracture, such as a joint, vein, or dike. Pick locations on this figure, not associated with the capital letters annotating it, to identify the following small elastic strains: *extension*, *contraction*, and *shearing*. Use your choices to describe the heterogeneous strain field around the opening fracture.

4.19. Vectors overlying material lines in the undeformed and deformed states are used in **Figure 4.32** to derive kinematic relationships between displacements and the small normal and shear strains. Use this figure to describe the *approximations* employed to define the small strains.

4.20. In the context of Newtonian mechanics, stress *causes* strain. One might suppose that each element of the stress matrix $[\sigma]$ causes only the corresponding element of the small strain matrix $[\varepsilon]$. However, Hooke's law relates the normal strain in the y-direction to all three normal stresses: $\varepsilon_{yy} = [\sigma_{yy} - \nu\,(\sigma_{zz} + \sigma_{xx})]/E$. Explain how this comes about and describe the two elastic constants in this equation and their SI units.

4.21. One of Cauchy's first laws of motion is quoted in Section 4.11 of this chapter as:

$$\rho\frac{Dv_z}{Dt} = \frac{\partial\sigma_{xz}}{\partial x} + \frac{\partial\sigma_{yz}}{\partial y} + \frac{\partial\sigma_{zz}}{\partial z} + \rho g_z^*$$

Name the *dependent* and *independent* variables and the *constants* in this equation, along with their SI units. Explain why these equations cannot be solved without introducing *constitutive laws*. In doing so, indicate what a constitutive law is.

4.22. Describe in words how the equation in Question 4.21 is modified to find the corresponding Navier Displacement Equation of Motion for the linear elastic solid. Name the *dependent* and *independent* variables and the *constants* in this equation along with their SI units.

MATLAB EXERCISES FOR CHAPTER 4: ELASTIC BRITTLE DEFORMATION

Linear elastic behavior dominates the deformation of rock in Earth's lithosphere. In the upper 15 to 20 km elastic deformation is accompanied by localized brittle fracturing or faulting. Exercises for Chapter 4 explore the concepts of rock stiffness, compressibility, and strength in the context of elastic deformation. They also demonstrate that MATLAB is particularly adept at solving the classic eigenvalue problem for the principal values and principal orientations of the stress and small strain tensors, key physical quantities in the analysis of brittle deformation. Other exercises use data from laboratory tests on various rock types to confirm that the limit on elastic behavior, set by the compressive strength, increases with depth in Earth's lithosphere. Exercises on the kinematics of elastic deformation relate small strains to displacement gradients. A series of exercises show how to use the elastic constitutive equations to find Navier's equations of motion, which can be solved for the stress, strain, and displacement fields. To demonstrate one important solution, students model the displacement field associated with slip on the Tanna fault in Japan using MATLAB. These exercises help students understand why linear elasticity is such a powerful tool for structural geologists investigating rock deformation in Earth's upper lithosphere. www.cambridge/SGAQI

FURTHER READING

For citations in figure captions see the reference list at the end of the book.

Barber, J. R., 2010. *Elasticity*. Springer, New York.
This textbook for engineers provides a very approachable coverage of many problems in elasticity theory, emphasizing the analytical approach to such problems and providing excellent background material for the structural geologist.

Bieniawski, Z. T., 1984. *Rock Mechanics Design in Mining and Tunneling*. A. A. Balkema, Rotterdam.
This textbook for mining and civil engineers focuses on applications of rock mechanics to the design process for mining and tunneling.

Cosgrove, J. W., and Hudson, J. A., 2016. *Structural Geology and Rock Engineering*. World Scientific Publishing Company, Inc, London.
This textbook brings the disciplines of structural geology and rock engineering together and offers case study examples of how to integrate the concepts and principles of the two disciplines to achieve safe and sustainable engineering projects using rock.

Crouch, S. L., and Starfield, A. M., 1983. *Boundary Element Methods in Solid Mechanics (with Applications in Rock Mechanics and Geological Engineering)*. George Allen & Unwin, London.
The elastic solution for the plane of displacement discontinuity (4.70) is derived and applied in this book, which lays out the theory for boundary elements, one of the standard techniques for solving elastic boundary-value problems using numerical methods.

Gordon, J. E., 1976. *The New Science of Strong Materials, or Why You Don't Fall Through the Floor*, 2nd edition. Princeton University Press, Princeton, NJ.

Written for the layperson, this treatment of the history of strength of materials provides a useful perspective and is an enjoyable read.

Hoek, E., and Martin, C., 2014. Fracture initiation and propagation in intact rock–a review. *Journal of Rock Mechanics and Geotechnical Engineering* 6, 287–300.
The authors of this paper review the rock mechanics literature on laboratory testing of intact rock and show that brittle failure at modest confining pressures is a process that begins with tensile fracturing and is followed by coalescing fractures, strain localization, and sample-scale shear failure.

Jaeger, J. C., Cook, N. G. W., and Zimmerman, R. W., 2007. *Fundamentals of Rock Mechanics*, 4th edition. Blackwell Publishing, Oxford.
This is the fourth edition of a classic book on rock mechanics. Most topics of engineering rock mechanics addressed in this book focus on brittle elastic deformation.

Kwasniewski, M., Li, X., and Takahashi, M., 2013. True triaxial testing of rocks, in: Kwasniewski, M. (Ed.) *Geomechanics Research Series*, Volume 4. CRC Press/Balkema, Leiden, The Netherlands.
The challenging problems that should be addressed in undertaking true triaxial tests, and the interesting data taken from tests, are described and discussed in this edited volume.

Lockner, D. A., 1995. Rock failure, in: Ahrens, T. J. (Ed.) *Rock Physics & Phase Relations: A Handbook of Physical Constants*. American Geophysical Union, Washington, DC, pp. 127–147.
Brittle failure mechanisms are described in this review that includes a helpful overview, an extensive reference list, and summary plots of data on compressive and tensile strength for unconfined samples, and for differential strength as a function of confining pressure.

Lockner, D. A., and Beeler, N. M., 2003. Rock failure and earthquakes, Chapter 32, in: Lee, W. H. K., Kanamori, H., Jennings, P. D., and Kisslinger, C. (Eds.) *International Handbook of Earthquake and Engineering Seismology*. Academic Press, San Diego, CA, pp. 505–537.
This review considers earthquakes in the brittle crust and reviews laboratory tests that address fault strength and stability as well as theoretical concepts related to fracture of intact rock and frictional sliding on existing faults.

Paterson, M. S., and Wong, T-f., 2005. *Experimental Rock Deformation – The Brittle Field*. Springer-Verlag, Berlin.
The brittle properties of rock as measured in laboratory tests and the physics of rock deformation in the brittle field are comprehensively reviewed in this book, which includes an extensive reference list covering the relevant literature.

Segall, P., 2010, *Earthquake and Volcano Deformation*. Princeton University Press, Princeton, NJ.
This advanced textbook builds upon the elementary mechanics and structural geology covered here, with a focus on active deformation related to faulting.

Terzaghi, K, 1943. *Theoretical Soil Mechanics*. John Wiley and Sons, New York.
This classic book covers the concept the author conceived called effective stress, and the impact of pore fluid pressure on strength of permeable materials.

Timoshenko, S. P., and Goodier, J. N., 1970. *Theory of Elasticity*, 3rd edition, McGraw-Hill Book Company, New York.
This third edition of a classic book on elasticity covers most aspects of two- and three-dimensional theory with applications to a host of engineering problems, some of which are relevant to rock deformation.

Wong, T.-f., and Baud, P., 2012. The brittle-ductile transition in porous rock: a review. *Journal of Structural Geology* 44, 25–53.
This paper reviews the laboratory behavior of porous rock as a function of confining pressure and identifies a transition from brittle faulting with dilatant failure to delocalized cataclasis with shear-enhanced compaction and strain hardening.

Chapter 5
Elastic–Ductile Deformation

Introduction

Chapter 5 introduces elastic–ductile deformation of rock using field observations of geologic structures, laboratory tests of mechanical behaviors, and theoretical concepts relevant to strength and ductility. We show that ductile deformation takes over from brittle deformation as confining pressure and temperature increase, and as the rate of deformation decreases, so ductile behavior is characteristic of rock deformation in the deeper levels of Earth's lithosphere. To quantify ductile deformation, we introduce the idealized elastic–plastic solid, a mathematical construct in which strain is linearly proportional to stress and is recoverable up to the yield point. Thereafter, the stress remains constant while permanent strain accumulates. We review the mechanisms of plastic deformation that involve the motion of dislocations in the crystal lattice. The dependence of plastic deformation on time motivates consideration of flow laws that relate strain rate to the state of stress. We conclude by showing how Newton's laws, embodied in Cauchy's equations of motion, model ductile deformation of salt flowing toward a rising diapir.

Ductile deformation of rock can occur as *distributed* strain throughout a large volume of rock, but also may be spatially limited to narrow, tabular regions called ductile shear zones. Shear zones are examples of *localized* ductile strain. Unlike the displacement discontinuity across fractures that open, or across faults that slip in brittle rock, the displacement fields across ductile shear zones are continuous. Because the stress in the rock surrounding a shear zone did not reach the yield strength, the deformation is dominantly elastic. In the lower parts of Earth's lithosphere, and throughout the asthenosphere, the mechanical behavior of rock usually is elastic and ductile, hence the chapter title: elastic–ductile deformation.

While the drama of major earthquakes along the boundaries of tectonic plates highlights brittle deformation of rock in the upper parts of Earth's lithosphere, ductile deformation prevails in the lower parts. At depths greater than about 10 to 15 km, earthquakes are scarce, yet the relative plate motion continues, presumably accompanied by ductile deformation. Whether this deformation is distributed or localized is an open question. Even at Earth's surface some faults display continuous but very slow relative motion, suggestive of *localized* ductile deformation. In exhumed metamorphic rocks the occurrence of highly contorted folds with little fracturing is a clear signal that the stress state reached the yield strength, and that ductile deformation led to large irrecoverable strains *distributed* throughout the folded rock mass. Interest in the tectonic history of metamorphic rock, and in the coupling of brittle and ductile deformation along plate boundaries, motivate structural geologists to understand elastic–ductile deformation.

5.1 IDEALIZED PLASTIC SOLIDS

If you place a sphere of modeling clay on a table (**Figure 5.1a**) and move a *light* weight on and off the top of the clay, a sensitive strain gauge would detect a very small recoverable deformation as the sphere slightly deforms and then returns to its original diameter. These observations indicate that the clay is elastic: the applied stress produces a *distributed* and *recoverable* elastic strain, which we introduced in Section 4.1. Young's modulus of elasticity for clay soils ranges from about 0.5 to perhaps 50 MPa, so clay is much softer than most rocks, which range from about 1 to 100 GPa in elastic stiffness (refer to **Table 4.1**). On the other hand, most clay is stiffer than the rubber band described in Section 4.1, which has a Young's modulus of about 0.5 MPa.

When a heavy weight is placed on the clay sphere (**Figure 5.1b**), the clay flows outward in all directions, and the sphere flattens to a thick disk without fracturing. Upon removing the heavy weight (**Figure 5.1c**) a small elastic strain is recovered, but the clay remains flattened and does not return to its original spherical shape. From this behavior we understand that an applied stress equal to the yield strength produces a *distributed* and *irrecoverable* strain that is much greater than the elastic strain. When the same experiment is performed on a rubber racquetball, the result is quite different. The ball is spherical before loading (**Figure 5.1d**); flattens to a thick disk without fracturing under the heavy weight (**Figure 5.1e**); but springs back to its original spherical shape upon unloading (**Figure 5.1f**). The deformation is *distributed* and *recoverable* for this elastic material.

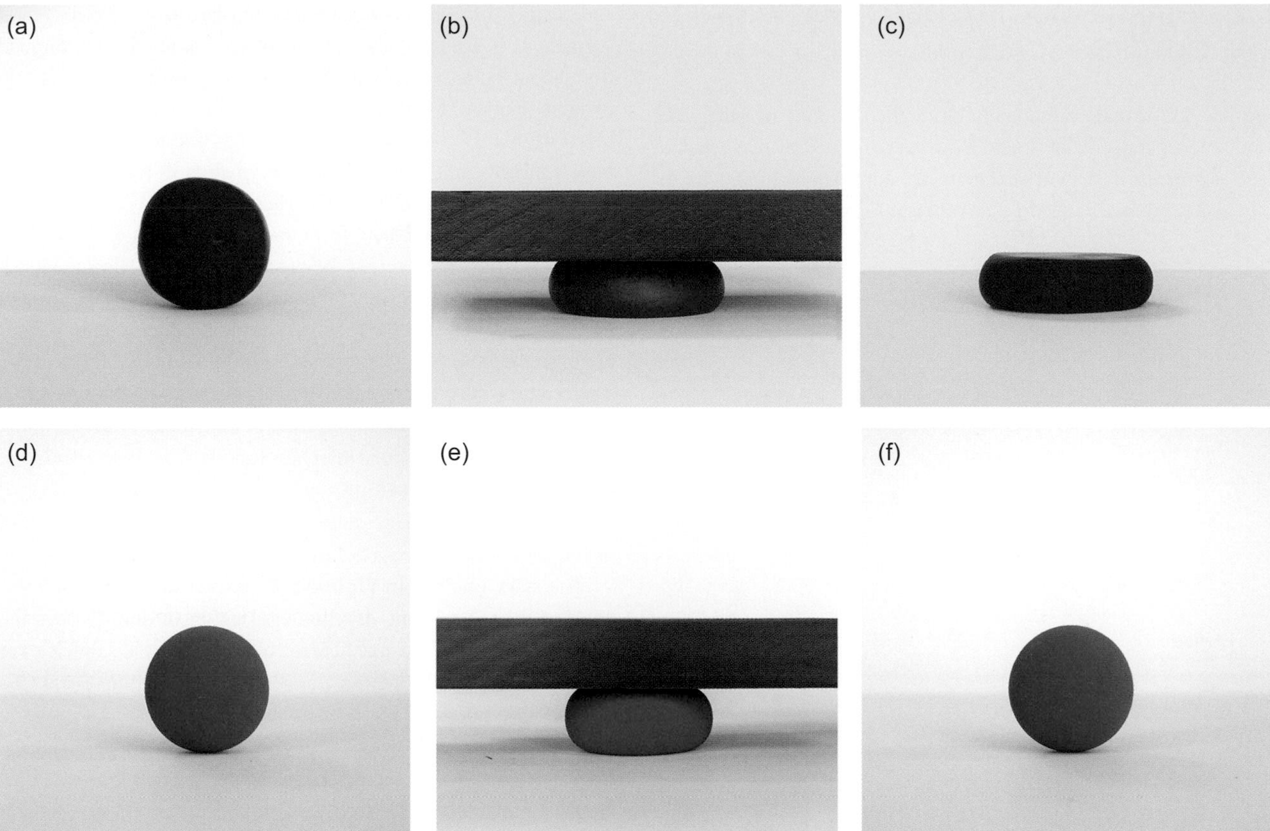

Figure 5.1 Three stages in the deformation of a modeling clay sphere (red) and a rubber racquetball (blue). (a) Unloaded sphere rests on tabletop. (b) Weight (gray) imposes stress equal to the yield strength of the clay and it deforms to a flattened disk. (c) Weight removed; clay remains a flattened disk. (d) Unloaded rubber ball rests on tabletop. (e) Weight (gray) imposes stress and ball flattens. (f) Weight removed; ball springs back to its original shape. Photography by Richard Stultz.

The behavior of the modeling clay pictured in **Figure 5.1a–c** is typical of materials that we characterize broadly as ductile. This behavior is in sharp contrast to what we described in Section 4.4 as brittle. Under sufficient stress ductile solids deform by *flowing*, such that significant strain accumulates, without the prominent *fracturing* that is characteristic of brittle solids. Left for many hours, the unloaded sphere and disk (**Figure 5.1a,c**) do not permanently deform, but retain their shapes. Therefore, the clay has *strength*: it can stand up under its own weight without significant deformation. In this sense the clay is solid, and we contrast that with materials described in Chapter 6 that are liquid. Liquids must be contained, for example in a cup or bottle, or they will flow under their own weight.

In Section 4.1 we introduced the idealized elastic solid that was conceived by Robert Hooke. For a cylindrical bar of that material, the axial strain is proportional to the axial stress, and Young's modulus measures the uniform slope of the stress versus strain graph (**Figure 4.2**). For the idealized elastic solid no limit was placed on the magnitude of the stress or strain, but laboratory tests reviewed in Section 4.4 and schematically summarized in **Figure 4.7** indicate that rock has a limiting stress, called the strength. Those tests also indicate that the linear relationship between stress and strain changes to a non-linear relationship at a particular stress called the yield point (point B in **Figure 4.7**). In this section we focus on an idealized plastic solid, which has a well-defined yield point and a specific relationship between stress and strain.

Consider a cylindrical bar loaded by an axial stress, σ_a, that responds to this loading with an elastic axial strain, ε_a, and a linear stress versus strain relationship with a slope equal to Young's modulus, E (**Figure 5.2**). A tensile stress induces an extension (quadrant 1) and a compressive stress induces a contraction (quadrant 3). The deformation is linear and elastic (recoverable) as long as the stress magnitude is less than the yield strength: $|\sigma_a| < \sigma_{ys}$. Upon reaching the yield point, yp, the axial strain continues to accumulate while the stress remains constant: $|\sigma_a| = \sigma_{ys}$. If the bar is unloaded while deforming plastically, the elastic strain is recovered along a straight line (dashed) with slope equal to Young's modulus. When unloaded to zero stress, the permanent plastic strain is equal to ε_p, the amount of strain accumulated during the plastic yielding. Upon reloading, the dashed (elastic) line is followed and additional plastic strain is accumulated when the stress equals the yield strength.

The plastic solid is a special case of the more broadly defined category of ductile materials, all of which are solids that are capable of flowing under sufficient stress. Although some rocks approximate the *bilinear* stress–strain relations shown in **Figure 5.2** for the idealized elastic–plastic solid, others have non-linear relations that include both ductile and brittle behaviors. We review these using selected laboratory results in Section 5.4.

5.2 ELASTIC–DUCTILE DEFORMATION AT THE OUTCROP SCALE

Figure 5.3 reveals remarkably different behavior of otherwise similar leucocratic dikes, A and B, when deformed by two echelon fault segments. The fault segments are approximately parallel to one another, but separated by a 3 cm right step. Dike A is broken and sharply offset by about 6 cm in a left-lateral sense across fault segment 1. In contrast, dike B was rotated about 75°, and thinned from about 4 cm to 1 cm within the right step between the right tip of fault segment 1 and the left tip of fault segment 2. The total offset of dike B is about 11 cm and this is due to slip on the two fault segments *and* to distributed deformation in the ductile shear zone between the fault segments. There, the highly deformed dike may be traced continuously from one fault segment to the other.

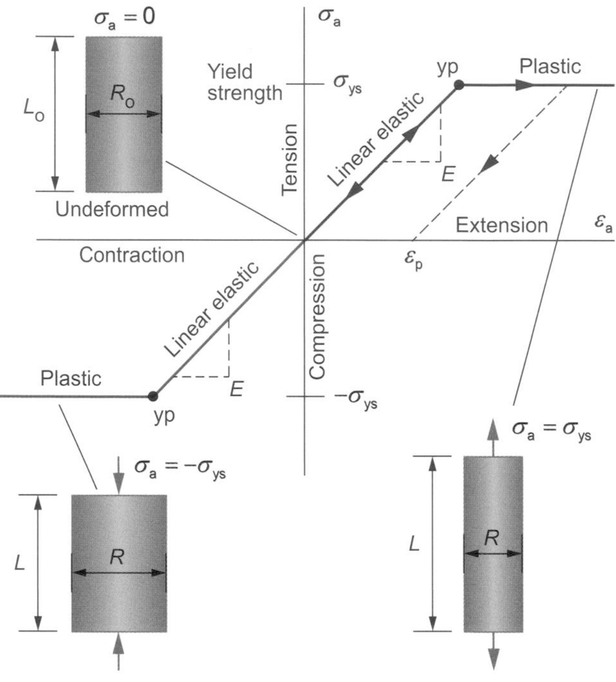

Figure 5.2 Axial stress, σ_a, is plotted versus axial strain, ε_a, for a cylindrical bar of elastic–plastic material. The slope of the linear elastic line is Young's modulus, E. Tension tests (red) plot in the first quadrant and compression tests (blue) plot in the third quadrant. The stress first reaches the strength at the yield point, yp, and thereafter strain accumulates at constant stress equal to the yield strength. See Jaeger et al. (2007).

Outside the small region between the two fault segments in **Figure 5.3** evidence is scant for ductile deformation. At the scale of the entire fault the offset is about 10 cm and the trace length is about 10 m, so the ratio of offset to length is approximately 1:100 (1%). This is a measure of the average strain associated with faulting and is close to or within the limit of elastic deformation. In contrast, the ratio of deformed to original thickness for dike B is about 1:4 (25%), corresponding to a much greater strain, one that is indicative of ductile deformation. This outcrop, therefore, captures a beautiful example of elastic–ductile deformation, the subject of this chapter.

The stark contrast in **Figure 5.3** between the brittle deformation of dike A and the ductile deformation of dike B begs the question: why did the mechanical behavior vary so much over distances of a few decimeters during the same faulting event? To address this question we need to understand how deformation depends upon conditions such as pressure, temperature, and the state of stress. These topics are taken up in Section 5.4 of this chapter, but first, in the next section, we describe evidence for ductile deformation at the scale of Earth's lithosphere.

5.3 ELASTIC–DUCTILE DEFORMATION AT THE CRUSTAL SCALE

In Section 4.2 we introduced Earth tides to demonstrate the elastic deformation of the lithosphere. Over the 24 hour time scale of Earth's rotation, and the 29.5 day lunar cycle, the Arbuckle dolomite, buried 840 m below the surface in central Oklahoma, moves up and down about 30 cm in a remarkably repeatable fashion

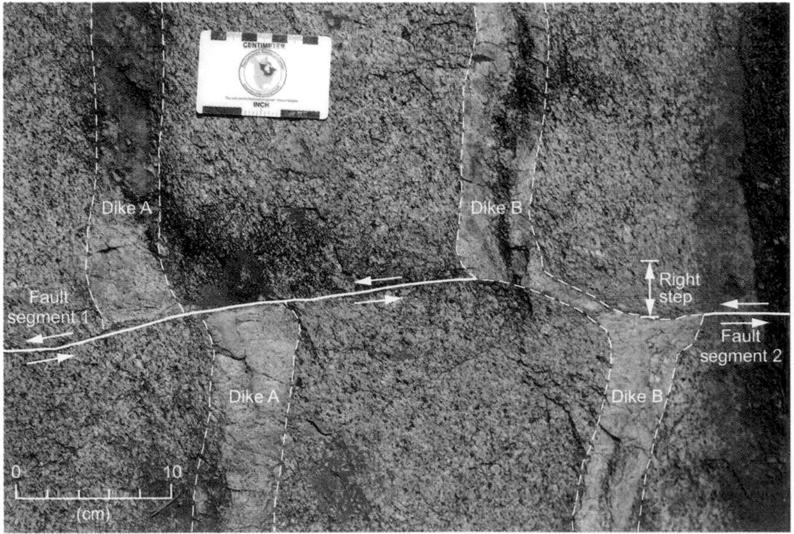

Figure 5.3 Two echelon segments of a left-lateral fault in Lake Edison Granodiorite, Sierra Nevada, CA with two offset dikes. Dike A is broken and offset about 6 cm along fault segment 1, but little deformed. Dike B is offset about 11 cm and is rotated, stretched and thinned in the right step between fault segments 1 and 2. Photograph by J. M. Nevitt.

(**Figure 4.3**). Although these data represent the motion at a single point, it is clear from global geodetic observations that the entire lithosphere is in motion due to the gravitational interaction of Earth and Moon, and that this deformation is recoverable and repeatable: it is *elastic*. However, if the time scale is lengthened to years or decades, a variety of geophysical and geodetic observations demonstrate that elasticity alone is not sufficient to describe and understand deformation of Earth's lithosphere and asthenosphere. In this section we use the time-dependent displacement field at Earth's surface after a large earthquake to demonstrate the inelastic *ductile* deformation of the asthenosphere.

Large strike-slip earthquakes in California are associated with slip on faults that extend to depths of 10 to 15 km in Earth's lithosphere (**Figure 5.4**). This slip causes deformation throughout the nearby lithosphere and asthenosphere that is associated with a measurable displacement field at Earth's surface.

Geodetic surveys using the Global Positioning System (GPS) after the 1999 M_w 7.1 Hector Mine earthquake (**Figure 5.5**) demonstrate that Earth's surface continues to displace after these earthquakes on a multi-year time scale. Apparently, the stress field in the region surrounding the faults is perturbed by the elastic deformation that accompanies fault slip, and this change in stress induces time-dependent flow in the underlying asthenosphere. That continuing inelastic deformation in the asthenosphere, in turn, deforms the overlying lithosphere, leading to displacements that can be measured at Earth's surface by GPS surveys. This conceptual model requires inelastic behavior for the asthenosphere and raises the questions: what is the constitutive equation for this behavior, and what are the physical mechanisms for this deformation?

Laboratory creep tests on olivine aggregates, described in more detail in Section 5.4.6, provide important constraints on the mechanical behavior of rock in the asthenosphere. These tests suggest a power-law relationships between the rate of axial strain, $\dot{\varepsilon}$, and the differential stress, $\Delta\sigma$:

$$\dot{\varepsilon} = A(\Delta\sigma)^n \exp(-Q/RT) \tag{5.1}$$

This relationship is called a flow law. The laboratory tests are designed to determine the constant A, the power-law exponent, n, and the activation energy, Q. This is a thermally activated deformation, which means the strain rate is a function of the absolute temperature, T. In this case as the temperature increases, the strain rate increases. The testing equipment measures the differential stress, axial strain rate, and absolute temperature, and the universal gas constant is $R = 8.3145\ \mathrm{J\,mol^{-1}\,K^{-1}}$. If n has a value close to one, the strain rate is approximately linearly related to the differential stress, and the deformation is referred to as Newtonian flow, the subject of Chapter 6. On the other hand, if n is greater than one, the deformation is referred to as power-law flow, or non-linear flow, one of the topics of this chapter. A power-law material with $n = 3.5$ provides a good match to the GPS displacement fields illustrated in **Figure 5.5**.

The length and time scale of the laboratory creep tests, and the conditions of confining pressure and temperature in those tests, may not match those at appropriate depths in the asthenosphere. On the other hand, one can investigate models of post-earthquake displacements at Earth's surface that constrain the value of n while utilizing *natural* length scales and appropriate conditions of confining pressure, temperature, and strain rate. The results of such a modeling effort are summarized in **Figure 5.6**, which is a time series for the horizontal displacement in millimeters versus time in years for the geodetic station OPCX in **Figure 5.5**. Three flow laws overlay the displacement data: two are Newtonian, $n = 1$, and have viscosities of 2.5×10^{18} Pa s and 2.5×10^{19} Pa s; the third is power law with $n = 3.5$. Although the two Newtonian models adequately match the slope of the data at earlier and later times respectively, the power-law model satisfies the data over the entire time.

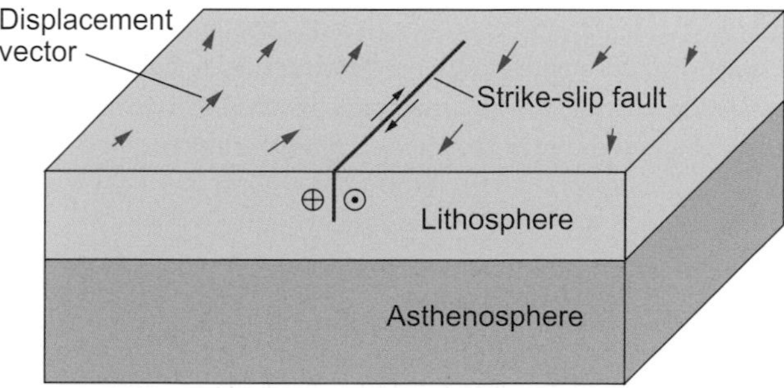

Figure 5.4 Sketch of a crustal scale strike-slip fault. The pair of arrows along the fault indicate the sense of slip (right lateral), but not the magnitude, which is much less than the height or length of the fault. Blue arrows are schematic displacement vectors at Earth's surface, again with exaggerated magnitudes.

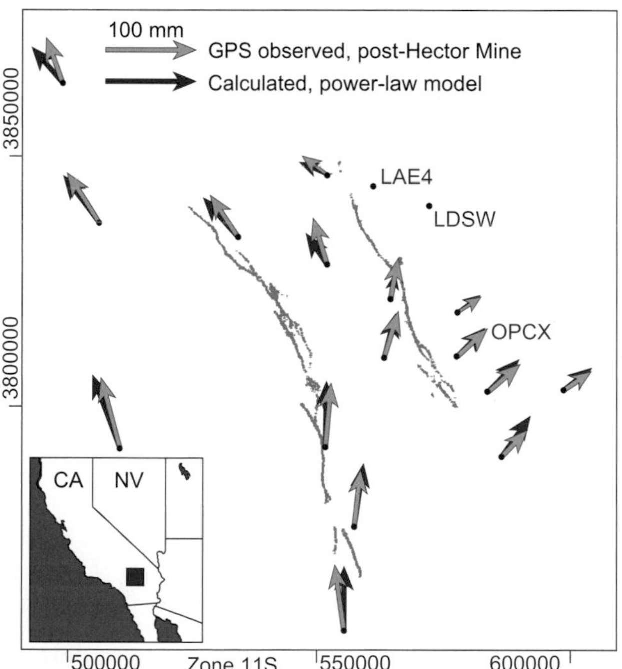

Figure 5.5 Observed GPS horizontal displacement vectors after the 1999 Hector Mine earthquake (October 1999–December 2002, green), compared to model displacement vectors (red) using a power-law material with $n = 3.5$ for the asthenosphere. Modified from Freed and Bürgmann (2004), Figure 1.

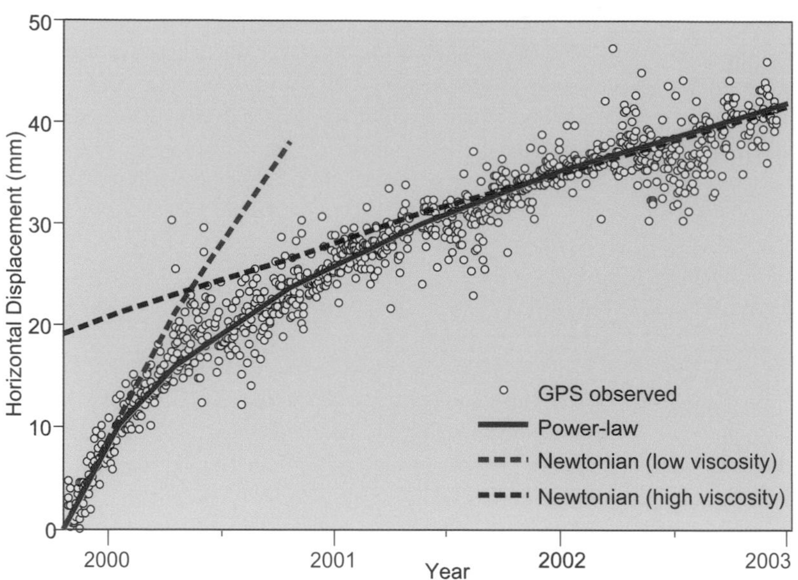

Figure 5.6 Horizontal displacements obtained from GPS data for station OPCX (**Figure 5.5**) plotted versus time since the 1999 Hector Mine earthquake. The high viscosity line is shifted upward from the origin to match the slope of the data. Modified from Freed and Bürgmann (2004), Figure 3.

The mechanism responsible for deformation in the laboratory creep tests on olivine aggregates under conditions that yield a power-law with $n = 3.5$ has been identified as dislocation creep. Dislocations are line defects at the nanometer scale in the crystal lattice that break and restore the bonds between atoms as the dislocation moves. With this mechanism individual crystals can undergo large deformations without fracturing. In other words, this is a physical mechanism for ductile deformation. We devote Section 5.6 to a detailed description of dislocations and show how their motion distorts the crystal lattice. In the next section we describe laboratory criteria for ductility and laboratory data that help to distinguish conditions of confining pressure, temperature, pore pressure, and strain rate that produce ductile deformation.

5.4 ELASTIC–DUCTILE DEFORMATION IN THE LABORATORY

Important concepts about ductility and strength, derived from conventional triaxial compression tests, are defined in **Figure 5.7**. This is a plot of differential stress, $\Delta\sigma$, versus axial strain, ε_a. The schematic curves plot in the third quadrant because the differential stress is compressive and axial strain is a contraction (both are negative quantities). The differential stress at the point on the stress–strain curve that separates elastic from ductile states is defined as the yield point. From the curves OA and OC the yield point is not easy to identify. However, for the idealized *elastic–plastic* material (**Figure 5.2**), represented by the bilinear curve OB, the slope changes from positive to zero at the yield point. After the yield point, inelastic strain accumulates for this idealized material at constant differential stress equal to the differential strength $D_c(\text{OB}) = |\min(\Delta\sigma)|$.

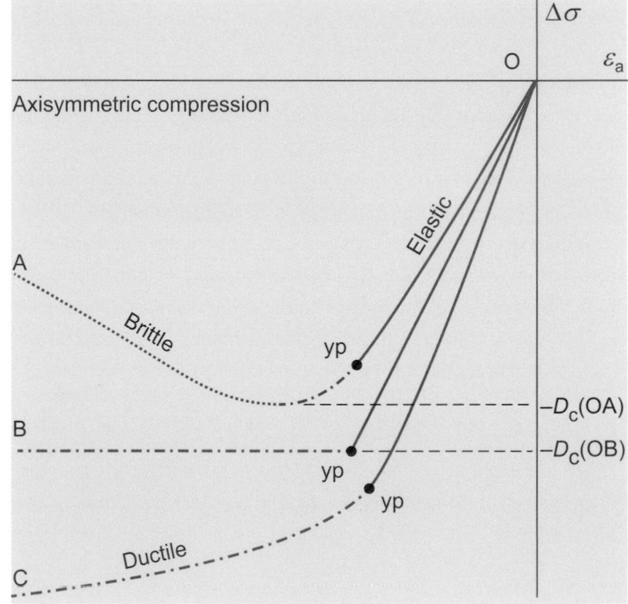

Figure 5.7 Schematic plot of differential stress versus axial strain for axisymmetric (conventional triaxial) compression tests exhibiting elastic (solid line), ductile (dash–dot curve), and brittle (dotted curve) deformation. Yield point is yp; differential strength is D_c. See Jaeger et al. (2007).

What happens after yielding can be quite varied, but curves OA and OC on **Figure 5.7** serve to bracket typical laboratory behavior. Curve OA passes from an elastic state (solid line) to a ductile state (dash–dot curve) with a concave upward curve and decreasing slope until the slope goes to zero, marking a well-defined differential strength, $D_c(\text{OB}) = |\min(\Delta\sigma)|$. Thereafter, the material is in a brittle state (dotted curve): the slope is negative and the differential stress and load carrying capacity decrease as strain accumulates. In contrast, curve OC passes from an elastic state (solid line) to a ductile state (dash–dot curve) at the yield point, but the curve does not reach a minimum thereafter, so the differential strength is not defined. The slope of the curve OC remains positive and the load carrying capacity continues to increase until the test is terminated.

Using the slope of the stress–strain curve to distinguish brittle from ductile deformation, as in **Figure 5.7**, provides an easily identified criterion for laboratory data, but this is not readily applicable to outcrop data because the slope of the relevant stress–strain curve usually is unknown. Another criterion, also developed during rock mechanics testing, distinguishes ductile and brittle deformation based on arbitrary ranges of strain accumulated before failure. In this laboratory context failure denotes significant loss of load carrying capacity, which usually causes the test to be terminated. If the axial strain at failure is less than 0.03 (< 3%) the behavior is called brittle; between 0.03 and 0.05 (3–5%) it is transitional; and greater than 0.05 (> 5%) it is called ductile. This concept of ductility also is not easily applicable to outcrop observations because *failure* in laboratory tests depends upon the stiffness of the testing machine and the design of the pressure vessel, neither of which are factors in nature.

Referring to the outcrop pictured in **Figure 5.3**, the sharp offset of dike A by slip on fault segment 1 is associated with a very small strain, estimated as the ratio of fault slip to trace length, here about 1%. According to the criterion just described, that small strain and the displacement discontinuity across the fault put this deformation in the *brittle* regime. On the other hand, the stretching and thinning of dike B between fault segments 1 and 2 is associated with about 25% strain. The substantial strain and the lack of fractures between fault segments 1 and 2, put this deformation in the *ductile* regime. Estimates of strain and observations of fractures (or the lack thereof) provide information to distinguish brittle from ductile deformation at outcrop.

In the late 1950s a series of rock mechanics tests were undertaken to investigate the effects of confining pressure, temperature, and pore fluid pressure on strength and the transition from brittle to ductile behavior for the Solnhofen limestone. This limestone, of late Jurassic age, is quarried near the village of Solnhofen in Bavaria, Germany, and sold for architectural and decorative purposes. It was chosen for the rock mechanics tests because of its homogeneity, small grain size relative to the laboratory sample size (~½ by 1 inch cylinders), and mechanical isotropy. At the time the test results were reported, 1960, the transition from brittle to ductile behavior was recognized to occur at the confining pressures and temperatures found deep in Earth's lithosphere. Researchers at the time acknowledged that the presence of hydrothermal solutions and very slow strain rates probably were conducive to ductile deformation and the accumulation of large strains.

Testing Solnhofen limestone has continued for more than 50 years with new discoveries being made as different conditions were prescribed and the testing equipment was improved. We focus on these results in order to understand how changes in the prescribed variables (confining pressure, temperature, pore fluid pressure, and strain rate) affect the mechanical behavior. Clastic sedimentary rocks, metamorphic rocks, and igneous rocks have mechanical behaviors that are different *quantitatively*, but most have similar behavior to Solnhofen limestone from a *qualitative* point of view. For example, different rock types pass from brittle to ductile behavior across different ranges of increasing temperature, but most are brittle at low temperatures and ductile at elevated temperatures. We use Solnhofen limestone as a well-studied example, and recommend that readers use the literature to explore the particular mechanical behaviors of other rock types.

5.4.1 Strength and Ductility Variations with Confining Pressure

The conventional triaxial tests on Solnhofen limestone reported in 1960 covered a confining pressure range from 0.1 to 550 MPa (atmospheric pressure to that at about 22 km depth) and a temperature range from 25 to 600 °C (room temperature to that at about 20 km depth). The pore fluid pressures ranged from 0.1 MPa (atmospheric pressure) to the value of the confining pressure. For all experiments the testing machine imposed an axial strain rate of $10^{-4}\,s^{-1}$. It should be noted that individual tests did not necessarily correspond to the conditions expected in Earth's lithosphere, where the lithostatic pressure increases at about 25 MPa/km, temperature increases at about 30 °C/km, and hydrostatic pressure increases at about 10 MPa/km. Rather, a wide range of pressures and temperatures were tested to understand the mechanical behavior. For example, tests were conducted at atmospheric pressure and 500 °C, and also at 550 MPa confining pressure and 25 °C, even though neither condition is likely to be common in Earth's lithosphere.

Conventional triaxial compression test results showing the effect of confining pressure on strength and the transition from brittle to ductile behavior for dry Solnhofen limestone are shown in **Figure 5.8** where differential stress is plotted versus axial strain. Confining pressures range from 0.1 to 552 MPa, while

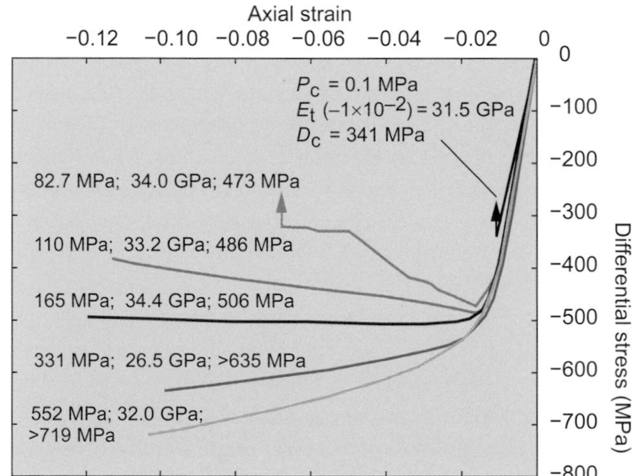

Figure 5.8 Differential stress versus axial strain for triaxial compression tests of Solnhofen limestone. Temperature = 25 °C; strain rate = $10^{-4}\,s^{-1}$. Triplets of numbers refer to confining pressure, P_c, tangent Young's modulus, E', and differential strength, D_C. Arrow at end of curve indicates test terminated due to fracture; no arrow indicates test terminated. Modified from Heard (1960), Figure 3a.

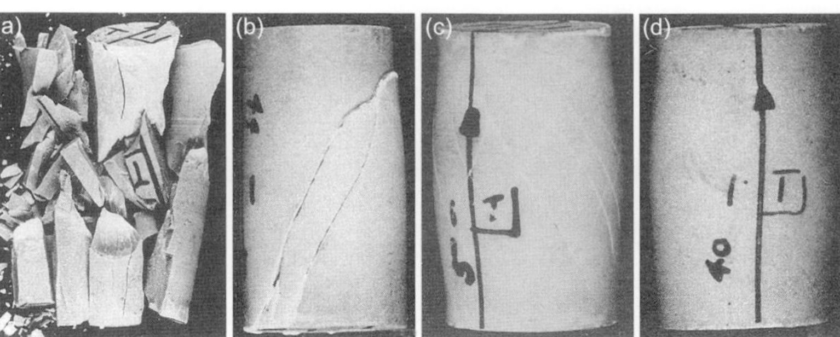

Figure 5.9 Specimens after conventional triaxial compression tests of Solnhofen Limestone. Temperature = 25 °C; strain rate = $10^{-4}\,s^{-1}$. Confining pressure (left to right): P_c = 0.1, 82.7, 110, 165 MPa. From Heard (1960), Plate 1.

the temperature is held constant at 25 °C. For all samples the initial deformation was nearly linear and elastic, with Young's modulus at a contractional strain of –0.01 varying *non-systematically* from 26.5 to 34.4 GPa. On the other hand, the differential strength in compression, D_c, increased *systematically* from 341 MPa to greater than 719 MPa as confining pressure increased. At the lowest confining pressure (0.1 MPa), the specimen fractured immediately upon reaching the minimum differential stress. At a confining pressure of 82.7 MPa, a sharp change in slope occurs at the minimum value of differential stress, followed by a negative slope and jagged stress–strain curve, leading ultimately to fracture at about –0.07 axial strain. Above 165 MPa confining pressure, the differential stress did not reach a minimum value and the test at this confining pressure is approximately at the transition to ductile behavior as defined by the nearly zero slope of the stress–strain curve. Tests were terminated at axial strains of about –0.1 (10%).

Representative specimens from these compression tests are shown in **Figure 5.9** to illustrate the range of deformation styles. At 0.1 MPa confining pressure (**Figure 5.9a**), the specimens shatter with wedge and splitting fractures. At 82.7 MPa confining pressure (**Figure 5.9b**), a couple of shear fractures cross the specimen obliquely. Note that this specimen is no longer a straight-sided cylinder: the sides bulge outward, a phenomenon known as barreling. This is evidence for inelastic deformation, not associated with macroscopic shear fracture, and presumably associated with the more than 5% axial strain accumulated before fracture. At 110 MPa confining pressure (**Figure 5.9c**), barreling is readily apparent and multiple oblique deformation bands cross the sample. Both of these mechanisms for inelastic deformation in Solnhofen limestone are associated with the more than 11% axial strain that accumulated before the test was terminated, and before any macroscopic fracture occurred. At 165 MPa confining pressure (**Figure 5.9d**), the only observable sign of deformation resulting from nearly 12% axial strain is the pronounced barreling of the specimen sides.

5.4.2 Strength and Ductility Variations with Temperature

Conventional triaxial compression tests on Solnhofen limestone at 44 MPa confining pressure reveal a *systematic* decrease in strength as the temperature is increased from 300 to 600°C (**Figure 5.10**).

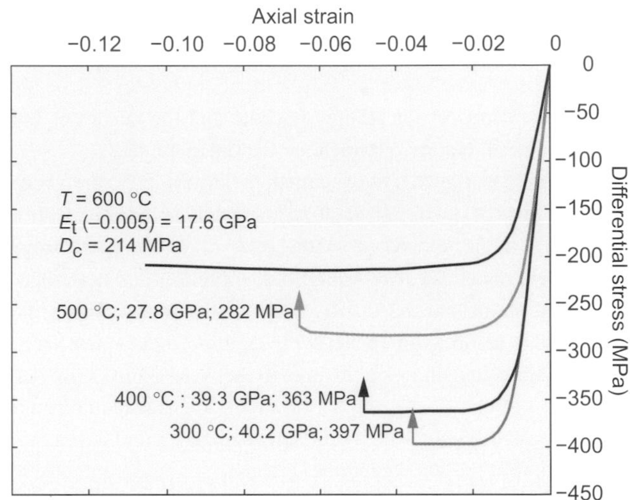

Figure 5.10 Differential stress versus axial strain for triaxial compression tests of Solnhofen limestone. Confining pressure = 44.1 MPa; strain rate = $10^{-4}\,s^{-1}$. Triplets of numbers next to each curve refer to temperature, *T*, tangent Young's modulus, E_t, and differential strength, D_c. Arrow at end of curve indicates test terminated due to fracture; no arrow indicates test terminated. Modified from Heard (1960), Figure 4b.

All of the tests begin with nearly linear elastic behavior, but Young's modulus (evaluated at a contractional strain of −0.005) decreases systematically from 40.2 to 17.6 GPa as the temperature increases. The form of the stress–strain curves is similar over this range of temperatures with linear elastic deformation giving way to a smoothly decreasing slope that approaches zero at a few percent axial strain and continues with a nearly zero slope to fracture. In addition, the differential compressive strengths decrease systematically from 397 to 214 MPa, and the axial strain at fracture increases systematically from less than 4% to greater than 10% as the temperature increases from 300 to 600 °C.

5.4.3 Strength and Ductility Variations with Pore Fluid Pressure

Conventional triaxial compression tests also were conducted on Solnhofen limestone at 121 MPa confining pressure and 150 °C

(**Figure 5.11**) with pore fluid pressures varying from 67.3 to 104 MPa. These tests reveal a more-or-less *systematic* decrease in strength and decrease in the axial strain at fracture as the pore pressure is increased. All of the tests begin with nearly linear elastic behavior, but Young's modulus (evaluated at a contractional strain of –0.005) varies non-systematically over a range from 29.8 to 42.3 GPa. The differential strengths vary from 416 MPa at the least pore pressure to 362 MPa at the greatest pore pressure.

The inset photograph (**Figure 5.11**) shows the specimen after the test at 72.8 MPa pore pressure. Barreling of the specimen sides presumably was associated with the inelastic accumulation of strain to about 8%. The test terminated when a single oblique shear fracture broke the specimen into two parts. Under these conditions initial elastic deformation gives way to ductile deformation at nearly constant differential stress, but the ultimate failure of the specimen is by shear fracturing.

5.4.4 Strength and Ductility Variations with Strain Rate

To appreciate the significant effect of strain rate on mechanical properties we describe conventional triaxial (axisymmetric) tests carried out on Solnhofen limestone. The tests shown in **Figure 5.12** were at 400 °C and the total axial strain ranged up to 12%. The confining pressure was fixed at 165 MPa, and the pore fluid pressure was fixed at 15.0 MPa, so the effective confining pressure was 150 MPa. The estimated value for the tangent elastic modulus, $E_t(-0.005)$ = 18 GPa, is somewhat softer than most values taken from **Figure 5.8**, **Figure 5.10**, and **Figure 5.11**.

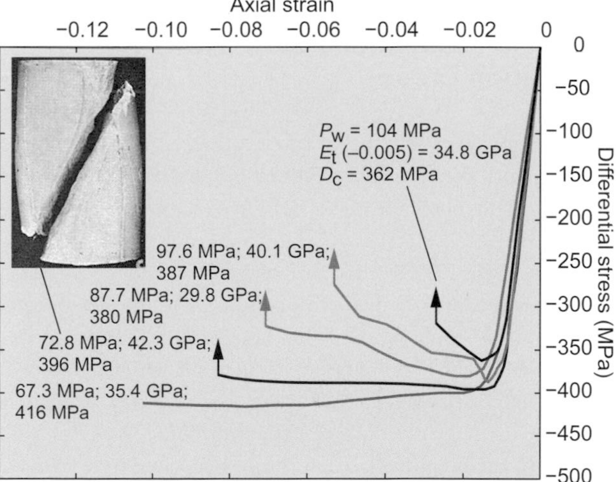

Figure 5.11 Differential stress versus axial strain for triaxial compression tests of Solnhofen Limestone. Confining pressure = 121 MPa; temperature = 150 °C; strain rate = $10^{-4}\,s^{-1}$. Triplets of numbers next to each curve refer to interstitial fluid pressure, P_w, tangent Young's modulus, E_t, and differential strength, D_c. Arrow at end of curve indicates test terminated due to fracture; no arrow indicates test terminated. Modified from Heard (1960), Figure 11b.

The yield point is poorly defined for Solnhofen limestone under the conditions illustrated in **Figure 5.12**, but the curves have positive slopes to the strain at which the tests were terminated. Significant strain accumulates with increasing stress, so the rock is ductile under these conditions. The limestone lacks a well-defined differential strength, but the differential stress when the tests terminated changes systematically from about –190 MPa to about –360 MPa as the strain rate changes from 10^{-8} to $10^{-4}\,s^{-1}$. Under dry conditions at the same effective confining pressure and temperature, the differential stress when the tests terminated changes systematically from about –280 MPa to –360 MPa as the strain rate changes. Thus, water and elevated pore pressure cause a significant decrease in the load carrying capacity at the smaller strain rates.

Most laboratory rock mechanics tests are conducted at greater strain rates than those that typically occur in Earth's lithosphere and asthenosphere. Accordingly, the strengths inferred from the laboratory, such as those illustrated in **Figure 5.12**, are likely to be greater than those in nature.

5.4.5 Transitions in Mechanical Behavior with Depth

Laboratory data can be used to define significant transitions in the mechanical behavior of rock with depth in Earth's lithosphere by increasing temperature and confining pressure simultaneously at

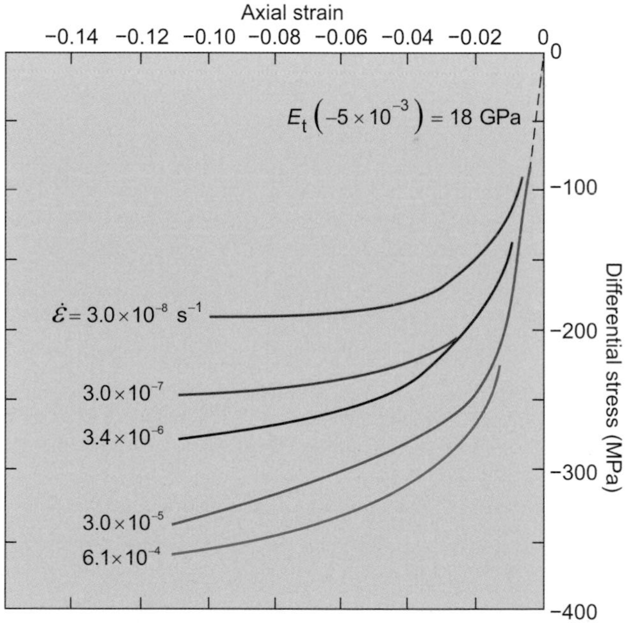

Figure 5.12 Differential stress versus axial strain for triaxial compression tests of wet Solnhofen Limestone. Confining pressure = 165 MPa; pore fluid pressure = 15 MPa; temperature = 400 °C. Numbers next to each curve refer to axial strain rate. Slope of dashed line through origin gives tangent Young's modulus of 18 GPa at an axial strain of –0.005. Tests terminated between 10 and 12% axial strain. Modified from Rutter (1974), Figure 1.

rates that are similar to those found in the Earth. In the laboratory context, specimens that lose their load-carrying capacity at lesser axial strain (usually by fracturing) are termed brittle, whereas those that continue to sustain their load-carrying capacity to greater axial strain are termed ductile. In this sense a transition may be defined as conditions pass from those that induce brittle deformation to those that induce ductile deformation. Recall that a different definition of such a transition, one based on the *slope* of the stress–strain curve after reaching the yield point, was used with **Figure 5.7**. Clearly the context and the definition must be clearly enunciated when discussing a transition from brittle to ductile behavior. Other important distinctions may be made based on the *mechanisms* of deformation, and some of these also will be highlighted in this section.

Mechanical behaviors at the specimen scale referred to as *brittle* include wedge and splitting fractures (**Figure 5.9a**), and shearing fractures (**Figure 5.9b**). At the grain scale, the deformation mechanism associated with the wedge and shearing fractures is cataclasis: fracture of individual grains, frictional sliding along grain boundaries, and wholesale rotation of grain fragments. Behaviors at the specimen scale referred to as *ductile* include conjugate deformation bands (**Figure 5.9c**) and barreling (**Figure 5.9d**). At the grain scale, the deformation mechanisms include cataclasis within the deformation bands, and intracrystalline plasticity: distortion of mineral grains without fracture caused by the motion of crystal-lattice defects (see Section 5.6).

To focus on conditions in Earth's lithosphere, and to emphasize transitions in the grain-scale deformation mechanisms, consider data from conventional triaxial compression tests on Solnhofen limestone at a constant strain rate of $10^{-5}\,s^{-1}$ (**Figure 5.13**). The conditions of confining pressure and temperature were restricted to those due to "normal" gradients in Earth's lithosphere (25 MPa/km and 30 to 35 °C/km). One curve and associated data points are for wet samples with a ratio of pore water pressure to confining pressure of 0.1, whereas the other curve and associated data points are for oven-dried samples. The transition (thick gray line) at an equivalent depth of 2 to 3 km is from dominantly localized cataclasis to dominantly distributed cataclasis, whereas the transition at an equivalent depth of about 7 km is to dominantly intracrystalline plasticity. Neither of these transitions is sharply defined, but the change in deformation mechanism clearly is a function of depth, and therefore of confining pressure and temperature.

The strength of Solnhofen limestone (**Figure 5.13**) increases with depth through the transition from localized to distributed cataclasis, reaches a peak somewhat after the transition to intracrystalline plasticity, and then decreases with further increases in depth. The changing slope of this plot of strength versus depth may be understood by noting that cataclasis is accompanied by the development of opening cracks, which cause the volume of the sample to increase. Therefore, this mechanism of deformation is inhibited by increasing the confining pressure, so the rock strength increases with depth where cataclasis is the dominant mechanism. On the other hand, intracrystalline plasticity is accompanied by the motion of dislocations, which is enhanced by increases in temperature, but involves little volume change.

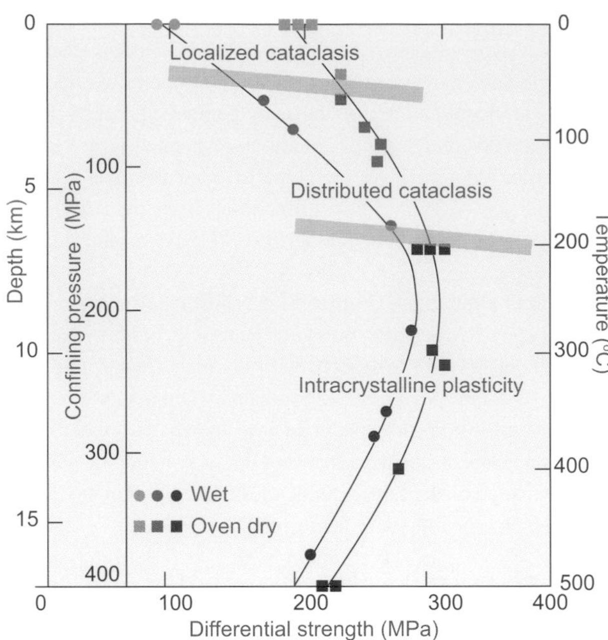

Figure 5.13 Strength of Solnhofen limestone in conventional triaxial tests under conditions of temperature and confining pressure designed to follow the lithostatic and geothermal gradients in Earth's lithosphere. Transitions are marked by thick gray lines between localized cataclasis, distributed cataclasis, and intracrystalline plasticity. Modified from Rutter (1986), Figure 2.

Thus, this mechanism is not inhibited by increases in confining pressure, so the rock strength decreases with depth where dislocation motion is the dominant mechanism.

5.4.6 Non-linear Solid-State Flow: Dislocation Creep

From the elastic–ductile behaviors idealized in **Figure 5.7** and the laboratory data plotted in **Figure 5.8** and **Figure 5.10**, we gain no insights about the rate of deformation, because the testing machine imposed a *fixed* axial strain rate of $10^{-4}\,s^{-1}$. Plastic deformation ensues when the differential stress equals the yield strength, and continues until the stress drops below the strength, but the rate at which plastic strain accumulates is fixed. In this section we describe laboratory experiments called creep tests in which a fixed differential stress that exceeds the yield strength is maintained, and the rate of axial strain is recorded. We begin with a brief description of the laboratory apparatus used to investigate this solid-state flow.

The laboratory experiments use equipment that is broadly similar to the conventional triaxial apparatus described in Section 4.7. A uniaxial testing machine (**Figure 4.6**) provides an axial load and a containment vessel (**Figure 4.16**) provides a confining pressure and elevated temperature. The cylindrical samples are subjected to an axial stress, σ_a, and a radial stress equal to the negative of the confining pressure, $\sigma_r = -P_c$. The difference between the axial stress and the radial stress is the differential stress, $\Delta\sigma = \sigma_a - \sigma_r$, and this is held constant during the creep test. For the different creep tests described here the differential stresses ranged from 20 to 300 MPa and the confining pressures ranged from 100 to 450 MPa. Argon gas provided the confining pressure, so this equipment is referred to as a gas-medium apparatus. Temperatures as great as about 1,500 K are provided by an internal furnace. The extreme conditions of these tests require specialized equipment and sample preparation techniques that are documented in the contributions listed under Further Reading at the end of this chapter.

For a perfectly elastic material the sample length would not vary under a constant differential stress, but for a material that includes solid-state flow the length changes with time by *creeping* deformation. This is monitored using a displacement transducer attached to the moving platen. This displacement is associated with a change in length of the sample, $L - L_o$, where L is the current length and L_o is the original length. These lengths are used to calculate the axial strain: $\varepsilon_a = (L - L_o)/L_o$. The displacement is recorded as a function of time, so on a graph of axial strain versus time the slope is a scalar quantity called the strain rate, $\dot{\varepsilon} = d\varepsilon_a/dt$, and this is what typically is reported as data from a creep test. Because strain is a ratio of lengths and therefore without dimensions and units, the strain rate has SI units of reciprocal seconds, $\dot{\varepsilon}\,[=]\,s^{-1}$. For the tests described here the strain rates ranged from 7×10^{-7} to $1.2 \times 10^{-4}\,s^{-1}$.

As one example we describe samples that are encased in thin malleable capsules to provide the desired chemical conditions and containment (**Figure 5.14**). These capsules are placed inside the confining pressure vessel between the platens that are driven inward by the testing machine. The samples range from 14 to 19 mm long and 6.5 to 9 mm in diameter and are enclosed in a nickel foil can that is only 0.2 mm thick. The talc sleeve is 1.1 mm thick and dehydrates at the elevated temperatures of these tests to provide enough water to saturate the sample. For example, in the tests described here the water fugacity ranged from ~85 to 520 MPa. This is an effective pressure that measures the chemical potential of the water. While necessary to control the conditions of the test, the sample enclosure does support some fraction of the axial force and confining pressure. This support must be taken into account in calculating the axial and radial stress acting on the sample.

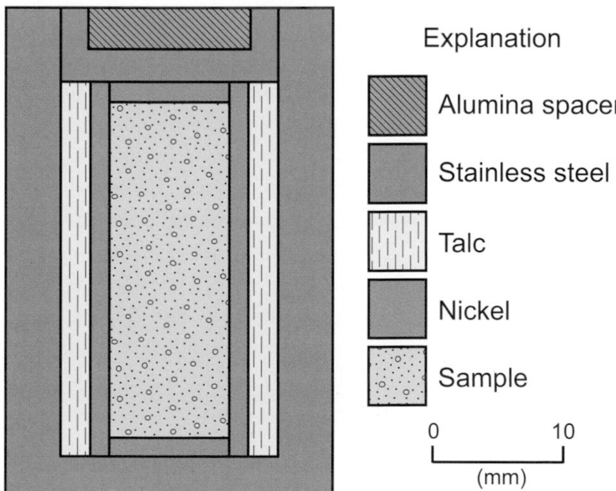

Figure 5.14 Schematic drawing of sample enclosure for conventional triaxial test of olivine aggregate under hydrous conditions. Thickness of alumina, stainless steel, talc, and nickel are exaggerated relative to the sample size. Modified from Mei and Kohlstedt (2000a), Figure 3.

For the tests described here each sample was prepared by crushing hand-picked grains of San Carlos olivine, initially 3 to 6 mm in size, down to a powder with particle sizes ranging from 2 to 18 μm. This powder was dried and cold-pressed at 200 MPa into the nickel can and then hot-pressed at 300 MPa and 1523 K to eliminate porosity. The small grain size facilitates chemical diffusion within grains and along grain boundaries. Thus, by reducing the grain size, a process that takes place on a geologic time scale can be studied on a laboratory time scale.

The analysis of data from creep tests involves the determination of constants in an empirical relationship between strain rate and differential stress called a flow law. Depending upon the minerals being tested and the test conditions, the strain rate may be a function of the differential stress, $\Delta\sigma$, the grain size, d, the fugacity of water, f_{H_2O}, the absolute temperature, T, and the pressure, P, here equated to the confining pressure:

$$\dot{\varepsilon} = A(\Delta\sigma)^n d^{-p} f^r_{H_2O} \exp\left[-\frac{Q + PV}{RT}\right] \quad (5.2)$$

The constants to be determined in this relationship are a pre-exponential factor, A, the differential stress exponent, n, the grain size exponent, p, the water fugacity exponent, r, the activation energy, Q, and the activation volume, V. The pre-exponential factor, A, carries units necessary to leave the right side of (5.2) with units of reciprocal seconds. The three exponents (n, p, r) are unit-less numbers. Note that the strain rate is directly proportional to differential stress and water fugacity, but inversely proportional to grain size, all raised to their respective powers. The activation energy and activation volume have the SI units $Q\,[=]\,J\,mol^{-1}$ and $V\,[=]\,m^3\,mol^{-1}$. The two terms in square brackets in (5.2) provide, respectively, measures of the sensitivity of the strain rate to temperature and to pressure. The pressure times the activation volume has the SI units $PV\,[=]\,Pa\,m^3\,mol^{-1} = N\,m\,mol^{-1} = J\,mol^{-1}$, so the two terms in square brackets have the same units. The gas constant has the value and SI units $R = 8.3145\,J\,mol^{-1}\,K^{-1}$.

Experimental results on olivine aggregates where the differential stress was greater than about 50 to 100 MPa are illustrated in **Figure 5.15**. At these differential stresses there was little dependence of strain rate on grain size: that is $p = 0$, so $d^{-p} = 1$ and that quantity drops out of (5.2). Empirical evaluation of these data yielded an activation energy of $Q = 470 \times 10^3\,J\,mol^{-1}$, and an activation volume near zero (i.e. the strain rate was not very sensitive to pressure). Setting $V = 0\,m^3\,mol^{-1}$, the sensitivity of the strain rate to temperature from the last term in (5.2) evaluates as 7.60×10^{-17}. A statistical analysis of the data at the three different confining pressures determined a stress exponent $n \approx$

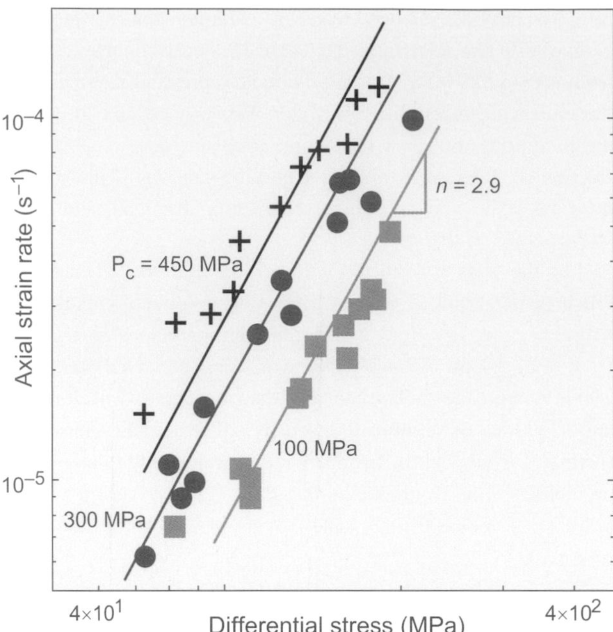

Figure 5.15 Plot of strain rate versus differential stress on log–log graph for water-saturated olivine aggregates at 1523 K and three different confining pressures, P_c. Best fit straight lines suggest a non-linear relationship with stress exponent near 3. At each confining pressure three different samples were tested, yielding similar stress exponents. Modified from Mei and Kohlstedt (2000b), Figure 4.

2.9. Including the empirically determined exponents (n = 3, r = 0.7), the flow law (5.2) is written:

$$\dot{\varepsilon} = \left(4.6 \times 10^3 \text{MPa}^{-3.7}\text{s}^{-1}\right)(\Delta\sigma)^3 f_{\text{H}_2\text{O}}^{0.7}\left(7.60 \times 10^{-17}\right)$$

The distinctly non-linear relationship between strain rate and differential stress is interpreted as deformation caused by the motion of *dislocations* within the olivine grains. Thus, this mechanism of solid-state deformation is dislocation creep, which we describe in more detail in Section 5.6.

Experiments similar to those illustrated in **Figure 5.15**, but using anhydrous conditions, yielded stress exponents that varied from n = 2.9 to 3.1. For these experiments on similarly prepared olivine aggregates, tested at comparable differential stress and temperature, but without water present, the strain rates were about *five* to *six* times less than those for hydrous conditions. The presence of water reduces the resistance to deformation and these differences are great enough to have significant implications for solid-state deformation in Earth's upper mantle.

The confining pressures (100 to 450 MPa) used in the tests on olivine aggregates (**Figure 5.15**) are representative of depths from 4 to 18 km, but the results from these laboratory creep tests are extrapolated to the lower crust and upper mantle at depths up to 120 km. Confining pressures necessary to replicate conditions at these depths would range up to 3 GPa, well beyond the capability of conventional triaxial testing machines and pressure containment vessels. Extrapolation of experimental results to these greater confining pressures rely on the efficacy of flow laws such as (5.2), the abundance of data, and understanding of the physical mechanisms for solid-state flow. The lack of direct access to the lower crust and upper mantle of Earth, and the inability of laboratory equipment to replicate the conditions there, provide many challenges for those investigating deformation in Earth's upper mantle. Confidence in the required extrapolations can be gained through the interpretation of geophysical data using crustal scale mechanical models of flow in the asthenosphere (e.g. see Section 5.3).

5.5 DEFORMATION AND STRAIN FOR THE DUCTILE SOLID

The kinematic equations for small strains, introduced in Section 4.9.2, describe elastic deformation, but they may not be adequate to describe ductile deformation. Gradients in displacement for ductile deformation can be so large that the approximations leading to the definitions of small strains introduce substantial errors. If that is the case, the description of ductile deformation requires a more robust kinematics.

Here we take a second look at the deformation of an infinitesimal material line in the neighborhood of a particle and introduce the deformation gradient tensor, **F**, that accounts for stretching, shearing, and rotation of large magnitudes. Then, the finite strain tensor, **E**, is defined in terms of the gradient in displacement using no approximations. Finally, we compare the finite strain components to their small strain counterparts and show how one decides which kinematic relations to employ in analyzing geologic structures. All of this is done in two dimensions to simplify the presentation, and in Section 5.5.7 we generalize to three dimensions without adding new concepts.

5.5.1 Deformation Gradient Tensor

To characterize the deformation in the neighborhood of any particle in a rock mass viewed as a material continuum, we compare two states using a referential description of motion. For convenience, we refer to these as the initial state and final state, but they could be any two moments in time during a complicated and protracted history of tectonic deformation. To ease the presentation we take a two-dimensional perspective, but generalize this to three dimensions in Section 5.5.7 with no new concepts. We use the Cartesian coordinates (X, Y) for the initial state, and the coordinates (x, y) for the final state (**Figure 5.16**). These two sets of coordinates share the same origin and their respective axes are coincident.

In the initial state (**Figure 5.16**) an arbitrary particle is at a point defined by the position vector $\mathbf{X} = \langle X, Y \rangle$. From the initial to the final state the particle moves to the position $\mathbf{x} = \langle x, y \rangle$ through a displacement $\mathbf{u}(\mathbf{X})$, so the coordinates change as a function of the initial position: $\mathbf{x} = \mathbf{X} + \mathbf{u}(\mathbf{X})$. An arbitrarily oriented material line of infinitesimal length extends from the particle, and the vector $\text{d}\mathbf{X} = \langle \text{d}X, \text{d}Y \rangle$ with its tail at $\mathbf{X}$ overlies this material line. The material line is stretched and rotated, such that the overlying vector in the final state is $\text{d}\mathbf{x} = \langle \text{d}x, \text{d}y \rangle$. To quantify

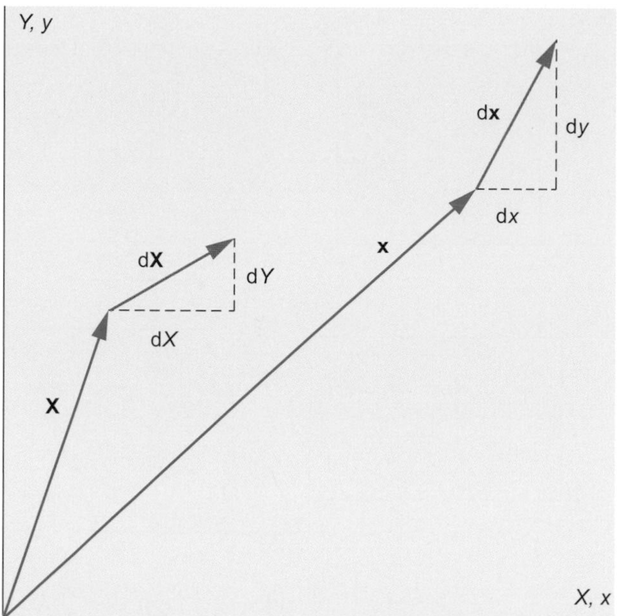

Figure 5.16 Vectors used to define deformation in the infinitesimal neighborhood of a particle at position **X** in the initial state and **x** in the final state. Vector d**X** represents the material line at **X** and d**x** represents the same material line after stretching, rotation, and translation to **x**. Lengths of differential vectors are greatly exaggerated relative to lengths of position vectors.

the deformation for the particle initially at **X**, we focus attention on the vector d**x**, and ask how this vector changes as we consider all the possible material lines emanating from the particle.

The material line represented by the vector d**X** in **Figure 5.16** is taken as infinitesimal because, if it were of finite length, it could be curved in the final state and represented poorly by the vector d**x**. Straight material lines of finite length in the initial state are straight in the final state only if the deformation does not vary over the length of that line. This is homogeneous deformation at the length scale of the line. However, geologic structures involve deformation that does vary with position: it is heterogeneous deformation. For example, folded sedimentary strata and folded metamorphic foliations are composed of surfaces that changed from planar in the initial state to curved in the final state. In cross section, material lines within these folded surfaces once were straight, but now are curved. Understanding geologic structures requires a kinematic theory that admits heterogeneous deformation. This necessitates consideration of infinitesimal material lines represented by differential vectors, d**X** and d**x**, as in **Figure 5.16**.

Use of the infinitesimal material line also is consistent with the objective to describe the deformation in the neighborhood of a particle (i.e. at a point in the material continuum). With this objective in mind, we relate the components of d**x** to the components of d**X** using the total differentials for a vector function of two independent variables (see Section 2.5.3):

$$\mathrm{d}x = \frac{\partial x}{\partial X}\mathrm{d}X + \frac{\partial x}{\partial Y}\mathrm{d}Y$$
$$\mathrm{d}y = \frac{\partial y}{\partial X}\mathrm{d}X + \frac{\partial y}{\partial Y}\mathrm{d}Y \quad (5.3)$$

These total differentials are measures of the spatial variation in length and orientation of the infinitesimal material lines emanating from the particle at **X**. Note that the partial derivatives in (5.3) are taken with respect to the coordinates in the initial state, X and Y. Each partial derivative is a dimensionless number (a ratio of two lengths) that has a unique value at **X** in the initial state. Thus, each component of the vector d**x** in the final state is composed of dimensionless numbers times the respective components of the vector in the initial state, so dx and dy depend *linearly* on dX and dY. These components of d**X** define the orientation of the material line in the initial state.

Recall from Section 2.6.1 that the gradient of a vector field is a tensor field. Here the vector is d**x**, so the partial derivatives in (5.3) are components of a tensor, **F**, called the deformation gradient tensor. Using matrix notation (5.3) is:

$$\begin{bmatrix}\mathrm{d}x\\ \mathrm{d}y\end{bmatrix} = \begin{bmatrix}F_{xx} & F_{xy}\\ F_{yx} & F_{yy}\end{bmatrix}\begin{bmatrix}\mathrm{d}X\\ \mathrm{d}Y\end{bmatrix},$$
$$0 < F_{xx}, F_{yy} < +\infty, \quad -\infty < F_{xy}, F_{yx} < +\infty \quad (5.4)$$

The two subscripts on each F indicate the coordinates in the numerator and denominator of the partial derivatives in (5.3). The elements of [F] on the primary diagonal must be positive numbers. This *mathematical* constraint avoids the infinitesimal material line represented by d**X** in **Figure 5.17** shortening to a zero length. For example, ignoring shearing due to F_{xy} in (5.4), we have dx = F_{xx}dX. Setting $F_{xx} = 0$ means that the material line of initial length, dX, shortens such that its final length, dx, is zero. A similar condition constrains F_{yy}. The elements of [F] on the secondary diagonal impose shearing on the material line, and this can be negative, zero, or positive. Also, the elements on the secondary diagonal are not necessarily equal, so [F] is not necessarily symmetric.

Although just one vector d**X** is shown in **Figure 5.16** representing an infinitesimal material line in the initial state, (5.3) relates vectors with all possible orientations at the point **X** to the corresponding vector d**x** at the point **x**. The components of d**X** and d**x** determine the directions of vectors in the initial and final states:

$$\mathrm{d}X = |\mathrm{d}\mathbf{X}|\cos\Theta, \mathrm{d}Y = |\mathrm{d}\mathbf{X}|\sin\Theta$$
$$\mathrm{d}x = |\mathrm{d}\mathbf{x}|\cos\theta, \mathrm{d}y = |\mathrm{d}\mathbf{x}|\sin\theta \quad (5.5)$$

Here Θ and θ are direction angles for the corresponding vectors (**Figure 5.17**). Suppose |d**X**| is taken as the radius of a circle swept out by the heads of the vectors d**X** as the direction angle Θ ranges from 0 to 2π. In the final state the vector d**x** varies with the direction angle θ and the vector heads sweep out an ellipse as θ ranges from 0 to 2π. In Section 2.5.4 we use the semi-axes of these ellipses to characterize the magnitude of the deformation.

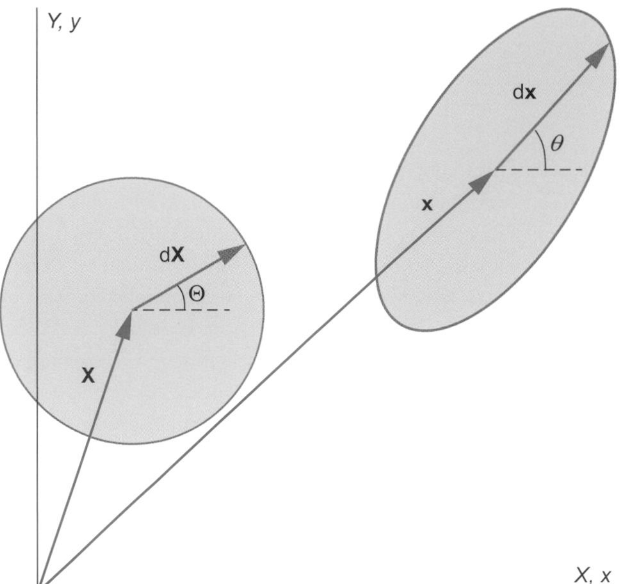

Figure 5.17 Deformation in the infinitesimal neighborhood of a particle at position **X** in the initial state. The vector d**X** represents a material line in the initial state at **X**; d**x** represents the stretched and rotated material line in the final state at **x**. All possible initial and final vectors trace out a circle in the initial state and an ellipse in the final state. Lengths of differential vectors are greatly exaggerated relative to lengths of position vectors.

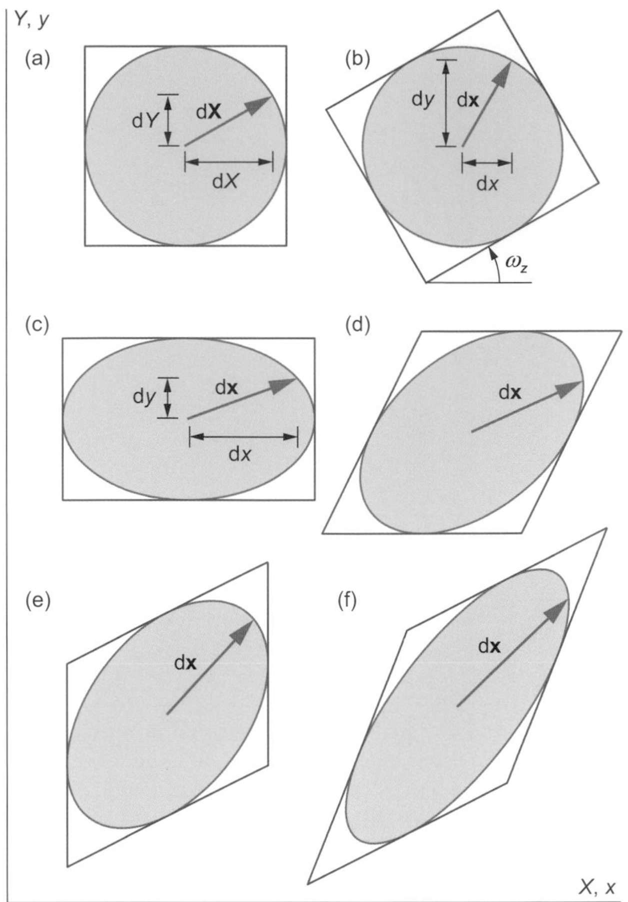

Figure 5.18 Illustrations of the deformation of a line, circle, and square using the deformation gradient tensor. (a) The *initial* shapes and orientations (no deformation). (b) Pure rotation. (c) Pure shear. (d) Simple shear parallel to X. (e) Simple shear parallel to Y. (f) Simple shear in both X and Y. No scale is marked on the coordinate axes because these shapes are infinitesimal in size and represent deformation at a point in the material continuum.

The nature of the deformation specified by the deformation gradient tensor can be visualized using regular geometric shapes such as lines, circles, and squares. In **Figure 5.18a** an arbitrary material line in the initial state is represented by the infinitesimal vector d**X**, which also defines the radius of a circle and half the side length of a square. These geometric shapes are used to represent the deformation at a point, but are enlarged for the sake of visibility. The initial shapes in **Figure 5.18a** are deformed using the following deformation gradient tensors (in matrix form) to give the shapes shown in **Figure 5.18a–f**, respectively, for the following values of the elements of [F]:

$$[\mathrm{F(a)}]=\begin{bmatrix}1 & 0\\0 & 1\end{bmatrix},[\mathrm{F(b)}]=\begin{bmatrix}0.866 & -0.5\\0.5 & 0.866\end{bmatrix},[\mathrm{F(c)}]=\begin{bmatrix}1.25 & 0\\0 & 0.8\end{bmatrix},$$
$$[\mathrm{F(d)}]=\begin{bmatrix}1 & 0.5\\0 & 1\end{bmatrix},[\mathrm{F(e)}]=\begin{bmatrix}1 & 0\\0.5 & 1\end{bmatrix},[\mathrm{F(f)}]=\begin{bmatrix}1 & 0.5\\0.5 & 1.25\end{bmatrix} \quad (5.6)$$

Two features common to the deformations described by the matrices in (5.6) are that circles transform to ellipses (or other circles), and squares transform to parallelograms (or other squares).

In (5.6) the matrix [F(a)], with ones on the primary diagonal and zeros on the secondary diagonal, results in no deformation: d**x** = d**X** and the vector, circle, and square shown in **Figure 5.18a** are unchanged. The matrix [F(b)] produces a counter-clockwise rotation, $\omega_z = 30°$, with no stretching of any radial material lines, and no shearing of any orthogonal material lines (**Figure 5.18b**). This pure rotation about the z-axis is associated with the two-dimensional version of the deformation gradient tensor introduced in Section 2.6.4:

$$[\mathrm{F}] = \begin{bmatrix}\cos\omega_z & -\sin\omega_z\\ \sin\omega_z & \cos\omega_z\end{bmatrix} \quad (5.7)$$

Notice that the secondary diagonal elements are equal in magnitude and opposite in sign, and the primary diagonal elements are equal and have the same sign. All four elements are related to the rotation angle through the trigonometric functions indicated here. Any matrix that provides a pure rotation is an orthogonal matrix (Section 2.6.4), for which material lines maintain their lengths and angular relationships to one another. A pure rotation sometimes is called a *rigid* rotation, which implies the material is incapable of deforming. However, we are describing a *local rotation* at a point, and the deformation at nearby points may include stretching and shearing, in which case the material is *not* rigid.

The first two matrices in (5.6) demonstrate that [F] *can* be symmetric, but it is not necessarily symmetric, because that property requires equal elements on the secondary diagonal. This contrasts with the matrices used to describe small strain, which *always* are symmetric. The first two matrices in (5.6) also demonstrate that [F] is *not* an appropriate measure of finite strain, because the elements do not all go to zero when the lengths of all radial lines are unchanged (no stretching), and the angular relation between all pairs of orthogonal radial lines is unchanged (no shearing). The finite strain tensor, introduced in Section 5.5.3, does have the property that all of its elements go to zero for no stretching and no shearing.

The matrix [F(c)] in (5.6) produces a stretching in the X-direction and a shortening in the Y-direction and results in no area change in the (X, Y)-plane. The matrix has zeros on the secondary diagonal, and the primary diagonal elements are reciprocals, that is $F_{yx} = F_{xy} = 0$ and $F_{xx} = 1/F_{yy}$. The traditional name for this style of deformation is pure shear strain. With these conditions, the ellipse of **Figure 5.18c** has the same area as the original circle, and the rectangle has the same area as the original square. Because the secondary diagonal elements are equal, the matrix is symmetric and does not include any pure rotation. Note that material lines originally parallel to the coordinate axes change length, but do not change orientation, so stretching, but no shearing occurs along the coordinate axes. This always is true if the secondary diagonal elements are zero. The name "pure shear" is a misnomer in the sense that stretching occurs for most material lines. Most material lines that are not originally parallel to the coordinate axes change both length and orientation. Pairs of orthogonal lines not parallel to the coordinate axes in the initial state are not orthogonal in the final state, so shearing occurs for those orientations.

The matrix [F(d)] in (5.6) involves shearing because the perpendicular sides of the square form acute and obtuse angles in the final state (**Figure 5.18d**): the square becomes a parallelogram. The traditional name for this style of deformation is simple shear strain. Notice that material lines parallel to the X-axis do not change length or orientation. However, the radius vector shown in the figure stretches and a radius vector at right angles to this shortens. Thus, the name should not imply that there only is shearing, because most material lines change length. Also, because the matrix is not symmetric, this deformation includes some pure rotation. The name simple shear strain is a misnomer in the sense that this deformation involves stretching, shearing, and rotation. Structural geologists commonly invoke simple shear kinematics for rock within fault zones and shear zones, and typically suppose the rock on either side of such a zone is rigid and moves in opposite directions, parallel to the zone. The authors of this textbook are skeptical that fault zones and shear zones are so simple.

To characterize the deformation associated with [F(d)] in (5.6), notice that the material line forming the top of the parallelogram in **Figure 5.18d** appears to have slid to the right relative to the material line forming the bottom. This is similar to the behavior of a deck of cards, laid on a table and sheared to the right, so each successive card slides a bit more than the card below (**Figure 5.22**). The cards do not change length, and no relative motion occurs perpendicular to the cards, so the deck of cards does not change cross-sectional area as it shears. The boards forming the wall of the cabin (**Figure 5.22**) play a similar role to the cards. The ellipse and parallelogram of **Figure 5.18d** have the same area as the original circle and square, respectively. The plane of the cards, and the (X, Z)-plane of **Figure 5.18d**, are referred to as the *shear plane*. The analogy to the card deck is not perfect because no discontinuities in displacement exist between the (X, Z)-planes of the continuum, whereas the cards clearly slip over one another. The matrix [F(e)] in (5.6) also produces simple shear, but the (Y, Z)-plane is the shear plane (**Figure 5.18e**).

The matrix [F(f)] in (5.6) is a particular combination of [F(d)] and [F(e)] that produces shearing (a square becomes a parallelogram), stretching (the long axis of the ellipse is greater than the circle radius), and shortening (the short axis of the ellipse is less than the circle radius) (**Figure 5.18f**). The sequence of deformation begins with [F(d)], simple shear on (X, Z)-planes (**Figure 5.18d**), followed by [F(e)], simple shear on (Y, Z)-planes (**Figure 5.18e**). To implement this, [F(d)] is the premultiplier and [F(e)] is the postmultiplier (Section 2.6.3), and this product defines a new deformation gradient tensor:

$$[\mathrm{F(e)}][\mathrm{F(d)}] = \begin{bmatrix} 1 & 0 \\ 0.5 & 1 \end{bmatrix} \begin{bmatrix} 1 & 0.5 \\ 0 & 1 \end{bmatrix} = \begin{bmatrix} 1 & 0.5 \\ 0.5 & 1.25 \end{bmatrix} = [\mathrm{F(f)}]$$

Changing the order of multiplication yields a different deformation gradient matrix:

$$[\mathrm{F(d)}][\mathrm{F(e)}] = \begin{bmatrix} 1.25 & 0.5 \\ 0.5 & 1 \end{bmatrix} = [\mathrm{F(g)}]$$

Changing the order of multiplication results in the same semi-axial lengths for the ellipse, but different axial orientations. In both cases the resulting deformation gradient matrix is *symmetric*, so includes no pure rotation. Matrix multiplication is *not commutative* in most instances, and here it clearly is not. From a geological perspective, this means that the order of tectonic events can play an important role in determining the resulting deformation.

Figure 5.19 Card deck shearing approximates simple shearing with the (X, Z)-plane as the shear plane and ϕ as the angle of shear.

5.5.2 Pure Stretch–Shear Tensor and Pure Rotation Tensor

The two-dimensional examples quantified as matrices in (5.6), and illustrated using lines, circles, and squares in **Figure 5.18**, make the point that the deformation gradient tensor, **F**, can produce deformation that involves stretching and shearing, but *no* rotation, and the corresponding matrices, [F], are symmetric. In contrast, **F** can produce deformation that *only* involves rotation, and the corresponding matrices, [R], are orthogonal. We introduced symmetric and orthogonal matrices in Sections 2.6.2 and 2.6.4. For many choices of [F] the matrix is not symmetric and not orthogonal, so it involves stretching, shearing, and rotation. Simple shearing, quantified as [F(d)] in (5.6) and illustrated in **Figure 5.18d**, is an example. Here we explore the general case and ask: can [F] be decomposed into a symmetric and an orthogonal matrix?

We begin by composing a symmetric deformation gradient matrix. Consider pure shearing, similar to what we illustrated in **Figure 5.18d**, and write the matrix as:

$$[\mathrm{U}] = \begin{bmatrix} 1.8 & 0 \\ 0 & 0.5556 \end{bmatrix}$$

Inspecting the elements on the secondary diagonal shows that this matrix is symmetric. To construct **Figure 5.20a** we translate the origin of coordinates to the particle in question and use a set of 2 by 1 column vectors $[\mathrm{X_c}]$ that define a unit circle:

$$[\mathrm{X_c}] = \begin{bmatrix} \cos\Theta \\ \sin\Theta \end{bmatrix}, \quad 0 \le \Theta \le 2\pi$$

For the sake of this illustration, we assume the deformation is homogeneous at this scale, but the concepts apply at the particle scale for heterogeneous deformation. Then, we use [U] to premultiply these unit vectors:

$$\underset{2\times1}{[\mathrm{X_e}]} = \underset{2\times2}{[\mathrm{U}]}\,\underset{2\times1}{[\mathrm{X_c}]}$$

This creates a new set of 2 by 1 column vectors $[\mathrm{X_e}]$ that define the perimeter of the stretch ellipse (**Figure 5.20a**).

Recall from Section 2.6.6 that symmetric tensors yield principal values and principal orientations. Because the secondary diagonal elements of [U] are zero, the elements on the primary diagonal are the principal values and the principal orientations correspond to the coordinate axes.

$$S_1 = 1.8, \hat{\mathbf{u}}_1 = \langle 1, 0 \rangle$$
$$S_2 = 0.5556, \hat{\mathbf{u}}_2 = \langle 0, 1 \rangle$$

Here S_1 and S_2 are the greatest and least principal stretches. The principal orientation vectors are the unit vectors $\hat{\mathbf{u}}_1$ and $\hat{\mathbf{u}}_2$. These unit vectors refer to directions in the initial state of material lines that acquire the principal stretches in the deformed state (**Figure 5.20a**).

Next consider an orthogonal deformation gradient matrix, similar to what we called pure rotation in **Figure 5.18b**, and write the matrix as:

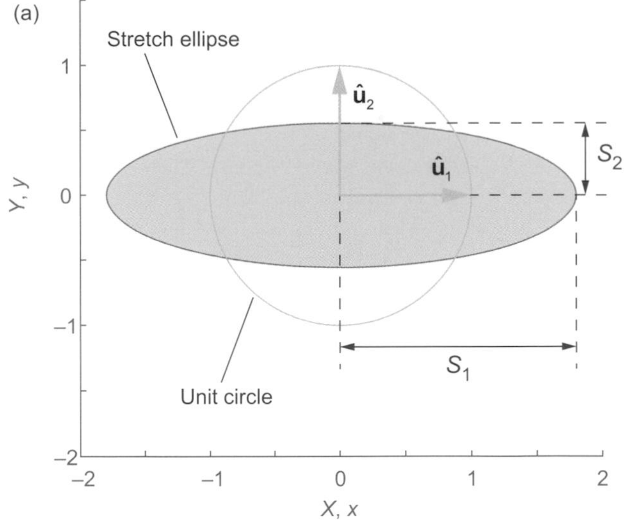

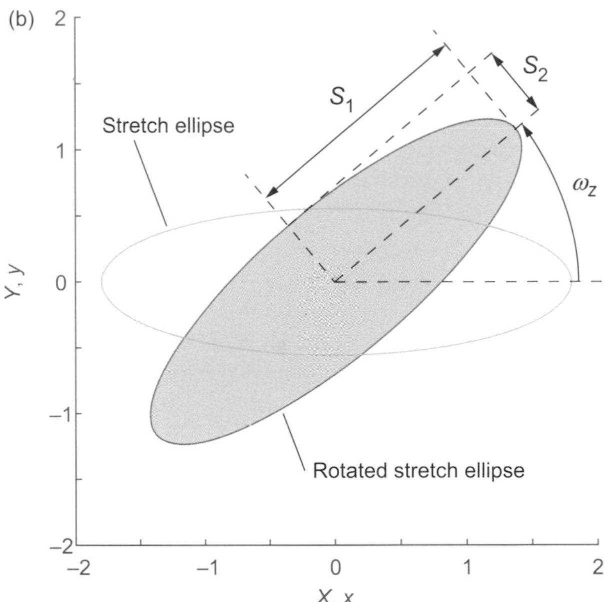

Figure 5.20 Illustration of the decomposition of the deformation gradient [F] = [R][U] into symmetric and orthogonal matrices. (a) Unit circle transforms to stretch ellipse due to [U]. (b) Stretch ellipse rotates through angle ω_z due to [R].

$$[\mathrm{R}] = \begin{bmatrix} 0.7660 & -0.6428 \\ 0.6428 & 0.7660 \end{bmatrix}$$

Confirming that $[\mathrm{R}'][\mathrm{R}] = [\mathrm{I}]$, shows that this matrix is orthogonal. The rotation angle about the *Z*-axis is $\omega_z = 40^\circ$, and the elements of this matrix are calculated using (5.7), the two-dimensional version of the rotation matrix defined in Section 2.6.4. To construct **Figure 5.20b** we use [R] to premultiply the set of column vectors $[\mathrm{X_e}]$ that define the stretch ellipse:

$$\underset{2\times1}{[\mathrm{X_r}]} = \underset{2\times2}{[\mathrm{R}]}\,\underset{2\times1}{[\mathrm{X_e}]}$$

This creates a new set of vectors $[\mathrm{X_r}]$ that define the perimeter of the rotated stretch ellipse.

To compose the general deformation gradient matrix [F], we use [R] to premultiply [U]:

$$[\mathrm{F}] = [\mathrm{R}][\mathrm{U}] = \begin{bmatrix} 1.3789 & -0.3571 \\ 1.1570 & 0.4256 \end{bmatrix}$$

This matrix is not symmetric by inspection, and it is not orthogonal because [F′][F] does not equal [I]. However, starting with the same set of column vectors $[\mathrm{X_c}]$ that define a unit circle, and transforming them by premultiplying by this [F], we recreate the set of column vectors $[\mathrm{X_r}]$ that define the perimeter of the rotated stretch ellipse (**Figure 5.20b**):

$$\underset{2\times 1}{[\mathrm{X_r}]} = \underset{2\times 2}{[\mathrm{F}]}\ \underset{2\times 1}{[\mathrm{X_c}]}$$

This deformation gradient matrix [F] includes stretching, shearing, and rotation.

Given a general deformation gradient matrix [F], how does one separate the stretching and shearing contained in the symmetric matrix [U] from the rotation contained in the orthogonal matrix [R]? Recall that the product of the transpose of a square matrix and the matrix itself is symmetric, and consider the following products (Section 2.6.4 introduces the matrix operations employed here):

$$[\mathrm{F'}][\mathrm{F}] = ([\mathrm{R}][\mathrm{U}])'([\mathrm{R}][\mathrm{U}])$$

Because the transpose of the product of two matrices is equal to the product of the transpose of those matrices in the reverse order, the right side is equal to [U′][R′][R][U]. For an orthogonal matrix, the product of the transpose and the matrix itself is the identity matrix, so [R′][R] = [I], which drops out. Finally, the transpose of a symmetric matrix is equal to the matrix itself, so $[\mathrm{U'}][\mathrm{U}] = [\mathrm{U}^2]$ and the displayed equation reduces to $[\mathrm{F'}][\mathrm{F}] = [\mathrm{U}^2]$. Taking the square root of both sides, we have:

$$[\mathrm{U}] = ([\mathrm{F'}][\mathrm{F}])^{1/2} \qquad (5.8)$$

The elements of [U] are the components of the pure stretch and shearing tensor **U**.

Given [U] and the general relationship [F] = [R][U], we postmultiply both sides by the inverse of [U] to find $[\mathrm{F}][\mathrm{U}^{-1}] = [\mathrm{R}][\mathrm{U}][\mathrm{U}^{-1}]$. The product of a square matrix and the inverse of that matrix is the identity matrix, so $[\mathrm{U}][\mathrm{U}^{-1}] = [\mathrm{I}]$ and:

$$[\mathrm{R}] = [\mathrm{F}][\mathrm{U}^{-1}] \qquad (5.9)$$

This operation depends upon the symmetric matrix [U] having an inverse. The elements of [R] are the components of the pure rotation tensor **R**.

The relationships summarized in (5.8) and (5.9) follow from the Polar Decomposition Theorem. This theorem proves that any deformation gradient tensor that is made up of real components (not complex numbers), and that is invertible (the inverse exists), is composed of the product [R][U] where [R] is orthogonal and a pure rotation, and [U] is symmetric and a pure stretch and shear.

5.5.3 Finite Strain Tensor

Because the deformation gradient tensor, **F**, quantifies any combination of stretching, shearing, and rotation, one might suppose this is the only kinematic quantity required to study deformation. However, three reasons compel us to introduce another tensor quantity, the finite strain tensor, **E**.

First, we want to divorce pure rotation from the deformation, because a pure rotation involves no stretching and no shearing, and we naturally think of those as the essential features of deformation. To remove the pure rotation from [F] we transform it into a symmetric matrix. A symmetric matrix can be constructed from [F] using the fact that the product of the transpose of a square matrix with the matrix itself is symmetric (see Section 2.6.4). Thus, [E] is constructed from [F], in part, using [F′][F]. The symmetry of [E] means the strain has real principal values and principal axes, which can be found by solving the *eigenvalue* problem.

Second, it would be intuitively more satisfying if all the elements of [F] go to zero when the stretching and shearing go to zero. However, the examples given in (5.6) and illustrated in **Figure 5.18** demonstrate that the elements on the primary diagonal of [F] go to one for no stretching and shearing. This problem is overcome by subtracting the identity matrix [I] when defining [E]. Recall that [I] is a square matrix with ones on the primary diagonal (Section 2.6.4).

Third, we seek a measure of deformation that can be compared to the small strain tensor [ε], introduced in Section 4.9.2, so we can assess when that tensor is an inadequate measure of deformation. By introducing a factor of ½ in the definition of [E], it reduces to [ε] for small deformations.

To satisfy the three requirements mentioned above, the finite strain tensor is defined as:

$$[\mathrm{E}] = \tfrac{1}{2}([\mathrm{F'}][\mathrm{F}] - [\mathrm{I}]) \qquad (5.10)$$

The elements of [E] are squares and products of the elements of [F]. It is not obvious what the geometric interpretations of the finite strain matrix elements might be, but we return to this important question shortly. First, however, because the small strain tensor [ε] is a function of the displacement gradient components (Section 4.9.2), we need to introduce the displacement in the expression for the finite strain (5.10). This is done be observing in **Figure 5.21** that the particle at **X** displaces to **x** by the displacement vector **u** such that:

$$\begin{aligned} x &= X + u_x(X, Y) \\ y &= Y + u_y(X, Y) \end{aligned} \qquad (5.11)$$

Taking the partial derivatives of (5.11) with respect to those coordinates, we have:

$$\begin{aligned} \frac{\partial x}{\partial X} &= 1 + \frac{\partial u_x}{\partial X}, \quad \frac{\partial x}{\partial Y} = \frac{\partial u_x}{\partial Y} \\ \frac{\partial y}{\partial X} &= \frac{\partial u_y}{\partial X}, \quad \frac{\partial y}{\partial Y} = 1 + \frac{\partial u_y}{\partial Y} \end{aligned} \qquad (5.12)$$

Substituting (5.12) into (5.10) for the appropriate elements of [F], we find the elements of [E] in terms of derivatives of the displacement components.

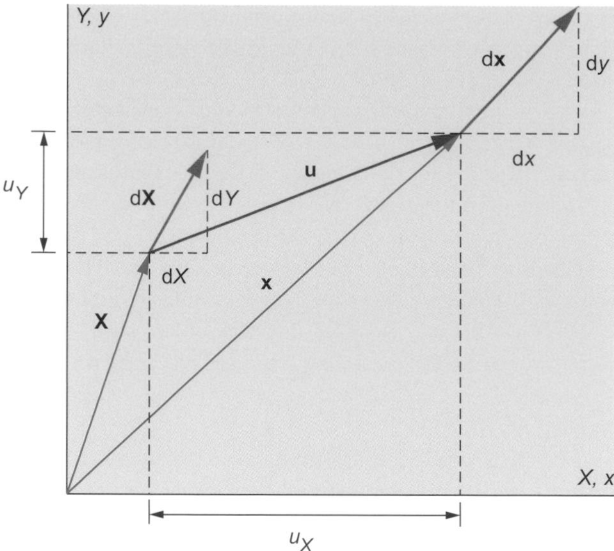

Figure 5.21 Vectors used to define the strain tensor for a particle at position **X** in the undeformed state and **x** in the deformed state. The particle displacement is **u**. Vector d**X** represents a material line at **X**, and d**x** represents the stretched and rotated material line at **x**. The lengths of the differential vectors are greatly exaggerated. See Malvern (1969).

$$E_{xx} = \frac{\partial u_x}{\partial X} + \frac{1}{2}\left[\left(\frac{\partial u_x}{\partial X}\right)^2 + \left(\frac{\partial u_y}{\partial X}\right)^2\right]$$
$$E_{xy} = \frac{1}{2}\left(\frac{\partial u_x}{\partial Y} + \frac{\partial u_y}{\partial X}\right) + \frac{1}{2}\left(\frac{\partial u_x}{\partial X}\frac{\partial u_x}{\partial Y} + \frac{\partial u_y}{\partial X}\frac{\partial u_y}{\partial Y}\right) = E_{yx}$$
$$E_{yy} = \frac{\partial u_y}{\partial Y} + \frac{1}{2}\left[\left(\frac{\partial u_x}{\partial Y}\right)^2 + \left(\frac{\partial u_y}{\partial Y}\right)^2\right] \quad (5.13)$$

These are the two-dimensional components of the Lagrangian strain tensor, **E**. The finite strain components include squares and products of the displacement gradient components. In this way they are clearly different from the small strain components (Section 4.9.2), which only contain displacement gradient components to the first power. From the perspective of dimensional analysis, the finite strain components are composed of ratios of lengths and squared ratios of lengths, so they are dimensionless numbers.

Each deformation described by the deformation gradient tensors in (5.6) and illustrated in **Figure 5.18** corresponds to a particular finite strain tensor. For those cases where the deformation involves a pure rotation, that rotation is not included in the corresponding finite strain. To show how to convert elements of [F] to the corresponding elements of [E] we use the example of simple shear [F(d)] from (5.6) where:

$$F_{xx} = \frac{\partial x}{\partial X} = 1, F_{xy} = \frac{\partial x}{\partial Y} = 0.5, F_{yx} = \frac{\partial y}{\partial X} = 0, F_{yy} = \frac{\partial y}{\partial Y} = 1$$

Using (5.12) the displacement gradients are:

$$\frac{\partial u_x}{\partial X} = \frac{\partial x}{\partial X} - 1 = 0, \quad \frac{\partial u_x}{\partial Y} = \frac{\partial x}{\partial Y} = 0.5, \quad \frac{\partial u_y}{\partial X} = \frac{\partial y}{\partial X} = 0,$$
$$\frac{\partial u_y}{\partial Y} = \frac{\partial y}{\partial Y} - 1 = 0$$

Substituting the displacement gradient terms into (5.13), yields the finite strain components:

$$E_{xx} = 0, \ E_{xy} = 0.25, \ E_{yx} = 0.25, \ E_{yy} = 0.125$$

This finite strain satisfies the condition that $E_{xy} = E_{yx}$, so the matrix is symmetric.

Following the procedure just summarized, the finite strain matrices corresponding to the deformation gradient matrices in (5.6) are:

$$[\mathrm{E(a)}] = \begin{bmatrix} 0 & 0 \\ 0 & 0 \end{bmatrix}, [\mathrm{E(b)}] = \begin{bmatrix} 0 & 0 \\ 0 & 0 \end{bmatrix}, [\mathrm{E(c)}] = \begin{bmatrix} 0.28125 & 0 \\ 0 & -0.18 \end{bmatrix},$$
$$[\mathrm{E(d)}] = \begin{bmatrix} 0 & 0.25 \\ 0.25 & 0.125 \end{bmatrix}, [\mathrm{E(e)}] = \begin{bmatrix} 0.125 & 0.25 \\ 0.25 & 0 \end{bmatrix},$$
$$[\mathrm{E(f)}] = \begin{bmatrix} 0.125 & 0.5625 \\ 0.5625 & 0.40625 \end{bmatrix} \quad (5.14)$$

Note that all of these matrices are symmetric, guaranteeing that [E] includes no pure rotation. Furthermore, the elements of the finite strain matrix all vanish for [E(a)] and for [E(b)], consistent with no stretching and no shearing for these two examples.

The geometric interpretation of the elements of the strain matrices in (5.14) may be understood, in part, from their relationship to the change in length of infinitesimal material lines parallel to X and Y. To illustrate this we use the deformation gradient tensor [F(c)] from (5.6), which is a particular case of pure shear. This produces an elongation ($F_{xx} > 1$) of the material line parallel to X and shortening ($0 < F_{yy} < 1$) of the line parallel to Y (**Figure 5.18**). The corresponding elements of the finite stain from [E(c)] in (5.14), are $E_{xx} = 0.28$ and $E_{yy} = -0.18$. The positive and negative signs of these finite strain elements signal elongation and shortening, respectively, but the values do not simply reflect the changes in length of these material lines. To evaluate these changes we must consider a different kinematic quantity called the stretch.

The stretch of the infinitesimal material lines parallel to X and Y provides a direct measure of elongation and shortening for the particular case of pure shear described in the preceding paragraph. The stretch is the final length divided by the initial length. For this example the material lines parallel to X and Y in the initial state are parallel to x and y in the final state, so they do not change orientation. Therefore, $\partial u_x/\partial Y = 0 = \partial u_y/\partial X$, and the stretch of these lines is found using (5.12) and (5.13) as follows:

$$S_{XX} = \frac{dx}{dX} = 1 + \frac{du_x}{dX} = \sqrt{1 + 2E_{xx}} = 1.25$$
$$S_{YY} = \frac{dy}{dY} = 1 + \frac{du_y}{dY} = \sqrt{1 + 2E_{yy}} = 0.8$$

Compared to their initial lengths, the infinitesimal line parallel to X lengthens by 25%, and the line parallel to Y shortens by 20%.

The same physical constraint that we used to determine the ranges for the partial derivatives in (5.3) limit the ranges of these stretches to positive numbers: $0 < S_{xx}, S_{yy} < +\infty$. Values between 0 and 1 correspond to shortening, and values greater than 1 correspond to elongation.

In matrix form, the two-dimensional finite strain is written:

$$[\mathrm{E}] = \begin{bmatrix} E_{xx} & E_{xy} \\ E_{yx} & E_{yy} \end{bmatrix}, -\tfrac{1}{2} < E_{xx}, E_{yy} < +\infty, -\infty < E_{xy}, E_{yx} < +\infty \quad (5.15)$$

The range of the elements on the primary diagonal of the finite strain [E] is found by squaring the stretch: $(S_{xx})^2 = 1 + 2E_{xx} > 0$. Values greater than –½ and less than 0 correspond to shortening, values of 0 indicate no change in length, and values greater than 0 correspond to elongations. The range of the elements on the secondary diagonal includes all the negative and positive numbers.

5.5.4 Principal Values and Orientations of Finite Strain

The components of finite strain (5.13) form a symmetric matrix (5.15), just like the components of small strain introduced in Section 4.9.2. This means the finite strain tensor has real principal values and principal axes, and the axes are orthogonal. The principal strain values and principal strain axes are found by solving the eigenvalue problem (Section 2.6.4) for the finite strain matrix [E] (5.15). For a two-dimensional example, consider *simple shear* strain described by the following deformation gradient tensor:

$$[\mathrm{F(h)}] = \begin{bmatrix} 1 & 1.5 \\ 0 & 1 \end{bmatrix}$$

Recall from the discussion of the matrix [F(d)] in (5.6) that the shear planes are in the (*X*, *Z*)-plane, like the deck of cards in **Figure 5.19**. The corresponding finite strain tensor is:

$$[\mathrm{E(h)}] = \begin{bmatrix} 0 & 0.75 \\ 0.75 & 1.125 \end{bmatrix}$$

This was found using (5.12) and (5.13) in the procedure described in Section 5.5.3.

The principal strain values (eigenvalues) and the principal strain axes (eigenvectors) are:

$$E_1 = 1.5, \hat{\mathbf{X}}_1 = \langle 0.4472, 0.8944 \rangle$$
$$E_2 = -0.375, \hat{\mathbf{X}}_2 = \langle -0.89440.4472 \rangle$$

Here the greatest principal strain is positive (elongation) and the least principal strain is negative (shortening). For a general deformation, all principal values may be positive ([E] is positive definite), or all may be negative ([E] is negative definite), or they may have a mix of signs as seen here for two-dimensional simple shear. By definition the principal values are ordered so $E_1 \geq E_2$. The principal orientation vectors, $\hat{\mathbf{X}}_1$ and $\hat{\mathbf{X}}_2$, are shown in **Figure 5.22** where their heads coincide with the unit circle: they are *unit vectors*. Because these vectors are orthogonal, their scalar product must be zero: $\hat{\mathbf{X}}_1 \cdot \hat{\mathbf{X}}_2 = 0$. They are written using capital letters because they refer to directions in the *initial state* of material lines that are deformed by the principal strains.

To illustrate the simple shear deformation in **Figure 5.22**, we relax the condition that the material lines in the initial state are *infinitesimal* in length, for example d**X** in **Figure 5.21**. Instead, we consider the two unit direction vectors, $\hat{\mathbf{X}}_1$ and $\hat{\mathbf{X}}_2$, for the principal strains as representing material lines in the initial state, and find the two vectors, $\mathbf{x}_1$ and $\mathbf{x}_2$, corresponding to these

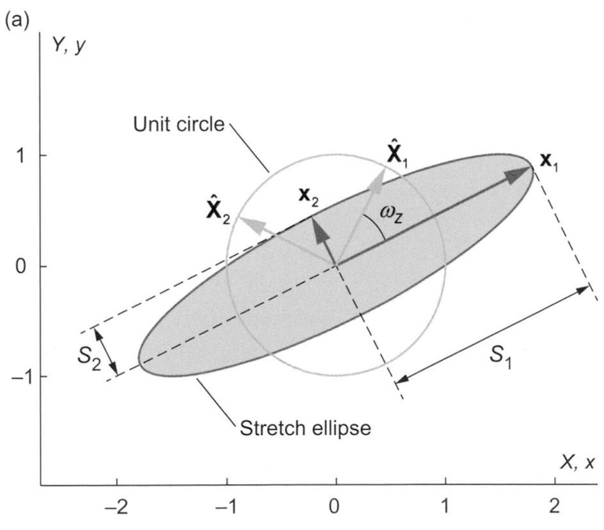

Figure 5.22 (a) Illustration of the initial state (circle) and final state (ellipse) where the deformation is simple shear. Vectors in the initial and final state are parallel to material lines that undergo the principal stretches. The unit circle deforms into the stretch ellipse with semi-axial lengths S_1 and S_2. (b) Cabin deformed (approximately) by simple shear.

material lines in the final state that have experienced the principal finite strains. This means that we have tacitly assumed the deformation is homogeneous at the length scale of the unit circle. For example, the unit direction vector, $\hat{\mathbf{X}}_1$, is deformed using the deformation gradient tensor [F(h)] to find $\mathbf{x}_1$:

$$[\mathrm{F(h)}][\hat{\mathrm{X}}_1] = [\mathrm{x}_1], \begin{bmatrix} 1 & 1.5 \\ 0 & 1 \end{bmatrix} \begin{bmatrix} 0.4472 \\ 0.8944 \end{bmatrix} = \begin{bmatrix} 1.7889 \\ 0.8944 \end{bmatrix}$$

This, and a similar operation for $\hat{\mathbf{X}}_2$, determines the components of the two vectors in the final state:

$$\mathbf{x}_1 = \langle 1.7889, 0.8944 \rangle, \ \mathbf{x}_2 = \langle -0.2236, 0.4472 \rangle$$

The vectors $\mathbf{x}_1$ and $\mathbf{x}_2$ are shown in **Figure 5.22** where they coincide with the semi-major and semi-minor axes of the ellipse that is the deformed unit circle. These vectors are orthogonal, so their scalar product is zero: $\mathbf{x}_1 \cdot \mathbf{x}_2 = 0$.

Because $\hat{\mathbf{X}}_1$ and $\hat{\mathbf{X}}_2$ are unit vectors, the magnitudes of $\mathbf{x}_1$ and $\mathbf{x}_2$ (**Figure 5.22**) are equal to the principal stretches:

$$S_1 = |\mathbf{x}_1| = \sqrt{1 + 2E_1} = 2.0$$
$$S_2 = |\mathbf{x}_2| = \sqrt{1 + 2E_2} = 0.5$$

Because the lengths of the semi-major and semi-minor axes of the ellipse in **Figure 5.22** are equal to the principal stretches, this is the stretch ellipse.

The finite strain tensor [E] does not necessarily yield an evocative graphical image, like the stretch ellipse, because it is not necessarily *positive definite*. Because the principal strain values, as in the previous example of simple shear (**Figure 5.22**), may be positive (elongation) and negative (shortening), they cannot collectively represent the lengths of the semi-axes of an ellipse. Despite this fact, the terms "strain ellipse" and "strain ellipsoid" are common in the literature of structural geology. These expressions should be avoided when referring to the Lagrangian finite strain tensor, so the definition (5.13) and permissible ranges (5.15) of the finite strain elements are not obscured.

5.5.5 Coaxial and Non-coaxial Deformation

The deformation gradient tensor (5.3) includes stretching, shearing, and pure rotation of infinitesimal material lines, whereas the finite strain tensor (5.13) includes only stretching and shearing. For two-dimensional deformation, such as the simple shear considered in Section 5.5.3, the pure rotation at a point in the (*X, Y*)-plane is the angle between the infinitesimal material line in the initial state that *will* experience the greatest stretch, and that same material line in the final state that *has* experienced the greatest stretch. If the vector that represents the material line with the greatest stretch does not change orientation from the initial to the final state, the deformation is called coaxial. If that vector does change orientation, the deformation is called non-coaxial.

Referring to **Figure 5.22** for the sake of an example, the pure rotation about the *z*-axis, ω_Z, for this case of simple shear is the angle between $\hat{\mathbf{X}}_1$ and $\mathbf{x}_1$. In order to calculate this angle we use the scalar product of the two vectors. Recall from Section 2.2.2 that the scalar product of vectors $\mathbf{v}$ and $\mathbf{w}$ is $\mathbf{v} \cdot \mathbf{w} = |\mathbf{v}||\mathbf{w}|\cos\theta_{\mathbf{vw}}$ where $\theta_{\mathbf{vw}}$ is the smaller angle between the vectors. To implement this relationship we use $\hat{\mathbf{x}}_1 = \mathbf{x}_1/S_1$, which is the unit direction vector in the direction of $\mathbf{x}_1$. Then the scalar product reduces to $\hat{\mathbf{X}}_1 \cdot \hat{\mathbf{x}}_1 = \cos\omega_Z$. Solving for the pure rotation we have:

$$\omega_Z = \cos^{-1}\left(\hat{\mathbf{X}}_1 \cdot \hat{\mathbf{x}}_1\right) = 36.87^{\circ}$$

This demonstrates that simple shear is *non-coaxial* (**Figure 5.22**). On the other hand, pure shear is *coaxial* (**Figure 5.18c**). An infinite number of other deformation gradient matrices and corresponding finite strain matrices produce coaxial or non-coaxial deformation. Because simple and pure shear are very special cases, they are unlikely to occur in nature, but they are instructive examples for a textbook.

5.5.6 Comparing Small Strain and Finite Strain: Error Analysis

The small strain components in two dimensions were derived in Section 4.9.2 and they are repeated here so we can compare them directly to the finite strain components (5.13):

$$\varepsilon_{xx} = \frac{\partial u_x}{\partial X}, \varepsilon_{xy} = \frac{1}{2}\left(\frac{\partial u_x}{\partial Y} + \frac{\partial u_y}{\partial X}\right) = \varepsilon_{yx}, \varepsilon_{yy} = \frac{\partial u_y}{\partial Y} \quad \textbf{(5.16)}$$

Note that these are the leading (linear) terms in the finite strain components (5.13): the second terms all are non-linear in the displacement gradient components. To evaluate the error introduced by ignoring the non-linear terms, consider the deformation prescribed using [F(f)] in (5.6). Components of the displacement gradients are found using (5.12):

$$\frac{\partial u_x}{\partial X} = 0, \ \frac{\partial u_x}{\partial Y} = 0.5, \ \frac{\partial u_y}{\partial X} = 0.5, \ \frac{\partial u_y}{\partial Y} = 0.25$$

These are introduced into (5.13) to evaluate the finite strain components:

$$E_{xx} = 0 + 0.125, \ E_{xy} = 0.5 + 0.0625 = E_{yx},$$
$$E_{yy} = 0.25 + 0.15625$$

Using $[(E - \varepsilon)/E] \times 100$ to estimate the percent error introduced by ignoring the non-linear terms we have 100%, 11%, 11%, and 38%, respectively.

Whether or not one can tolerate such errors depends upon the application. In some engineering applications errors greater than 10% may not be tolerable, while in some geologic applications such errors may not be large compared to others introduced by poor data quality due to lack of exposure. A considerable incentive exists to approximate strain using small strain theory, because the equations are much simpler, and because the theory of linear elasticity is so powerful. However, such a decision should be justified with an error analysis, and a discussion of the tolerance for error in the particular application.

5.5.7 Three-Dimensional Deformation and Finite Strain

The deformation gradient tensor, **F**, and the finite strain tensor, **E**, are defined here in three dimensions without introducing new concepts. The elements of the deformation gradient tensor in matrix form are:

$$[\mathrm{F}] = \begin{bmatrix} F_{xx} & F_{xy} & F_{xz} \\ F_{yx} & F_{yy} & F_{yz} \\ F_{zx} & F_{zy} & F_{zz} \end{bmatrix} = \begin{bmatrix} \frac{\partial x}{\partial X} & \frac{\partial x}{\partial Y} & \frac{\partial x}{\partial Z} \\ \frac{\partial y}{\partial X} & \frac{\partial y}{\partial Y} & \frac{\partial y}{\partial Z} \\ \frac{\partial z}{\partial X} & \frac{\partial z}{\partial Y} & \frac{\partial z}{\partial Z} \end{bmatrix} \quad (5.17)$$

The deformation gradient tensor describes the stretch, shear, and rotation of infinitesimal material lines of any orientation at a point in the material continuum. An infinitesimal spherical object of unit radius is transformed into an infinitesimal ellipsoidal object by the deformation gradient tensor. This ellipsoid has three orthogonal semi-axes with lengths that are equal to the principal values of the stretch at that point.

In general each component of the deformation gradient tensor is a function of the three spatial coordinates and time: $F_{xx} = F_{xx}(X, Y, Z, t)$, $F_{xy} = F_{xy}(X, Y, Z, t)$, etc. As such they describe deformation that is heterogeneous (varies in space), and not constant (varies in time). In rare cases deformation in Earth's lithosphere may be shown to be approximately homogeneous over a particular length scale, and approximately time independent over a particular time scale. In this case each component of the deformation gradient tensor is taken as uniform in space and constant in time.

The elements of the finite strain tensor in three dimensions in matrix form are:

$$[\mathrm{E}] = \begin{bmatrix} E_{xx} & E_{xy} & E_{xz} \\ E_{yx} & E_{yy} & E_{yz} \\ E_{zx} & E_{zy} & E_{zz} \end{bmatrix} \quad (5.18)$$

The primary diagonal elements of the finite strain tensor are related to the spatial derivatives of the displacement components as:

$$\begin{aligned} E_{xx} &= \frac{\partial u_x}{\partial X} + \frac{1}{2}\left[\left(\frac{\partial u_x}{\partial X}\right)^2 + \left(\frac{\partial u_y}{\partial X}\right)^2 + \left(\frac{\partial u_z}{\partial X}\right)^2\right] \\ E_{yy} &= \frac{\partial u_y}{\partial Y} + \frac{1}{2}\left[\left(\frac{\partial u_x}{\partial Y}\right)^2 + \left(\frac{\partial u_y}{\partial Y}\right)^2 + \left(\frac{\partial u_z}{\partial Y}\right)^2\right] \\ E_{zz} &= \frac{\partial u_z}{\partial Z} + \frac{1}{2}\left[\left(\frac{\partial u_x}{\partial Z}\right)^2 + \left(\frac{\partial u_y}{\partial Z}\right)^2 + \left(\frac{\partial u_z}{\partial Z}\right)^2\right] \end{aligned} \quad (5.19)$$

The finite strain tensor is symmetric, so the respective secondary diagonal elements are equal, and they are related to the spatial derivatives of the displacement components as:

$$\begin{aligned} E_{xy} &= \frac{1}{2}\left(\frac{\partial u_x}{\partial Y} + \frac{\partial u_y}{\partial X}\right) + \frac{1}{2}\left(\frac{\partial u_x}{\partial X}\frac{\partial u_x}{\partial Y} + \frac{\partial u_y}{\partial X}\frac{\partial u_y}{\partial Y} + \frac{\partial u_z}{\partial X}\frac{\partial u_z}{\partial Y}\right) = E_{yx} \\ E_{yz} &= \frac{1}{2}\left(\frac{\partial u_y}{\partial Z} + \frac{\partial u_z}{\partial Y}\right) + \frac{1}{2}\left(\frac{\partial u_x}{\partial Y}\frac{\partial u_x}{\partial Z} + \frac{\partial u_y}{\partial Y}\frac{\partial u_y}{\partial Z} + \frac{\partial u_z}{\partial Y}\frac{\partial u_z}{\partial Z}\right) = E_{zy} \\ E_{zx} &= \frac{1}{2}\left(\frac{\partial u_z}{\partial X} + \frac{\partial u_x}{\partial Z}\right) + \frac{1}{2}\left(\frac{\partial u_x}{\partial Z}\frac{\partial u_x}{\partial X} + \frac{\partial u_y}{\partial Z}\frac{\partial u_y}{\partial X} + \frac{\partial u_z}{\partial Z}\frac{\partial u_z}{\partial X}\right) = E_{xz} \end{aligned} \quad (5.20)$$

For each finite strain component in (5.19) and (5.20) the leading term is equal to the respective component of the small strain tensor, for example see (5.16). The second term for each component is non-linear in the displacement derivatives and provides the necessary correction to make the small strain approximation exact. The finite strain components describe the non-rotational part of the deformation, including stretching and shearing, at a point in a deforming rock mass with no approximations.

Given the components of finite strain tensor referred to an arbitrary coordinate system, it always is possible to solve the eigenvalue and eigenvector problem for the principal values and axes. By convention the principal values are ordered such that $E_1 \geq E_2 \geq E_3$. Principal values in the range $-½ \leq (E_1, E_2, E_3) < 0$ represent a shortening of the respective material lines, and those in the range $0 < (E_1, E_2, E_3) < \infty$ represent an elongation. The finite strain is zero if all three principal strains are zero.

5.6 MECHANISMS OF DUCTILE DEFORMATION

At the scale of mineral grains, we distinguish three different mechanisms of ductile deformation. The first mechanism is cataclasis, which involves fragmentation of grains by brittle fracture. Cataclastic flow is the inelastic deformation of fragmented rock by frictional sliding on fragment contacts, distortion of pores, and rotation of fragments. Dominantly a physical mechanism, cataclasis is sensitive to the mean compressive stress, which inhibits both fracture and frictional sliding.

The second mechanism is pressure solution, which involves the dissolution of soluble minerals at points of stress concentration. Diffusion of the dissolved constituents in pore fluids, or transport of these constituents with the flowing fluid to a different location, and the precipitation of new minerals, result in permanent deformation of the rock mass. Dominantly a chemical mechanism, pressure solution is sensitive to the solubility of the mineral species and the chemistry of the groundwater.

The third mechanism, called intracrystalline plasticity, involves the motion of defects in the crystal lattice (e.g. dislocations and vacancies), and the resulting deformation of mineral grains without fracturing or dissolution. This is a physical mechanism operating entirely within mineral grains. Of the three mechanisms of ductile deformation introduced above, intracrystalline plasticity is, perhaps, the best understood. This mechanism is a major focus for engineers dealing with plasticity of manufactured materials and is a core subject for the discipline of materials science. In addition, the role of dislocations in plasticity is amenable to mechanical analysis that provides important insights about deformation at the crystal lattice scale. Finally, the mechanical models for dislocation motion have extensive applications at larger scales. For these reasons, we focus on dislocations in this section, and refer to other literature at the end of this chapter that deals with cataclasis and pressure solution in more detail.

Models utilizing dislocations have provided a deeper understanding of the physical processes that shape mountain ranges and continents, and they can be generalized to understand outcrop and crustal scale geologic structures, including those as diverse as igneous dikes and plate-bounding strike-slip faults. This versatility of the dislocation concept warrants a careful description at the introductory level. In this section, we describe the physical nature of dislocations at the scale of a mineral lattice.

5.6.1 The Edge and Screw Dislocation

Dislocations are line defects that move through a mineral grain, breaking bonds and offsetting the crystalline lattice, but these bonds are healed after the dislocation line passes, so the grain is deformed, but remains intact. The passage of an edge dislocation line through such a grain is shown schematically in **Figure 5.23** where the initial grain shape is taken as cubic to simplify the drawings. The four parts of the figure are like selected frames of a movie, used to illustrate discrete stages in the passage of the dislocation line through the mineral grain.

In **Figure 5.23a** orthogonal gray lines on the front left face of the cubic grain form a square grid, with each grid intersection representing the same atom in a periodic arrangement, and the gray lines representing the bonds between these atoms. The atoms are arranged periodically, with the same spacing in the third dimension, so this grain is taken as having a simple cubic crystalline lattice. Metallic Polonium (Po) is the only substance known to have such a simple lattice, but we use it here to provide a clear illustration of the effects of dislocation motion. Halite (NaCl) and pyrite (FeS_2) are common rock-forming minerals that have a somewhat more complex array of atoms in their cubic lattice, and this basic geometry is reflected in the familiar cubic shape of natural halite ("rock salt") and pyrite ("fool's gold") grains. We use the grid of gray lines to illustrate the distortion and offset of the cubic crystalline lattice in the subsequent stages of dislocation motion shown in **Figure 5.23b–d**. Although we refer only to atoms and bonds associated with the grid on the left front face, the behavior of each atom and bond is mimicked by the corresponding row of atoms and bonds parallel to the dislocation line in the third dimension.

Application of the shear stress pictured in the inset above **Figure 5.23a** causes the upper half of the grain to move toward the left rear, relative to the lower half. The blue line indicates where the dislocation line enters the grain, and two red circles indicate bonds that break for the dislocation line to advance toward the left rear of the grain. **Figure 5.23b** shows the lattice after those two bonds have broken. A small step that is one lattice dimension wide now offsets the right front face of the grain. The first broken bond has re-established, connecting the upper half of the right front face to the vertical interior lattice plane just inside the lower half of that face. The dislocation line intersects the left front face at the inverted "T," where the second broken bond creates a defect in the lattice. The inverted "T" is the standard symbol for an edge dislocation.

Three more lines of bonds, indicated by the red circles in **Figure 5.23b**, must break for the dislocation to advance to the center of the grain. Notice in **Figure 5.23c** that the vertical lattice plane above the dislocation line does not connect to another plane below that line. Again, the dislocation line marks a defect in the

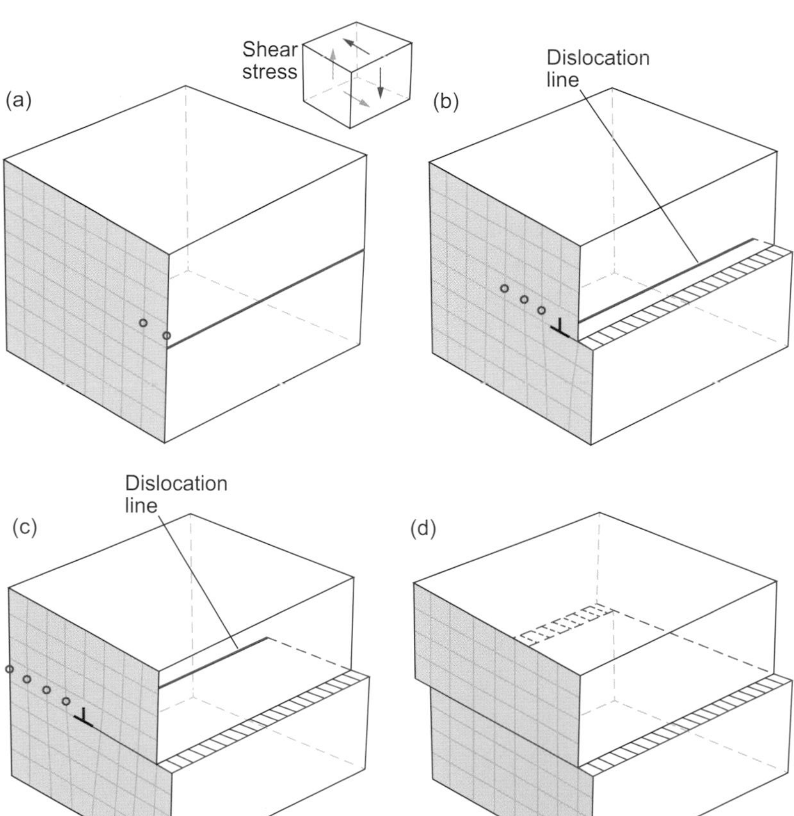

Figure 5.23 Four stages in the passage of an edge dislocation line through a mineral grain. Gray lines on front left face are edges of a cubic crystal lattice; red circles mark bonds that break as the dislocation advances to the next stage. Shear stress (inset) drives the motion of the dislocation line. (a) undistorted mineral grain; (b) dislocation line (blue) has moved into the grain interior leaving a step in the right front face; (c) dislocation line has moved farther into the grain interior; (d) dislocation line has emerged from the left rear face leaving a step. See Weertman and Weertman (1964).

crystal lattice. The truncated lattice plane above the dislocation line often is referred to as an extra half plane. It is *not* really extra, because the vertical lattice plane on the lower right front face also is an isolated half plane, which was left behind as the dislocation entered the grain. When the dislocation line emerges from the left rear face of the grain (**Figure 5.23d**), a step is created there that isolates a half plane in the upper part of the grain. Aside from these two isolated half planes, all other vertical lattice planes are bonded across the path that the dislocation line took through the grain. In this way the grain has changed shape *without fracturing*. Furthermore, this process required only one line of bonds to break for each incremental advance of the dislocation line.

We use **Figure 5.24** to identify some of the defining characteristics of the edge dislocation in a mineral grain with a cubic lattice. Again, the lattice planes are shown intersecting the left front face of the grain as gray lines and the dislocation line is marked on this face using an inverted "T." The linear defect in the lattice is identified as the dislocation line. As we illustrated in **Figure 5.23** the dislocation line moved perpendicular to itself, along the slip plane from the right front face of the grain to its current position. The orientation of the dislocation line and the sense of direction along that line are defined by a vector **t**, called the tangent vector or sometimes the sense vector. Above the dislocation line the lattice has a half plane of atoms that appears to push aside and distort the adjacent lattice planes. The dislocation line marks the lower *edge* of this half plane, which is evocative of the name *edge* dislocation. The horizontal part of the inverted "T" coincides with the slip plane, and the vertical part points to the half plane.

A second vector is needed to describe the dislocation completely. This vector is found by taking a circuit (red) around the dislocation line. In this case the circuit is shown on the left front face of the grain (**Figure 5.24**), but it could be anywhere within the grain. Starting at the arbitrary atom labeled "S," one proceeds down 6 atoms, then left 5, up 6, and right 5 to the atom labeled "F." The number of steps down and up must be equal, as must the number of steps left and right. For a perfect crystal lattice the atom labeled "F" would coincide with "S." However, because this circuit encloses an edge dislocation, a gap exists, which is measured using the vector **b** that extends from "S" to "F." This is called Burgers vector after the Dutch physicist J. M. Burgers who introduced the concept.

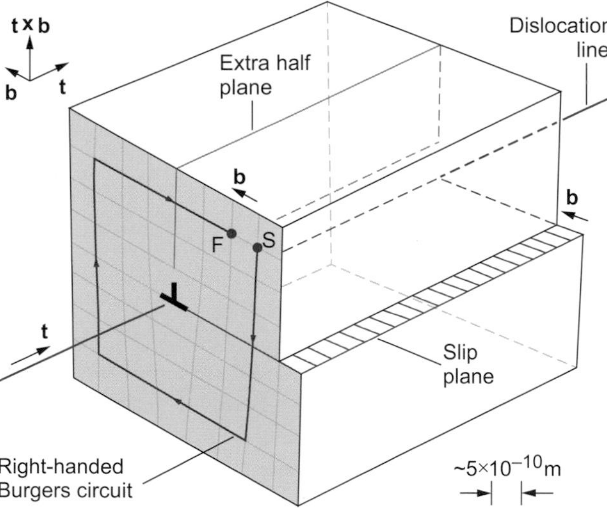

Figure 5.24 Schematic illustration of positive edge dislocation in cubic mineral grain with dislocation line and slip plane (blue), Burgers circuit (red), and "extra" half plane (gray). Inverted "T" is short-hand symbol indicating the slip plane and half plane. See Cai and Nix (2016).

Two conventions are used to define Burgers vector (**Figure 5.24**). The first provides the direction of the Burgers circuit by pointing the thumb of your right hand in the direction of the tangent vector, **t**. The circuit is taken in the direction your fingers curl. The second convention chooses Burgers vector, **b**, to be directed from "S" to "F." These constitute the RH/SF conventions, meaning *right-handed* and *start-finish*. Using the RH/SF conventions, the vector $\mathbf{t} \times \mathbf{b}$ always is directed toward the "extra" half plane of atoms, regardless of the direction chosen for **t**. For all edge dislocations Burgers vector, **b**, lies in the slip plane and is perpendicular to the dislocation line. The magnitude of Burgers vector in this case is $\approx 5 \times 10^{-10}$ m, which is equal to the spacing of the cubic lattice planes and the width of the step in the side of the grain. A fundamental feature of deformation associated with an edge dislocation is that the displacement field is discontinuous across the slip plane. In this example, the atoms above the slip plane are displaced to the left relative to those below. The magnitude of the displacement discontinuity is $|\mathbf{b}|$, and the slip is directed perpendicular to the dislocation line.

Figure 5.25 is a schematic illustration of a cubic crystal lattice containing the second type of dislocation, a screw dislocation. Again, the linear defect in the lattice is identified as the dislocation line. This dislocation line moved perpendicular to itself, along the slip plane from the right front face of the block to its current position. The orientation of the dislocation line, and the sense of direction along that line are defined by the tangent vector, **t**. The name *screw* dislocation comes from the fact that the shift of atoms creates a spiral ramp about the dislocation line. Starting at the atom labeled "S," the right-handed Burgers circuit takes one down the ramp to the left, circling around the dislocation line, and proceeding down the other ramp to the right. The atoms along this part of the circuit are located on the next lattice plane toward the right rear of the block. Continuing to turn right one arrives at the atom labeled "F." Again, the number of steps between atoms down and up must be equal, as must the number of steps left and right. If the lattice were undistorted, "S" and "F" would be the same atom. The fact that the circuit fails to close reveals the presence of the screw dislocation.

The Burgers vector, **b**, for the screw dislocation points from "S" to "F" (**Figure 5.25**), lies in the slip plane, and is parallel to the dislocation line. Using the RH/SF convention, the scalar product $\mathbf{t} \cdot \mathbf{b}$ always is positive (the two vectors point in the same direction), regardless of the direction chosen for **t**. The magnitude of Burgers vector in this case is $\approx 5 \times 10^{-10}$ m, which is the spacing of the cubic lattice planes and the width of the step on the left front face of the grain. A fundamental feature of

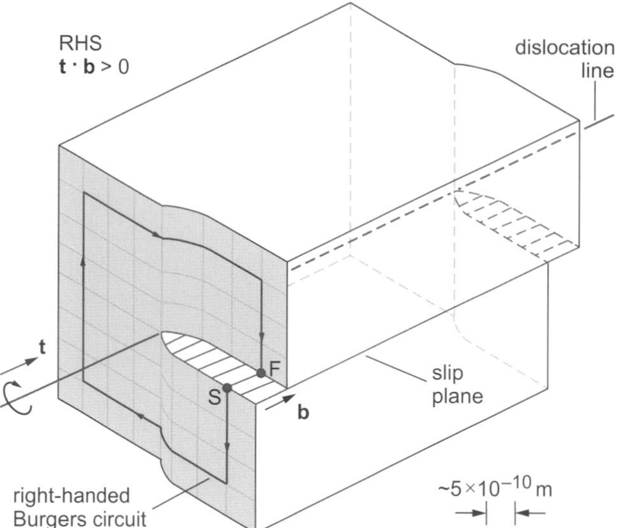

Figure 5.25 Schematic illustration of right-handed screw dislocation in cubic mineral with dislocation line and slip plane (blue), and Burgers circuit (red). The tangent vector, **t**, and Burgers vector, **b**, define the dislocation. Circular arrow wrapping around the dislocation line is a short-hand symbol for the screw dislocation and indicates the direction of Burgers circuit. See Cai and Nix (2016).

deformation associated with a screw dislocation is that the displacement field is discontinuous across the slip plane. In this example, the atoms above the slip plane are displaced to the right rear relative to those below. The magnitude of the displacement discontinuity is |**b**|, and the direction of slip is parallel to the dislocation line.

Comparing the edge and screw dislocations (**Figure 5.24** and **Figure 5.25**) note that the dislocation lines as drawn here are parallel and move in the same direction, but the resulting steps in the grain boundaries are on adjacent faces of the cubic grain. The slip directions for these edge and screw dislocations are orthogonal. The shift in atoms across the slip plane is *perpendicular* to the edge dislocation line and *parallel* to the screw dislocation line.

If 10^6 edge and screw dislocations moved on the slip planes of the mineral grains similar to those illustrated in **Figure 5.24** and **Figure 5.25**, steps of width 0.5 mm would form on the faces. The grains would be permanently deformed, but the atoms across the slip planes would be perfectly bonded. Repeating such dislocation motion throughout the grain would lead to large deformation without fracturing. In this way, the motion of edge and screw dislocations is an important mechanism for plastic deformation. This mechanism applied to manufactured materials is one of the central topics of Materials Science (**Box 5.1**).

Box 5.1 Materials Science and Engineering

Materials science is the study of the physical and chemical properties of manufactured materials such as metals, ceramics, semiconductors, polymers, biomaterials, nanomaterials, and composites. Materials engineering is the design of materials based on their physical and chemical properties to meet certain specifications and applications. Many universities have a department that combines these objectives as *materials science and engineering*. Research in these departments includes the characterization of existing materials and the creation and testing of new materials (Callister and Rethwisch, 2014).

One of two broad and overlapping areas of study in materials science and engineering focuses on the dependence of the load bearing capacity and other mechanical properties of materials on their microstructure and the history of processing. The mechanical properties of solids depend upon point defects, line defects, and surface defects at the crystal lattice and grain scale (**Figure B5.1**), so much of this area of study focuses on those imperfections (Cai and Nix, 2016). The other area of study focuses on the electronic, electrochemical, magnetic, and optical properties of materials. Applications of materials science and engineering research include improvements to the manufacturing process, engineering design using novel materials, energy storage and production, information technology, and biomedical engineering.

Modern materials science and engineering grew out of previously disparate investigations in disciplines such as metallurgy, mechanical engineering, solid-state physics, and chemistry during the middle of the twentieth century. For example, the journal *Materials Science and Engineering* began in 1966, and the discipline grew so rapidly that in 1988 the journal bifurcated into two journals, MS&E A and B. These two journals divide the subject matter based on the two broad areas of study mentioned in the previous paragraph. Topics most closely related to structural geology appear in MS&E A as articles on how strength depends on microstructure and the history of processing. Of course, for structural geologists, the materials are rocks and minerals; the history is geologic history; and the processing is tectonic.

Materials science intersects structural geology, as presented in this introductory textbook, in three prominent ways. In the laboratory, some of the methods for testing the mechanical properties of rock during brittle (Sections 4.4 and 4.7) and ductile (Section 5.4) deformation are similar to those used on engineered materials. On the theoretical side, engineering fracture mechanics applies elasticity theory to understand the initiation and propagation of fractures in brittle engineered materials. These concepts are similar to those applied to rock to understand the development of joints, dikes, and veins (Sections 7.3 and 7.4). Finally, investigations of dislocations by materials scientists as a mechanism for deformation in metals and other ductile engineered materials (e.g. Cai and Nix, 2016) have many parallels to investigations of ductile deformation in rock (Section 5.6) (Poirier, 1985).

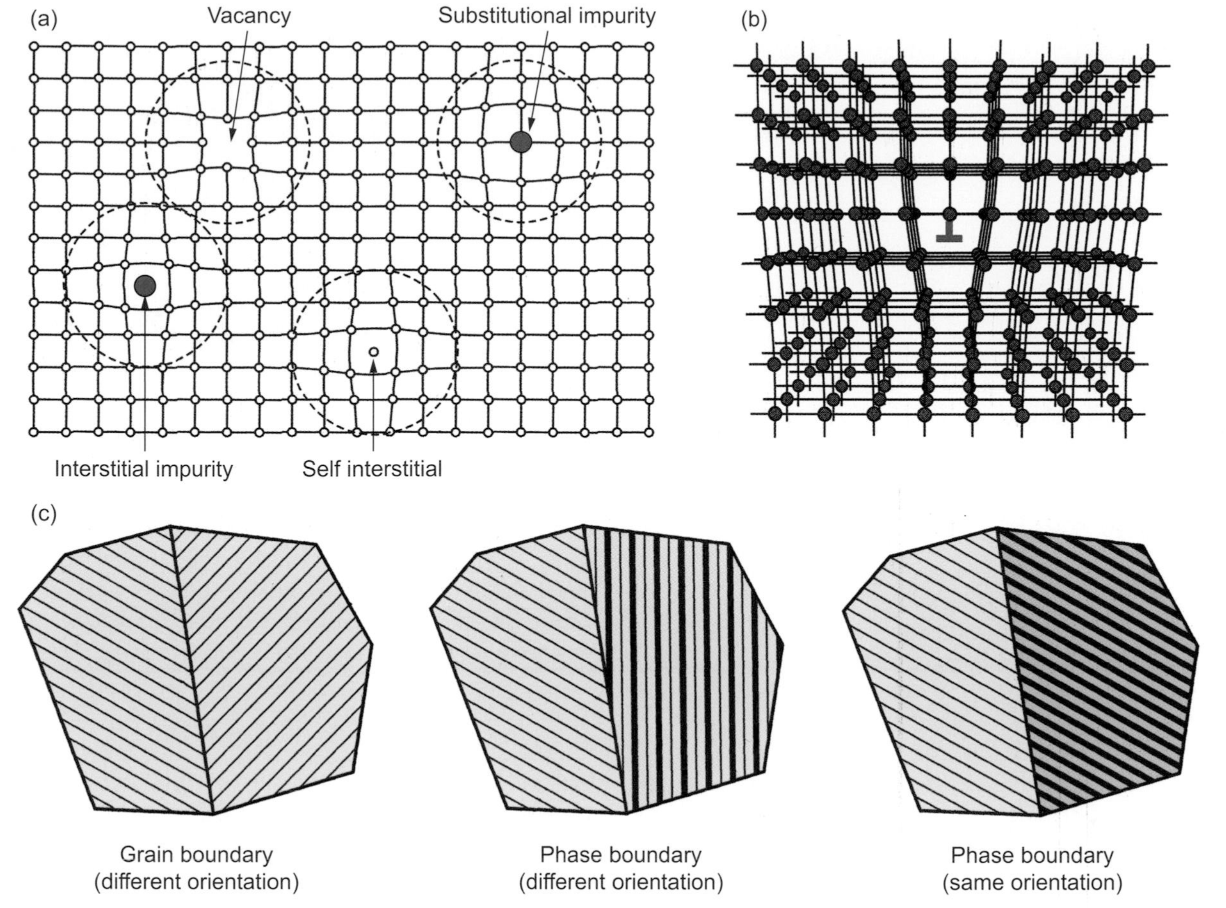

Figure B5.1 Three classes of defects in crystalline solids. (a) Zero-dimensional or *point* defects in a crystal lattice. (b) One-dimensional or *line* defects. Here an edge dislocation resides at the inverted red "T" in this cubic crystal lattice (Section 5.6.1). (c) Two-dimensional or *surface* defects. Parallel lines indicate lattice orientations and background colors represent different phases. Note that (a) and (b) are at the atomic scale, and (c) is at the grain scale. From Cai and Nix (2016).

5.6.2 The Dislocation Loop

In the previous section edge and screw dislocations were described as linear defects in the mineral lattice, and the two ends of the straight dislocation line terminated at opposite grain boundaries (**Figure 5.24** and **Figure 5.25**). This linearity of the dislocation line simplified the drawings and the explanations, because it reduced the geometry to two dimensions. However, most dislocations are three-dimensional structures and dislocation lines often are curves that form a closed dislocation loop entirely within a mineral grain (**Figure 5.26**). Notice that the tangent vector, **t**, is consistently directed around the entire loop, and that Burgers vector is the same everywhere on the loop. The shear stress causing the dislocation line to advance is σ_{zx} (upper left inset). Studying this dislocation loop helps to clarify the relationships between edge and screw dislocations. Also, it helps to define the sign of the edge dislocation and the handedness of the screw dislocation.

Traveling along the dislocation loop shown in **Figure 5.26** in the direction specified by the tangent vector, **t**, one encounters successively a negative edge (–E), right-handed screw (RHS), positive edge (+E), and left-handed screw (LHS) dislocation. Between these pure forms are dislocations that are mixtures of edge and screw dislocations. For the Cartesian coordinates used in this figure, the vector $\mathbf{t} \times \mathbf{b}$ has only a z-component, and it is negative for the edge dislocation labeled "–E." In contrast, the vector $\mathbf{t} \times \mathbf{b}$ is positive for the edge dislocation labeled "+E." Looking in the direction of **t**, the screw dislocation labeled "RHS" has a right-handed Burgers circuit that advances clockwise, and the scalar product $\mathbf{t} \cdot \mathbf{b}$ is positive, whereas the screw dislocation labeled "LHS" has a left-handed Burgers circuit that advances counter-clockwise, and the scalar product $\mathbf{t} \cdot \mathbf{b}$ is negative.

The sense of slip across the slip plane of the dislocation loop (**Figure 5.26**) is top to the right rear. With sufficient shear stress, σ_{zx}, the dislocation loop propagates outward in the gray (x, y)-plane, eventually passing through all four faces of this cubic mineral grain. After passage of the dislocation loop, steps are created on the faces that are parallel to the (y, z)-plane (inset lower right), providing evidence for the plastic deformation. On the other hand, those faces parallel to the (x, z)-plane are not

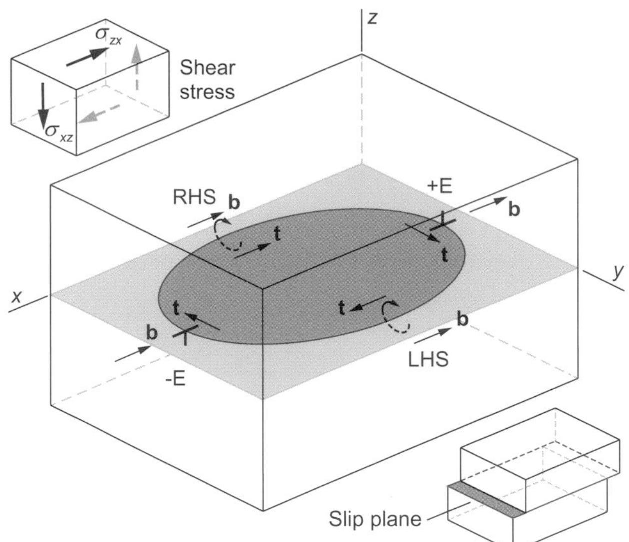

Figure 5.26 Schematic illustration of dislocation loop composed of positive (+E) and negative (–E) edge dislocations, right (RHS) and left (LHS) handed screw dislocations, and mixed dislocations between these. Each dislocation is viewed in the direction of the tangent vector, **t**. The loop encloses the slip plane (blue) and slip is top to the right rear. Shear stress driving dislocation motion (upper left inset). Offset of mineral faces (lower right inset). See Cai and Nix (2016).

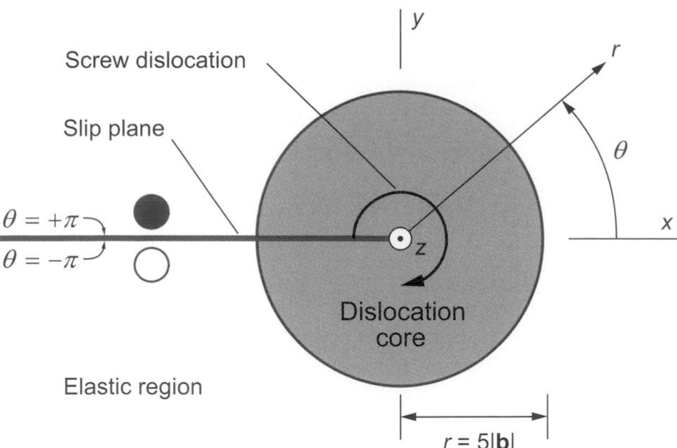

Figure 5.27 Cartesian (x, y) and polar (r, θ) coordinates for the screw dislocation. Positive z is toward the observer and the dislocation line lies along the z-axis. Tangent vector, **t**, and Burgers vector, **b**, are directed away from observer along the dislocation line. Positive displacement (filled blue circle) and negative displacement (filled white circle) are directed toward positive and negative z, respectively, and thick blue line is the slip plane.

affected by passage of the dislocation and the mineral grain appears undamaged on these faces.

5.6.3 Displacement and Stress Fields for the Screw Dislocation

The discussion of dislocations to this point has been schematic, portraying the edge and screw dislocation using the distortion of a cubic crystal lattice (**Figure 5.24** and **Figure 5.25**), or symbols on a block diagram to suggest relative displacements associated with a dislocation loop (**Figure 5.26**). Both edge and screw dislocations deform a crystal lattice in ways that can be quantified quite accurately using elastic theory, and the solutions to these elastic boundary-value problems provide important insights. Although the deformation associated with the screw dislocation is, perhaps, more difficult to visualize than that for the edge dislocation, the governing equations and resulting distributions of displacement and stress are mathematically simpler, so we focus first on the screw dislocation. Navier's equations of motion (Section 4.11) must be solved for the appropriate boundary conditions to find the elastic fields in the material surrounding the screw dislocation.

As suggested by **Figure 5.25**, the displacement vectors for the pure screw dislocation are parallel to the dislocation line. We use a right-handed Cartesian coordinate system with the z-axis parallel to the dislocation line (**Figure 5.27**) and consider the variation of the displacement component u_z in the (x, y)-plane:

$$u_x = 0, \;\; u_y = 0, \;\; u_z = f(x, y)$$

Notice in **Figure 5.27** that the positive z-axis is directed outward, toward the observer. In contrast, the tangent vector, $\mathbf{t} = \langle 0, 0, t_z \rangle$, is directed inward, in the look direction of the observer, so t_z is negative. Employing the RH/SF convention, Burgers vector, $\mathbf{b} = \langle 0, 0, b_z \rangle$, also is directed inward, so b_z is negative.

The first two Navier equations are repeated here, so we can assess all the terms in light of the restrictions on the displacements:

$$\begin{aligned}\rho\frac{\partial^2 u_x}{\partial t^2} &= G\left(\frac{\partial^2 u_x}{\partial x^2} + \frac{\partial^2 u_x}{\partial y^2} + \frac{\partial^2 u_x}{\partial z^2}\right) \\ &\quad + (G+\lambda)\left(\frac{\partial^2 u_x}{\partial x^2} + \frac{\partial^2 u_y}{\partial x \partial y} + \frac{\partial^2 u_z}{\partial x \partial z}\right) + \rho g_x^* \\ \rho\frac{\partial^2 u_y}{\partial t^2} &= G\left(\frac{\partial^2 u_y}{\partial x^2} + \frac{\partial^2 u_y}{\partial y^2} + \frac{\partial^2 u_y}{\partial z^2}\right) \\ &\quad + (G+\lambda)\left(\frac{\partial^2 u_x}{\partial x \partial y} + \frac{\partial^2 u_y}{\partial y^2} + \frac{\partial^2 u_z}{\partial y \partial z}\right) + \rho g_y^*\end{aligned} \tag{5.21}$$

The partial derivatives on the left sides and the first five partial derivatives on the right sides are zero because u_x and u_y are zero. The last partial derivative on each right side is zero because u_z is not a function of z. As a consequence, the body force (per unit volume) term on each far right side must be zero. Thus, the conditions we imposed on the displacements for the screw dislocation eliminates the first two Navier equations from further consideration.

The third Navier equation is:

$$\begin{aligned}\rho\frac{\partial^2 u_z}{\partial t^2} &= G\left(\frac{\partial^2 u_z}{\partial x^2} + \frac{\partial^2 u_z}{\partial y^2} + \frac{\partial^2 u_z}{\partial z^2}\right) \\ &\quad + (G+\lambda)\left(\frac{\partial^2 u_x}{\partial x \partial z} + \frac{\partial^2 u_y}{\partial y \partial z} + \frac{\partial^2 u_z}{\partial z^2}\right) + \rho g_z^*\end{aligned} \tag{5.22}$$

The solution we seek is for quasi-static deformation, where the term for mass (per unit volume) times acceleration on the left side is very small compared to the forces (per unit volume) on the right side. Therefore, the term on the left side is set to zero. At the very small length scale of a dislocation, the body force per unit volume (last term on right side) also is set to zero. This is justified by recalling that variations in stress due to gravity in Earth's lithosphere are of order 25 MPa/km, or 25×10^{-12} MPa/nm. This is inconsequential compared to stress variations due to the dislocation, which we show below are of order 25 MPa/nm. Finally, because we stipulated that u_x and u_y are zero, and that u_z is a function of x and y, but not a function of z, only the first two partial derivatives on the right side survive. Dividing both sides of the reduced equation by G, we find that the elastic problem for the screw dislocation is governed by:

$$\frac{\partial^2 u_z}{\partial x^2} + \frac{\partial^2 u_z}{\partial y^2} = 0 \tag{5.23}$$

This is the two-dimensional case of the famous partial differential equation called Laplace's equation, which has many particular solutions with different applications across a wide range of physical problems. The one dependent variable is the displacement component, u_z; the two independent variables are the coordinates, x and y.

The boundary conditions for the screw dislocation specify the displacement u_z on either side of the slip plane:

$$u_z = -\frac{1}{2} b_z \text{ on } x \leq 0, y = 0^+, \quad u_z = \frac{1}{2} b_z \text{ on } x \leq 0, y = 0^-$$

Because b_z is negative, the material above the slip plane moves in the positive z-direction (toward the observer of **Figure 5.27**). Below the slip plane the material moves in the negative z-direction (away from the observer). The solution to Laplace's equation for these boundary conditions yields the following displacement component (see Weertman and Weertman, 1964, in Further Reading):

$$u_z = \frac{-b_z}{2\pi} \tan^{-1}\left(\frac{y}{x}\right) \tag{5.24}$$

This may be verified by substituting (5.24) into (5.23) and differentiating. Note that the displacement is proportional to Burgers vector and it is independent of the elastic moduli.

Despite satisfying the governing equation, it is not obvious how the inverse tangent function in (5.24) solves the problem of the screw dislocation. Changing from Cartesian to cylindrical coordinates (r, θ) clearly reveals the discontinuity in displacement, and the required boundary values for the displacement. From **Figure 5.27** note that $x = r\cos\theta$ and $y = r\sin\theta$. Substituting for x and y in (5.24), the displacement equation is rewritten:

$$u_z = \frac{-b_z}{2\pi}\left[\tan^{-1}\left(\frac{r\sin\theta}{r\cos\theta}\right)\right] = \frac{-b_z}{2\pi}\left[\tan^{-1}(\tan\theta)\right]$$
$$= \frac{-b_z\theta}{2\pi} \tag{5.25}$$

The z component of displacement is independent of the radial distance, r, from the dislocation, and varies linearly around the dislocation with angular position, θ. Approaching the slip plane (**Figure 5.27**) in the second quadrant, $\theta \rightarrow +\pi$, whereas approaching the slip plane in the third quadrant, $\theta \rightarrow -\pi$. A discontinuous jump in θ of 2π occurs across the slip plane, and this corresponds to a jump in the displacement from $u_z = b_z/2$ in the second quadrant to $u_z = b_z/2$ in the third quadrant. The displacement discontinuity is $\Delta u_z = u_z(r, -\pi) - u_z(r, +\pi) = b_z$.

The distribution of displacements in the (x, y)-plane near the screw dislocation is illustrated in **Figure 5.28**. The slip plane (thick blue line) extends from the screw dislocation at the origin to the left along the negative x-axis to $x = -\infty$. Displacements in a small region around the dislocation (inside dashed black circle) are not computed, because the elastic solution breaks down there (see discussion below about the dislocation core). The magnitude of u_z is indicated by the diameter of the colored circles, and the direction of u_z is indicated by the fill color: blue is positive (directed toward the observer); white is negative (directed away from the observer). Along the extension of the slip plane ($x > 0$, $y = 0$), the displacement u_z is zero. The displacement is uniform along any radial line from the dislocation. For example, along the gray radial line extending toward the lower right corner of the figure, all the open circles have the same diameter. On the other hand, the displacement decreases along orthogonal lines away

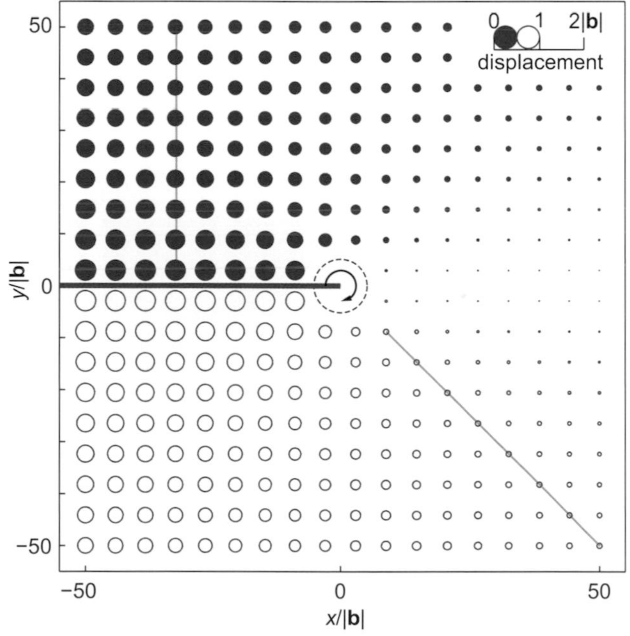

Figure 5.28 Displacement component u_z for the screw dislocation. Positive displacement (blue circles) and negative displacement (white circles) are directed toward and away from the observer, respectively; displacement magnitude is proportional to circle diameter. Thick blue line is slip plane; circular arrow indicates screw dislocation. Distance is normalized by Burgers vector magnitude, |**b**|, and displacement magnitudes are exaggerated relative to the length scale. Gray lines are described in text.

from the slip plane. For example, along the gray line near the upper left side of the figure, the solid circles decrease in diameter.

The only non-zero small strain component for the screw dislocation is found from the cylindrical form of the small strain kinematic equations as (see Weertman and Weertman, 1964, in Further Reading):

$$\varepsilon_{\theta z} = \frac{1}{r}\frac{\partial u_z}{\partial \theta} = \frac{b_z}{2\pi r}$$

Using the cylindrical form of Hooke's law, the only non-zero stress component is:

$$\sigma_{\theta z} = G\varepsilon_{\theta z} = \frac{Gb_z}{2\pi r} \quad (5.26)$$

This shear stress is independent of the angle θ, and proportional both to Burgers vector and the elastic shear modulus. Perhaps the most noteworthy fact from inspection of (5.26) is that the shear stress is inversely proportional to radial distance from the dislocation line. This creates a stress (and strain) concentration near the screw dislocation that increases without bounds as $r \to 0$. To avoid stresses that exceed the yield strength, and strains that exceed those of small strain kinematics, one must avoid a small region surrounding the dislocation (red circular region in **Figure 5.27**), called the dislocation core, where elastic theory is not applicable.

The distribution of shear stress (5.26) in the plane perpendicular to the screw dislocation line is illustrated in **Figure 5.29**. The circular contours emphasize the conclusions reached in the previous paragraph that the shear stress is independent of angle θ. The range of colors from cool to warm as the dislocation is approached is consistent with the concentration of shear stress at the dislocation. From the dislocation core boundary to the outer contour is about 25 nm and the shear stress varies from about 850 to 250 MPa, for a gradient of about 25 MPa/nm. This intense gradient and the shear stress concentration presumably lead to breakage of the nearby bonds and propagation of the dislocation line. Because the shear stress is independent of the angle, θ, continued propagation parallel to the slip plane is not necessarily preferred, so any such tendency must come from the anisotropy of the mineral grain.

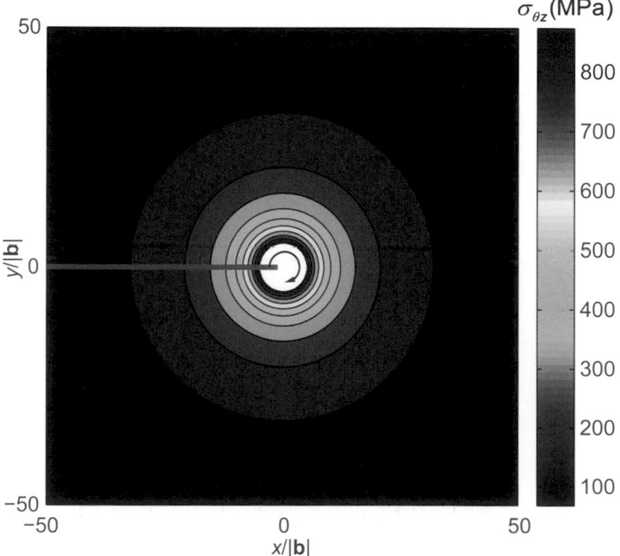

Figure 5.29 Shear stress (MPa) in the vicinity of a right-handed screw dislocation. View is along dislocation line in direction of tangent vector. Thick blue line is slip plane; small white circle is dislocation core; distance is normalized by the Burgers vector magnitude, |**b**|.

5.6.4 Displacement and Stress Fields for the Edge Dislocation

Displacement vectors for the pure edge dislocation are perpendicular to the dislocation line, and vary in any plane perpendicular to that line, but do not vary with position along that line. To address this symmetry we choose a right-handed Cartesian coordinate system with the z-axis parallel to the dislocation line (**Figure 5.30**), and consider the variation of the displacement components u_x and u_y in the (x, y)-plane:

$$u_x = f_1(x, y), \quad u_y = f_2(x, y), \quad u_z = 0$$

In **Figure 5.30** the positive z-axis is directed outward, toward the observer. In contrast, the tangent vector, $\mathbf{t} = \langle 0, 0, t_z \rangle$, is directed inward, in the look direction of the observer, so t_z is negative. For this choice of coordinates, and using the RH/SF convention, Burgers vector, $\mathbf{b} = \langle b_x, 0, 0 \rangle$ is directed to the left, so b_x is negative.

Because we take the dislocation line as straight and very long compared to Burgers vector, the problem is one of plane strain (Section 4.11.1) in which the displacement in the z-direction (along the dislocation line) is zero everywhere. Navier's three displacement equations of motion, (5.21) and (5.22), must be

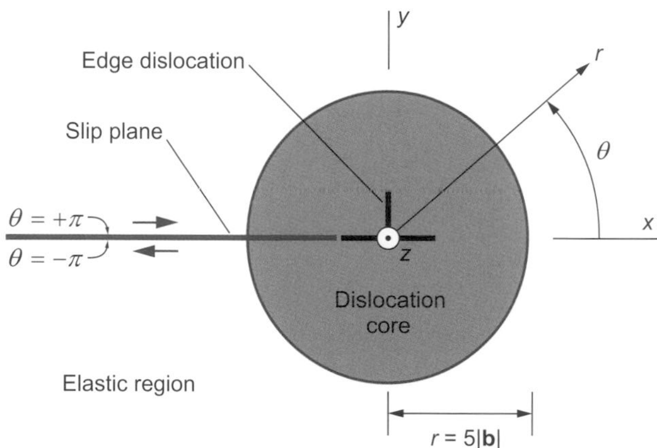

Figure 5.30 Cartesian (*x*, *y*) and polar (*r*, *θ*) coordinates for the edge dislocation. Positive *z* is toward the observer and the dislocation line lies along the *z*-axis. The slip plane is blue and dislocation core is red. Positive *z* is toward observer. Tangent vector, **t**, is directed away from observer along the dislocation line and Burgers vector, **b**, is directed in negative *x*. Blue arrows indicate displacement direction, not magnitude.

solved using the appropriate boundary conditions to find the elastic fields in the material surrounding the edge dislocation. As justified in Section 5.6.3 for the screw dislocation, the left sides of Navier's equations of motion and the body force terms on the right sides are set to zero. Given the constraints on the displacement components in the previous paragraph, all the partial derivatives in the third Navier equation are zero, and the first two of Navier's equations of motion reduce to:

$$\begin{aligned} (2G+\lambda)\frac{\partial^2 u_x}{\partial x^2} + G\frac{\partial^2 u_x}{\partial y^2} + (G+\lambda)\frac{\partial^2 u_y}{\partial x \partial y} &= 0 \\ G\frac{\partial^2 u_y}{\partial x^2} + (2G+\lambda)\frac{\partial^2 u_y}{\partial y^2} + (G+\lambda)\frac{\partial^2 u_x}{\partial x \partial y} &= 0 \end{aligned} \tag{5.27}$$

The two dependent variables are the displacement components, u_x and u_y; the two independent variables are the coordinates, x and y; and the two elastic constants are G and λ. Unlike the governing equations for the screw dislocation (5.23), the governing equations for the edge dislocation depend upon the elastic constants.

The solution to (5.27) yields the two displacement components in the (x, y)-plane (see Weertman and Weertman, 1964, in Further Reading) for the edge dislocation:

$$\begin{aligned} u_x &= -\frac{b_x}{2\pi}\left[\tan^{-1}\left(\frac{y}{x}\right) + \left(\frac{G+\lambda}{2G+\lambda}\right)\left(\frac{xy}{x^2+y^2}\right)\right] \\ u_y &= -\frac{b_x}{2\pi}\left[\left(\frac{G}{2(2G+\lambda)}\right)\ln\left(\frac{x^2+y^2}{C}\right) - \left(\frac{G+\lambda}{2G+\lambda}\right)\left(\frac{y^2}{x^2+y^2}\right)\right] \end{aligned} \tag{5.28}$$

The displacement components are proportional to the Burgers vector, which in this case means they are proportional to b_x. The first term in square brackets for u_x is the same inverse tangent function that defined the displacement distribution for the screw dislocation (5.24). Making substitutions for the cylindrical coordinates, we find $u_x = -b_x\theta/2\pi$. A discontinuous jump in θ of 2π occurs across the slip plane, and this corresponds to a jump in the displacement from $u_x = -b_x/2$ in the second quadrant to $u_x = b_x/2$ in the third quadrant. The displacement discontinuity is $\Delta u_x = u_x(r, -\pi) - u_x(r, +\pi) = b_x$. The arbitrary constant C in (5.28) has dimensions of length squared, so the argument of the logarithmic term is dimensionless. This constant (assigned a unit value) provides a uniform translation parallel to the y-axis that has no effect on the strain or stress fields, which depend on spatial derivatives of the displacements.

Here we focus on u_x, because that provides the displacement discontinuity associated with the slip plane of the edge dislocation. The displacement component u_x from (5.28), normalized by the magnitude of Burgers vector, is plotted in **Figure 5.31**, where the slip plane (thick blue line) extends from the edge dislocation at the origin to the left along the negative x-axis to $x = -\infty$. Displacements in a small region (dashed circle) around the dislocation are not computed, because the elastic solution breaks down inside the dislocation core. The magnitude of u_x is constant along the slip plane, and it is positive just above and negative just below the slip plane. To the right of the dislocation line, ($x > 0$, $y = 0$), u_x is zero, and it varies continuously in the y-direction across the positive x-axis, so the displacements are continuous. The displacement component u_x decreases along orthogonal lines away from the slip plane, for example, along the gray line near the upper left side of the slip plane. The displacement is uniform along any radial line from the dislocation. For example, along the gray radial line extending toward the lower right corner of the figure, all the displacement vectors have the same length.

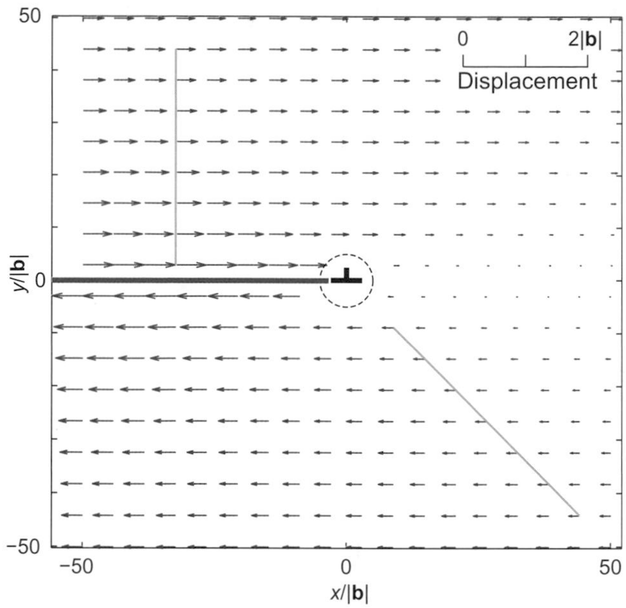

Figure 5.31 Displacement component u_x for the edge dislocation normalized by Burgers vector magnitude, |**b**|. Thick blue line is the slip plane; inverted "T" is the edge dislocation. Vector arrow lengths are exaggerated for visibility. Distance is normalized by Burgers vector magnitude, |**b**|. Gray lines are described in text.

The small strain kinematic equations (5.16) are used to determine the strain components from the displacement components by differentiating (5.28). Then, the equations for Hooke's law for plane strain conditions are used to determine the stress components from the strain components. Here we examine the shear stress component acting parallel to the slip plane (see Weertman and Weertman, 1964, in Further Reading):

$$\sigma_{yx} = -\left(\frac{b_x}{2\pi}\right)\left[\frac{2G(G+\lambda)x(x^2-y^2)}{(2G+\lambda)(x^2+y^2)^2}\right] \tag{5.29}$$

Note that the shear stress is proportional to Burgers vector. Referring to the cylindrical coordinates (r, θ) shown in **Figure 5.30**, the denominator in this equation is proportional to r^4 and the numerator is proportional to r^3. Thus, the shear stress is proportional to b_x/r, so it goes to an infinite value as $r \to 0$, i.e. at the dislocation line. To avoid stresses that exceed the yield strength, and strains that exceed those of small strain kinematics, we must avoid the small region surrounding the dislocation (red circle in **Figure 5.30**), called the dislocation core, where elastic theory is not applicable.

The radius of the dislocation core is estimated using the theoretical shear strength, σ_t, which has an approximate range for solids related to the elastic shear modulus as $G/30 \leq \sigma_t \leq G/3$. For a shear modulus $G = 3 \times 10^4$ MPa, the range of theoretical strengths would be $10^3 \leq \sigma_t \leq 10^4$ MPa. The shear stress near the dislocation is approximated using (5.29) as $|\sigma_{yx}| \approx G|\mathbf{b}|/2\pi r \leq \sigma_t$. Conservatively, taking the lower end of the range for strength and setting the radius to that of the core, $r = r_c$, we estimate $r_c \approx 5|\mathbf{b}|$. Outside the dislocation core, and over length scales greater than the radius of the core, the continuum concept is valid and the assumptions of linear elasticity are not violated. This has been verified by direct observation of the displacement field near an edge dislocation in silica (see Section 3.4.2).

The distribution of shear stress, σ_{yx}, near the edge dislocation is illustrated in **Figure 5.32** as a contour map for a region that is $100b_x$ on a side, omitting the dislocation core, and using $G = \lambda = 3 \times 10^4$ MPa. The view is in the direction of the tangent vector. In general, the shear stress decreases in magnitude away from the dislocation. Two main lobes of shear stress are aligned with, and symmetric about, the x-axis. The shear stress lobes are negative along the slipped portion, $x < 0$, and positive along the unslipped portion, $x > 0$ of the x-axis. To the right of the dislocation, and just outside the core, the concentration of positive shear stress is about 1,000 MPa. This is about $G/30$, at the lower end of the range of theoretical shear strengths necessary to break the next bond, so the dislocation line should advance to the right, parallel to the slip plane. This is in contrast to the stress distribution for the screw dislocation, which lacks a preferred direction for propagation.

5.7 CONSTITUTIVE EQUATIONS FOR ELASTIC–DUCTILE DEFORMATION

Conventional triaxial tests of rock at elevated confining pressures (e.g. **Figure 5.8**) and elevated temperatures (e.g. **Figure 5.10**) demonstrate that elastic behavior is followed at greater differential stress by yielding and the accumulation of significant non-recoverable strain. In Section 5.7.1 we introduce relationships that limit the maximum shear stress to the yield strength in ways that approximate stress–strain behaviors documented in some laboratory tests. The rate of strain is not specified for these idealized behaviors, and the deformation is treated as independent of time. While some geological structures can be analyzed successfully with time-independent plasticity, others require a constitutive law that explicitly relates strain rate to stress, or rate of deformation to stress. Time dependent deformation is addressed in Section 5.7.2 by considering the strain rate associated with dislocation creep.

5.7.1 Rigid Plastic and Elastic–Plastic Behaviors

The stress–strain behavior for an idealized rigid plastic material is illustrated in **Figure 5.33a**. On the ordinate we plot the maximum shear stress. For a three-dimensional state of stress with unequal extreme principal stresses, $\sigma_1 > \sigma_3$, the maximum shear stress acts on two conjugate planes (**Figure 5.33b**). Those two planes contain the intermediate principal stress axis, and make angles of 45° with the axes of the extreme principal stresses. The magnitude of the maximum shear stress is one half the difference between the greatest and least principal stress:

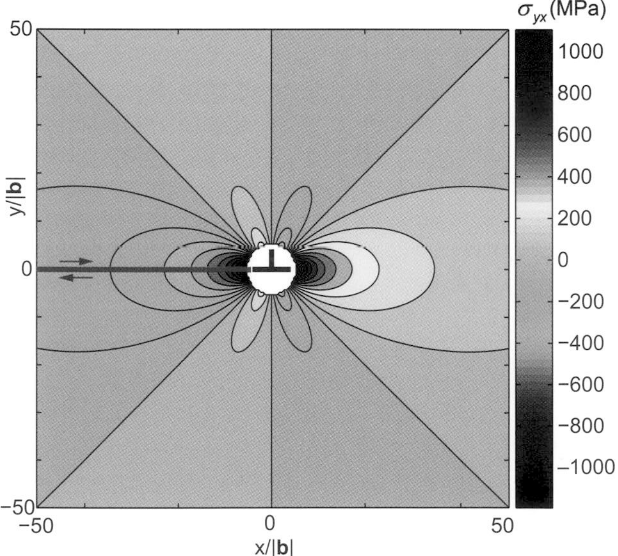

Figure 5.32 Shear stress (MPa) in the vicinity of a positive edge dislocation. Thick blue line is the slip plane; small white circle is dislocation core; inverted "T" is edge dislocation; distance is normalized by Burgers vector magnitude, |**b**|. Blue arrows give sense of slip, not magnitude.

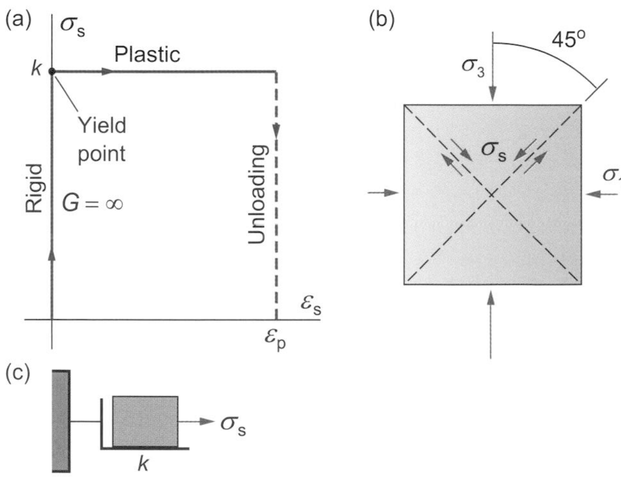

Figure 5.33 (a) Idealized rigid plastic behavior illustrated by plotting maximum shear stress, σ_s, versus shear strain, ε_s, with an infinite elastic shear modulus, G, and finite yield strength, k. (b) Principal stresses acting on a small element and conjugate planes carrying the maximum shear stress. (c) One-dimensional phenomenological model for rigid plastic behavior. Modified from Jaeger et al. (2007).

$$\sigma_s = \frac{1}{2}(\sigma_1 - \sigma_3) \quad (5.30)$$

Because plasticity is a shearing phenomenon, we anticipate that the rigid plastic material yields when the maximum shear stress equals some critical value.

The idealized rigid plastic material, illustrated in **Figure 5.33a**, begins to deform plastically when the maximum shear stress equals the yield strength, k:

$$\sigma_s = k \quad (5.31)$$

This is known as Tresca's yield criterion. Recall from the discussion of laboratory testing in Section 4.7.2 that the differential stress applied during a conventional triaxial compression test is the difference between the axial stress and the radial stress. This, in turn, is the difference between the least and greatest principal stresses:

$$\Delta\sigma = \sigma_a - \sigma_r = \sigma_3 - \sigma_1 = -2\sigma_s \quad (5.32)$$

Thus, the differential stress is the negative of twice the maximum shear stress.

On the abscissa of **Figure 5.33a** we plot the maximum shear strain, ε_s. This is the shear strain on the conjugate planes of **Figure 5.33b**, so it is directly associated with the maximum shear stress. According to Hooke's law for an elastic material (Section 4.10), the maximum shear strain, ε_s, accumulates linearly with increased shear stress and constant shear modulus, G, such that $\sigma_s = 2G\varepsilon_s$. However, for the idealized rigid plastic behavior, the shear modulus has an infinite value, so the shear stress increases without any elastic shear strain. At the yield point the maximum shear stress is equal to the yield strength (5.31). Thereafter, the maximum shear stress is constant and non-recoverable strain accumulates indefinitely as the material deforms plastically. Note that the rate at which strain accumulates is not specified, so this is a form of time-independent plasticity.

When the shear stress is decreased, unloading occurs along a straight line with an infinite slope (**Figure 5.33a**). Thus, when unloaded to zero shear stress, the shear strain does not return to zero, but to a finite value, ε_p, the non-recoverable plastic shear strain. Rock does not have an infinite shear modulus, but if the magnitude of plastic strain far exceeds the elastic strain, rigid plastic behavior may be an adequate approximation.

Rigid plastic behavior is characterized schematically using a one-dimensional phenomenological model in **Figure 5.33c**. Here, plasticity is represented by a rigid rectangular block, called the *slider*, resting on a fixed plate. As shear stress, σ_s, is applied to the slider it remains stuck on the plate and does not move. When the shear stress is equal to the yield strength, k, the rigid block slides along the plate until the shear stress drops below the yield strength, and the motion stops. Again, the rate of sliding is not specified. Although the relative motion is localized to the interface between the slider and the fixed plate, it should be understood that shear strain accumulates everywhere in the rigid plastic material where the maximum shear stress is equal to the yield strength. Thus, in **Figure 5.33b**, all conjugate planes throughout the block that are parallel to the two dashed lines, would yield and accumulate shear strain: the homogeneous state of stress induces a homogeneous plastic strain. Such distributed plastic deformation appears to be a reasonable approximation for the deformation near the center of the Solnhofen limestone sample in **Figure 5.9d**.

At a confining pressure of 165 MPa and a temperature of 25 °C (**Figure 5.8**), or at a confining pressure of 44 MPa and a temperature of 600 °C (**Figure 5.10**), the deformation of Solnhofen limestone proceeds at nearly constant differential stress after yielding. However, the behavior is not rigid before yielding, but includes elastic deformation. This behavior is approximated reasonably well by the idealized elastic–plastic deformation (**Figure 5.34**). Using Hooke's law for an elastic material (Section 4.10), the maximum shear strain accumulates linearly with increased shear stress and constant shear modulus, G, such that $\sigma_s = 2G\varepsilon_s$. At the yield point, the maximum shear stress is equal to the yield strength, k. Thereafter, the maximum shear stress is constant and non-recoverable strain accumulates indefinitely as the material deforms plastically. Because the rate at which strain accumulates is not specified, this is another form of time-independent plasticity. If the shear stress is decreased, unloading occurs along a straight line with a slope equal to twice the shear modulus. When unloaded to zero shear stress, the shear strain does not return to zero, but to a finite value, ε_p, the non-recoverable plastic shear strain.

The one-dimensional phenomenological model for elastic–plastic behavior has a *spring* in series with a slider (**Figure 5.34c**). The slider remains stuck and elastic strain accumulates due to the spring until the shear stress is equal to the yield strength, k. Then, the slider moves along the plate until the shear stress drops below the yield strength and elastic unloading occurs. Again, the shear strain accumulates everywhere in the elastic–plastic material where the maximum shear stress is equal to the yield strength. The most obvious deviation from this

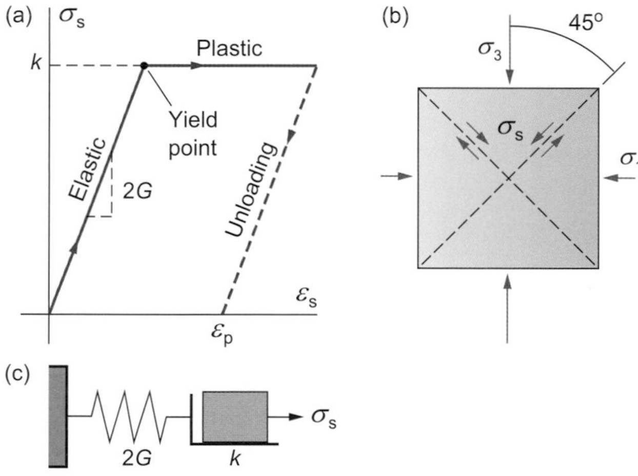

Figure 5.34 (a) Idealized rigid plastic behavior illustrated by plotting maximum shear stress, σ_s, versus shear strain, ε_s, with finite elastic shear modulus, G, and yield strength, k. (b) Principal stresses and planes carrying the maximum shear stress. (c) One-dimensional phenomenological model for elastic–plastic behavior. See Jaeger et al. (2007).

idealization for the Solnhofen limestone at elevated temperatures (**Figure 5.10**) is near the transition from elastic to plastic deformation where the laboratory stress–strain plots are curved rather than having a sharp discontinuity in slope. The curvature makes it difficult to pick an exact yield point, but for these particular tests the elastic moduli and yield strength at large plastic strain are well defined and approximately constant.

5.7.2 Orowan's Equation and Dislocation Creep

Solid-state deformation that depends upon the motion of dislocations within the crystal lattice is called dislocation creep. Recall that edge and screw dislocations were introduced in Section 5.6 to explain plastic deformation of mineral grains without fracture, but the rate of deformation and the velocity of a dislocation were not specified. To understand dislocation creep we first must establish a kinematic relationship between strain rate and dislocation motion. We do this using the edge dislocation as an example, but the resulting relationship also applies for screw dislocations.

Consider a cubic mineral grain of width W and height H (**Figure 5.35**) that contains an edge dislocation line of length L (perpendicular to the plane viewed here), with a Burgers vector $\mathbf{b}$ of magnitude $|\mathbf{b}|$. The dislocation spans the entire grain, so the grain has length L perpendicular to this view. When the dislocation line has moved by dislocation glide a width ΔW (**Figure 5.35a**), the average angle of shear is $\phi = \tan^{-1}(|\mathbf{b}|/H)(\Delta W/W)$ and the average shear strain is $\varepsilon = \tan\phi = (|\mathbf{b}|/H)(\Delta W/W)$. When the dislocation line has moved across the entire grain, the average angle of shear is $\phi = \tan^{-1}(|\mathbf{b}|/H)$ and the average shear strain is $\varepsilon = (|\mathbf{b}|/H)$. In other words, the average shear strain is reduced by the fractional width of the dislocation, $\Delta W/W$. We say *average* shear strain because the deformation actually is localized at the glide plane of the dislocation and not uniformly distributed over the height of the grain.

Now suppose N parallel edge dislocations contribute to the shear strain. Taking $\Delta W/W$ as the average fractional width of these dislocations, the total shear strain is $\varepsilon = N\,|\mathbf{b}|\Delta W/HW)$. As the number of dislocations increases, and their distribution becomes more homogeneous, the strain becomes more uniformly distributed over the height of the grain. Dislocation density, ρ_d, is used to quantify dislocations, and this is defined as the total length of all dislocation lines within a grain, divided by the grain volume: $\rho_d = NL/LWH$. Thus, the number of dislocations can be written in terms of this density as $N = \rho_d WH$, and the accumulated shear strain becomes $\varepsilon = \rho_d|\mathbf{b}|\Delta W$. Both dislocation density and average width may be functions of time, so the strain rate is:

$$\dot{\varepsilon} = \frac{\mathrm{d}\varepsilon}{\mathrm{d}t} = |\mathbf{b}|\left(\rho_d\frac{\mathrm{d}\Delta W}{\mathrm{d}t} + \Delta W\frac{\mathrm{d}\rho_d}{\mathrm{d}t}\right) \approx |\mathbf{b}|\left(\rho_d\frac{\mathrm{d}\Delta W}{\mathrm{d}t}\right)$$

For the last step it is postulated that the dislocation density does not change significantly with time.

The time rate of change of average dislocation width is associated with the average dislocation velocity, $\mathrm{d}\Delta W/\mathrm{d}t = v$, so the strain rate is written:

$$\dot{\varepsilon} \approx \rho_d|\mathbf{b}|v \quad (5.33)$$

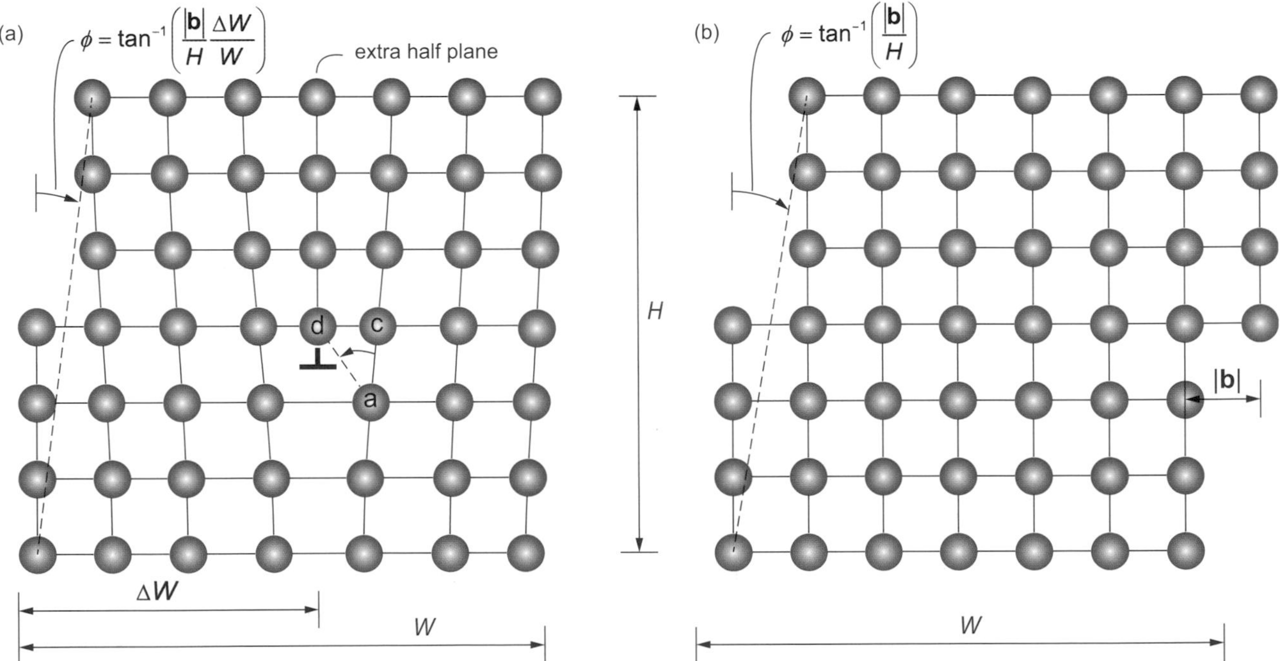

Figure 5.35 Schematic illustration of dislocation glide in a cubic grain of width W and height H. (a) Edge dislocation with Burgers vector of magnitude $|\mathbf{b}|$ moved ΔW from the left edge to its current position, leaving a step in the left edge. (b) The edge dislocation moved across the entire grain width. Average angle of shear is ϕ. See Poirier (1985) in Further Reading.

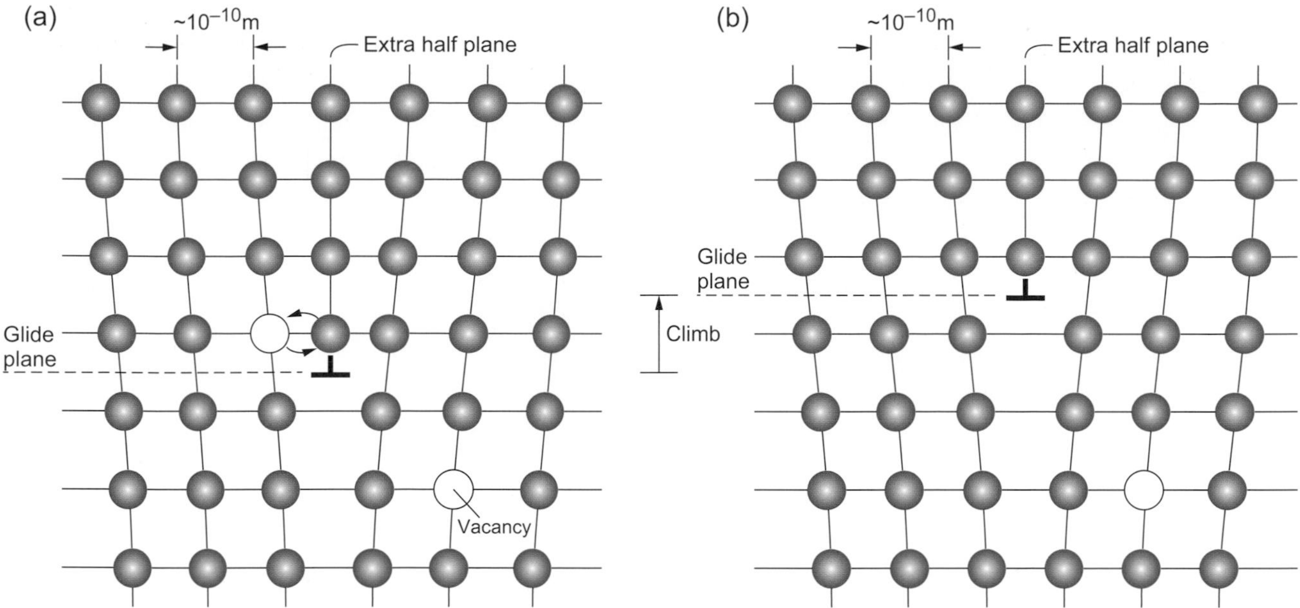

Figure 5.36 Schematic illustration of dislocation climb. The atom at the base of the extra half plane jumps to the adjacent vacancy, so the dislocation line climbs up one lattice dimension. See Poirier (1985) in Further Reading.

This kinematic relationship was proposed by Egon Orowan (1902–1989) and now is known as Orowan's equation. The strain rate is proportional to the dislocation density, the magnitude of Burgers vector, and the dislocation velocity.

Orowan's equation applies both to dislocations that move along the glide plane (**Figure 5.35**) and also to those that move along the extra half plane, perpendicular to the glide plane (**Figure 5.36**). The latter movement is called dislocation climb. Climb in the direction of the half plane occurs when a vacancy replaces the atom at the base of the half plane. Climb in the direction away from the half plane occurs when a vacancy just below the dislocation line is replaced by an adjacent atom.

For dislocation climb Orowan's equation (5.33) is written $\dot{\varepsilon} \approx \rho|\mathbf{b}|v_c$, where v_c is the climb velocity. Climb velocity depends on the flux of vacancies to the dislocation core, which is proportional to the applied stress: $v_c \propto \sigma$. This relationship is described in Section 6.4.4 where the topic of diffusion creep is introduced. Because the dislocation density is proportional to the stress squared, $\rho \propto \sigma^2$, the strain rate associated with dislocation climb is proportional to stress cubed: $\dot{\varepsilon} \propto \sigma^3$. Recall from the discussion of the laboratory data presented in **Figure 5.15** that the strain rate in these experiments was proportional to the differential stress to the third power. This is consistent with the mechanism for deformation being dislocation motion, either climb or glide, so the deformation is referred to as dislocation creep.

5.8 EQUATIONS OF MOTION FOR RIGID PLASTIC DEFORMATION

Rock mechanics tests typically are designed to induce a uniform stress throughout the cylindrical specimen. When this is achieved the stress–strain relations are representative of the entire specimen, which is elastic throughout at small strains and ductile throughout at greater strains. The "barrel" shape of deformed samples in **Figure 5.9c** and **d** demonstrate the uniformity of strain is achieved only approximately in these conventional triaxial tests. As geologic structures develop in Earth's lithosphere, elastic deformation and small recoverable strains may occur in some parts, whereas large non-recoverable plastic strains may occur at the same time in other parts. The boundaries between the differently behaving parts are difficult to predict, so the formulation of mechanical models is challenging.

One productive methodology is to simplify the models by idealizing the mechanical behavior. Two idealizations that have proved very useful are: (1) postulate the material is *incompressible*; and (2) postulate the elastic strain is so small compared to the plastic strain that it can be ignored. We exploit these postulates to investigate a horizontal layer of salt flowing toward a rising diapir using a *rigid plastic* constitutive law (**Figure 5.33**). In these models the entire layer of salt is at the yield stress and therefore flowing, while the surrounding sedimentary rock is rigid. This is called fully-developed plastic flow, and we consider an example related to the flow of salt in this section.

5.8.1 Salt Flow Toward an Ascending Diapir

Clastic sediments near Earth's surface typically are less dense than salt, so shallow salt layers are gravitationally stable. However, compaction during burial eventually makes these sediments denser than salt, so the salt is buoyant. This density inversion usually occurs at burial depths between 500 m and 2 km. At greater depths, the buoyancy of salt causes it to flow upward and the rising salt body is called a salt diapir (**Figure 5.37**). The upward flow of salt in a diapir is fed by lateral flow in the salt layer. We describe and model a rising diapir in Section 11.6, and

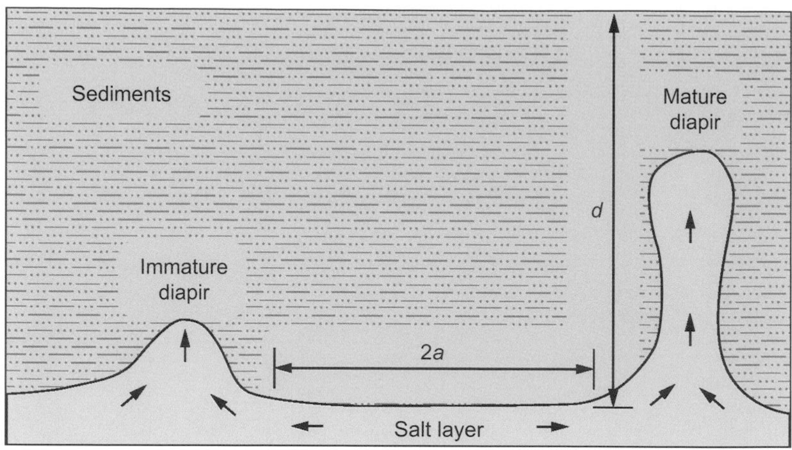

Figure 5.37 Vertical section in sedimentary basin showing lateral flow of salt within a salt layer toward nearby rising salt diapirs. Depth to top of salt is *d* and width of salt layer between diapirs is 2*a*. Modified from Jackson and Talbot (1986), Figure 10.

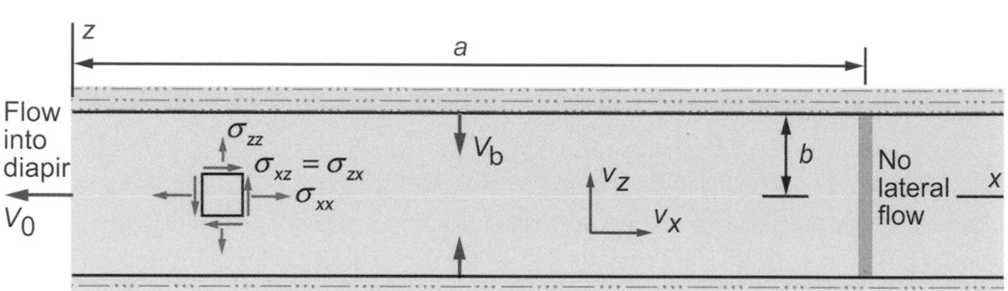

Figure 5.38 Model of lateral flow within salt layer toward rising diapir on left. Layer half-thickness is *b* and half-length is *a*. No lateral flow occurs along the dark gray boundary, so symmetry allows consideration of only the left half of the layer. Inward velocity of layer boundaries is V_b, and net flow into the diapir at x = 0 is V_0. See **Figure 5.39** for velocity vectors computed using this model. See Kachanov (1974) in Further Reading.

here we focus only on the lateral flow. Because the salt can flow out of the layer and into the growing diapir, it is not completely constrained by the surrounding rock mass. Thus, the entire salt layer may be slowly deforming and the plastic strains there may greatly exceed the elastic strains.

Although some salt diapirs are approximately axisymmetric, others develop from long salt anticlines (e.g. see **Figure 11.33**). That geometry is consistent with a two-dimensional model where the plane of interest is perpendicular to the anticline, and the geometry and loading conditions do not change significantly along the anticline. We take the (x, z)-plane as the plane of interest (**Figure 5.38**) and postulate conditions of plane strain (Section 4.11.1) where the displacement component in the y-direction is zero: $u_y = 0$. We suppose that the layer of salt is feeding two salt diapirs, one to the left and one to the right, with a line of symmetry and no lateral flow at $x = a$. The three stress components in the (x, z)-plane are taken as independent variables, and each may be a function of x and z:

$$\sigma_{xx} = \sigma_{xx}(x, z),\ \sigma_{xz} = \sigma_{xz}(x, z),\ \sigma_{zz} = \sigma_{zz}(x, z) \quad (5.34)$$

For plane strain conditions the shear stresses acting in the y-direction are zero: $\sigma_{xy} = 0 = \sigma_{zy}$. Also, the normal stress in y for plane strain conditions is related to the other normal stresses as $\sigma_{yy} = \nu\,(\sigma_{xx} + \sigma_{zz})$. Here ν is Poisson's ratio, and because the model salt is *incompressible*, we specify $\nu = ½$.

Suppose the entire salt layer to the left of the symmetry line at $x = a$ is flowing laterally toward the diapir and has accumulated plastic strains that far exceed the elastic strains. We use the criterion for flow introduced in Section 5.7.1 and known as Tresca's yield criterion (5.31). The maximum shear stress magnitude is related to the stress components in the (x, z)-plane for this plane strain problem as:

$$\sigma_s = \frac{1}{4}\left[(\sigma_{xx} - \sigma_{zz})^2 + 4\sigma_{xz}^2\right]^{1/2} \quad (5.35)$$

Setting the maximum shear stress equal to the yield strength to satisfy the Tresca's yield criterion (5.31), the stress components in the salt layer must satisfy:

$$(\sigma_{xx} - \sigma_{zz})^2 + 4\sigma_{xz}^2 = 4k^2 \quad (5.36)$$

This is the *first* of three governing equations for the three unknown stress components. Because the entire layer has reached the yield stress, this is referred to as fully-developed plastic flow.

5.8.2 Equations of Motion for Fully-Developed Flow

Flow of the model salt toward the diapir is governed by Cauchy's laws of motion (Section 3.6.1), but many of the terms in these equations can be set to zero for the geometry in **Figure 5.38** and plane strain conditions. We repeat the first equation here to assess the terms:

$$\rho\frac{Dv_x}{Dt} - \left(\frac{\partial}{\partial x}\sigma_{xx} + \frac{\partial}{\partial y}\sigma_{yx} + \frac{\partial}{\partial z}\sigma_{zx}\right) + \rho g_x$$

The solution we seek is for slow steady flow of the model salt toward the diapir, so the material time derivative of velocity on the left side is taken as very small compared to the stress gradients on the right side. For plane strain, the shear stress, σ_{yx}, is zero. Because the x-axis is horizontal, the component of gravitational acceleration, g_x, is zero. Under these conditions Cauchy's first equation reduces to:

$$\frac{\partial \sigma_{xx}}{\partial x} + \frac{\partial \sigma_{zx}}{\partial z} = 0 \tag{5.37}$$

This is the *second* governing equation for the unknown stress components.

Every term in the second Cauchy equation of motion is zero. The third Cauchy equation of motion is:

$$\rho \frac{Dv_z}{Dt} = \left(\frac{\partial}{\partial x}\sigma_{xz} + \frac{\partial}{\partial y}\sigma_{yz} + \frac{\partial}{\partial z}\sigma_{zz} \right) + \rho g_z$$

The left side is set to zero by the same argument used above. However, here the gravitational body force per unit volume is not zero. For a salt density of 2200 kg m^{-3} and gravitational acceleration of 9.8 m s^{-2} this term is $\rho g_z = 2.2 \times 10^4$ N m^{-3}. This is equivalent to a gradient in stress of 0.02 MPa m^{-1}. We show below for a yield strength of 1 to 10 MPa and a layer half thickness of 5 m, the stress gradients would be 0.2 to 2 MPa m^{-1}. The weight of the overlying sediments drives the plastic flow of salt, but the stress gradient due to weight is not significant compared to the stress gradient due to plastic flow, so we neglect the gravitational stress. The surviving terms of Cauchy's third equation are:

$$\frac{\partial \sigma_{xz}}{\partial x} + \frac{\partial \sigma_{zz}}{\partial z} = 0 \tag{5.38}$$

This is the *third* governing equation for the unknown stress components. Taken together, (5.37) and (5.38) are referred to as the equilibrium equations for two-dimensional problems because they guarantee the equilibrium (balance) of forces per unit volume.

In addition to the three stress components (5.34), the two velocity components in the (x, z)-plane are independent variables and each may be a function of x and z:

$$v_x = v_x(x, z), \; v_z = v_z(x, z) \tag{5.39}$$

We have already mentioned that the model salt is incompressible, and this imposes the following condition on the velocity components (Section 3.5.3):

$$\frac{\partial v_x}{\partial x} + \frac{\partial v_z}{\partial z} = 0 \tag{5.40}$$

One final condition relates the velocity components to the stress components by requiring the planes carrying the greatest shear stress to experience the greatest rates of deformation:

$$\frac{\sigma_{xx} - \sigma_{zz}}{2\sigma_{xz}} = \left(\frac{\partial v_x}{\partial x} - \frac{\partial v_z}{\partial z} \right) \Big/ \left(\frac{\partial v_x}{\partial z} + \frac{\partial v_z}{\partial x} \right) \tag{5.41}$$

The derivation of this equation is beyond the scope of this textbook, but see Kachanov (1974) in Further Reading. The equations (5.40) and (5.41) are the fourth and fifth governing equations for this problem of fully-developed flow of a rigid plastic material. These five equations are sufficient to determine the three stress and two velocity components.

5.8.3 Stress and Velocity in the Flowing Salt

The three stress components satisfying the governing equations within the flowing layer are distributed as:

$$\sigma_{xx} = -p - k\left[\frac{x}{b} - 2\left(1 - \frac{z^2}{b^2}\right)^{1/2} \right], \sigma_{xz} = k\left(\frac{z}{b}\right) = \sigma_{zx},$$
$$\sigma_{zz} = -p - k\left(\frac{x}{b}\right) \tag{5.42}$$

Here p is an arbitrary and constant normal stress that is determined using the boundary conditions. Substitution of these stress components into (5.36) through (5.38) demonstrates that they satisfy the condition for fully-developed plastic flow and the two equilibrium equations. Note that the horizontal normal stress has a linear distribution along the layer and a parabolic distribution across the layer. The shear stress has a linear distribution across the layer and the vertical normal stress varies linearly along the layer.

From (5.42) we note that the shear stress, σ_{zx}, is equal to the yield stress along the top and bottom of the layer, $z = \pm b$, so the overlying and underlying material is subjected to a shear stress of this magnitude. The yield strength of the adjacent material in this model is presumed to be greater than that of salt, and it has an infinite elastic modulus, so it remains perfectly rigid under this shear stress. Because a vertical plane of symmetry exists at $x = a$, the shear stress there should be zero, but it only is zero at the midline of the layer where $z = 0$, and it increases to equal the yield strength at $z = \pm b$. This is a shortcoming of the model, but one that can be ignored if the length of the salt layer is much greater than the thickness.

The two velocity components are distributed within the flowing layer as:

$$v_x = V_0 + V_b\left[\frac{x}{b} - 2\left(1 - \frac{z^2}{b^2}\right)^{1/2} \right], v_z = -V_b\frac{z}{b} \tag{5.43}$$

V_0 and V_b are constant velocities to be determined by the boundary conditions. Substitution of the velocity components into (5.40) and (5.41) demonstrates that incompressibility is satisfied and the greatest shear stress and rate of deformation are related as specified. Along the top and bottom of the flowing layer, where $z = \pm b$, the velocity component in the z-direction is $v_z = \mp V_b$, so these boundaries are moving vertically towards each other with a constant velocity V_b that is independent of position along the layer. Thus, the layer changes thickness uniformly over its entire length. This is consistent with a perfectly rigid overlying and underlying material.

The constant velocity, V_0, in (5.43) is determined using the fact that the model salt is incompressible. At the left end of the layer, $x = 0$, the quantity of material flowing out must equal that displaced by the inward moving boundaries at $z = \pm b$. Taking advantage of symmetry, this condition is evaluated by considering just the upper half of the layer and the inward moving top boundary:

$$\int_0^b v_x(x=0)\mathrm{d}z = v_z(z=b)a = -V_b a$$

Substituting for the lateral velocity $v_x(x = 0)$ using (5.43) and evaluating the integral, we find $V_0 b - (V_b b\pi/2) = -V_b a$. Solving for V_0 gives $V_0 = V_b[\pi/2 - a/b]$. If the layer is very long compared to its thickness, then $V_0 \approx -V_b(a/b)$ and $V_0 b \approx -V_b a$. In other words, the quantity of model salt swept out by the inward moving layer top approximates the quantity carried into the diapir. This condition is met if $a/b >> \pi/2$.

Velocity vectors are illustrated in **Figure 5.39** using (5.43) for the layer with $a/b = 10$. The flow is dominantly to the left, with increasing velocity toward the exit into the diapir at $x = 0$. The model salt flows inward at the top and bottom boundaries, but also flows laterally. Also, note that the symmetry condition of no lateral flow at $x/b = a/b = 10$ is only approximated. For example, the position with no lateral flow along $z = 0$ is obtained by substituting for V_0 in (5.43) and solving for $v_x = 0$ to find $x/a = 1 + (b/a)(2 - \pi/2)$. For the position of no lateral flow to be approximately at $x = a$, we must have $b/a << 2 - \pi/2$. In other words, the layer must be very thin compared to its length.

What stress due to the weight of overlying strata would be sufficient to cause fully-developed plastic flow within the model salt layer? To address this question we need to determine the constant p in the equations for the normal stresses (5.42). This is accomplished using the boundary condition that the net horizontal force is zero where the model salt exits into the diapir. Using the distribution of normal stress, σ_{xx}, from (5.42):

$$\begin{aligned} & f_x(x=0, -b \le z \le +b) \\ &= w\int_{-b}^{+b} \sigma_{xx}(x=0)\mathrm{d}z = w\int_{-b}^{+b}\left[-p + (2k/b)\sqrt{b^2 - z^2}\right]dz \\ &= -2pwb + \pi kb = 0 \end{aligned}$$

Here w is the width of the layer in the y-direction. Solving for the constant, we find $p = \pi k/2$.

The next step is to evaluate the net force in the z-direction on the top of the layer. Using the distribution of normal stress, σ_{zz}, from (5.42) and the relation for p just found, we have:

$$\begin{aligned} f_z(0 \le x \le a, z=b) &= w\int_0^a \sigma_{zz}\mathrm{d}x = -wk\int_0^a \left[\frac{\pi}{2} + \frac{x}{b}\right]dx \\ &= -wk\left(\frac{\pi a}{2} + \frac{a^2}{2b}\right) \end{aligned}$$

The average normal stress on the top of the layer necessary for yielding throughout the layer is found by dividing the net force by the area, wa:

$$\sigma_{zz}(\text{ave}) = -\frac{k}{2}\left(\pi + \frac{a}{b}\right) \approx -\tfrac{1}{2}\,k(a/b) \qquad \textbf{(5.44)}$$

The approximation depends upon very long layers, $a/b >> \pi$. Taking a representative range of yield strengths we have $k = 1$ to 10 MPa, and using a = 100 m and $b = 10$ m we find σ_{zz} (ave) $\approx$ 5 to 50 MPa. This range of vertical stress would be found at depths from a few hundred meters to a few kilometers, and these are representative of depths where salt diapirs initiate.

Although the idealized model of lateral flow provides important insights about the relationships among the physical quantities (e.g. stress and velocity components, layer thickness and length, and yield strength), the lateral flow of salt in nature likely involves more complicated velocity distributions. If the salt layer is not homogeneous, but is itself layered with impurities, these internal layers may be sheared and folded during lateral flow in a layer with irregular boundaries (**Figure 5.40**). The sense of shearing also changes from near the floor to near the roof of the layer and this would be reflected in the asymmetry of the folds.

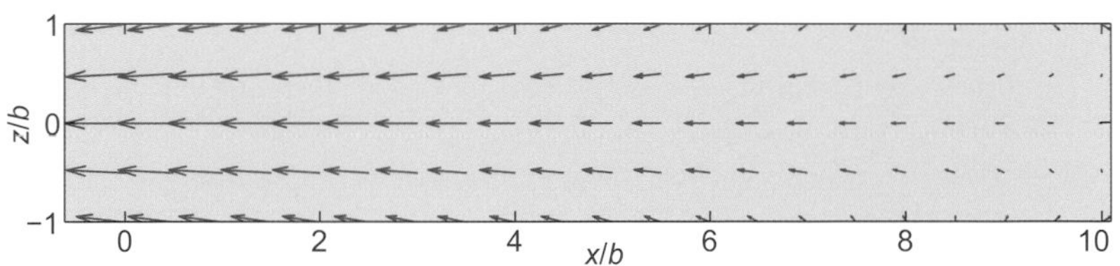

Figure 5.39 Velocity vectors for flow of a rigid plastic layer according to (5.43). The greatest shear stress equals the yield strength, k, throughout the layer. See Kachanov (1974) in Further Reading.

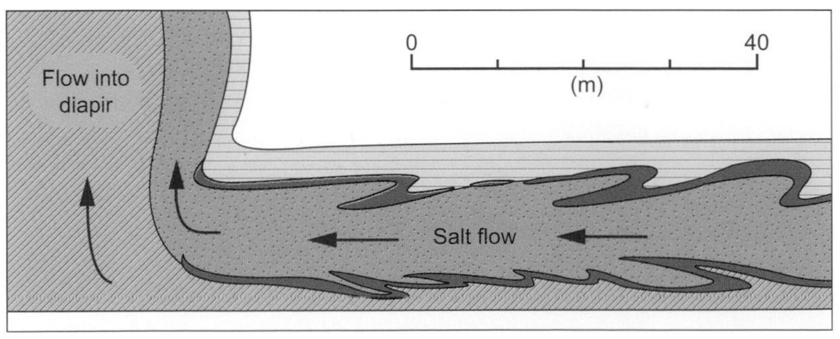

Figure 5.40 Lateral flow of salt into diapir with folding of layers near the boundaries. Sense of shearing reverses from floor to roof zone. Heterogeneity in material properties due to layering is not part of the rigid plastic model for lateral flow. Modified from Talbot and Jackson (1987), Figure 12.

This is in contrast to the uniform sense of shearing across a typical shear zone, such as that illustrated in **Figure 10.27**. Near the floor the folds would take on shapes like a flattened letter "S," while near the roof the shapes would look like a flattened letter "Z." From (5.42) we noted that the shear stress, σ_{yx}, is equal to the yield stress along the top and bottom of the layer, so the overlying and underlying sedimentary units are subjected to a shear stress of this magnitude and may deform.

Recapitulation

We began this chapter by describing the well-known ductile behavior of modeling clay, which flows readily and accumulates large strains without fracturing when the stress equals the yield strength. However, modeling clay deforms like an elastic solid with small recoverable strains at lesser stress. Commonly observed structures in metamorphic rock, including folds and shear zones, are indicative of ductile behavior because the strains are large and distributed throughout the structures without significant fracturing. In most cases a region of ductile deformation is wholly contained within a rock mass that has deformed outside this region only by small elastic strains, thus emphasizing the coupling of elastic and ductile deformation.

During *conventional* triaxial compression tests rock may exhibit *brittle* behavior, in which the stress–strain curve acquires a negative slope after reaching the differential strength. Or, it may exhibit *ductile* behavior, in which the stress–strain curve maintains a positive slope. A useful idealization of some laboratory tests is the perfectly elastic–plastic solid that progresses from linear elastic behavior to a yield point, followed by accumulation of large strain at constant stress. As confining pressure and temperature are increased most rock passes from brittle to ductile behavior in conventional triaxial tests. Strength increases with increased confining pressure, but decreases with increased temperature. For brittle deformation, the stress–strain curve is terminated by fracture or localized shearing at small strains, whereas strains greater than about 5% accumulate in the ductile state and specimens deform by distributed shearing without fracturing. Elevated pore fluid pressure is associated with brittle deformation and fracture at small strains. As strain rate decreases, most rock becomes more ductile and strength decreases.

The kinematics of ductile deformation requires consideration of large heterogeneous strains. In this context, the *deformation gradient tensor* completely characterizes the change in length and orientation of an infinitesimal material line at a point in a deforming rock mass. Infinitesimal material lines representing all possible radii of a sphere at that point deform such that the sphere becomes an ellipsoid, the *stretch ellipsoid*. Simple shear is an example of two-dimensional deformation that includes stretching, shearing, and pure rotation, whereas pure shear includes only stretching and shearing. The *polar decomposition theorem* proves that any deformation gradient tensor that is made up of real components (not complex numbers), and that is invertible (the inverse exists), is composed of the product [R][U] where [R] is orthogonal and a pure rotation, and [U] is symmetric and a pure stretch.

The *finite strain tensor* characterizes deformation that does not include any pure rotation. This tensor is useful to compare large and small strains. The displacement gradient components as well as squares and products of these components define the components of finite strain with no approximations. In contrast, the small strain tensor approximates the strain by excluding these squares and products. Understanding this difference enables one to assess quantitatively whether the small strain theory is adequate to describe the kinematics of deformation. The finite strain tensor has three principal values that include the greatest and least strain. The directions of the three principal strains are orthogonal. For *coaxial* deformation, the material lines that experience the principal strains do not change orientation, whereas for *non-coaxial* deformation those material lines rotate.

Edge and screw dislocations are *line defects* in minerals that can move through the mineral lattice by breaking and healing bonds between atoms along the dislocation line. In this way mineral grains can be distorted without fracturing, so the motion of dislocations is a mechanism for plastic deformation. Surprisingly, the deformation around these line defects is adequately quantified using elastic theory. The elastic displacement field around the edge dislocation has been imaged using electron microscopy, thereby validating the use of elastic theory to within several nanometers of the dislocation.

The rigid plastic material is a useful idealization for modeling a region of ductile deformation that is not completely contained by the surrounding elastic material. In such cases the small elastic strains are neglected and one focuses on the large plastic strains. The stress is at the *yield strength* everywhere in the ductile region, so the plastic deformation is *fully developed*, and the surrounding material is treated as rigid. As an example, we describe a two-dimensional plane strain model for the lateral flow of salt in a horizontal layer toward a rising salt diapir. Five governing equations are solved analytically for the three stress and two velocity components within the layer. The vertical squeezing of the horizontal layer boundaries pushes salt into the diapir. Given yield strengths of 1 to 10 MPa, the necessary thickness of overburden for fully-developed plastic flow is a few hundred meters to a few kilometers, consistent with the known depth of initiation of salt domes.

REVIEW QUESTIONS

The following questions are designed to highlight the expected *learning outcomes* for this chapter. Each question is taken directly from the material in the chapter and, for the most part, in the same sequence that it appears in the chapter. If an answer is not forthcoming, students are advised to read the relevant section of the chapter and discover the answer.

5.1. Using words and two familiar materials, compare and contrast *elastic–brittle* behavior and *elastic–ductile* behavior. In doing so, define what is meant by the *fracture* strength and the *yield* strength.

5.2. On a graph of axial stress versus axial strain for uniaxial tests in compression (use quadrant three) and in tension (use quadrant one) plot results for a material that exhibits *linear elastic deformation*, followed by *perfectly plastic deformation*. Identify Young's modulus for the initial loading phase. Also, identify the permanent plastic strain developed upon unloading.

5.3. On a graph of differential stress versus axial strain for axisymmetric (conventional triaxial) compression tests plot a representative curve (use quadrant three) that progresses from an *elastic* state to a *ductile* state to a *brittle* state. Define the differential stress, and identify the *yield point* and the *differential strength* in compression on the graph.

5.4. Laboratory data from axisymmetric (conventional triaxial) compression tests show how the differential strength of rock specimens vary with *confining pressure* and *temperature*. Describe how these variations compete with one another for tests that represent the mechanical behavior of rock with increasing depth in Earth's lithosphere. In doing so indicate typical *gradients* in confining pressure and temperature with depth.

5.5. Laboratory data from axisymmetric (conventional triaxial) compression tests show how the differential strength and ductility of rock specimens vary with *pore pressure* and *axial strain rate*. Describe typical variations, and indicate why the strengths inferred from laboratory tests are likely to be greater than those in nature.

5.6. *Creep tests* on rock specimens in the laboratory are compared to a *flow law* such as $\dot{\varepsilon} = A(\Delta\sigma)^n \exp(-Q/RT)$, where $\dot{\varepsilon}$ is the rate of axial strain, A is a constant, $\Delta\sigma$ is the differential stress, n is the power-law exponent, Q is the activation energy, R is the universal gas constant, and T is the absolute temperature. Evaluate the power-law exponent for Newtonian *viscous flow* and for *dislocation creep*. Explain why the dislocation creep mechanism is more consistent with GPS data after the Hector Mine earthquake, as displayed in **Figure 5.6**.

5.7. Before deformation an arbitrarily oriented material line segment of infinitesimal length is overlain by the vector $\mathrm{d}\mathbf{X} = \langle \mathrm{dX}, \mathrm{dY} \rangle$ with its tail at the particle located by the position vector $\mathbf{X} = \langle X, Y \rangle$. After deformation the particle is displaced to the position $\mathbf{x} = \langle x, y \rangle$, and the material line is stretched and rotated, such that the overlying vector is $\mathrm{d}\mathbf{x} = \langle \mathrm{dx}, \mathrm{dy} \rangle$ (see **Figure 5.16**). The components of $\mathrm{d}\mathbf{x}$ are linearly related to the components of $\mathrm{d}\mathbf{X}$ using:

$$\mathrm{d}x = \frac{\partial x}{\partial X}\mathrm{d}X + \frac{\partial x}{\partial Y}\mathrm{d}Y$$

$$\mathrm{d}y = \frac{\partial y}{\partial X}\mathrm{d}X + \frac{\partial y}{\partial Y}\mathrm{d}Y$$

Explain why this kinematic description admits *heterogeneous deformation*. Also explain why heterogeneous deformation is necessary to understand the kinematics, for example, of folded sedimentary strata.

5.8. The four partial derivatives in the two indented equations in Question 5.7 are components of the *deformation gradient tensor* and elements of its matrix representation, [F]. Give an example matrix for each of the following deformation styles: *pure rotation*; *pure shear*; *simple shear*. Illustrate each style using a square as the initial geometric figure.

5.9. Describe how the deformation gradient tensor, [F], is transformed into the *finite strain tensor*, [E], in three steps that: (1) make the matrix symmetric; (2) make all elements go to zero for no stretching and no shearing; and (3) introduce a numerical factor, so the finite strain tensor reduces to the small strain tensor, if the deformation is small.

5.10. The *principal strain values* (eigenvalues) and *principal strain directions* (eigenvectors) are found from the symmetric matrix, [E], representing the finite strain. Despite the fact that the term "strain ellipse" is common in the literature of structural geology, explain why the principal values of finite strain do not necessarily represent the semi-axes of an ellipse. Also, explain why the principal stretches are suitably represented by the semi-axes of an ellipse. Hint: consider the ranges of the components of finite strain and stretch.

5.11. Draw a sketch of a cubic lattice containing an *edge dislocation* and identify the dislocation line, the slip plane, the tangent vector, and Burgers vector. Explain the RH/SF conventions used to define Burgers vector.

5.12. The solution to the elastic boundary-value problem for a *screw dislocation* with Burgers vector b_z gives the displacement component, u_z, as:

$$u_z = \frac{-b_z}{2\pi}\tan^{-1}\left(\frac{y}{x}\right)$$

Explain how this function provides the requisite displacement discontinuity across the slip plane illustrated in **Figure 5.28**.

5.13. *Dislocation creep* is solid-state deformation that depends upon the motion of dislocations within the crystal lattice. Use **Figure 5.35** to explain how Orowan's equation, $\dot{\varepsilon} \approx \rho_d b v$, relates the rate of strain, $\dot{\varepsilon}$, to dislocation density, ρ_d, Burgers vector magnitude, b, and dislocation velocity, v.

MATLAB EXERCISES FOR CHAPTER 5: ELASTIC DUCTILE DEFORMATION

In the shallow lithosphere, elastic deformation gives way to localized fracturing and faulting when the stress equals the rock strength. At greater depths, with greater pressure and temperature, ductile shear zones and distributed solid state flow develop where the stress state reaches the elastic limit. MATLAB simulations of the deformation of fossil brachiopods promote an understanding of ductile shearing based on the change in orientation of originally orthogonal material lines. Special cases including pure shear, pure dilation, and simple shear help to build intuition about the orientations and magnitudes of the principal stretches as estimated from collections of deformed fossils. We employ MATLAB to calculate the principal strain orientations and magnitudes. These exercises also elucidate the distinction between coaxial and non-coaxial deformation, and they provide a procedure for evaluating when one must use the non-linear terms in the finite strain equations. We offer exercises on deformation associated with edge dislocations to address problems of plasticity due to the motion of linear defects in a crystal lattice. These exercises help students appreciate the basic concepts of ductile deformation and gain working knowledge of the tools for investigating rocks and minerals that have experienced finite strains without fracturing. www.cambridge/SGAQI

FURTHER READING

For citations in figure captions see the reference list at the end of the book.

Blenkinsop, T. G., 2000. *Deformation Microstructures and Mechanisms in Minerals and Rocks*. Kluwer Academic Publishers, Dordrecht, The Netherlands.
This is an abundantly and beautifully illustrated guidebook for the study and interpretation of deformation features in rocks and minerals using thin sections and a petrographic microscope.

Cai, W., and Nix, W. D., 2016. *Imperfections in Crystalline Solids*. Cambridge University Press, Cambridge.
The principles of mechanics and thermodynamics are used in this textbook to introduce students to the behavior of defects in crystalline solids, which can significantly affect the bulk properties of the host crystals.

Callister, W., and Rethwisch, D., 2014. *Materials Science and Engineering*. John Wiley & Wiley, Inc., New York.
This popular textbook introduces both materials science and materials engineering at the undergraduate level with abundant teaching aids including simulations and animations.

Christensen, R. M., 2013. *The Theory of Materials Failure*. Oxford University Press, Oxford.
This comprehensive coverage of failure from a theoretical point of view considers homogeneous materials that are isotropic or anisotropic with respect to material properties, and range from brittle to ductile.

Duba, A. G., Durham, W. B., Handin, J. W., and Wand, H. F. (Eds.), 1990. *The Brittle-Ductile Transition in Rocks (The Heard Volume)*. Geophysical Monograph 56. American Geophysical Union, Washington, DC.

This monograph is dedicated to one of the pioneers in laboratory studies of rock deformation, Hugh Heard, who provided the data on Solnhofen limestone reviewed in Section 5.4. Both testing equipment and test results are described in this monograph that bear on the transition from brittle to ductile behavior.

Evans, B., and Kohlstedt, D. L., 1995. Rheology of rocks, in: Ahrens, T. J. (Ed.) *Rock Physics & Phase Relations: A Handbook of Physical Constants*. American Geophysical Union, Washington, DC, pp. 148–165.

This review paper covers laboratory data for diffusion creep, dislocation creep, and pressure solution with useful tables of evaluated parameters for flow laws.

Gratier, J-P, Dysthe, D. K., and Renard, F., 2013. The role of pressure solution creep in the ductility of the Earth's upper crust. *Advances in Geophysics*, 54, 47–179.

This review paper combines evidence from geological observations, laboratory experiments, and theory to describe and evaluate the role of pressure solution creep as a mechanism for ductile deformation in the Earth's upper crust.

Griggs, D. T., and Handin, J. (Eds.), 1960. *Rock Deformation*. Memoir 79, The Geological Society of America, New York.

This memoir contains papers from a symposium attended by researchers in theoretical and experimental rock deformation in the United States in 1960. The data for Solnhofen limestone used for **Figure 5.8**, **Figure 5.9**, and **Figure 5.10** come from this memoir.

Hambrey, M. J., and Lawson, W., 2000. Structural styles and deformation fields in glaciers: a review, in: Maltman, A., Hubbard, B., and Hambrey, M. (Eds.) *Deformation of Glacial Materials*. Geological Society, London, pp. 59–83.

This review paper considers structures such as folds, foliations, boudins, and shear zones that develop in glaciers on short time scales, yet may be good models for ductile deformation in rock.

Hobbs, B. E., and Ord, A., 2015. *Structural Geology: The Mechanics of Deforming Metamorphic Rocks*. Elsevier, Amsterdam.

This monograph covers the physical and chemical processes leading to the structures that develop in metamorphic rocks including deformation, mineral reactions, fluid flow, heat transport, and microstructural adjustments.

Kachanov, L. M., 1974. *Fundamentals of the Theory of Plasticity*, M. Konyaeva (trans.), MIR Publishers, Moscow.

This classic book on the theory of time-independent plastic deformation has many example problems, including the one described in Section 5.8 for lateral flow of salt toward a diapir.

Karato, S., 2008. *Deformation of Earth Materials: An Introduction to the Rheology of Solid Earth*. Cambridge University Press, Cambridge.

This is an advanced textbook for geologists and geophysicists on the materials science of rock deformation at the grain scale including elasticity, plasticity, diffusion, and dislocation creep with applications at scales ranging from mineral defects to plate tectonics.

Kohlstedt D. L., and Hansen, L. N., 2015. *Constitutive Equations, Rheological Behavior, and Viscosity of Rocks*. Treatise on Geophysics, 2nd edition, Elsevier, pp. 441–472.

The micromechanics of deformation and the relevant constitutive laws are reviewed in this book chapter and applied to how laboratory flow laws relate to the viscosity structure of Earth's upper mantle determined from geophysical observations.

Passchier, C. W., and Trouw, R. A. J., 1996. *Microtectonics*. Springer-Verlag, Berlin.

This is a comprehensive summary of microscopic fabrics of rock with abundant illustrations and thin section photographs in black and white of different fabrics and their kinematic interpretation.

Poirier, J.-P., 1985. *Creep of Crystals, High-temperature Deformation Processes in Metals, Ceramics and Minerals*. Cambridge University Press, London.

This book is written, in part, for geologists and geophysicists to introduce the methods and concepts of materials science relevant to the plastic deformation of minerals and rocks.

Segall, P., 2010. *Earthquake and Volcano Deformation*. Princeton University Press, Princeton, NJ.
This advanced textbook goes beyond the elementary mechanics and structural geology covered here, with a focus on geophysical data related to active deformation associated with earthquake hazards.

Shewmon, P. (Ed.), 2016. *Diffusion in Solids*. The Minerals, Metals & Materials Series, Springer, Switzerland.
Although focused on diffusion in alloys for materials engineers, this book offers theoretical background and experiments results on diffusion of interest to structural geologists.

Weertman, J., and Weertman, J. R., 1964. *Elementary Dislocation Theory*. The Macmillan Company, New York.
This very readable textbook covers the fundamental properties of dislocations in crystals and their role in providing mechanisms for ductile deformation.

Wong, T.-f., and Baud, P., 2012. The brittle-ductile transition in porous rock: a review. *Journal of Structural Geology* 44, 25–53.
This paper reviews the laboratory behavior of porous rock as a function of confining pressure and identifies a transition from brittle faulting with dilatant failure to delocalized cataclasis with shear-enhanced compaction and strain hardening.

Chapter 6
Elastic–Viscous Deformation

Introduction

In this chapter we describe the laboratory and field observations that provide conceptual models for flow of viscous liquids and viscoelastic solids. We review laboratory and field methods for measuring the apparent viscosity of liquid and solid rock under a variety of temperature and pressure conditions. We also describe the ideal behavior of a linear (Newtonian) viscous liquid and show that measured behaviors approximate idealized behaviors under some conditions. We formalize the relationship between strain rate and stress in a solid-state flow law, discuss the concept of pressure in viscous liquids, and explore the nature of the stress state during flow. After introducing the constitutive equations that link stress to rate of deformation, we develop the equations of motion for viscous flow. Next, we derive the scaling relations that define the transition from laminar flow to turbulent flow and use them to classify flow regimes. Finally, we present a solution to the equations of motion and apply it to flow in a sill.

To contribute to the mitigation of volcanic hazards, structural geologists must understand the physical behavior of viscous liquids and be able to calculate rates of deformation and flow velocities. Furthermore, the formation of ore deposits often is closely related to magmatic intrusions, so developing strategies for exploration and production of mineral resources also requires an understanding of the viscous flow of magma. Given the fact that salt tectonics may be intimately involved in trapping hydrocarbons in sedimentary basins, the flow of viscoelastic solids and their gravitational instabilities are high priorities for students of geology contemplating a career in the hydrocarbon industry. Finally, that grandest expression of the dynamic Earth, plate tectonics, is driven by the solid-state flow of rock in Earth's asthenosphere.

6.1 VISCOUS AND VISCOELASTIC LIQUIDS

A few moments after you partly fill a beaker with water and place it on a table, the upper surface of the water will be horizontal and motionless (**Figure 6.1a**). The normal vector to this surface is parallel to the local acceleration of gravity, **g**, introduced in Section 1.1.2, which defines what we mean by the direction *down*. If the beaker is tilted, the surface of the water becomes oblique to the downward directed gravitational force (**Figure 6.1b**), and this subjects the water to a shearing stress. As a consequence, the water flows until the surface is horizontal and again perpendicular to **g** (**Figure 6.1c**). Having a horizontal surface is a characteristic of static viscous liquids. Even a very small tilt produces flow and adjustment of the surface to horizontal. Thus, water does not have a significant *yield strength* like the plastic material (modeling clay) described in Section 5.1. Because of its strength, the clay sphere is capable of resisting permanent deformation under gravitational loading, but the water is not.

If you incline the table top and pour the water onto the table, it flows in the direction of inclination. The greater the inclination, the greater the flow velocity. On the other hand, if you restore the table top to horizontal, the water does not flow back to the position where it was first poured out. Unlike the elastic material described in Section 4.1 the deformation is not recoverable when the applied stress is removed. These are characteristics of viscous liquids: an applied shear stress produces a *distributed* and *irrecoverable* deformation without fracturing, and the rate of deformation depends on the shear stress magnitude.

Lava erupting from a volcano can behave like a viscous liquid (Section 1.2.3): it flows down slopes and the rate of flow increases as the slope increases. Also, the lava fills depressions and comes to rest with a nearly horizontal surface. If the flow rate of lava is not as rapid as water down the same slope, we say the lava has a greater viscosity than the water. Molten rock, whether it is lava at Earth's surface or magma deep in Earth's crust, can have more complex mechanical behavior than viscous liquids, but we begin this chapter by considering those Earth materials that are described adequately as viscous liquids.

Another familiar material that flows readily is silly putty, a toy product based on silicone polymers. On time scales much less than a second, this material behaves like an elastic solid: a small sphere bounces like a rubber ball. On longer time scales silly putty behaves like a viscous liquid (**Figure 6.2**): loaded only by its own weight the sphere flows with a *distributed* and *irrecoverable* deformation. No fracturing occurs and there appears to be no significant yield strength. This combination of behaviors, elastic

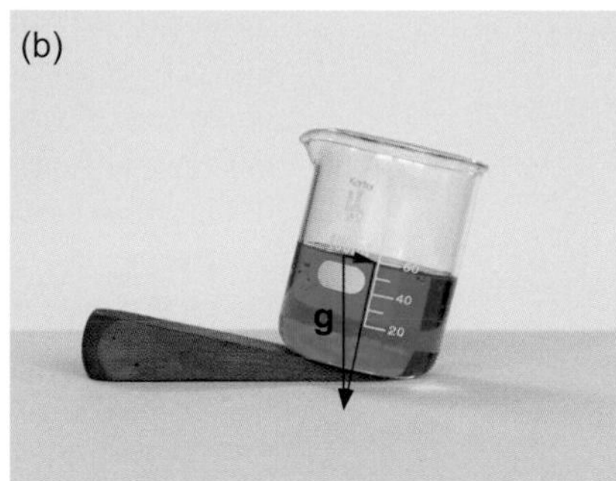

Figure 6.1 (a) Beaker partly filled with water (dyed blue) with surface perpendicular to acceleration of gravity, **g**. (b) In first instant after tilting the beaker the surface is not perpendicular to **g**, so a component of **g** is parallel to the surface. Refer to Section 6.5 for analysis of flow. (c) Tilting induces flow such that **g** is again perpendicular to surface as flow ceases. Photography by Richard Stultz.

Figure 6.2 (a) Silly putty sphere in first instant that it is placed on table. (b) After a few minutes loaded only by gravity, silly putty has flowed outward like a viscous liquid. (c) Later image shows continued outward flow. However, when a sphere of silly putty is dropped, it bounces like an elastic solid. Photography by Richard Stultz.

and viscous, is documented for rock using laboratory experiments, and is inferred from geophysical data gathered to study rock deformation at a crustal scale, hence the chapter title: elastic–viscous deformation.

6.1.1 Newtonian Viscous Liquid

In his seminal work, *The Principia*, the English natural scientist and mathematician Sir Isaac Newton (1642–1727) considered the relationship between the resistance to flow and the relative motion of fluid particles. To do this he focused on what he called the *circular* motion of fluids in a geometry very much like that of the Couette–Hatschek viscometer described in Section 6.4.1. Here we examine his hypothesis about this relationship and describe what now is known as the Newtonian liquid. The modern discipline that was founded by Newton is called fluid dynamics and this deals with both gases and liquids.

Although not the circular motion envisioned by Newton, a somewhat simpler flow is depicted in **Figure 6.3** that is relevant to Newton's hypothesis. A slab of liquid of thickness H, length L, and width, W, perpendicular to the plane of view, rests on the planar and horizontal surface of a stationary and rigid plate. The lengths L and W are much greater than H, so conditions at the ends of the slab will not have a significant effect on the flow in the interior. A rigid plate is placed on top of the liquid, and it is capable of moving laterally in the x-direction under the action of an applied force, F_x, in that direction. We refer to this rectilinear flow as a viscous shear zone. We ignore the weight of the upper plate and the liquid and focus on the shear stress imparted to the viscous liquid by motion of the upper plate.

Actually performing an experiment like this would create a mess, because the liquid is not contained and eventually would run out between the ends of the plates. This behavior can be ignored in the *thought* experiment. We suppose that the surfaces of both plates are sufficiently rough that the liquid adheres to them, so the velocity of the liquid at the interface matches that of the adjacent plate:

$$\text{at } z=0,\ v_x=0;\ \text{at } z=H,\ v_x=V \qquad (6.1)$$

The velocity of the upper plate is V, once it achieves a steady motion due to the applied force.

We digress to record what Newton wrote regarding this hypothesis in *Philosophiae Naturalis Principia Mathematica*, or *Mathematical Principles of Natural Philosophy*, which he published first in 1687, and as a third edition in 1726. Book 2 of that monumental tome is titled *The Motion of Bodies*, and preceding Proposition 51 of Book 2 the following hypothesis is written in the original Latin text (see Langlois, 1964, p. 46, in Further Reading):

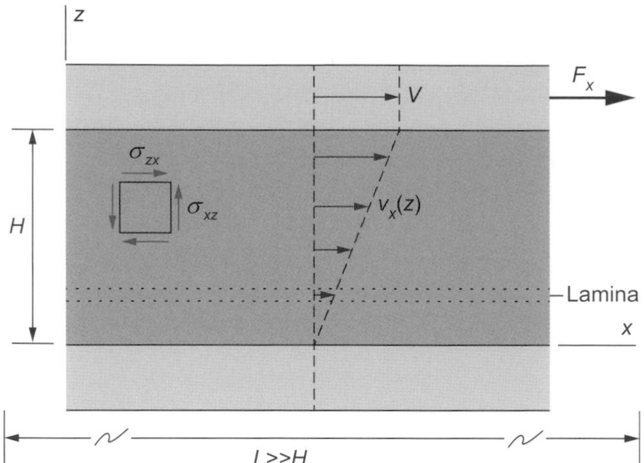

Figure 6.3 A schematic interpretation of Newton's hypothesis. Idealized flow of a slab of Newtonian viscous material of thickness H and length L between two parallel rigid plates. The lower plate is fixed; the upper plate is driven to the right by the force F_x. The velocity distribution, $v_x(z)$, is linear in z and the shear stress, $\sigma_{zx} = \sigma_{xz}$, is constant across the height of the liquid. See Newton et al. (1999) in Further Reading.

> *Resistentiam quae oritur ex defectu lubricitatis partium fluidi, caeteris paribus, proportionatem esse velocitati, qua partes fluidi separator ab invicem.*

A modern translation of the Latin reads as follows (see Newton et al., 1999, p. 779, in Further Reading):

> The resistance that arises from the lack of lubricity [or lack of slipperiness] of the parts of a fluid is, other things being equal, proportional to the velocity with which the parts of the fluid are separated from one another.

Langlois (1964, see Further Reading) points out that empirical evidence, gathered over the subsequent nearly three centuries, demonstrates that Newton's hypothesis adequately describes the behavior of many liquids over a wide range of conditions.

Returning to the viscous shear zone depicted in Figure 6.3, and a modern understanding of the phenomenon, suppose you apply a horizontal force, F_x, to the upper plate. After a brief *startup* period during which the motion of the upper plate is transferred downward through the liquid, a steady state is achieved with the upper plate moving horizontally to the right with velocity V. Because the rigid plate spreads the force uniformly over the surface of the liquid, the shear stress acting on horizontal planes in the liquid and in the direction of the applied force is:

$$\sigma_{zx} = F_x/LW \tag{6.2}$$

Recall that the stress tensor is symmetric, so the conjugate shear stresses are equal: $\sigma_{zx} = \sigma_{xz}$. In other words, the horizontal force also induces a shear stress on vertical planes within the liquid. At steady state, the shear stresses σ_{zx} and σ_{xz} are constant and uniform throughout the slab, and the shearing liquid exerts a traction on the bottom plate in the x-direction. The resultant force that must be applied to keep the bottom plate fixed is equal and opposite to F_x. Although the concept of the stress tensor was unknown to Newton, we take the shear stress σ_{zx} as the measure of his "*resistance that arises from lack of lubricity*."

Imagine that thin laminae of liquid (Figure 6.3) occupy planes that are parallel to the rigid plates, and therefore parallel to the (x, y)-plane. Each successive lamina in the z-direction moves only in the x-direction under the applied force, sliding over its underlying neighbor. Thus, the only non-zero component of velocity is v_x and "*the velocity with which the parts of the fluid are separated from one another*" is the relative velocity of adjacent laminae. This velocity of separation is measured by the velocity gradient, $\partial v_x/\partial z$. Under the conditions so imagined, and certainly for a liquid with homogeneous properties, it seems reasonable to assert that the velocity is distributed *linearly* from zero at the stationary lower plate to V at the upper plate. Thus, the velocity gradient throughout the liquid is *uniform* and *constant*.

Newton hypothesized that the *resistance* is proportional to the *velocity of separation*:

$$\sigma_{zx} = \eta \frac{\partial v_x}{\partial z} = 2\eta D_{zx} \tag{6.3}$$

The proportionality constant, η, now is called the Newtonian viscosity. Because it will be used throughout this chapter, we also introduce the quantity D_{zx}, which is one component of the rate of deformation tensor. This component is defined using gradients in velocity as:

$$D_{zx} = \tfrac{1}{2}\left(\frac{\partial v_z}{\partial x} + \frac{\partial v_x}{\partial z}\right)$$

However, $v_z = 0$ for the viscous shear zone illustrated in Figure 6.3, so the first term on the right side is zero. We define and describe the entire rate of deformation tensor in Section 6.6.2. For now, it suffices to understand that each component of shear stress is linearly related to the corresponding component of the rate of deformation for the Newtonian liquid.

Rearranging (6.3) to solve for viscosity we have $\eta = \sigma_{zx}/(\partial v_x/\partial z)$. From this we note that the dimensions of viscosity are those of stress divided by a velocity gradient: $(ML^{-1}T^{-2})/(LT^{-1}L^{-1}) = ML^{-1}T^{-2}T$. Thus, the dimensions of viscosity are those of stress multiplied by time. The corresponding SI units are Pa s. The viscosity of water is about 10^{-3} Pa s; that of Hawaiian lava is about 10^{3} Pa s; that of glacial ice is about 10^{13} Pa s; and that of Earth's asthenosphere is about 10^{20} Pa s. If these substances were flowing down a steep slope, you would not be able to outrun the water; you could jog faster than the Hawaiian lava; you would have to wait weeks or months to see significant flow of the glacial ice; and rock from the asthenosphere would appear solid on a human time scale. It is important to recognize that viscosities of Earth materials span at least 23 orders of magnitude! Unlike the elastic stiffness for Earth materials described in Section 4.4.2, which span only 3 orders of magnitude, the range of viscous resistance is almost incomprehensible.

Based on (6.3) values of the applied shear stress, σ_{zx}, are plotted versus twice the corresponding rate of deformation, $2D_{zx}$, as straight lines on **Figure 6.4** to represent the loading and unloading of the viscous shear zone in **Figure 6.3**. The slope of the straight line is the Newtonian viscosity. Lines with greater slope represent *more* viscous liquids; those with lesser slope are *less* viscous. Force applied to the upper plate in the positive x-direction induces positive shear stress and positive rate of deformation in the liquid, which plot in the first quadrant. Force applied in the negative x-direction induces negative shear stress and rate of deformation, which plot in the third quadrant. As the applied force decreases to zero, the rate of deformation decreases to zero.

Unlike the linear elastic solid (Section 4.1), the linear viscous liquid does not undeform as the applied force and the *rate* of deformation go to zero. The liquid simply stops flowing and whatever deformation has accumulated remains set in the static liquid. This behavior obviates any need to refer to a *former* state of the liquid, and focuses attention only on the current state. This is referred to as a spatial description of motion, which we describe in more detail in Section 6.6.1. Non-linear relationships between shear stress and shearing rate of deformation (e.g. the gray curve in **Figure 6.4**) are diagnostic of a non-Newtonian liquid. We describe the mechanical behavior of liquids that have a power-law relationship between stress and rate of deformation in Section 11.2.3.

Recall from the discussion of Hooke's law for linear elasticity in Section 4.10 that the shear stress is proportional to the displacement gradient and small shear strain:

$$\sigma_{zx} = G\frac{\partial u_x}{\partial Z} = 2G\varepsilon_{zx} \tag{6.4}$$

Here the proportionality constant, G, is the elastic shear modulus. Note that both this equation for the elastic solid and equation (6.3) for the viscous liquid are *linear* relationships between shear stress and a kinematic quantity. However, the kinematic quantities are quite different. For the elastic solid it is the derivative of displacement with respect to the initial coordinate, and for the viscous liquid it is the derivative of velocity with respect to the current coordinate. Hooke and Newton were contemporaries, both were leading English natural philosophers of their time, and they were great rivals. They provided future generations with these useful and insightful descriptions of the mechanical behavior of elastic solids and viscous liquids.

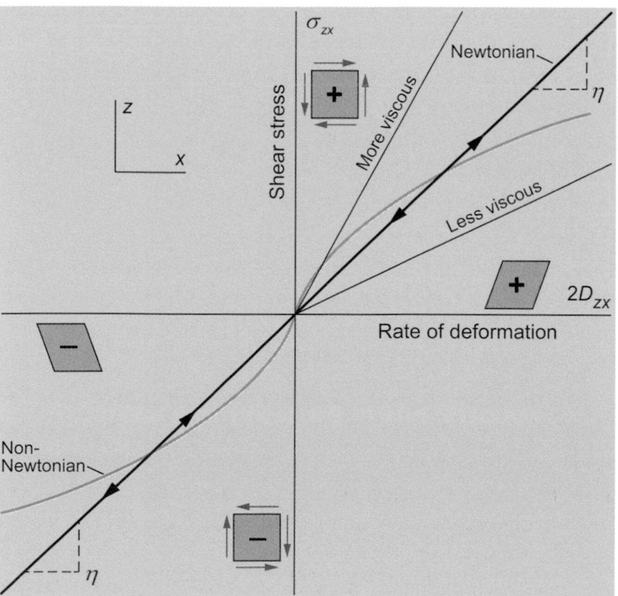

Figure 6.4 Plot of shear stress versus rate of deformation for a linear viscous liquid (heavy black line). The slope of this line is the Newtonian viscosity, η. Gray curve represents a non-Newtonian (non-linear) liquid.

6.2 VISCOUS DEFORMATION AT THE OUTCROP SCALE

In late June of 1952 an eruption began that continued until mid-November inside a crater called Halemaumau (**Figure 6.5**) near the summit of Kilauea Volcano. Volcanic craters are steep-sided depressions that result from the withdrawal of magma from subsurface chambers, and the subsequent collapse of the overlying rock. Halemaumau crater was about 1 km in diameter and approximately 250 m deep at the beginning of the 1952 eruption. The early phase of the eruption lasted from June 27 to July 5, 1952 and was distinguished by the development of a lava lake in the crater. The upper surface of the lava lake cooled and solidified in a matter of hours to form a roughly horizontal crust over a molten interior. The horizontal surfaces of successive flows seen in this cross section attest to the liquid nature of the lava. Measuring the shape of the crater before an eruption, and recording the depth of filling, provides an estimate of the volume of lava erupted. Over the duration of the 1952 eruption approximately 49×10^6 m^3 of new lava partly filled the crater, reducing the depth by about 95 m.

The spectacular rivers of lava flowing down the flanks of a cinder cone (**Figure 6.6**) provide compelling visual evidence that molten rock behaves like a familiar viscous liquid. Between July 5 and August 9 of 1952 cinder cones developed on the solidified crust of the lava lake in Halemaumau crater and rivers of lava flowed down the sides of these cones. Such flows were used to estimate the lava viscosity during the final phase of the eruption,

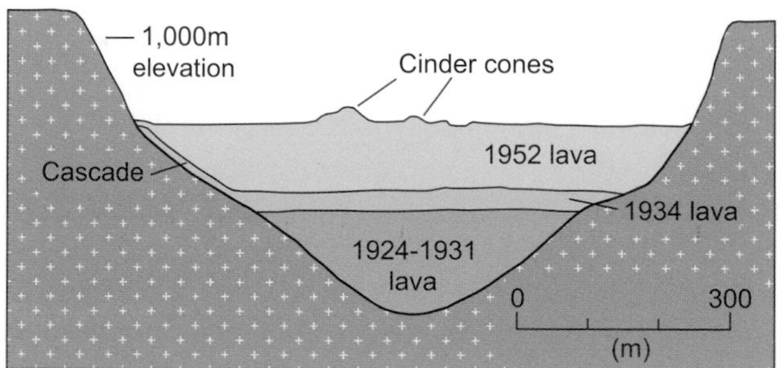

Figure 6.5 Cross section of Halemaumau Crater in June 1954. Successive lava flows have nearly horizontal surfaces, except for small cinder cones. UTM: 5 Q 260200.84 m E, 2147408.22 m N. Modified from Macdonald and Abbott (1970).

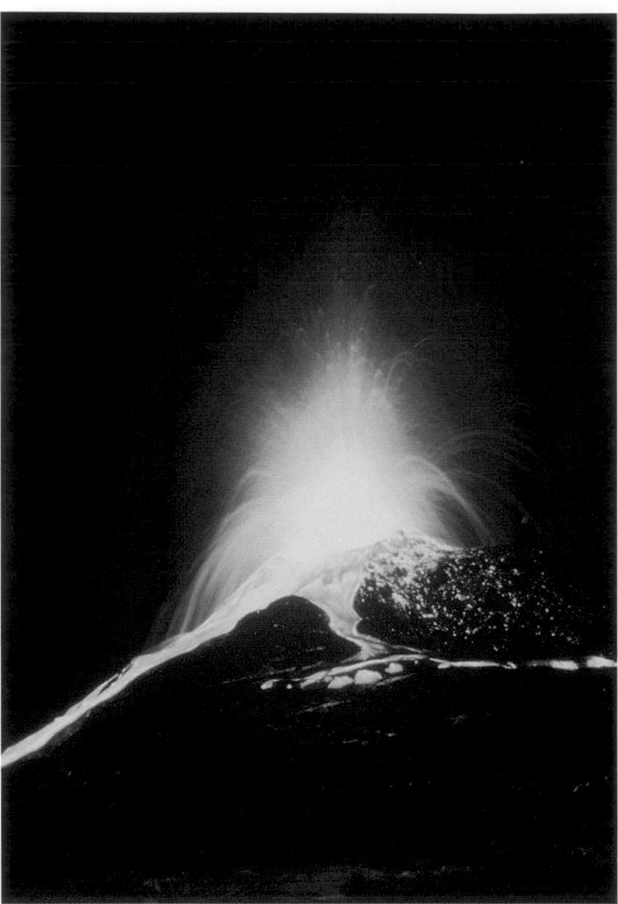

Figure 6.6 Night view (time exposure) of fountain of lava ~20 m high from the summit of a small cinder cone during the 1972–74 Mauna Ulu eruption of Kilauea Volcano, Hawaii. Photograph by R. T. Holcomb.

from August 9 through November 10. In Section 6.5 we explain how the viscosity of the liquid lava was estimated.

Ice and salt are crystalline solids, not liquids, but they form *glaciers*: rivers of ice and salt that flow slowly down gentle slopes (**Figure 6.7**). Ice and salt crystals have a well-defined elastic stiffness, so small strains are completely recovered if the loading is applied and removed on a short time scale (minutes to hours). However, under modest temperatures and pressures at shallow depths in Earth's crust, salt flows on a human time scale (tens of years) to close tunnels in salt mines. Also, on a relatively short geologic time scale (a few million years) salt domes rise to Earth's surface from kilometer depths in sedimentary basins. Thus, in this chapter we also consider the flow of materials that are adequately described as viscoelastic solids.

Rock in Earth's asthenosphere undergoes solid-state flow as part of the slow convection associated with tectonic plate motions (Section 1.1). In the next section we describe evidence that Earth's asthenosphere behaves approximately as a viscous liquid, with much greater viscosity than water or magma, over time scales of tens of thousands of years. On time scales of seismic waves the asthenosphere is elastic, so its behavior appears to be that of a viscoelastic solid.

6.3 VISCOUS DEFORMATION AT THE CRUSTAL SCALE

Earth tides demonstrate elastic (recoverable) deformation of rock at global length scales and at time scales of hours to days (Section 4.2). In Section 5.3, the time-dependent displacement field after a major earthquake showed that plastic deformation (dislocation creep) is an important component of rock behavior at length scales of tens of kilometers and time scales of a few years. Here we use the uplift of shorelines around glacial lakes to demonstrate the viscous deformation of rock in Earth's asthenosphere at length scales of hundreds of kilometers and time scales of tens of thousands of years. Each of the different mechanical behaviors (elastic, plastic, and viscous) plays an important role in the development of geologic structures. In some cases one behavior dominates the deformation, but in other cases combinations of these behaviors must be considered. Deformation at these length and time scales is a topic of interest in both structural geology and geodynamics (**Box 6.1**).

About 20 to 25 ka, ice glaciers spread southward across much of North America and their weight depressed Earth's lithosphere (**Figure 6.8a**). Although the elastic lithosphere would respond to such changes in vertical loading immediately, the viscosity of the underlying asthenosphere, which had to flow laterally to accommodate the downward motion of the lithosphere, governed the rate of depression (Watts, 2001; see Further Reading). The elasticity of the lithosphere caused the depression to spread beyond the leading edge of the ice sheet, and this low-lying topography was filled by water, forming large glacial lakes. Wave action cut benches in the hill slopes around the shoreline of these lakes. About 16 ka, the ice sheet retreated northward, and the water of the glacial lake drained away or evaporated. The lithosphere displaced upward in response to this unloading (**Figure 6.8b**). Again, the *rate* of vertical motion of the lithosphere depended upon the viscosity of the asthenosphere, which must flow laterally to return to its former location. The uplifted wave-cut benches now sit well above the lake level and make terraces in the hill slopes.

In Utah the former glacial lake, known as Lake Bonneville (**Figure 6.9a**), covered an area of about 50,000 km^2. Today, the Great Salt Lake and Sevier Lake are the only remnants of Lake Bonneville. However, evidence for the former shoreline is found in many localities, including just east of Logan, UT (**Figure 6.9b**). There, one can see terraces stepping up the hill slope that are the wave-cut benches from the former lake. The highest terrace corresponds to the highest stand of the lake, just before the glacial retreat at 16 ka. The lake level apparently went down in stages, with each stage cutting a new bench.

By dating the terraces and surveying their current elevations, one obtains a record of the crustal uplift due to retreat of the ice sheet and lowering of the lake as a function of distance from the center of the lake (**Figure 6.10**). Along arms of Lake Bonneville that extended the greatest distance from the lake center, the initial depression and later uplift was the least, while around islands near the lake center the initial depression and later uplift was the

Figure 6.7 Two salt glaciers (gray) flow out of the Zagros Mountains into the adjacent valley. The tongue-shaped glaciers are about 5 km long. UTM: 40 R 254840.70 m E, 3047656.06 m N. Mountain chains are folded sedimentary rock in the form of anticlines. The Google Earth image data are from CNES/Airbus.

Box 6.1 Geodynamics

Geodynamics is the study of the motion and deformation of Earth materials, from the global scale to the plate tectonic scale, using geophysical data and models based on continuum mechanical principles. Structures at the global scale that are formed by these motions include the tectonic plates, plate-boundary fault zones, mantle plumes and convection cells, mid-ocean ridges, subduction zones, and transform faults (Watts, 2001; Turcotte and Schubert, 2002; see Further Reading). Geophysical data used to constrain geodynamic models come mostly from seismicity, heat flow, gravity, geochronology, paleomagnetism, and geodesy. The continuum mechanical principles are identical to those used in structural geology (Chapter 3). Modern geodynamics gained recognition as a subdiscipline of geophysics with the acceptance of plate tectonic theory towards the end of the twentieth century.

For the most part, the motions associated with the growth of global-scale structures are very slow, with velocities of centimeters per year or less. However, the effects of plate tectonic motions also include earthquakes and volcanic eruptions (Segall, 2010; see Further Reading), which can involve velocities of kilometers per second. Thus, the time scales of interest to the geodynamicist extend from those relevant to an earthquake rupture on a plate-boundary fault to the age of Earth. Because of the length, depth, and time scales inherent to global-scale structures, the range of material properties used in geodynamic models (e.g. Hirth and Kohlstedt, 2003; Karato, 2008; see Further Reading) are similar to those used in modeling geologic structures, including elastic (Chapter 4), ductile (Chapter 5), and viscous (Chapter 6) behaviors. Given the similar range of time scales and the similar material properties used in geodynamic and structural models, many studies overlap between these two disciplines. Although most of the geologic structures documented on photographs and maps in this textbook are at the meter to kilometer scale, structural geologist deal with a range of length scales that extends from microscopic to that of tectonic plates.

Because geodynamic models typically address large-scale structures, they can provide a global or tectonic plate scale framework for consideration of the structures investigated by structural geologists at smaller length scales. The structures studied by structural geologists usually are restricted to the lithosphere and upper asthenosphere. In many cases, the insights and results from geodynamic research provide the boundary and initial conditions for modeling structures within the lithosphere (**Figure B6.1**). Recall, for example, that surface forces (tractions) on the base of the lithosphere due to mantle convection (Section 1.1.1), and body forces due to gravity and buoyancy (Section 1.1.2) cause the motions and

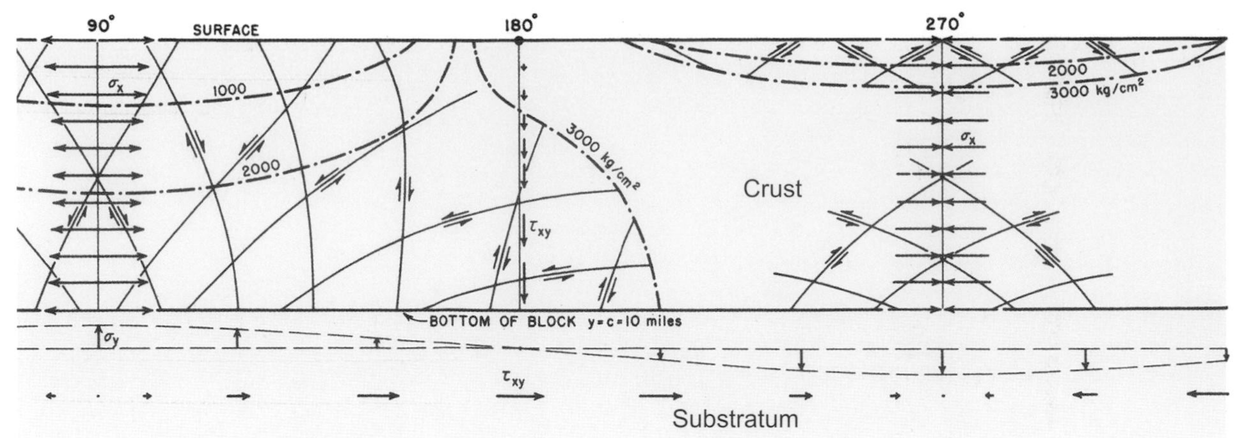

Figure B6.1 Historic geodynamic model of deformation in Earth's crust due to sinusoidal distributions of normal stress (σ_y) and shear stress (τ_{xy}) on the base of the crust caused by convection currents in the substratum. See text for explanation. Modified from Hafner (1951).

relative motions in these outermost shells of Earth. Reversing that perspective, the insights and results from structural geology research help to constrain geodynamic concepts and models.

Figure 1.3 illustrates the conceptual model for one of the earliest contributions to geodynamic analyses. In the middle of the twentieth century, geophysicists suggested that large-scale convection currents in the weak substratum underlying Earth's **crust** could provide the primary forces for mountain building and crustal deformation. Hafner (1951) investigated this concept by solving the boundary-value problem for an elastic plate with a stress free upper surface and a sinusoidal distribution of normal and shear stresses along the lower surface (Figure B6.1). The unit "kg" on this figure refers to the archaic unit kilogram force, and 1 kg/cm^2 = 0.1 MPa. The dash-dot contours bound regions where the stress given by a linear failure criterion exceeds the strength indicated by the contour value. Thus, inside the 1000 kg/cm^2 contour, the failure stress exceeds 100 MPa. Also shown on Figure B6.1 are two conjugate sets of potential fault surfaces (solid curves) oriented at 30° to the maximum compressive stress trajectories (not shown). The failure criterion for the conjugate faults is similar to the Coulomb shear strength criterion introduced in Section 4.8.2.

greatest. For example, around a former island, about 50 km from the center of the lake, the uplift is about 64 m. The furthest terrace is about 340 km from the lake center and that is taken as the reference elevation.

The data provided in Figure 6.10 have been modeled using an elastic plate for the lithosphere and a viscous substrate for the asthenosphere. The thickness of the elastic plate effects the amplitude and breadth of the depression and uplift. In Section 9.6 we derive the governing equations for the bending of an elastic plate, but here we simply point out that the uplift data are consistent with effective thickness between 30 km and 40 km. For the models shown in Figure 6.10 we use an effective thickness of T_e = 30 km. Inspection of this figure indicates that viscosities for the asthenosphere between 5×10^{19} Pa s and 5×10^{20} Pa s adequately bound most of the data. Some outliers at distances between 150 and 300 km require further evaluation and modeling, but we take 1×10^{20} Pa s as a good estimate of the viscosity.

In the next section we turn to laboratory experiments, exploring how viscosity is measured in molten rock, and how it depends upon chemical composition, temperature, and water content. We also describe laboratory creep experiments that document the mechanisms for viscous flow in the solid state.

6.4 VISCOUS DEFORMATION IN THE LABORATORY

Laboratory measurements of viscosity are relatively simple for liquids like water and oil, but the special requirements for molten rock can make these tests very challenging. Temperatures in the range of 500 to 1500 °C are necessary to melt the rock, and these elevated temperatures introduce the possibility of chemical reaction with the equipment and the atmosphere. In addition, the possible presence of crystals and gas bubbles in the liquid can introduce non-linear mechanical behavior (Figure 6.4) that is quite different from that used to define Newtonian viscosity (6.3). We begin by describing the mechanical principles for a laboratory viscometer, and leave the details of the experimental equipment to documents listed in Further Reading at the end of this chapter.

6.4.1 A Laboratory Viscometer

We consider viscometric tests for materials in the *liquid state*, and assume the liquid is isotropic with respect to viscosity. This means the resistance to shearing deformation does not depend on orientation of the applied shear stress. Also, we assume, as in

(6.3), that the relationship between shear stress and shearing rate of deformation is *linear* (**Figure 6.4**). The objective of the viscometric test is to measure the shear stress and associated rate of shearing deformation for a given liquid under prescribed conditions. If the ratio of these two quantities is constant over a given range of conditions, that ratio defines the Newtonian viscosity.

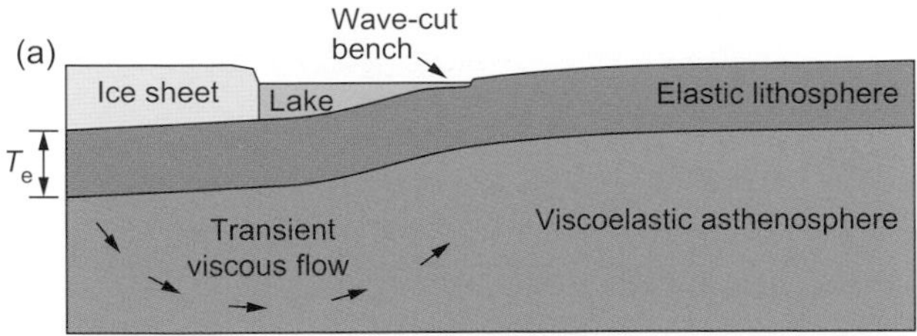

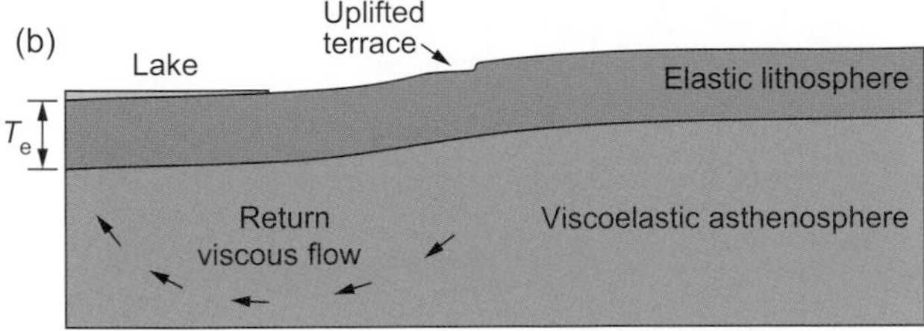

Figure 6.8 Schematic illustration (not to scale) of consequences of continental glacier advance and retreat for depression and uplift of lithosphere. (a) Weight of glacier depresses elastic lithosphere with effective thickness T_e and creates glacial lake. Waves cut a bench along the shoreline. (b) Retreat of glacier causes uplift of lithosphere and leaves terrace etched in the landscape well above current lake. Time scale for depression and uplift depends upon viscosity of asthenosphere. See Watts (2001) in Further Reading.

The Couette–Hatschek viscometer (**Figure 6.11**) is a classic testing apparatus for viscous liquids. It is composed of a cylindrical crucible that contains the liquid, and a centered cylindrical rod of length, L. The annulus between the rod and crucible has an inner radius, aR, and an outer radius R. Given the cylindrical geometry of the apparatus, we use a cylindrical coordinate system, (r, θ, z) to describe the motion. The objective of this test is to achieve a *steady flow*, in which the velocity does not change with time. Also, the test is designed to achieve a *two-dimensional* and *laminar* distribution of velocity:

$$v_r = 0,\ v_\theta = v_\theta(r),\ v_z = 0$$

Despite its name, the viscometer does not measure the viscosity directly, but relies on a calculation that only is valid if the basic assumptions of the test are met.

For the prescribed conditions within the Couette–Hatschek viscometer, the circumferential velocity distribution is found by solving the equations of motion (see Section 6.8), written for a linear viscous liquid using a cylindrical coordinate system (see Bird et al., 2007, in Further Reading):

$$v_\theta = \omega R\left[\frac{(aR/r) - (r/aR)}{a - (1/a)}\right]$$

Here ω is the angular velocity of the crucible, which is controlled by a variable speed motor. This velocity is expressed as a dimensionless angle per unit time, $\omega\,\{=\}\,\mathrm{T}^{-1}$, so the

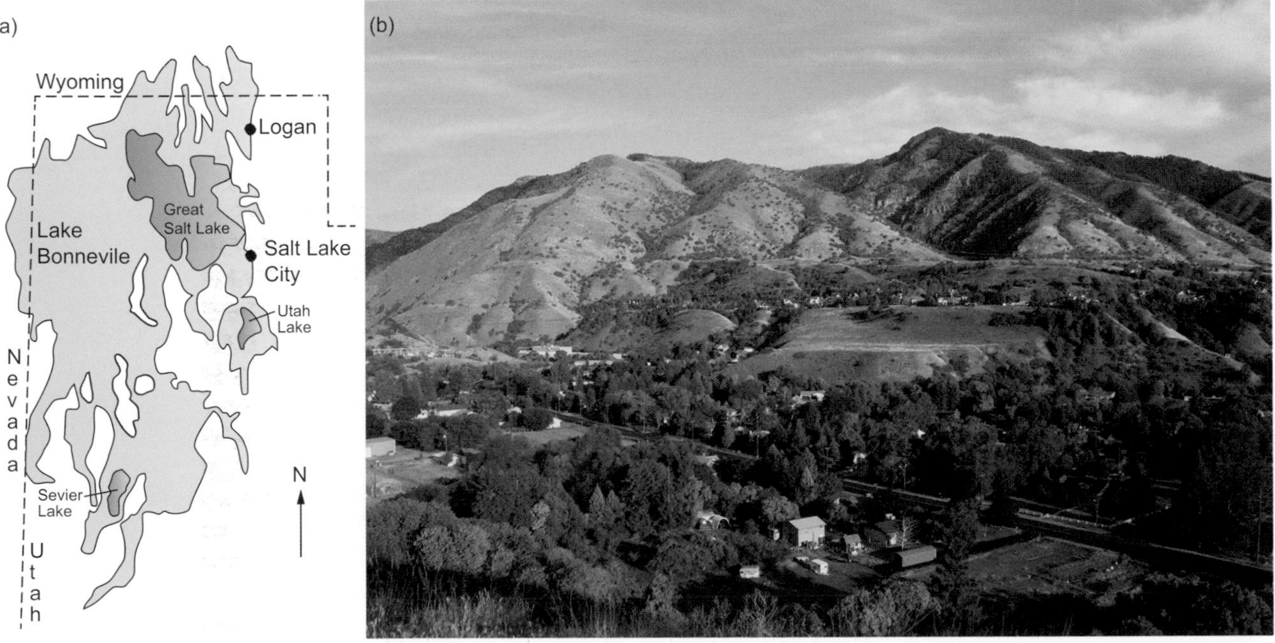

Figure 6.9 (a) Outline of Lake Bonneville in northwest Utah at about 16 ka. The modern Great Salt Lake also shown. Modified from Utah Geological Survey. (b) Lake Bonneville terraces in Logan, UT. Upper terrace was raised by the uplift that began about 16 ka as the ice sheet retreated northward. Photograph by J. S. Horsburgh.

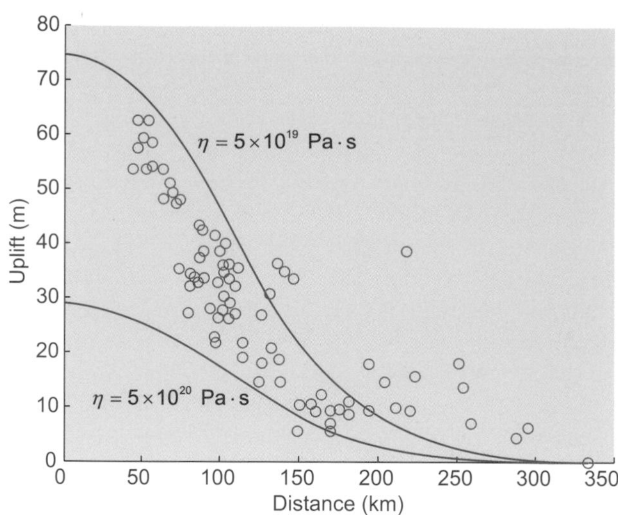

Figure 6.10 Plot of Lake Bonneville shoreline uplift since retreat of the last ice sheet at 16 ka versus distance from the lake center. Thickness of elastic lithosphere is 30 km. Red circles are data from raised terraces. Blue curves are model results for two viscosities. Modified from Iwasaki and Matsu'ura (1982). See also Watts (2001) in Further Reading.

units are inverse seconds, $\omega\,[=]\,s^{-1}$. The circumferential velocity in the liquid increases from $v_\theta = 0$ at $r = aR$, at the fixed inner rod, to a maximum $v_\theta = \omega R$ at $r = R$, the inner edge of the crucible.

The net driving force on the liquid is the torque, τ, required to turn the cylindrical crucible. This torque is the product of the radius, R, and the applied force, which is equal to the shear stress at the outer edge of the annulus multiplied by the area of the crucible wall, $2\pi RL$:

$$\tau = (R)(\sigma_{r\theta}2\pi RL) \tag{6.5}$$

The dimensions of torque are the same as a force multiplied by a length, that is $\tau\,\{=\}\,MLT^{-2}L$, so the units are $\tau[=]$N m. The shear stress at the outer edge of the annulus, $r = R$, also is found from the equations of motion:

$$\sigma_{r\theta} = 2\eta\omega\left[\frac{a^2}{1-a^2}\right] \tag{6.6}$$

Substituting (6.6) for the shear stress in (6.5), and rearranging we find the viscosity as:

$$\eta = \frac{\tau}{4\pi R^2 L\omega}\left[\frac{1-a^2}{a^2}\right] \tag{6.7}$$

All the terms on the right side of (6.7) are known for a given laboratory test. The geometry of the rod and crucible give L, R, and a; the torque, τ, is measured with a transducer on the drive shaft of the apparatus; and the angular velocity, ω, can be adjusted by controlling the speed of the motor driving the crucible.

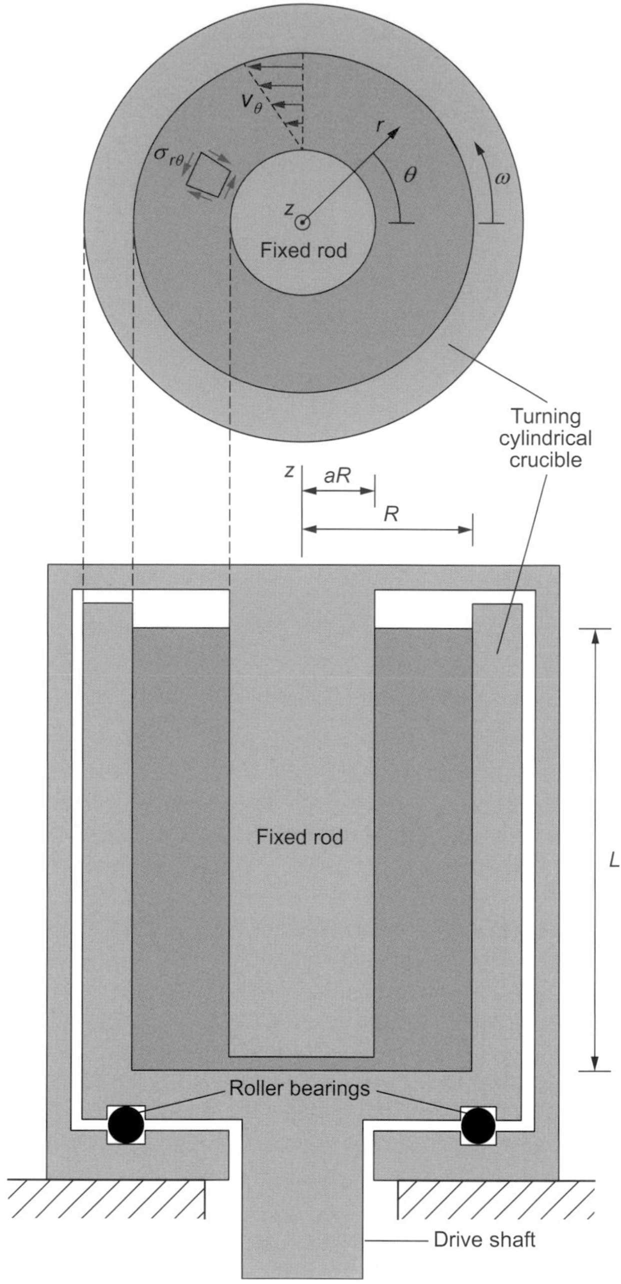

Figure 6.11 Schematic cross sections of a Couette–Hatschek viscometer with liquid sample (red). Not shown are parts of the apparatus necessary to maintain a constant temperature and prevent chemical reactions with the atmosphere. See Bird et al. (2007) in Further Reading.

6.4.2 Liquid Viscosity Variations with Composition, Temperature, and Water Content

The viscosity of four different volcanic rock melts at 1,200 °C (Table 6.1) demonstrate that this material property has a wide range as a function of composition. At this temperature the basaltic lavas have viscosities between 10^1 and 10^2 Pa s, whereas the viscosity of the rhyolite lava is between 10^5 and 10^6 Pa s. Note that viscosity increases systematically with silica content. The rhyolite is more viscous than the andesite, and the Columbia River basalt is more viscous than the Galápagos olivine basalt. The viscosity varies with the other chemical constituents of the melt, but the variation with silica content is pronounced.

Because the petrographic names are related to silica content, we can group these volcanic rocks by their names (rhyolite, andesite, basalt) and expect a correlation to their viscosities in a molten state. Also, a strong correlation exists between the petrographic names and the form of the volcanic flows. Lava with high silica content (e.g. rhyolite and dacite), and therefore greater viscosity forms thick, bulbous flows that are reminiscent of ice or salt glaciers (**Figure 6.7**) in their shape and speed (**Figure 6.12**). In contrast, lava with low silica content (e.g. basalt), and therefore lesser viscosity, forms thin flows that spread rapidly and move quickly down the slope of a volcano (**Figure 6.13**).

Table 6.1 Viscosity at 1,200 °C

Rock name and type	Weight % SiO_2	Viscosity (Pa s)
Newberry rhyolite	73%	5×10^5
Mount Hood andesite	61%	2×10^3
Columbia River basalt	51%	8×10^1
Galápagos olivine basalt	46%	1×10^1

Data from Murase and McBirney (1973)

The viscosity of most liquids is strongly dependent upon temperature. This is illustrated for rock melts in **Figure 6.14** where the ordinate is the base 10 logarithm of viscosity and the abscissa is the temperature. Note that viscosities span almost fifteen orders of magnitude. As described with reference to **Table 6.1**, the dependence on silica content separates the basalts from the andesite and the rhyolite. Also, it is clear that viscosity increases dramatically as these molten rocks cool. For example, the viscosity for the Newberry rhyolite (NRO) increases from about 5×10^5 Pa s at 1,200 °C to about 8×10^{11} Pa s at 700 °C, more than a million times greater viscosity.

Water is another chemical component that can play an important role in determining the apparent viscosity of molten rock. In (**Figure 6.15**) the apparent viscosity is plotted versus temperature

Figure 6.12 Dacite dome at Mt. Saint Helens, Washington. Geologists are measuring the changing shape of the growing dome using a survey instrument. 10 T 562671.70 m E, 5117295.07 m N.

Figure 6.13 Basaltic lava flow on Kilauea Volcano, Hawaii. Backpack height ~ 40 cm. See Pollard et al. (1983). UTM: 5 Q 259686.71 m E, 2144647.44 m N.

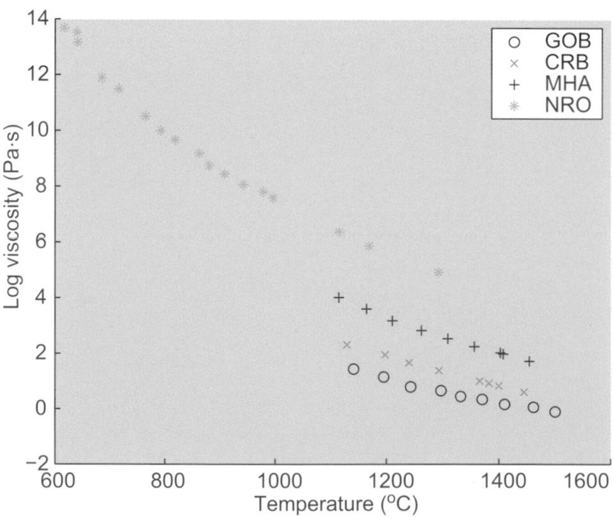

Figure 6.14 Plot of the log of apparent viscosity versus temperature for four different lavas. The names NRO, MHA, CRB, and GOB refer to the Newberry rhyolite, the Mount Hood andesite, the Columbia River basalt, and the Galápagos olivine basalt, respectively. Modified from Murase and McBirney (1973).

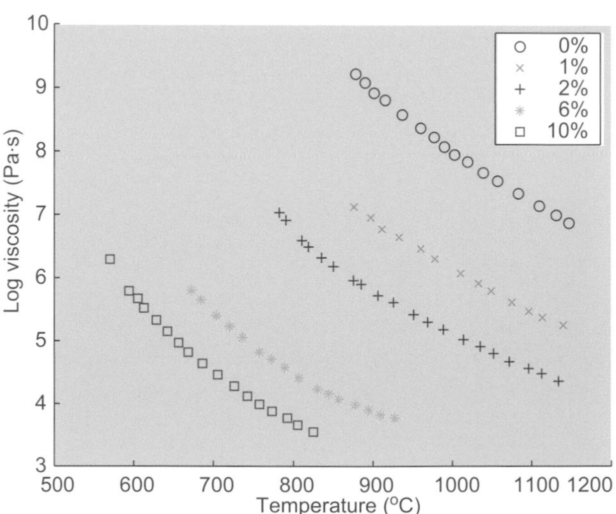

Figure 6.15 Plot of the log of apparent viscosity versus temperature for rhyolite melt with different weight percentages of water. Modified from Shaw (1963).

for rhyolites of the same composition, except for water. As the weight percent of water in this rhyolite melt is increased, the viscosity decreases dramatically. For example at 900 °C increasing the water content from 0 to 10% causes a decrease in apparent viscosity by five orders of magnitude. From these data we conclude the apparent viscosity of molten rocks decreases markedly with decreasing silica content and with increasing water content; for a particular composition the apparent viscosity decreases markedly with increasing temperature.

Unlike material properties such as rock stiffness and strength, which typically vary over less than two orders of magnitude (see Chapter 4), the viscosity of lava and magma can vary over more than ten orders of magnitude. This makes it difficult to estimate the viscosity of rock melts without knowledge of the temperature and composition. To make matters even more challenging, rock can exhibit solid-state flow like a viscous material, and the associated viscosities can range up to ten orders of magnitude greater than those for rock melts. In the next section we describe the acquisition and interpretation of laboratory data on solid-state flow.

6.4.3 Laboratory Viscosity for Solid-State Flow

Laboratory experiments designed to investigate the solid-state flow of crystalline materials at high temperatures and pressures use equipment and procedures described in Section 5.4.6. A uniaxial testing machine provides an axial load and a containment vessel provides the confining pressure and elevated temperature. Cylindrical samples are subjected to an axial stress, σ_a, and a radial stress, $\sigma_r = -P_c$, where P_c is the confining pressure, usually provided by Argon gas. Temperatures up to about 1,500 K are achieved using an internal furnace.

To determine the flow behavior, the samples are subjected to a creep test, in which the differential stress, $\Delta\sigma = \sigma_a - \sigma_r$, is held constant. A displacement transducer measures the change in length of the sample, $L - L_o$, where L is the current length and L_o is the original length. These lengths are used to calculate the axial strain: $\varepsilon_a = (L - L_o)/L_o$, which is recorded as a function of time, and the strain rate, $\dot{\varepsilon} = d\varepsilon_a/dt$, is reported as data for each creep test at a constant differential stress. We note that the choice here of a kinematic variable, a component of the strain rate tensor, is not necessarily consistent with the components of the rate of deformation tensor, adopted as the kinematic variables in fluid dynamics. This distinction can be ignored if the displacement gradient components are small compared to unity (see Malvern, 1969, in Further Reading), but that condition rarely is acknowledged in reporting laboratory data, or in applying such data to fluid dynamical models of tectonic processes.

A primary objective of the creep test is to determine the constants in the empirical relationship between axial strain rate and differential stress, called the flow law. For the tests described here the strain rate is a function of the differential stress, $\Delta\sigma$, grain size, d, fugacity of water, f_{H_2O}, and absolute temperature, T:

$$\dot{\varepsilon} = A(\Delta\sigma)^n d^{-p} f_{H_2O}^r \exp(-Q/RT) \qquad (6.8)$$

The constants to be determined by the testing are the pre-exponential factor, A, the differential stress exponent, n, the grain size exponent, p, the water fugacity exponent, r, and the activation energy, Q. The last term in (6.8) measures the sensitivity of the strain rate to temperature in this thermally activated process. The activation energy has the SI units Q [=]J mol^{-1} and the gas constant has the value in SI units R = 8.3145 J mol^{-1} K^{-1}. The three exponents (n, p, r) are unit-less numbers. Note that the strain rate is directly proportional to differential stress and water fugacity, but inversely proportional to grain size, all raised to their respective powers.

Data from creep tests on olivine aggregates at 1523 K and at three different confining pressures are plotted in **Figure 6.16**, a log–log plot of strain rate versus differential stress. Empirical evaluation of these data yield an activation energy of $Q = 295 \times 10^3$ J mol^{-1}, and the three exponents are evaluated as $n = 1.1$, $p = 3$, and $r = 0.7$. Because the units of differential stress and water fugacity are MPa, and their exponents are 1.1 and 0.7, respectively, the units of the pre-exponential factor must include the unit MPa$^{-1.8}$. Similarly, because the units of grain size are μm and the exponent is –3, the pre-exponential factor must include the unit μm^{-3}. The pre-exponential factor also must include the units for strain rate, s^{-1}. With these units the flow law is dimensionally homogeneous.

Of special importance in these experimental results are the exponents for the differential stress and for grain size. A statistical analysis of data on **Figure 6.16**, and other data not shown at the three different confining pressures, determined the stress exponent, $n = 1.1 \pm 0.2$. This nearly linear relationship between strain rate and differential stress is suggestive of deformation caused by the *diffusion* of chemical species from regions of greater compression to regions of lesser compression. Thus, this mechanism of solid-state viscous deformation is called diffusion

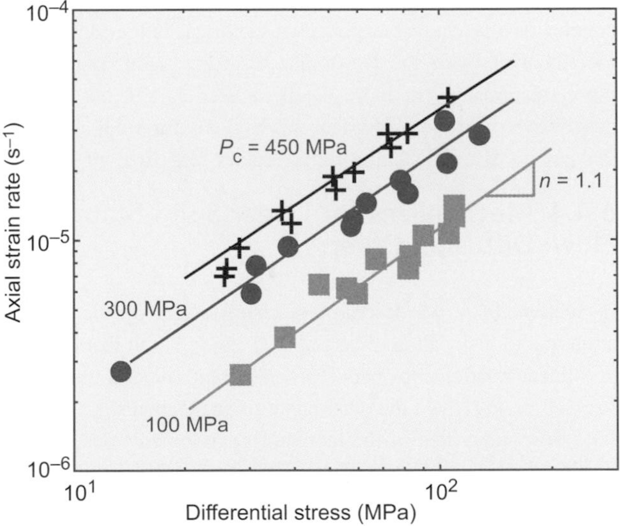

Figure 6.16 Plot of axial strain rate versus differential stress on log–log graph for water saturated olivine aggregates at 1523 K and three different confining pressures, P_c. Best fit straight lines suggest a nearly linear relationship with stress exponent about 1. At each confining pressure, three different samples were tested, yielding similar stress exponents. Modified from Mei and Kohlstedt (2000a).

creep, which we describe in more detail in Section 6.4.4. The locus of the diffusion, through the grain interiors or along grain boundaries, is identified by the grain size exponent. Again, statistical analysis of the data on **Figure 6.16** determined the grain size exponent, $p = 2.8$. If the chemical species diffuse along grain boundaries, the mechanism is referred to as Coble creep and the anticipated grain size exponent is $p = 3$. These data suggest that the dominant diffusion in the olivine aggregates was along grain boundaries.

Flow laws such as (5.2) address data from conventional triaxial experiments and as such, quantify the relationship between axial strain rate and differential stress. Subsuming quantities related to the activation energy, water fugacity, and grain size into the pre-exponential factor, A, and considering only a linear relationship between strain rate and differential stress, the flow law is written:

$$\dot{\varepsilon} = A\Delta\sigma \tag{6.9}$$

The lack of subscripts on $\dot{\varepsilon}$ and $\Delta\sigma$ reminds us that these are *scalar* quantities. Here, the units of the pre-exponential factor are $A[=]\mathrm{MPa}^{-1}\,\mathrm{s}^{-1}$. Now recall from Section 6.1 that Newton's thought experiment defined viscosity in equation (6.3), where a component of the shearing rate of deformation is related to the corresponding component of the shear stress as:

$$D_{zx} = (1/2\eta)\sigma_{zx} \tag{6.10}$$

The double subscripts on D_{zx} and σ_{zx} indicates these are components of *tensor* quantities. Here the units of the term in parentheses are $(1/2\eta)[=]\mathrm{MPa}^{-1}\,\mathrm{s}^{-1}$, exactly the same as those for the pre-exponential factor in (6.9). However, the kinematic quantities (left sides) and the stress quantities (on the right sides) are different in (6.9) and (6.10), so one should not jump to the conclusion that the pre-exponential factor, A, is the same as the reciprocal of twice the Newtonian viscosity, $1/2\eta$. We explore these relationships in more detail in Section 6.9, but first we describe two different physical mechanisms for solid-state flow that can be distinguished using laboratory test data.

6.4.4 Mechanisms for Linear Solid-State Flow: Diffusion Creep

In Section 5.4.6, we described mechanisms for ductile deformation due to dislocation glide and dislocation climb that involved non-linear relationships between axial strain rate and differential stress: $\dot{\varepsilon} \propto (\Delta\sigma)^3$. On the other hand, some mechanisms for solid-state flow, identified in the laboratory experiments described in Section 6.4.3, result in linear relationships: $\dot{\varepsilon} \propto \Delta\sigma$. Here we describe two such mechanisms that operate at the scale of the crystal lattice and involve chemical diffusion, so the associated deformation is referred to as diffusion creep. With a linear relationship between strain rate and differential stress these mechanisms are suggestive of viscous deformation. However, because the deforming material is a solid rather than a liquid, this is referred to as solid-state flow.

In very general terms, a chemical substance diffuses in the opposite direction to a vacancy in the crystal lattice: when a given atom, molecule, or ion jumps to an adjacent vacant site in the lattice, the vacancy jumps to the former site of the chemical substance. The first mechanism we consider depends upon *diffusion* of chemical substances (e.g. atoms, molecules, or ions) in one direction and of point defects (vacancies) in the opposite direction *within* the crystal lattice. The seminal papers calling attention to this mechanism and providing quantitative relationships were published by F. R. N. Nabarro and C. Herring (see Poirier, 1985, in Further Reading), so the mechanism is referred to as Nabarro–Herring creep.

For a cubic mineral grain with side length d at a fixed temperature T with no applied stress, the equilibrium concentration of vacancies within the grain is C_0. However, suppose a tension, $\sigma_1 = \sigma$, exists in one direction and a compression of the same magnitude, $\sigma_3 = -\sigma$, in a perpendicular direction. The sides of the grain perpendicular to the tension develop a greater concentration of vacancies, $C_T > C_0$, because the tension extends the lattice, so less work is required to form a vacancy. Sides perpendicular to the compression develop a lesser concentration, $C_C < C_0$, because the compression contracts the lattice and more work is required to form a vacancy.

The mineral grain elongates in the direction of tension and shortens in the direction of compression as diffusion brings atoms or molecules toward the sides carrying the tension, and sends vacancies toward the sides carrying the compression. In this way the non-hydrostatic stress drives diffusion that changes the shape of the grain. The strain rate, $\dot{\varepsilon}$, is estimated as:

$$\dot{\varepsilon} = \frac{2\alpha D_V C_0 \sigma b^6}{d^2 kT} \tag{6.11}$$

In this equation α is a numerical factor that accounts for the geometry of the grain and the diffusion path, D_v is the vacancy diffusion coefficient, the vacancy volume is b^3, and $k = 1.38 \times 10^{-23}\,\mathrm{m}^2\,\mathrm{kg}\,\mathrm{s}^{-2}\,\mathrm{K}^{-1}$ is the Boltzmann constant. The strain rate is inversely proportional to the square of the grain size, so this mechanism is enhanced by smaller grain size. The relationship between strain rate and stress is linear, so deformation dominated by Nabarro–Herring creep is referred to as *linear* solid-state flow. Note that the change of shape is not due to elastic or plastic stretching and shortening in response to the applied stress, but rather to the transport of vacancies and matter. In this respect diffusion creep is a kind of chemical deformation rather than a mechanical deformation related to the motion of dislocations.

The second mechanism that depends upon *diffusion* for solid-state flow considers the motion of chemical substances (e.g. atoms, molecules, or ions) along *grain boundaries*, rather than through the crystal lattice. The paper calling attention to this mechanism and providing quantitative relationships was published by R. L. Coble, so the mechanism is referred to as Coble creep (see Poirier, 1985, in Further Reading). In this case vacancies do not play a prominent role, and the relevant diffusion coefficient is that for flux along the grain boundaries, D_{GB}. The relationship between strain rate and stress is:

$$\dot{\varepsilon} = \frac{\alpha D_{GB} C_0 \sigma \delta b^3}{d^3 kT} \tag{6.12}$$

Here δ is the width of the grain boundary. Again, the relationship between strain rate and stress is linear. However, strain rate is inversely proportional to the *cube* of the grain size in (6.12), whereas for Nabarro–Herring creep, strain rate is inversely proportional to the *square* of the grain size (6.11). Thus, the enhancement of diffusion creep by smaller grain size is greater for Coble creep than for Nabarro–Herring creep. Because diffusion occurs more readily along grain boundaries than through crystal lattices, Coble creep tends to be the dominant mechanism for creep at lower temperatures.

6.4.5 Comparing Dislocation and Diffusion Creep

Recall from Chapter 5 that we described a non-linear relationship, $\dot{\varepsilon} \propto (\Delta\sigma)^3$, between strain rate and differential stress for dislocation creep, in which the gliding or climbing motion of dislocations was responsible for the deformation. This non-linear relationship is distinctly different from the linear relationship, $\dot{\varepsilon} \propto (\Delta\sigma)^1$, described in the preceding section for diffusion creep. However, at a fixed temperature, confining pressure, and grain size, results from creep tests reveal a transition between diffusion and dislocation creep as a function of the differential stress (**Figure 6.17**). For each creep test represented by a red circle on this graph the differential stress is held constant and the corresponding axial strain rate is recorded and plotted.

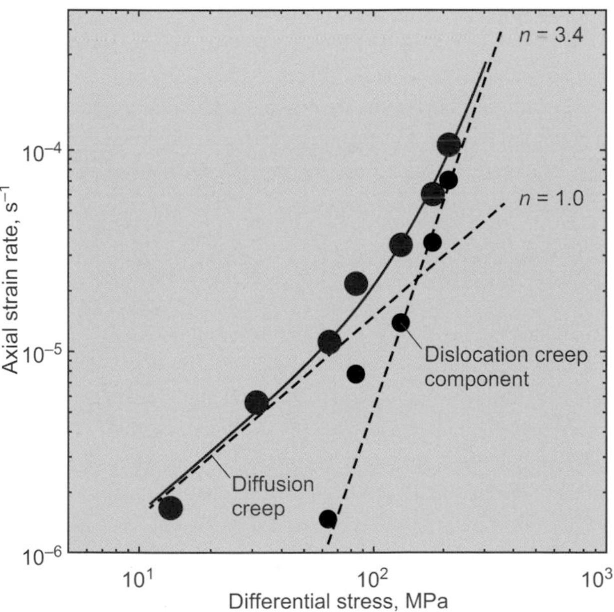

Figure 6.17 Plot of axial strain rate versus differential stress on a log–log graph for wet saturated olivine aggregates at absolute temperature 1523 K and confining pressure, P_c = 300 MPa. Black symbols, labeled dislocation creep component, are obtained by subtracting the diffusion creep component from the data. Modified from Hirth and Kohlstedt (2003).

For creep tests where the differential stress is less than about 100 MPa (**Figure 6.17**), the relationship between strain rate and differential stress is approximately linear, that is $n \approx 1$. For greater differential stress the relationship is distinctly non-linear with a stress exponent $n \approx 3.4$. Thus, these data span the transition from creep dominated by diffusion mechanisms to creep dominated by dislocation mechanisms as the applied differential stress is increased. Within the transition, both mechanisms operate.

We have described laboratory test data that evaluate both the viscosity of silicate liquids and the creeping behavior of rock in the solid state. While data from these tests provide valuable constraints on the mechanical properties of magma and rock, field tests described in the next section provide complementary constraints at length and time scales not necessarily achievable in the laboratory.

6.5 FIELD ESTIMATES OF LAVA VISCOSITY

Field observations from Kilauea Volcano demonstrate the liquid nature of basaltic lava – it flows down slopes under its own weight (**Figure 6.6**), and it adopts a level surface when contained in a lava lake (**Figure 6.5**). Here we use a field test for estimating the viscosity of lava by postulating that it behaves like a viscous liquid and using a solution to the equations of motion for such a liquid flowing in a thin sheet down a uniform slope.

First we postulate that the only non-zero component of velocity is in the x-direction, parallel to the slope (**Figure 6.18**), and this velocity component is only a function of the z coordinate:

$$v_x = v_x(z),\ v_y = 0,\ v_z = 0$$

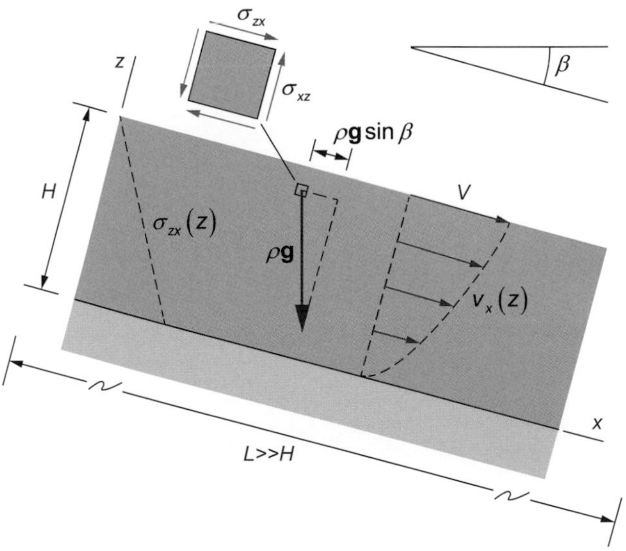

Figure 6.18 Idealized flow of a sheet of viscous liquid (red) of thickness H and length L down a slope with uniform inclination β. The velocity distribution, $v_x(z)$, is parabolic and the shear stress distribution $\sigma_{zx}(z)$ is linear. The mass density is ρ, and the gravitational acceleration is $\mathbf{g}$.

Under these restrictions the flow is perfectly *laminar*: the velocity varies only with distance from the base of the flow like a deck of cards sliding over one another (Section 5.5.1) with each card representing one lamina of liquid. In Section 6.8.2 of this chapter we define the conditions for laminar flow explicitly, but suffice it to say that these conditions usually are met by lava flows. At the base of the flow the velocity is zero, because the lava sticks to the rock or soil it is flowing over, and on the upper surface the lava attains the maximum velocity, V.

For a Newtonian viscous liquid the maximum velocity of a thin sheet on a uniform slope (**Figure 6.18**) is found from the equations of motion (see Section 6.8):

$$v_x(z = H) = V = \frac{H^2 \rho g \sin\beta}{2\eta} \tag{6.13}$$

The driving force for this flow is the component of gravitation force acting parallel to the slope, $\rho g \sin\beta$, and the resistance to flow is provided by the viscosity, η. Solving (6.13) for the viscosity we have:

$$\eta = \frac{H^2 \rho g \sin\beta}{2V} \tag{6.14}$$

To implement (6.14), mass density, $\rho = 2.65 \times 10^3$ kg m^{-3}, and acceleration of gravity, $g = 9.8$ m s^{-2}, are taken from laboratory and geophysical measurements representative of Hawaiian volcanoes. The flow depth, H, was taken from observations of drained channels, and the slope, β, was measured in the field using surveying instruments. The velocity was calculated by measuring the travel time down the channel for fragments of the chilled lava crust near the middle of the stream. **Table 6.2** summarizes average values for different lava streams and field days on Kilauea Volcano, Hawaii. The results are systematic in that the greater velocities are associated with the greater slopes.

A number of uncertainties suggest that the estimates given in **Table 6.2** may be no more precise than an order of magnitude. The depths of the lava streams were not observed directly and H enters the equation to the second power, so an error by a factor of two in H would produce a four-fold error in the viscosity. The channels were postulated to be flat and two-dimensional, but in fact they had finite widths and irregular bottoms. Furthermore, although viscosity was postulated to be uniform and constant throughout the flow, loss of heat to the surroundings would cause a temperature drop in the lava and subsequent increases in viscosity both near the top and the bottom of the flow, and with time. Despite these uncertainties, order-of-magnitude estimates such as these are quite instructive.

Table 6.2 Field estimates of apparent viscosity

Date	Number of readings	Velocity (m/s)	Slope (degrees)	Viscosity (Pa s)
August 10	20	3.00	20	2.9×10^3
August 12	55	5.88	29	2.2×10^3
August 13	20	5.49	25	1.9×10^3
August 13	?	2.00	18.5	3.8×10^3

Data from Macdonald (1955)

To this point in Chapter 6 we have used well-known kinematic quantities like the velocity vector and the scalar strain rate, but the most important kinematic quantity for viscous flow is the rate of deformation tensor. One component of this tensor quantity was used in (6.3), but in the next section we define all of the tensor components and other relevant quantities.

6.6 KINEMATICS OF FLOW

Recall from Section 4.9 that we used a referential description of motion with the *original* positions (X, Y, Z) of particles as the reference for elastic deformation. We compared the later positions of those particles to their original positions to define the displacement. We also compared the lengths and orientations of material lines in the deformed state to their lengths and orientations in the original state, in order to calculate the small strain and rotation.

For example, before the dike in **Figure 6.19** opened, the Mancos Shale on the left side of the dike was in contact with the same bed of Mancos Shale on the right side. Originally adjacent sedimentary particles, located at what now is the center of the dike, were displaced about 1.5 m in opposite directions as the dike opened. Because of the symmetry of this opening, we place the coordinate origin at the center of the dike with the Y-axis perpendicular to the dike contact and the Z-axis up. The displacements of two adjacent particles originally at the dike center are related as $\mathbf{u}(X, Y = 0^-, Z) = -\mathbf{u}(X, Y = 0^+, Z)$. We know both the original and the final positions of these particles because the dike has simply opened. In contrast, we can only speculate about the original position of a particle of the magma that formed the igneous rock (basalt) exposed at this outcrop. Therefore, we need a different description of motion to address questions about the flowing magma.

6.6.1 Spatial Description of Motion

To understand the flow of magma through the dike at Ship Rock (**Figure 6.19**) we adopt a spatial description of motion. As the name implies, we focus on particular positions in space (x, y, z), rather than particular particles. For example, a small portion of the dike contact is shown in **Figure 6.20** with Mancos Shale in contact with basalt. The basaltic magma may have flowed to this location in the dike from many kilometers away, and individual particles may have experienced a large range of velocities as they flowed from that source region to this location and beyond. Instead of referring to the original positions of magma particles, we ask: at a given time when the magma was flowing, what was the velocity of the particle at *this position* in space?

To address the question raised in the preceding paragraph we use the components of the position vector, $\mathbf{x}$, which are the coordinates for the designated position in space (**Figure 6.20**). The velocity vector, $\mathbf{v}$, is the dependent variable at that position, and the coordinates and time are the independent variables:

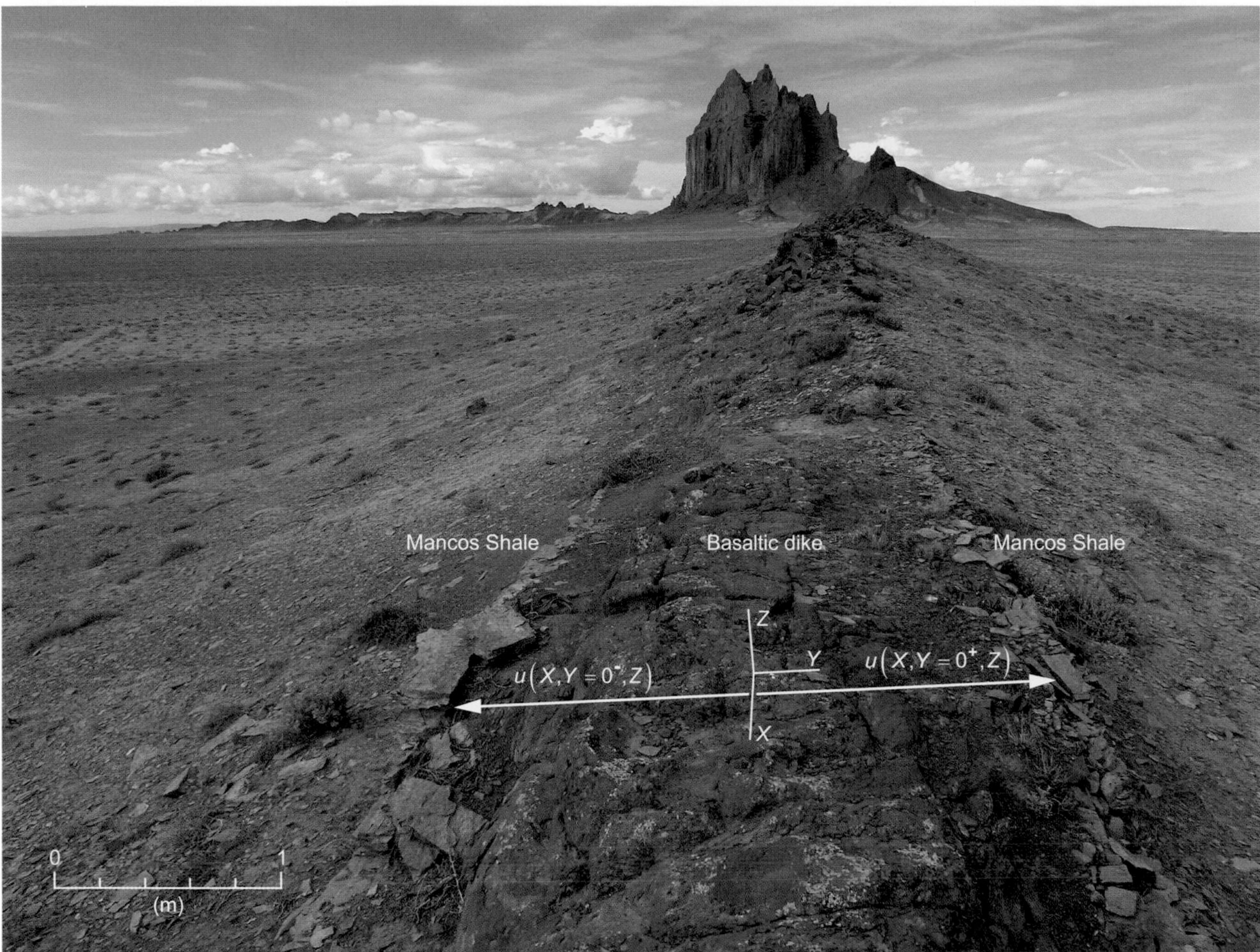

Figure 6.19 Northeastern dike (foreground) at Ship Rock (background), NM. Black basalt and tan Mancos Shale are in contact along two sides of the dike and are about three meters apart. A referential description of motion with the coordinate system (*X*, *Y*, *Z*) is used to define the displacement, **u**, of particles of Mancos Shale that originally were adjacent at what now is the dike center. See Delaney and Pollard (1981). UTM: 12 S 695445.22 m E, 4064253.78 m N.

$$\mathbf{v} = \mathbf{v}(\mathbf{x}, t) = \mathbf{v}(x, y, z, t) \tag{6.15}$$

This quantity is called the local velocity, because it is the velocity at the designated position and not the velocity of a particular particle. At earlier or later times different particles of magma occupied that position and could have had different velocities. In general, the flow of viscous liquids is investigated by focusing attention on the current state of the flowing magma and ignoring the initial or reference state. By *current* we don't necessarily mean today, but rather a specified time, t, after the flow began at this position, and before the magma stopped flowing. In the case of the dike at Ship Rock, that time was about 25 million years ago, when the Navajo Volcanic Field was active.

6.6.2 Rate of Deformation and Spin

To quantify the kinematics of flow for viscous liquids like magma, or viscous rock in Earth's lower lithosphere or asthenosphere, we use the spatial derivatives of velocity arranged in two tensors, the rate of deformation tensor and the rate of spin tensor. The first tensor describes the rate of stretching of infinitesimal material line segments, and the second tensor describes the rate of rotation of those line segments. These tensors are parallel in concept, respectively, to the pure stretch tensor, and the pure rotation tensor defined in Section 5.5.2 for material line segments in the ductile solid. However, we describe significant differences here.

We adopt the spatial description of motion described in the previous section and a two-dimensional perspective to ease the presentation (**Figure 6.21**). Then, we focus attention on an arbitrary location p in the flow defined by the position vector, $\mathbf{x}$. At a designated time, the local velocity vector, $\mathbf{v}$, defined in (6.15) is tangent to the particle path that passes through p. At that same time, and at an arbitrary neighboring location q, defined by the position vector $\mathbf{x} + \mathrm{d}\mathbf{x}$, the local velocity vector, $\mathbf{v} + \mathrm{d}\mathbf{v}$, is tangent to the particle path through that position. The objective is to determine the velocity of the particle at q, *relative* to the particle at p, at the designated time. This relative velocity

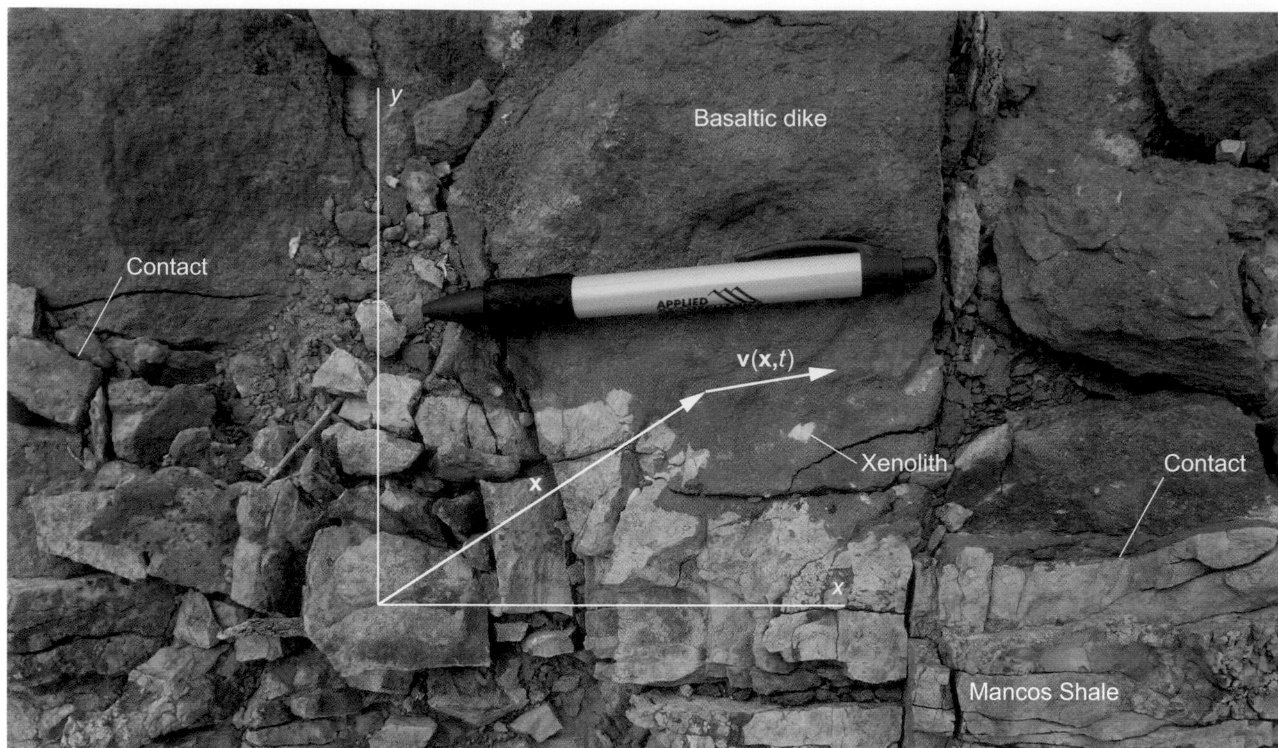

Figure 6.20 Outcrop of the contact for the northeastern dike at Ship Rock, NM. A spatial description of motion with the coordinate system (x, y, z) is used to define the local velocity, $\mathbf{v}(\mathbf{x}, t)$. See Figure 6.19 for location.

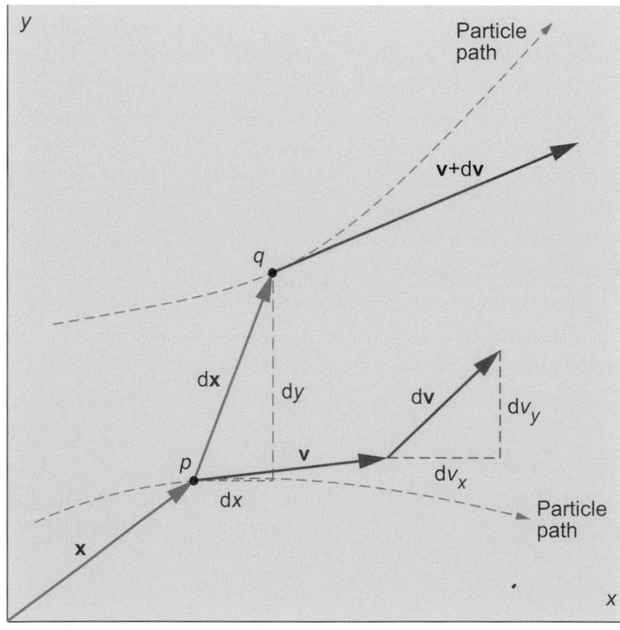

Figure 6.21 Using a spatial description of motion, the velocity at $\mathbf{x}$ is compared to the velocity at the neighboring position $\mathbf{x} + \mathrm{d}\mathbf{x}$ to determine the spatial derivatives of velocity. Brown dashed curves are particle paths. The finite and differential vectors and distances are not drawn to scale. Modified from Malvern (1969).

provides a measure of the rate of stretching (rate of deformation) of the infinitesimal material line, represented by $\mathrm{d}\mathbf{x}$, between the two particles.

We overlay the material line between the particle at p and the particle at q in Figure 6.21 using the vector $\mathrm{d}\mathbf{x}$ with its tail at $\mathbf{x}$. The material line is taken as infinitesimal in length because we want to evaluate the kinematics at p. Therefore, we are interested only in the changes in velocity from that position, $\mathbf{x}$, to a neighboring position, $\mathbf{x} + \mathrm{d}\mathbf{x}$, an infinitesimal distance away. However, the local velocities at p and q are not restricted to being small.

The total differentials (see Section 2.5.3) for the components of the relative velocity, $\mathrm{d}\mathbf{v}$, of the particle at q, relative to the particle at p, are (Figure 6.21):

$$\begin{aligned} \mathrm{d}v_x &= \frac{\partial v_x}{\partial x}\mathrm{d}x + \frac{\partial v_x}{\partial y}\mathrm{d}y \\ \mathrm{d}v_y &= \frac{\partial v_y}{\partial x}\mathrm{d}x + \frac{\partial v_y}{\partial y}\mathrm{d}y \end{aligned} \tag{6.16}$$

These total differentials are measures of the spatial variation of the velocity components in the neighborhood of a point in the velocity field. The coefficients on the right sides of (6.16) are partial derivatives of the local velocity with respect to the spatial coordinates. Note that the partial derivatives in (6.16) are taken with respect to the coordinates in the *current state*, x and y.

The matrix form of (6.16) is:

$$\begin{bmatrix} dv_x \\ dv_y \end{bmatrix} = \begin{bmatrix} L_{xx} & L_{xy} \\ L_{yx} & L_{yy} \end{bmatrix} \begin{bmatrix} dx \\ dy \end{bmatrix}, \quad -\infty < L_{xx}, L_{xy}, L_{yx}, L_{yy} < +\infty \tag{6.17}$$

The elements of the square matrix [L] are the two-dimensional components of the spatial gradient of velocity tensor, **L**. Recall from Section 2.6.1 that the gradient of a vector field is a tensor field. Here, the vector is the relative velocity and the partial derivatives of the velocity components in (6.16) are the components of the gradient vectors. The first subscript on each element of [L] indicates the velocity component, and the second subscript indicates the coordinate with respect to which the partial derivative is taken. The components of this tensor have dimensions of reciprocal time, T^{-1}, and this tensor is not necessarily symmetric, so L_{xy} might not equal L_{yx}.

The spatial gradient of velocity tensor, **L**, plays a parallel role for flowing liquids to the role played by the spatial gradient of deformation tensor, **F**, for ductile solids (see Section 5.5.1). A notable differences is that [F] compares initial and final states using a referential description of motion, whereas [L] is focused on the current state using a spatial description of motion. Also, the differential vector quantity used to formulate [F] is d**x**, whereas the differential vector used to formulate [L] is d**v**. Furthermore, while the elements of [F] on the primary diagonal must be positive numbers, so the material line does not shrink to zero length, those on the primary diagonal of [L] are not restricted. That is, L_{xx} and L_{yy} can be negative, zero, or positive numbers.

To understand the role of the spatial gradient of velocity, [L], in the kinematics of flow, it is helpful to separate it into a symmetric matrix, [D], and a skew-symmetric matrix, [W], such that [L] = [D] + [W]. Recall from Section 2.6.4 that a square matrix converts to a symmetric matrix by adding the matrix to its transpose, so:

$$[D] = \tfrac{1}{2}\left(\begin{bmatrix} L_{xx} & L_{xy} \\ L_{yx} & L_{yy} \end{bmatrix} + \begin{bmatrix} L_{xx} & L_{yx} \\ L_{xy} & L_{yy} \end{bmatrix} \right) \tag{6.18}$$

Substituting the partial derivatives of velocity for the elements of [L] using (6.16) and (6.17):

$$[D] = \begin{bmatrix} D_{xx} & D_{xy} \\ D_{yx} & D_{yy} \end{bmatrix} = \begin{bmatrix} \dfrac{\partial v_x}{\partial x} & \dfrac{1}{2}\left(\dfrac{\partial v_x}{\partial y} + \dfrac{\partial v_y}{\partial x}\right) \\ \dfrac{1}{2}\left(\dfrac{\partial v_y}{\partial x} + \dfrac{\partial v_x}{\partial y}\right) & \dfrac{\partial v_y}{\partial y} \end{bmatrix} \tag{6.19}$$

Here, [D] is the two-dimensional matrix form of the rate of deformation tensor, **D**, which also is called the rate of stretch and shearing tensor. [D] is a symmetric matrix because $D_{xy} = D_{yx}$, and it is a measure of the rate of stretching and shearing of infinitesimal material lines in all orientations at a given point in the flowing material.

For the sake of an example consider the following spatial gradient of velocity:

$$[L] = \begin{bmatrix} 1.3 & 0 \\ 0.6 & 1.1 \end{bmatrix} \tag{6.20}$$

Note that this matrix is not symmetric because $L_{xy} \neq L_{yx}$. Using (6.18) the corresponding symmetric rate of deformation matrix is:

$$[D] = \begin{bmatrix} 1.3 & 0.3 \\ 0.3 & 1.1 \end{bmatrix}$$

To illustrate the consequences of this stretching matrix we choose one vector d**x** (**Figure 6.22**), representing the infinitesimal material line connecting the particle at p and its neighbor at q. The components of the corresponding relative velocity vector, d**v**, are:

$$\begin{bmatrix} dv_x \\ dv_y \end{bmatrix} = \begin{bmatrix} D_{xx} & D_{xy} \\ D_{yx} & D_{yy} \end{bmatrix} \begin{bmatrix} dx \\ dy \end{bmatrix} \tag{6.21}$$

Placing this vector with its tail at p in **Figure 6.22**, it represents the spatial change in velocity from the particle at p to that at q. The actual velocity vectors at p and q are not shown, and usually would *not* be differential quantities. Also, note that the change in velocity from p to q is not necessarily directed along the material line between those positions.

Although only one pair of vectors, d**x** and d**v**, are shown in **Figure 6.22**, the rate of deformation tensor determines the relative velocity of all possible neighboring particles to the particle at p using (6.21). To locate those particles, note that the components of d**x** are dx = |d**x**|cosθ and dy = |d**x**|sinθ (**Figure 6.22**, inset). Thus, for a fixed magnitude, |d**x**|, the components dx and dy change as the angle θ varies from 0 to 2π, and the heads of the vectors d**x** sweep out a *circle*. As d**x** varies in this way, the components of d**v** change according to (6.17) and their heads sweep out an *ellipse* (**Figure 6.22**).

The ellipse in **Figure 6.22** illustrates the variation in *relative* velocities for all particles neighboring the particle at p. This

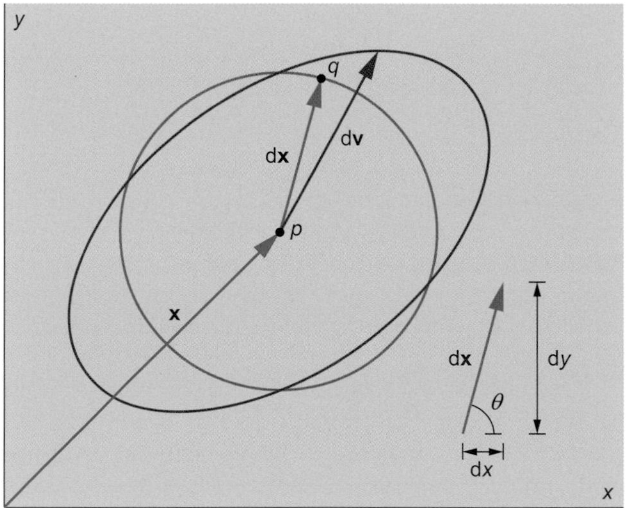

Figure 6.22 Neighborhood of the particle at p located by the position vector **x** at a designated time (see **Figure 6.21**). A neighboring particle at q is located by the differential vector d**x**, and is moving with the relative velocity d**v**.

ellipse is *not* the deformed shape of an originally circular object, such as the ooliths that we described in Section 2.1 using a referential description of motion. The ooliths are finite in size, not infinitesimal, and their shape reflects the displacements from an initial shape, not relative velocities. For similar reasons, the ellipse in **Figure 6.22** is *not* the same as the stretch ellipse for ductile deformation that we described in Section 5.5.1, again using a referential description of motion.

The skew-symmetric matrix [W] is found from the spatial gradient of velocity, [L], by subtracting the transposed matrix:

$$[\mathrm{W}] = \tfrac{1}{2}\left(\begin{bmatrix} L_{xx} & L_{xy} \\ L_{yx} & L_{yy} \end{bmatrix} - \begin{bmatrix} L_{xx} & L_{yx} \\ L_{xy} & L_{yy} \end{bmatrix}\right) \quad (6.22)$$

Substituting the partial derivatives of velocity for the elements of [L] using (6.16) and (6.17):

$$[\mathrm{W}] = \begin{bmatrix} 0 & W_{xy} \\ W_{yx} & 0 \end{bmatrix} = \begin{bmatrix} 0 & \frac{1}{2}\left(\frac{\partial v_x}{\partial y} - \frac{\partial v_y}{\partial x}\right) \\ \frac{1}{2}\left(\frac{\partial v_y}{\partial x} - \frac{\partial v_x}{\partial y}\right) & 0 \end{bmatrix} \quad (6.23)$$

This is the two-dimensional rate of spin tensor, **W**, which also is called the vorticity tensor. [W] is a skew-symmetric matrix because the elements on the principal diagonal are zero, and those on the secondary diagonal are equal in magnitude and opposite in sign, $W_{xy} = -W_{yx}$.

To illustrate the rate of spin, we substitute the elements of the same spatial gradient of velocity matrix (6.20) into (6.22) to find the following skew-symmetric matrix:

$$[\mathrm{W}] = \begin{bmatrix} 0 & -0.3 \\ 0.3 & 0 \end{bmatrix}$$

Then, the components of the relative velocity vector, **dv**, are calculated using:

$$\begin{bmatrix} \mathrm{d}v_x \\ \mathrm{d}v_y \end{bmatrix} = \begin{bmatrix} W_{xx} & W_{xy} \\ W_{yx} & W_{yy} \end{bmatrix} \begin{bmatrix} \mathrm{d}x \\ \mathrm{d}y \end{bmatrix} \quad (6.24)$$

The vectors **dv** are plotted with their tails at the respective particle q, that is at the distal end of the material line between p and q. Each relative velocity vector, **dv**, is *perpendicular* to the vector, **dx**, representing the infinitesimal material line between p and q, so these material lines do not stretch. The motion due to the rate of spin, [W], in the neighborhood of the particle at p is a pure rotation of all the neighboring points about the point p.

It is important to understand that the actual velocities of the particles at p and q are not illustrated in **Figure 6.22** and **Figure 6.23**. Those velocities usually would be finite in magnitude, and usually would have different directions than the relative velocity, as can be seen in **Figure 6.21**. What is illustrated in **Figure 6.22** and **Figure 6.23** are the changes in velocity from the particle at p to the particle at q. Because the distance between these particles is infinitesimal, these relative velocities are the differential quantities, **dv**.

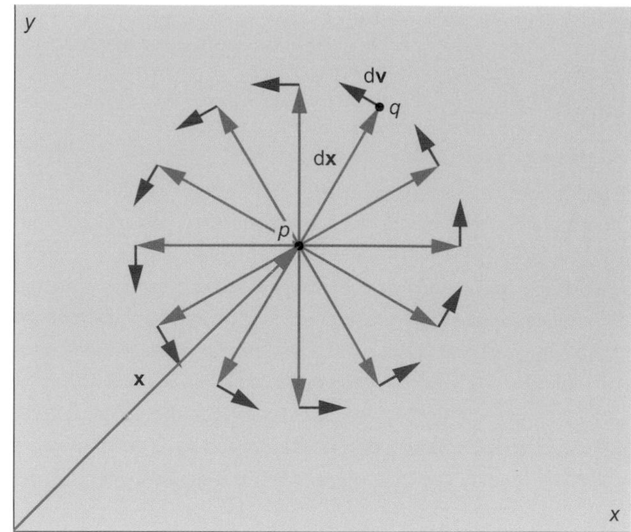

Figure 6.23 Representation of [W], the *rate of spin* in the neighborhood of the particle at *p*, located by the position vector **x** (see **Figure 6.21**). A neighboring particle at *q* is located by the vector **x** + **dx**, and is moving with the relative velocity **dv**.

6.6.3 Rate of Deformation and Spin in Three Dimensions

The two-dimensional rate of deformation and rate of spin introduced in the preceding section may be generalized to three dimensions without introducing new concepts. We use the spatial description of motion introduced in Section 6.6.1, so the coordinates (x, y, z) refer to the *current position* in space. The velocity is the *local velocity*, which is the dependent variable at that position, and the spatial coordinates and time are the independent variables (6.15).

Generalizing (6.16) and (6.17) to three dimensions, the spatial gradient of velocity tensor is written in matrix form as:

$$[\mathrm{L}] = \begin{bmatrix} L_{xx} & L_{xy} & L_{xz} \\ L_{yx} & L_{yy} & L_{yz} \\ L_{zx} & L_{zy} & L_{zz} \end{bmatrix} = \begin{bmatrix} \frac{\partial v_x}{\partial x} & \frac{\partial v_x}{\partial y} & \frac{\partial v_x}{\partial z} \\ \frac{\partial v_y}{\partial x} & \frac{\partial v_y}{\partial y} & \frac{\partial v_y}{\partial z} \\ \frac{\partial v_z}{\partial x} & \frac{\partial v_z}{\partial y} & \frac{\partial v_z}{\partial z} \end{bmatrix} \quad (6.25)$$

The first subscript on each element of [L] indicates the velocity component, and the second subscript indicates the coordinate with respect to which the partial derivative is taken.

The symmetric part of [L] is found by adding the transpose to the matrix itself, [D] = [L] + [L′], to define the components of the rate of deformation tensor:

$$D_{xx} = \frac{\partial v_x}{\partial x}, D_{yy} = \frac{\partial v_y}{\partial y}, D_{zz} = \frac{\partial v_z}{\partial z}$$
$$D_{xy} = \tfrac{1}{2}\left(\frac{\partial v_x}{\partial y} + \frac{\partial v_y}{\partial x}\right) = D_{yx}, D_{yz} = \tfrac{1}{2}\left(\frac{\partial v_y}{\partial z} + \frac{\partial v_z}{\partial y}\right) = D_{zy},$$
$$D_{zx} = \tfrac{1}{2}\left(\frac{\partial v_z}{\partial x} + \frac{\partial v_x}{\partial z}\right) = D_{xz} \quad (6.26)$$

Arranging the components from (6.26) in a square matrix, we have:

$$[\mathrm{D}] = \begin{bmatrix} D_{xx} & D_{xy} & D_{xz} \\ D_{yx} & D_{yy} & D_{yz} \\ D_{zx} & D_{zy} & D_{zz} \end{bmatrix} \quad (6.27)$$

From the second line of (6.26) note that pairs of the off-diagonal elements of [D] are equal, so this is a *symmetric* matrix.

The rate of spin tensor (vorticity) is the skew-symmetric part of [L], which is defined as [W] = [L] – [L′]:

$$W_{xx} = 0, W_{yy} = 0, W_{zz} = 0$$
$$W_{xy} = \frac{1}{2}\left(\frac{\partial v_x}{\partial y} - \frac{\partial v_y}{\partial x}\right) = -W_{yx}, W_{xz} = \frac{1}{2}\left(\frac{\partial v_z}{\partial x} - \frac{\partial v_x}{\partial z}\right) = -W_{zx},$$
$$W_{yz} = \frac{1}{2}\left(\frac{\partial v_y}{\partial z} - \frac{\partial v_z}{\partial y}\right) = -W_{zy} \quad (6.28)$$

In matrix form, the rate of spin is:

$$[\mathrm{W}] = \begin{bmatrix} 0 & \mathrm{W}_{xy} & \mathrm{W}_{xz} \\ \mathrm{W}_{yx} & 0 & \mathrm{W}_{yz} \\ \mathrm{W}_{zx} & \mathrm{W}_{zy} & 0 \end{bmatrix} \quad (6.29)$$

Because elements on the principal diagonal are zero, and elements in symmetric positions across the principal diagonal are equal in magnitude and opposite in sign, the rate of spin is a *skew-symmetric* matrix.

6.7 CONSTITUTIVE EQUATIONS FOR LINEAR VISCOUS MATERIALS

The viscous properties of a material relate the components of stress and the components of rate of deformation to define constitutive equations. These equations describe the mechanical behavior that results from the internal *constitution* of the material, hence the name *constitutive*. The constitutive equations for the viscous liquid also include a term that represents the pressure in the static and flowing liquid, so we define pressure in this context and show how it is related to stress. Then, we introduce the linear equations that relate rate of deformation, pressure, and stress. In the context of Newtonian mechanics one takes a given state of stress as the *cause* of flow, and the rate of deformation as the *effect* of those stresses.

6.7.1 Stress and Pressure in the Viscous Liquid

We begin by describing the general state of stress in a flowing viscous liquid. Recall from Section 4.6 that the components of the stress tensor populate the elements of the 3 by 3 *matrix* as follows:

$$[\sigma] = \begin{bmatrix} \sigma_{xx} & \sigma_{xy} & \sigma_{xz} \\ \sigma_{yx} & \sigma_{yy} & \sigma_{yz} \\ \sigma_{zx} & \sigma_{zy} & \sigma_{zz} \end{bmatrix} \quad (6.30)$$

Also, Cauchy's second law of motion renders the conjugate shear stresses equal ($\sigma_{xy} = \sigma_{yx}$, $\sigma_{yz} = \sigma_{zy}$, $\sigma_{zx} = \sigma_{xz}$), so this matrix is symmetric and has only six independent elements. If the normal stresses on the diagonal of (6.30) are not all equal, extreme values of these stresses, called principal stresses, can be defined such that $\sigma_1 > \sigma_3$. The principal stresses are orthogonal and no shear stresses act on planes perpendicular to the principal stresses. However, shear stresses act on planes oblique to these principal planes. Recall from Section 5.7.1 that the maximum shear stress is one-half the difference of the extreme principal stresses:

$$\sigma_s = \tfrac{1}{2}(\sigma_1 - \sigma_3) \geq 0 \quad (6.31)$$

This shear stress acts on conjugate planes that contain the intermediate principal stress, σ_2. The normals to these two conjugate planes are at 45 degrees to the extreme principal stress axes.

If a Newtonian viscous liquid is not flowing (*static*), the shear stresses are zero on planes of all orientations. The normal stresses are compressive, due to the weight of the overlying liquid, and are of equal magnitude (**Figure 6.24a**). Under these conditions the definition of pressure is straightforward: the static pressure, $\bar{p}_o$, is defined using mechanical principles as the negative of the compressive normal stress in any one of the three coordinate directions:

$$\bar{p}_o = -\sigma_{xx} = -\sigma_{yy} = -\sigma_{zz}, \text{static} \quad (6.32)$$

Because the orientation of the coordinate system is arbitrary, the normal stresses in *all* orientations are equal, and this is called a *spherical* or isotropic stress. Under these special conditions the principal stresses are *not* defined.

The mechanical behavior of viscous liquids sometimes is described by stating that such a liquid is "incapable of supporting shear stress." This does not mean that the greatest shear stress (6.31) always is zero in a viscous liquid, but rather it means that

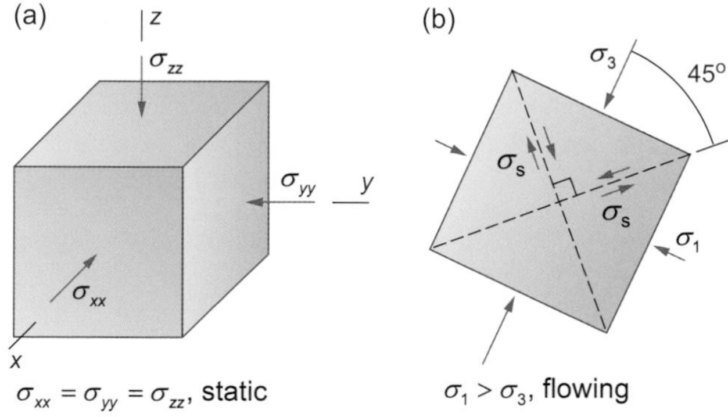

Figure 6.24 (a) small volume element with isotropic normal stress ($\sigma_{xx} = \sigma_{yy} = \sigma_{zz}$) and no shear stress, representative of a *static* liquid. Orientation of coordinate axes is arbitrary. (b) Cross section perpendicular to σ_2 through a small volume element in a *flowing* liquid. The maximum shear stress, σ_s, acts on orthogonal planes at 45 degrees to the extreme principal stress orientations.

$\sigma_s = 0$ when the liquid is static (*not flowing*). The greatest shear stress also is zero when the body of liquid is moving with a *uniform velocity*, because that implies no shearing throughout.

For the viscous liquid in *non-uniform motion*, shear stresses act on all planes except those perpendicular to the principal stress orientations (**Figure 6.24b**). Under these conditions the principal stresses are defined, and the concept of pressure loses the conventional meaning expressed in (6.32) because the normal stresses are not equal in all directions. However, the mean normal pressure, $\bar{p}$, can be defined as the negative of the mean value of the three normal stresses acting in orthogonal directions:

$$\begin{aligned}\bar{p} &= -\tfrac{1}{3}\left(\sigma_{xx}+\sigma_{yy}+\sigma_{zz}\right)\\ &= -\tfrac{1}{3}(\sigma_1+\sigma_2+\sigma_3),\ \tfrac{1}{2}(\sigma_1-\sigma_3)>0,\text{flowing}\end{aligned} \quad (6.33)$$

The sum of the normal stresses acting in any three orthogonal directions is the same. This sum does not depend upon the choice of coordinates: it is an *invariant* for a given state of stress at a point. Thus, the mean normal pressure also can be defined as the negative of the mean value of the principal stresses (6.33).

It should be understood that the mean normal pressure, $\bar{p}$, does *not* represent a spherical state of normal stress in the flowing liquid, because the normal stress components are different in different directions. To emphasize this point we consider a typical laboratory "pressure" gauge, which actually is designed to measure the magnitude of the normal traction (normal stress) acting on the sensing *surface* of the gauge. This small flat surface will have a particular orientation in the liquid depending upon how the gauge is held. In a static liquid the gauge will measure the same value, the static pressure $\bar{p}_o$, as the sensing surface is turned in different orientations at a given point, because the state of stress is *spherical* (6.32). However, in a flowing liquid the gauge will measure different values in different orientations at a given point. The values measured in three *orthogonal* directions can be used to calculate the mean normal pressure, $\bar{p}$, using (6.33). As the flowing liquid comes to rest, the mean normal pressure goes to the *static* pressure: $\bar{p} \to \bar{p}_o$.

The static pressure, $\bar{p}_o$, is defined using mechanical principles in terms of the normal stress components in (6.32), but it also may be defined using thermodynamic principles. In this context for a pure liquid at thermodynamic equilibrium, the static pressure is a function of the mass density and absolute temperature:

$$\bar{p}_o = f(\rho, T),\ \text{static} \quad (6.34)$$

This is a thermodynamic equation of state, and the liquid density, ρ, and absolute temperature, T, are the state variables. Defined in this way, we refer to $\bar{p}_o$ as the thermostatic pressure, to emphasize that it is determined by an equation of state, rather than the normal stress components.

The definition of the thermostatic pressure in (6.34) begs the question: can pressure be defined in the flowing liquid using thermodynamic principles? The answer is yes, but the definition is based on the *assumption* that such a thermodynamic pressure, p, employs the same function, $f(\rho, T)$, used for the static liquid:

$$p = f(\rho, T),\ \text{flowing} \quad (6.35)$$

Here, the mass density and absolute temperature are those in the flowing liquid, and the stress state (6.30) is taken as a combination of the thermodynamic pressure and the viscous stresses due to flow. In the next section we explicitly define this combination using constitutive equations. Here we focus on the fact that as the flow stops, the shear stresses go to zero, and the normal stresses equal the negative of the thermodynamic pressure:

$$[\sigma] = \begin{bmatrix} -p & 0 & 0 \\ 0 & -p & 0 \\ 0 & 0 & -p \end{bmatrix},\ \text{static} \quad (6.36)$$

Thus, in the static liquid, the static pressure defined in terms of the normal stresses (6.32) is equal to the thermodynamic pressure defined by the equation of state (6.34). That is, $\bar{p}_o = p$ for the static liquid. This provides the linkage between the thermodynamic concept of pressure, and the mechanical concept of pressure in a viscous liquid.

We have *not* demonstrated that the mean normal pressure, $\bar{p}$, in the flowing liquid (6.33) is equal to the thermodynamic pressure, p, in that flowing liquid (6.35). This is true only for special conditions that we define after introducing the constitutive equations in the next section.

6.7.2 Linear and Isotropic Viscous Properties

In the Newtonian context it would be tempting to think that each element of the stress matrix $[\sigma]$ in (6.30) causes only the corresponding element of the rate of deformation matrix [D] in (6.27), but that is not the case. A constitutive law for the viscous liquid continuum was developed by the Irish mathematician and hydrodynamicist George Stokes (1819–1903). He proposed that the state of stress is related to a combination of the thermodynamic pressure (6.35) and a *linear* function of the rate of deformation components (6.27). Any liquid obeying this general linear form, or simplifications of it, is referred to as a Newtonian liquid, because of Newton's insightful investigations of viscous flow.

The general linear form of the constitutive equations are simplified here for an isotropic liquid: one in which the viscous material properties are not dependent upon direction. For such a liquid the stress–rate of deformation relationships are:

$$\begin{aligned}\sigma_{xx} &= -p + 2\eta D_{xx} + \Lambda\left(D_{xx}+D_{yy}+D_{zz}\right)\\ \sigma_{yy} &= -p + 2\eta D_{yy} + \Lambda\left(D_{xx}+D_{yy}+D_{zz}\right)\\ \sigma_{zz} &= -p + 2\eta D_{zz} + \Lambda\left(D_{xx}+D_{yy}+D_{zz}\right)\\ \sigma_{xy} &= 2\eta D_{xy}, \sigma_{yz} = 2\eta D_{yz}, \sigma_{zx} = 2\eta D_{zx}\end{aligned} \quad (6.37)$$

Here p is the thermodynamic pressure (6.35), and the two material constants that characterize the viscosity of the liquid are η and Λ. Note that the constant 2η links the particular rate of deformation components to the corresponding stress component for both normal and shear stresses. The constant Λ links the sum of the normal rates of deformation to each normal stress. For dimensional homogeneity both η and Λ must have dimensions $ML^{-1}T^{-2}T$, the same as stress times time.

Important similarities and important differences exist between the constitutive equations for the isotropic liquid (6.37) and the constitutive equations for the isotropic solid that were introduced in Section 4.10. We repeat the equations for the elastic solid here, so those comparisons can be made:

$$\begin{aligned}\sigma_{xx} &= 2G\varepsilon_{xx} + \lambda(\varepsilon_{xx} + \varepsilon_{yy} + \varepsilon_{zz})\\ \sigma_{yy} &= 2G\varepsilon_{yy} + \lambda(\varepsilon_{xx} + \varepsilon_{yy} + \varepsilon_{zz})\\ \sigma_{zz} &= 2G\varepsilon_{zz} + \lambda(\varepsilon_{xx} + \varepsilon_{yy} + \varepsilon_{zz})\\ \sigma_{xy} &= 2G\varepsilon_{xy},\ \ \sigma_{yz} = 2G\varepsilon_{yz},\ \ \sigma_{zx} = 2G\varepsilon_{zx}\end{aligned} \tag{6.38}$$

For the normal stresses the last two terms on the right sides of (6.37) are similar in form to those on the right sides of (6.38), but the kinematic variables are the rate of deformation components instead of the small strain components. For the normal stresses an additional term appears on the right sides of (6.37), which is the *thermodynamic pressure*, p. For the viscous liquid at rest (zero rate of deformation) the normal stress components are equal to the negative of the thermodynamic pressure, but for the elastic solid at rest (zero strain) the normal stresses are zero. For the shear stresses the proportionality constant is $2G$ for the elastic solid (6.38) and 2η for the viscous liquid (6.37). The dimensions of the elastic constants are $ML^{-1}T^{-2}$ (the same as stress), but the dimensions of the viscous constants are $ML^{-1}T^{-2}T$ (the same as stress times time).

Perhaps the most important difference between the constitutive equations for the viscous liquid (6.37) and those of the elastic solid (6.38) is the recognition that the kinematics of viscous flow and the kinematics of elastic strain are based on different *descriptions of motion*. Viscous deformation is based on a spatial description of motion (Section 6.6.1), which focuses on particular *positions* in the flowing liquid and describes the local velocity at those positions at particular times. In contrast, the kinematics of elastic deformation is based on a referential description of motion (Section 4.9), which focuses on particular *particles* in the straining solid, and describes the displacement of those particles from their original positions to their positions at some later time. These two descriptions of motion are compared in Malvern (1969), which is listed in Further Reading.

In the next section we examine those special conditions in the flowing viscous liquid under which the mean normal pressure, $\bar{p}$, is equal to the thermodynamic pressure, p. These conditions play a prominent role in solutions for viscous flow using the linear constitutive equations (6.37).

6.7.3 Incompressible Deformation of the Linear Viscous Liquid

In the analysis of geologic structures formed by viscous deformation it commonly is postulated that the material is *incompressible*. Recall from Section 3.5.2 that the equation of continuity guarantees the conservation of mass as follows:

$$\frac{\partial\rho}{\partial t} = -\rho\left(\frac{\partial v_x}{\partial x} + \frac{\partial v_y}{\partial y} + \frac{\partial v_z}{\partial z}\right) - \left(v_x\frac{\partial\rho}{\partial x} + v_y\frac{\partial\rho}{\partial y} + v_z\frac{\partial\rho}{\partial z}\right)$$

The time rate of change of mass density at any point (left side) is determined by the *stretching* deformation of the material (terms in first parentheses on right side), and by the *flow* of material with a different density through that point (terms in second parentheses on the right side). For a material that is *homogeneous* with respect to mass density, all the terms in the second parentheses are zero. If the stretching in one coordinate direction is *exactly* compensated by shortening in the other coordinate directions, the terms in the first parentheses sum to zero, and the volume does not change. Under these conditions the time rate of change of density is zero.

The incompressible liquid behaves such that the volume at any point does not change with time. Based on the discussion in the preceding paragraph, this behavior is defined in terms of the spatial gradient of velocity components as follows:

$$\frac{\partial v_x}{\partial x} + \frac{\partial v_y}{\partial y} + \frac{\partial v_z}{\partial z} = D_{xx} + D_{yy} + D_{zz} = 0 \tag{6.39}$$

Here, the spatial gradient of velocity components are identified with the corresponding components of rate of deformation using the first three equations of (6.26). The special case (6.39) of the equation of continuity is one of the equations that governs the flow of viscous incompressible liquids: the other equations of motion are introduced in Section 6.8.

To understand the consequences of incompressible material behavior for the pressure in a *flowing* viscous liquid we calculate the mean normal pressure as defined in (6.33) using the stress components for the linear and isotropic viscous liquid (6.37):

$$\begin{aligned}\bar{p} &= -\tfrac{1}{3}(\sigma_{xx} + \sigma_{yy} + \sigma_{zz})\\ &= p - \left(\tfrac{2}{3}\eta + \Lambda\right)(D_{xx} + D_{yy} + D_{zz})\end{aligned} \tag{6.40}$$

Thus, the pressure, p, defined using thermodynamic principles (6.35), and the mean normal pressure, $\bar{p}$, defined using mechanical principles (6.33), are equal if the flowing liquid is *incompressible*:

$$\bar{p} = p, \text{if } D_{xx} + D_{yy} + D_{zz} = 0 \tag{6.41}$$

If the flow stops, or proceeds with a uniform velocity, the static pressure defined in terms of normal stresses (6.32) equals the thermostatic pressure defined by the equation of state (6.34). In this way the thermodynamic understanding of pressure is rationalized with the mechanical understanding of pressure in static and flowing liquids.

Using the *incompressible* constraint (6.39), the constitutive equations for the *linear* and *isotropic viscous liquid* (6.37) are simplified because the material constant Λ drops out and the thermodynamic pressure is replaced by the mean normal pressure:

$$\begin{aligned}&\sigma_{xx} = -\bar{p} + 2\eta D_{xx}, \sigma_{yy} = -\bar{p} + 2\eta D_{yy}, \sigma_{zz} = -\bar{p} + 2\eta D_{zz}\\ &\sigma_{xy} = 2\eta D_{xy}, \sigma_{yz} = 2\eta D_{yz}, \sigma_{zx} = 2\eta D_{zx}\end{aligned} \tag{6.42}$$

The stress components are linearly related to the rate of deformation components and the Newtonian viscosity, η, is the only material property. Note that the normal stress components revert to the negative of the mean normal pressure if $D_{xx} = D_{yy} = D_{zz} = 0$, that is in the absence of any normal rate of deformation. The shear stresses are proportional to the respective shearing rate of deformation, and the proportionality constant is twice the Newtonian viscosity, as suggested by Newton's hypothesis (6.3).

Before proceeding to the equations of motion for the Newtonian viscous liquid, we quantify the pressure in a static groundwater system, and in a static column of magma residing, for example, in a vertical dike. These pressures are evaluated and compared to the mean normal pressure in rock due to the weight of the overburden.

6.7.4 Hydrostatic, Lithostatic, and Magmastatic Pressure

Both water and magma play important roles in structural geology, and the principles of fluid statics determine the pressure in these liquids when they are not flowing. For example, consider a porous and permeable sedimentary rock saturated with water below the water table (**Figure 6.25**). The water pressure at depth D in this static groundwater system that extends up to the water table at depth W is calculated as:

$$\bar{p}_o = p_{atm} - \int_{z=-W}^{z=-D} \rho g \, dz \tag{6.43}$$

Here, the coordinate origin is at Earth's surface with z vertical and positive upward, p_{atm} is the pressure due to the atmosphere, ρ is the density of water, and g is the acceleration of gravity. This pressure is referred to as the hydrostatic pressure. At the water table the pressure ($\approx$ 0.1 MPa) due to the weight of the atmosphere ($\approx$ 10 N) usually is ignored, because it is small relative to the hydrostatic pressure at depths below the water table greater than a few tens of meters.

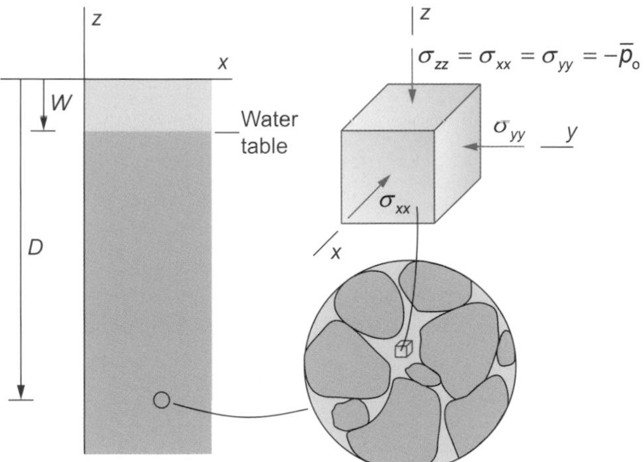

Figure 6.25 Permeable rock column saturated with water below the water table at depth W. All normal stresses in the water at depth D are equal to the negative of the hydrostatic pressure, $\bar{p}_o$.

Both the density of water and the acceleration of gravity in (6.43) vary with depth, but a useful estimate of the hydrostatic pressure may be obtained using average values and ignoring these variations. For example, the mass density of water may be taken as $\rho = 10^3$ kg m^{-3}, and the acceleration of gravity as $g = 9.8$ m s^{-2}. Then, from (6.43) the hydrostatic pressure evaluates as $\bar{p}_o \approx \rho g(D - W)$. Furthermore, if the water table nearly coincides with Earth's surface, the hydrostatic pressure is $\bar{p}_o \approx \rho g D$. Given the average near-surface values for density and acceleration of gravity mentioned above, the unit weight of water is $\rho g \approx 10^4$ N m^{-3}. Thus, at one kilometer depth with a near-surface water table, the hydrostatic pressure is $\bar{p}_o \approx 10^7$N m$^{-2} \approx 10$ MPa. We conclude that a typical gradient in hydrostatic pressure is 10 MPa per kilometer depth.

For comparison, recall that the gradient in lithostatic pressure (that due to the weight of the overlying rock) is about 25 MPa per kilometer depth. The concept of a *lithostatic* pressure assumes that the stress state at a given depth in the lithosphere is spherical (6.32), so all normal stresses in the solid rock are equal. Then, the mean normal pressure is equal to the normal stress. The unit weights of sedimentary, metamorphic, and igneous rocks only range from about 1.4×10^4 to 3.4×10^4 N m^{-3} (see Clark, 1966, in Further Reading), so the lithostatic pressure gradient is unlikely to be less than 14 MPa km^{-1}, or greater than 34 MPa km^{-1}. A gradient of 25 MPa km^{-1} is a good first estimate, but (6.43) should be used if variations in density and acceleration of gravity are known and significant.

The gradient in static pressure in a vertical, magma-filled dike that extends to Earth's surface also would be about 25 MPa km^{-1}. In keeping with the nomenclature used in this section we refer to this as the magmastatic pressure.

6.8 EQUATIONS OF MOTION FOR LINEAR VISCOUS MATERIALS

Cauchy's equations of motion for the material continuum were derived in Chapter 3. These follow from the fundamental physical laws of conservation of linear and angular momentum. Cauchy's first law (Section 3.6.1) is recalled here:

$$\begin{aligned}\rho\frac{Dv_x}{Dt} &= \frac{\partial\sigma_{xx}}{\partial x} + \frac{\partial\sigma_{yx}}{\partial y} + \frac{\partial\sigma_{zx}}{\partial z} + \rho g_x \\ \rho\frac{Dv_y}{Dt} &= \frac{\partial\sigma_{xy}}{\partial x} + \frac{\partial\sigma_{yy}}{\partial y} + \frac{\partial\sigma_{zy}}{\partial z} + \rho g_y \\ \rho\frac{Dv_z}{Dt} &= \frac{\partial\sigma_{xz}}{\partial x} + \frac{\partial\sigma_{yz}}{\partial y} + \frac{\partial\sigma_{zz}}{\partial z} + \rho g_z\end{aligned} \tag{6.44}$$

The left sides are products of mass density and acceleration, so they are forces per unit volume according to Newton's second law, $m\mathbf{a} = \mathbf{F}$. On the right sides of (6.44) we have the sum of the surface forces per unit volume due to stress, and the body forces per unit volume due to gravity. Cauchy's second law (Section 3.7.1) equates the conjugate shear stresses:

$$\sigma_{xy} = \sigma_{yx}, \; \sigma_{yz} = \sigma_{zy}, \; \sigma_{zx} = \sigma_{xz} \tag{6.45}$$

Thus, only six of the stress components are independent. However, the six stress and the three velocity components are too many dependent variables for the number of equations. To address this problem, constitutive equations (6.42) are used to write the stresses in terms of the velocities in the next section, and thus eliminate the stresses from (6.44). This leaves three dependent variables, the velocity components, and three equations.

6.8.1 Navier–Stokes Equations of Motion

The constitutive equations for the *linear*, *isotropic*, and *incompressible* material are used to derive equations of motion for the viscous liquid. Employing (6.42) and making the substitutions for the stress components on the right side of the first of (6.44) we have:

$$\frac{\partial}{\partial x}(-\bar{p}+2\eta D_{xx})+\frac{\partial}{\partial y}(2\eta D_{yx})+\frac{\partial}{\partial z}(2\eta D_{zx})+\rho g_x=$$
$$-\frac{\partial \bar{p}}{\partial x}+\eta\left[2\frac{\partial}{\partial x}\left(\frac{\partial v_x}{\partial x}\right)+\frac{\partial}{\partial y}\left(\frac{\partial v_x}{\partial y}+\frac{\partial v_y}{\partial x}\right)+\frac{\partial}{\partial z}\left(\frac{\partial v_z}{\partial x}+\frac{\partial v_x}{\partial z}\right)\right]+\rho g_x$$

On the second line, the spatial derivatives of the velocity components from (6.26) are substituted for the components of the rate of deformation. Carrying out the partial derivatives in the second line we have:

$$-\frac{\partial \bar{p}}{\partial x}+\eta\left[2\frac{\partial^2 v_x}{\partial x^2}+\left(\frac{\partial^2 v_x}{\partial y^2}+\frac{\partial^2 v_y}{\partial x \partial y}\right)+\left(\frac{\partial^2 v_z}{\partial x \partial z}+\frac{\partial^2 v_x}{\partial z^2}\right)\right]+\rho g_x$$

Three of the partial derivatives in the square brackets may be eliminated by noting that:

$$\frac{\partial^2 v_x}{\partial x^2}+\frac{\partial^2 v_y}{\partial x \partial y}+\frac{\partial^2 v_z}{\partial x \partial z}=\frac{\partial}{\partial x}\left(\frac{\partial v_x}{\partial x}+\frac{\partial v_y}{\partial y}+\frac{\partial v_z}{\partial z}\right)=0$$

Because the material is incompressible, (6.39) indicates that the term in parentheses is zero. This leaves the right side of the first of (6.44) as:

$$-\frac{\partial \bar{p}}{\partial x}+\eta\left[\frac{\partial^2 v_x}{\partial x^2}+\frac{\partial^2 v_x}{\partial y^2}+\frac{\partial^2 v_x}{\partial z^2}\right]+\rho g_x$$

This eliminates the stress components on the right side of the first of Cauchy's equations of motion (6.44).

Following this procedure for the other equations in (6.44), we have:

$$\begin{aligned}
\rho\frac{\mathrm{D}v_x}{\mathrm{D}t} &= -\frac{\partial \bar{p}}{\partial x}+\eta\left(\frac{\partial^2 v_x}{\partial x^2}+\frac{\partial^2 v_x}{\partial y^2}+\frac{\partial^2 v_x}{\partial z^2}\right)+\rho g_x \\
\rho\frac{\mathrm{D}v_y}{\mathrm{D}t} &= -\frac{\partial \bar{p}}{\partial y}+\eta\left(\frac{\partial^2 v_y}{\partial x^2}+\frac{\partial^2 v_y}{\partial y^2}+\frac{\partial^2 v_y}{\partial z^2}\right)+\rho g_y \\
\rho\frac{\mathrm{D}v_z}{\mathrm{D}t} &= -\frac{\partial \bar{p}}{\partial z}+\eta\left(\frac{\partial^2 v_z}{\partial x^2}+\frac{\partial^2 v_z}{\partial y^2}+\frac{\partial^2 v_z}{\partial z^2}\right)+\rho g_z
\end{aligned} \qquad \textbf{(6.46)}$$

Equations (6.46) are the celebrated Navier–Stokes equations for flow of a linear, isotropic, and incompressible viscous liquid with uniform mass density. Recall that we describe the material time derivatives of velocity, D()/Dt, on the left sides of (6.46) in Section 2.5.5. The Navier–Stokes equations are fundamental to the discipline of fluid mechanics (**Box 6.2**).

Each distinct term in (6.46) must have the same dimensions to satisfy the condition of dimensional homogeneity (Section 3.3.1). On the left sides, we have mass density times temporal derivatives of velocity, so the dimensions are $(\mathrm{ML}^{-3})(\mathrm{LT}^{-2}) = \mathrm{MLT}^{-2}\mathrm{L}^{-3}$. On the right sides, the first terms are spatial derivatives of the mean pressure. These have the same dimensions as (force/area)/length, so the dimensions are $(\mathrm{MLT}^{-2})(\mathrm{L}^{-2})(\mathrm{L}^{-1}) = \mathrm{MLT}^{-2}\mathrm{L}^{-3}$. The next terms are products of viscosity and second spatial derivatives of velocity. The dimensions of viscosity are the same as (force/area) × time, that is $(\mathrm{MLT}^{-2})(\mathrm{L}^{-2})(\mathrm{T})$. The second spatial derivatives of velocity have dimensions $(\mathrm{LT}^{-1})(\mathrm{L}^{-2})$. Thus, the dimensions of the products are $\mathrm{MLT}^{-2}\mathrm{L}^{-3}$. The last term is mass per unit volume times acceleration of gravity, which has dimensions $\mathrm{ML}^{-3}\mathrm{LT}^{-2} = \mathrm{MLT}^{-2}\mathrm{L}^{-3}$. Thus, all of the distinct terms have the dimensions of force per unit volume, so equations (6.46) are dimensionally homogeneous.

The dimensions of each term in the Navier–Stokes equations are those of mass times acceleration divided by volume, or force divided by volume. Using these words to reconstruct these equations in physical terms, we have:

$$\frac{\text{mass} \times \text{acceleration}}{\text{volume}}=\frac{\text{pressure force}}{\text{volume}}+\frac{\text{viscous force}}{\text{volume}}+\frac{\text{body force}}{\text{volume}} \qquad \textbf{(6.47)}$$

With this interpretation the Navier–Stokes equations reduce to Newton's second law, $m\mathbf{a} = \mathbf{F}$, as applied to the *viscous continuum*.

The four independent variables of the Navier–Stokes equations (6.46) are three spatial coordinates and time (x, y, z, t). The four dependent variables are the three velocity components and the mean pressure $(v_x,\ v_y,\ v_z,\ \bar{p})$. The four equations for the four unknowns are the Navier–Stokes equations (6.46) and the equation of continuity (6.39). These equations are solved for viscous flow with a prescribed geometry subject to boundary and initial conditions on the three velocity components and mean pressure. The velocity components and mean pressure everywhere in the flowing liquid, which are functions of the spatial coordinates and time, are the solution to this boundary and initial value problem:

$$v_x = v_x(x,y,z,t),\ v_y = v_y(x,y,z,t),\ v_z = v_z(x,y,z,t),$$
$$\bar{p} = \bar{p}(x,y,z,t) \qquad \textbf{(6.48)}$$

Once the velocity components and mean pressure are determined, the kinematic equations (6.26) are used to calculate the rate of deformation components. Then, the constitutive equations (6.42) are used to calculate the stress components. In this way all of the relevant physical variables (velocity, mean pressure, rate of deformation, and stress) are accounted for as functions of the spatial coordinates and time.

The Navier–Stokes equations (6.46) are the subject of classic textbooks in hydrodynamics and fluid mechanics (see Langlois,

Box 6.2 Fluid Mechanics

Fluid mechanics is the study of liquids, gases, and plasmas in terms of their physical properties, such as density and viscosity, when treated as a material continuum (Section 3.4) subject to the fundamental constraints of mass, momentum, and energy conservation (Sections 3.5 through 3.8). Thus, fluid mechanics is a sub-discipline of continuum mechanics. Fluid mechanics is broadly divided into fluid statics (Section 6.7.4), which considers fluids at rest, and fluid dynamics (Section 6.8), which considers the action of forces on fluids, the flow of fluids, and their resistance to flow. Viscous fluid flows are characterized as laminar or turbulent (Section 6.8.2). Applications of fluid mechanics occur in many disciplines, including civil, chemical, aeronautical, and biomedical engineering, physics, biology, geology, and geophysics.

Structural geologists, who use concepts from fluid mechanics, typically focus on static liquids and viscous liquids in laminar flow. For example, we introduced the concepts of gravitational force and buoyant force in Section 1.1.2 to understand why a body of salt is static at a level of neutral buoyancy in a sedimentary basin. The dynamics of viscous liquids has applications ranging from the flow of debris and lava on Earth's surface, to the flow of magma in volcanic conduits (Section 6.8.3), to the folding of rock strata in the lithosphere (Section 9.7), to the solid-state flow of rock in the Earth's lower lithosphere and asthenosphere (Section 6.4.4). These applications indicate the relevance of fluid mechanics for addressing many important problems in structural geology, and make the case for learning the basic concepts and tools of fluid mechanics during an introductory course in structural geology.

The canonical model for Chapter 11 on intrusions is the rise of a sphere of dense, viscous fluid in a surrounding fluid of greater density and different viscosity (Section 11.6). The streamlines inside the sphere (**Figure B6.2**) illustrate the flow pattern due to viscous drag against the surrounding fluid. The solution to the Navier–Stokes equations provides these streamlines (**Figure B6.2a**), and laboratory experiments (**Figure B6.2b**) corroborate their form.

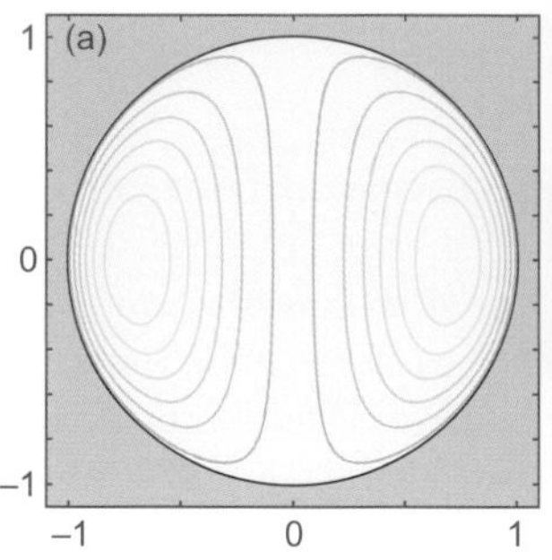

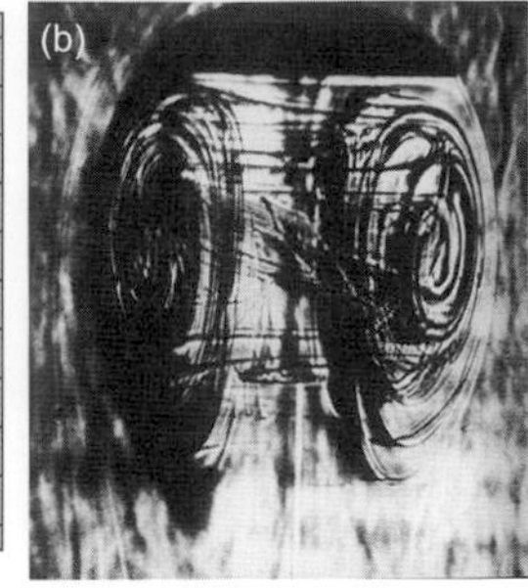

Figure B6.2 Meridional view of a rising sphere of viscous fluid in viscous surroundings. (a) Calculated streamlines using Navier–Stokes equations of motion. (b) Visualized streamlines in laboratory experiment. From Van Dyke (1982).

The founders of fluid mechanics were physicists, engineers, and applied mathematicians. For example, Newton (Newton, Cohen, and Whitman, 1999), who proposed the linear relationship between applied force (shear stress) and velocity gradient (rate of deformation) (Section 6.1) was a mathematician and natural philosopher. Navier and Stokes, whose names are associated with the foundational equations of viscous fluid dynamics (Section 6.8.1) were, respectively, a mechanical engineer and a physicist. This discipline is associated with so many fields of science and engineering today that the authors of modern textbooks on fluid mechanics come from many different disciplines (e.g. Happel and Brenner, 1965; Turcotte and Schubert, 2002; Bird et al., 2007; see Further Reading). For structural geologists the challenge in applying fluid mechanics lies, in part, in sorting through this vast literature to find those concepts and solutions that are relevant to their particular problem.

1964; Happel and Brenner, 1965; Bird et al., 2007, in Further Reading). They are the basis for a host of applications in physics, engineering, and the earth sciences that deal with the flow of viscous materials. Especially important in the context of structural geology is slow viscous flow, which also is referred to as *creeping flow* or *low Reynolds number flow*. For slow viscous flow, the products of mass density and the material time derivative of velocity on the left sides of (6.46) are considered negligible compared to the terms on the right sides. Therefore, the left sides are eliminated, leaving the equations of motion for slow viscous deformation:

$$\begin{aligned} 0 &= -\frac{\partial \bar{p}}{\partial x} + \eta\left(\frac{\partial^2 v_x}{\partial x^2} + \frac{\partial^2 v_x}{\partial y^2} + \frac{\partial^2 v_x}{\partial z^2}\right) + \rho g_x \\ 0 &= -\frac{\partial \bar{p}}{\partial y} + \eta\left(\frac{\partial^2 v_y}{\partial x^2} + \frac{\partial^2 v_y}{\partial y^2} + \frac{\partial^2 v_y}{\partial z^2}\right) + \rho g_y \\ 0 &= -\frac{\partial \bar{p}}{\partial z} + \eta\left(\frac{\partial^2 v_z}{\partial x^2} + \frac{\partial^2 v_z}{\partial y^2} + \frac{\partial^2 v_z}{\partial z^2}\right) + \rho g_z \end{aligned} \quad (6.49)$$

Dimensional analysis, as carried out for (6.46), emphasizes that each term on the right sides of equations (6.49) is equivalent to a force per unit volume. The first terms are forces due to gradients in the mean pressure; the second terms are forces due to the viscous resistance to flow; and the third terms are forces due to gravity. These forces must sum to zero in each coordinate direction to maintain quasi-static equilibrium.

Applications of (6.49) include the folding of stratified sedimentary rock and foliated metamorphic rock (Chapter 9), the flow of magma in igneous intrusions, and the flow of salt in diapirs (Chapter 11). In these applications mass density, ρ, acceleration of gravity, $\mathbf{g}$, and viscosity, η, commonly are taken as given by laboratory or field data. These viscous flow problems have solutions that yield distributions of velocity in rather simple patterns. However, with increased velocity, flowing viscous liquids also can devolve into extremely complex patterns. The transition from simple to complex patterns of flow is the subject of the next section.

6.8.2 Laminar and Turbulent Flow

Osborne Reynolds published the defining experiments in 1883 that identified the transition from simple to complex patterns of flow of viscous fluids. In these experiments Reynolds monitored the flow of a viscous liquid through a cylindrical tube and injected a very narrow stream of black dye at different positions in the tube to identify the nature of the flow patterns (**Figure 6.26**). The liquid density and viscosity were controlled to be uniform and constant throughout the tube, which had a uniform internal diameter, and the velocity at the center of the tube was adjustable by changing the pressure in a large reservoir that fed liquid into the tube.

At relatively low velocity, the dye stream was perfectly straight throughout the length of the tube for a given injection point (**Figure 6.26a**). The dye moved more rapidly if injected nearer to the center of the tube, but it always moved along a straight line. This pattern is defined as laminar flow because the liquid moved as though in narrow laminae parallel to the wall of the tube, and never mixed with adjacent laminae. However, when Reynolds increased the flow velocity the straight dye stream began to waiver (**Figure 6.26b**), and then the waves began to curl over. This marked the beginning of the transition to turbulent flow. As the velocity increased the stream of dye took on a very convoluted geometry with intricate eddies made visible by the dye (**Figure 6.26c**). Eventually the dye mixed with the viscous liquid in a very complicated pattern from one side of the tube to the other.

For a geological example of the flow of viscous magma in a nearly parallel-sided conduit we refer to the sills at Shonkin Sag, Montana (**Figure 6.27**). These intrusions are thin horizontal sheets of nearly constant thickness that emanate from a larger intrusion called the Shonkin Sag laccolith. That laccolith presumably played a role similar to the large reservoir of liquid used by Reynolds in the laboratory experiment (**Figure 6.26**). Each sill played a role similar to the tube into which Reynolds injected the dye. No markers reveal the flow pattern in the sills, but we use dimensional analysis and the results of Reynolds experiments to evaluate whether the flow in the Shonkin sills was laminar or turbulent.

The Navier–Stokes equations (6.46) are the governing equations for viscous flow. The conceptual model for flow in the Shonkin sills is a parallel-sided conduit of half-height, H, filled with a viscous liquid (**Figure 6.28**). The Newtonian viscosity, η, measures the resistance to flow and is postulated to be uniform in space and constant in time. The mass density of the liquid, ρ, also is taken as uniform and constant. The length of the conduit, L, parallel to the x-axis, and the width of the conduit, W, parallel to the y-axis, are very great compared to the height. The pressure, p, is the second dependent variable of this problem and it can vary with position, x, along the direction of flow, and with time. The pressure drop, $p_1 > p_2$, drives the flow from left to right.

Following Reynolds observations in **Figure 6.26a**, we postulate that the only non-zero component of velocity, v_x, is directed along the length of the conduit (**Figure 6.28**), so $v_y = 0 = v_z$. This

Figure 6.26 Images of the experiments used to define the transition from laminar to turbulent flow of a constant property viscous liquid flowing in a cylindrical tube. Velocity at the tube center increases from (a) to (b) to (c). From O. Reynolds (1883).

Figure 6.27 Photograph of Shonkin Sag laccolith (left) with sills (right) that intruded laterally into the horizontal sedimentary rock. See Pollard et al. (1975). Approximate UTM: 12 T 554410.37 m E, 5265511.19 m N.

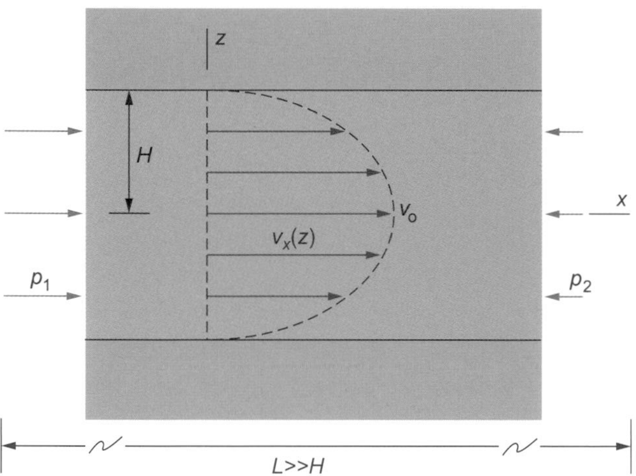

Figure 6.28 Conceptual model for flow of magma in a sill of uniform half-height, H, length parallel to x, $L >> H$, and width parallel to y, $W >> H$, due to a pressure drop, $p_1 > p_2$. For the Newtonian liquid the velocity distribution, $v(z)$, is parabolic with a maximum at the center $v_x(z{=}0) = v_o$. See Bird et al. (2007).

is consistent with the laminar flow regime observed at low velocities, but clearly will not describe the turbulent flow regime in **Figure 6.26c**. If the velocity components in y and z are zero, the continuity equation (6.39) reduces to:

$$\frac{\partial v_x}{\partial x} = D_{xx} = 0 \tag{6.50}$$

Thus, the velocity in the x-direction is not changing along the conduit. Writing out these conditions on the velocity we have:

$$v_x = v_x(z, t)\,\text{only}, \quad v_y = 0 = v_z \tag{6.51}$$

The velocity component, v_x, is one of the dependent variables in the parallel-sided conduit used to model magma flow in a sill.

The conditions of zero velocity in y and z eliminate the left sides and all the terms in parentheses on the right sides of the second and third Navier–Stokes equations (6.46). The condition that the x-component of velocity is only a function of z and t, eliminates the first two terms in parentheses on the right side of the first equation (6.46). Recall from Section 3.6 that the material time derivative for the x-component of velocity is expanded as:

$$\frac{Dv_x}{Dt} = \frac{\partial v_x}{\partial t} + v_x\frac{\partial v_x}{\partial x} + v_y\frac{\partial v_x}{\partial y} + v_z\frac{\partial v_x}{\partial z}$$

The conditions on this velocity component in (6.51) eliminate all but the first term on the right side. Finally, the gravitational force acts only in the z-direction, so the body force per unit volume is eliminated from the first and second of (6.46), and $g_z = -g$, the gravitational acceleration. Thus, the Navier–Stokes equations reduce from (6.46) to the following three equations:

$$\rho\frac{\partial v_x}{\partial t}=-\frac{\partial\bar{p}}{\partial x}+\eta\frac{\partial^2 v_x}{\partial z^2},0=-\frac{\partial\bar{p}}{\partial y},0=-\frac{\partial\bar{p}}{\partial z}-\rho g \quad (6.52)$$

These three equations and the continuity equation (6.50) govern flow of the viscous liquid in the horizontal, parallel-sided conduit (**Figure 6.28**).

In the first of (6.52), the velocity component in x and mean normal pressure defined in (6.40) are the two dependent variables. The three independent variables are the x- and z-coordinates and time, and the two constants are mass density and viscosity. For the dimensional analysis each of the variables must be normalized by a physical quantity that shares the same dimensions. It is customary in fluid dynamics to select a characteristic distance and a characteristic velocity for this purpose. Here the characteristic distance is the half-height of the conduit, H. We choose the velocity, v_o, at the center of the conduit to be the characteristic velocity. This is the maximum velocity, but the selection is arbitrary, so we could have selected the average velocity. The normalized independent variables are:

$$x^*=\frac{x}{H},\quad z^*=\frac{z}{H},\quad t^*=\frac{v_o}{H}t$$

The characteristic velocity and distance are used in the ratio v_o/H to define a dimensionless time. The normalized independent variables are:

$$v_x^*=\frac{v_x}{v_o},\quad p^*=\frac{\bar{p}-p'}{\rho v_o^2}$$

A reference pressure, p', is subtracted from the pressure, and $\rho(v_o)^2$ is used to normalize this reduced pressure. The reference pressure could be that at the entrance to the channel. The differential operators are normalized as follows:

$$\frac{\partial}{\partial t^*}=\frac{H}{v_o}\frac{\partial}{\partial t},\quad \frac{\partial}{\partial x^*}=H\frac{\partial}{\partial x},\quad \frac{\partial^2}{\partial(z^*)^2}=H^2\frac{\partial^2}{\partial z^2}$$

The normalized variables and differential operators are substituted into the governing equation (6.52) to find:

$$\rho\frac{v_o}{H}\frac{\partial}{\partial t^*}(v_o v_x^*)=-\frac{1}{H}\frac{\partial}{\partial x^*}(\rho v_o^2 p^*+p')+\eta\frac{1}{H^2}\frac{\partial^2}{\partial(z^*)^2}(v_o v_x^*)$$

Bringing the constants outside the derivatives and eliminating the derivative of the constant reference pressure, the normalized equation of motion becomes:

$$\frac{\rho v_o^2}{H}\frac{\partial v_x^*}{\partial t^*}=-\frac{\rho v_o^2}{H}\frac{\partial p^*}{\partial x^*}+\frac{\eta v_o}{H^2}\frac{\partial^2 v_x^*}{\partial(z^*)^2} \quad (6.53)$$

The two different ratios of physical constants that multiply the partial derivatives in (6.53) have dimensions of force per unit volume:

$$\frac{\rho v_o^2}{H}\{=\}\mathrm{MLT^{-2}L^{-3}},\ \text{inertial force per unit volume} \quad (6.54)$$

$$\frac{\eta v_o}{H^2}\{=\}\mathrm{MLT^{-2}L^{-3}},\ \text{viscous force per unit volume} \quad (6.55)$$

Each ratio is associated with the magnitude of a different force.

The dimensional ratios identified in (6.54) and (6.55) are called scale factors, because they indicate how the flow depends upon the constants. For example, the viscous forces are related to the product of the viscosity and velocity divided by the square of the characteristic distance. If the viscosity doubles, but the conduit half-height and maximum velocity are unchanged, we would expect the viscous forces to double. If the height of the conduit is cut in half, but the viscosity and velocity are unchanged, we would expect the viscous forces to increase by a factor of four. Similarly, if the velocity is doubled, the inertial forces would increase by a factor of four, but the viscous forces would only double.

Dividing each term in (6.53) by $\rho v_o^2/H$, a single dimensionless group is identified:

$$\frac{\partial v_x^*}{\partial t^*}=-\frac{\partial p^*}{\partial x^*}+\left[\frac{\eta v_o/H^2}{\rho v_o^2/H}\right]\frac{\partial^2 v_x^*}{\partial(z^*)^2}$$

This group is a ratio of the viscous force (6.55) to the inertial force (6.54).This dimensionless group usually is written as the reciprocal of the form given here, and is called the Reynolds number:

$$\text{Reynolds number}=\mathrm{Re}=\frac{Hv_o\rho}{\eta},\quad \frac{\text{inertial force}}{\text{viscous force}} \quad (6.56)$$

The magnitude of the Reynolds number can be used to characterize the transition in style of flow between two dramatically different flows as demonstrated in the classic experiments by Reynolds (**Figure 6.26**). The transition from laminar to turbulent flow in a pipe occurs at Reynolds numbers ranging from 2,000 to 13,000, depending upon the roughness of the pipe and the geometry of the entrance. When the product of the diameter, velocity and density, divided by the viscosity is less than 2,000, viscous forces dominate over inertial forces and the flow is laminar flow. In the literature of fluid dynamics this is referred to as *low Reynolds number flow* (see Happel and Brenner, 1965).

The great viscosity of magma, relative to typical products of conduit width, velocity, and density, usually puts magmatic intrusions in the category of low Reynolds number flow. For example, the basaltic lava in fissure eruptions on Kilauea Volcano has a viscosity of η ~ 1,300 Pa s at 1135 °C. The mass density of the lava is estimated as ρ ~ 2.7×10^3 kg m^{-3}, a typical velocity for dike propagation is v_o ~ 0.1 m s^{-1}, and a representative dike half thickness is H ~ 1 m. Calculating Reynolds number we find:

$$\mathrm{Re}\sim\frac{Hv_o\rho}{\eta}\sim\frac{(1\,\mathrm{m})(0.1\,\mathrm{m\,s^{-1}})(2.7\times10^3\,\mathrm{kg\,m^{-3}})}{1.3\times10^3\,\mathrm{kg\,m^{-1}\,s^{-1}}}\sim0.2$$

For the values of the parameters used here the flow of this lava in a dike feeding a fissure eruption should be laminar (Re << 2,000). However, very dynamic flow conditions at Earth's surface during an eruption could involve turbulence.

Figure 6.29 Xenoliths of Mancos Shale and basalt near contact of northeastern dike at Ship Rock, NM. See **Figure 6.19** for location. Dike contact is parallel to bottom edge of photograph. Alignment of long dimension of xenoliths is suggestive of laminar (low Reynolds number) flow approximately parallel to the contact.

Direct evidence for laminar flow in volcanic and igneous rock includes the alignment of phenocrysts, individual mineral grains that crystallized early in the history of the magma and were carried to their current location by flow of the magma. The regular patterns and alignments of xenoliths, fragments or blocks of the host rock that were caught up in the flowing magma, also is evidence for laminar flow (**Figure 6.29**). In still other cases layers of different composition within a magma body reveal the organized patterns of laminar flow. In the absence of these fabrics, the inference that magma flow was laminar usually is based on calculations of Reynolds number, which requires measurements or estimates of a characteristic length and velocity, as well as mass density and viscosity (6.56).

6.8.3 Steady Laminar Flow in a Sill

All of the velocity fields that are quantified using the equations of motion in this chapter, including that for Newton's hypothesis (**Figure 6.3**), and for the Couette–Hatschek viscometer (**Figure 6.11**), and for flow in the sills at Shonkin Sag (**Figure 6.27**), are examples of steady viscous flow. By steady we mean that the *local velocity* at any fixed position, as given by the spatial description of motion, does not change with time. Note that this condition constrains the local velocity and not necessarily the velocity of individual particles of liquid, which may vary with time, as we showed when introducing the material time derivative in Section 2.5.5.

Steady flow is a condition that facilitates solving the equations of motion, and makes it easier to understand the relationships among the independent and dependent variables. This provides a strong incentive to employ the steady flow condition, but one should understand that this is a special case. For example, before the viscometer illustrated in **Figure 6.11** is turned on, the liquid is at rest. Once the motor is turned on, a *starting* phase begins as the motion of the cylindrical crucible is transferred to the liquid, which adheres to the wall of the crucible. The motion of the cylindrical lamina of fluid next to the crucible is transferred to the next adjacent lamina, and so on, until the entire liquid is in motion, except for the inner-most lamina that adheres to the fixed rod. During this starting phase, at any given radius within the liquid, the velocity increases with time, so this is *not* steady flow. Eventually, when the velocity at any particular radius stops changing with time, steady flow is achieved. When the motor is turned off, a *stopping* phase occurs as the velocity decreases to zero throughout the crucible.

Adopting the steady flow condition, the left side of the first Navier–Stokes equation in (6.52) goes to zero, so we have:

$$\frac{\partial \bar{p}}{\partial x} = \eta \frac{\partial^2 v_x}{\partial z^2}, \ \frac{\partial \bar{p}}{\partial y} = 0, \ \frac{\partial \bar{p}}{\partial z} = -\rho g \qquad \textbf{(6.57)}$$

These are the governing equations for steady laminar flow in one horizontal direction, x, through a parallel-sided conduit (**Figure 6.28**) of a linear viscous fluid with uniform and constant mass density, ρ, and viscosity, η, subject to a uniform and constant acceleration of gravity, g. The dependent variables are the velocity, v_x, which is only a function of z, and the pressure, $\bar{p}$, which is a function of x and z. The independent variables are the two coordinates, x and z.

We seek an explicit relationship that, for example, enables a structural geologist to use field measurements of sill thickness, along with laboratory measurements of viscosity and density, to estimate the magma velocity for a given pressure gradient in x, and to understand how the velocity and pressure are distributed throughout the sill. Such estimates are important for assessing how quickly magma might flow through the brittle part of Earth's lithosphere and for the mitigation of volcanic hazards.

Noting that mean pressure is a function of both x and z, the third of (6.57) is rearranged and integrated with respect to z to find:

$$\bar{p} = -\rho g z + f_1(x) \qquad \textbf{(6.58)}$$

Integration of (6.57) introduces $f_1(x)$, a function only of x, which is the unknown distribution of pressure along the middle of the conduit where $z = 0$. Taking the derivative of this function with respect to x and substituting into the first of (6.57) we have:

$$\frac{\mathrm{d}f_1(x)}{\mathrm{d}x} = \eta\left(\frac{\partial^2 v_x}{\partial z^2}\right)$$

Integrating with respect to z:

$$z\frac{\mathrm{d}f_1(x)}{\mathrm{d}x} + f_2(x) = \eta\left(\frac{\partial v_x}{\partial z}\right)$$

This integration introduces $f_2(x)$, a second unknown function of x. At the middle of the conduit, symmetry requires the change of velocity with respect to z to be zero, and this determines $f_2(x)$:

$$\text{at } z = 0, \frac{\partial v_x}{\partial z} = 0, \text{ so } f_2(x) = 0$$

Integrating once more with respect to z, we introduce $f_3(x)$, a third unknown function of x:

$$\tfrac{1}{2}z^2\frac{\mathrm{d}f_1(x)}{\mathrm{d}x} + f_3(x) = \eta v_x \quad (6.59)$$

We find a relationship between $f_1(x)$ and $f_3(x)$ by taking the derivative of (6.59) with respect to x, and recalling from the continuity equation (6.50) that $\partial v_x/\partial x = 0$:

$$\tfrac{1}{2}z^2\frac{\mathrm{d}^2 f_1(x)}{\mathrm{d}x^2} + \frac{\mathrm{d}f_3(x)}{\mathrm{d}x} = \eta\frac{\partial v_x}{\partial x} = 0$$

Rearranging, we have:

$$\tfrac{1}{2}z^2\frac{\mathrm{d}^2 f_1(x)}{\mathrm{d}x^2} = -\frac{\mathrm{d}f_3(x)}{\mathrm{d}x}$$

This equality holds for any choice of x and z within the conduit, only if both derivatives are zero, which implies:

$$\frac{\mathrm{d}^2 f_1(x)}{\mathrm{d}x^2} = \mathrm{C}_1 \text{ and } f_3(x) = \mathrm{C}_2$$

Here C_1 and C_2 are constants, which are substituted into the equation for velocity (6.59) to find:

$$\tfrac{1}{2}z^2\mathrm{C}_1 + \mathrm{C}_2 = \eta v_x \quad (6.60)$$

Taking the second derivative of (6.60) with respect to z, and using the first of (6.57), identifies C_1 as the gradient in pressure with respect to x:

$$\mathrm{C}_1 = \eta\frac{\partial^2 v_x}{\partial z^2} = \frac{\partial \bar{p}}{\partial x} \quad (6.61)$$

The conventional boundary condition for viscous fluids at a stationary boundary is that the velocity is zero, because the fluid adheres to the solid surface. This is called the no slip boundary condition. We invoke this boundary condition at both surfaces of the conduit and use (6.60) and (6.61) to find:

$$\text{at } z = \pm H, v_x = 0, \text{ so } \mathrm{C}_2 = -\tfrac{1}{2}H^2\mathrm{C}_1 = -\tfrac{1}{2}H^2\frac{\partial \bar{p}}{\partial x} \quad (6.62)$$

Substituting for the two constants in (6.60) using (6.61) and (6.62), the velocity distribution throughout the model sill is:

$$v_x = \frac{1}{2\eta}\frac{\partial \bar{p}}{\partial x}\left(z^2 - H^2\right) \quad (6.63)$$

Note that the velocity is directly proportional to the pressure gradient in x and inversely proportional to the viscosity. The velocity is positive if the mean pressure gradient is negative, that is this pressure decreases in the positive x-direction. The velocity falls to zero at the conduit surfaces ($z = \pm H$) as required by the *no slip* boundary condition. Gravitational forces play no role in driving magma through the model sill, because it is horizontal, so flow is entirely due to the pressure gradient along the middle of the sill.

Returning to the pressure distribution (6.58), we now know from (6.61) that the gradient in pressure with respect to x is a constant, so the unknown function $f_1(x)$ is linear in x: $f_1(x) = \mathrm{C}_1 x + \mathrm{C}_0$. Thus, the mean pressure distribution throughout the model sill is:

$$\bar{p} = -\rho g z + \frac{\mathrm{d}\bar{p}}{\mathrm{d}x}x + \mathrm{C}_0 \quad (6.64)$$

The constant C_0 must be fixed at some arbitrary position, but the velocity of the fluid is only dependent on the gradient in pressure, so the velocity does not depend on C_0.

The velocity has a *parabolic* distribution, with a maximum at mid-height where:

$$\max(v_x) = v_x(z = 0) = v_o = -\frac{H^2}{2\eta}\frac{\partial \bar{p}}{\partial x} \quad (6.65)$$

For example, consider a sill 2 m thick and 1 km long containing a basaltic magma with viscosity 10^3 Pa s. Suppose the magma pressure at the feeder dike was 2 MPa and fell to zero at the tip-line, over a distance of 1 km. The maximum velocity would be:

$$v_o = -\frac{(1\,\mathrm{m})^2}{2(10^3\,\mathrm{Pa\,s})}\frac{-(2\times 10^6\,\mathrm{Pa})}{(10^3\,\mathrm{m})} = 1\,\mathrm{m\,s^{-1}} \quad (6.66)$$

This is the order of magnitude for the velocity of magma in propagating dikes on Kilauea Volcano. Given a mass density for basaltic magma of 2.6×10^3 kg m^{-3}, Reynolds number for flow in the model sill is:

$$\mathrm{Re} = \frac{2Hv_o\rho}{\eta} = \frac{(2\,\mathrm{m})(1\,\mathrm{m\,s^{-1}})(2.6\times 10^3\,\mathrm{kg\,m^{-3}})}{1\times 10^3\,\mathrm{Pa\,s}} \approx 5 \quad (6.67)$$

This is more than two orders of magnitude less than the nominal transition to turbulent flow. Considering that few sills or dikes exceed several tens of meters in thickness and magma viscosities rarely are much less than 10^3 Pa s, laminar flow is likely for sheet intrusions.

Normalizing the velocity distribution (6.63) by the maximum velocity (6.65) we have a dimensionless velocity to plot versus a dimensionless distance across the model sill:

$$v_x/v_o = 1 - (z/H)^2 \quad (6.68)$$

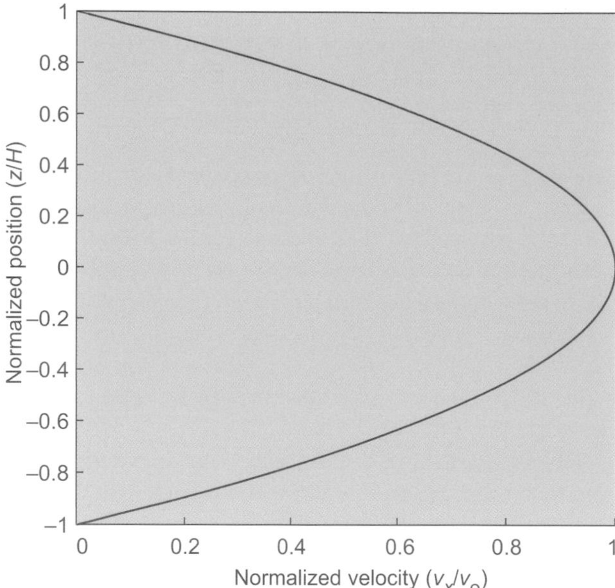

Figure 6.30 Position across a model sill normalized by the half height, *H*, plotted versus velocity normalized by the velocity at the center. The distribution is parabolic. Modified from Malvern (1969).

The parabolic velocity distribution is illustrated in **Figure 6.30**. The normalized velocity is greatest at the center of the model sill where $z/H = 0$, and falls to zero at both sill contacts where $z/H = \pm 1$.

The velocity gradient in the model sill is found by taking the derivative of (6.63) with respect to z:

$$\frac{dv_x}{dz} = \frac{z}{\eta}\frac{\partial \bar{p}}{\partial x} \tag{6.69}$$

At the middle of the model sill where $z = 0$ the velocity gradient is zero. At the conduit surfaces where $z = \pm H$, the velocity gradient is proportional to the height and the pressure gradient, and inversely proportional to the viscosity.

The volume rate of flow, Q, through the model sill is found by integrating the velocity over the half-height, and then multiplying by 2 and by a constant width, W, perpendicular to the (x, z)-plane:

$$\begin{aligned} Q &= 2W\int_0^H v_x dz = \frac{W}{\eta}\frac{\partial \bar{p}}{\partial x}\int_0^H \left(z^2 - H^2\right)dz \\ &= \frac{W}{\eta}\frac{\partial p}{\partial x}\left(\frac{1}{3}z^3 - H^2 z\right)\Big|_0^H = -\frac{2WH^3}{3\eta}\frac{\partial p}{\partial x} \end{aligned} \tag{6.70}$$

Equation (6.70) provides the scaling relations for volumetric flow rate. The volumetric flow rate is proportional to the half-height cubed, the conduit width, and the pressure gradient. It is inversely proportional to the viscosity. Using the values of the quantities considered above for Kilauea Volcano, and considering the volumetric flow rate over a width in the y-direction of 1 m, we have:

$$Q = \frac{-2(1\,\mathrm{m})(1\,\mathrm{m})^3}{3(10^3\,\mathrm{Pa\,s})}\cdot\frac{-2\times 10^6\,\mathrm{Pa}}{10^3\,\mathrm{m}} = \tfrac{4}{3}\mathrm{m}^3\,\mathrm{s}^{-1} \tag{6.71}$$

In other words, the feeder dike would deliver just over one cubic meter of magma to this one meter wide part of the sill every second.

Despite the success of Newton's hypothesis that established the general linear relationship between stress components and rate of deformation components for the isotropic viscous material, modern experimental evidence suggests that consideration of non-linear relationships may be important. To address this we consider the flow of a power-law magma in a sill in Section 11.2.3.

What we have derived in this chapter considers *isothermal* flow, so the temperature is constant in time and uniform throughout the model sill. This postulate is appropriate if the time scale for intrusion is less than that for significant heat loss to the host rock. In Section 11.1.5 we address this loss of heat by conduction from a model dike. Conditions will be established there under which changes in magma viscosity due to temperature changes may be ignored, and those under which heat transfer and variable viscosity must be included in the analysis.

Recapitulation

Viscous liquids flow down slopes, fill depressions, and maintain a horizontal upper surface when at rest. That molten rock behaves approximately as a *viscous liquid* is confirmed by direct observations on volcanoes where basaltic lava flows satisfy these criteria. Other Earth materials, for example salt and ice, are crystalline and behave like an elastic solid on short time scales, but flow like viscous liquids on long time scales. These materials are *viscoelastic solids*. For example, indirect evidence, such as the elevated shorelines around former glacial lakes, indicate that Earth's asthenosphere behaves like a viscoelastic solid with a viscosity of about 10^{20} Pa s on a time scale of several thousand years. For comparison, the viscosity of glacial ice is about 10^{13} Pa s, that of Hawaiian lava is about 10^3 Pa s, and that of water is about 10^{-3} Pa s. This huge range of viscosities makes the viscous properties of Earth materials both interesting and challenging for structural geologists.

Laboratory measurements of viscosity for molten rock demonstrate that composition in terms of weight percent silica plays an important role, with basalts being orders of magnitude less viscous than rhyolites. Viscosity also is very sensitive to temperature, decreasing by orders of magnitude as temperature increases by several hundred degrees Celsius. Dissolved water, ranging from zero to 10% by weight, causes a decrease in viscosity by orders of magnitude. As lava and magma cool, minerals crystallize and volatiles form bubbles, both of which increase the viscosity and make the mechanical properties more complex than that of a purely viscous liquid. Laboratory measurements define the parameters of flow laws for solid-state flow that relate strain rate to differential stress, grain size, and temperature. If the stress exponent is approximately one (Newtonian) the mechanism for solid-state flow is *diffusion creep*, either through the crystal lattice (Nabarro–Herring creep) or along grain boundaries (Coble creep).

Quantification of the concept of viscosity is attributed to Newton who proposed that the resistance to flow (now measured by the shear stress) is proportional to the velocity with which adjacent laminae of liquid are separated (now measured by the shearing rate of deformation). For the Newtonian liquid at rest the static pressure is equal to the spherical normal stress, and when that liquid is flowing, a proxy for pressure is the mean normal pressure. The kinematics of flow are described using a *spatial* description of motion that focuses on particular positions and keeps track of the local velocity at those positions. The spatial gradient of velocity is used to calculate the rate of deformation (stretching) and the rate of spin (vorticity) in the flowing liquid.

The constitutive equations for the *Newtonian viscous liquid* relate the stress tensor to the rate of deformation tensor. The rate of spin tensor plays no role in these equations because it represents a rigid rate of rotation. The linear relationships between components of the stress tensor and the rate of deformation tensor usually are simplified by postulating that the liquid is incompressible, in which case the only material property is the Newtonian viscosity, and this is not a function of orientation, so the liquid is isotropic. For the incompressible liquid, the pressure defined using thermodynamic principles is equal to the mean normal pressure in the flowing liquid. Using this incompressible and isotropic form of the constitutive equations, Cauchy's first law of motion reduces to the legendary *Navier–Stokes equations* of motion. Because the viscosity of molten rock is very great, the special case of the Navier–Stokes equations applicable to creeping flow usually is adopted and the flow is considered *laminar* (low Reynolds number flow), not turbulent.

As an example of a solution to the Navier–Stokes equations, we consider magma flow in a sill. The model magma is postulated to be a Newtonian viscous liquid with uniform and constant density and viscosity. If the viscosity is sufficiently great, or the velocity sufficiently small, the flow is laminar. For a horizontal sill with parallel and planar contacts, gravity effects the pressure distribution with height, but it is the imposed pressure gradient along the length of the sill that drives the flow. We consider steady flow, so the local velocity does not change with time. For this problem the distribution of velocity across the sill is parabolic. The maximum velocity is proportional to the pressure drop and the square of the sill thickness, and it is inversely proportional to the viscosity.

REVIEW QUESTIONS

The following questions are designed to highlight the expected *learning outcomes* for this chapter. Each question is taken directly from the material in the chapter and, for the most part, in the same sequence that it appears in the chapter. If an answer is not forthcoming, students are advised to read the relevant section of the chapter and discover the answer.

6.1. Describe the thought experiment in **Figure 6.3** that was used to introduce Newtonian viscosity, η. In particular, discuss the physical quantities in the linear relationship, $\sigma_{zx} = 2\eta D_{zx}$, their relationship to the illustrated experiment, and Newton's concepts of "lack of lubricity" and "the velocity with which the parts of the fluid are separated from one another."

6.2. Hooke's law of elasticity and Newton's law of viscosity may be written for one shear component of the stress tensor as:

$$\sigma_{zx} = G\frac{\partial u_x}{\partial Z} \text{ and } \sigma_{zx} = \eta\frac{\partial v_x}{\partial z}$$

Compare these two classic relationships by describing each variable and each constant, including their SI units. Contrast the referential and spatial descriptions of motion in the context of these two relationships.

6.3. The following Google Earth image is taken from the file: Question 06 03 Zagros Mountains salt glacier.kmz. Open the file; click on the Placemark for this question to open the popup box; copy and paste the questions into your answer document; and address the assigned questions. The Google Earth image data are from CNES / Airbus.

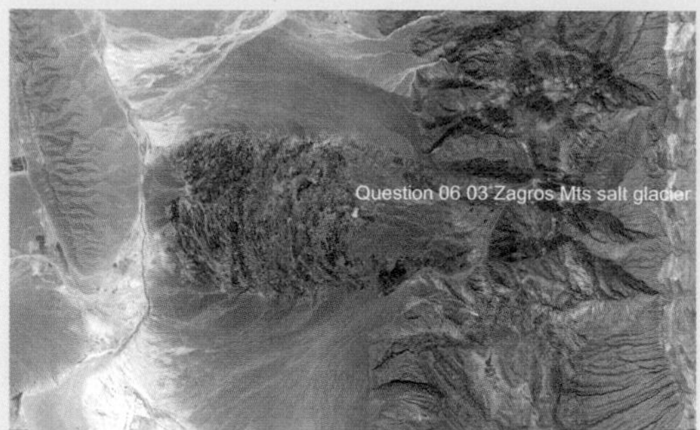

6.4. The Couette–Hatschek viscometer (**Figure 6.11**) is a laboratory instrument used to determine the viscosity of liquids from the relationship:

$$\eta = \frac{\tau}{4\pi R^2 L\omega}\left[\frac{1-a^2}{a^2}\right]$$

Describe each physical quantity in this equation, along with its SI units. Show how the units on the right side are resolved to become the units of viscosity.

6.5. Referring to **Figure 6.14**, describe how the viscosity of Newberry rhyolite and Columbia River basalt vary with temperature and weight % SiO_2.

6.6. Mechanisms for ductile deformation include dislocation motion (glide and climb) and chemical diffusion (diffusion creep). Both mechanisms result in power-law relationships between rate of strain and differential stress: $\dot{\varepsilon} \propto (\Delta\sigma)^n$. What are representative values for the exponent n for these two mechanisms? Why is this deformation referred to as *solid-state* flow?

6.7. Lava flowing down the side of a cinder cone has been idealized as viscous flow down a smooth slope (**Figure 6.18**) to determine the viscosity from the relationship:

$$\eta = \frac{H^2 \rho g \sin\beta}{2V}$$

Describe each physical quantity in this equation along with its SI units. Show how the units on the right side are resolved to become the units of viscosity.

6.8. For the spatial description of motion, which underlies the kinematics of viscous flow, the velocity vector, $\mathbf{v}$, is the dependent variable and the position vector, $\mathbf{x}$, (coordinates) and time are the independent variables:

$$\mathbf{v} = \mathbf{v}(\mathbf{x}, t) = \mathbf{v}(x, y, z, t)$$

Explain why $\mathbf{v}$ in this context is called the *local* velocity and not the *particle* velocity.

6.9. Using a spatial description of motion, the local velocity at $\mathbf{x}$ is compared to the local velocity at a neighboring position, $\mathbf{x} + \mathrm{d}\mathbf{x}$, where $\mathrm{d}\mathbf{x} = \langle \mathrm{d}x, \mathrm{d}y \rangle$ is the vector that overlies the infinitesimal material line between these two positions (see **Figure 6.21**). The components of the relative velocity, $\mathrm{d}\mathbf{v} = \langle \mathrm{d}v_x, \mathrm{d}v_y \rangle$, for these two positions are linearly related to the components of $\mathrm{d}\mathbf{x}$ as:

$$\mathrm{d}v_x = \frac{\partial v_x}{\partial x}\mathrm{d}x + \frac{\partial v_x}{\partial y}\mathrm{d}y$$

$$\mathrm{d}v_y = \frac{\partial v_y}{\partial x}\mathrm{d}x + \frac{\partial v_y}{\partial y}\mathrm{d}y$$

Explain why the material line is taken as infinitesimal, but the velocities at the two positions need not be small. Also explain why these constraints are necessary to understand the kinematics, for example, of lava flow down the side of a cinder cone.

6.10. The four partial derivatives in the two indented equations in Question 6.9 are components of the spatial gradient of velocity tensor and elements of its matrix representation, [L]. Show how this matrix is composed of a symmetric matrix, [D], and a skew-symmetric matrix, [W], such that [L] = [D] + [W]. Name the two matrices on the right side of this equation and describe in words how they characterize the deformation and motion of infinitesimal material lines at the position of interest in the flowing material.

6.11. Compare and contrast the principal stresses in a Newtonian viscous liquid that is static and one that is flowing. In doing so define the static pressure, $\bar{p}_o$, and describe the orientation and magnitude of the maximum shear stress, σ_S. Under what circumstances is the viscous liquid capable of supporting shear stress?

6.12. For a viscous liquid in non-uniform motion, a proxy for pressure is the mean normal pressure, $\bar{p}$. Define this quantity in terms of the stress components, and in terms of the principal stresses. Contrast the mean normal pressure with the static pressure.

6.13. The thermodynamic pressure is defined using an equation of state: $p = f(\rho, T)$ where the liquid density, ρ, and absolute temperature, T, are the state variables. Describe conditions under which the static pressure equals the thermodynamic pressure, $\bar{p}_o = p$. Describe conditions under which the mean normal pressure equals the thermodynamic pressure, $\bar{p} = p$.

6.14. Give a single range of unit weights that covers typical sedimentary, metamorphic, and igneous rocks. Use this to define the range of gradients with respect to depth for the lithostatic pressure in Earth's lithosphere. Extract from this range a value of the gradient that is useful for "back of the envelope" calculations.

6.15. Starting with Cauchy's laws of motion, (6.44) and (6.45), describe in words the assumptions and constraints necessary to derive the Navier–Stokes Equations of Motion (6.46). What are the independent and dependent variables in the Navier–Stokes equations?

6.16. A dimensional analysis of viscous flow in a sill using the Navier–Stokes equations identifies two-dimensional ratios with dimensions of force per unit volume that compete to determine the nature of the flow. Describe these ratios and use them to derive Reynolds number. What are the characteristics of low Reynolds number flow?

MATLAB EXERCISES FOR CHAPTER 6: ELASTIC VISCOUS DEFORMATION

Laboratory experiments confirm that a combination of elastic and viscous behaviors occur for rock samples at elevated temperatures and pressures. In addition, geophysical data are consistent with elastic viscous behavior for deformation in Earths asthenosphere. The exercises for this chapter help distinguish the characteristic features of viscous deformation from elastic or ductile deformation, and they define linear viscous behavior in terms of shear stress and rate of deformation. Calculations with MATLAB provide insights about how fast a person would have to run to stay ahead of lava flowing down the side of a volcano. Another exercise highlights the need to use the spatial description of motion, rather than the referential description of motion. Velocity is the primary kinematic variable, and the spatial gradient of velocity tensor plays a parallel role to the spatial gradient of deformation tensor for ductile solids. MATLAB helps to visualize how the rate of deformation tensor quantifies the stretching of material lines and how the rate of spin tensor quantifies their rotation in a flowing liquid. These exercises help students understand the constitutive law proposed by Stokes that led to the celebrated Navier–Stokes equations of motion. www.cambridge/SGAQI

FURTHER READING

For citations in figure captions see the reference list at the end of the book.

Bird, R. B., Stewart, W. E., and Lightfoot, E. N., 2007, *Transport Phenomena*. John Wiley & Sons, Inc., New York.
This engineering textbook provides a thorough coverage of momentum and heat transfer in fluids, including many examples of solutions to worked problems concerning the flow of viscous materials, some of which have direct applications to magma flow and viscous creep.

Clark, S. P., (Ed.), 1966. *Handbook of Physical Constants*. Memoir 97. The Geological Society of America, Inc., New York.

Decker, R. W., Wright, T. L., and Stauffer, P. H. (Eds.), 1987. *Volcanism in Hawaii*. US Geological Survey, Professional Paper 1350.

The contents of this volume should be compared to that of Lipman and Mullineaux (1981), which reviews the explosive eruption of Mount St. Helens Volcano.

Flugge, W., 1967. *Viscoelasticity*. Blaisdell Publishing Co., Waltham, MA.
This book is intended for an introductory course in viscoelasticity, and also for self-study of the subject: it stays focused on the theoretical aspects.

Happel, J., and Brenner, H., 1965. *Low Reynolds Number Hydrodynamics with Special Applications to Particulate Media*. Prentice-Hall, Inc., Englewood Cliffs, NJ.
Written by two chemical engineers, this textbook explores the slow flow of a viscous fluid in concert with solid particles, a topic with wide industrial as well as geological applications. The authors develop the theoretical aspects of this subject from first principles.

Hirth, G., and Kohlstedt, D., 2003. Rheology of the upper mantle and the mantle wedge: a view from the experimentalists, in Eiler, J. (Ed.) *Inside the Subduction Factory*. American Geophysical Union, Washington, DC, pp. 83–105.
This review paper on the viscosity of the upper mantle is based on laboratory experimental data for olivine aggregates and single crystals deformed under diffusion creep and dislocation creep regimes.

Jiang, D., 2007. Numerical modeling of the motion of deformable ellipsoidal objects in slow viscous flows. *Journal of Structural Geology*, 29, 435–452.
This paper provides the equations and modeling tools to investigate the deformation of a viscous ellipsoidal object embedded in a viscous fluid during laminar flow.

Karato, S., 2008. *Deformation of Earth Materials: An Introduction to the Rheology of Solid Earth*. Cambridge University Press, Cambridge.
This advanced textbook provides a thorough description and evaluation of the materials science of deformation at the mineral grain scale with applications to geophysical and geological phenomena including solid-state viscous creep.

Kohlstedt D. L., and Hansen, L. N., 2015. Constitutive equations, rheological behavior, and viscosity of rocks, in: *Treatise on Geophysics*, 2nd edition. Elsevier, pp. 441–472.
The micromechanics of deformation and the relevant constitutive laws are reviewed in this book chapter and applied to how laboratory flow laws relate to the viscosity structure of Earth's upper mantle determined from geophysical observations.

Lai, W. M., Rubin, D. H., and Krempl, E., 2010. *Introduction to Continuum Mechanics*. Elsevier, New York.
Presented as a first course in continuum mechanics for undergraduate students of engineering, this textbook provides chapters on Newtonian viscous flow, and both linear and non-linear viscoelastic flow.

Langlois, W. E., 1964. *Slow Viscous Flow*. The Macmillan Co., New York.
This is a scholarly book on the dynamics of slow viscous flow written by an applied mathematician with a deep understanding of the mechanics.

Lipman, P. W., and Mullineaux, D. R. (Eds.), 1981. *The 1980 Eruptions of Mount St. Helens*. US Geological Survey Professional Paper 1250, Washington.
The contents of this volume should be compared to that of Decker et al., (1987), which reviews the non-explosive eruptions of Hawaiian volcanoes.

Malvern, L. E., 1969. *Introduction to the Mechanics of a Continuous Medium*. Prentice-Hall, Inc., Englewood Cliffs, NJ.
This authoritative book on continuum mechanics provides chapters on constitutive equations, including linear viscous and linear viscoelastic responses, and on fluid mechanics.

Newton, Isaac, Cohen, I. B., and Whitman, A., 1999. *The Principia, Mathematical Principles of Natural Philosophy, a New Translation by I. Bernard Cohen and Anne Whitman*. University of California Press, Berkeley, CA.
This modern translation of Newton's monumental tome provides the fundamental ideas that led to the concept of the viscous liquid and the linear relationship between shear stress and shearing rate of deformation.

Poirier, J.-P., 1985. *Creep of Crystals: High-temperature Deformation Processes in Metals, Ceramics and Minerals*. Cambridge University Press, Cambridge.
This book is written, in part, for geologists and geophysicists in order to introduce the methods and concepts of materials science relevant to the plastic and viscous deformation of minerals.

Segall, P., 2010. *Earthquake and Volcano Deformation*. Princeton University Press, Princeton, NJ.
This advanced textbook goes beyond the elementary mechanics and structural geology covered here, with a focus on geophysical data related to active deformation associated with volcanic and earthquake hazards.

Tilling, R.I., Heliker, C., and Swanson D.A., 2010. *Eruptions of Hawaiian volcanoes; past, present, and future*, US Geological Survey General Information Product 117, online at http://pubs.usgs.gov/gip/117/
The material covered in this online document focuses on the Hawaiian volcanoes and is meant to compliment Lipman and Mullineaux (1981), on Mount St. Helens Volcano, thus comparing shield and composite volcanoes.

Turcotte, D. L., and Schubert, G., 2002. *Geodynamics*, 2nd edition. Cambridge University Press, Cambridge.
Geodynamics approaches many of the same topics covered here, but typically uses geophysical data to constrain mechanical models, rather than geological data.

Van Dyke, M., 1982. *An Album of Fluid Motion*. The Parabolic Press, Stanford, CA.
A collection of classic and beautiful images of fluid flow.

Watts, A. B., 2001. *Isostasy and Flexure of the Lithosphere*. Cambridge University Press, Cambridge.
This graduate textbook covers the history of the concept of isostasy and derives the theory of flexure of elastic plates, and applies both to understand deformation of Earth's lithosphere in the ocean basins and continental interiors.

Zoback, M. D., 2010. *Reservoir Geomechanics*. Cambridge University Press, Cambridge.
Insights from geophysics, rock mechanics, structural geology, and petroleum geology are integrated in this textbook with abundant data and useful applications to the hydrocarbon industry.

PART IV

Part IV (Chapters 7–11) covers five categories of geologic structures: fractures, faults, folds, fabrics, and intrusions. For each category, we use outcrop photographs, detailed maps, and thin sections to document the geometry of these structures and constrain the mechanics of their formation. For example:

- Chapter 7 uses a map of the transition zone between two fracture domains in the Entrada Sandstone at Arches National Park, Utah, to reveal their relative age;
- Chapter 8 uses a map of the Frog fault and the Lone Mountain monocline in the Western Grand Canyon, Arizona, to document how faulting and folding are related;
- Chapter 9 uses Airborne Laser Swath Mapping (ALSM) at Sheep Mountain anticline, Wyoming, to prepare a structure contour map and quantify the 3D form of this fold;
- Chapter 10 uses thin sections from a right step in a fault from the Sierra Nevada, California, to study ductile fabrics that are characteristic of a shear zone; and
- Chapter 11 uses the geologic map of Mt Ellsworth, Henry Mountains, Utah, to investigate three stages in the growth of laccolithic intrusions.

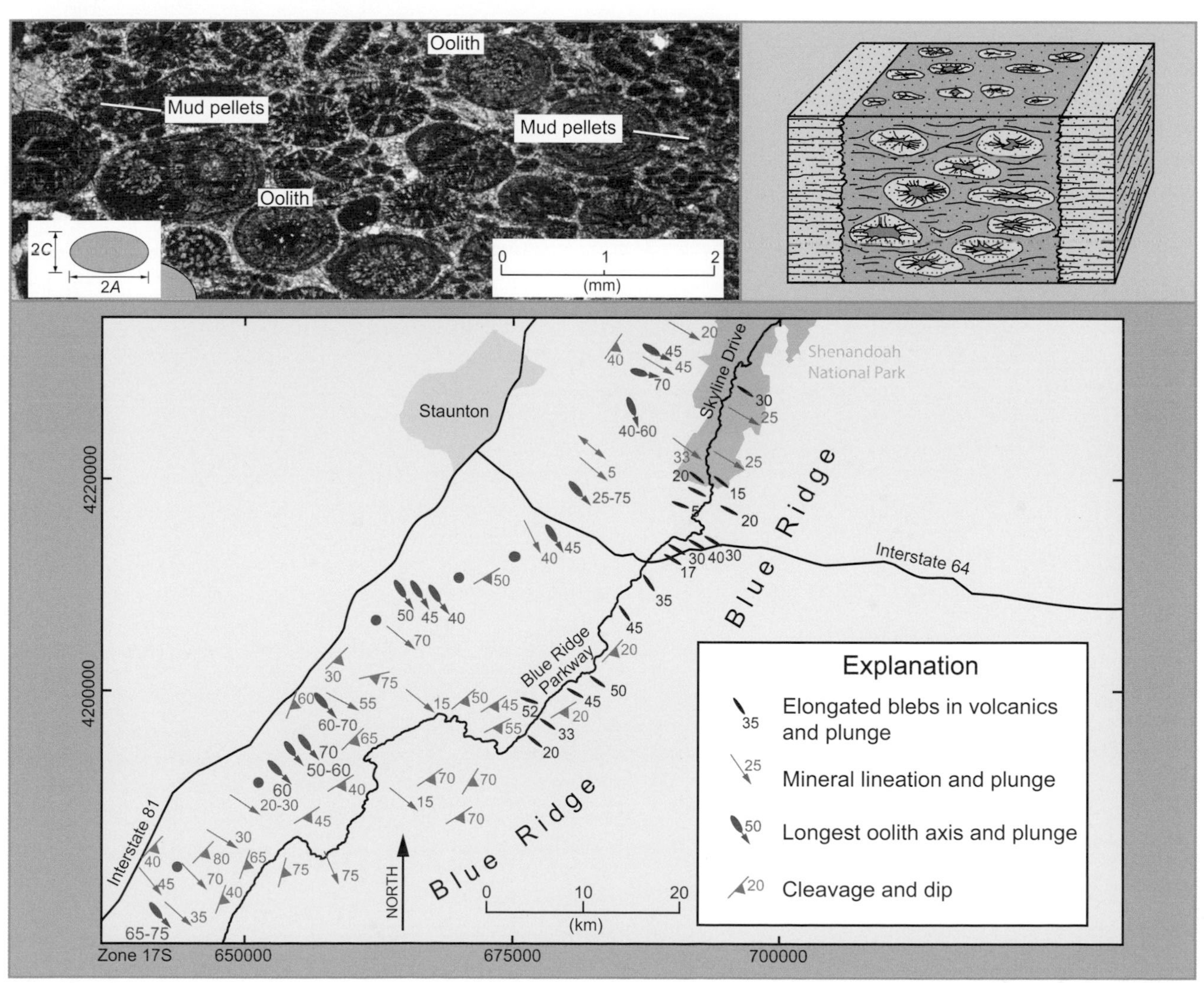

Chapter 7
Fractures

Introduction

Chapter 7 focuses on fractures in rock by describing outcrops of joints, veins, and dikes, introducing a canonical model for opening fractures, and considering fracture initiation and propagation using linear elastic fracture mechanics. The outcrop descriptions serve to highlight the characteristic geometric features of these structures, and provides the background necessary to build a conceptual model for opening fractures in rock. The canonical fracture model is based on a pure opening fracture in an elastic rock mass. With attention focused on the fracture tips, we explore the stress concentration there that leads to fracture propagation, and identify the three modes of fracture tip deformation. We explain how fractures initiate at flaws in rock and propagate when the stress intensity in the near-tip region reaches a critical value called the fracture toughness. Although many opening fractures are approximately planar, we describe how minor amounts of shearing can alter the propagation path and lead to kinked or echelon fractures. These interesting geometries provide evidence for interpreting the state of stress at the time the fractures formed.

Although the canonical fracture model presented in this chapter is for *pure opening*, many of these fractures exhibit minor amounts of shearing. In some cases, that shearing occurs when the fracture is open and propagating. This shearing may be caused, for example, by heterogeneities in rock properties such as elastic stiffness, or by mechanical interactions with other nearby fractures. To acknowledge this minor shearing, yet keep these fractures clearly distinguished from shear fractures and faults, we refer to them as *dominantly opening* fractures. In some cases shearing occurs long after an opening fracture formed, and if the fracture surfaces are in contact, *slickenlines* may develop to bear witness to this shearing. If the open fractures were mineralized, for example by precipitates from hydrothermal fluids, those minerals will be deformed by the later shearing. To acknowledge this sequence of events (first opening, then shearing) we refer to these structures as *sheared* joints or *sheared* veins.

The study of joints and veins has a practical side because they play an important role in providing pathways for groundwater flow through aquifers, for ore-forming fluids in mineral deposits, and for hydrocarbon flow within oil and gas reservoirs. In addition, joints affect the strength of rock near Earth's surface, and thus must be considered carefully by engineering geologists working on major construction projects such as dams and bridges, and on nuclear waste repositories. These opening fractures also contribute to the permeability of geothermal reservoirs. Although faulting and earthquake generation is primarily a process of shearing rather than opening, outcrops exposing the damage zone around faults provide abundant evidence that opening fractures also play a prominent role in fault development.

7.1 DESCRIPTIONS OF JOINTS, VEINS, AND DIKES

Opening fractures are the most ubiquitous structures in Earth's lithosphere, occurring in sedimentary, metamorphic, and igneous rocks in all tectonic environments. Those fractures that are barren, or contain only traces of deposited minerals on the fracture surfaces, are called joints. Those that are filled with minerals deposited from aqueous fluids are called veins or filled joints. Those filled with minerals crystallized from a magma are sills if they are interleaved with sedimentary strata or metamorphic foliation. They are dikes if they crosscut sedimentary strata, metamorphic foliation, or larger bodies of igneous rock. Because Chapter 11 covers intrusions of magma into Earth's lithosphere, most material related to dikes and sills is found there. It should be understood, however, that mechanical models for opening fractures introduced here apply to dikes and sills, as well as joints and veins. While the mineral fillings, or lack thereof, provide useful information about the geologic history and origins of these structures, the underlying physical commonality is that these fractures *opened* as they formed.

7.1.1 Map and Outcrop Patterns

Joints can strongly affect the physiography of landforms from coastlines to stream networks to mountain peaks to desert mesas.

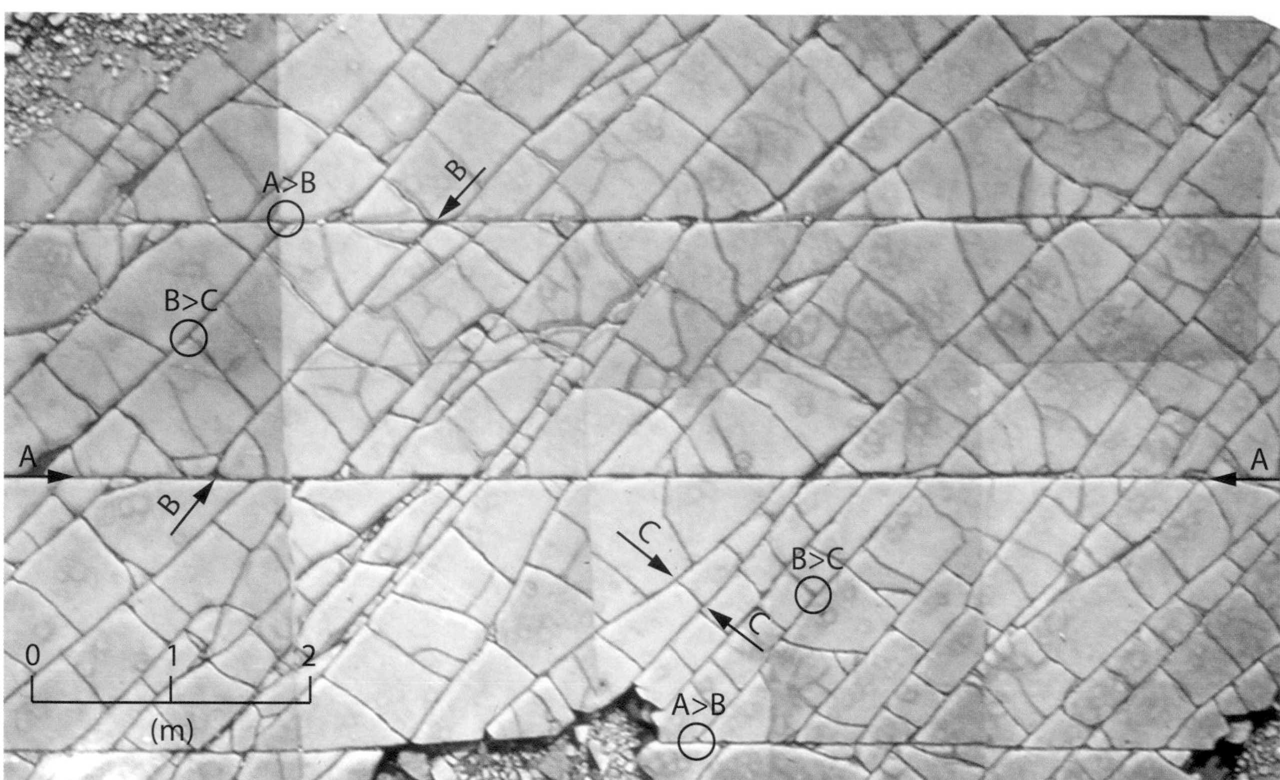

Figure 7.1 Three sets of joints (A, B, and C) in a limestone bed at Lavernook Point, Bristol Channel Basin, England. Terminating relationships, for example within the small circles, indicate set A is older than set B (A > B), and set B is older than set C (B > C). Modified from Rawnsley et al. (1998), Figure 8a. UTM: 30 U 488029.02 m E, 5694972.16 m N.

In **Figure 1.14** the horizontal Cedar Mesa Sandstone is cut by nearly orthogonal vertical joints that divide the rock unit into rectangular prismatic blocks. A joint set is composed of individual fractures that are roughly planar and sub-parallel to other members of the set. In this photograph members of the joint set that strike parallel to the view direction are more widely spaced than members of the set that strike perpendicular to the view direction. Other than orientation and spacing, the two joint sets do not have distinctive differences when viewed from this vantage point. The two joint sets appear to be, for the most part, mutually crosscutting, so their relative ages are difficult to determine from this photograph.

If members of one joint set systematically terminate against members of another set, we would interpret the terminating set as younger. For example, on the photomosaic of a limestone bedding surface exposed on a wave-cut bench along the Bristol Channel, England (**Figure 7.1**) joint sets in three different orientations have systematic terminating relationships. Most joints of set B terminate against members of set A, and most joints of set C terminate against members of set B. These relationships lead to the interpretation that set A is older than set B, and that set B is older than set C. Apparently the set C joints stopped propagating when they intersected the set B joints. An alternative explanation is that set C joints initiated at set B joints and propagated away from them. These explanations can be distinguished using textures on the joint surfaces, but in both cases set C joints are younger than set B joints.

The interpreted relative ages of the three joint sets in **Figure 7.1** also are supported by the facts that set A joints are the straightest, most widely spaced, and the most regularly spaced joints. These geometric qualities are consistent with those joints propagating through the limestone bed with no interference from other joint sets. In sharp contrast, many members of set C are curved, less widely spaced, and less regularly spaced. These qualities are consistent with those joints propagating through the limestone bed that already contained sets A and B. The inference that joint sets formed at different times in different orientations leads to the conclusion that they formed under *different* tectonic stress. Recall from Section 4.8.1 that opening fractures form perpendicular to σ_1, the greatest tensile (or least compressive) stress. Based on that concept one would infer that the principal stresses were directed differently when the different joint sets formed.

Calcite-filled veins in a limestone outcrop exposed near Crackington Haven are shown in **Figure 7.2**. This outcrop is about one meter across and the veins vary in thickness from less than a millimeter to about two centimeters. At first glance, one might conclude that two sets of veins exist, one with traces crossing the outcrop from left to right, and the other with traces crossing from bottom to top. However, note that vein B is crosscut by vein C at location #1, so B must be older than C. On the other hand, vein A is crosscut by B at location #2, so A must be older than B. From these observations we deduce that vein A is older than vein C, and that they are not members of the same set. After the formation of vein A, the tectonic stress apparently changed, and vein B formed in a perpendicular orientation. Then, the stress apparently changed again, and vein C formed

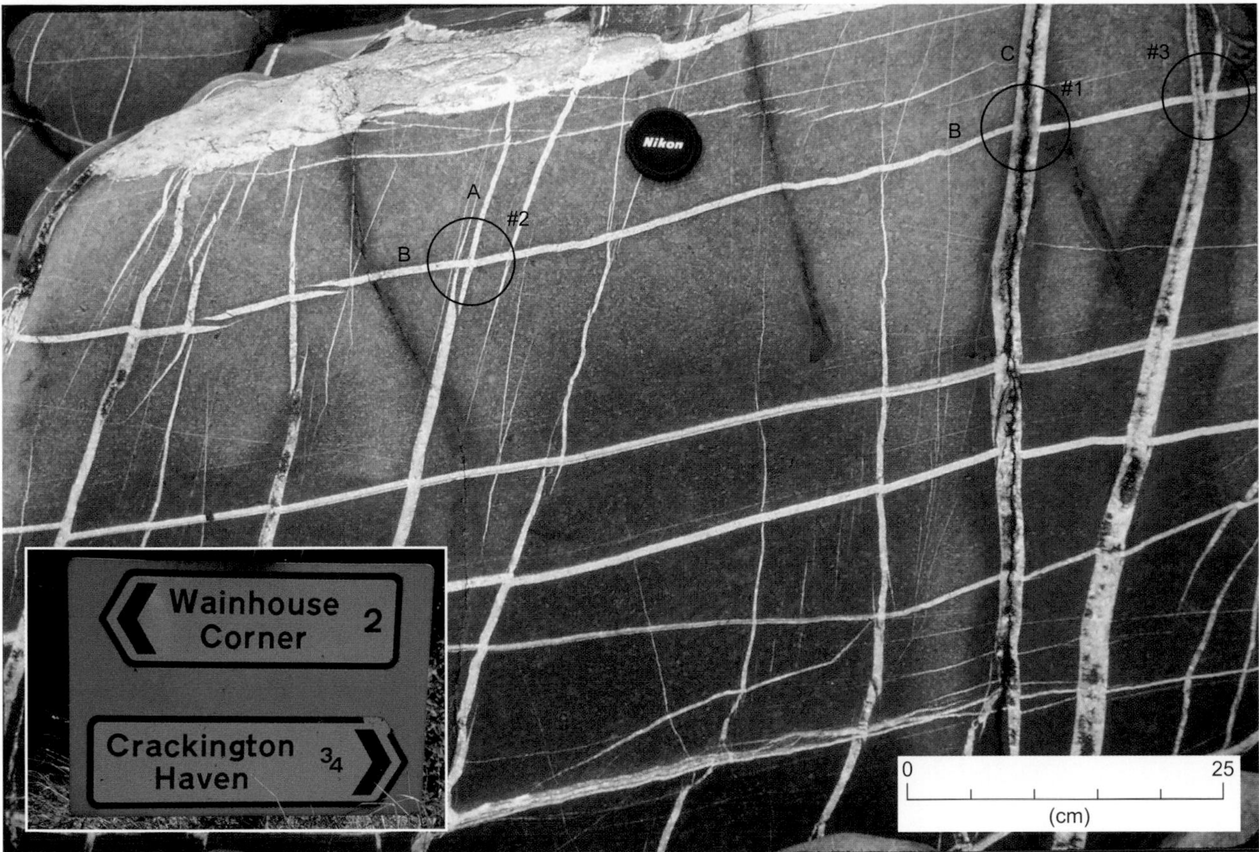

Figure 7.2 Calcite (white) filled veins in limestone (gray) from Crackington Haven, Cornwall, England. Note crosscutting relationships that indicate relative ages of veins. Vein segments labeled A, B, and C at numbered locations are discussed in the text. UTM: 30 U 384651.99 m E, 5622463.77 m N.

parallel to vein A. Following this logic, one would conclude that at least three sets of veins occur in this outcrop. The fact that vein C is not completely filled (black central region is open), but vein A is completely filled, lends support to the interpretation that they formed at different times, and under different hydrologic conditions. Can you explain the age relationships at location #3?

To explore the kinematics of opening fractures consider the schematic example in Figure 7.3, which depicts parts of two veins, G and H. For the sake of this example, the displacements related to the formation of the veins are directed in the plane of this illustration. Also, the geometry shown in these figures continues in and out of this plane without variation, so it is truly two-dimensional. Vein G is older, because it is cut by vein H, and veins G and H are orthogonal in Figure 7.3a and b. Two particles (two small black circles) are located at the intersections of the top surface of vein G, and the left and right surfaces of vein H. These were neighboring particles that now are separated by the development of vein H. We use the distance between these particles as a measure of the *opening* and *shearing* as vein H developed. The opening, o, is measured perpendicular to the plane of vein H on the outcrop. The shearing, s, is measured parallel to the plane of H. In Figure 7.3a the relative motion is pure opening ($o > 0$, $s = 0$). In Figure 7.3b the relative motion includes opening and shearing ($o > 0$, $s > 0$). Geometric concepts and techniques such as this one are explored in greater detail in the text by Ragan (2009; see Further Reading).

In Figure 7.3c and d the younger vein H cuts the older vein G at about 45°. In Figure 7.3c the plane of vein G appears to have been sheared: if the top surface of G is extended across vein H (see dashed gray line), it does not align with the top surface of G on the other side. However, this does not mean that the formation of vein H involved any shearing. Inspection of the originally neighboring particles demonstrates that the younger vein only opened ($o > 0$, $s = 0$). In Figure 7.3d the top surface of vein G is in alignment across vein H (see dashed gray line), suggesting that there was no shearing. However, the separation of the two particles demonstrate that the younger vein has both opened and sheared ($o > 0$, $s > 0$).

The particles describe in the preceding two paragraphs are located at what is called a piercing point in the geometry of these structures. This name comes from the fact that in three dimensions the upper surface of vein G and the coincident surfaces of vein H before it opens intersect along a line that *pierces* the plane of view in Figure 7.3. Particles lying along that line are separated by the opening of vein H and they displace exactly as the two surfaces of vein H displace. Thus, the current positions of the two particles facilitates the measurement of the relative displacement

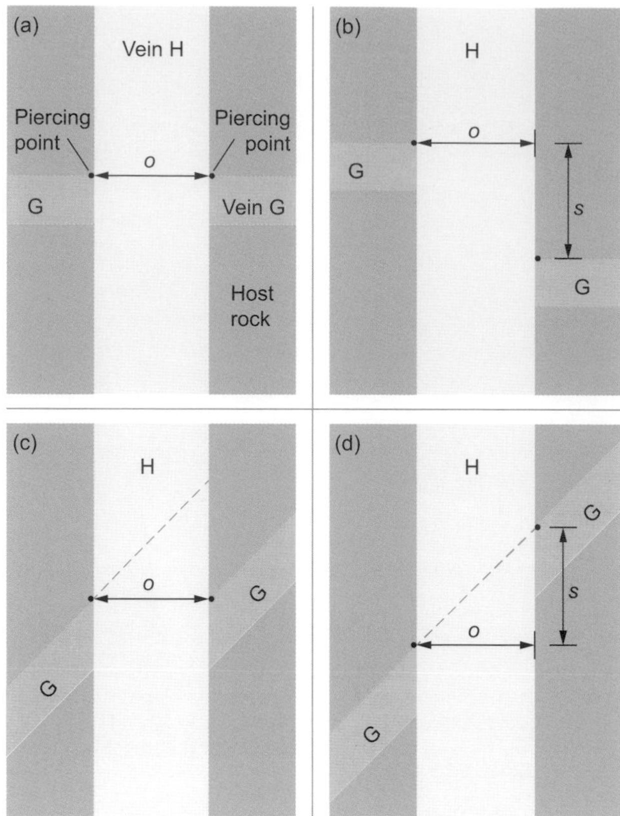

Figure 7.3 Two veins, G (gray) and H (light gray), cut the host rock in these schematic outcrops. The planes of both veins are perpendicular to the page, and the displacements are parallel to the page. Vein G is older than vein H. Vein H is perpendicular to vein G in (a) and (b); vein H is 45° to vein G in (c) and (d). Small black dots locate *piercing points*, which are used to measure opening, *o*, and shearing, *s*. Vein H opens in (a) and (c), but opens and shears in (b) and (d).

as vein H formed. These schematic examples serve to highlight the need for careful inspection of fractures to identify piercing points and deduce the kinematics of fracture formation.

An example of the kinematic relationships illustrated in **Figure 7.3c** is shown at location #1 of the limestone outcrop in **Figure 7.2**. The plane of vein B has a small apparent offset, but this can be explained by pure opening of vein A, because vein A and B are not orthogonal. Most of the veins in this outcrop have dominantly opened, but a case can be made for shearing at some intersections. Can you find such an intersection?

The kinematic relationships illustrated in **Figure 7.3d** are helpful in interpreting **Figure 7.4**, an outcrop in the Twin Lakes pluton, from the Sierra Nevada, CA, where a mafic dike from the Independence dike swarm cuts older felsic dikes. For the sake of this example, we assume all of these dikes are vertical, but their actual orientations should be confirmed. Felsic dike A makes an acute angle of about 70° with the Independence dike, and felsic dike B makes an angle of about 90°. The contact of dike A, marked with a dashed line, is in nearly perfect alignment across the Independence dike, suggesting that there was no shearing. However, for pure opening this contact should crop out at the left end of the double-headed arrow labeled *o*. Instead this contact apparently has sheared about 7 cm in a left-lateral sense across the Independence dike, which also apparently has opened about 20 cm. This magnitude and sense of shearing is consistent with the kinematics of felsic dike B. The dike contact, marked with a dashed line, is offset parallel to the Independence dike, and this offset is about 7 cm in a left-lateral sense.

The field examples displayed in **Figures 7.1**, **7.2**, and **7.4** show that joints, veins, and dikes can vary widely as to mineral filling, thickness, length, segmentation, orientation, spacing, and many other geometric characteristics. However, all are dominantly opening fractures, a characteristic we will exploit when we develop a mechanical model for these structures in Section 7.2.

7.1.2 Fracture Tips and Tipline

In the early 1920s the English mechanical engineer A. A. Griffith proposed that fractures *initiate* at flaws in solids where stress is concentrated, and *propagate* away from these flaws because of the stress concentration associated with the leading edge of the fracture. Eventually, for reasons we explore later in this chapter, the stress concentration is reduced and fractures *arrest*; they cease propagating. The conceptual model of Griffith envisions a physical *process* with a beginning and an end, and focuses attention on the leading edge of the fracture, the location of the mechanical action responsible for propagation and arrest. The story of Griffith's discovery and its place in the history of fracture mechanics is told in the very readable book by Gordon (1976), listed in Further Reading.

For fractures exposed in rock outcrops we refer to the leading edge, where the two fracture surfaces join, as the fracture tip. The tip is a point on the outcrop surface, but fractures are three-dimensional structures, with two surfaces that join along a curved line, called the fracture tipline. The intersection of a fracture tipline and an outcrop surface defines the point that is the exposed fracture tip. In **Figure 7.5** a dike segment tapers in thickness from several decimeters on the horizon to a tip in the immediate foreground. The dike is more resistant to erosion, so it stands up a few decimeters above the surrounding shale. The tip is rounded and has a radius of curvature of a few centimeters. The shale near the tip has been weathered to a loose soil, so structures within the host rock are not exposed.

Different styles of opening fracture tips occur, as revealed by calcite-filled vein segments at the decimeter scale in the outcrop photograph of **Figure 7.6**. Some of the veins taper gradually to sharp tips (e.g. at location #1), but others, in echelon arrangements with neighboring vein segments, have blunt tips (e.g. at location #2). Some of the veins have very blunt tips (e.g. at location #3) within the vein that cuts across the upper right corner of the photograph. At location #4 two veins overlap and thin abruptly, but then extend outside the marker circle as very thin veins to a sharp tip. Each of these styles provides clues about the growth of the vein segments, which we return to after introducing the mechanical model for opening fracture propagation.

Figure 7.4 Kinematic relationships between a member of the Independence dike swarm and older felsic dikes in the Twin Lakes pluton, Sierra Nevada. See text for discussion of apparent opening, $o \sim 20$ cm, and apparent left-lateral shearing, $s \sim 7$ cm, estimated here assuming all dikes are vertical. Photograph by S. J. Martel. UTM: 11 S 375251.79 m E, 4085132.50 m N.

7.1.3 Textures on Opening Fracture Surfaces

Woodworth published one of the earliest and most thorough descriptions of joint surface textures in 1896. In Woodworth's drawings, families of fine curved lines, called hackle, bound parts of a fracture surface with slight differential relief (Figure 7.7, part 4). Here two different fracture surfaces, each decorated with hackle, depart from a common plane and intersect. A family of hackle also can form a pattern that looks like a *feather*, or half of a feather, called plumose structure (Figure 7.7, parts 3 and 10). Woodworth referred to joints with these surface textures as feather fractures. In some cases, a full plume develops with an axis near the middle of a layer and parallel to bedding. For example, the joint surfaces at mid-height in Figure 7.8 display plumose structure that is more-or-less symmetric about an axis along the mid-plane of the sandstone beds. In other cases, only half a plume is developed and the axis is located near the bedding top or bottom.

Hackle may be almost imperceptible near the middle of a fracture surface (Figure 7.7, part 10a), but can grow in height near the distal margins, or fringe of a joint (Figure 7.7, parts 5 and 10b). Woodworth calls the individual joint segments, bounded by more pronounced hackle, cross fractures. These gradually or abruptly twist about an axis that is parallel to the bounding hackle (Figure 7.7, part 5). The joint segments form an echelon pattern when viewed in cross section (Figure 7.7, parts 6, 7, and 8), and they too may have plumose structure on their surfaces (Figure 7.7, part 9). A single two-dimensional view of a three-dimensional structure can be very informative, but it also is incomplete. For example, the cross sections in Figure 7.7, parts 6–8, are different representations of a structure that is similar to that shown in part 5.

On some joint surfaces curved lines or swaths, with some differential relief, are roughly perpendicular to hackle (Figure 7.7, part 10d). These rib marks usually are arcuate in shape and may form a nearly complete oval or ellipse that is concentric to neighboring rib marks. The joint origin is the focal point of the hackle (Figure 7.7, part 10a), and concentric rib marks enclose this focal point. A heterogeneity, which acted as the initiation point of the fracture, may be located at the origin.

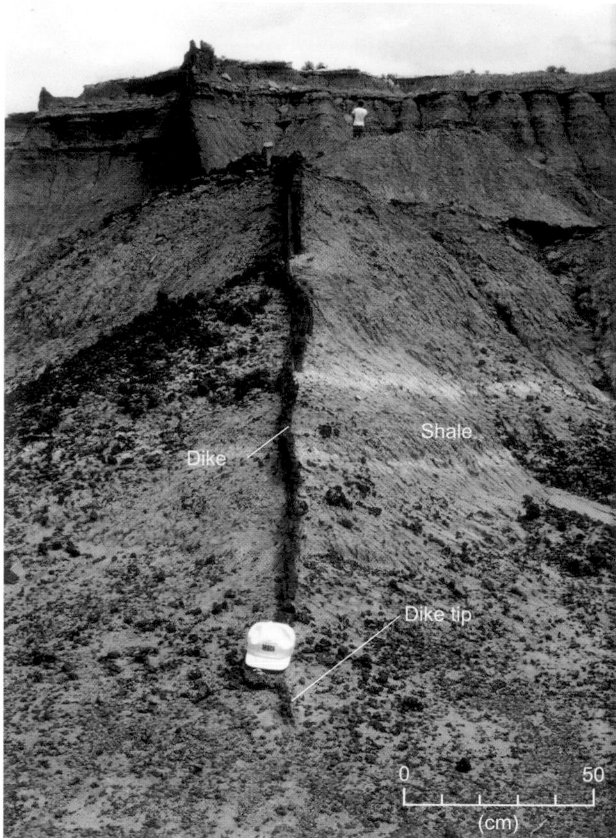

Figure 7.5 Dike segment from the San Rafael swarm, UT, with exposed dike tip in foreground. See Delaney and Gartner (1997) for maps and descriptions of dikes and sills. General location of San Rafael swarm is UTM 12 S 490047.35 m E, 4271300.94 m N.

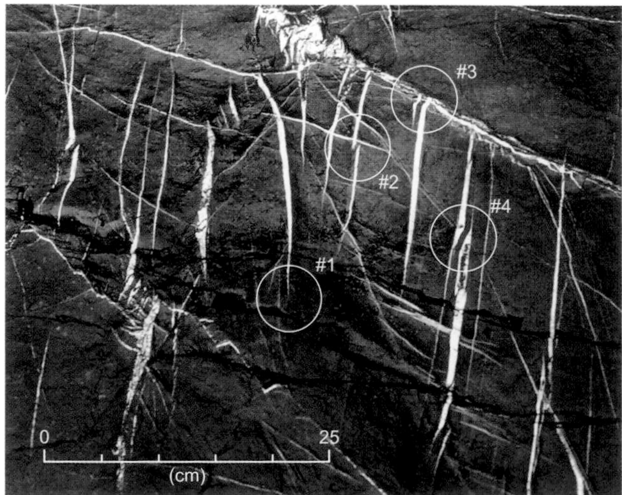

Figure 7.6 Calcite-filled veins in Carboniferous turbidites near Millook Haven, Cornwall, England. See Figure 1.16 for approximate location. Numbered sites and circles highlight vein tips discussed in the text.

Hackle and rib marks support Woodworth's conjecture that joints are opening fractures, because shearing of the two fracture surfaces in frictional contact would destroy these delicate textures. The symmetry of hackle and rib marks provides evidence concerning the symmetry of the loading (Figure 7.9).

Plumose structure on fractures in lava reveal the sequential development of columnar joints (Figure 7.10). These joints bound columns of lava that form as the lava cools. Often the columns have hexagonal cross sections. Fractures initiate at the top and bottom surfaces of the lava flow, and grow toward the middle of the flow. The textures on columnar joint surfaces reveal a very interesting growth history and the direction of propagation of the individual fracturing events. Figure 7.10a illustrates the composite nature of the fractures forming one side of a columnar joint, which is divided into horizontal *bands*, each about 1 m wide and 10 to 25 cm high. Each band has one plumose structure and represents one fracturing event. Apparently, these columnar joints developed by the successive addition of new bands, each initiating from the top of the previous band as the column grew up from the base of the flow. For example, the plume axis for segment 1 (Figure 7.10a, lower) extends along the bottom of that band away from the lower left corner (red dot), which is interpreted as the origin. The fracture propagated dominantly from left to right, but also upward.

7.1.4 Sheared Joints and Wing Cracks

If shearing occurs after a joint formed, and if the surfaces are in contact, slickenlines may develop to bear witness to this shearing. Textures on joint surfaces due to opening, such as hackle and rib marks, are likely to be obliterated by later shearing. Also, minerals deposited in veins, for example as precipitates from a hydrothermal fluid, will be deformed by later shearing. Where evidence for a sequence of initial opening and later shearing is found in outcrop, these structures are called sheared joints or sheared veins.

The sequence from initial opening of fractures to later shearing of the fracture surfaces was recognized by Woodworth (1896). He wrote:

> A cursory examination of the joint-planes in the Mystic River quarries shows that these secondary divisional planes differ as regards the form of their surface. . . . Such joints frequently show slickensides, as noted by B. F. Becker (1894), but I believe the rubbing to be largely secondary. . . . The preservation of feather-fracture depends upon the joint blocks moving away from each other in a direction transverse to the surface, so the feather-fractures usually gape.

Here Woodworth uses the term *feather fracture* for what we now call *plumose structure*. The sequence from initial opening to later shearing is common, because joints are weak surfaces within the rock mass that can slip when subjected to sufficient shear stress, without having to break the rock again.

Sheared joints can be mistaken for shear fractures, because slickenlines or offset older structures are diagnostic of shearing.

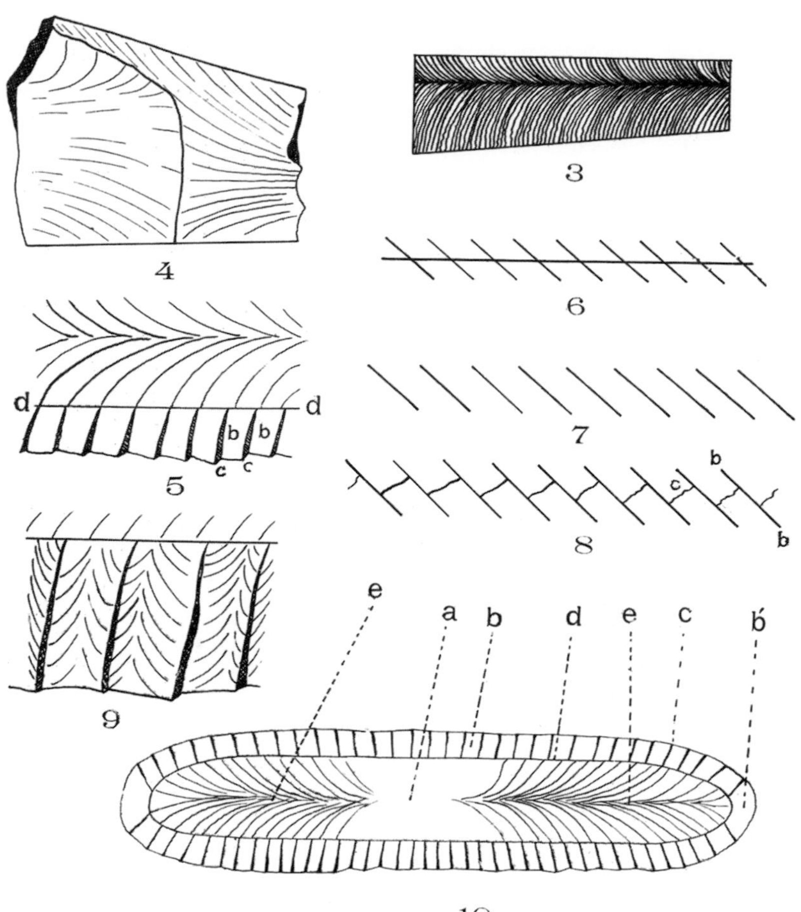

Figure 7.7 Drawings of surface markings and profiles of joints from Mystic River, Massachusetts: plumose structure (3); hackle (4); fringe region and echelon segments (5); profiles through echelon segments (6–8) with cross fractures (c); plumose structure on fringe segments (9); entire joint surface (10) with origin (a), fringe (b), cross fracture (c), rib mark (d), and plume axis (e). Modified from Woodworth (1896).

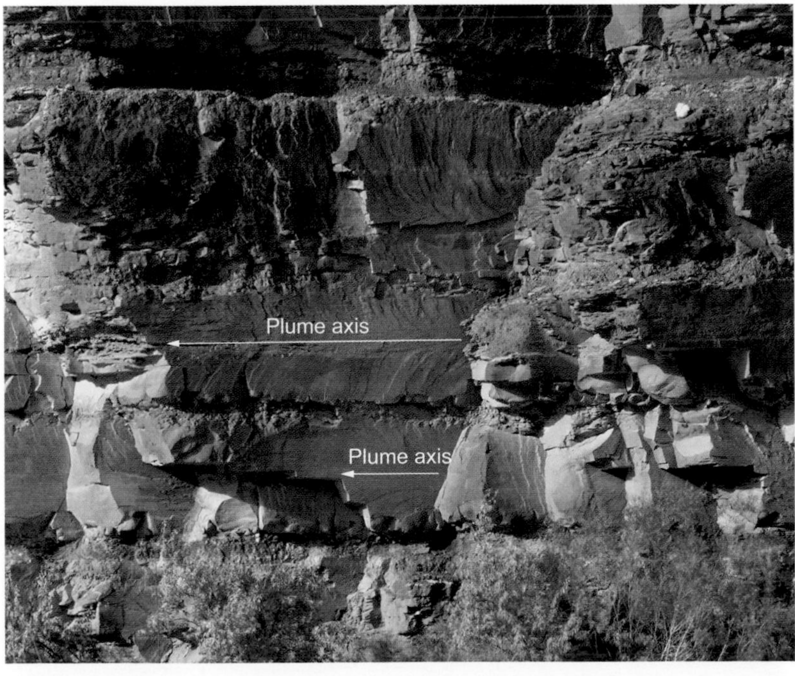

Figure 7.8 Plumose structure on joint surfaces in the Rico formation, Raplee Ridge anticline, UT. Two examples of plume axes are marked: hackle spread upward and downward from these axes. Approximate location on San Juan River at UTM: 12 S 603485.93 m E, 4114337.90 m N.

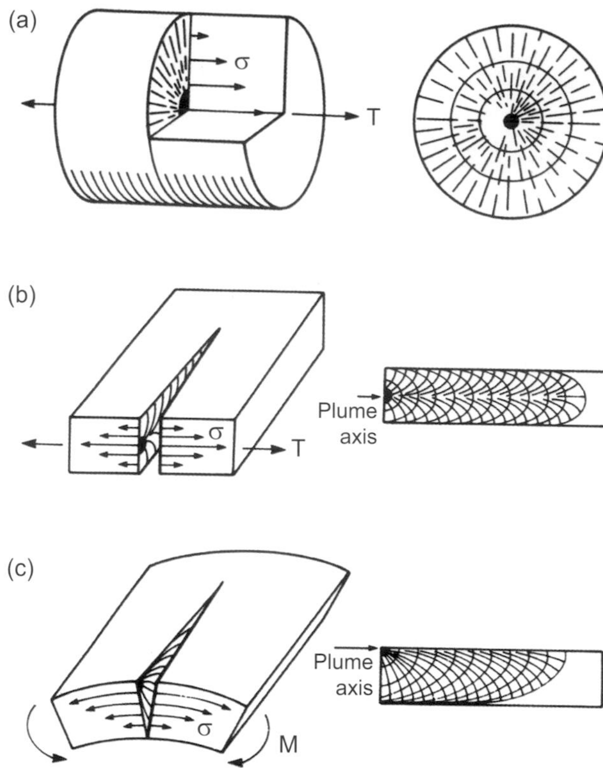

Figure 7.9 Possible relationships between plumose structure and loading. (a) Axisymmetric loading in tension, T, is associated with radial plumose pattern. (b) Loading in tension, T, that is symmetric about the mid-plane of a layer is associated with a central plume axis. (c) Loading by a bending moment, M, is associated with a plume axis along the top of the layer and the development of a half-plume. In all cases, σ represents the tensile stress before growth of the fracture. Modified from Pollard and Aydin (1988), Figure 24. See also Bankwitz (1965).

The *key question* to ask at the outcrop is: did that shearing occur as the fracture propagated, or was it a later event, induced by a different stress field? In many cases later shearing of joints produces wing cracks (also called *splay* cracks), often near the tips of the sheared fracture. The sheared joints in the Lake Edison Granodiorite (**Figure 7.11**) are members of a filled joint set that probably formed as the pluton was cooling. Some members display no shearing and the hydrothermal minerals within those fractures are undeformed, so they are *filled joints*. Those members that offset older aplite veins in a left-lateral sense contain deformed hydrothermal minerals and most have wing cracks; they are *sheared joints*.

Wing cracks extend from the sheared joint at acute angles that can range from a few tens of degrees up to nearly ninety degrees. Apertures of wing cracks typically are greatest where they intersect the sheared joint, and they taper away from that intersection. If wing cracks are gently curved, their concave side typically faces the sheared joint or vein. Most wing cracks are quite short relative to the trace length of the sheared joint or vein. The sense of shearing is readily deduced from the geometry of wing cracks: they form *only* on the side of a sheared joint that moved away from the nearby joint tip (see **Figure 7.11**, inset).

Wing cracks at a larger scale and in a sedimentary rock are displayed on a structure map of the Entrada Sandstone at Arches National Park, UT (**Figure 7.12**). Each line on this map represents the trace of a joint, or the trace of echelon and parallel joints forming a narrow *zone* of joints. Two sets of joints are mapped: a NE-striking set is exposed alone in the eastern part of the map, and a NW-striking set is exposed alone in the western part. Near the middle of the map both sets are exposed in a 300 m wide *transition zone*. The NE-striking set of joints are interpreted as older, because some members of the NW-striking set occur as *wing cracks* that extend from the NE-striking joints, commonly near the tips of these joints.

The development of the wing cracks striking to the northwest on the NE-striking joints in **Figure 7.12** is consistent with left-lateral shearing of these joints. This shear sense is associated with a counter-clockwise rotation of the regional horizontal principal stresses. Recall from Section 4.8.1 that opening fractures form in a plane that is perpendicular to σ_1, the greatest tensile stress, or least compressive stress in the presence of pore fluid pressure (**Figure 7.13a**). Counter-clockwise rotation of the principal stresses results in a left-lateral shear stress resolved on the opening fracture. Under this new stress the fracture shears and new stresses are concentrated near the tip, which lead to the opening of wing cracks that trend in a direction perpendicular to the current greatest tensile stress (**Figure 7.13b**).

7.2 A CANONICAL MODEL FOR OPENING FRACTURES

A canonical model is one that reduces a physical process to the simplest form possible, without loss of generality. The canonical model introduced here is defined in the context of a material continuum governed by Cauchy's laws of motion (see Sections 3.6.1 and 3.7.1). In other words, linear and angular momentum are conserved. In order to reduce Cauchy's equations to equations that can be solved, choices must be made about the *constitutive properties*, the *geometry of the structure*, and the *boundary conditions*. Thus, we are motivated to ask three questions:

(1) What is the simplest constitutive law relevant to rock fracture?
(2) What is the simplest geometry of the fracture and the surrounding rock mass?
(3) What are the simplest boundary conditions required to open the fracture?

The answers to these questions, provided in Sections 7.2.1 and 7.2.2, lead us directly to a well-known boundary-value problem of continuum mechanics, which is the canonical model.

7.2.1 Constitutive Law and Model Geometry

The discussion of laboratory testing for brittle rock in Section 4.4 suggests Hooke's law is the appropriate choice for constitutive behavior. Typically, in these tests the strain is linearly related to

Figure 7.10 Columnar joints in a Snake River Basalt flow along the Boise River near Lucky Peak Dam, ID. (a) Plumose patterns confined to fracture bands on the side of a columnar joint. Scale is graduated in inches. Discrete fracture events are represented by a single plumose pattern with an origin (red dot) on the edge of the previous band. Arrows indicate local fracture propagation direction. Modified from DeGraff and Aydin (1987), Figure 5b. (b) Columnar joints at the same site with well-developed fracture bands on each column side. Photograph by A. Aydin. Approximate location at UTM: 11 T 576656.19 m E, 4820037.69 m N.

the stress, and it is almost completely recoverable if the strains are small enough. Strains associated with opening fractures are small, because the ratio of fracture opening to length is small. For example, for the dikes of the San Rafael swarm (**Figure 7.5**) the median dike thickness (opening) is 1.1 m and the median dike outcrop length is 1,090 m, so this ratio is about 1/1,000. That is small enough to be within the elastic limit for most rocks. Some of the Millock Haven calcite-filled veins (**Figure 7.6**) are a few millimeters thick (opening) and a few decimeters to a few meters long, so the ratio is within the elastic limit. However, other veins of similar length in that outcrop are a few centimeters thick, so the ratio is about 1/10. For those veins we anticipate an inelastic constitutive law would be more appropriate.

The descriptions of joints, veins, and dikes earlier in this chapter suggest that idealizing their surfaces as *planar* is appropriate. For example, look again at the Cedar Mesa joints (**Figure 1.14**) and the calcite-filled veins at Crackington Haven (**Figure 7.2**). If one pulls back from examining the interesting small-scale textures on fracture surfaces, such as those on the joints in the Rico formation at Raplee anticline (**Figure 7.8**), the overall geometry of these fractures is nearly planar. We consider a fracture that is very long and unchanging in one dimension to take advantage of a two-dimensional geometry. Some natural fractures closely approximate this two-dimensional idealization, but many others do not. For example, the rib marks on the joint surface shown in **Figure 7.9** indicate this fracture was approximately elliptical in shape. We acknowledge that more can be learned by extending the geometry to three-dimensional shapes, but accept the two-dimensional idealization here to maintain simplicity. A single fracture of length $2a$ is centered at the origin of Cartesian coordinates for this two-dimensional problem (**Figure 7.14**).

The simplest option for the geometry of the body containing the fracture is to let it extend to an infinite distance in all directions. This avoids having to specify the exterior shape of the body and the location of the exterior surface. Furthermore, the mathematical effort required to solve the boundary-value problem for a finite elastic body is formidable and, when accomplished, usually does not negate what we will learn from the infinite body solution.

The practical constraint for applying the infinite body solution to natural fractures is that the actual size of the rock mass must be large compared to the length of the fracture. That certainly is the case for the veins at Crackington Haven (**Figure 7.2**), but one might wonder if the dikes in the San Rafael swarm (**Figure 7.5**) were shallow enough, relative to their heights, to interact mechanically with Earth's traction-free surface. If so, one would have to include Earth's surface as an exterior boundary for the model.

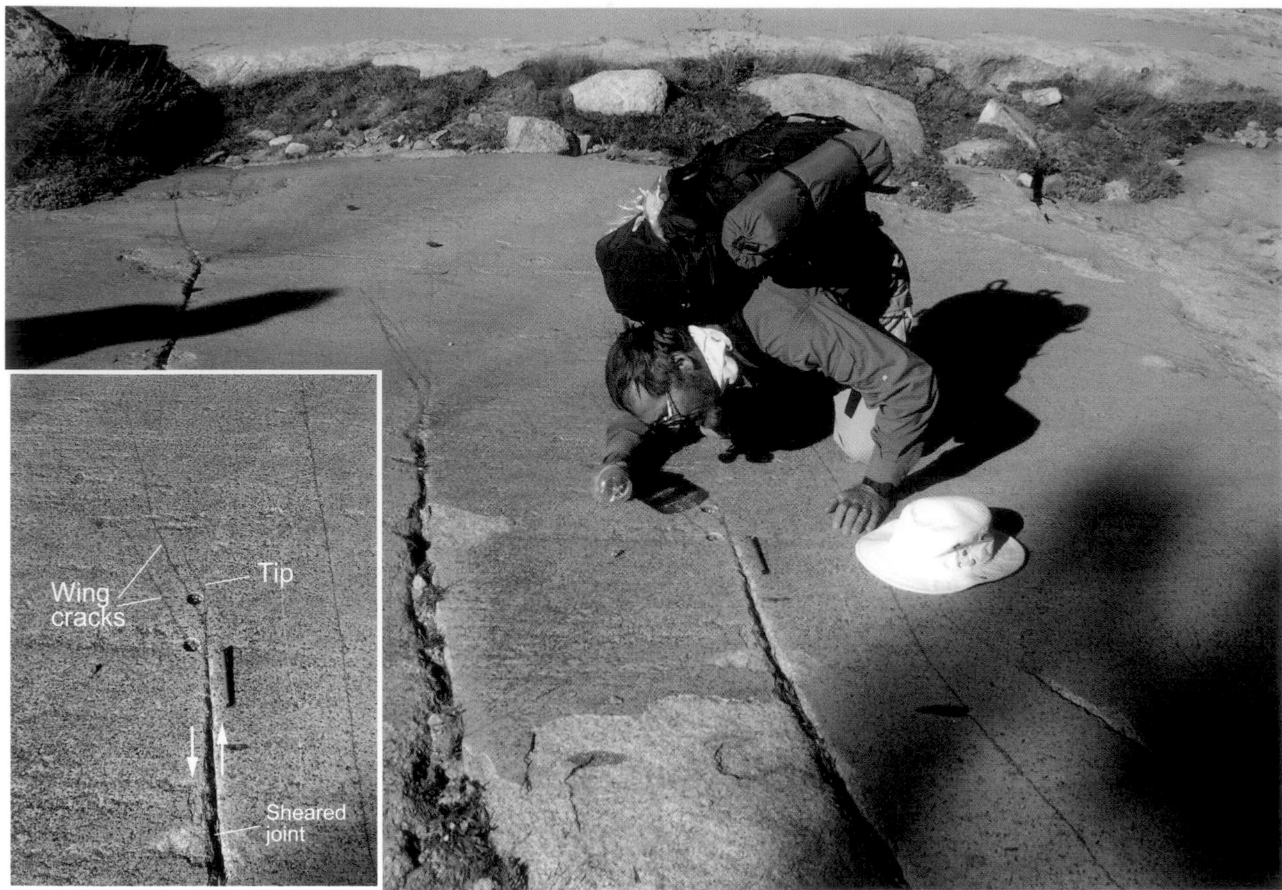

Figure 7.11 Steve Martel examining wing cracks near the ends of sheared joints in the Lake Edison Granodiorite, Sierra Nevada, CA. A second and third example are seen to the left, and the inset photograph displays the geometry of the wing cracks under Martel's right arm. The sense of shearing on the filled joints is inferred to be left lateral. See Segall and Pollard (1983). Outcrop location is UTM: 11 S 333373.00 m E, 4136564.00 m N.

7.2.2 Boundary Conditions and Cauchy's Formula

Boundary conditions constrain either *tractions* or *displacements* acting on the boundaries of the elastic body, so one must ask: what is known about these vector quantities, and their distributions, on the surfaces of the fracture and at great distances from the fracture? Because one can measure the thickness of a dike or vein where exposed in outcrop (e.g. Figure 7.5 or Figure 7.6), specifying the opening displacement (half the thickness) for both fracture surfaces seems attractive. However, evidence often is absent for what the tangential component of displacement might be, and one must specify both components of the displacement vector to define the displacement boundary conditions. We will learn shortly that choosing zero for the tangential displacement is not a good idea, so this lack of knowledge is a deterrent to using displacement as a boundary condition.

Because the surfaces of dikes and veins are pushed open by the pressure of the invading magma or hydrothermal fluid, a traction boundary condition is attractive. In this case the normal component of the traction would have the magnitude of the pressure, and the tangential component would be zero, because a static fluid imparts no shear traction to the fracture surfaces (Figure 7.14b). We write the boundary conditions on these surfaces as follows:

$$\begin{aligned} &\text{on } |x| \leq a, y = 0^+, \quad t_x = 0, t_y = +P \\ &\text{on } |x| \leq a, y = 0^-, \quad t_x = 0, t_y = -P \end{aligned} \tag{7.1}$$

These conditions apply over the entire fracture in the z-direction: $-\infty \leq z \leq +\infty$. The sign of the normal traction component, t_y, is positive on the $y = 0^+$ surface and negative on the $y = 0^-$ surface to account for the fact that this traction *pushes* against both surfaces. A problem defined in this way is a *traction* boundary-value problem.

Given the traction acting on the model fracture surface (Figure 7.14b), it is reasonable to ask: what is the stress state on a small element adjacent to those surfaces? The general three-dimensional relationship between the traction vector acting on a surface and the stress tensor acting on a small volume element adjacent to that surface is known as Cauchy's formula (Section 2.6.5). Because of its importance to many facets of structural

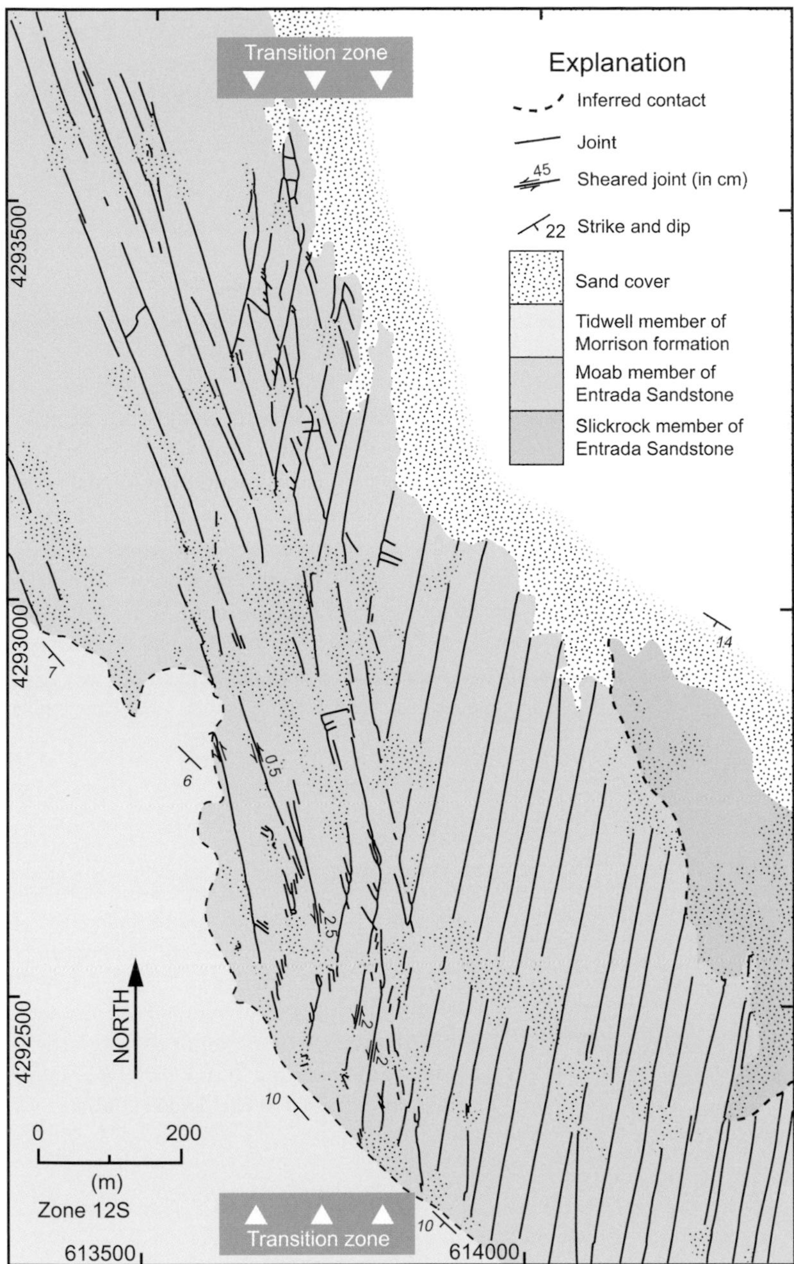

Figure 7.12 Structure map of the transition zone between two joint domains in Entrada Sandstone at Arches National Park, UT. Some NW-striking joints initiated as wing cracks near tips of NE-striking joints. Modified from Cruikshank and Aydin (1995), Figure 9. Outcrop location at UTM: 12 S 613963.61 m E, 4292665.01 m N.

geology, we consider this formula in its two-dimensional form here, and then apply it to the two-dimensional model of the opening fracture.

To understand the two-dimensional relationship between the traction vector and the stress tensor, we consider a small triangular element with a traction acting on the oblique face and stress components acting on the coordinate faces (**Figure 7.15**). The area of the oblique face is ΔA and the outward directed unit normal is $\hat{\mathbf{n}} = \langle n_x, n_y \rangle$. In the limit as $\Delta A \to 0$ the traction and stresses act at the same point. The components of a unit vector are equal to the respective direction cosines: $\hat{\mathbf{n}} = \left\langle \cos\theta_{\hat{\mathbf{n}}\hat{\mathbf{i}}}, \cos\theta_{\hat{\mathbf{n}}\hat{\mathbf{j}}} \right\rangle$. In what follows we transform the stress components into forces by multiplying them by the area of the face on which they act. These areas are found by projecting the oblique face onto the coordinate planes. For example, the area of the face perpendicular to the y-axis is $\Delta A_y = \Delta A \cos\theta_{\hat{\mathbf{n}}\hat{\mathbf{j}}} = \Delta A n_y$.

The physical principle that Cauchy employed was Newton's second law, which requires that the sum of all forces acting in each coordinate direction must be zero for quasi-static conditions. Taking the y-direction as an example, we write:

$$\sum f_y = t_y \Delta A - \sigma_{xy} \Delta A_x - \sigma_{yy} \Delta A_y = 0$$

Here the traction component, t_y, acts on the oblique face in the positive y-direction, and the two stress components, σ_{xy} and σ_{yy}, act on the x- and y-coordinate planes in the negative y-direction. The area ΔA_y is eliminated from this equation using the projection relation ($\Delta A_y = \Delta A n_y$) and the other areas are similarly eliminated. Then, we divide both sides by ΔA and solve for the traction component. Following a similar derivation for the other traction components, we arrive at the two-dimensional Cauchy's formula:

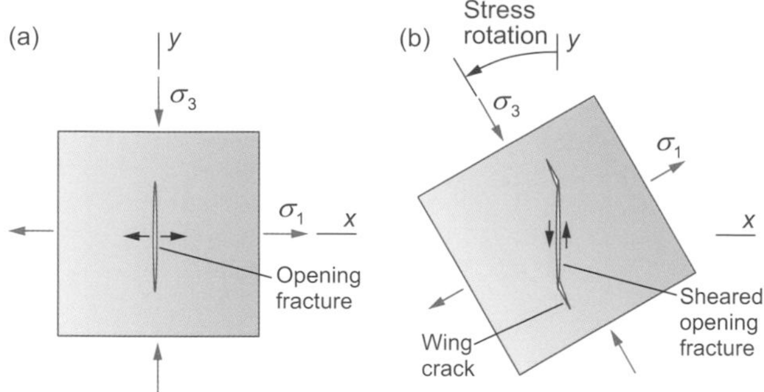

Figure 7.13 Formation of wing cracks. (a) Opening fracture (joint) forms perpendicular to least compressive or greatest tensile stress. (b) Counter-clockwise rotation of principal stresses induces left-lateral shearing on the joint, and the near-tip stress concentration promotes wing cracks that veer to the left when looking toward the tip. Modified from Segall and Pollard (1983).

$$t_x = \sigma_{xx} n_x + \sigma_{yx} n_y$$
$$t_y = \sigma_{xy} n_x + \sigma_{yy} n_y \qquad (7.2)$$

In the (x, y)-plane $n_x = \cos\theta_{\hat{i}\hat{n}}$, $n_y = \cos\theta_{\hat{j}\hat{n}} = \cos\left(\frac{\pi}{2} - \theta_{\hat{i}\hat{n}}\right) = \sin\theta_{\hat{i}\hat{n}}$. Therefore, Cauchy's formula may be written:

$$t_x = \sigma_{xx}\cos\theta_{\hat{i}\hat{n}} + \sigma_{yx}\sin\theta_{\hat{i}\hat{n}}$$
$$t_y = \sigma_{xy}\cos\theta_{\hat{i}\hat{n}} + \sigma_{yy}\sin\theta_{\hat{i}\hat{n}} \qquad (7.3)$$

Here $\theta_{\hat{i}\hat{n}}$ is the counter-clockwise angle measured from the positive x-axis to the outward unit normal vector, $\hat{\mathbf{n}}$, for the plane on which the traction vector, $\mathbf{t}$, acts (**Figure 7.15**).

Returning now to the model of the opening fracture (**Figure 7.14b**), note that the outward unit normal is in the negative y-direction for the fracture surface $y = 0^+$, so $n_y = -1$ and $n_x = 0$. Thus, the second Cauchy Formula in (7.2) reduces to $t_y = -\sigma_{yy}$, but the boundary condition (7.1) assigns $t_y = +P$, so the applied traction induces a compressive normal stress $\sigma_{yy} = -P$. For the fracture surface $y = 0^-$, the outward unit normal is in the positive y-direction, so $n_y = +1$ and $n_x = 0$. Thus, the second Cauchy Formula in (7.2) reduces to $t_y = \sigma_{yy}$ and the boundary condition (7.1) assigns $t_y = -P$, so the applied traction again induces a compressive normal stress $\sigma_{yy} = -P$. The applied

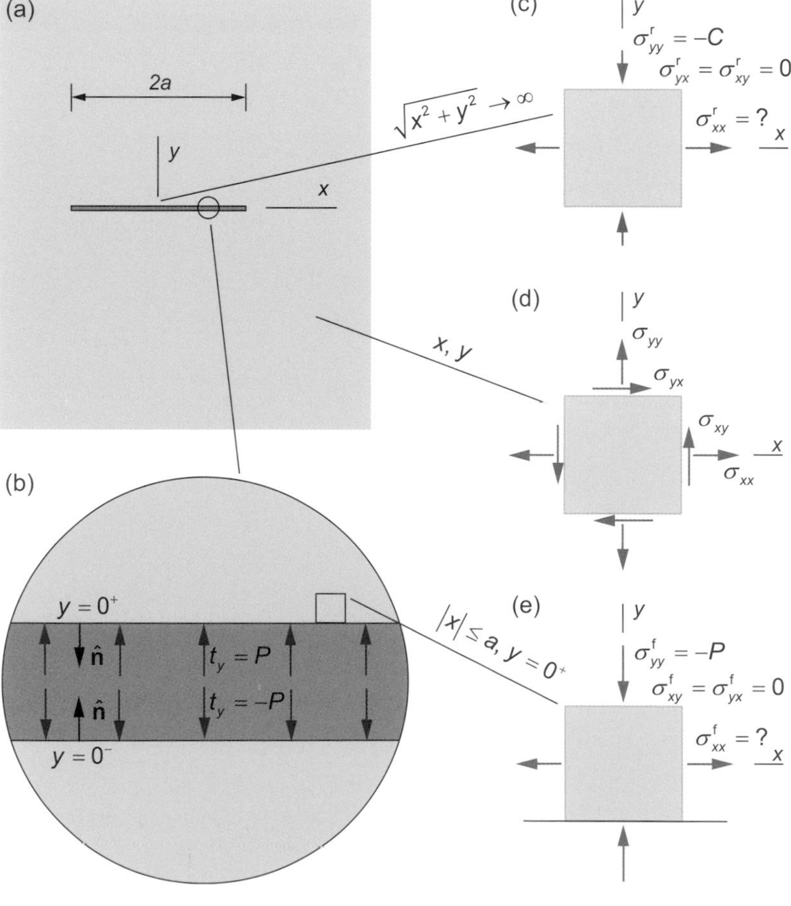

Figure 7.14 A canonical model for opening fractures. (a) Fracture of length $2a$ in an infinite elastic body. (b) Traction acting on fracture surfaces opened by fluid pressure P. (c) Remote stress state at an infinite distance. (d) Stress state in the interior of the elastic body. (e) Stress state at one surface of the fracture. See Pollard and Segall (1987).

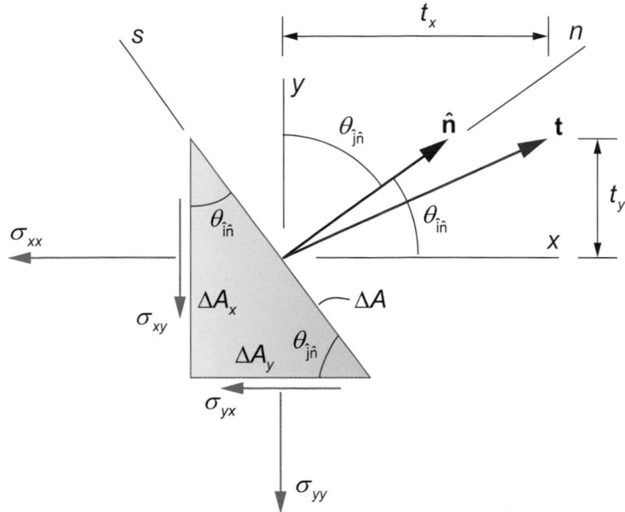

Figure 7.15 Two-dimensional version of the Cauchy tetrahedron with traction vector **t** acting on the oblique face with outward unit normal $\hat{\mathbf{n}}$, and stress components, σ_{xx}, $\sigma_{xy} = \sigma_{yx}$, and σ_{yy}, acting on the coordinate planes. Modified from Malvern (1969).

tangential traction, t_x, on both fracture surfaces is zero and $n_x = 0$, so the first Cauchy Formula in (7.2) reduces to $\sigma_{yx} = 0$.

Summarizing the boundary conditions described in the last paragraph we have:

$$\begin{aligned} &\text{at } |x| \leq a, y = 0^+, \quad \sigma^{\mathrm{f}}_{yy} = -P, \sigma^{\mathrm{f}}_{yx} = 0 \\ &\text{at } |x| \leq a, y = 0^-, \quad \sigma^{\mathrm{f}}_{yy} = -P, \sigma^{\mathrm{f}}_{yx} = 0 \end{aligned} \tag{7.4}$$

The superscript f indicates that the stress components are acting on a small element at the *fracture* surface (**Figure 7.14e**). Cauchy's second law requires the conjugate shear stress, σ^{f}_{xy}, to be zero also. The normal stress in the x-direction, σ^{f}_{xx}, is *unconstrained* by the boundary condition, which only specifies the *traction* acting on the boundary, and not the complete state of stress. The value of σ^{f}_{xx} is found from the *solution* for the elastic boundary-value problem, rather than being prescribed as a boundary condition.

The remote boundary conditions are conceptually problematic owing to difficulties imagining where infinity is, and how one would specify a boundary there. Instead of trying to define the location and orientation of a boundary at infinity, one specifies limiting values on the stress components as the radial distance from the origin goes to an infinite value. If, for example, the fracture is oriented perpendicular to the remote least compressive principal stress (**Figure 7.13a**), the normal stress acting on remote planes that are parallel to the fracture is σ^{r}_1, and the shear stress on those planes is zero (**Figure 7.14c**). On remote planes that are perpendicular to the fracture, the normal stress is σ^{r}_3 and the shear stress is zero. The superscript r indicates that these are the *remote* stresses acting at a great distance from the fracture. Thus, the *remote* "boundary" condition is:

$$\text{as } \sqrt{x^2 + y^2} \to \infty, \quad \begin{cases} \sigma_{xx} \to \sigma^{\mathrm{r}}_3 \\ \sigma_{xy} = \sigma_{yx} \to 0 \\ \sigma_{yy} \to \sigma^{\mathrm{r}}_1 \end{cases} \tag{7.5}$$

The square root term is the radial distance from the origin, so this implies that distance goes to infinity in all directions. In the absence of a fracture, (7.5) would define a homogeneous stress state throughout the elastic body. The fracture perturbs that homogeneous stress state and the solution to the elastic boundary-value problem provides the stress state at any point (**Figure 7.14d**). The literature commonly refers to this as a *stress* boundary-value problem.

At most locations in the elastic body surrounding the fracture all three independent components of the stress tensor are nonzero (**Figure 7.14d**). Near the fracture surfaces (**Figure 7.14e**) the stress components approach the values given in (7.4). At a great distance from the origin, the stress components approach the values given in (7.5). The remarkable property of the boundary-value problem is that the solution provides a *unique* distribution of the stress anywhere within the elastic body by specifying only the tractions on the internal boundaries and the stresses in the remote field.

7.2.3 Equations of Motion, Equilibrium, and Compatibility

Cauchy's laws of motion, derived in Sections 3.6.1 and 3.7.1, are the basic physical laws that underlie the canonical model for opening fractures. They assure that the model obeys conservation of linear and angular momentum for the material continuum. Restating Cauchy's equations:

$$\begin{aligned} \rho\frac{Dv_x}{Dt} &= \frac{\partial\sigma_{xx}}{\partial x} + \frac{\partial\sigma_{yx}}{\partial y} + \frac{\partial\sigma_{zx}}{\partial z} + \rho g^*_x \\ \rho\frac{Dv_y}{Dt} &= \frac{\partial\sigma_{xy}}{\partial x} + \frac{\partial\sigma_{yy}}{\partial y} + \frac{\partial\sigma_{zy}}{\partial z} + \rho g^*_y \\ \rho\frac{Dv_z}{Dt} &= \frac{\partial\sigma_{xz}}{\partial x} + \frac{\partial\sigma_{yz}}{\partial y} + \frac{\partial\sigma_{zz}}{\partial z} + \rho g^*_z \end{aligned} \tag{7.6}$$

$$\sigma_{xy} = \sigma_{yx}, \; \sigma_{yz} = \sigma_{zy}, \; \sigma_{zx} = \sigma_{xz} \tag{7.7}$$

The left sides of (7.6) are the product of mass per unit volume and the material time derivative of velocity (in other words, the particle acceleration) in each coordinate direction. The first three terms on the right sides of (7.6) are the spatial derivatives of stress components (that is surface forces per unit volume), and the last terms are the body forces per unit volume due to gravity in each coordinate direction. In other words, these are a statement of Newton's second law, $m\mathbf{a} = \mathbf{F}$, for a material continuum. Equations (7.7) state that the conjugate shear stress components are equal.

The development of opening fractures in Earth's brittle crust may occur under conditions in which the product of mass and acceleration are small compared to the surface and body forces per unit volume. For these cases the left sides of (7.6) are set to zero and the problem is reduced to one of quasi-static equilibrium. Taking the z-axis as vertical, $\rho g_x = 0 = \rho g_y$, and $\rho g_z \approx 2.6 \times 10^4 \text{ N m}^{-3}$ in Earth's lithosphere. This gravitational body force produces a gradient in stress that is approximately $2.6 \times 10^4 \text{ MPa m}^{-1}$. If the gradients in stress due to fracture opening are much greater than this gradient due to the rock

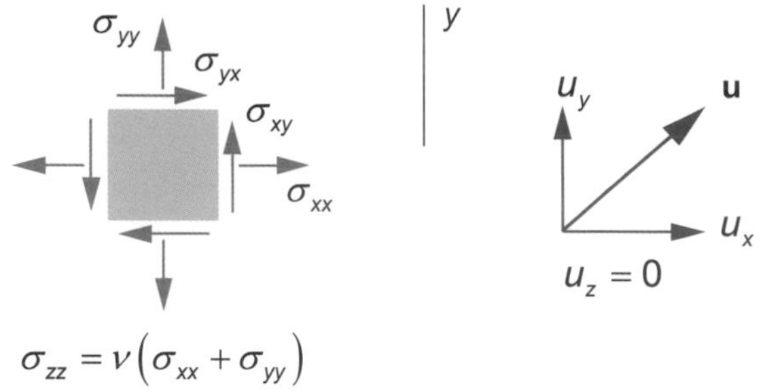

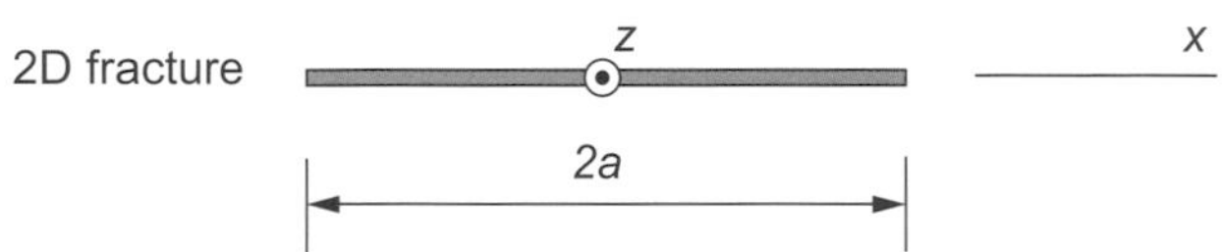

Figure 7.16 Two-dimensional, plane strain conditions ($u_z = 0$) for the canonical opening fracture. The fracture is very long in z compared to the uniform length, $2a$, in x. The stress components ($\sigma_{xx}, \sigma_{xy} = \sigma_{yx}, \sigma_{yy}, \sigma_{zz}$) and displacement components (u_x, u_y) do not vary with z.

weight, it can be neglected. Then the equilibrium equations without body forces are:

$$\frac{\partial\sigma_{xx}}{\partial x} + \frac{\partial\sigma_{yx}}{\partial y} + \frac{\partial\sigma_{zx}}{\partial z} = 0$$
$$\frac{\partial\sigma_{xy}}{\partial x} + \frac{\partial\sigma_{yy}}{\partial y} + \frac{\partial\sigma_{zy}}{\partial z} = 0 \tag{7.8}$$
$$\frac{\partial\sigma_{xz}}{\partial x} + \frac{\partial\sigma_{yz}}{\partial y} + \frac{\partial\sigma_{zz}}{\partial z} = 0$$

We consider an opening fracture (**Figure 7.16**) that is very long in the z-direction compared to its length, $2a$, in the (x, y)-plane. If the geometry and loading do not change significantly along the z-axis, the structure may be treated as a two-dimensional cylindrical structure, and this special case of deformation is called plane strain, which is defined using the displacement components:

$$u_x = u_x(x, y),\ u_y = u_y(x, y),\ u_z = 0 \tag{7.9}$$

In other words, the out-of-plane displacement component, u_z, is zero and the two in-plane displacement components are functions of x and y only.

The constraints on the displacement components from (7.9) have consequences for some of the strain components, which are apparent by recalling the kinematic equations for small strains from Section 4.9.2:

$$\varepsilon_{zz} = \frac{\partial u_z}{\partial z} = 0, \varepsilon_{yz} = \frac{1}{2}\left(\frac{\partial u_y}{\partial z} + \frac{\partial u_z}{\partial y}\right) = 0 = \varepsilon_{zy}, \varepsilon_{zx}$$
$$= \frac{1}{2}\left(\frac{\partial u_z}{\partial x} + \frac{\partial u_x}{\partial z}\right) = 0 = \varepsilon_{xz}$$

Here, in keeping with most of the literature on linear elasticity theory, we ignore the difference between derivatives of displacement with respect to the *initial* coordinates (X, Y, Z) employed to derive the small strains, and those with respect to the *final* coordinates (x, y, z) used here.

Recalling Hooke's law for the isotropic elastic material from Section 4.10, and noting that $\varepsilon_{zz} = [\sigma_{zz} - \nu(\sigma_{xx} + \sigma_{yy})]/E = 0$, the normal stress in the z-direction is:

$$\sigma_{zz} = \nu(\sigma_{xx} + \sigma_{yy}) \tag{7.10}$$

Although the displacement and normal strain components in z are zero, the normal stress in z is not. It varies in the (x, y)-plane, but does not vary with z. This non-zero normal stress is required to prevent displacement in the z-direction. If Poisson's ratio is zero, the normal stress σ_{zz} is zero everywhere in the (x, y)-plane. Hooke's law requires each shear stress to be proportional to the corresponding shear strain, so the following shear stress components are zero: $\sigma_{yz} = 0 = \sigma_{zy}$, $\sigma_{zx} = 0 = \sigma_{xz}$.

Using the constraints on the stress components described in the previous paragraph, and (7.8), the two-dimensional equilibrium equations are:

$$\frac{\partial\sigma_{xx}}{\partial x} + \frac{\partial\sigma_{yx}}{\partial y} = 0, \quad \frac{\partial\sigma_{xy}}{\partial x} + \frac{\partial\sigma_{yy}}{\partial y} = 0 \tag{7.11}$$

For example, suppose the normal stress, σ_{xx}, *increases* at a spatial rate of 1 MPa/m in the x-direction. The only stress component that can serve to balance the force due to the changing normal stress would be the shear stress σ_{yx}, which also acts in the x-direction. For equilibrium, the first of (7.11) requires σ_{yx} to *decrease* at a spatial rate of 1 MPa/m in the y-direction. The equilibrium equations are fundamental to the deformation of solids and the discipline of solid mechanics (**Box 7.1**).

Hooke's law from Section 4.10 is reduced using (7.10) for plane strain deformation, such that the surviving strain components in the (x, y)-plane are related to the stress components as:

$$\varepsilon_{xx} = \frac{1+\nu}{E}\left[(1-\nu)\sigma_{xx} - \nu\sigma_{yy}\right]$$
$$\varepsilon_{yy} = \frac{1+\nu}{E}\left[(1-\nu)\sigma_{yy} - \nu\sigma_{xx}\right] \tag{7.12}$$
$$\varepsilon_{xy} = \frac{1+\nu}{E}\sigma_{xy}$$

where E is Young's modulus and ν is Poisson's ratio. These are the strain–stress relationships for plane strain deformation. The solution to the traction (or stress) boundary-value problem provides the stress components in the (x, y)-plane, and (7.12) is used to compute the associated strain components given the two elastic moduli.

Because a complete solution to the elastic boundary-value problem includes the three independent small strain components ($\varepsilon_{xx}, \varepsilon_{xy}, \varepsilon_{yy}$), and the two displacement components (u_x, u_y), the interrelations of these quantities must be considered, and that reveals an additional constraint on solutions to these problems. Again, recalling the kinematic equations for small strains from Section 4.9.2, the kinematic equations relating small strain and displacement components for the two-dimensional plane strain problem are:

$$\varepsilon_{xx} = \frac{\partial u_x}{\partial x}, \varepsilon_{xy} = \frac{1}{2}\left(\frac{\partial u_x}{\partial y} + \frac{\partial u_y}{\partial x}\right), \varepsilon_{yy} = \frac{\partial u_y}{\partial y} \tag{7.13}$$

Box 7.1 Solid Mechanics

Solid mechanics is the study of solids in terms of their physical properties, such as density and stiffness, when treated as a material continuum (Section 3.4) subject to the fundamental constraints of mass, momentum, and energy conservation (Sections 3.5 through 3.8). Thus, solid mechanics is a subdiscipline of continuum mechanics. Here we focus on linear elasticity (Timoshenko and Goodier, 1970; Barber, 2010; see Further Reading), the branch of solid mechanics that employs Hooke's law to reduce Cauchy's equations of motion to Navier's displacement equations of motion (Section 4.11), or to the biharmonic equation for the Airy stress function (Section 7.2.4). Elasticity theory is the most fully developed and widely applicable branch of solid mechanics. It is divided broadly into quasi-static elasticity, which we employ throughout this textbook, and dynamic elasticity, which finds applications in seismology and engineering rock mechanics. Applications of elasticity theory occur in many other disciplines, including civil, mechanical, and aeronautical engineering, physics, and geodesy.

The founders of solid mechanics were physicists, engineers, and applied mathematicians. For example, Hooke (Section 4.1), who proposed the linear relationship between applied force (stress) and displacement gradient (strain) was a mathematician and natural philosopher. Navier (Section 4.11), whose name is associated with one of the sets of foundational equations of elasticity theory, was a mechanical engineer. Airy (Section 7.2.4), whose name is associated with the other set of foundational equations, was an astronomer. Mechanical engineers (Timoshenko and Goodier, 1970) and an applied mathematician (Muskhelishvili, 1975) wrote two of the classic textbooks on elasticity theory in the second half of the twentieth century. These books, along with modern textbooks (e.g. Barber, 2010) provide many analytical solutions to the foundational equations of elasticity, and are listed in Further Reading. For example, two of the most widely applied solutions are the stress distributions around a circular hole (**Figure B7.1a**) and around a crack (**Figure B7.1b**).

Structural geologists apply solutions from elasticity to a variety of geological problems, including fracturing (Section 7.2) and faulting (Section 8.4). They use concepts from elasticity theory to interpret the geometry of geological structures in terms of rock stiffness at the kilometer scale (Section 4.5), and to understand and quantify the results of laboratory experiments at the centimeter scale that document brittle strength changes associated with confining pressure and pore fluid pressure variations (Section 4.7). They also use elasticity theory to interpret the early stages of deformation in laboratory experiments that document strength and ductility changes associated with variations in confining pressure, temperature, pore fluid pressure, and strain rate (Section 5.4). These laboratory tests demonstrate that most rock passes through a nearly linear and elastic stage of deformation before proceeding to brittle fracture or ductile shearing. Because field observations document that inelastic deformation of rock usually localizes at propagating fracture tips (Section 7.3), or within ductile shear zones (Section 10.5), most of the surrounding rock mass continues to deform as a linear elastic solid. These applications indicate the relevance of solid mechanics in general, and elasticity theory in particular, for addressing important problems in structural geology. They make the case for learning the basic concepts and tools of solid mechanics during an introductory course in structural geology.

Figure B7.1 Photoelastic images using white light: color bands are contours of maximum shear stress. (a) Circular hole in thin plate loaded by vertical uniaxial tension. (b) Crack-like slot in thin plate with vertical uniaxial tension. Although contour patterns are similar, the slot has a much greater stress concentration than the hole when subject to the same remote loading.

Now suppose one has functions for the three stress components that satisfy equilibrium (7.11). These may be substituted into Hooke's law (7.12) to give the functions for the strain components without ambiguity. However, the *three* functions describing the spatial variations of the strain components *over-determine* the *two* functions needed to describe variations of the displacement components using (7.13). There must be a relationship among the strain components that settles this ambiguity.

To find the relationship conceptualized in the previous paragraph, we differentiate the first of (7.13) twice with respect to y; the second once each with respect to x and y; and the third twice with respect to x:

$$\frac{\partial^2 \varepsilon_{xx}}{\partial y^2} = \frac{\partial^3 u_x}{\partial x \partial y^2}, \frac{\partial^2 \varepsilon_{xy}}{\partial x \partial y} = \frac{1}{2}\left(\frac{\partial^3 u_x}{\partial x \partial y^2} + \frac{\partial^3 u_y}{\partial x^2 \partial y}\right),$$

$$\frac{\partial^2 \varepsilon_{yy}}{\partial x^2} = \frac{\partial^3 u_y}{\partial x^2 \partial y}$$

Noting the common terms in these equations we find:

$$\frac{\partial^2 \varepsilon_{xx}}{\partial y^2} - 2\frac{\partial^2 \varepsilon_{xy}}{\partial x \partial y} + \frac{\partial^2 \varepsilon_{yy}}{\partial x^2} = 0 \tag{7.14}$$

This is called the compatibility equation for small strains. The compatibility equation is a necessary mathematical constraint on the three strain components to assure that one can integrate the kinematic equations (7.13) and find consistent distributions for the two displacement components. By *consistent* we mean that at every point in the continuum a single value exists for each displacement component. For this reason (7.14) also is called the integrability condition.

To be useful for solving traction (or stress) boundary-value problems, (7.14) must be written in terms of the stress components. Substituting stresses for strains using Hooke's law in the form of (7.12), and then using (7.11) to eliminate the shear stress we find:

$$\left(\frac{\partial^2}{\partial x^2} + \frac{\partial^2}{\partial y^2}\right)(\sigma_{xx} + \sigma_{yy}) = 0 \tag{7.15}$$

This is the compatibility equation in terms of the stress components. The governing equations for the two-dimensional, quasi-static elastic problem of plane strain deformation are the two equilibrium equations (7.11) and one compatibility equation (7.15). Together these three equations govern the spatial variation of the three stress components in the (x, y)-plane: σ_{xx}, σ_{xy}, and σ_{yy}.

7.2.4 Solving the Elastic Boundary-Value Problem

A mathematical method to solve the governing equations for the plane strain problem was proposed by G. B. Airy in 1862 (see Malvern, 1969, in Further Reading). He defined the following relationships between the stress components and a scalar function, $\Phi(x, y)$, that now is called the Airy stress function:

$$\sigma_{xx} = \frac{\partial^2 \Phi}{\partial y^2}, \sigma_{xy} = -\frac{\partial^2 \Phi}{\partial x \partial y}, \sigma_{yy} = \frac{\partial^2 \Phi}{\partial x^2} \tag{7.16}$$

Substituting these equations for the stress components in the equilibrium equations (7.11), they both are satisfied. Substituting for the stress components in the compatibility equation (7.15), the single governing equation is:

$$\frac{\partial^4 \Phi}{\partial x^4} + 2\frac{\partial^4 \Phi}{\partial x^2 \partial y^2} + \frac{\partial^4 \Phi}{\partial y^4} = 0 \tag{7.17}$$

In mathematical terms this is called the biharmonic equation. A scalar function $\Phi(x, y)$ is a solution to a traction (or stress) boundary-value problem in elasticity if it satisfies the biharmonic equation throughout the region of interest, and gives the prescribed stresses at infinity using (7.16), and gives the prescribed tractions on all interior boundaries using (7.16) and Cauchy's formula (7.2).

For the opening fracture subject to uniform tractions on the fracture surfaces described by (7.4), and a homogeneous stress state at infinity described by (7.5), the Airy stress function was found by N. I. Muskhelishvili in 1919 and is described in his 1975 textbook (see Further Reading). The method of solution involves functions of complex variables, a mathematical subject not considered a pre-requisite for this textbook, so we do not delve further into the method. The solution itself is described in the paper by Pollard and Segall (1987), also listed in Further Reading. In the next section we evaluate the displacement components found from the Airy stress function for this canonical model in order to gain physical insight about opening fractures in Earth's lithosphere.

7.2.5 Displacement Field

The remotely applied stress state (7.5), in absence of a fracture, results in the following displacement field throughout the elastic body:

$$\left.\begin{aligned} u_x &= (x/2G)\left[(1-\nu)\sigma^{\mathrm{r}}_{xx} - \nu\sigma^{\mathrm{r}}_{yy}\right] \\ u_y &= (y/2G)\left[(1-\nu)\sigma^{\mathrm{r}}_{yy} - \nu\sigma^{\mathrm{r}}_{xx}\right] \end{aligned}\right\}, \text{ no fracture} \tag{7.18}$$

Here G is the elastic shear modulus and ν is Poisson's ratio. The displacement magnitudes are inversely proportional to the elastic shear modulus: the greater the elastic stiffness, the lesser the displacements for the same remote stress state. According to (7.18) the displacement at the origin ($x = 0 = y$) is zero. This is the *reference point* for this displacement field: all displacements due to application of the remote stresses are evaluated relative to a zero value at the origin. Selection of the reference point is arbitrary, and it is chosen for convenience to coincide with the origin of the coordinate system.

A displacement field from (7.18) is shown in **Figure 7.17** for a compression acting parallel to the x-axis and a tension acting parallel to the y-axis, with no fracture. The compressive remote stress induces displacements that are directed toward the origin, and the tensile remote stress induces displacements directed away

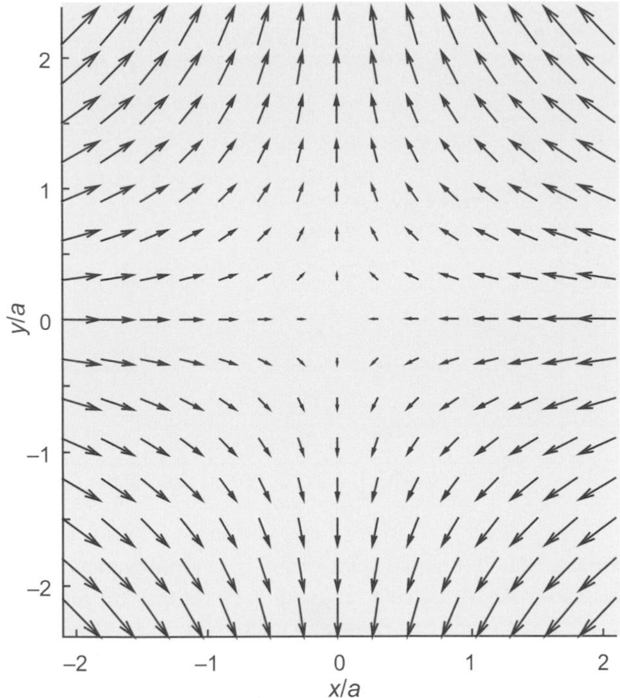

Figure 7.17 Displacement field due to a remote biaxial stress state using (7.18) with $\sigma^{\mathrm{r}}_{xx} = -1\,\mathrm{MPa}$, $\sigma^{\mathrm{r}}_{yy} = 1\,\mathrm{MPa}$, $G = 3{,}000\,\mathrm{MPa}$, and $\nu = 1/4$. The greatest displacement components in this field of view have a magnitude of 0.35×10^{-3} m. Displacement magnitude is exaggerated, so vectors can be visualized. See Pollard and Segall (1987) in Further Reading.

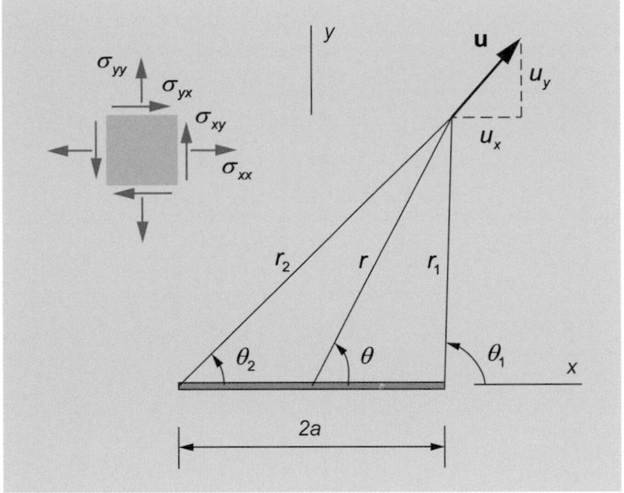

Figure 7.18 Tri-polar coordinate system used for opening fracture model: (r, θ), (r_1, θ_1), and (r_2, θ_2). For every point specified by these coordinates the solution to the boundary-value problem provides the displacement vector components, u_x and u_y, and the stress components, σ_{xx}, $\sigma_{xy} = \sigma_{yx}$, and σ_{yy}.

from the origin. The displacement magnitudes increase with distance from the reference point at the origin, and the greatest displacement components in this field of view have a magnitude that is approximately the ratio of the applied stress to the elastic shear modulus, for example $\sigma^{\mathrm{r}}_{xx}/G$.

The displacement components in absence of the fracture, (7.18), vary linearly with distance from the origin. This occurs because Hooke's law requires the strains due to a uniform state of stress (7.5) to be uniform, and integrating a uniform strain yields a displacement that increases proportionally with distance from the reference point. Looking at the kinematics from the point of view of the strains, substitute the first of (7.18) into the first of (7.13) and take the partial derivative, to find the small normal strain $\varepsilon_{xx} = \sigma^{\mathrm{r}}_{xx}(1-\nu)/2G$. This strain is uniform throughout the infinite elastic body; it is proportional to the remote normal stress in the x-direction and to $(1 - \nu)$; and it is inversely proportional to the elastic shear modulus.

The remote stresses usually are treated as constant during the initiation, propagation, and arrest of an opening fracture. This can be justified if any changes in the remote stresses occur on a time scale that is much greater than that of the fracturing. For example, rates of propagation of basaltic dikes are of order one decimeter per second, whereas creep rates on tectonic plate boundaries that might lead to regional changes in the stress state are of order one centimeter per year. In such cases the displacement field (7.18) associated with any changes in the remote stress are ignored, and we focus on the displacement field due to the opening of the fracture.

The equations for the displacement field for the canonical model (**Figure 7.14**) are rather complicated in appearance, but we write them out because it is important to appreciate that the solution provides a unique displacement field with no approximations. To write out these equations we introduce a tri-polar coordinate system (**Figure 7.18**): the origin for (r, θ) is at the Cartesian origin; the origin for (r_1, θ_1) is at the right tip of the fracture ($x = +a$, $y = 0$); and the origin for (r_2, θ_2) is at the left tip ($x = -a$, $y = 0$). We group the coordinates for the two fracture tip coordinate systems as $R = (r_1 r_2)^{1/2}$ and $\Theta = (\theta_1 + \theta_2)/2$, because these quantities appear in the displacement equations.

Fracture opening is directly proportional to the mode I driving stress, $\Delta\sigma_{\mathrm{I}}$, defined in terms of the normal stress acting perpendicular to the fracture in the both the remote field, σ^{r}_{yy}, and adjacent to the fracture surfaces, σ^{f}_{yy}:

$$\Delta\sigma_{\mathrm{I}} = \sigma^{\mathrm{r}}_{yy} - \sigma^{\mathrm{f}}_{yy} \tag{7.19}$$

The subscript I indicates this is the driving stress for *opening* fractures. In Section 7.3 we introduce the *modes* of fracture and opening fractures are termed mode I. Recall from the discussion of boundary conditions for this model (**Figure 7.14**) that $\sigma^{\mathrm{r}}_{yy} = \sigma^{\mathrm{r}}_{1}$, that is the most tensile or least compressive stress. This principal stress acts throughout the elastic body before the fracture is introduced. Also, $\sigma^{\mathrm{f}}_{yy} = -P$, where P is the fluid pressure inside the fracture. If the fluid pressure exceeds the magnitude of the least compressive stress, then the driving stress is *positive* and the fracture will *open*. For no fluid pressure in the fracture, it will

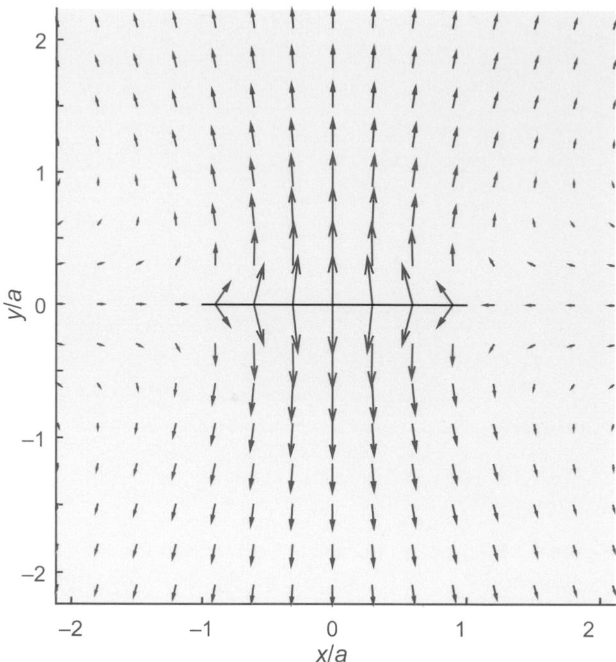

Figure 7.19 Displacement field for the opening fracture in an elastic solid. Using, $\Delta\sigma_I \approx 1$ MPa, and elastic moduli $G = 3{,}000$ MPa and $\nu = 1/4$, the greatest displacement has a value 0.25×10^{-3} m. Displacement magnitude is greatly exaggerated, so vectors can be visualized. Calculation based on Pollard and Segall (1987).

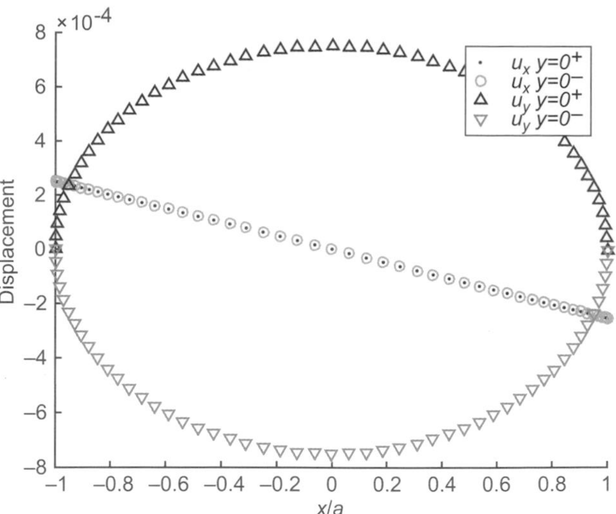

Figure 7.20 Opening displacement, u_y, and tangential displacement, u_x, calculated using (7.21) for the two surfaces, $y = 0^{\pm}$, of a mode I fracture with uniform normal traction, $\Delta\sigma_I \approx 1$ MPa, and elastic moduli $G = 1$ GPa and $\nu = 1/4$.

open only if the remote stress is tensile, $\sigma^r_{yy} > 0$. In this case the remote stress acting normal to the fracture plane pulls the two fracture surfaces apart.

The two components for the displacement field around the opening fracture (**Figure 7.18**) act in the x- and y-directions (see Pollard and Segall, 1987, in Further Reading):

$$2Gu_x = \Delta\sigma_I\{(1-2\nu)(R\cos\Theta - r\cos\theta) - r\sin\theta\left[rR^{-1}\cos(\theta-\Theta) - 1\right]\}$$
$$2Gu_y = \Delta\sigma_I\{2(1-\nu)(R\sin\Theta - r\sin\theta) - r\sin\theta\left[rR^{-1}\cos(\theta-\Theta)\right]\} \quad (7.20)$$

Each displacement component is proportional to the mode I driving stress (7.19), and therefore proportional to $\sigma^r_{yy} - \sigma^f_{yy}$. These two stresses cannot be distinguished from knowledge of the displacement field. That is, the same fracture opening can be achieved by any combination of σ^r_{yy} and σ^f_{yy} that yields the same driving stress. Each displacement component is inversely proportional to the elastic *shear modulus*, G. Thus, for a given driving stress, the softer the rock, the greater the fracture opening and the greater the displacements around the fracture.

The displacements from (7.20) due to fracture opening (not including those due to the remotely applied stresses) are illustrated in **Figure 7.19**. Fracture opening results in displacements directed toward the tips along the x-axis, and directed away from the fracture along the y-axis. Each coordinate plane is a plane of symmetry. Four "circulation" patterns of displacement vectors occur, one in each quadrant. The displacements are greatest on the fracture surfaces and decrease in magnitude with distance from the fracture. The greatest displacement has a value 0.25×10^{-3} m in this figure, which is approximately equal to the ratio of driving stress to elastic shear modulus, $\Delta\sigma_I/G$.

The displacement components on the fracture surfaces are:

$$\left.\begin{aligned} u_x &= -\Delta\sigma_I[(1-2\nu)x]/2G \\ u_y &= \pm\Delta\sigma_I\left[(1-\nu)\left(a^2-x^2\right)^{1/2}\right]/G \end{aligned}\right\} \text{ for } |x| \le a, y = 0^{\pm} \quad (7.21)$$

The distributions of the two components on the two fracture surfaces are shown in **Figure 7.20**. It is apparent from the first of (7.21) and from the graph, that the tangential displacement component, u_x, is identical for the two opposing fracture surfaces. In other words, this component has *no* displacement discontinuity between the two surfaces. Also, from the first of (7.21) and from the graph, the tangential displacement component, u_x, varies linearly with x, and is negative (directed toward the origin) for positive x, and is positive (also directed toward the origin) for negative x. In other words, the fracture appears to *shorten* as it opens. Because a field geologist typically will not know the original length of a fracture as it opens, this shortening usually is not detectable in outcrops.

Inspection of **Figure 7.20** and the second of (7.21) shows that the opening displacement component, u_y, has the same magnitude, but opposite sign, for initially neighboring particles on the two fracture surfaces. In other words, this component has a displacement discontinuity. Further inspection of **Figure 7.20** shows that the opening displacement component, u_y, varies such that the deformed surfaces have the shape of a very eccentric ellipse. The term $(a^2 - x^2)^{1/2}$ in the second of (7.21) provides this *elliptical* distribution. Note that the opening is greatly

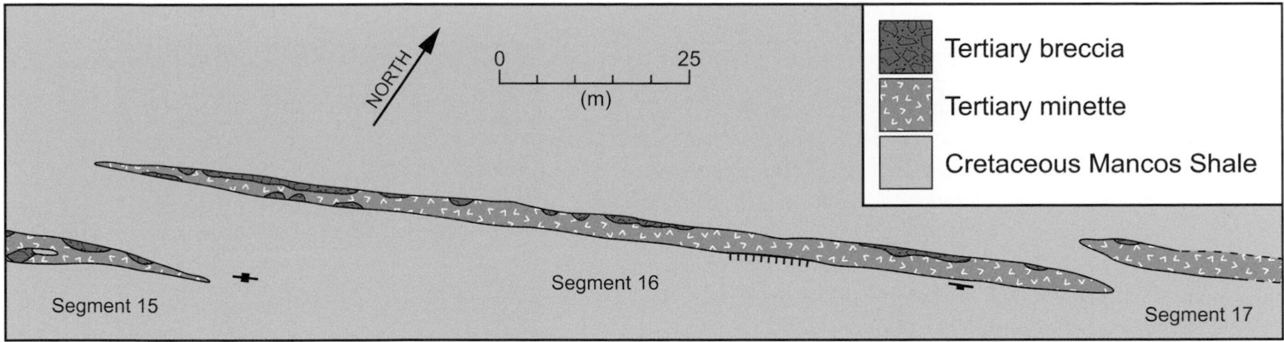

Figure 7.21 Geologic map of dike segments from Ship Rock, NM. Km – Cretaceous Mancos Shale; Tmn – Tertiary minette; Thb – Tertiary breccia. Modified from Delaney and Pollard (1981), Plate 1. UTM: 12 S 694843.47 m E, 4063820.22 m N.

exaggerated in **Figure 7.20** where the numbers on the ordinate are of order 10^{-4} and those on the abscissa are of order one.

The displacement components on the fracture surfaces given in (7.21) are for a two-dimensional geometry in which the fracture is infinitely long perpendicular to the (x, y)-plane. Thus, the two parallel tipline are infinitely long. No natural fractures have this geometry, but the mechanical behavior of fractures with constrained tipline is qualitatively similar and only differs quantitatively by a modest numerical factor. For example, if the fracture has a *circular* tipline with radius a, the opening displacements are (see Tada et al., 2000, in Further Reading):

$$u_y = \pm 2\Delta\sigma_{\mathrm{I}}[(1-\nu)(a^2-r^2)^{1/2}]/\pi G \quad \text{for } r \leq a,\ y = 0^{\pm} \tag{7.22}$$

This displacement distribution also is elliptical, and differs from that for the two-dimensional fracture (7.21) by the factor $2/\pi$.

To apply the opening displacement distribution (7.21) for the canonical model we use field data on the geological map (**Figure 7.21**) from the northeastern dike at Ship Rock, NM. The host rock is the upper part of the Cretaceous Mancos Shale (Km), and the igneous rock is Tertiary minette (Tmn). In addition to the minette, the dike contains pods of breccia (Thb) composed of fragments of shale and minette and most of these are found along the dike contact with the shale. The northeastern dike is 2.9 km in outcrop length and is aligned with an azimuth of N56°E. It is composed of 35 segments, but here we focus on segment 16, which is exposed between 1075 m and 1255 m, measured from the southwest tip. That segment is 136 m in outcrop length and has a maximum thickness of 3.4 m. We have 118 measurements of dike thickness that include the igneous rock (minette, Tms) and the breccia (Thb) with estimated errors of ±0.2 m. The average thickness is 2.5 m, so the measurement error is about 10% of the average.

The thickness of dike segment 16 is plotted versus distance along strike from the middle of the segment in **Figure 7.22**. Also plotted is a curve (blue) for an elliptical distribution of *opening* displacement discontinuity found from the second of (7.21):

$$\Delta u_y = \frac{2\Delta\sigma_{\mathrm{I}}(1-\nu)}{G}\left(a^2 - x^2\right)^{1/2} \tag{7.23}$$

The leading quotient in this equation has a value of about 0.04 for the fitted curve. Ignoring the minor contribution of Poisson's ratio

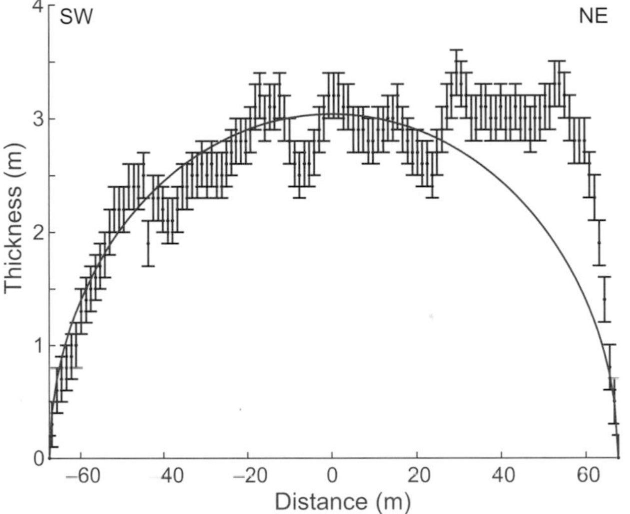

Figure 7.22 Thickness of dike segment 16 with red error bars plotted versus distance along the outcrop of the Northeastern dike at Ship Rock, NM. See **Figure 7.21** for location. Blue curve is approximate fit to data using elliptical opening displacement discontinuity from (7.23). Data from Delaney and Pollard (1981).

compared to one, this is consistent with a driving pressure $\Delta\sigma_{\mathrm{I}} \approx$ 2 MPa and a shear modulus $G \approx 100$ MPa. This highlights the *first benefit* of understanding and applying a canonical model to geologic structures: one can use the quantitative relationships found in the solution to estimate the values of key physical quantities.

The estimated shear modulus, $G \approx 100$ MPa, is an order of magnitude (or more) softer than values from laboratory tests. Why might this occur? In part, this is due to the 100 m scale of the deformation associated with the dike segment. Many heterogeneities (fractures, bedding surfaces, etc.) at the field scale soften the elastic response. Perhaps of more importance is the presence of the other 34 segments of the northeastern dike at Ship Rock. The length of the entire dike is 2,901 m, and this much greater length provides a greater leverage for the given magma pressure acting on the elastic host rock. Taking a maximum thickness $\Delta u_y(x = 0) \approx$ 3 m, a half-length for the entire dike $a \approx 1{,}450$ m, and a driving pressure $\Delta\sigma_{\mathrm{I}} \approx 2$ MPa, we rearrange (7.23) to find $G \approx 2$ GPa. This is more in keeping with laboratory results for the elastic shear modulus.

The model curve (blue) in **Figure 7.22** is a good fit to the thickness data for the SW half of dike segment 16, but significantly underestimates the thickness near the NE tip of this segment. A *second benefit* of applying a canonical model to a geologic structure is that discrepancies serve to identify interesting aspects of the structure that might otherwise have gone unappreciated. Having identified that segment 16 exceeds the model thickness near its NE tip we return to the map (**Figure 7.21**) to look for explanations of this behavior. One possible explanation is the presence of segment 17, which is offset by only 6 m and overlaps segment 16 by 5 m. This close spacing and small overlap suggests that the two segments merge a short distance below the current outcrop, so the mechanical behavior of the two segments should be similar to that of a single segment with a length equal to their combined lengths. The two segments, although clearly separated at the current level of erosion, apparently opened more than they would have, if they had been farther apart.

A more subtle discrepancy between dike thickness and model fracture opening (**Figure 7.22**) is related to the presence of breccia between the igneous rock and the Mancos Shale. The dike thickness includes the breccia, but the breccia is composed, in part, of fragments of Mancos Shale. Thus, *two mechanisms* led to the thickness of dike segment 16. The first mechanism, and the one that is addressed by the canonical model described above, is the opening of a 136 m long fracture in the Mancos Shale due to the invading magma. The second mechanism is the local *fragmentation* of the Mancos Shale at the centimeter to meter scale to form the breccia. The solution to the elastic boundary-value problem that leads to the opening displacement discontinuity evaluated in (7.23) does not address the issue of breccia formation, so one should not expect the model to match the field data where a significant part of the dike is composed of breccia. Some of the fluctuations in thickness apparent in **Figure 7.22** are ascribed to the breccia. This example illustrates a *third benefit* of applying a canonical model to a geologic structure: discrepancies between the model and data may bring to light new mechanisms that require a completely different modeling approach.

7.2.6 Stress Field

We evaluate the stress field for the canonical model of an opening fracture in an elastic solid to gain further insight about fractures in rock. Again, we employ tri-polar coordinates (**Figure 7.18**) to write out the analytical equations for the three stress components in the (x, y)-plane around the model fracture (**Figure 7.18**) (see Pollard and Segall, 1987, in Further Reading):

$$\sigma_{xx} = \sigma^{\mathrm{r}}_{xx} + \Delta\sigma_{\mathrm{I}}\left[rR^{-1}\cos(\theta-\Theta) - 1 - a^2rR^{-3}\sin\theta\sin 3\Theta\right]$$
$$\sigma_{xy} = \sigma^{\mathrm{r}}_{xy} + \Delta\sigma_{\mathrm{I}}\left[a^2rR^{-3}\sin\theta\cos 3\Theta\right]$$
$$\sigma_{yy} = \sigma^{\mathrm{r}}_{yy} + \Delta\sigma_{\mathrm{I}}\left[rR^{-1}\cos(\theta-\Theta) - 1 + a^2rR^{-3}\sin\theta\sin 3\Theta\right] \quad (7.24)$$

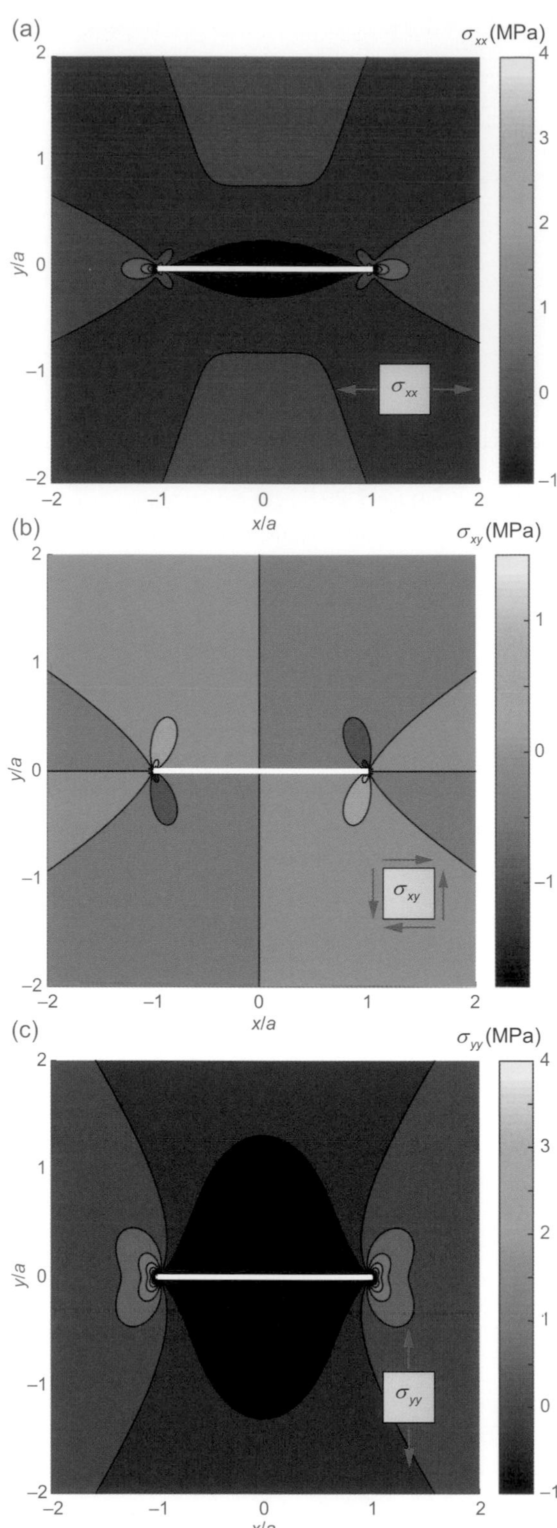

Figure 7.23 Stress components for an opening fracture (white line) in a linear elastic material under plane strain conditions subject to a 1 MPa pressure within the fracture and zero stresses at infinity. Tensile stress is positive and the stress magnitudes are in units MPa. (a) σ_{xx}, (b) σ_{xy}, (c) σ_{yy}.

The first term in each equation is the corresponding remote stress component that is present and homogeneously distributed before the fracture opens. The second term accounts for the *perturbation* due to fracture opening. The perturbed stress is proportional to the mode I driving stress and *independent* of the elastic moduli. For two-dimensional problems with traction boundary conditions the stress state at any point is not a function of the elastic properties. However, for displacement boundary conditions the stress components are functions of the elastic moduli. Examples of displacement boundary-value problems are given in Section 5.6.3 and Section 5.6.4 for the screw and edge dislocations.

The stress components are distributed around the fracture (and out to infinity) so that equilibrium (7.11) and compatibility (7.15) are satisfied. Examples of these distributions are illustrated in **Figure 7.23** for the case of a unit pressure within the fracture and zero stress at infinity. Each stress component contoured in **Figure 7.23** increases in magnitude toward the two fracture tips, so we say the stresses are *concentrated* at the tips. Each stress component decreases with distance from the fracture and tends toward zero at infinity.

The two normal stress components (**Figure 7.23a, c**) are distributed symmetrically with respect to the coordinate axes and with respect to the fracture plane. Both normal stress components are compressive (negative) to either side of the fracture, and this is referred to as a stress shadow. Opening of the fracture induces this compression, which would tend to prevent other parallel fractures from opening nearby. Both normal stress components are tensile (positive) ahead of the fracture tips, and increase without bounds toward the tips. This is referred to as a stress concentration. Propagation of opening fractures is attributed to this elevated tension, for example in the stress component σ_{yy} (**Figure 7.23c**). In the next section we zoom in and explore these near-tip stress distributions. The shear stress component has an anti-symmetric distribution with respect to the coordinate axes and the fracture plane (**Figure 7.23b**). Taking a clockwise circuit around the right fracture tip starting at the $y = 0^+$ surface, the sign of the shear stress changes from negative to positive, and then to negative and positive. The shear stress also increases without bounds toward the tips.

7.3 FRACTURE MODES AND THE NEAR-TIP FIELDS

The fracture modes characterize the deformation near a fracture tip, where a subsequent increment of propagation is likely to occur. Therefore, to study the initiation, propagation, and arrest of joints, veins, and dikes we need to understand the fracture modes, which depend on the displacement discontinuity between the two fracture surfaces (see Anderson, 1995, in Further Reading).

7.3.1 Displacement Discontinuity

A displacement discontinuity develops between the two fracture surfaces, one of which initially lies along $|x| \leq a$, $y = 0^+$ and the other lies along $|x| \leq a$, $y = 0^-$, as depicted in **Figure 7.24** in two

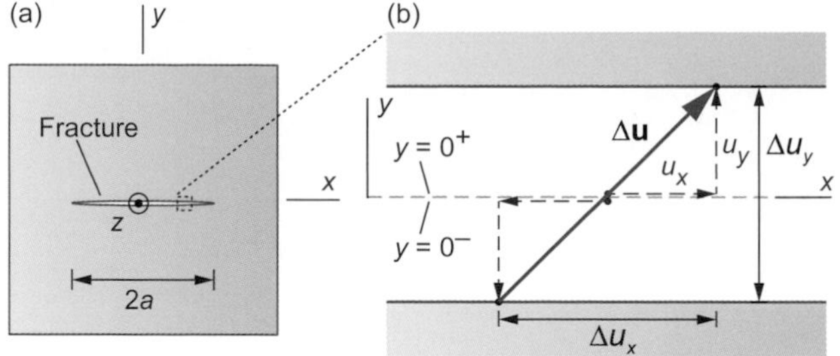

Figure 7.24 Displacement discontinuity for a fracture, shown here in two dimensions only. (a) Schematic fracture of length 2*a*. The *z*-axis points toward the viewer. (b) Fracture surfaces initially lie along the dashed blue line. Two neighboring particles (black circles) are cut apart by the fracture and displaced in opposite directions. The displacement discontinuity vector, $\Delta\mathbf{u} = \langle \Delta u_x, \Delta u_y \rangle$, is the difference between these particle displacements as defined in (7.25).

dimensions. Two neighboring particles are located, respectively, on what will become the two deformed surfaces of the fracture. The displacement of the particle located on $y = 0^+$ is $\mathbf{u} = \langle u_x, u_y, u_z \rangle$, and that of the neighboring particle on $y = 0^-$ is of equal magnitude, but opposite direction. The displacement discontinuity, $\Delta\mathbf{u} = \langle \Delta u_x, \Delta u_y, \Delta u_z \rangle$, is the difference between these two displacements. This vector quantity has three components:

$$\text{on } |x| \leq a \begin{cases} \Delta u_x = u_x(y = 0^+) - u_x(y = 0^-) \\ \Delta u_y = u_y(y = 0^+) - u_y(y = 0^-) \\ \Delta u_z = u_z(y = 0^+) - u_z(y = 0^-) \end{cases} \quad (7.25)$$

The displacement discontinuity is defined over the entire fracture, from the left tip to the right tip, and by definition it goes to zero at the tips: $\Delta\mathbf{u} = 0$ for $|x| = a$. The displacement discontinuity measures a *jump* in the displacement across the fracture. For the pure opening fracture the tangential displacements (7.21) are identical for initially neighboring particles on the two fracture surfaces, so $\Delta u_x = 0$. However, the perpendicular displacements are equal and opposite, so $\Delta u_y > 0$.

Determining the displacement discontinuity from geological observations in the field requires a piercing point in order to identify initially neighboring particles (**Figure 7.3**). Given such a piercing point the displacement discontinuity vector is directed from one particle to the other, and the magnitude of this vector is the measured distance between the two particles, as illustrated in the example of the Independence dike from the Sierra Nevada (**Figure 7.4**).

7.3.2 Fracture Modes

Conceptualizing fractures as two surfaces joined at a tipline leads to a simple geometric idealization of the fracture tip (**Figure 7.25**). Here the distance along the fracture from the point where the displacement discontinuity is measured to the fracture tip is symbolized as Δx, and it is understood that this distance is much less than the half length of the fracture, $\Delta x << a$ (**Figure 7.24a**). We define a small neighborhood surrounding

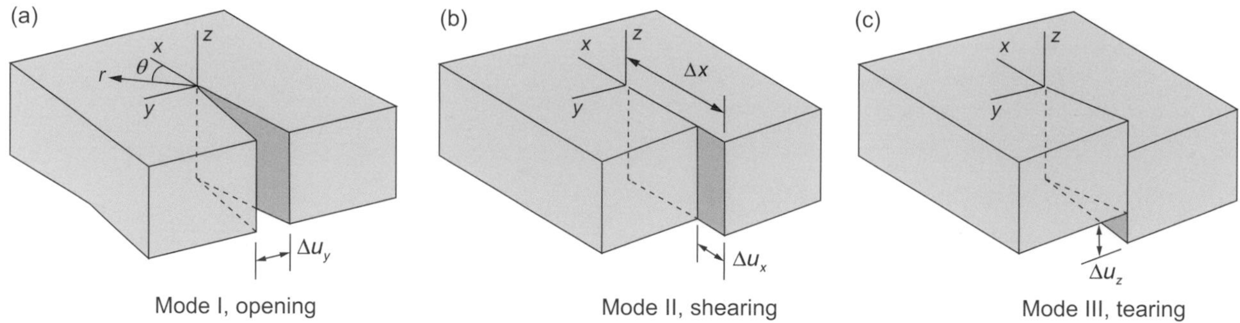

Figure 7.25 Fracture modes based on the components of the displacement discontinuity, $\Delta\mathbf{u}$, very near the fracture tip. (a) Δu_y, mode I opening. (b) Δu_x, mode II shearing. (c) Δu_z, mode III tearing. The displacement directions for the two fracture surfaces are accurately drawn, but the magnitudes, relative to the distance, Δx, from the tip is exaggerated for clarity.

the tip by restricting the radial distance from the tip, r, such that $r < 0.01a$. In this neighborhood the stress and displacement distributions are highly dependent on the local geometry of the fracture tip. This *near-tip* perspective has been exploited in the discipline of fracture mechanics (**Box 7.2**) to characterize the mode of fracture and to understand fracture propagation.

In the near-tip area as depicted in **Figure 7.25** we suppose the tipline of the fracture is adequately described as a *straight line*, and associate that with the z-axis. Furthermore, we suppose the fracture surfaces are adequately described as *planar* before the deformation, and take the normal to that plane as the y-axis.

We identify the three components of the displacement discontinuity (7.25) very near the fracture tipline (**Figure 7.25**), and use these components to define the fracture modes. For mode I fracture the displacement discontinuity is parallel to the y-axis: $\Delta\mathbf{u} = \langle 0, \Delta u_y, 0\rangle$. Thus, the fracture surfaces move apart and this is referred to as an *opening* fracture tip. For mode II fracture the discontinuity is parallel to the x-axis: $\Delta\mathbf{u} = \langle \Delta u_x, 0, 0\rangle$. The fracture surfaces remain in contact and this is referred to as a *shearing* fracture tip. For mode III fracture the discontinuity is parallel to the z-axis: $\Delta\mathbf{u} = \langle 0, 0, \Delta u_z\rangle$. The fracture surfaces again remain in contact and this is referred to as a *tearing* fracture tip. Fractures that include combinations of these pure modes are referred to as mixed-mode fractures. In Earth's lithosphere the abundance of fractures, the prevalence of heterogeneities in material properties, and the common spatial variations in the state of stress suggest that fracture tips subject only to a pure mode are rare.

Although the modes are introduced here using a geometric motivation (**Figure 7.25**), each mode is associated with a unique local stress field, a different criterion for propagation, and different directions of propagation. Thus, the modes are intimately related to key mechanical aspects of fracture. In the next section we describe the local stress field associate with each mode of fracture.

7.3.3 Near-Tip Stress Distributions

To investigate the stress distributions in the vicinity of a fracture tip, the polar coordinate system (r, θ, z) with origin at the tip is convenient (**Figure 7.25a**). To be within the vicinity of the fracture tip, we restrict the radial distance as $r < 0.01a$, where $2a$ is a characteristic length of the fracture (**Figure 7.24a**). This restriction is necessary to assure that the equations for the near-tip stresses adequately approximate the actual stresses. For pure mode I fracture tips (**Figure 7.25a**) the stress components in the (x, y)-plane are approximated as (see Lawn, 1993, in Further Reading):

$$\begin{Bmatrix} \sigma_{xx} \\ \sigma_{xy} \\ \sigma_{yy} \end{Bmatrix} \approx \frac{K_\mathrm{I}}{\sqrt{2\pi r}} \begin{Bmatrix} \cos(\theta/2)[1 - \sin(\theta/2)\sin(3\theta/2)] \\ \sin(\theta/2)\cos(\theta/2)\cos(3\theta/2) \\ \cos(\theta/2)[1 + \sin(\theta/2)\sin(3\theta/2)] \end{Bmatrix} \quad (7.26)$$

This is the near-tip stress field for the mode I fracture.

The quantity K_I in (7.26) is called the mode I stress intensity, and it is a measure of the stress magnitude in the near-tip region. For consistent units in this equation the stress intensity must have units of stress times the square root of a length: K_I [=]MPa m$^{1/2}$. The stress intensity depends upon the gross geometry of the fracture and the configuration of the loads that are causing the fracture tip to open. For example, the mode I stress intensity at the tips of the canonical opening fracture (**Figure 7.14**) is:

$$K_\mathrm{I} = \left(\sigma^\mathrm{r}_{yy} - \sigma^\mathrm{f}_{yy}\right)\sqrt{\pi a} \quad (7.27)$$

The stress intensity is proportional to the mode I driving stress. Engineering handbooks (e.g. Tada et al., 2000, in Further Reading) compile hundreds of examples of stress intensities for a wide variety of fracture geometries and loads.

To visualize an example of the near-tip stress field, we consider the normal stress, σ_{yy}, for the canonical mode I fracture and use the second of (7.26) to produce the contours in **Figure 7.26**. For this example K_I = 1 MPa m$^{1/2}$. Note that the field of view is centered on the right fracture tip and extends only a distance $\pm 0.02a$ in the x- and y-directions from this tip.

The equations for the near-tip stresses, (7.26), are particularly useful because they separate *stress magnitude*, which is proportional to the appropriate stress intensity, from *stress distribution*, which is proportional to $1/(r)^{1/2}$ and to trigonometric functions of the angle θ. These distributions are independent of how the

Box 7.2 Fracture Mechanics

Fracture mechanics is the study of materials in terms of their resistance to the initiation, propagation, and termination of fractures, when treated as a material continuum (Section 3.4) subject to the fundamental constraints of mass, momentum, and energy conservation (Sections 3.5 through 3.8). Thus, fracture mechanics (Anderson, 1995; see Further Reading) is a sub-discipline of continuum mechanics. Fractures commonly are associated with elastic–brittle deformation of rock (Chapter 4), but also occur during ductile and viscous deformation. However, in this textbook, we focus on *linear elastic* fracture mechanics (LEFM), which restricts attention to materials and conditions where the length-scale of inelastic yielding, associated with the stress concentration at the fracture tip, is very small compared to the fracture length. This enables us to introduce the basic concepts of rock fracture mechanics (Atkinson, 1987; Gudmundsson, 2011; see Further Reading) with many applications to structural geology (**Figure B7.2**). Fracture mechanics also finds applications in engineering rock mechanics, civil and mechanical engineering, and solid earth geophysics.

The founders of fracture mechanics in the first half of the twentieth century were engineers dealing with the problems of designing ships, airplanes, and critical structures such as bridges and dams, that were susceptible to failure by fracture (Gordon, 1976; Anderson, 1995; see Further Reading). One of the founders, C. E. Inglis, solved the elastic boundary-value problem for an elliptical hole in an elastic plate in 1913 and learned that the **stress concentration** at the end of the long axis of that hole (Section 7.2.6) increased in proportion to the ratio of the length of the long axis to that of the short axis. For a very eccentric elliptical hole that resembles a crack (**Figure B7.1b**), this ratio is very large. The second seminal insight came from A. A. Griffith in 1921, who proposed and demonstrated that fractures initiate at **flaws** (Section 7.4.1), where the remotely applied stress is concentrated. The third seminal insight came from G. R. Irwin in 1957, who showed that the stress concentration is separable into three independent **fracture modes** (Section 7.3.2) that depend on the orientation of the displacement discontinuity near the fracture tip.

Recognition in the second half of the twentieth century that fracture growth in an engineered structure could lead to catastrophic failure, with significant loss of property and life, prompted the rapid growth of engineering fracture mechanics. The *International Journal of Fracture* began publishing in 1965, and *Engineering Fracture Mechanics* began publishing in 1968. Today, fracture mechanics is a recognized sub-discipline in the engineering schools of major universities. Joints, veins, and dikes are examples of dominantly mode I (opening) fractures in rock (Section 7.1). Sheared joints, shear fractures, and faults are examples of dominantly mode II and mode III fractures in rock (Section 7.1.4 and Chapter 8). Thus, opportunities abound for a structural geologist to take advantage of the basic concepts and results of ongoing research in engineering fracture mechanics.

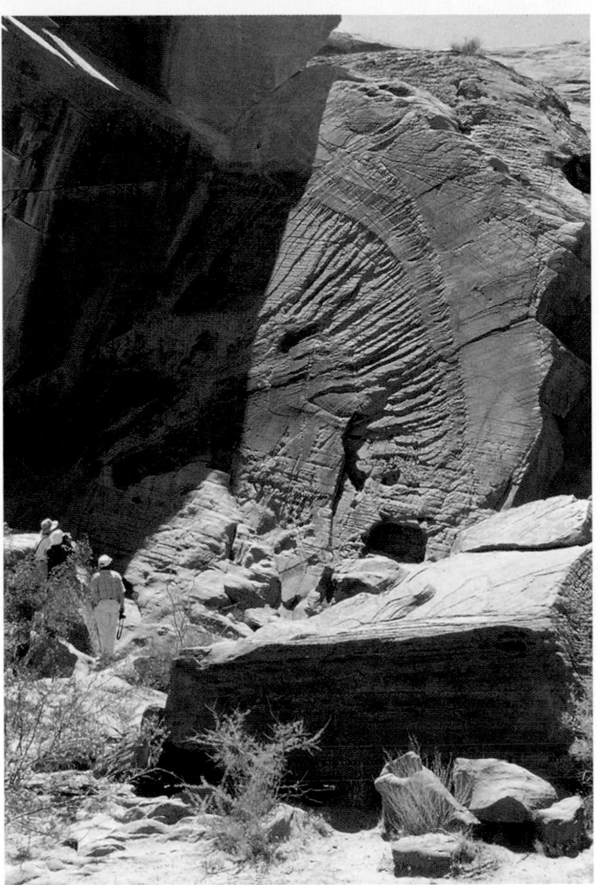

Figure B7.2 Structural geologists admire plumose structure and echelon segments on a joint surface at the Valley of Fire, NV. See **Figure 7.7** for naming conventions and **Figure 7.35b** for a mechanical interpretation.

fracture is loaded to produce the opening, and are the same for all opening fracture tips in homogeneous and isotropic linear elastic materials. These dependencies do not change, for example, for mode I fractures that are opened by an internal pressure, or by a remote tension, or by point forces, or by bending moments, or by any loading that produces opening near the tip. Therefore, we can investigate all opening fractures in such materials using the distribution of stress given by (7.26).

The dependence of the near-tip stresses on $1/(r)^{1/2}$ deserves special consideration, because the stresses go to an infinite value as the radial distance goes to zero. Thus, the stress state from the elastic solution is non-physical near the tip. To amend the solution one has to change the constitutive properties of the material and require an upper bound on the stress, a requirement that is not enforced in elasticity theory because the idealized elastic materials have infinite strength. One choice is to include inelastic deformation that limits the stress by plastic yielding, as we discussed in Section 5.1. If the region of plastic deformation is very small compared to the near-tip region, then this plastic deformation can be ignored. This condition is called small-scale yielding, and we invoke this condition in what follows. Small-scale yielding implies that the material is *brittle*, so the fracture propagates rather than sustaining more wide spread plastic deformation near the fracture tip.

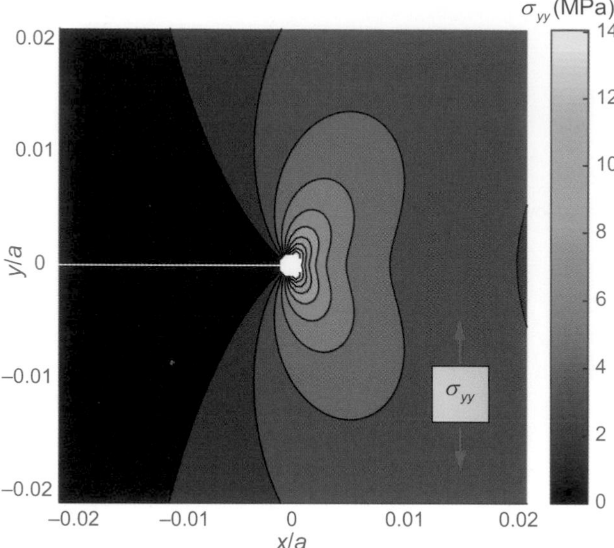

Figure 7.26 Normal stress σ_{yy} near the tip of an opening fracture (white line) of unit half-length in a linear elastic material under plane strain conditions subject to a 1 MPa pressure within the fracture. The remote stresses are zero. Tensile stress is positive and the stress magnitudes are in units MPa.

Important features of the near-tip stress distribution are illuminated by plotting the normalized stress, $\sigma\sqrt{2\pi r}/K_{\mathrm{I}}$, versus the angle θ using equations (7.26). We consider the range $-\pi \leq \theta \leq +\pi$ from one fracture surface around to the other as illustrated by the dashed circle in **Figure 7.27a**. The distributions of the three normalized stress components are plotted in **Figure 7.27b**.

Referring to **Figure 7.23c**, we suggested that propagation of opening fractures might be attributed to elevated tension in the component σ_{yy} just ahead of the fracture tip. The stress σ_{yy} is indeed positive (tensile) everywhere around the fracture tip (**Figure 7.27b**). However, this stress does *not* reach a maximum directly ahead of the fracture tip. Instead, the distribution is bimodal, reaching two equal maxima to either side of the tip, where $\theta = \pm 60°$. This distribution suggests that opening fracture propagation may not be explained simply by referring to one stress component immediately ahead of the fracture tip. Where $\theta = 0°$, the two orthogonal normal stresses are equal and the shear stress is zero. Thus, the normal stresses in all directions are equal along $\theta = 0°$: a state of *isotropic tension* exists directly ahead of the fracture.

While conceding that the near-tip stress distribution does not lead to a simple explanation for opening fracture propagation, the

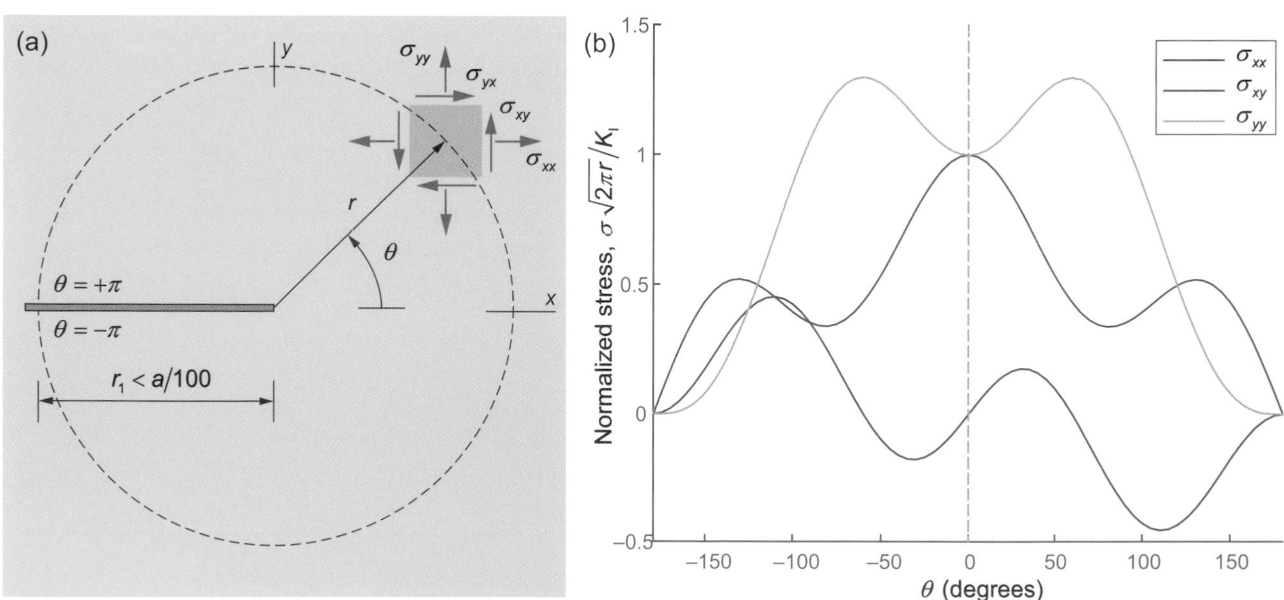

Figure 7.27 (a) Near-tip region for the right fracture tip. Polar coordinates (r, θ) are centered at the fracture tip. Stress components are those defined in (7.26) for mode I. (b) Distributions of normalized stress components versus angle θ using the trigonometric functions in (7.26). Modified from Lawn and Wilshaw (1975).

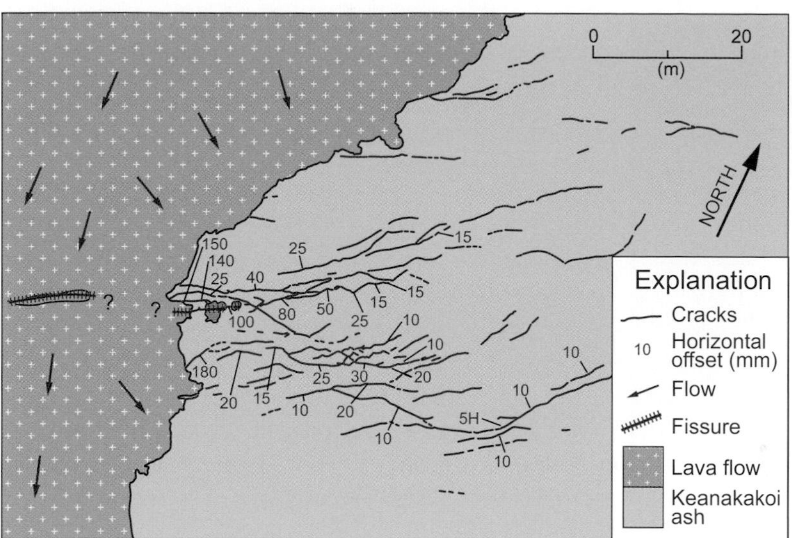

Figure 7.28 Structure map of open cracks in the Keanakakoi ash deposit from the 1974 eruption on the southwestern rift of Kilauea Volcano. Horizontal crack opening "H" in millimeters. The array of eruptive fissures extends 775 m to the southwest. Modified from Pollard et al. (1983), Figure 23. UTM: 5 Q 259524.42 m E, 2144350.27 m N.

distributions of stress (**Figure 7.27b**) do lead to important insights about secondary cracks that develop in the host rock to either side of a fracture tip. We take up this application of equations (7.26) in the next section.

7.3.4 Secondary Cracks Near Fracture Tips

A dike breeched the surface of the southwest rift zone of Kilauea Volcano on December 31, 1974 and fed a small lava flow. Eruptions that are fed by dikes are referred to as fissure eruptions and the gapping conduit through which the lava flows onto the surface is referred to as a *fissure*. Near the northeast end of the fissure eruption several meters of an ash deposit, called the Keanakakoi Formation, provided a nearly unbroken and smooth cover that was ruptured by the upward propagating dike to form the fissures. The Keanakakoi ash also is broken by much smaller vertical cracks that are exposed near the ends of the longer eruptive fissures. A structure map (**Figure 7.28**), constructed after the eruption near the northeast end of a 775 meter-long fissure, documents the open vertical cracks in the ash deposit. Typical examples of these secondary cracks are 5 to 20 m long and they gap open by a few centimeters at the surface. The cracks are approximately parallel to the closest fissure and they cluster into two groups, one to either side of the projection of that fissure to the northeast. This set of cracks is about 50 m wide and extends about 50 m to the northeast beyond the nearest fissure.

The timing of the vertical cracks (**Figure 7.28**) is corroborated by a dramatic nighttime photograph taken during the fissure eruption (**Figure 7.29a**). The photographer saw the set of cracks open as the fissure eruption advanced, but most were quickly covered by the lava. However, when the dike stopped propagating some cracks were preserved off the end of each long array of fissures. We interpret the formation of these cracks as due to the dual maxima (**Figure 7.27b**) in the normal stress component, σ_{yy}. In this case the upward propagating dike is the mode I fracture, and this normal stress component would be parallel to the ground surface and approximately perpendicular to the dike and the vertical cracks. The maxima are at $\theta = \pm 60^\circ$ according to the stress analysis. As the dike propagated toward the surface, the distance between the two maxima decreased, so additional cracks formed closer to the dike and were more closely spaced. Those cracks closest to the dike tip emitted steam, presumably from groundwater brought to a boil by the hot lava.

The model stress distribution (7.26) does not include the effect of the traction-free boundary condition, but a model that does account for this condition gives a similar dual maximum for σ_{yy}. The development of a set of secondary cracks, smaller in scale than the primary fracture, and clustered near the primary fracture trace, is a common occurrence. Such crack sets have been documented in association with deeply buried dikes, and with propagating opening mode fractures at the centimeter scale in laboratory experiments. This zone of cracking is referred to as the process zone of the mode I fracture.

7.3.5 A Small Primer on Stress Intensity

The basic premise of linear-elastic fracture mechanics is that propagation of an opening fracture depends on the near-tip stress magnitude, which is determined by the mode I stress intensity. Therefore, knowledge of the stress intensity for a fracture with a given geometry and loading is of paramount importance. In order to build intuition about the mechanical behavior of fractures we review the stress intensity for three classic fracture geometries. For each case, the stress intensity was found by solving the corresponding elastic boundary-value problem. Hundreds of solutions are available in the literature of engineering mechanics, and these are documented in the handbook by Tada et al. (2000) in Further Reading.

The two-dimensional fracture with uniform remote tension, $\sigma^r_{yy} = T$, is the canonical opening fracture model (**Figure 7.14**). We refer to this geometry as the 2D blade-shaped fracture

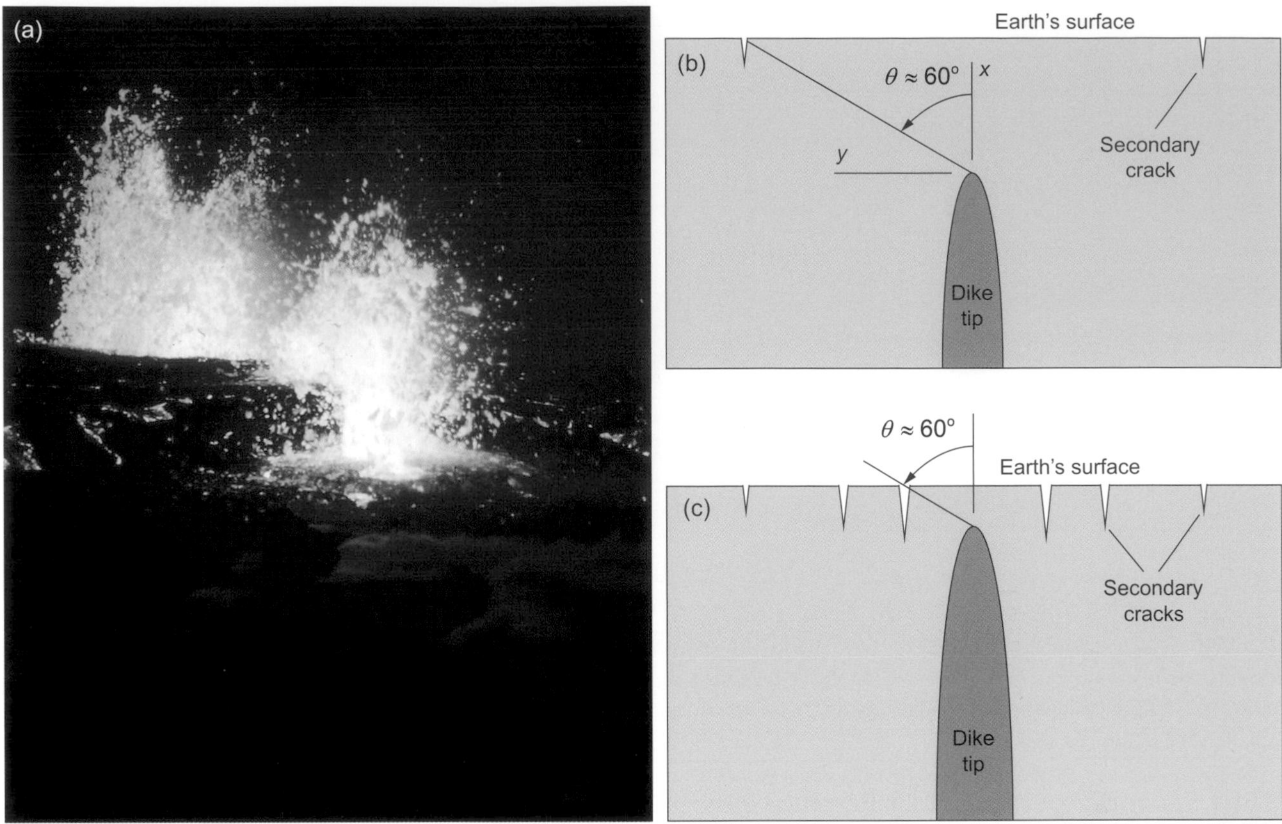

Figure 7.29 (a) Photograph of lava erupting from fissure that breeches the surface in the background. In the foreground cracks have opened in the Keanakakoi ash and steam is flowing out of them. Photograph by R. Holcomb. (b) Schematic vertical cross section through the dike with two cracks forming at the surface. (c) As dike tip propagates upward, more secondary cracks form, and they are closer to the dike. See Pollard et al. (1983) and Figure 7.28 for location.

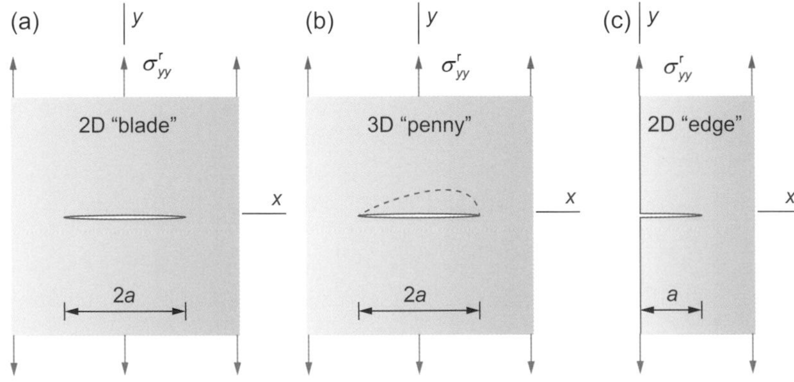

Figure 7.30 Effects of different fracture geometry, but the same uniform remote tension, on the mode I (opening) stress intensity. (a) 2D blade-shaped fracture. (b) 3D penny-shaped fracture. (c) 2D edge fracture.

(**Figure 7.30a**) because the fracture is very long (infinite) in the z-direction with a uniform length, $2a$, in the x-direction, so it resembles the blade of a long sword. For this geometry and loading the stress intensity is:

$$K_I = \sigma^r_{yy}\sqrt{\pi a}, \text{ blade-shaped fracture} \tag{7.28}$$

The stress intensity is proportional to the remotely applied tensile stress and to the square root of the fracture half length. The remote stress must be positive (tensile) for the fracture to open, so the stress intensity is positive.

One might suppose that changing the geometry of the fracture tipline from straight to circular would change the stress intensity dramatically, but it does not. We refer to this geometry as the 3D penny-shaped fracture (**Figure 7.30b**), because the fracture resembles a very flat penny with a uniform diameter, $2a$, in the (x, z)-plane. The stress intensity is:

Figure 7.31 Joint initiation at a small cavity in siltstone. Hackle radiate from the hole, which apparently was the source of the stress concentration that induced the fracture. Photograph by A. Aydin. See Pollard and Aydin (1988) in Further Reading.

$$K_{\mathrm{I}} = \left(\frac{2}{\pi}\right)\sigma^{\mathrm{r}}_{yy}\sqrt{\pi a}, \text{ penny-shaped fracture} \quad (7.29)$$

The stress intensity for the penny-shaped fracture is less than that for the blade-shaped fracture by a factor $2/\pi$. It is intuitive that the stress intensity for the penny shape is less, because the surface area of the fracture is less, and therefore the fracture is exposed to less total force. It is not intuitive that the stress intensity is 64% of that for the infinitely long fracture. What seems a radical change in geometry, results in only a modest change in stress intensity.

Perhaps an even more radical change in geometry is the 2D edge fracture (**Figure 7.30c**). The edge fracture applies to fractures that extend downward from Earth's surface, such as those associated with fissure eruptions (**Figure 7.28**). The surface $x = 0$ is traction-free, so for this case the entire infinite half-space $x < 0$ is removed, leaving a fracture that has a length a in the remaining half-space. With a shorter fracture length, one might suppose the stress intensity would be less, but removing half of the elastic body means that half of the resistance to fracture opening provided by the elastic stiffness is removed. How these two competing effects play out is not obvious, but the stress intensity for the 2D edge fracture is:

$$K_{\mathrm{I}} = 1.1215\sigma^{\mathrm{r}}_{yy}\sqrt{\pi a}, \text{ edge fracture} \quad (7.30)$$

An exact solution for this problem is not available, but numerical methods show that the leading numerical factor is correct within one unit in the last digit. Thus, the stress intensity is greater than that for the fracture in the whole elastic body by about 12%. The competition between fracture length and elastic stiffness nearly ends in a draw, with the lesser resistance to opening due to the removed half-space having a slightly greater effect.

7.4 FRACTURE INITIATION AND PROPAGATION

Here we describe typical flaws in rock that serve as stress concentration points necessary to get fractures started. Criteria for failure in a homogeneous stress state, such as the tensile strength criterion that we introduced in Section 4.8.1 inform us about the limiting stress conditions at failure, but do not explicitly include the structure (e.g. the fracture) that is associated with the *process* of failure. By combining solutions to elastic boundary-value problems with principles of fracture mechanics, one can explore the process of fracturing in rock from the *initiation* stage through a stage of *propagation* to the eventual *arrest* of the fracture as it attains the size and configuration we observe in outcrop today. In this section we introduce the concepts and laboratory data necessary to quantify the propagation of fractures that are dominantly mode I, opening fractures.

7.4.1 Fracture Initiation at Flaws

Figure 7.31 shows one surface of a joint decorated with hackle that radiate from a small hole to form a plumose pattern. Because the hole is near the bottom of the siltstone bed, the plumose pattern extends upward more than downward, but it focuses on the small hole. We use this pattern to infer the *fracture kinematics*: the joint initiated at the hole and propagated away from it in all directions. We introduce the mechanics of fracture initiation by showing how the stress field in an elastic body is perturbed by a flaw, such as this hole, to create a local tensile stress that exceeds the tensile strength of the rock. The equations given for the stress state are derived by solving the governing equations of elasticity theory, (7.11) and (7.15).

Fossils, mineral grains, and other objects with different elastic moduli than the surrounding rock can concentrate a remote stress. For example, consider a circular heterogeneity (**Figure 7.32a**) that is bonded to the surrounding material so the interface does not delaminate. The elastic shear modulus of the heterogeneity is G_{h}, and for the surrounding rock mass the shear modulus is G_{s}, so the ratio of these moduli is $k = G_{\mathrm{h}}/G_{\mathrm{s}}$. We take Poisson's ratio for the heterogeneity and surrounding to be equal: $\nu_{\mathrm{h}} = 1/4 = \nu_{\mathrm{s}}$. The radius of the circular heterogeneity is R, and the polar coordinates of the point where the stress is calculated are (r, θ). Suppose the loading is a remote tensile stress, $\sigma^{\mathrm{r}}_{yy} > 0$ acting in the y-direction

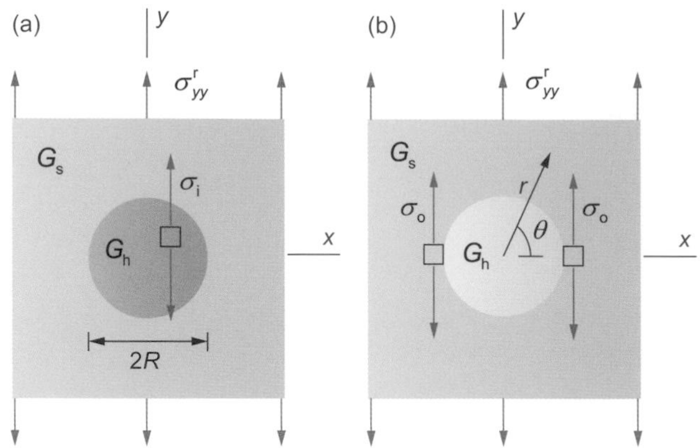

Figure 7.32 Opening fracture initiation mechanisms based on stress concentration. The remotely applied tension is σ^{r}_{yy} and the shear modulus is G. (a) Circular heterogeneity of radius R that is stiffer than the surroundings, $G_h > G_s$. (b) Circular heterogeneity that is softer than the surroundings, $G_h < G_s$. See text for equations describing the concentrated stress. See Jaeger et al. (2007) in Further Reading.

(**Figure 7.32a**). This remote tension induces a *uniform* tension throughout the heterogeneity and this tension is only a function of the ratio of shear moduli:

$$\sigma_i(r \le R, \theta) = \sigma^{r}_{yy}[3k/(2k+1)] \tag{7.31}$$

If the heterogeneity is stiffer than the surrounding rock, $G_h > G_s$, so $k > 1$, the tension inside the heterogeneity exceeds the applied tension. For $k >> 1$, the stress concentration approaches a factor of 3/2 or 150%. Thus, we would expect new fractures to initiate inside the stiffer heterogeneity.

Suppose the circular heterogeneity is softer than the surrounding rock, $G_h < G_s$, so $k < 1$ (**Figure 7.32b**). Again, the loading is a remote tensile stress, acting in the y-direction. For the softer heterogeneity the tension inside is less than the applied tension, and this stress goes to zero as the stiffness of the heterogeneity goes to zero (open hole). However, just outside the heterogeneity, along the sides that are parallel to the direction of the remotely applied tension, the local stress is concentrated as:

$$\sigma_o(r = R^+, \theta = 0, \pi) = \sigma^{r}_{yy}[3/(2k+1)] \tag{7.32}$$

For a much softer heterogeneity ($k << 1$) the normal stress at these two points is enhanced by a factor of 3.0 (open hole). We would expect new fractures to initiate at the boundary of the heterogeneity where the stress is most concentrated, and propagate away from it.

7.4.2 Mode I Fracture Propagation

Using the coordinates defined in **Figure 7.25**, the near-tip stress components in the (x, y)-plane for a mode I fracture (opening) are given using (7.26) as:

$$\begin{Bmatrix} \sigma_{xx} \\ \sigma_{yy} \\ \sigma_{xy} \end{Bmatrix} \approx \frac{K_I}{\sqrt{2\pi r}} \begin{Bmatrix} f_{xx}(\theta) \\ f_{yy}(\theta) \\ f_{xy}(\theta) \end{Bmatrix} \tag{7.33}$$

Table 7.1 Fracture toughness (MPa m$^{1/2}$)

Rock type	From	To
Granite	1.66	3.52
Basalt	0.99	3.75
Quartzite	1.31	2.10
Marble	0.87	1.49
Limestone	0.86	1.65
Sandstone	0.34	2.66
Shale	0.17	2.61

Data from Atkinson and Meredith (1987)

This description of the near-tip stress field neatly separates the stress *magnitude*, which is proportional to the mode I stress intensity, K_I, from the stress *distribution*, which is proportional to $1/(r)^{1/2}$, and to the trigonometric functions of the angle θ. Because these distributions are the same for all fracture tips in homogeneous and isotropic linear elastic materials, we can investigate the propagation of *all* opening fractures in these materials using the same distribution of stress. In this way the distribution of stress is taken "off the table" and we focus on the *stress intensity*.

Because the stress state everywhere in the near-tip region (7.33) is proportional to the stress intensity, we formulate a fracture propagation criterion based only on the mode I stress intensity increasing to some *critical* value:

$$K_I = K_{IC} \text{ at propagation} \tag{7.34}$$

Here K_{IC} is the critical stress intensity for mode I propagation, otherwise known as the mode I fracture toughness. Textbooks on linear elastic fracture mechanics, such as Anderson (1995), explore all aspects of the fracture toughness of materials (see Further Reading).

The underlying assumption of linear elastic fracture mechanics is that the fracture toughness is a *property* of the material that can be measured in the laboratory and then used to predict the conditions of propagation. The prerequisites for a laboratory measurement of stress intensity are that a fracture of known geometry resides in a sample, also of known geometry, and that loads can be applied to the sample that cause pure mode I opening of the fracture. The testing equipment records the load when the fracture tip starts to propagate. The stress intensity at that moment is computed and set equal to the fracture toughness (7.34). For example, if the sample were a large plate with a single fracture of length $2a$ loaded by a uniform tensile stress σ^{r}_{yy}, the stress intensity would be calculated using (7.28) as $K_I = \sigma^{r}_{yy}\sqrt{\pi a}$.

Values of fracture toughness for rock samples under room temperature and atmospheric pressure are recorded in **Table 7.1**. These values are taken from a compilation by Atkinson and Meredith (1987), see Further Reading. Although the igneous rocks tend to have greater fracture toughness than clastic sedimentary rocks, no distinct trends appear in the data. We draw the following general conclusions: *laboratory specimens under room temperature and pressure have fracture toughness that ranges from about* 0.1 to 4 MPa m$^{1/2}$, *with a typical value of* 1 MPa m$^{1/2}$.

The propagation criterion (7.34) avoids having to be specific about *which* stress component in the near-tip region is related to propagation, and *where* one should calculate that component. We know, for example, that we can't use the value of any of the stress components in (7.33) exactly at the fracture tip, because the $1/(r)^{1/2}$ distribution leads to an infinite value. Also, we have learned (**Figure 7.27**) that if we select the stress component σ_{yy}, some small distance directly ahead of the tip where $\theta = 0°$, that is not the location of the maximum value, which occurs where $\theta = \pm 60°$.

As the name implies, the mode I fracture toughness is a measure of the resistance of a given material, under a given set of conditions, to the propagation of an opening fracture. Despite the fact that samples in the rock mechanics laboratory fail in uniaxial tension by the propagation of one or more opening fractures (Section 4.4.2), fracture toughness is not the same physical quantity as tensile strength. Recall that the uniaxial tensile strength criterion was written $\sigma_1 = T_u$. The left side of this equation is the greatest principal stress with units MPa. This stress is believed to be a representative value throughout the sample before it fails. The left side of (7.34) is the stress intensity with units MPa m$^{1/2}$. This is a measure of the stress magnitude throughout the near-tip stress field of an existing fracture. Typical values of tensile strength for rock are of order 10 MPa, and typical values of fracture toughness for the same rock are of order 1 MPa m$^{1/2}$.

7.4.3 Mixed-Mode Fracture Propagation

Joints, veins, and dikes are fractures that dominantly *open*, so the deformation near the tips of these fractures should be dominantly mode I (**Figure 7.25a**). **Figure 7.33a** shows the near-tip region of a pure mode I fracture with a dashed plane that represents the next increment of fracture growth extending from the tipline in the (x, z)-plane. If the remote least compressive stress is perpendicular to this plane, so is the local principal stress, σ_1. Recall from Section 4.8.1, that opening fractures are expected to form perpendicular to this principal stress. It should be no surprise then, that many joints, veins, and dikes have very straight traces in outcrop (e.g. **Figure 7.2** and **Figure 7.5**). And, when the three-dimensional geometry of these fractures is exposed, their surfaces are nominally *planar*.

With the addition of mode II deformation at the fracture tip (**Figure 7.33b**), some shear stress is resolved on the next increment of growth in the fracture plane, so the normal stress acting across that increment is not a principal stress. The plane carrying the least compression, or most tension, is *oblique* to the existing fracture plane. This obliquity is measured by the propagation angle, θ_o, which is related to the ratio of mode II to mode I stress intensity:

$$\theta_o = \sin^{-1}\left[\frac{K_{II}}{K_I}\cos\left(\tan^{-1}\frac{3K_{II}}{K_I}\right)\right] - \tan^{-1}\left(\frac{3K_{II}}{K_I}\right) \tag{7.35}$$

If the mode II shearing is right lateral as viewed on the top of **Figure 7.33b**, the predicted increment veers to the right. Left-lateral shearing is associated with a veering to the left. In this way the geometry of fracture traces in outcrop may be used to infer the presence of shearing and the sign of the mode II stress intensity during fracture propagation.

Examples of curved fracture paths that demonstrate mixed-mode I–II deformation are illustrated for a variety of length scales for echelon fracture pairs in **Figure 7.34**. Traces of cracks in glass at a 25 μm scale, of hydrothermal veins in granitic rock at a 25 cm scale, of basaltic dikes in shale at a 250 m scale, and of oceanic ridges along the East Pacific Rise at a 2.5 km scale all have hook-shaped paths that curve toward the neighboring fracture. These paths are characteristic of opening fracture propagation under conditions of significant mechanical interaction between echelon segments. Each fracture segment induces a shear stress on its neighbor that is associated with mode II deformation and is of the proper sense to veer the fracture tip toward the neighbor.

The addition of mode III tearing to a dominantly mode I fracture induces a change in the direction of propagation that causes fracture breakdown into echelon segments that follow a *twisted* path. The cross sections of Woodworth (**Figure 7.7**) provide good examples of these echelon geometries.

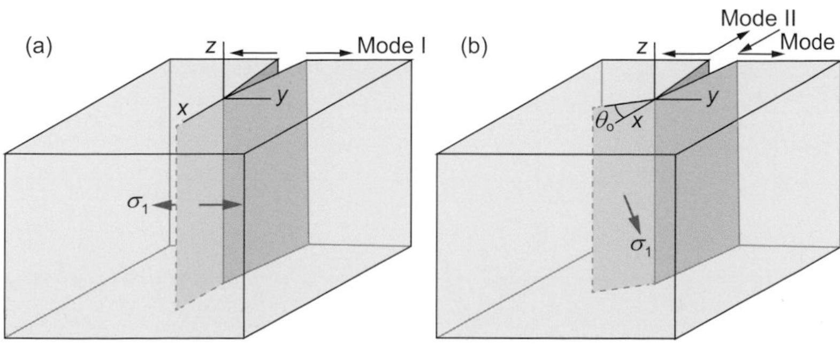

Figure 7.33 Propagation paths for dominantly opening mode fractures. (a) Pure mode I opening: σ_1 is perpendicular to next increment (dashed), which is in the (x, z)-plane. (b) Mode I opening plus mode II shearing: σ_1 is perpendicular to oblique increment (dashed) that makes fracture angle θ_o with x-axis. Modified from Thomas and Pollard (1993).

To understand this phenomenon consider the near-tip region of a pure mode I fracture (**Figure 7.35a**). Each increment of fracture propagation produces new fracture surface that aligns with the plane of the existing fracture. However, if some mode III deformation is introduced (**Figure 7.35b**) the near-tip stress field is perturbed such that some shear stress is resolved on the extension of the fracture plane just ahead of the tip. This means that the normal stress acting across that plane is not a principal stress, so the plane that carries the greatest tension must be *oblique* to the existing fracture plane. This direction defines a set of planes that contain the x-axis, and are rotated about this axis through a propagation angle ϕ_o (**Figure 7.35b**).

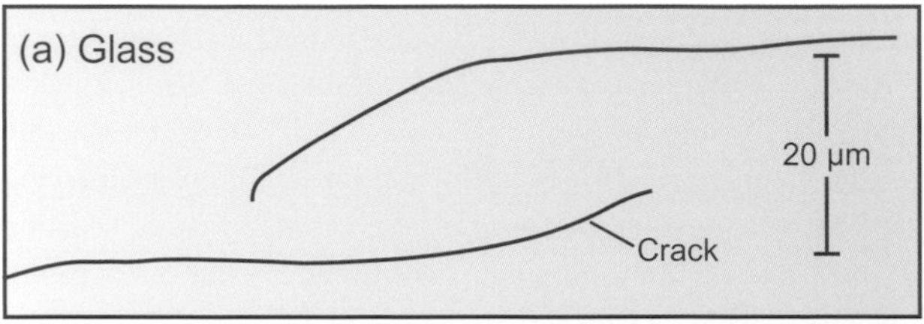

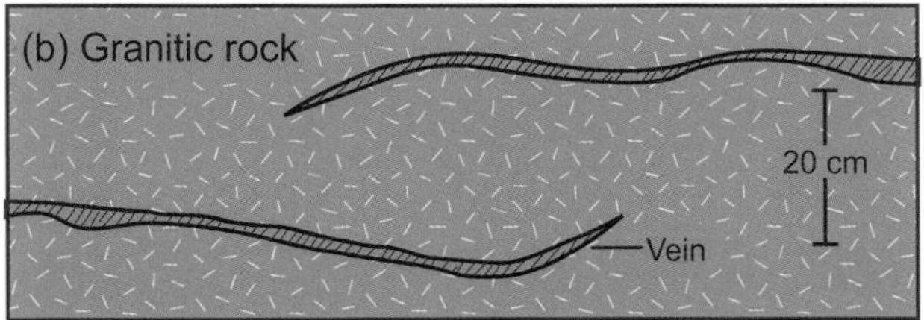

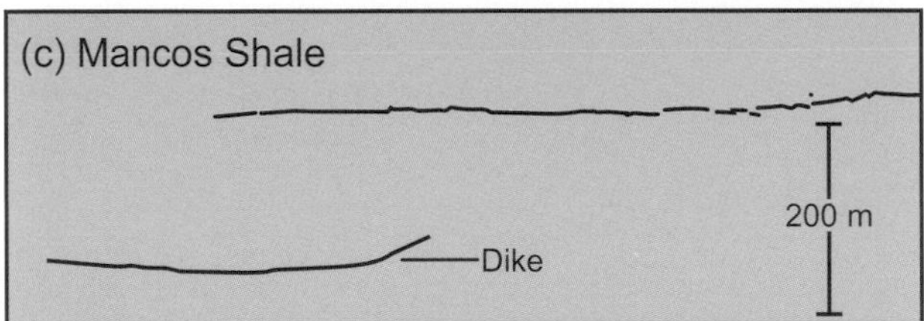

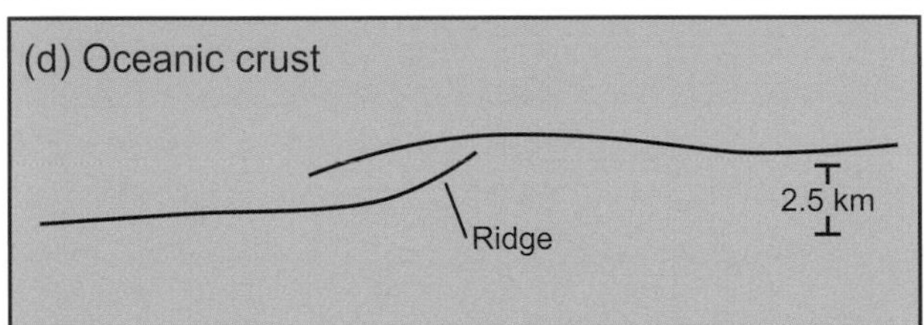

Figure 7.34 Examples of curved paths of opening fractures that responded to the interaction of a neighboring fracture and turned toward it. (a) Laboratory-induced cracks in glass at 20 micron scale. (b) Veins in granitic rock at 20 cm scale. (c) Dikes in Mancos Shale at 200 m scale. (d) Mid-ocean ridges in oceanic crust at 2.5 km scale. Modified from Pollard and Aydin (1984), Figure 6.

The propagation angle ϕ_o is related to the ratio of mode III stress intensity to the mode I stress intensity:

$$\phi_o = \frac{1}{2}\tan^{-1}\left[\frac{K_{\mathrm{III}}}{K_{\mathrm{I}}\left(\frac{1}{2} - \nu\right)}\right] \tag{7.36}$$

Here ν is Poisson's ratio. For a positive mode III deformation, the fracture angle ϕ_o is positive and the local greatest principal stress rotates in a clockwise sense as viewed in the positive x-direction (**Figure 7.35b**). With continued propagation the fracture breaks down into a set of echelon fractures that extend from the tipline where the mode III loading was introduced.

In **Figure 7.36a** segments 9 to 19 of the northeastern dike at Ship Rock are organized into a left-stepping echelon pattern. The segments are approximately parallel, with an average strike of 062°, whereas the strike of the entire dike is 056°. In **Figure 7.36b** we construct a three-dimensional illustration of a dike that is continuous near the bottom of the drawing, but as the tipline ascends to the position marked by the dotted curve, it breaks down into five echelon segments. These segments twist about the propagation axis in a clockwise sense to form the echelon pattern at the top of the drawing. If the ratio of mode III to mode I stress intensity factors is negative the segments twist in the opposite sense. The key concept is that a continuous opening fracture breaks into many shorter opening fractures that attain a different orientation.

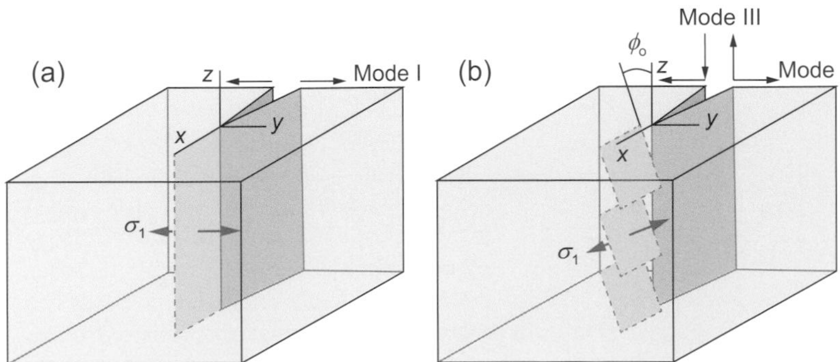

Figure 7.35 Propagation paths for dominantly opening mode fractures. (a) Pure mode I opening: straight path. (b) Mode I opening plus mode III tearing: twisted path. Modified from Pollard et al. (1982).

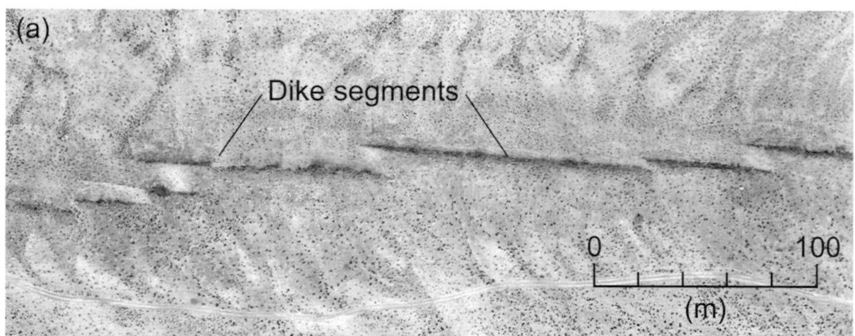

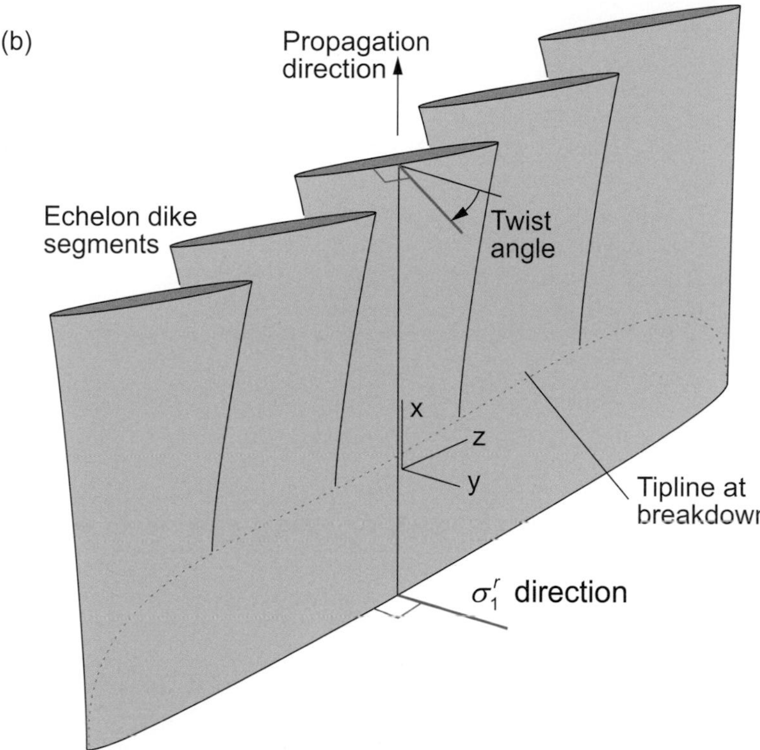

Figure 7.36 (a) Aerial photograph of dike segments 9 through 19 for northeastern dike at Ship Rock, NM. See **Figure 7.21** for location and map. (b) Echelon dike segments formed because the direction of the greatest principal remote stress rotates in a clockwise sense along the vertical propagation direction. Modified from Delaney and Pollard (1981), Figure 29.

Recapitulation

The abundance of fractures exposed in outcrops at a wide variety of length scales and in most tectonic settings provides many opportunities to study fracture geometry and patterns. In this chapter we limit attention to fractures that dominantly open, and include igneous *dikes*, *joints*, and *veins* as natural examples. We introduce the concept of a fracture set that contains many members with similar orientations, and use the terminating relationships between members of different sets to infer their relative ages: fractures that systematically terminate at members of another set are younger. We also show how intersecting fractures provide piercing points that enable one to measure the amount of opening and shearing as the younger fracture developed.

The concept that opening fractures *propagate* through the rock mass brings attention to their tips, where stresses are concentrated that break the rock. In some cases fracture surfaces have plumose structure and rib marks, imprinted as the fracture propagated, that can be used to interpret the kinematic history of fracturing. We point out that some joints and veins are sheared after they formed, because they are weaker than the surrounding intact rock. These sheared joints and sheared veins often are misinterpreted as faults, but they initiated as opening fractures.

Fractures are composed of two sub-parallel and roughly mirror-image surfaces that are joined at a tipline. These surfaces dominantly move apart from one another, but may exhibit minor shearing. The opening displacement discontinuity is very small compared to the dimensions of a fracture measured in its plane. The canonical model for opening fractures is defined using a *linear elastic* constitutive law, a planar and two-dimensional fracture, and boundary conditions of an internal pressure that exceeds the magnitude of the remote least compressive stress. The two equilibrium equations and the compatibility equation form the set of three equations governing the spatial distribution of the three stress components. The three governing equations reduce to one biharmonic equation for the *Airy stress function*, which we use to solve the canonical model, and explore the displacement and stress fields for the opening fracture.

Fractures initiate at flaws in rock where tensile stresses are concentrated. Once initiated, fractures propagate because of the *stress concentration* near their tips. Near the fracture tip, the displacement discontinuity vector has three components that correspond to opening, shearing, and tearing, and these are defined as the three *modes* of fracture: mode I, II, and III. For each mode a unique stress field exists near the fracture tip that neatly separates the distribution of stress from the magnitude of stress. The stress magnitude is proportional to the stress intensity, which depends upon the detailed geometry of the fracture and the loading conditions.

The criterion for propagation of opening fractures is that the mode I stress intensity reaches a critical value called the *fracture toughness*. Laboratory values for fracture toughness are of order 1 $MPa \cdot m^{1/2}$. Dominantly opening fractures in isotropic rock tend to propagate along straight paths and produce planar fractures. Introduction of minor amounts of mode II shearing causes the fracture to turn sharply producing a kinked path, or turn gradually producing a curved path. Minor amounts of mode III tearing cause the fracture to break down into segments that twist about the propagation direction producing echelon arrays of fractures.

REVIEW QUESTIONS

The following questions are designed to highlight the expected *learning outcomes* for this chapter. Each question is taken directly from the material in the chapter and, for the most part, in the same sequence that it appears in the chapter. If an answer is not forthcoming, students are advised to read the relevant section of the chapter and discover the answer.

7.1. Describe the characteristics of a *set* of joints. If two joint sets occur in an outcrop what observations help one to determine their *relative age*? Can you point out geometric relationships among the joint sets in **Figure 7.1** that cast doubts upon the interpretation that set A is older than set B, and that set B is older than set C? How might these doubts be resolved?

7.2. Identify a possible *piercing point* in **Figure 7.4** that leads to an estimate of shearing and opening for the Independence dike. Describe the geometric properties of a piercing point and explain why the offset of a single planar structure by a fault is not sufficient to estimate the slip direction or magnitude.

7.3. Describe how the *rib marks* and *hackle* on the joint surface in **Figure 7.9** are used to estimate the location of the *origin*, which is the point where fracture propagation initiated.

7.4. The *wing cracks* displayed in **Figure 7.11** have been used to interpret the fractures from which they emanate as *sheared joints*. Explain how the sense of shearing is related to the geometry of the wing cracks. What diagnostic feature of the minerals within the sheared joints distinguishes them from members of the same set of fractures that have not been sheared?

7.5. The following Google Earth image is taken from the file: Question 07 05 Arches Park joints.kmz. Open the file; click on the Placemark for this question to open the popup box; copy and paste the questions into your answer document; and address the assigned questions. The Google Earth image data are from Landsat / Copernicus.

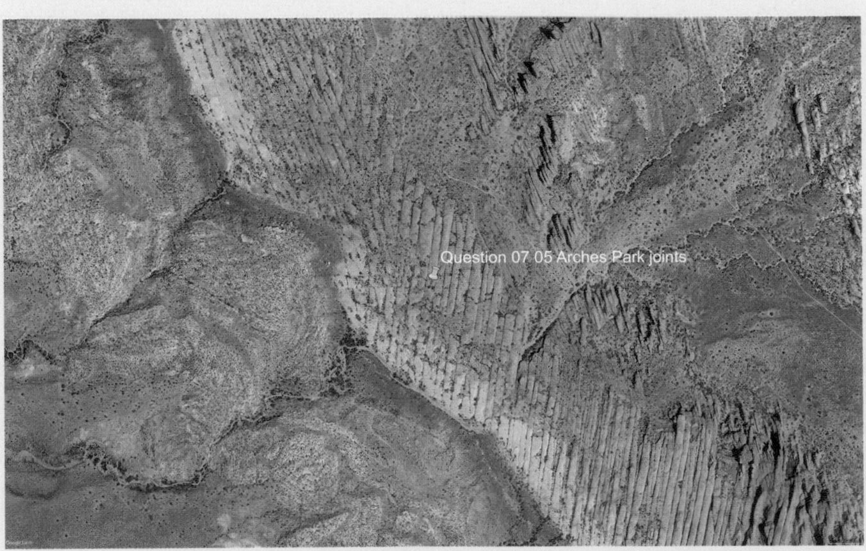

7.6. Describe the geometry, boundary conditions, and constitutive law for the *canonical model* for opening fractures (**Figure 7.14**). What aspect of fracture geometry suggests that using small strain kinematics is likely to be satisfactory?

7.7. Draw Cauchy's tetrahedron (**Figure 7.15**) loaded by stress components on the coordinate faces and an arbitrary traction on the inclined face. Derive the two-dimensional version of *Cauchy's formula* using your sketch to find equation (7.2).

7.8. Show how the tractions on the two surfaces of the canonical fracture (**Figure 7.14**) are related to the applied fluid pressure, *P*, and show that this pressure induces a compressive stress in the host rock using Cauchy's formula.

7.9. Cauchy's laws, (7.6) and (7.7), are reduced to two *equilibrium equations* (7.11) and one *compatibility equation* (7.15) for the canonical fracture problem (**Figure 7.14**). Describe in words the assumptions and conditions used to derive these three equations, and then list the independent and dependent variables. The solution for this problem provides equations for the three stress components, σ_{xx}, $\sigma_{xy}=\sigma_{yx}$, and σ_{yy}. Explain how the strain components are calculated from these stresses and identify the form of Hooke's law appropriate for two-dimensional *plane strain* problems.

7.10. The *opening displacements* for the canonical fracture surfaces (**Figure 7.19**) given in equation (7.21). Describe the physical quantities in this equation and give representative values for the elastic moduli and driving pressure for the dike segments at Ship Rock (**Figure 7.21**). Using these values, what is the maximum opening displacement for a dike segment that is 100 m long?

7.11. The thickness (opening displacement) for dike segment 16 of the Northeastern dike at Ship Rock is plotted versus distance along the outcrop in **Figure 7.22**. Compared to a best-fitting elliptical opening displacement, the dike thickness has an *asymmetric* distribution: thinner near the SW tip and thicker near the NE tip. Explain this asymmetry based on the map (**Figure 7.21**) of segment 16 and the adjacent segments.

7.12. When a fracture opens, the stress is *diminished* (the stress shadow) in two regions, and the stress is concentrated in two other regions. Using the contour maps of the stress components in **Figure 7.23**, describe where these regions are located with respect to the fracture, and discuss the implications of these perturbations in the stress field for: (1) propagation of the opening fracture; and (2) opening of a nearby fracture that is parallel to the given fracture.

7.13. The propagation direction of fractures is closely related to the relative displacements (displacement discontinuity) of the two fracture surfaces very near the fracture tip. These displacement discontinuities are classified into three modes, described by the words *opening*, *shearing*, and *tearing*. Sketch three fracture tips, each one of which illustrates a different mode. For each mode describe the orientation of displacement discontinuity relative to the orientation of the fracture tipline and the fracture plane.

7.14. The nature of the stress concentration near a fracture tip enables one to separate the local stress *magnitude* from the local stress *distribution*. Use the equations (7.26) for the stress components for the mode I fracture to describe this separation, and thereby highlight the importance of the stress intensity, K_I. Write down the equation for the stress intensity for the canonical opening fracture and describe each physical quantity in this equation along with the SI units.

7.15. Opening fractures propagate when the mode I stress intensity equals the *fracture toughness*: $K_I=K_{IC}$. Describe with words and a sketch how the fracture toughness would be measured in a laboratory experiment based on the canonical fracture problem. What are typical values for fracture toughness using SI units?

7.16. Use the two mechanisms illustrated in **Figure 7.32** to discuss the *initiation* of opening fractures at heterogeneities in a rock mass. Illustrate the location of initiation and the orientation of the fracture(s).

7.17. The pairs of opening fractures illustrated in **Figure 7.34** appear to have propagated along nearly straight parallel paths until their tips passed one another. Use the concepts of *mixed-mode* fracture mechanics to explain why the fractures then turned and propagated toward one another along gently curving paths.

7.18. Use the outcrop photograph and illustrative sketch in **Figure 7.36** of part of the northeastern dike at Ship Rock, NM, to describe the origin of the *echelon* geometry of the dike segments. In particular identify which fracture modes played a role in the change of orientation of the segments as they propagated upward.

MATLAB EXERCISES FOR CHAPTER 7: FRACTURES

The identification of fractures, the documentation of their geometric and kinematic features, and the determination of their relative ages takes place at the outcrop. These field observations are the focus of the first two exercises for this chapter and they provide motivations and constraints for subsequent mechanical models. The canonical model for opening fractures employs the simplest geometry, the simplest constitutive law, and the simplest boundary conditions to solve Cauchy's equations of motion. The computational and graphical powers of MATLAB are employed to investigate the displacement field for the canonical opening fracture model. MATLAB also helps to visualize the stress field surrounding the fracture, with both stress shadows and stress concentrations. Very close to the fracture tip, the stresses simplify to the mode I near-tip field, which is utilized in an exercise to understand secondary fractures near an igneous dike. In a separate exercise, the mixed mode I–II near-tip stress field explains the kinked fracture geometry of a sheared joint in sandstone. The exercises for this chapter combine basic field observations with linear elastic theory to devise a canonical model for brittle fracture, which students apply to interpret joints, dikes, and sheared joints in a variety of geological settings. www.cambridge/SGAQI

FURTHER READING

For citations in figure captions see the reference list at the end of the book.

Anderson, T. L., 1995. *Fracture Mechanics: Fundamentals and Applications*. CRC Press, Boca Raton, FL.
The field of engineering fracture mechanics, which started in the middle of the twentieth century, reached a level of maturity by the end of the century that is broadly covered in this textbook.

Atkinson, B. K. (Ed.), 1987. *Fracture Mechanics of Rock*. Academic Press, London.
This edited book contains eleven chapters on field, experimental, and theoretical aspects of rock fractures.

Barber, J. R., 2010. *Elasticity*. Springer, New York.
This textbook for engineers provides approachable coverage of elasticity theory applied to fractures and fracture tip stress concentrations.

Gordon, J. E., 1976, *The New Science of Strong Materials, or Why You Don't Fall Through the Floor*, 2nd edition. Princeton University Press, Princeton, NJ.
Written for the layperson, this treatment of the history of strength of materials provides a useful perspective and is an enjoyable read.

Gudmundsson, A., 2011. *Rock Fractures in Geological Processes*. Cambridge University Press, Cambridge.
This textbook integrates abundant illustrations and photographs of fractures in the field with basic mechanical results to explain the initiation and propagation of fractures in rock.

Lawn, B. R., 1993. *Fracture of Brittle Solids*, 2nd edition. Cambridge University Press, New York.
This is a revised and expanded version of the 1975 book with the same title authored by Lawn and Wilshaw, which introduced the concepts of linear elastic fracture mechanics starting with the seminal papers of A. A. Griffith.

Mandl, G., 2005. *Rock Joints – The Mechanical Genesis*. Springer, Berlin.
To facilitate the engineering evaluation of rock joints, this book adopts the premise that the origin of fractures is a mechanical process, and employs the graphical method of Mohr's stress circle to aid in the explanation and interpretation of fracturing.

Muskhelishvili, N. I., 1975. *Some Basic Problems of the Mathematical Theory of Elasticity*. Noordhoff International Publishing, Leyden.
This textbook introduces a method to solve two-dimensional boundary-value problems in linear elastic theory using functions of complex variables and describes many such solutions in detail.

Olson, J. E., 2004. *Predicting Fracture Swarms – The Influence of Subcritical Crack Growth and The Crack-tip Process Zone on Joint Spacing in Rock.* Geological Society, London, Special Publications 231, pp. 73–88.

This paper investigates the effects of propagation velocity on the patterns and spacings of layer-confined fractures (joints) in sedimentary rock.

Pollard, D. D., and Aydin, A., 1988. Progress in understanding jointing over the past century. *Geological Society of America Bulletin*, 100, 1181–1204.
In this review paper the authors survey research by geologists on jointing in rock from throughout the twentieth century, and they provide a prescription for future research.

Pollard, D. D., and Segall, P., 1987. Theoretical displacements and stresses near fractures in rock: with applications to faults, joints, veins, dikes, and solution surfaces, Chapter 8, in: Atkinson, B. K. (Ed.) *Fracture Mechanics of Rock*. Academic Press, London, pp. 277–349.
The elastic theory and analytical solutions for the canonical problems of two-dimensional fractures with uniform loading are derived and applied to geologic structures.

Price, N. J., 1966. *Fault and Joint Development in Brittle and Semi-brittle Rock*. Pergamon Press, Oxford.
This classic book on brittle and semi-brittle rock deformation uses laboratory test results, elasticity theory, and fracture mechanics to interpret fractures, with special attention to joints.

Rubin, A. M., 1995. Propagation of magma-filled cracks. *Annual Review of Earth and Planetary Sciences* 23(1), 287–336.
This review paper on dike propagation covers dike initiation, dike propagation relevant to the ascent of granitic magmas, and the association of dikes and earthquakes.

Tada, H., Paris, P. C., and Irwin, G. R., 2000. *The Stress Analysis of Cracks Handbook*, 3rd edition. The American Society of Mechanical Engineers Press, New York.
The authors have compiled a remarkable number and variety of solutions for crack problems and provide the crack tip stress intensity factors, as well as a useful reference list.

Chapter 8
Faults

Introduction

Chapter 8 focuses on faults by reviewing terminology, describing faults at the outcrop and crustal scales, building a canonical model for faulting, and discussing relationships between earthquakes and faults. We begin with conventional fault terminology, and then offer detailed descriptions of faults in granite and in sandstone at the outcrop scale. At the crustal scale we describe four sets of normal faults at Chimney Rock, UT; curved thrust faults in the Elk Hills oil field, CA; seismically active strike-slip faults in the Imperial Valley, CA; and the association of faults and folds in the Western Grand Canyon. The canonical model for an idealized fault provides the basis for exploring the mechanics of faulting using the displacement field and stress field due to fault slip in a linear elastic rock mass. Then, we review the kinematics of faulting using the small strain and small rotation fields. Finally, we define earthquake moment and magnitude, along with slip rate and rupture tip velocity, and investigate rock melted by frictional heating using data from the Sierra Nevada, CA.

Faults are roughly planar to tabular structures that occur in sedimentary, metamorphic, and igneous rocks in all tectonic environments. Faults are not as common in Earth's lithosphere as the opening fractures described in Chapter 7, but they are abundant, and in many cases they affect the surrounding rock more than joints, veins, and dikes. They are distinguished from opening fractures because the relative motion of rock on either side of the fault is *dominantly shearing*: the displacement discontinuity is directed approximately parallel to the fault plane. We say dominantly shearing because many faults provide evidence for minor opening or closing displacement discontinuities, and these should not be overlooked.

One could make the argument that faults are simply another class of fractures that sheared rather than opened. If that argument were persuasive we could have included faults in Chapter 7 as a type of fracture to be distinguished from joints and veins by the kinematics of the fracture surfaces. Following that train of thought, and exploiting what we have already learned about fracture mechanics in Section 7.3.2 we would anticipate using *mode II* and *mode III* stress fields to understand the deformation near fault tips, and we would expect to conceptualize fault growth in terms of the in-plane propagation of a shearing or tearing mode fracture tip through otherwise unbroken rock. The challenge for structural geologists seeking to understand faults, and the process of faulting, is that this straightforward fracture mechanics interpretation rarely is adequate.

We hinted at this inadequacy in Section 7.1.4 when we described *sheared* joints and veins. For these structures the initial phase of growth was as an opening fracture, and slip occurred later along that pre-existing weakness. Although sheared joints and veins have demonstrable shearing displacement discontinuities (offset piercing points) these structures did not develop by the propagation of a shearing or tearing mode fracture tip through intact rock. Many examples of this multi-stage process of fault development have been published, so the assumption that a particular fault originated as a shear fracture must be carefully evaluated. In this chapter we describe field examples of faults that provide a conceptual framework for how to evaluate fault development.

An understanding of faults and the process of faulting has many practical applications. The most obvious is that slip on faults generates earthquakes. Understanding whether a given fault is active, and its likelihood of generating earthquakes of a particular magnitude, are first-order matters that structural geologists can help to address, and thereby mitigate hazards to people and infrastructures. In this task structural geologists work closely with geophysicists and civil engineers. Also, the enhanced permeability of some faults offers *pathways* for groundwater flow through aquifers, for ore-forming fluids to the sites of mineral deposits, for hydrocarbon flow within oil and gas reservoirs, and for hot fluids in geothermal reservoirs. Some faults, or portions of faults, have a much reduced permeability due to gouge formation, and hence can divide a reservoir into hydraulically isolated compartments, each one of which would need to be drilled to tap the stored fluids.

8.1 FAULT TERMINOLOGY

In **Figure 8.1a** no offset structures are drawn that would provide a clue that an inclined fault crosses the block diagram. Perhaps the fault was identified by the juxtaposition of different rock types, or by crushed rock between the two fault surfaces. In any case the orientation of this fault would be determined by the *strike* and *dip* as illustrated here and described in Section 2.3.2.

The rock mass above the fault is referred to as the hanging wall, and that below is called the foot wall. These names came from the mining industry. Imagine you are in a tunnel that is bored along the fault, parallel to the strike direction (see dashed line). Your feet would be on rocks in the foot wall, and hanging over your head would be rocks in the hanging wall. A vertical fault would not have foot or hanging walls.

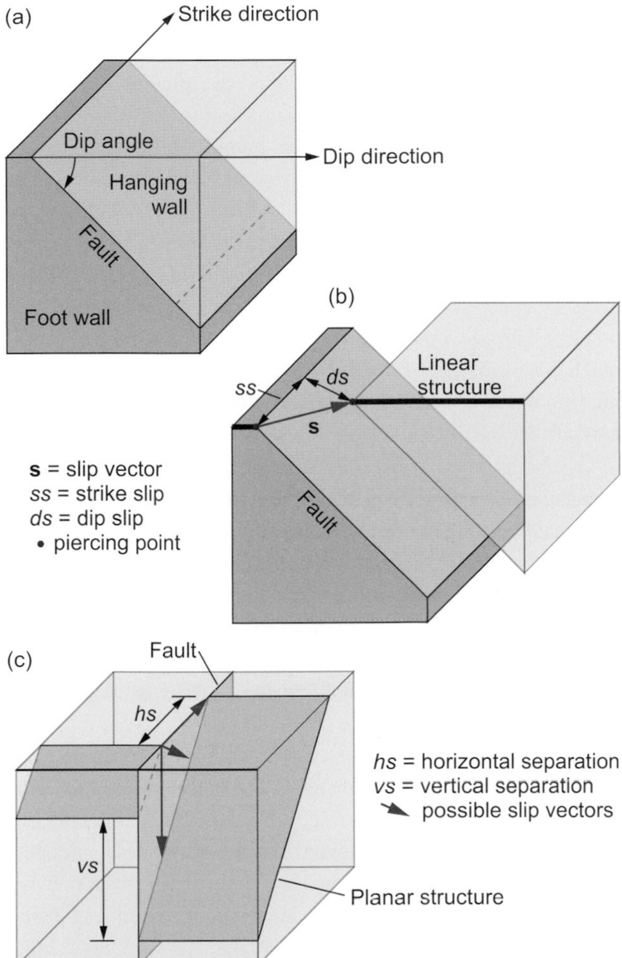

Figure 8.1 Fault terminology and kinematics illustrated using block diagrams. (a) Fault with no offset marker illustrating foot wall and hanging wall. (b) Fault with offset linear structure (heavy black line) that provides a piercing point (red) to measure slip vector. (c) Fault with offset planar structure and measures of horizontal and vertical separation. With no piercing point, an infinite number of possible slip vectors exist.

The heavy black line along the front upper edge of the block in **Figure 8.1b** represents a linear structure offset by slip on the fault. This could be, for example, the line of intersection of a horizontal sedimentary bed, coincident with the top of the block, and a vertical dike, coincident with the front face of the block. These two intersecting planar structures define a piercing point where their line of intersection meets the fault. Before the fault slipped, this lineation was continuous across the would-be fault. After slip, the lineation is cut by the fault and the truncated ends are piercing points on either fault surface. The distance between these piercing points is the slip magnitude, and the direction from one point to the other is the slip direction, so the slip is a vector quantity. Components of the slip vector in the strike and dip directions are called the strike slip and dip slip. Another name for the slip vector is the displacement discontinuity, introduced in Section 7.3.1, which is the difference in displacement vectors for originally neighboring points across the fault. The displacement discontinuity is a more general vector quantity, because it includes relative motion of the piercing points due to opening.

In **Figure 8.1c** a vertical fault cuts the block from front to back and offsets a planar structure. Unlike the linear structure depicted in **Figure 8.1b**, the single planar structure does *not* provide any piercing points, so the slip vector is indeterminate. In fact, an infinite number of slip vectors are consistent with the observed offset. Therefore, a different terminology is used to describe and quantify the distance between the offset segments of the planar structure. On the top surface of the block, which represents a map view, this distance is called the horizontal separation. On the front face of the block, which represents the view exposed on a vertical cliff or road cut, this distance is called the vertical separation. If the slickenlines on this fault are horizontal, the horizontal separation is a measure of the slip magnitude. If the slickenlines are vertical, the vertical separation equals the slip magnitude. Without slickenlines or piercing points the structural geologist must be content with measuring the *separation* of points on offset planar structures.

Faults may be classified according to the direction of slip, using strike slip (**Figure 8.2a**) and dip slip (**Figure 8.2b**) as end members and oblique slip for cases between these two. A steeply dipping strike-slip fault also is called a wrench fault. For a normal fault (**Figure 8.2b**), dip slip carries the hanging wall down

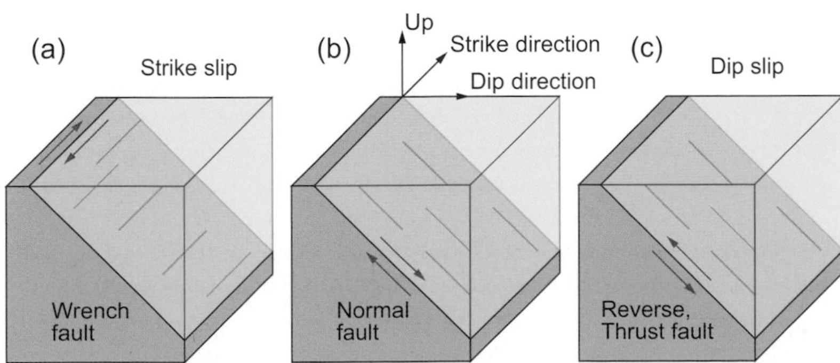

Figure 8.2 Block diagrams of fault types with slickenlines in red. (a) Wrench fault with strike slip. (b) Normal fault with dip slip: hanging wall moves down. (c) Reverse or thrust fault with dip slip: hanging wall move up.

relative to the foot wall, whereas the hanging wall moves up relative to the foot wall for a reverse fault (**Figure 8.2c**).

As drawn schematically in **Figure 8.1** and **Figure 8.2**, faults appear to be composed of two rock surfaces that have slipped in frictional contact with no material caught between these surfaces. In contrast, fault surfaces in nature usually are separated by so-called fault rock, which helps to identify the structure as a fault, and may be suggestive of the mechanism of faulting. For example, some faults contain material between the two fault surfaces that is broadly termed cataclasite (**Figure 8.3a**). This fault rock may be composed of breccia (larger fragments of host rock in a fine-grained matrix) or gouge (finely comminuted host rock). These two fault rocks may compose the fault core, which is bordered by fractured host rock in the damage zone of the fault. Cataclastic rock may lack a well-developed fabric, and the fragmented nature of the constituents suggests *brittle* deformation. Other faults contain material called mylonite (**Figure 8.3b**), where the minerals between the fault surfaces have been highly distorted by crystal-plastic deformation, and/or greatly reduced in size by dynamic recrystallization. Mylonites are characterized by well-developed fabrics that suggest *ductile* deformation. Still other faults contain pseudotachylyte (**Figure 8.3c**), a glassy material interpreted as forming due to frictional heating at high rates of slip, suggesting *dynamic* deformation.

Mapping cataclasite, mylonite, and pseudotachylyte associated with faults, provides important information about the geologic history and origins of the faults. However, it also is important to document the underlying physical commonality that faults are associated with a *shearing* displacement discontinuity. In this way, one distinguishes faults from the fractures we discussed in Chapter 7, including joints, veins, and dikes, which all are associated with an *opening* displacement discontinuity.

8.2 DESCRIPTIONS OF FAULTS AT THE OUTCROP SCALE

The identification of some faults and the documentation of their geometric and kinematic features can take place at outcrops. Other faults are of such a large scale, or are so hidden by surficial features, or only are detectable by sub-surface imaging technology, that outcrop characterization is not possible. In this section we focus on examples of faults that are exposed in outcrop, so they can be characterized by detailed mapping and photography. Because faults are wonderfully diverse we do not attempt to be encyclopedic, but rather choose a few well-characterized examples that illustrate some of this diversity.

8.2.1 Fault Development in Granitic Rock: Sierra Nevada, CA

The Bear Creek drainage along the John Muir Trail in the central Sierra Nevada, CA exposes numerous outcrops of faults at various stages of development. Using maps and photographs of these faults, and thin sections of rock within and near these faults, one can infer the sequence of events in their development, and document the geometry and constituents of the faults and related structures. The rock hosting these faults is the Lake Edison Granodiorite (Kle) of late Cretaceous age (88 ± 1 Ma), one of three prominent plutons that are elongate to the northwest, as seen on the geological map (**Figure 8.4**), and beautifully exposed in this glaciated terrain. The Lamarck Granodiorite (Kl) is intruded by the Lake Edison Granodiorite, which is intruded by the Mono Creek Granite (Kmc), so these Cretaceous plutons are successively younger to the northeast. The Lake Edison Granodiorite is composed primarily of plagioclase, quartz, alkali feldspar, and mafic minerals, with typical grain sizes from 1 to 5 mm. The granodiorite has a weak foliation, defined by the alignment of mafic minerals and flattened xenoliths that dip steeply and strike about 315, approximately parallel to the pluton contacts. Where dikes of aplite, pegmatite, or basalt cut the granodiorite they are no more than a few meters thick, and they have trace lengths of as much as several hundred meters. Although too small to appear on the geologic map (**Figure 8.4**), the dikes provide abundant *markers* for measuring the offset across the faults.

A structure map of the Kip Camp (KC) outcrop (**Figure 8.5**) reveals several aplite dikes cut by a set of steeply dipping fractures that strike to the northeast, from 050° to 070°. Following the map explanation, we use the general term fracture for these structures, and emphasize their common features: later we distinguish those that only opened from those that opened and then sheared (Section 7.1.4). The dikes and fractures were mapped on an aerial photograph at a scale of approximate 1:370. While some smaller fractures could not be identified on the photograph, most of those that are more than several meters long are mapped. The fractures are filled with a mineral assemblage including epidote, chlorite, quartz, calcite, and muscovite, so they served as conduits for flow of hydrothermal fluids that deposited these minerals. Fracture thicknesses vary from less than 0.1 mm to somewhat greater than 1 cm, and their trace lengths range from less than a meter to several tens of meters. The fracture spacing is quite irregular, ranging from less than one meter to more than ten meters. Most fractures are mapped as continuous, but many are composed of echelon segments with steps that are too small to resolve at this map scale.

At each location where a fracture cuts a dike at the Kip Camp outcrop (**Figure 8.5**) the structural relations were examined in the field. Those locations marked with a "0" have no discernable strike separation, but they are propped open as much as a few millimeters by the hydrothermal mineral assemblage. On the other hand, those fractures marked with numbers greater than zero offset older dikes and the number indicates the strike separation in centimeters. Typically, the fractures with demonstrable separation are as much as a centimeter thick. The outcrop photograph (**Figure 8.6**) is approximately 1 m across, and shows an aplite dike and several fractures, two of which do not appear to offset the dike and are interpreted as joints, fractures that dominantly opened. Two other fractures clearly offset the dike by a few to several centimeters. On the Kip Camp map (**Figure 8.5**) the offsets commonly are a few tens of centimeters and can range up to 2 meters, so these structures are interpreted as faults.

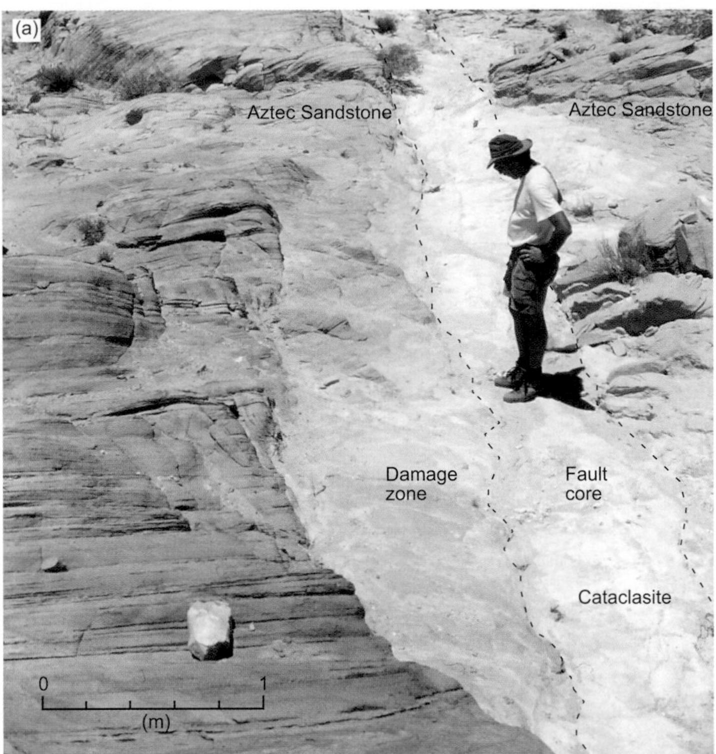

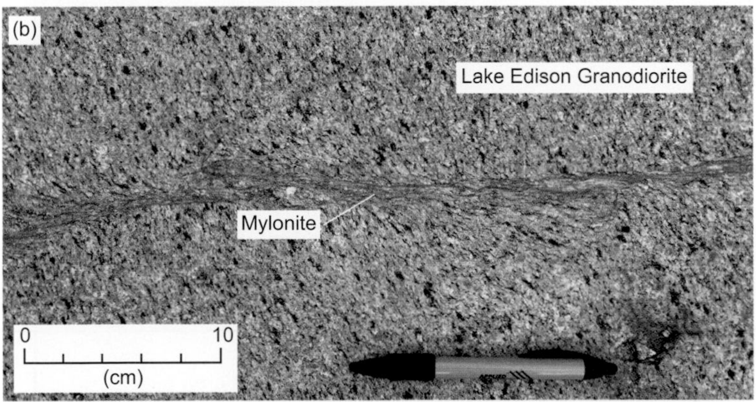

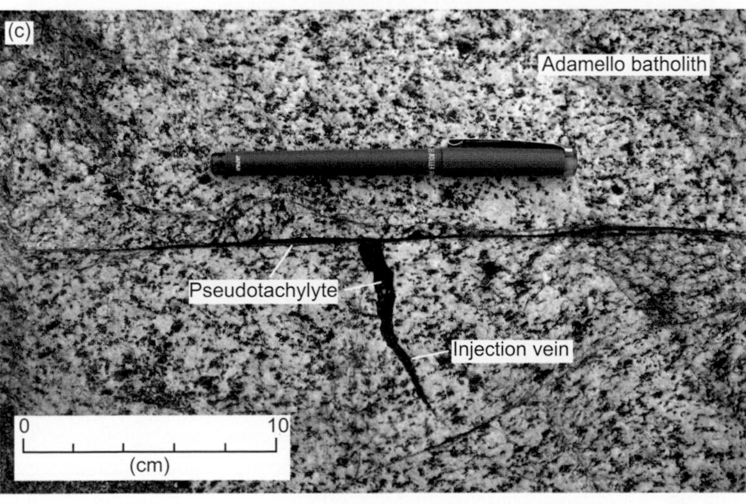

Figure 8.3 Faults and fault rocks. (a) Cataclasite in fault core cutting the Aztec Sandstone at the Valley of Fire, NV. See Myers and Aydin (2004). UTM: 11 S 722604.90 m E, 4036956.23 m N. (b) Mylonite in left-lateral shear zone cutting the Lake Edison Granodiorite, Sierra Nevada, CA. See Nevitt et al. (2017b). UTM: 11 S 337637.43 m E, 4132518.41 m N. (c) Pseudotachylyte (black) in fault cutting the Adamello batholith, Italy. Injection vein filled with pseudotachylyte extends from the fault. See Di Toro and Pennacchioni (2005). Approx. UTM: 32 T 622420 m E, 5114230 m N.

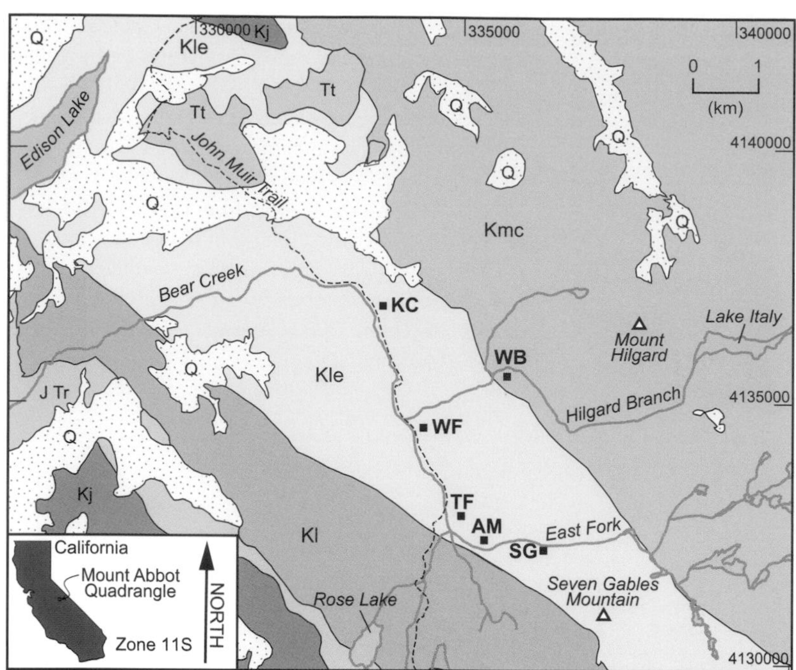

Figure 8.4 Location and geologic map for the Bear Creek field area in the southern portion of the Mount Abbot Quadrangle, Sierra Nevada, CA: Kmc – Mono Creek Granite; Kle – Lake Edison Granodiorite; Kl – Lamarck Granodiorite; Kj – granitic rocks of uncertain affinities; JTr – Metavolcanic rocks; Tt – olivine trachybasalt; Q – alluvium. Outcrops referred to in the text include Kip Camp (KC), Waterfall (WF), White Bark (WB), Trail Fork (TF), Ape Man (AM), and Seven Gables (SG). Modified from Nevitt et al. (2017a), Figure 1.

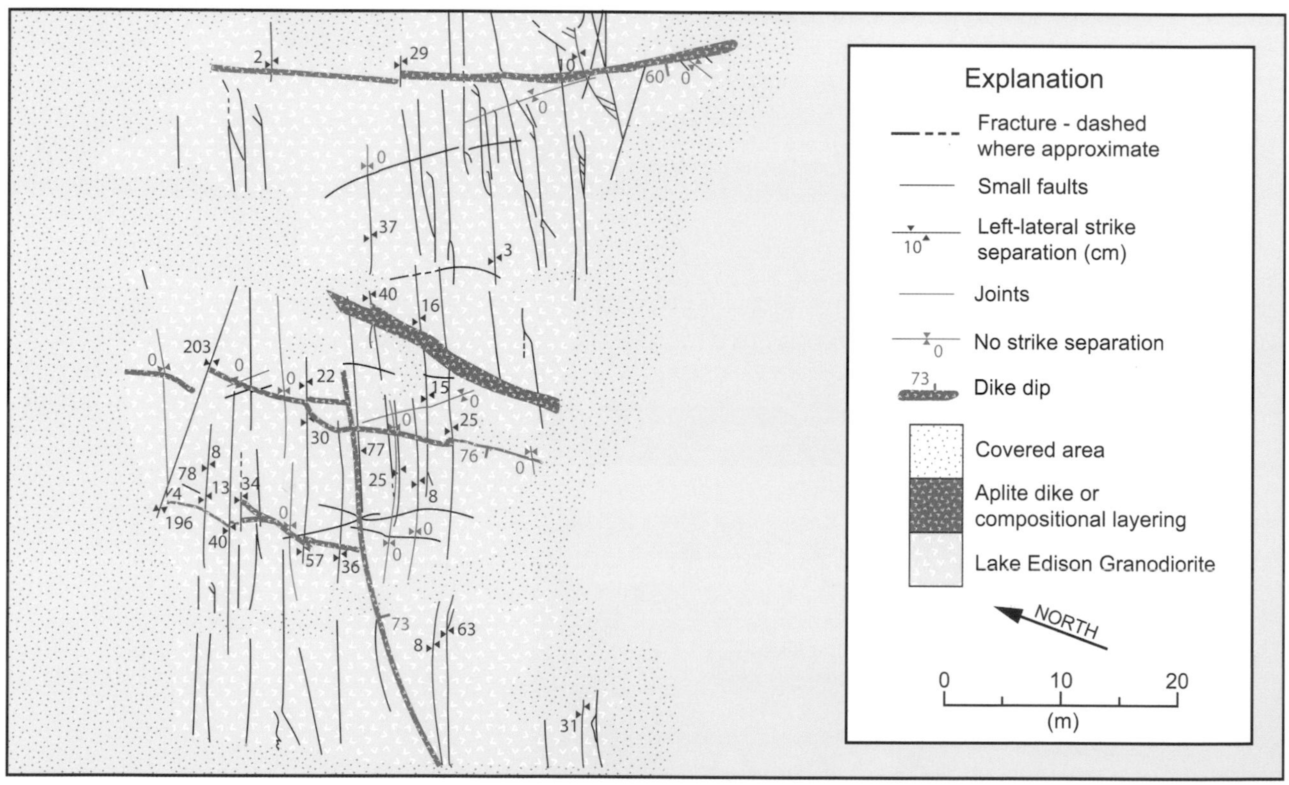

Figure 8.5 Structure map of Kip Camp (KC) outcrop showing joints and small faults (sheared joints) with measured strike separation. Modified from Segall and Pollard (1983), Figure 2. See Figure 8.4 for geologic map of region. Kip Camp UTM: 11 S 333362.00 m E, 4136556.00 m N.

Did the small faults at Kip Camp grow as shear fractures to the lengths shown on the map (Figure 8.5)? Shearing would require a shear stress resolved on the fault plane, but the faults are subparallel to the joints, which carried no shear stress because opening fractures form in a principal stress plane (Section 4.8.1). The shearing must have occurred under a different state of stress, and therefore at a different time, than the formation of the joints. Yet, the joints and the small faults share the same outcrop, are sub-parallel, and contain similar hydrothermal minerals. Apparently, the joints and small faults originated as opening fractures with an ENE strike, and were sealed by the same suite of hydrothermal minerals. After a change in the orientation of the principal stresses that resolved shear stress on this set of fractures, some slipped, offset the dikes, and thus were transformed into small faults.

In a close-up photograph (**Figure 8.7a**) showing the individual mineral grains of the granodiorite, a fracture trace cuts a feldspar grain, but does not discernably offset the boundaries of that grain. Thin section observations of several samples taken from such fractures demonstrate that relative displacements have parted individual mineral grains normal to the fracture surfaces, but have not discernably displaced the two parts tangential to the

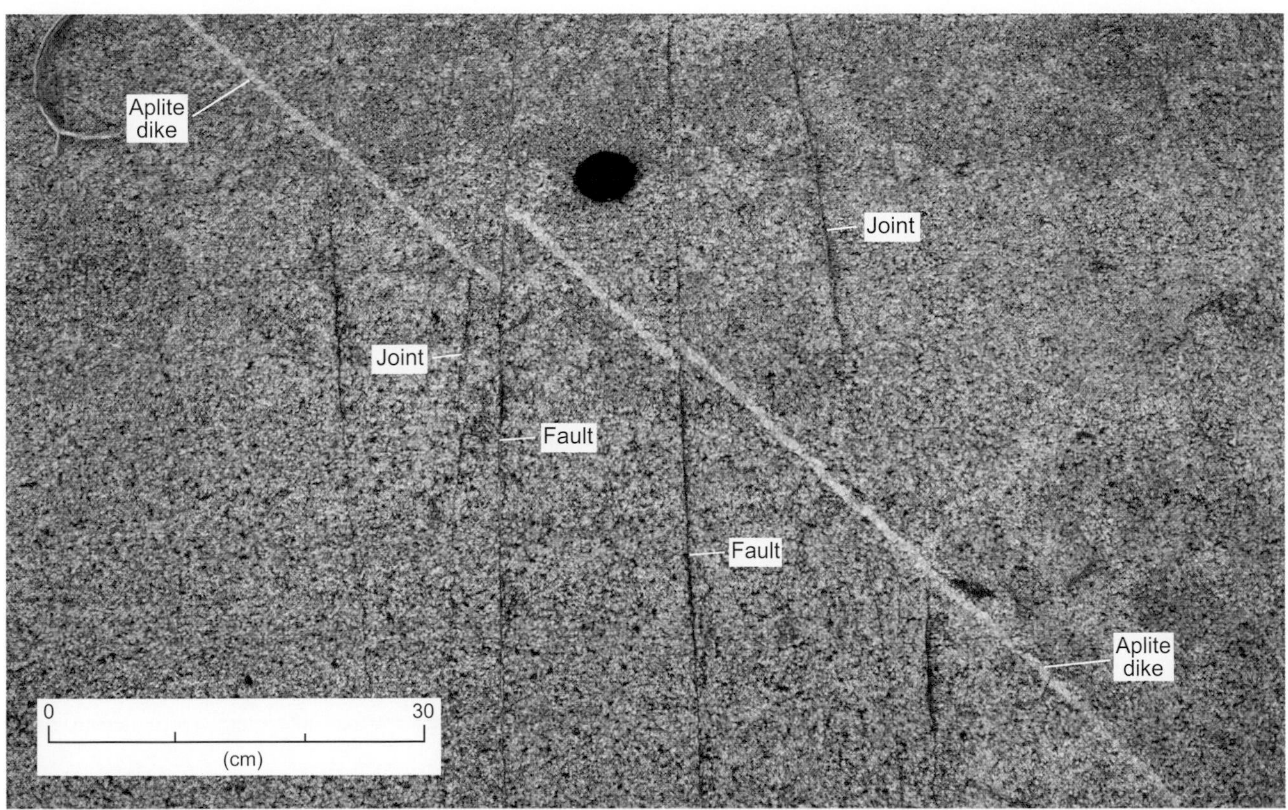

Figure 8.6 Aplite dike cut by joints and faults in the Lake Edison Granodiorite, Bear Creek, Sierra Nevada, CA. See **Figure 8.4** for geologic map of region.

Figure 8.7 (a) Close-up photograph of joint in granodiorite cutting feldspar grain without detectable shear offset. Photograph by P. Segall. (b) Photo of 1 cm thick left-lateral small fault (sheared joint) in Lake Edison Granodiorite. Deformed hydrothermal minerals define a foliation that is parallel to, or slightly rotated clockwise from, the strike of the fault surfaces. See **Figure 8.4** for geologic map of region.

fracture surfaces. Furthermore, thin section observations demonstrate that the hydrothermal mineral grains are *undeformed* inside those fractures that are interpreted as *joints*. In contrast, on a close-up photograph of a fracture, interpreted as a *small fault* because it has measurable strike separation (**Figure 8.7b**), the hydrothermal mineral assemblage is *deformed* and re-arranged into thin, platy aggregates, giving this material a distinct planar fabric or foliation. This foliation varies in strike from 0° to about 20° clockwise from the strike of the small faults, and developed after deposition of the hydrothermal minerals. We devote Chapter 10 to rock fabrics and describe the highly deformed fault rocks from this field area in more detail there.

Some small faults in the Lake Edison Granodiorite are composed of distinct segments arranged in echelon patterns, with left and right steps. The perpendicular distance between the parallel echelon segments usually is less than a meter, and the segments commonly overlap by as much as a few meters. For example, **Figure 8.8** shows a 7 cm left step between two segments that overlap by about 30 cm. An aplite dike just to the left of this step is offset by the fault segments in a left-lateral sense by about 7 cm. Several open fractures span the left step obliquely, connecting the two small fault segments. These open fractures project only to the left of a given fault segment as viewed toward the step. In this regard they are similar to the wing cracks described in Section 7.1.4. The granodiorite is bleached within the step and for a few centimeters around the step. Apparently, the open fractures served as conduits for hydrothermal fluids that chemically altered the granodiorite.

As wing cracks within a left step open (**Figure 8.8**), they transfer slip from one small fault segment to the other. This means that the two segments on either side of the step can behave more like a continuous fault with a length equal to the combined length of the two segments. This mechanism of *end-to-end linkage* explains how the small faults in the Lake Edison Granodiorite grew to lengths that greatly exceed that of single joints. Some of these faults offset dikes as much as 2 m.

Tabular structures (**Figure 8.9**) occur in the Lake Edison Granodiorite that are parallel to the joints and faults, but these structures offset aplite dikes by as much as 10 m. These narrow zones of deformation, mapped in **Figure 8.5** and pictured in **Figure 8.6**, are composed of side-by-side pairs of faults, each one of which is similar to the small fault segments shown in **Figure 8.8**. The spacing of the two faults bounding these zones varies from about 0.5 to 3 m, and these zones can be hundreds of meters to a few kilometers in length along strike. We refer to these structures as simple fault zones in order to distinguish them from the small faults described above, and from more complex fault zones that are composed of several such simple zones.

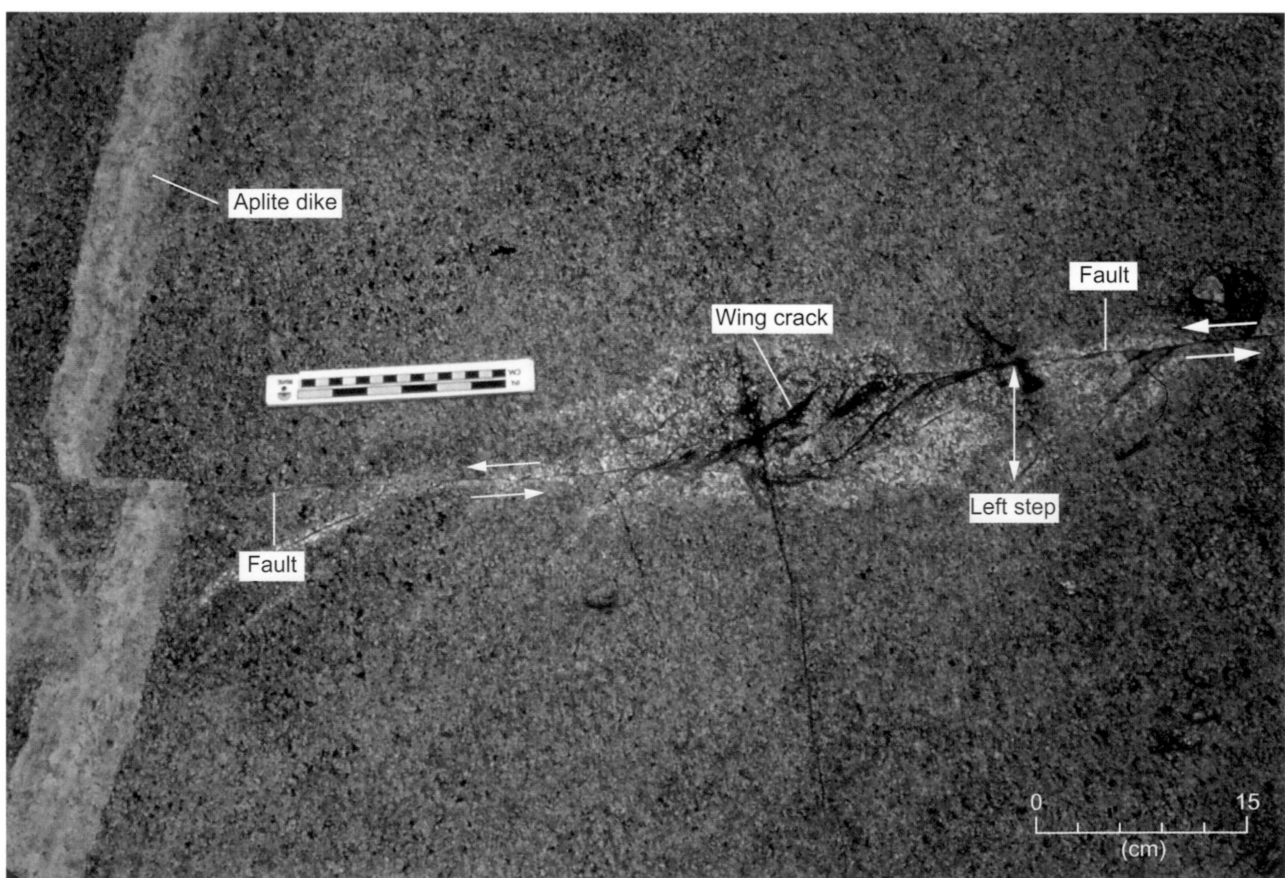

Figure 8.8 Photograph of left step with wing cracks between two echelon fault segments (sheared joints). Older aplite dike is offset about 7 cm in a left-lateral sense. See **Figure 8.4** for geologic map of region.

Between the boundary faults of a simple fault zone the granodiorite is cut by a multitude of oblique fractures (Figure 8.9). Most of these fractures are oriented counter-clockwise from the strike of the zone, and some resemble the wing cracks depicted in Figure 8.8. Immediately outside simple fault zones, as seen on the structure map in Figure 8.10, the granodiorite is cut only by the sparse set of joints and small faults, seen for example, on the map in Figure 8.5. The contrast in deformation from inside to outside the simple fault zone is dramatic. Because most of the fractures within a simple fault zone are oriented oblique to the strike of the zone, they do not play a prominent role in offsetting the aplite dikes. Instead, the aplite dikes are sharply offset at the

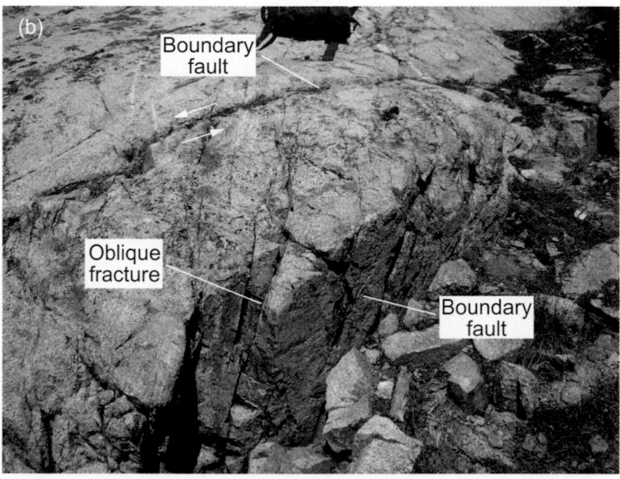

Figure 8.9 Two photographs of a simple fault zone at the Waterfall site (WF) composed of two boundary faults with left-lateral slip, and oblique fractures confined to the zone. This outcrop is mapped and described by Martel (1990). See Figure 8.4 for geologic map of the region. Waterfall site UTM: 11 S 334022.00 m E, 4134542.00 m N.

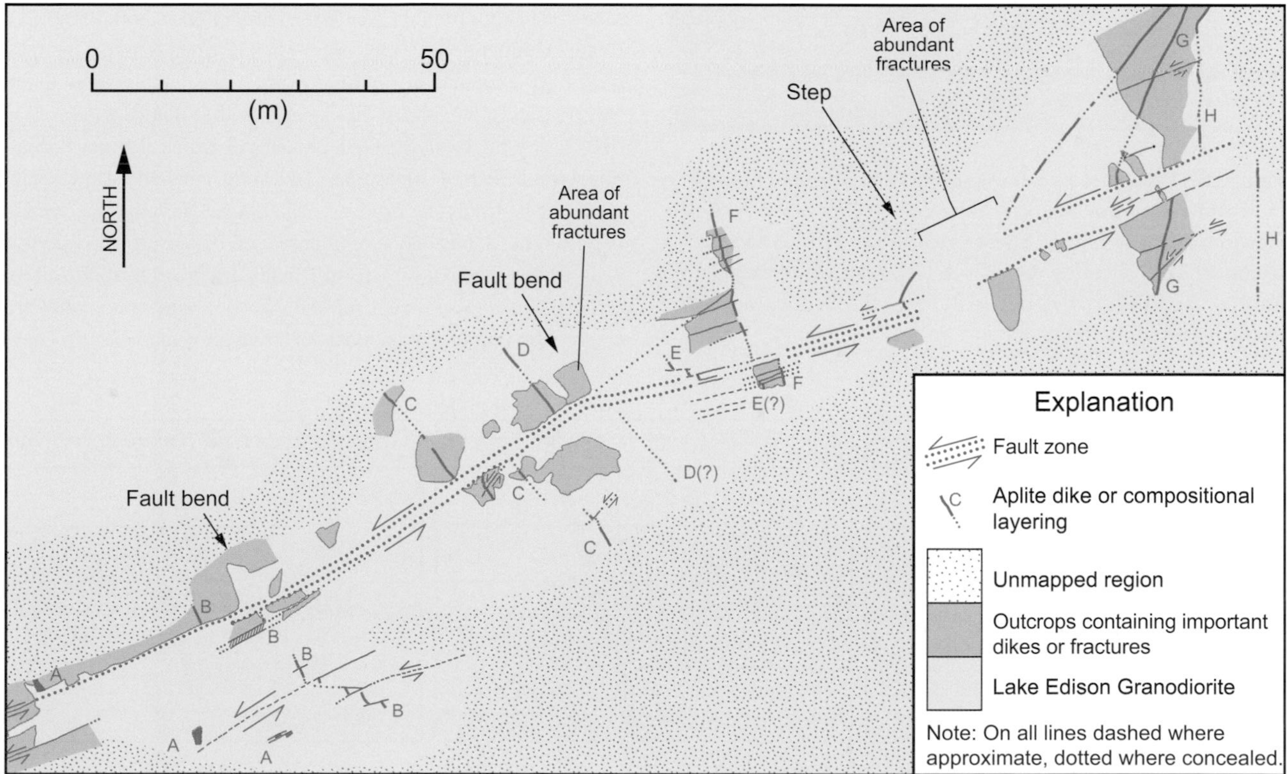

Figure 8.10 Map of simple fault zone at the Trail Fork (TF), Lake Edison Granodiorite. See Figure 8.4 for geologic map of region. A–H are planar markers used to estimate the separation across the fault zone. Modified from Martel et al. (1988), Figure 9b. UTM: 11 S 334866.05 m E, 4132867.93 m N.

boundary faults, where the majority of the strike separation is *localized*.

The mineral assemblage within the boundary faults is similar to that in the small faults, including epidote, chlorite, quartz, feldspar, calcite, and muscovite. However, the texture of part of the boundary fault material is *cataclastic* (**Figure 8.11**), in contrast to that of most small faults, which is dominantly *mylonitic*. The boundary fault material is composed of angular fragments, predominantly of quartz and feldspar, in a very fine-grained unfoliated matrix. The angular fragments are as much as a few millimeters in size, and the matrix grains are less than a micron. Relict mylonitic textures occur within some of the angular quartz fragments, suggesting that the cataclastic deformation followed and overprinted the mylonitic deformation.

The sequence of events leading to the development of simple fault zones in the Lake Edison Granodiorite is summarized in **Figure 8.12**. The granodiorite was emplaced and cut by aplite dikes. Then, both the granodiorite and the dikes were cut by a set of opening fractures (joints) that were sealed by hydrothermal minerals (**Figure 8.12a**). The joints formed in a principal stress plane with the least compressive stress perpendicular to that plane.

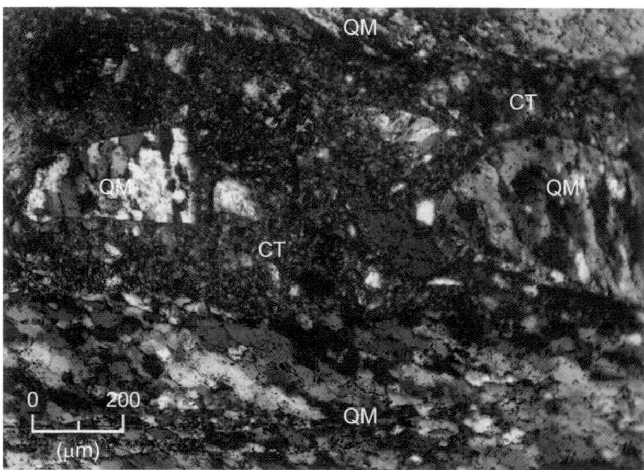

Figure 8.11 Microscopic image of fault rock from a boundary fault at the Waterfall site, WF, Lake Edison Granodiorite, Sierra Nevada, CA. See **Figure 8.4** for location of WF. Quartz mylonite (QM) is cut by cataclasite (CT) containing angular fragments of quartz mylonite with rotated foliation. Modified from Griffith et al. (2008), Figure 5a.

In the next stage of deformation in the stress state changed, so shear stress was resolved on the joint planes and some slipped in a *left-lateral* sense to form small faults (**Figure 8.12b**). The hydrothermal minerals within the small faults deformed predominantly by ductile mechanisms and a mylonitic foliation developed in the fault rock (Section 10.2.3). In some cases the slip was transferred across left steps by a new set of opening fractures called wing cracks that are oblique to the fault planes. This end-to-end linkage of small faults increased the slip, which can be as great as 2 m. The next stage of deformation involved pairs of closely spaced small faults (**Figure 8.12c**). Oblique fractures formed between the boundary faults of these *simple fault zones* and slip increased to as great as 10 m on the boundary faults. The minerals within the boundary faults deformed by brittle mechanisms and a *cataclastic* texture developed in the fault rock, which includes rotated angular fragments of mylonite.

8.2.2 Fault Development in Sandstone: San Rafael Desert, UT

To emphasize that faults can develop with very different mechanisms than those illustrated in **Figure 8.12**, we describe faults in sandstone exposed in the San Rafael Desert, just south of the San Rafael Monocline in southeastern Utah (**Figure 8.13**). The massive, buff-colored unit holding up the ridge in this view is the Navajo Sandstone, which dips steeply to the southeast (left). It is overlain by the thinly bedded, chocolate-colored Carmel Formation, composed of limestone, sandstone, siltstone, and shale, which make triangular *hogbacks* against the Navajo. The overlying Entrada Sandstone is composed of both buff and reddish beds, and is the host rock for the faults described in this section. Similar faults also occur in the Navajo Sandstone, and are common in porous sandstone units throughout southeastern Utah.

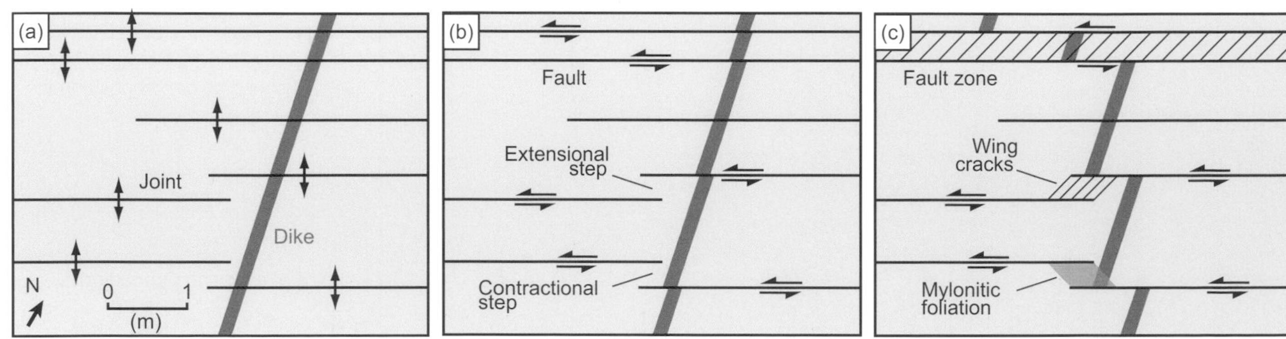

Figure 8.12 Summary diagram of fault development from Bear Creek sites. (a) Joints form as opening fractures. (b) Some joints slip to become small faults. (c) Simple fault zones developed with much greater slip on bounding faults. Wing cracks link small faults at extensional (left) steps. Shear zones with mylonitic foliation link small faults at contractional (right) steps. Modified from Nevitt et al. (2014), Figure 3.

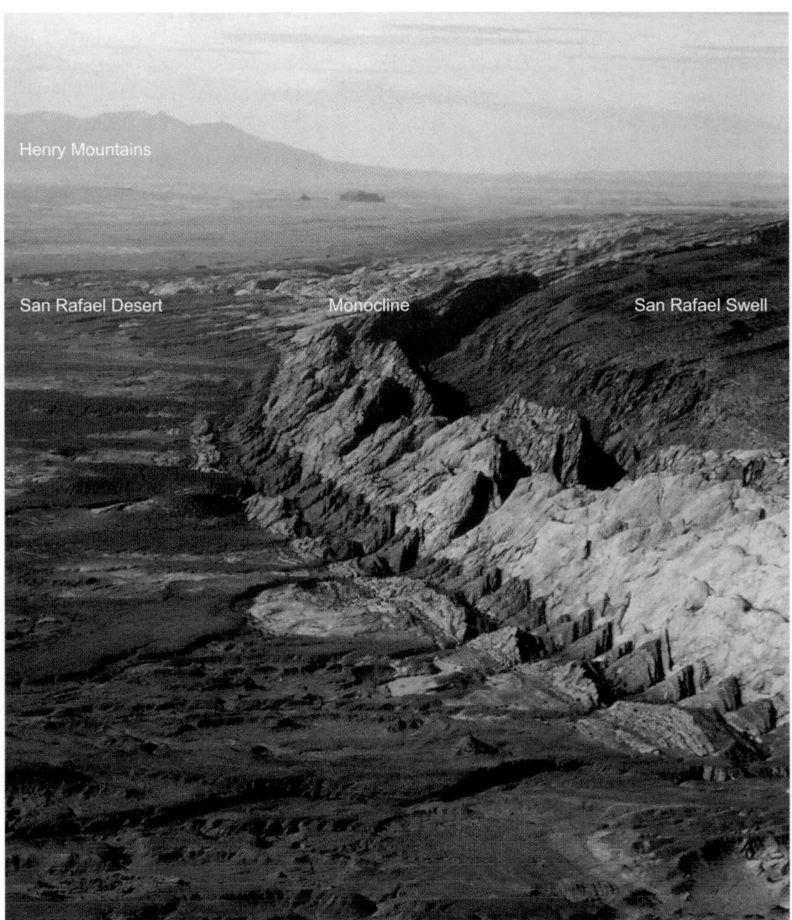

Figure 8.13 Aerial photograph of San Rafael Monocline looking to the southwest towards the Henry Mountains on the horizon. The prominent buff-colored unit in the limb of the monocline is the Navajo Sandstone. Approx. UTM: 12 S 547614.25 m E, 4301495.71 m N.

On the geological map of the San Rafael Desert (**Figure 8.14a**) note that the sedimentary formations exposed west of Highway U-24 increase in age toward the west, from Jurassic (J) to Triassic (T) and Permian (P). This is a classic example of a monoclinal fold, with gently dipping sedimentary rocks to the east and west, and a single limb that is steeply inclined to the east. **Figure 8.14b** is a cross section roughly perpendicular to the strike of the strata in the monocline. The uplift of strata on the San Rafael swell is about 1.2 km. This region was relatively stable from the Permian to the Cretaceous time as these sedimentary rocks accumulated, and the uplift is associated with the Laramide orogeny, a period of mountain building that started in the Late Cretaceous and affected much of western North America.

The Jurassic sedimentary rocks, including the Navajo Sandstone, Carmel Formation, and Entrada Sandstone, form the most steeply dipping portion of the San Rafael Monocline in **Figure 8.14b**. The Carmel Formation underlies much of the San Rafael Desert where it is crossed by Hwy 24, with occasional mesas held up by the overlying Entrada Sandstone. Localities with well-exposed deformation bands and faults in this desert terrain include Iron Wash, Molly's Castle, and Goblin Valley.

A typical example of the first stage in the development of faults in the Entrada Sandstone is shown in **Figure 8.15a**. Thin sedimentary laminae within the sandstone are offset about 8 mm along a light-colored band that stands up as a narrow rib, because it is more resistant to erosion than much of the sandstone. These deformation bands offset bedding from a few millimeters to a couple of centimeters, and they are about 1 mm thick and tens to hundreds of meters in trace length. In many outcrops deformation bands with more than one orientation are present and younger bands systematically offset older bands. Bands that offset bedding or older bands with dip-slip, oblique-slip, and strike-slip movement have the same characteristics, so they are not differentiated based on the direction of relative motion.

Petrographic investigation of thin section images (e.g. **Figure 8.15b**) show that the Entrada Sandstone is composed predominantly of quartz and feldspar grains cemented by calcite, with an average grain size of about 0.15 mm. The mineralogical composition in deformation bands is the same as that in the surrounding Entrada Sandstone, but the grains are about an order of magnitude smaller and have a broader range of sizes. In other words, grains in the bands are *poorly sorted*. The porosity of the undeformed sandstone averages about 25%, whereas that of the material in deformation bands averages about 10%, resulting in a substantial *reduction* in porosity. The mechanisms of deformation include collapse of pore space, rearrangement of grains by translation and rotation, and fracturing of grains. Surfaces of opening or shearing displacement discontinuity are not observed within these bands, but rather shearing appears to be distributed across the thickness of the band. Given typical offsets of bedding,

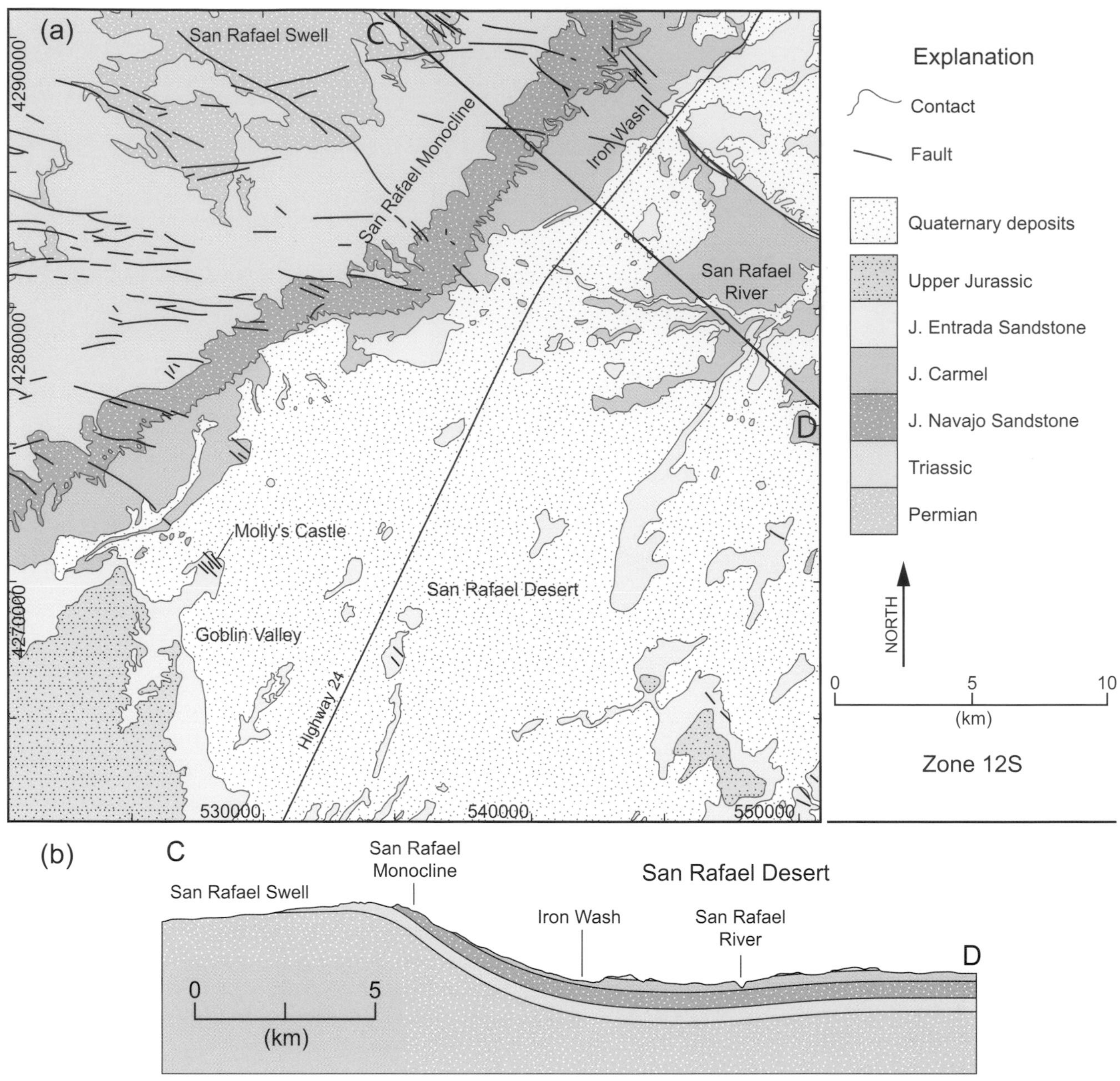

Figure 8.14 (a) Geologic map of a portion of the San Rafael Swell and Desert. Jurassic sedimentary rocks form the monocline on the eastern margin of the swell. (b) Schematic (vertical exaggeration) cross section from the San Rafael Swell (C) to the San Rafael River (D) through the monocline. Modified from Aydin (1977). Center of map UTM: 12 S 534058.32 m E, 4273139.32 m N.

$\Delta u \approx 1$ to 10 mm, and typical band thicknesses, $t \approx 1$ mm, estimates of angular shearing, $\phi \approx \tan^{-1}(\Delta u/t)$, range from 45° to 85°.

While single deformation bands are found in most outcrops of Entrada Sandstone in the San Rafael Desert, structures representing the second stage in fault development are more sparsely distributed, but easily recognized. Walls of sandstone a few decimeters thick (**Figure 8.16**) stand up to a few meters above the surrounding outcrop, or pile of wind-blown sand. These zones of deformation bands are composed of numerous (up to about 100) closely spaced single deformation bands with a common orientation. The spacing of single bands within the zone is not uniform. Thus, lenses of undeformed sandstone from less than a millimeter to a few centimeters thick remain between some individual bands within the wider zones.

Locally older single bands with a different orientation are crosscut and offset within a zone of deformation bands (**Figure 8.16**). These occurrences demonstrate that bands within a zone have similar offsets to isolated bands, that is $\Delta u \approx 1$ to 10 mm. The offset of older structures across a zone of deformation bands is equal to the sum of the offsets across each band within the zone. For example, a zone with 10 bands would be expected to have an offset of 1 to 10 cm. Even with as many as 100 bands, zones typically do not display offsets greater than 30 cm.

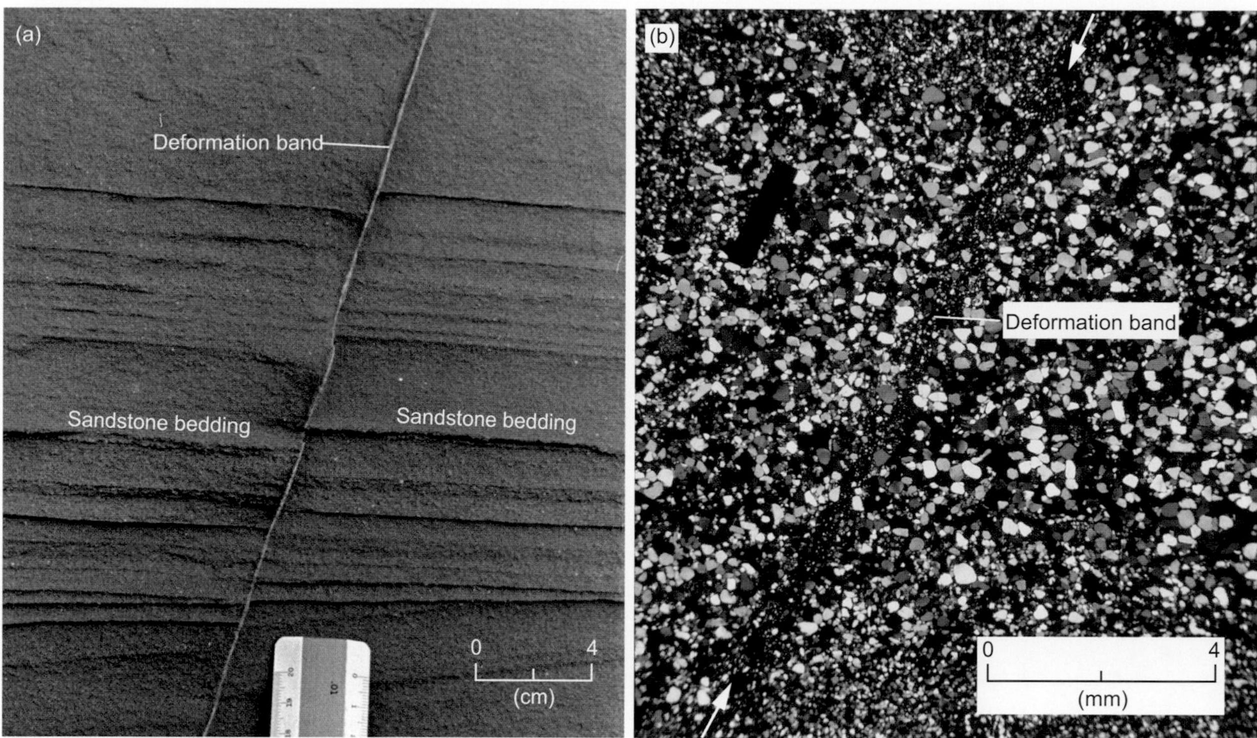

Figure 8.15 (a) Outcrop photograph of single deformation band cutting bedding in Entrada Sandstone, San Rafael Desert, UT. Deformation band offsets bedding about 8 mm. (b) Microscopic image of single deformation band in Entrada Sandstone at Buckskin Spring, San Rafael Desert, UT. Approx. UTM: 12 S 528353.93 m E, 4274604.43 m N. Photography by A. Aydin. Both figures modified from Aydin (1977).

The third stage in the development of faults in the Entrada Sandstone involves a very different mechanism from the first two stages, both of which depend upon the formation of deformation bands. On the left side of the zone shown in Figure 8.16, a thin gray layer resides adjacent to a concentration of deformation bands within the zone. The left-most edge of this layer is a surface that bounds the entire zone of deformation bands. That surface is polished to a high luster and contains slickenlines, both suggesting that it was a surface that slid in frictional contact against the adjacent sandstone. This is referred to as a slip surface. The grain size of material in the thin gray layer is reduced, relative to that within adjacent deformation bands, and the porosity is estimated to be about 1%. Offsets can be as great as 10 m on slip surfaces.

The three stages in the development of faults in Entrada Sandstone exposed in the San Rafael Desert are summarized in Figure 8.17. An isolated deformation band with up to 1 cm of slip is the first structure to form in these rocks (Figure 8.17a). They usually are less than a centimeter thick, contain crushed sandstone grains and have a reduced porosity. The deformation is related to distributed shearing across the band with no surfaces of displacement discontinuity. In the second stage (Figure 8.17b) multiple sub-parallel deformation bands congregate in a closely spaced zone of deformation bands that accommodates up to 30 cm of slip, the sum of the slip on each band. In the third and final stage (Figure 8.17c) a slip surface develops on the margin of a concentrated zone of deformation bands. These surfaces of displacement discontinuity accommodate as much as several meters of slip, they are polished, and they display slickenlines, indicative of frictional sliding.

8.3 DESCRIPTIONS OF FAULTS AT THE CRUSTAL SCALE

Outcrop photographs, maps, and thin sections are appropriate for characterizing the meter-scale faults in granite and sandstone described in the previous section, but here we consider three examples that illustrate some of the characteristics of faults at kilometer length scales, so other methods are required. For example, the normal faults at Chimney Rock, UT (Section 8.3.1), were mapped using the Global Positioning System, and the thrust faults in the Elk Hills oil field, CA (Section 8.3.2), were mapped using seismic imaging.

8.3.1 Normal Faults: Chimney Rock, UT

The Chimney Rock fault array crops out near the northern end of the San Rafael Swell in eastern Utah. The abundance of easily eroded sedimentary units (e.g. shale, calcareous sandstone, and friable limestone) in the Chimney Rock area means that mapping the faults depends upon identification of resistant beds that are likely to be exposed in the local stratigraphy, and that are laterally continuous. The relevant stratigraphy extends about 15 meters from the top of the Navajo Sandstone up through the blue-gray limestone near the base of the Carmel Formation (**Figure 8.18**). The top of the massive Navajo Sandstone is relatively resistant to erosion and therefore is found in prominent outcrops. The blue-gray limestone is only one meter thick, but also is resistant to erosion, laterally continuous, and forms distinct ledges and mesa tops. Thus, the Navajo Sandstone and the blue-gray limestone are good marker horizons for mapping.

Figure 8.16 Zone of deformation bands in Entrada Sandstone, Molly's Castle, San Rafael Desert, UT. Zone is about 30 cm thick (hand lens for scale). Modified from Aydin (1977). Approx. UTM: 12 S 527929.19 m E, 4270253.41 m N.

The rock units making up the local stratigraphic section (**Figure 8.18**) are shown in an outcrop photograph in **Figure 8.19**. The top of the Navajo Sandstone crops out near the bottom of the gully in the center and right foreground. About three-quarters of the way up the ridge on the right side of the photograph, the blue-gray limestone forms a prominent ledge. However, on the left side the blue-gray limestone caps the top of the mesa, several meters higher than the outcrop on the right. Although a fault is not exposed between these two outcrops, because the rock units are covered with soil and colluvium, an *inferred* normal fault is mapped with a dashed line on the photograph with a sense of movement given by the words Up and Down on the photograph.

The tops of the four ledge-forming units identified in the local stratigraphy (**Figure 8.18**), and the faults cutting these units at Chimney Rock were mapped using a portable GPS system. Points mapped on the Navajo Sandstone, the reddish limestone, and the gray limestone were extrapolated upward using the measured thickness of units to provide additional data at the level of the blue-gray limestone. From these data the structure contour map in **Figure 8.20** was constructed. The contour interval is 10 m and the range of elevations above mean sea level is 1,600 to 1,750m. Although the blue-gray limestone is cut and offset locally by many faults, the structure contours indicate that the regional dip is to the east with an inclination of about 2° to 3°.

On the Chimney Rock map (**Figure 8.20**) northerly-dipping faults (dip directions ~315° to 045°) are black, whereas southerly-dipping faults (dip directions ~135° to 225°) are red. Because both black and red fault traces trend in two different directions, four fault sets exist. Referring to these sets in an order based upon increasing strike of the fault plane, members of the first set strike about 070° and dip to the southeast; members of the second set strike about 110° and dip to the southwest; members of the third

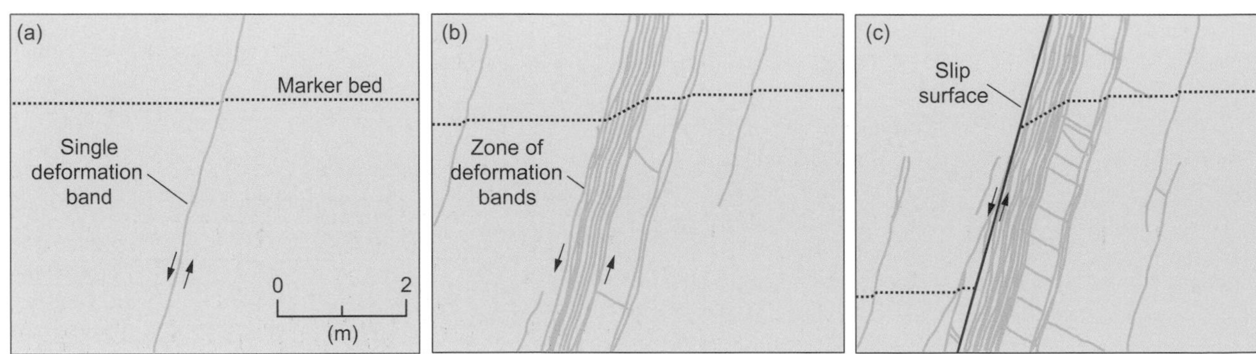

Figure 8.17 Stages in the development of a deformation band-based fault. (a) Single deformation band offsets a marker bed. (b) Zone of deformation bands with cumulative offset of a few decimeters. (c) Zone of deformation bands with much greater offset on a slip surface. Modified from Davatzes and Aydin (2003), Figure 1a.

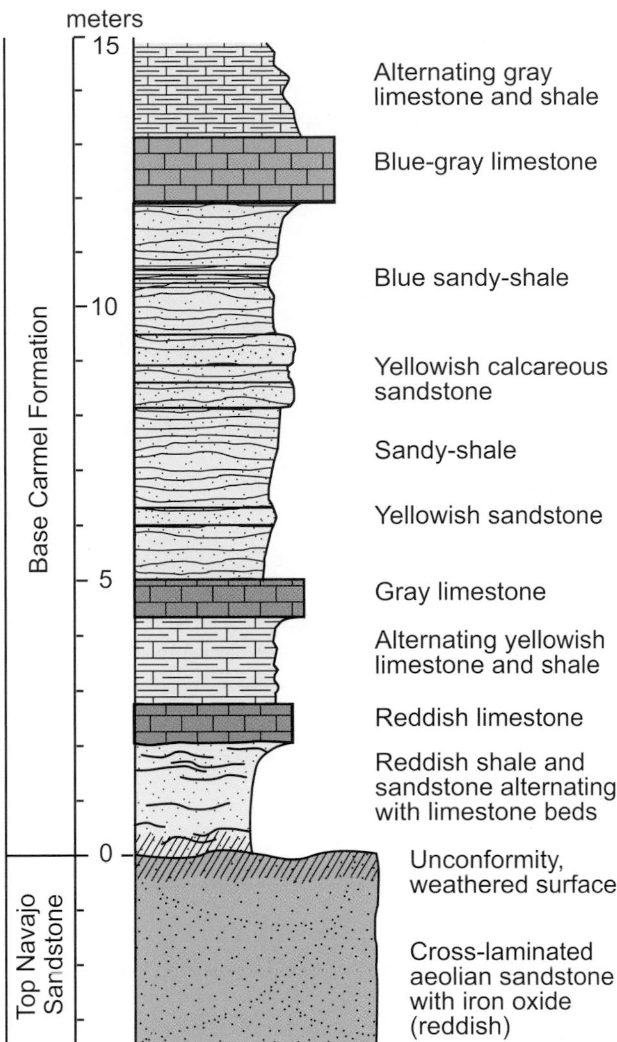

Figure 8.18 Detailed stratigraphic section for the top of the Jurassic Navajo Sandstone and base of the Carmel Formation in the Chimney Rock area of the San Rafael Swell, UT. The tops of the ledge-forming units were mapped using GPS techniques. Modified from Maerten et al. (2001), Figure 3.

set strike about 250° and dip to the northwest; and members of the fourth set strike about 290° and dip to the northeast.

8.3.2 Thrust Faults: Elk Hills Oil Field, CA

The Elk Hills oil field is located in the San Joaquin Valley, the southern portion of California's Great Valley. The petroleum potential of Elk Hills was identified in the early twentieth century and it was designated Naval Petroleum Reserve No. 1 in 1912. The oil field was turned over to commercial exploitation in 1976. Drilling and geological studies have documented the upper Tertiary stratigraphy in great detail, with particular focus on the Monterey formation, which has been identified as both a source and a reservoir for hydrocarbons. A stratigraphic section is shown in **Figure 8.21**. By the first decade of the twenty-first century more than 2,000 wells had produced more than 1.2 billion barrels of oil and 1.8 trillion cubic feet of natural gas. This production ranks Elk Hills as the seventh largest petroleum field in the continental United States. Understanding the geologic structures in this field has direct implications for sorting out the migration routes of the hydrocarbons from source to trap, for identifying and quantifying the potential reserves, and for formulating a drilling strategy to produce these reserves.

The major faults and associated folds at Elk Hills have been imaged using a three-dimensional seismic reflection data set acquired from 1999 to 2000 (**Box 8.1**). This seismic data, along with the abundant well data, provide an opportunity to unravel the evolution of these structures in space and time. Five strong reflecting surfaces (**Figure 8.21**, right side) were identifiable throughout the seismic volume and therefore were selected to map the faults and folds: the top of the McDonald formation, the base of the Reef Ridge formation, the Calitroleum horizon, the Wilhelm horizon, and the Mya 4A horizon. These surfaces extend in time from upper Miocene to upper Pliocene. Hundreds of locations on each of these surfaces were picked from the well data to calibrate the seismic volume.

The partitioning of seismic energy at geological interfaces into transmitted and reflected waves is of fundamental importance to a seismic reflection survey. To provide an elementary understanding of this phenomenon, consider a *P*-wave traveling through a rock unit (1) with mass density, ρ_1, and wave velocity, α_1 (Sheriff and Geldart, 1995; see Further Reading). Suppose this wave encounters the interface with a different rock unit (2) with mass density, ρ_2, and wave velocity, α_2, and that the wave is traveling perpendicular to the interface. For this special case, the angle of incidence is 0°, but what we present here is a good approximation for angles of incidence up to 15°. Because the distance between the seismic source and the seismometers at the surface usually is much less than the depth to a reflector, this special case is widely applicable. The product of mass density and wave velocity is the acoustic impedance, so $Z_1 = \rho_1\alpha_1$ and $Z_2 = \rho_2\alpha_2$. The fractions of transmitted energy, E_T, and reflected energy, E_R, at the interface in terms of the acoustic impedances are:

$$E_T = \frac{4Z_1Z_2}{(Z_2+Z_1)^2}, \quad E_R = \left(\frac{Z_2-Z_1}{Z_2+Z_1}\right)^2$$

The total wave energy is conserved, so $E_T + E_R = 1$. If the impedance contrast, $Z_2 - Z_1$, is zero, $E_R = 0$ and no energy is reflected. Greater impedance contrasts, correspond to greater reflected energy, but E_R does not depend upon which rock unit has the greater impedance.

Because the acoustic impedances for common rock types are very similar, the fractions of reflected energy are very small (**Table 8.1**). For an example with sandstone on top of limestone, only 4% of the *P*-wave energy reflects, and having that limestone on top of the sandstone does not change the fraction of reflected energy. The value of E_R in the first two rows of this table is about an upper bound for common rock types. Values in the last two rows are typical, because rock density contrasts and wave velocity contrasts usually are small.

Figure 8.19 Photograph of offset blue-gray limestone at Chimney Rock, UT, with inferred normal fault (dashed white line). Navajo Sandstone is exposed at the bottom of the gully. Scale approximate in foreground. See Maerten et al. (2001). UTM: 12 S 542831.16 m E, 4342868.19 m N.

To illustrate the folding and faulting at Elk Hills, structure contour maps of the five marker horizons were constructed from the seismic reflection survey and data from more than 700 wells. In **Figure 8.22** structure contours (152 m interval) on top of the McDonald formation reveal three elongate ridges in the western part of the oil field that are identified as the 29R anticline, the 31S anticline, and the NWS anticline. These folds crest from about 1,500 to 2,500 m deep and trend about 120°, roughly parallel to the traces of the major faults. Each anticline is bounded on its northeast flank by a fault that dips to the southwest.

The folds and faults in the western Elk Hills field are interpreted in the cross section A–A′ (**Figure 8.23**) that is roughly perpendicular to the strike of bedding. The colored parts of this cross section, which include the five marker horizons, are very well constrained by the abundant data. The 1R, 2R, and 3R structures are thrust faults that tip out between the Base Reef Ridge and the Calitroleum horizon, and they all have dips that increase upward in the section. The 2R fault is associated with the NWS anticline; the 3R fault is associated with the 31S anticline; and the 1R and 5R faults bound the 29R anticline. Below the McDonald horizon the seismic data do not offer the same image quality, but the thrust faults are interpreted as decreasing in dip toward zero in the formations of Oligocene age.

The stratigraphic interval between the middle Miocene McDonald horizon and the lower Pliocene Reef Ridge horizon on the cross section in **Figure 8.23** changes thickness across each anticline. This interval thins as each underlying fault is approached from SW to NE. This interval also thickens discontinuously when crossing the 1R, 3R, and 2R faults from SW to NE. These relationships inform the interpretation of the timing of faulting and deposition. Using concepts from stratigraphic analysis, and the geometry of the marker horizons from the structure contour maps, the faulting at Elk Hills is interpreted as syn-depositional, that is the faults were *growing* as the sedimentary layers were deposited. The sequence of fault growth began with 3R and was followed by 2R and then 1R. The last major fault to grow was 5R, which formed after deposition of the Calitroleum horizon and before the Wilhelm.

8.3.3 Strike-Slip Faults: Imperial Valley, CA

Traces of the Coyote Creek, Imperial, and Brawley faults in the Imperial Valley of southern California are arranged as discontinuous echelon segments along strike, but all have predominantly right-lateral slip. Here we focus on a left step between two segments of the Coyote Creek fault, and on a right step between the Imperial and Brawley faults. These two examples, shown on the structure maps in **Figure 8.24**, serve to highlight the very different styles of deformation associated with the two different senses of step for faults that have the same sense of strike slip. These examples also demonstrate that deformation at kilometer-scale steps is quite different than what was described at a meter-scale step in granite (**Figure 8.8**).

The northern and southern segments of the Coyote Creek fault (**Figure 8.24a**) are approximately parallel; each segment is about 10 km in length; and their traces trend about 135°. The left step between these two segments is about 5 km long and 2 km wide, and the Ocotillo Badlands is located within this step. Slip on the

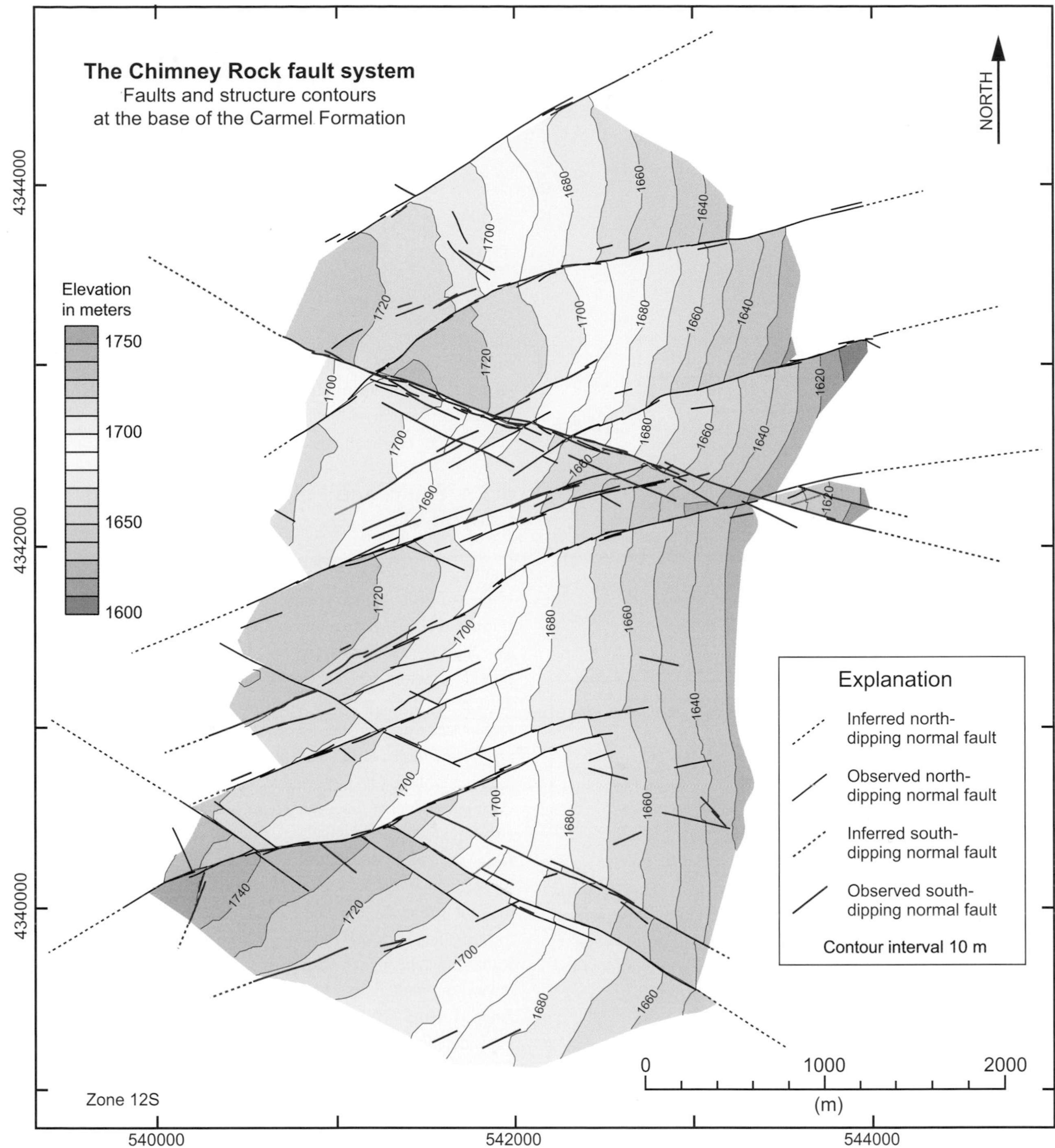

Figure 8.20 Structure contour map with 10 m intervals on the base of the Carmel Formation relative to mean sea level. Four sets of normal faults are mapped. Modified from Maerten et al. (2001), Figure 2. Approximate map center UTM: 12 S 542000 m E, 4342000 m N.

Coyote Creek fault during the 1968 Borrego Mountain earthquake was predominantly right lateral, however the local topography of the Ocotillo Badlands rises about 200 m above the surrounding desert, suggesting that there has been a significant component of dip slip on the overlapping portions of these fault segments in the past. Furthermore, the sedimentary strata within the left step are folded, with fold axes trending about 090°. Both the right-lateral slip on the fault segments and the trend of the fold axes are consistent with the greatest horizontal compression directed approximately north–south.

Table 8.1 Representative values for the range of physical quantities that affect seismic reflection

Interface	α_1 (m/s)	ρ_1 (kg/m^3)	α_2 (m/s)	ρ_1 (kg/m^3)	E_T	E_R
sandstone–limestone	2,000	2,400	3,000	2,400	0.96	0.04
limestone–sandstone	3,000	2,400	2,000	2,400	0.96	0.04
shallow interface	2,100	2,400	2,300	2,400	0.9979	0.0021
deep interface	4,300	2,400	4,500	2,400	0.9995	0.0005

From Sheriff and Geldart (1995), Table 3.1

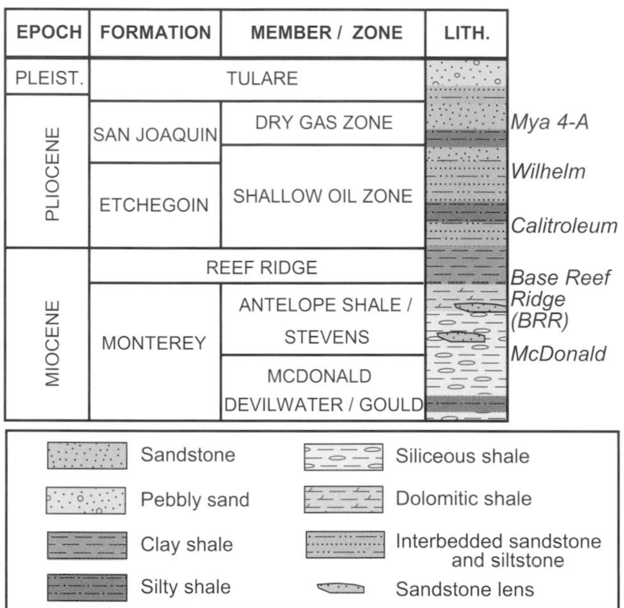

Figure 8.21 Simplified stratigraphy of the upper Miocene to Pleistocene at Elk Hills, CA. Five horizons used to analyze the seismic reflection and well data are the McDonald, Base Reef Ridge, Calitroleum, Wilhelm, and Mya 4-A. Lithologies are identified in the key and displayed along the right side of the column. Modified from Fiore et al. (2007), Figure 3.

The Brawley and Imperial faults are part of the right-lateral San Andreas Fault system in southern California. The southern tip of the 30 km long Brawley fault overlaps the northern tip of the 60 km long Imperial fault by several kilometers, and the step between these two echelon faults is a *right step* (**Figure 8.24b**). This step is about 15 km long, parallel to the strike of the segments, and about 10 km wide. The topography in this right step is depressed a few tens of meters below the surroundings, and the remnant of Mesquite Lake occupies the step.

The right step between the Brawley and Imperial faults is bounded by normal faults that indicate horizontal extension. These normal faults trend from the tips of the two strike-slip faults, form two sides of a rhomb-shaped depression, and dip toward the center of the depression. On October 15, 1979 a magnitude 6.5 earthquake nucleated about 30 km southeast of the step on the Imperial fault and propagated to the step. The rupture caused strike slip along much of the Imperial fault, but where this fault forms the southwestern border of the rhomb-shaped depression, the relative motion changed to dip slip, with about 20 cm of dip slip associated with the earthquake.

8.3.4 Faults and Folds: Western Grand Canyon, AZ

Faulting and folding can be intimately related to one another. Here we turn to exposures in the western Grand Canyon (**Figure 8.25a**) to explore an example of these relationships. The high desert terrain and the deep incisions of the canyons, provide outcrops revealing more than 1 km of the stratigraphic section. Quaternary basalt flows drape some of the faults in this region, and some of these flows are faulted, suggesting a young age for the faults. Historical seismicity indicates the area is active today, and the faults are considered to be younger than Miocene. Geologic evidence suggests there has been less than 500 m of erosion, other than in the deep canyons, since about 30 Ma, so the faults apparently developed at, or near, Earth's surface.

A schematic cross section (**Figure 8.25b**) shows that strata in the foot wall of the Frog fault bend upward approaching the fault, forming the lower half of a monoclinal fold. Strata in the hanging wall forms the upper half of a monocline with strata dipping toward the fault. These faults and the associated folds have been mapped using GPS instrumentation (see **Box 2.1**), which provides detailed quantitative data on the geometry of the faults and folded strata.

On the geological map (**Figure 8.26**) the Frog fault strikes to the southeast and dips about 70° to the southwest south of Paraschant Canyon. Slickenlines on fault surfaces trend nearly perpendicular to the fault strike and plunge at angles about equal to the fault dip, so the relative motion is dip slip. The vertical component of dip slip, called the fault throw, at the Permian Esplanade Formation is about 225 m, down to the southwest, so this is a normal fault. A smaller fault is exposed along a parallel trace about 0.5 km to the southwest of the main Frog fault. This fault has a strike and dip similar to the Frog fault, so it is a *synthetic* normal fault (**Figure 8.25b**). The throw on the synthetic fault is about 50 m, down to the southwest.

The upper hinge of a fold, called the Lone Mountain monocline, is parallel to the Frog fault and about 2 km to the southwest

Box 8.1 Seismic Surveying

Seismic surveys use elastic (seismic) waves to create images of rock structures in Earth's lithosphere (Sheriff and Geldart, 1995). The seismic waves are human-induced, not due to slip on a fault, which creates an earthquake. For example, detonating dynamite in a shallow well bore generates seismic waves, which travel from the source to a reflective interface in the subsurface, and then back to a set of seismometers (geophones) at the surface. For a two-dimensional survey, the geophones are located along a straight survey line extending from the well bore, and each geophone electronically records the time and the local ground motion due to passage of the seismic waves. The data from each geophone is a graph (trace) of ground motion versus time. A reflected wave that arrives at systematically different times for geophones at greater distances from the source is associated with a particular interface in the subsurface. Given the two-way travel time of that wave from the source to the geophone, and the velocity of the elastic wave, the distance to the interface is calculated. The compiled traces and processed data from all the geophones produce a cross section of the reflecting interfaces beneath the survey line.

The description of a seismic survey in the preceding paragraph greatly simplifies what usually is more complicated, thus requiring elaborate equipment and sophisticated data processing techniques (Sheriff and Geldart, 1995). For example, seismic waves may refract along a sedimentary bed with greater seismic velocity, before traveling back to the surface. In addition, sedimentary basins are composed of strata with differing seismic velocities, so the velocity structure under the survey line must be determined to compute the distance that a wave travels. Laboratory measurements provide seismic velocities for samples of rock extracted from well bores in the basin. Many survey lines, usually laid out in a rectangular grid, produce a three-dimensional survey of the subsurface interfaces. The volume of seismic data from such a survey, with suitable interpolation, leads to structure contour maps of those interfaces (**Figure 8.22**).

Seismic surveys are one of the primary geophysical methods for *exploration* of Earth's lithosphere for hydrocarbon and geothermal reservoirs, ore deposits, and groundwater aquifers. Typically, these surveys provide an image of geological structures such as folds, faults, and contacts between rocks with different mechanical properties, but they do not directly reveal hydrocarbons or valuable minerals (Sheriff and Geldart, 1995). Inferring the presence of hydrocarbon and mineral resources from the geometry of the structures requires *interpretation* of the seismic survey by geologists, who deduce the geological history of the region, and relate the origin and location of the resources to the geologic structures. In turn, the *production* of hydrocarbon reservoirs and ore deposits depends upon analysis by petroleum and mining engineers. Thus, the effective application of seismic surveying for resource recovery usually involves a team effort with geophysicists, geologists, and engineers contributing respectively to exploration, interpretation, and production.

To illustrate one of the many applications of seismic reflection for the hydrocarbon industry, we consider a three-dimensional survey from the North Sea (**Figure B8.1a**). About 130,000 data points on a 25 m grid spacing define the reflecting interface over an area that is about 10 km long and 8 km wide. Filtering undulations with wavelengths less than 2 km out of the data leaves a large central dome and smaller domes and basins separated by synformal saddles and antiformal saddles. The diameter of the central dome is about 6 km and its

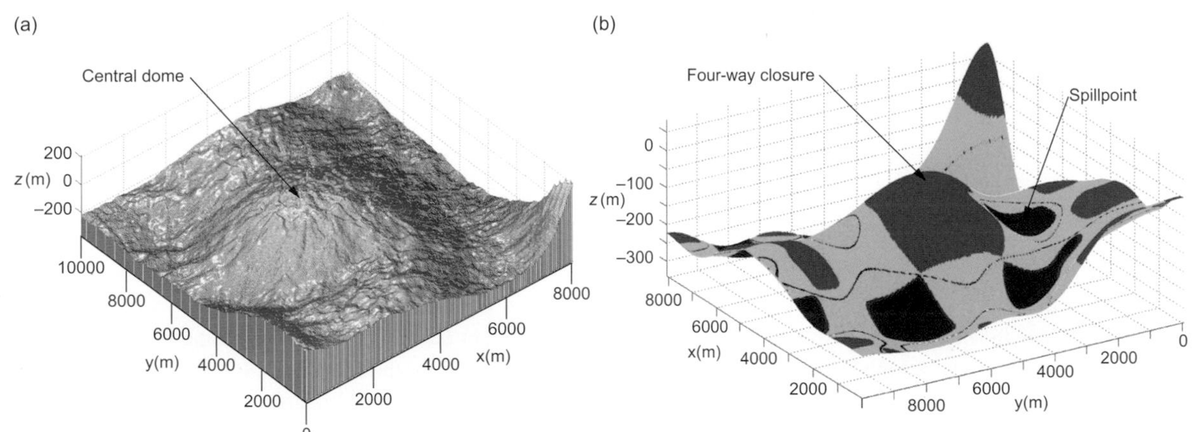

Figure B8.1 (a) Sedimentary interface from the North Sea imaged using a three-dimensional seismic reflection survey. From Bergbauer and Pollard (2003). (b) Smoothed interface (rotated counter-clockwise) with fault traces and survey noise removed to reveal large-scale structures identified using geological curvature classification (Section 9.5.4): dome (blue); antiformal saddle (green); synformal saddle (orange); basin (brown). From Mynatt et al. (2007).

amplitude is about 200 m. Over most of the central dome both principal curvatures are positive, and the Gaussian and mean curvatures are positive, so this structure meets the criteria for a dome according to the geological curvature classification (Section 9.5.4).

To be a potential trap for hydrocarbons the interface in a dome must have at least one horizontal point, and the trapping capacity is greater if the horizontal point is located near the middle of the dome, as it is for the central dome in **Figure B8.1b**. The term used in the hydrocarbon industry for this geometry is four-way closure. If impermeable formations overlie this interface, they would seal the dome and trap the hydrocarbons. Hydrocarbons can escape from a four-way closure at horizontal points in the synformal saddles that surround the dome. The industry term for these locations is spillpoint, although the fluid would be moving upward as it escaped the trap at such a point. Three possible spillpoints exist around the central dome, and the one with the highest elevation would "spill" first. The volume of the dome above this spillpoint, times the average porosity of the rock, provides an estimate for the trapped fluids.

Seismic surveys also play an important role in imaging the very shallow lithosphere for the design of tunnels, dams, nuclear waste repositories, and other engineered structures (Barton, 2007; see Further Reading). Again, the survey provides an image of the subsurface in terms of the geologic structures and contacts between rocks with different mechanical properties, but this image usually requires interpretation by geologists, and analysis by engineers, to understand the implications of those structures for rock quality relative to the specific objectives of the project. For example, the presence of fractured rock might be welcome if the project involved boring a large diameter tunnel (**Box 4.1**), but might be undesirable for the containment of nuclear waste, or as the footing of a large dam. Again, a multidisciplinary team involving geophysicists, geologists, and engineers is well suited to tackle big projects in regions with a complex geologic history.

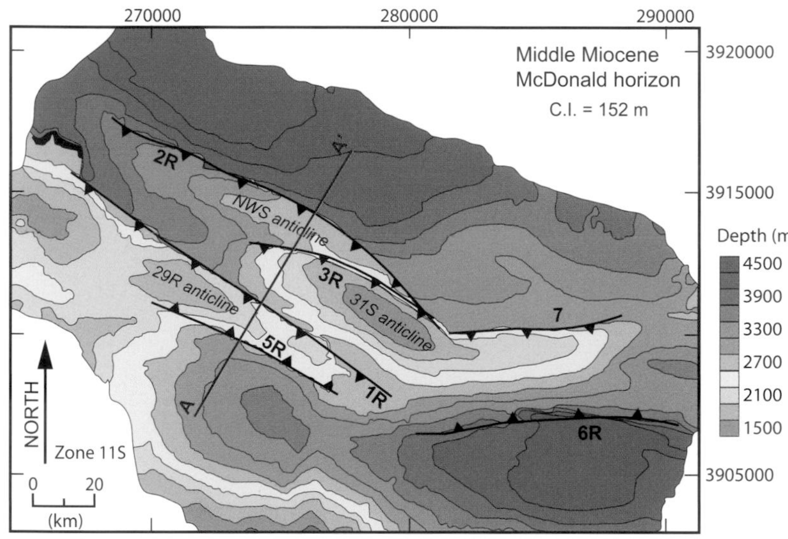

Figure 8.22 Structure contour maps for Elk Hills. Middle Miocene McDonald horizon with contour interval 152 m. Solid black lines mark thrust faults: teeth indicate dip direction. Cross section A–A′ shown in **Figure 8.23**. Modified from Fiore et al. (2007), Figure 2b.

of the fault (**Figure 8.26**). Northeast of this hinge line the Esplanade Formation typically dips toward the Frog fault, and the dips increase with proximity to the fault. Although called a monocline, this structure lacks a lower hinge, because it is truncated by the fault where the dips are steepest. In this regard it is a *half* monocline. To the northeast of the Frog fault the Esplanade Formation dips away from the fault, forming a lower half monocline. These structural data are consistent with the schematic cross section and aerial photograph shown in **Figure 8.25**.

To generate quantitative data on the faults and folds in the mapped area (**Figure 8.26**) a GPS survey was conducted on six prominent and well-exposed bedding surfaces of the upper Esplanade Formation. The bedding surfaces were walked out while recording the location and elevation of points every 5 seconds with an estimated horizontal and vertical precision of ~0.5 m and ~1.5 m, respectively. Errors were introduced because of difficulty in identifying the stratigraphic level of six different bedding surfaces within the upper Esplanade Formation. However, the combined GPS and stratigraphic error was only about 3.5 m, which is small compared to the 250 m range of elevations for the upper Esplanade Formation due to faulting and folding.

To construct a structure contour map for a surface in the upper Esplanade Formation (**Figure 8.27**), more than 24,000 GPS location points were extrapolated vertically up or down to a single surface, based on the measured thicknesses of the

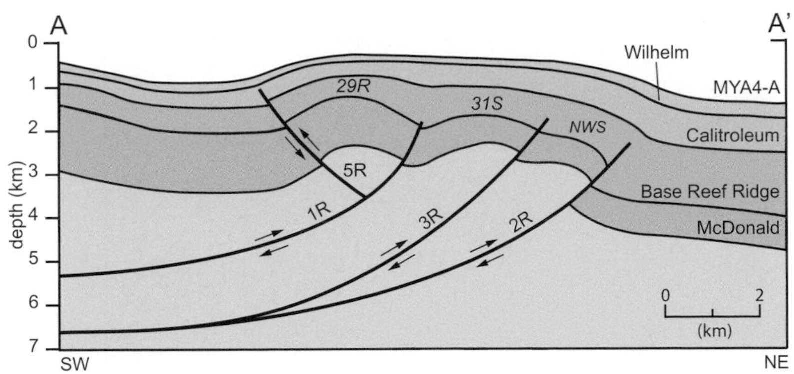

Figure 8.23 Cross section A–A′ with azimuth about 030°located in the western part of Elk Hills oil field (for location see Figure 8.22). Four major structure-bounding faults are 1R, 2R, 3R, and 5R; prominent anticlinal crests are 29R, 31S, and NWS; and five key seismic reflectors are McDonald, Base Reef Ridge, Calitroleum, Wilhelm, and MYA4-A. Modified from Fiore et al. (2007), Figure 2c.

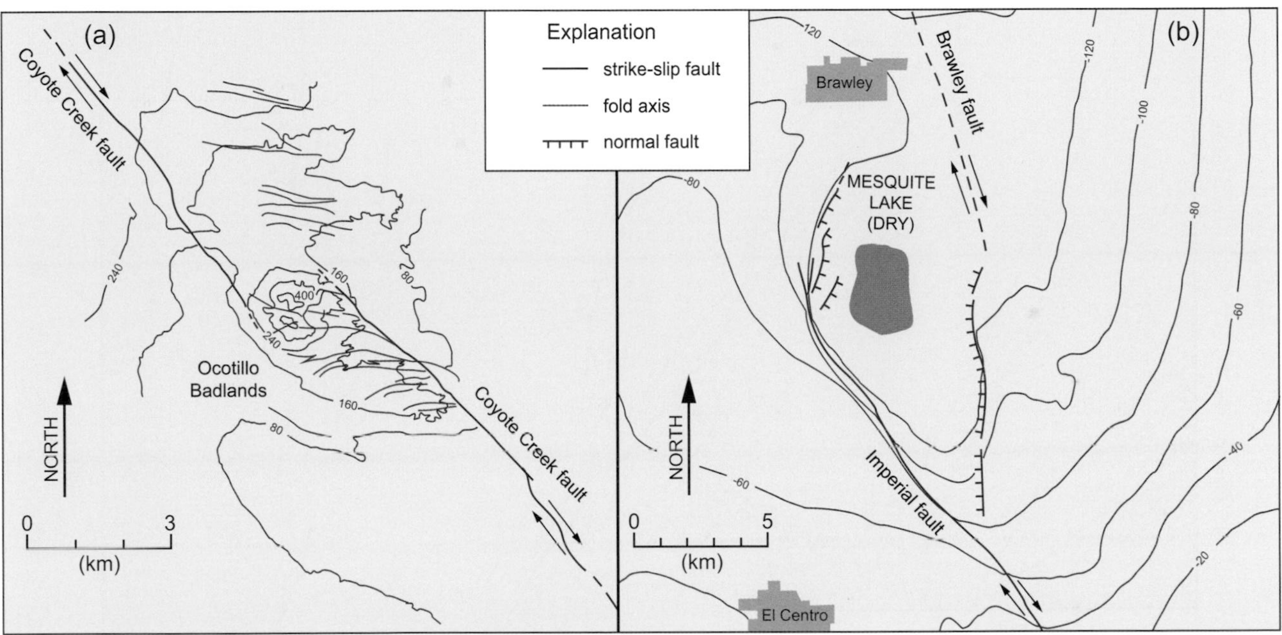

Figure 8.24 (a) Coyote Creek fault map with left step at Ocotillo Badlands. Map center UTM: 11 S 583814.02 m E, 3665730.57 m N. (b) Map of right step between Imperial and Brawley faults in the southern Imperial Valley, CA. Rhomb-shaped depression is bounded by normal faults. Approximate depression center UTM: 11 S 637906 m E, 3641335 m N. Modified from Segall and Pollard (1980), Figures 1 and 2.

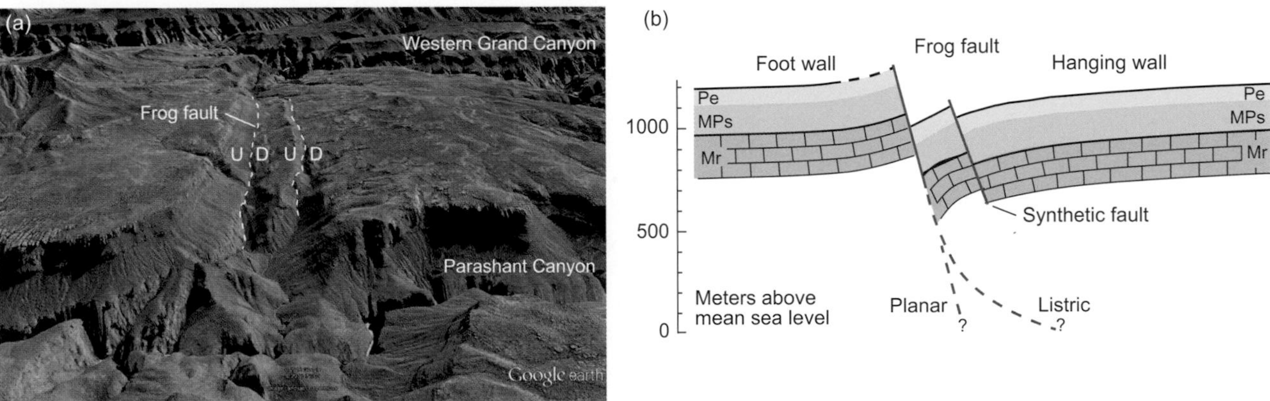

Figure 8.25 (a) Oblique Google Earth image looking southeast along strike of the Frog fault between Parashant Canyon and the Western Grand Canyon, AZ. Approximate UTM: 12 S 291315.00 m E, 4003819.00 m N. (b) Schematic cross section based on GPS mapping of the Permian Esplanade Formation showing the Frog fault and a synthetic fault. Fault geometry at depth is interpreted as either planar or listric. Modified from Resor (2008), Figure 3B.

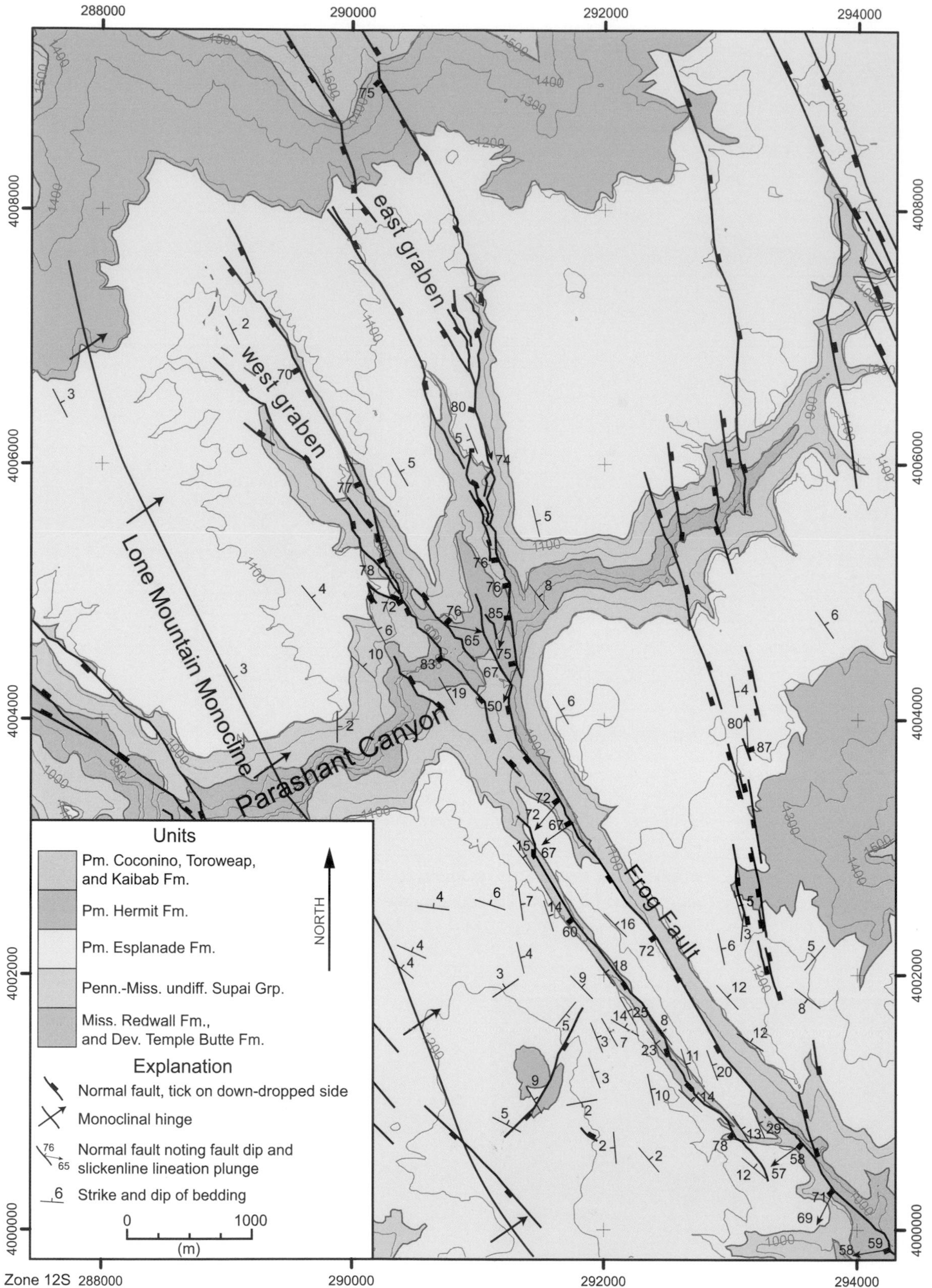

Figure 8.26 Geologic map of the Frog fault and Lone Mountain area near Parashant Canyon, AZ. WGS 1984 datum. Pm., Penn., Miss., and Dev. are Permian, Pennsylvanian, Mississippian, and Devonian strata. Approximate map center UTM: 12 S 291000.00 m E, 4005000.00 m N. Modified from Resor (2008), Figure 2.

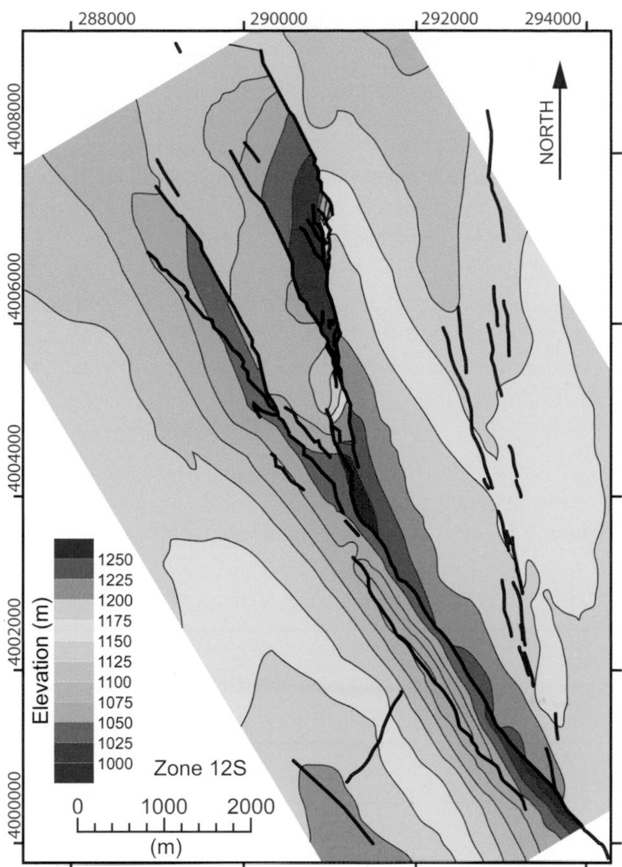

Figure 8.27 Structure contour map of the upper Esplanade Formation in the vicinity of the Frog fault and Lone Mountain monocline (Figure 8.26). More than 24,000 GPS location points were used to constrain elevations on this map. Modified from Resor (2008), Figure 6C.

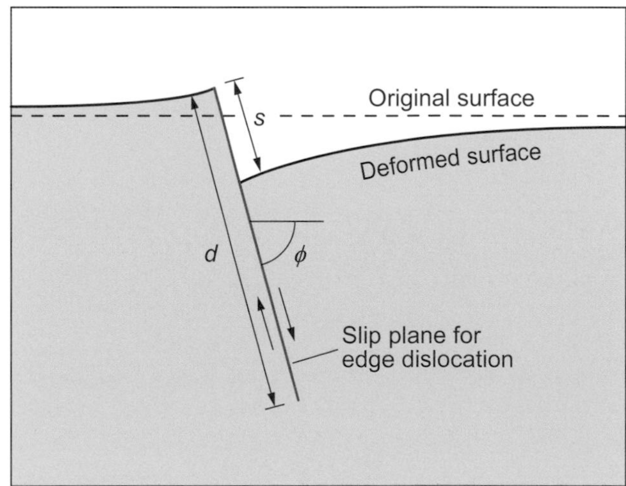

Figure 8.28 Mechanical model for normal fault using an edge dislocation in an elastic half-space. Refer to Section 5.6.4 for description of the edge dislocation. See Resor and Pollard (2012).

individual beds in this part of the stratigraphic section. Note the systematic decrease in elevation (warmer to cooler colors) approaching the Frog fault from southwest to northeast in the southern portion of this map. This defines the upper half monocline in the hanging wall of the fault. The structural discontinuity, based on the color change across the fault from blue to red, is about 225 m. Approaching the fault from northeast to southwest, the elevations gradually increase (cooler to warmer colors), defining the lower half monocline.

The geometry of the folds adjacent to the Frog fault (Figure 8.25b) can be compared to the solution for an elastic edge dislocation. Recall from Section 5.6.4 that dislocation models are based on solutions to *elastic* boundary-value problems. Because this is a two-dimensional model, we consider a cross section perpendicular to the fault and project the GPS data along the strike of the fault onto this cross section from distances as great as 2 km. The upper Esplanade Formation is assumed to have been horizontal before faulting, and at an elevation equal to the average elevation of that formation far from the fault today. The fault dip was fixed at 70° to the west, based on the field measurements. A range of down-dip fault heights, d, were considered, while searching for the best fitting slip, s, based on the lowest resulting sum of the squared differences between the model prediction and the data. A single edge dislocation was used, so the slip is constant over the entire fault (Figure 8.28).

The result shown in Figure 8.29a offers a relatively poor fit, especially for the foot wall data, but the fit improves significantly (Figure 8.29b) by removing a regional tilt of 1° to the east from the raw data. The most obvious discrepancies for this second model are near the synthetic fault in the hanging wall, where the dip of the surface from the GPS data exceed the model dip, with no discontinuity. However, including the synthetic fault as a second edge dislocation, also with a dip of 70°, improves the fit by about 30% (Figure 8.29c). For this model the Frog fault has 253 m of slip over a down-dip height of 1.1 km, and the synthetic fault has 40 m of slip over a down-dip height of 200 m. Only the GPS data between the main fault and the synthetic fault are not well-fit by this model. Perhaps the addition of inelastic deformation mechanisms such as jointing or bedding-plane faulting would improve the fit, but elastic deformation accounts for most of the observed folding in both the foot wall and hanging wall of the Frog fault.

8.4 A CANONICAL MODEL FOR FAULTING

A canonical model is one that reduces the physical process, in this cases faulting, to the simplest form possible without loss of generality. To maintain generality we consider a material continuum that is governed by Cauchy's laws of motion, introduced in Sections 3.6.1 and 3.7.1. To reduce Cauchy's equations to a practical problem that can be solved, a *constitutive law*, the *model geometry*, and the *boundary conditions* are defined here. We explore the solution to the boundary-value problem in terms of the displacement and stress fields near the model fault in some detail, owing to the importance of faulting in structural geology, and also to illustrate how much can be gleaned from this elementary solution.

8.4.1 Constitutive Law and Model Geometry

Following arguments similar to those of Section 7.2.1, we choose the constitutive law for a linear elastic material. This choice is justified by noting that the ratio of slip to length for single slip events on many faults is within the range $10^{-5} \leq (\Delta u_x/2a) \leq 10^{-3}$. In **Figure 8.30a** we suppose that the tangential displacement, u_x, near the fault middle decreases away from the fault along the y-axis over a distance that scales with the fault half-length, a. Thus, the ratio of displacement to length is a rough measure of angular shearing, ϕ_e, in the vicinity of the fault:

$$\tan \phi_e = \frac{u_x}{a} = \frac{\Delta u_x}{2a} \tag{8.1}$$

Typical angular shear strains calculated using (8.1) are within the range where laboratory samples exhibit recoverable deformation, and relations between axial stress and axial strain are linear. We use the subscript e to indicate this angular shearing usually is an elastic strain.

The conclusion that deformation accompanying slip on a fault is elastic has some notable exceptions. Within fault zones the deformation often is localized and intense (e.g. **Figure 8.9**, **Figure 8.16**). Structures may develop that indicate large irrecoverable strains that exceed the elastic limit, commonly by orders of magnitude (**Figure 8.30b**). There, estimates of angular shearing, ϕ_i, are based on the offset of older markers and the fault zone thickness, t, such that:

$$\tan \phi_i \approx \Delta u_x/t \tag{8.2}$$

For example, for the faults described in Section 8.2.1 (**Figure 8.5**) the ratio of slip to thickness usually is in the

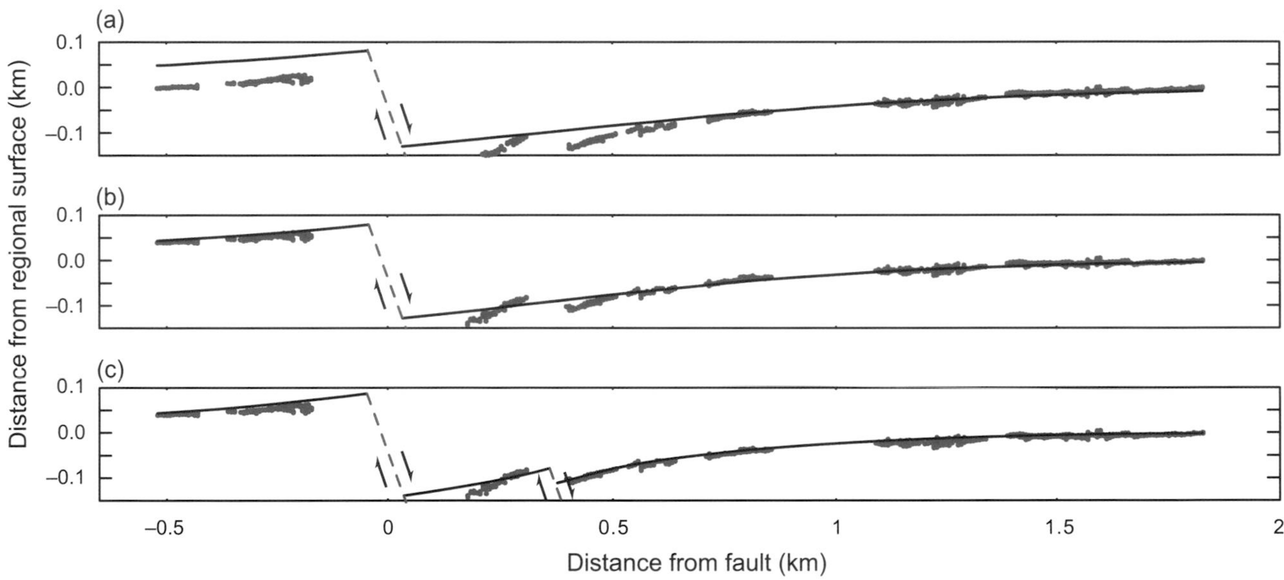

Figure 8.29 Elastic dislocation model displacements (black curve) compared to GPS data on structural elevations of upper Esplanade Formation (blue dots). (a) GPS data and single dislocation. (b) GPS data rotated 1° to account for regional tilt and single dislocation. (c) Rotated GPS data and two dislocations. Modified from Resor (2008), Figure 18.

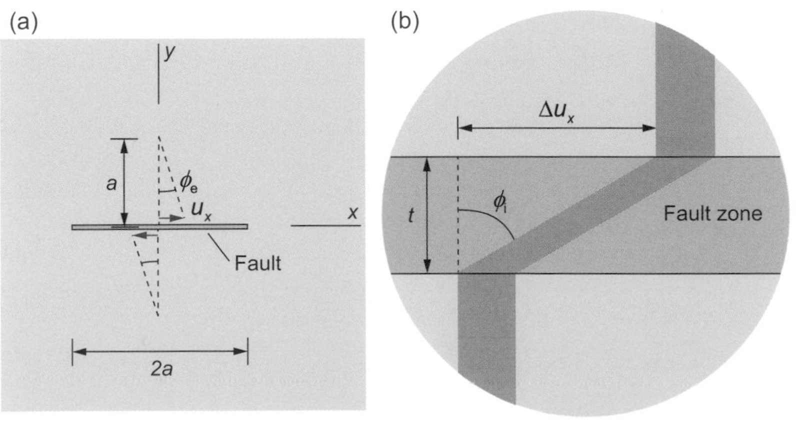

Figure 8.30 (a) Fault of length 2*a* slips with displacement u_x at middle of fault surface: $x = 0$, $y = 0^+$. ϕ_e is *elastic* angular shearing. (b) Shearing displacement discontinuity, Δu_x, determined by offset marker crossing fault zone. ϕ_i is *inelastic* angular shearing.

range $10^1 \leq \Delta u_x/t \leq 10^3$. These large strains are typical of those that are not recoverable, and are not linearly related to the applied stress. We use the subscript i for this angular shearing to indicate it usually is an inelastic strain. Near the tips of faults and adjacent to bends in faults (**Figure 8.10**), and at steps between echelon segments of faults (**Figure 8.8**, **Figure 8.24**), damage zones may exist that include fractures and secondary faults that would not heal if the slip were removed.

In some cases the inelastic deformation is isolated to regions that are very small compared to the fault length. For these cases of small scale yielding the elastic constitutive law is adequate for the vast majority of the rock mass. If the regions of yielding are not small, or if one wants accurate values of the stress and strain within these small regions, a more complicated constitutive law is required. Throughout this chapter we employ elasticity to understand the dominant behavior of faults in the brittle lithosphere.

The geometry of the idealized fault is motivated by the descriptions and images of faults reviewed earlier in this chapter: two planar surfaces that come together along a tipline. For the canonical model of faulting the geometry is taken as *two dimensional* and *planar*, with two infinitely long and perfectly straight tiplines parallel to the z-axis (**Figure 8.31**). The elastic body is in a state of plane strain, which means that the displacement components in the (x, y)-plane are only functions of x and y, and the displacement component in z is zero everywhere:

$$u_x = u_x(x, y),\ u_y = u_y(x, y),\ u_z = 0 \tag{8.3}$$

The length of the model fault in the x-direction is $2a$ and this length does not vary in z. Although the model fault is taken as infinite in the z-direction for mathematical simplicity, it need only be several times longer in z than in x for the stress and displacement fields to be adequately approximated by the two-dimensional solution. The distance between the two fault surfaces is negligible compared to $2a$ and the *surrounding* material is linear elastic and both *homogeneous* and *isotropic* with respect to shear modulus, G, and Poisson's ratio, ν.

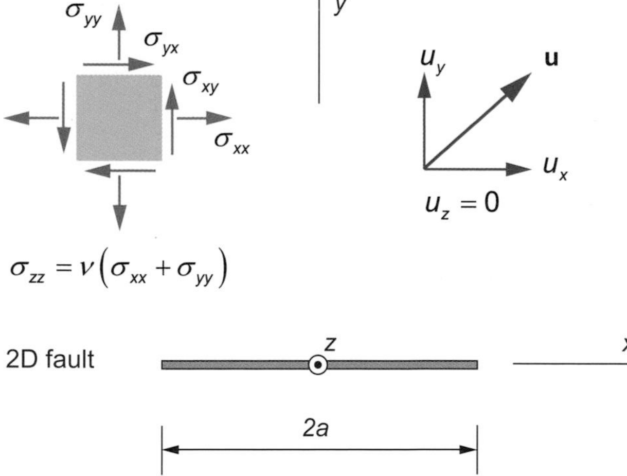

Figure 8.31 Two-dimensional, plane strain conditions for the canonical fault model. The fault is very long in z compared to the uniform length, 2a, in x. The non-zero stress components (σ_{xx}, $\sigma_{xy} = \sigma_{yx}$, σ_{yy}, σ_{zz}) and the non-zero displacement components (u_x, u_y) vary throughout the (x, y)-plane, but do not vary with z.

8.4.2 Boundary Conditions

Boundary conditions are specified using distributions of displacements or tractions on the internal and external boundaries of the model. The relative displacements *tangential* to fault surfaces commonly can be measured using offset markers, such as the aplite dike in **Figure 8.6**, or the bedding in **Figure 8.15**, but displacements normal to fault surfaces usually are unknown. The normal traction, however, can be related to the remote normal stress acting across the fault, and the shear traction can be related to the shear strength or frictional strength of the material in the fault zone. Therefore, in this section we formulate the canonical model in terms of traction boundary conditions on the fault surfaces and remote stress conditions.

The idealized two-dimensional fault is shown in cross section in **Figure 8.32a**. We take the remote normal stress acting perpendicular to the fault plane as a compression of magnitude C, which prevents the fault from opening (**Figure 8.32c**). The remote shear stress is given a magnitude S, and this is the stress that drives slip on the model fault. The remote normal stress acting parallel to the fault does not affect the tractions on the fault or the slip, so its value can be chosen arbitrarily. Thus, the remote stress boundary condition is written:

$$\text{as } \sqrt{x^2+y^2} \to \infty, \quad \begin{cases} \sigma^{\mathrm{r}}_{xx} \to ? \\ \sigma^{\mathrm{r}}_{xy} = \sigma^{\mathrm{r}}_{yx} \to S \\ \sigma^{\mathrm{r}}_{yy} \to -C \end{cases} \tag{8.4}$$

The superscript r indicates that these are the remote stresses, those that act at great distance from the model fault relative to the fault length, $2a$. Recall from Section 7.2.2 that the canonical model for the opening fracture specified remote stresses that are principal stresses, so the fracture was aligned with the principal stress planes. In **Figure 8.32a** the fault is not aligned with the principal stress planes, so the shear stress $\sigma^{\mathrm{r}}_{yx} = S$ is resolved on the fault from the remotely applied stress. In the absence of fault slip, (8.4) would define a homogeneous stress state throughout the elastic body. Fault slip perturbs that homogeneous stress state and the solution to the elastic boundary-value problem provides the stress state at any point (**Figure 8.32d**) for this boundary-value problem.

In **Figure 8.32b** we zoom in on the fault surfaces at $y = 0^+$ and $y = 0^-$, and ask: what are the traction components acting on those surfaces? The normal traction just balances the remotely applied compression, C, so no opening occurs. The shear traction provides the *resistance* to slip, which we prescribe as having a magnitude R. If the two fault surfaces are in frictional contact, R would be the frictional strength of that contact. If fault gouge occupied the very narrow space between the two fault surfaces, R would be the shear strength of that material. Thus, the boundary conditions on these surfaces are:

$$\mathrm{BC}: \begin{cases} \text{on } |x| \leq a, y = 0^+, & t_x = -R, t_y = C \\ \text{on } |x| \leq a, y = 0^-, & t_x = R, t_y = -C \end{cases} \tag{8.5}$$

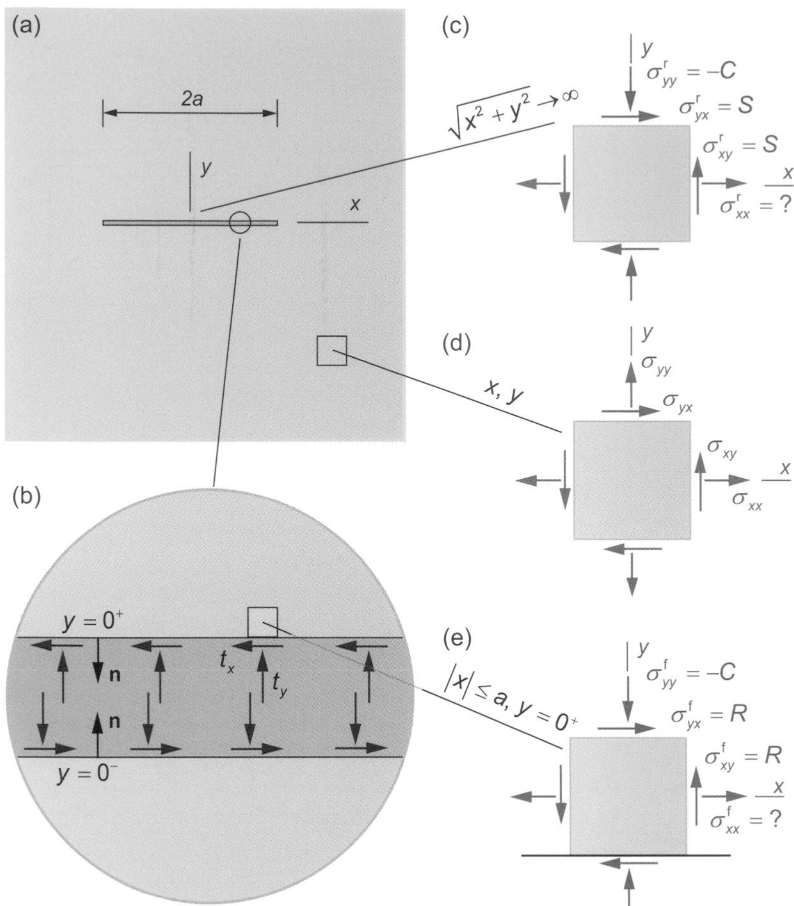

Figure 8.32 A canonical model for faults. (a) Fault of length 2*a* in an infinite elastic body. (b) Traction components, t_x and t_y, acting on fault surfaces. (c) Remote stress state at an infinite distance. (d) Stress state anywhere in the interior of the elastic body. (e) Stress state at one surface of the fault. See text for discussion of these tractions and stresses. See Pollard and Segall (1987).

These conditions apply over the entire model fault. The sign of the normal traction component, t_y, is positive on the $y = 0^+$ surface and negative on the $y = 0^-$ surface, because this traction *pushes* against both surfaces (**Figure 8.32b**). The sign of the shear traction component, t_x, is negative on the $y = 0^+$ surface and positive on the $y = 0^-$ surface, because this traction resists the slip due to the remote shear stress, $\sigma^{\mathrm{r}}_{yx} = S$, which drives the $y = 0^+$ surface to the *right* relative to the $y = 0^-$ surface.

Cauchy's formula, introduced in Section 7.2.2, relates the prescribed traction components acting on the fault surfaces (8.5) to the local stress components (**Figure 8.32b**, **e**). The general two-dimensional form of Cauchy's formula for the (x, y)-plane is:

$$\begin{aligned} t_x &= \sigma_{xx} n_x + \sigma_{yx} n_y \\ t_y &= \sigma_{xy} n_x + \sigma_{yy} n_y \end{aligned} \qquad (8.6)$$

Here n_x and n_y are components of the outward unit normal vector, $\hat{\mathbf{n}}$, for the plane on which the traction vector $\mathbf{t}$ acts. For the fault surface along $y = 0^+$ the components of $\hat{\mathbf{n}}$ are $n_x = 0$, $n_y = -1$, so $t_x = -\sigma^{\mathrm{f}}_{yx}$ and $t_y = -\sigma^{\mathrm{f}}_{yy}$. For the fault surface along $y = 0^-$ we have $n_x = 0$, $n_y = +1$, so $t_x = \sigma^{\mathrm{f}}_{yx}$ and $t_y = \sigma^{\mathrm{f}}_{yy}$. Using (8.5), the stress state at the fault surfaces is:

$$\begin{aligned} &\text{at } |x| \le a, y = 0^+, \quad \sigma^{\mathrm{f}}_{yy} = -C, \sigma^{\mathrm{f}}_{yx} = R \\ &\text{at } |x| \le a, y = 0^-, \quad \sigma^{\mathrm{f}}_{yy} = -C, \sigma^{\mathrm{f}}_{yx} = R \end{aligned} \qquad (8.7)$$

The normal stress σ^{f}_{xx} that is parallel to the surfaces is not constrained by the tractions according to (8.6), and must be found by solving the boundary-value problem.

In the elastic body surrounding the fault all three independent stress components usually are non-zero (**Figure 8.32d**). Near the fault surfaces (**Figure 8.32e**) the stress components approach the values given in the previous paragraph. At a great distance from the fault, the stress components approach the values given in (8.4). The solution to the boundary-value problem provides a *unique* distribution of the stress anywhere within the elastic body by specifying only the tractions on the internal boundaries and the stresses in the remote field.

8.4.3 Equations of Motion, Equilibrium, and Compatibility

Cauchy's laws of motion, derived in Sections 3.6.1 and 3.7.1, are the fundamental physical laws that underlie the canonical model for faults. They assure that the model obeys conservation of linear and angular momentum for the material continuum. The path from these equations of motion to the governing equations for the elastic boundary-value problem is exactly the same as the derivation in Section 7.2.3 for the opening fracture model. We recommend reviewing that section to prepare for what follows here in an abbreviated form.

For the two-dimensional, plane strain problem illustrated in **Figure 8.31** the *dependent* variables are the three stress components in the (x, y)-plane:

$$\sigma_{xx},\ \sigma_{xy} = \sigma_{yx},\ \sigma_{yy} \tag{8.8}$$

The two equilibrium equations that follow from the equations of motion are:

$$\frac{\partial \sigma_{xx}}{\partial x} + \frac{\partial \sigma_{yx}}{\partial y} = 0,\ \frac{\partial \sigma_{xy}}{\partial x} + \frac{\partial \sigma_{yy}}{\partial y} = 0 \tag{8.9}$$

The compatibility equation relates the small strain components to assure they correspond to single values for the two displacement components at every point. Written in terms of the stress components, the compatibility equation is:

$$\left(\frac{\partial^2}{\partial x^2} + \frac{\partial^2}{\partial y^2}\right)(\sigma_{xx} + \sigma_{yy}) = 0 \tag{8.10}$$

These three equations, (8.9) and (8.10), govern the spatial variation of the three stress components in the (x, y)-plane.

The mathematical approach used to solve the governing equations relates the stress components to a scalar function, $\Phi(x, y)$, called the Airy stress function:

$$\sigma_{xx} = \frac{\partial^2 \Phi}{\partial y^2},\ \sigma_{xy} = -\frac{\partial^2 \Phi}{\partial x \partial y},\ \sigma_{yy} = \frac{\partial^2 \Phi}{\partial x^2} \tag{8.11}$$

The equilibrium equations (8.9) are satisfied using (8.11), and substituting (8.11) for the stress components in (8.10) produces the biharmonic equation:

$$\frac{\partial^4 \Phi}{\partial x^4} + 2\frac{\partial^4 \Phi}{\partial x^2 \partial y^2} + \frac{\partial^4 \Phi}{\partial y^4} = 0 \tag{8.12}$$

A scalar function $\Phi(x, y)$ is a solution to a defined boundary-value problem if it satisfies (8.12) throughout the region of interest, gives the prescribed stresses at infinity using (8.11), and gives the prescribed tractions on the fault surfaces using (8.11) and Cauchy's formula (8.6). For faults subject to uniform tractions described by (8.5), and a homogeneous stress state at infinity described by (7.5), the Airy stress function is found using the mathematical method described in the textbook by Muskhelishvili (1975).

8.4.4 Displacement Field

The relative displacement of older structures (e.g. **Figure 8.6**), or cultural features at Earth's surface, provide directly measurable quantities, so we begin the description of the canonical model with the displacement field. The elastic body is loaded by uniform shear stresses, $\sigma^r_{xy} = \sigma^r_{yx} = S$, in the remote field that act on orthogonal planes with normals in the x- and y-coordinate directions (**Figure 8.32c**), respectively. In the absence of fault slip, these shear stresses are uniform throughout the entire elastic body and they *cause* a displacement field given by:

$$\left.\begin{aligned} u_x &= \sigma^r_{yx} y/2G \\ u_y &= \sigma^r_{xy} x/2G \end{aligned}\right\} \text{no fault slip} \tag{8.13}$$

The displacement magnitudes are inversely proportional to the elastic shear modulus, G: the greater the elastic stiffness, the lesser the displacements, for the same remote shear stress. The origin $(x = 0,\ y = 0)$ is the *reference point* for this displacement field (**Figure 8.32a**): all displacements due to application of the remote shear stresses are evaluated relative to a zero value at the origin. From the origin u_x increases linearly with distance in y, and u_y increases linearly with distance in x (**Figure 8.33**). The displacements are directed away from the origin in the first and third quadrants, and toward the origin in the second and fourth quadrants. This is the displacement field for uniform shear stress and uniform shear strain throughout the elastic body.

Recall from the discussion of small strains in Section 4.9.2 that the shear strain is related to gradients in displacement as $\varepsilon_{xy} = ½[(\partial u_x/\partial y) + (\partial u_y/\partial x)]$. Substituting (8.13) for the displacements and taking the partial derivatives, we find the *small* shear strain is:

$$\varepsilon_{xy} = \sigma^r_{xy}/2G$$

This is the shear strain–shear stress relationship from Hooke's law that we introduced in Section 4.10: the shear strain is proportional to the shear stress; and it is inversely proportional to twice the elastic shear modulus. In absence of perturbations caused by slip on the fault, the remote shear stress causes a uniform shear strain throughout the elastic body. Even though the displacement field is non-uniform, the shear strain field is uniform, because the displacement derivatives that contribute to the shear strain are uniform.

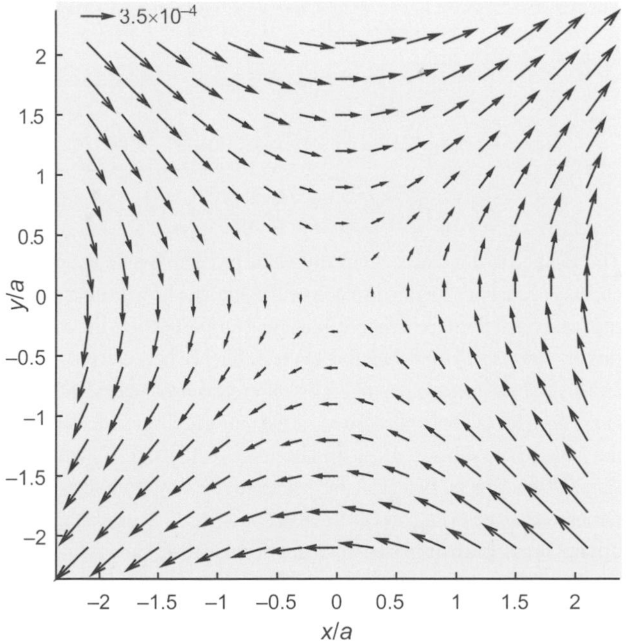

Figure 8.33 Normalized displacements, **u**/a, due to a remote shear stress using (8.13) with $\sigma^r_{yx} = 1 = \sigma^r_{xy}$ and G = 3,000 MPa. The greatest displacements in this field of view occur at the corners of the plot where $|u_x/a| = 3.5 \times 10^{-3} = |u_y/a|$. See Pollard and Segall (1987) in Further Reading.

The remote shear stress usually is treated as *constant* during fault slip. This can be justified if changes in the shear stress occur on a time scale that is much greater than that of the slip event. For example, slip rates on faults are of order meters per second, but creep rates on tectonic plate boundaries are of order centimeters per year, so they fit this prescription. In such cases the displacement field (8.13) associated with changes in the remote shear stress can be ignored and we focus attention on the displacement field due only to slip on the fault.

The local normal stress, σ^{f}_{yy}, acting perpendicular to the fault surfaces (**Figure 8.32e**) is taken as equal to the remote normal stress, σ^{r}_{yy}, in the same direction: $\sigma^{\mathrm{r}}_{yy} = \sigma^{\mathrm{f}}_{yy} = -C$. Recall from Section 7.2.5 that the driving stress for fracture opening is $\Delta\sigma_{\mathrm{I}} = \sigma^{\mathrm{r}}_{yy} - \sigma^{\mathrm{f}}_{yy}$, so this is zero for the boundary conditions on the model fault, and it does not open. On the other hand, the local shear stress, σ^{f}_{yx}, resisting slip at the fault surfaces (**Figure 8.32e**) is less in magnitude than the remote shear stress, σ^{r}_{yx}, and their difference is the driving stress for fault slip:

$$\Delta\sigma_{\mathrm{II}} = \sigma^{\mathrm{r}}_{yx} - \sigma^{\mathrm{f}}_{yx} \tag{8.14}$$

The subscript II indicates this is the driving stress for mode II fractures. Slip of the fault surface at $y = 0^+$ (**Figure 8.32b**) to the right, relative to the surface at $y = 0^-$, is associated with a positive driving stress (8.14).

To write down the displacements for the canonical fault model (**Figure 8.31**) we employ the same tri-polar coordinate system introduced in Section 7.2.5. The two displacement components due to slip on the model fault are (see Pollard and Segall, 1987, in Further Reading):

$$\begin{aligned} u_x &= (\Delta\sigma_{\mathrm{II}}/2G)\Big\{2(1-\nu)(R\sin\Theta - r\sin\theta) \\ &\quad + r\sin\theta\big[rR^{-1}\cos(\theta-\Theta) - 1\big]\Big\} \\ u_y &= -(\Delta\sigma_{\mathrm{II}}/2G)\Big\{(1-2\nu)(R\cos\Theta - r\cos\theta) \\ &\quad + r\sin\theta\big[rR^{-1}\sin(\theta-\Theta)\big]\Big\} \end{aligned} \tag{8.15}$$

The solution to this elastic boundary-value problem is unique, so these are exact equations for the displacement components, with no approximations. Each displacement component is proportional to the mode II driving stress (8.14) and therefore to the difference between the remote and local shear stress, $\sigma^{\mathrm{r}}_{yx} - \sigma^{\mathrm{f}}_{yx}$. This difference also is called the shear stress drop for faulting, because σ^{r}_{yx} is the uniform shear stress acting at the fault before slip and σ^{f}_{yx} is the shear stress after slip. Note that the same displacement field can be achieved for any combination of remote shear stress, σ^{r}_{yx}, and local shear stress, σ^{f}_{yx}, that yields the same driving stress. Each displacement component in (8.15) also is inversely proportional to the elastic shear modulus, G, so for the same driving stress, the displacements will be greater in a softer rock.

The displacement field from (8.15) is illustrated in **Figure 8.34** for a driving stress, $\Delta\sigma_{\mathrm{II}}$ = 1 MPa, shear modulus G = 3,000 MPa and Poisson's ratio ν = 0.25. This field is due only to fault slip; it does not include those displacements illustrated in **Figure 8.33** that arise from the remotely applied shear stress without slip. Slip on the model fault is associated with displacements directed away from the fault in quadrants 1 and 3, and toward the fault in quadrants 2 and 4. Two broad "circulation" patterns of displacement vectors exist, one in quadrants 1 and 2 and the other in quadrants 3 and 4. Similar circulation patterns have been measured using geodetic instruments after earthquake ruptures on active faults, such as the North Idu earthquake on the Tanna fault in Japan described in Section 4.11.3.

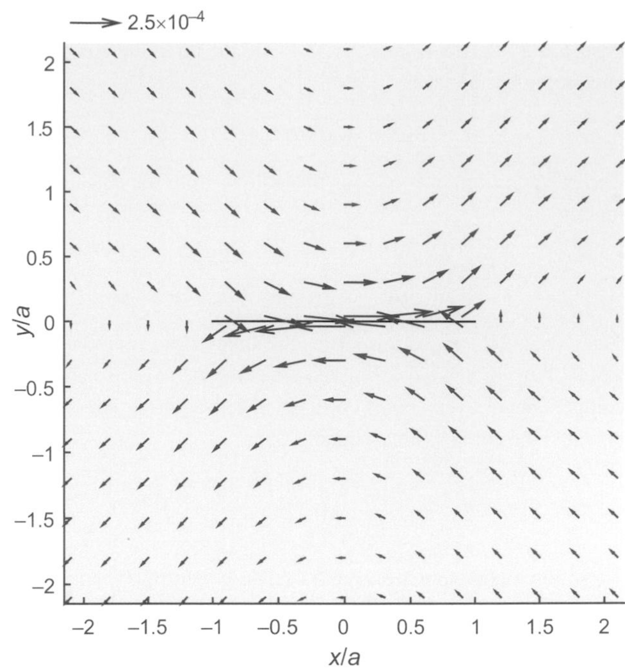

Figure 8.34 Normalized displacements, **u**/a, due to slip on the model fault using (8.15) with a positive driving stress, $\Delta\sigma_{\mathrm{II}}$ = 1 MPa, and elastic constants G = 3,000 MPa and ν = 0.25. The greatest displacements in this field of view occur at the middle of the model fault where $|u_x/a| = 2.5\times10^{-4}$. Calculation from Pollard and Segall (1987).

The displacement vectors illustrated in **Figure 8.34** are greatest on the fault surfaces and decrease in magnitude with distance from the fault. The greatest displacement is at the middle of the fault where $|u_x/a| = 2.5\times10^{-4}$, which is approximately equal to the ratio of driving stress (stress drop) to elastic shear modulus, $\Delta\sigma_{\mathrm{II}}/G = 3.3\times10^{-4}$. The displacements along the x-axis, and beyond the model fault ($|x| > a$, $y = 0$), are directed perpendicular to the fault: they are positive for positive x and negative for negative x, and they increase toward the fault tips. The displacements along the y-axis are directed parallel to the fault: they are positive for positive y and negative for negative y, and they increase toward the model fault.

Using (8.15) and focusing on the model fault surfaces, the displacements are:

$$\left.\begin{aligned} u_x &= \pm(\Delta\sigma_{\mathrm{II}}/G)\Big[(1-\nu)\big(a^2-x^2\big)^{1/2}\Big] \\ u_y &= (\Delta\sigma_{\mathrm{II}}/2G)[(1-2\nu)x] \end{aligned}\right\} \text{for } |x| \le a, y = 0^{\pm} \tag{8.16}$$

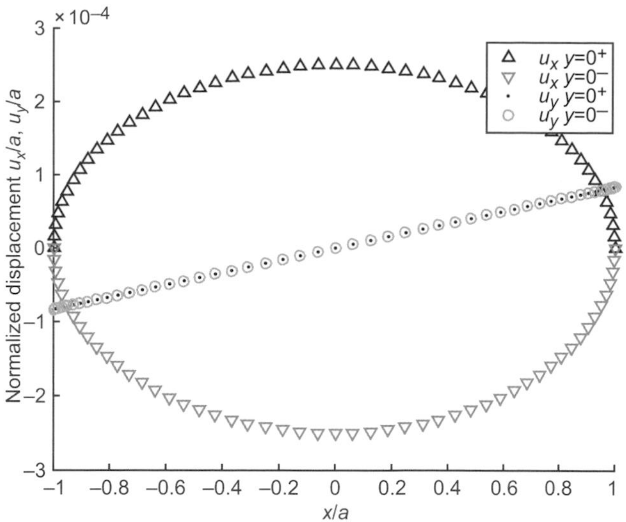

Figure 8.35 Normalized displacement components, u_x/a and u_y/a, for the surfaces, $y = 0^{\pm}$, of a model fault of half-length a, uniform driving stress, $\Delta\sigma_{II} = 1$ MPa, and elastic moduli $G = 3{,}000$ MPa and $\nu = 0.25$.

The $\pm$ sign for u_x refers to the two surfaces at $y = 0^{\pm}$. Therefore, this component is discontinuous from one surface to the other: a displacement discontinuity is associated with slip on the model fault (**Figure 8.35**). On both fault surfaces u_x has an elliptical distribution: for a positive driving stress this component is positive on $y = 0^+$ and negative on $y = 0^-$. In contrast, the component u_y is continuous from one surface to the other, and is linearly distributed along the fault. For a positive driving stress this displacement is positive (directed into the first quadrant) for $x > 0$, and negative (directed into the third quadrant) for $x < 0$, which produces a *counter-clockwise rotation* of the entire fault. The surrounding rock displaces *with* the fault surfaces, and this is seen in the vector field near the fault (**Figure 8.34**). The fault surfaces remain planar as they rotate. Because a field geologist typically will not know the original orientation of a fault before it slips, this rotation usually is not detectable in outcrops of ancient faults, but it is detectable using geodetic measurements on active faults.

The displacement discontinuity (slip) for the model fault is:

$$\Delta u_{II} = u_x(y=0^+) - u_x(y=0^-) = 2\Delta\sigma_{II}(1-\nu)(a^2 - x^2)^{1/2}/G \quad (8.17)$$

The slip goes to zero at the ends of the model fault, where $x = \pm a$, and is a maximum at the center where $x = 0$:

$$\max\Delta u_{II} = 2\Delta\sigma_{II}a(1-\nu)/G \quad (8.18)$$

The maximum slip is directly proportional to the shear stress drop, $\Delta\sigma_{II}$, and the fault length, $2a$, and it is inversely proportional to the elastic shear modulus, G.

To apply the displacement field for the canonical model, we turn to data from the November 24, 1979 earthquake in the Imperial Valley of Southern California, just south of the Salton Sea (**Figure 8.36**). This magnitude 6.6 earthquake was associated with right-lateral ruptures that broke the ground surface

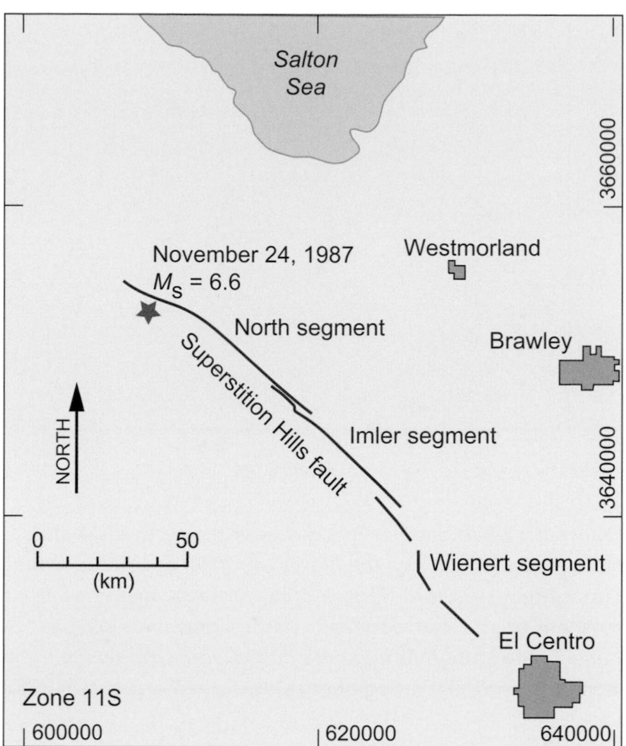

Figure 8.36 Location map for the November 24, 1987 magnitude 6.6 earthquake and associated right-lateral rupture on the Superstition Hills fault in southern California. Epicenter at red star. Note right steps in fault trace between North and Imler segments, and between Imler and Wienert segments. See Rymer (1989).

along the Superstition Hills fault zone for about 27 km to the southeast of the epicenter. The rupture trace is composed of the North, Imler, and Wienert segments that successively step to the right. The vertical and horizontal components of displacement discontinuity were measured at 296 sites along the Superstition Hills fault zone using displaced natural and cultural features. The measurement techniques and raw data are documented and described by Sharp et al. (1989). The fault continued to slip over the one year period during which data were recorded, but the general distribution of slip did not change appreciably.

For the purpose of comparison to the displacement discontinuity of the canonical model (8.17), we use the horizontal component of the slip vectors for the North and Imler segments measured on *day one* following the earthquake (**Figure 8.37**). The slip at individual sites is plotted versus distance from the northern tip of the North segment. The southern tip of that segment is at a distance of 14.86 km, and the overlapping northern tip of the Imler segment is at 11.23 km, both marked by red dots on the abscissa. The southern tip of the Imler segment is at 23.12 km. Ignoring the overlapped portions of these two fault segments for the moment, the horizontal slip generally increases from zero at the northern tip to about 45 cm near the middle of the fault zone, and then decreases back to zero at the southern tip of the Imler

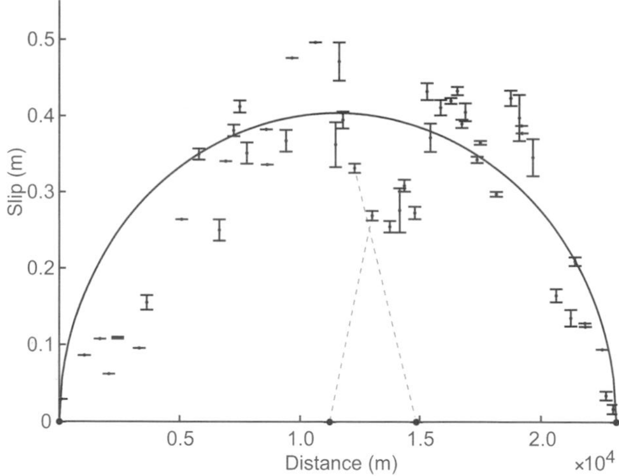

Figure 8.37 Slip data for first day after the earthquake on November 24, 1987 on the North and Imler segments of the Superstition Hills fault (**Figure 8.36**). Distance measured from northern tip of North segment. Dashed gray lines join penultimate station to tip station for overlapping tips at right step between the two segments. Data from Sharp et al. (1989).

segment. Within the overlapped portion, however, the slip decreases sharply and goes to zero (see gray dashed lines on **Figure 8.37**) at the overlapped tips.

For the canonical model the displacement discontinuity distribution is given in (8.17). We find a fit to the slip data illustrated in **Figure 8.37** (excluding data in the overlap portion) using a = 11.56 km and $2\Delta\sigma_{II}(1 - \nu)/G = 3.5 \times 10^{-5}$. This slip distribution is plotted in **Figure 8.37** as the blue curve, which is elliptical in shape with a maximum at the fault middle of about 0.4 m. This fit was found by minimizing the standard error of measured slip values relative to the model slip. However, inspection of **Figure 8.37** suggests this is not a particularly good representation of the data. The curve *overestimates* the data near the northern and southern tips of the combined segments, and misses the two inner tips at the overlapping portions of the North and Imler segments.

A better model for the slip along the Superstition Hills fault (**Figure 8.37**) would explicitly include the North and Imler segments, and address the mechanical interactions between them. For now we point out that the rough fit of the slip data to the single segment model enables one to calculate the quantity $2\Delta\sigma_{II}(1 - \nu)/G$ using (8.18). Estimating Poisson's ratio as ν = 0.25 and the elastic shear modulus as G = 10 GPa, we find a shear stress drop on the model fault, $\Delta\sigma_{II} \approx 0.2$ MPa, would approximately reproduce the measured slip distribution.

8.4.5 Stress Field

We evaluate the stress field for the canonical model to gain further insight about this style of deformation in Earth's crust (**Figure 8.38**). The three stress components in the (x, y)-plane around the model fault are (see Pollard and Segall, 1987, in Further Reading):

$$\begin{aligned}\sigma_{xx} &= \sigma^r_{xx} + \Delta\sigma_{II}\left[2rR^{-1}\sin(\theta - \Theta) - a^2 rR^{-3}\sin\theta\cos 3\Theta\right]\\ \sigma_{xy} &= \sigma^r_{xy} + \Delta\sigma_{II}\left[rR^{-1}\cos(\theta - \Theta) - 1 - a^2 rR^{-3}\sin\theta\sin 3\Theta\right]\\ \sigma_{yy} &= \sigma^r_{yy} + \Delta\sigma_{II}\left[a^2 rR^{-3}\sin\theta\cos 3\Theta\right]\end{aligned} \tag{8.19}$$

The solution to this elastic boundary-value problem is unique, so these are exact equations for the stress components with no approximations. The first term on the right side of each equation is the corresponding remote stress, which is found everywhere in the elastic body before the fault slips. The second terms are proportional to the mode II driving stress (8.14), and they are independent of the elastic moduli because this is a traction (stress) boundary-value problem.

Examples of the stress distributions are illustrated in **Figure 8.38** for the case of shear stress acting at infinity to drive slip ($\sigma^r_{xy} \rightarrow S = 1$ MPa), and zero tangential traction on the model fault surfaces to resist slip ($t_x = R = 0$). The normal traction on the fault surfaces and all other remote stress components are zero. These conditions produce a unit driving stress (stress drop) on the model fault ($\Delta\sigma_{II} = S - R = 1$ MPa). Note that slip on the model fault induces a stress concentration at the fault tips for all components. Slip also induces a stress shadow to either side of the model fault for the shear stress components.

The normal stress component σ_{xx} that acts parallel to the fault surfaces (**Figure 8.38a**) has particularly important effects on the mechanical behavior of the surrounding rock that are manifest as secondary structures. This component has an *antisymmetric* distribution: it has the same magnitude, but opposite sign, at neighboring points on the opposing fault surfaces. Thus, a stress discontinuity exists in this normal component across the fault. This stress component increases toward the fault tips without bounds, so the discontinuity in stress from one surface to the other can be very large.

We suggest that a discontinuity in σ_{xx}, like that illustrated in **Figure 8.38a**, produced the wing cracks between adjacent left-stepping faults in granite at the meter scale from the Sierra Nevada (**Figure 8.8**). The wing cracks are exclusively on the left side of a given fault as you face the tip. For these left-lateral faults, the signs of σ_{xx} in **Figure 8.38a** must be changed, so the stress is tensile in quadrants one and three, but the distribution remains the same. The discontinuity in σ_{xx} apparently played out in a different way in sedimentary rock at the ten kilometer scale in southern California (**Figure 8.24b**). The Imperial and Brawley faults are right lateral, so the signs of σ_{xx} in **Figure 8.38a** are appropriate. Tensile stress is induced within the right step between these faults, where the combined effects from the two faults complement each other. This apparently led to the development of normal faults, which in turn led to the development of the Mesquite Lake depression. The normal faults bounding the right step are interpreted as the large-scale analogue of the wing cracks on the small faults from the Sierra Nevada.

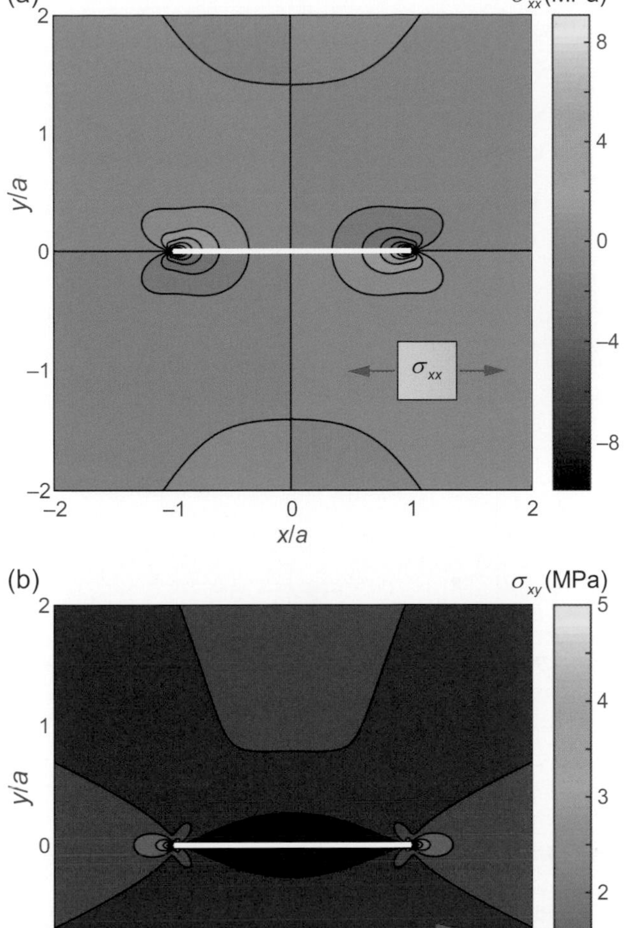

Figure 8.38 Stress distributions for the model fault (thin white line) in a linear elastic material under plane strain conditions subject to 1 MPa remote shear stress and zero resisting shear traction on the fault. Tensile stress is positive and the stress magnitudes are in units MPa. (a) Normal stress, σ_{xx}. (b) Shear stress, σ_{xy}.

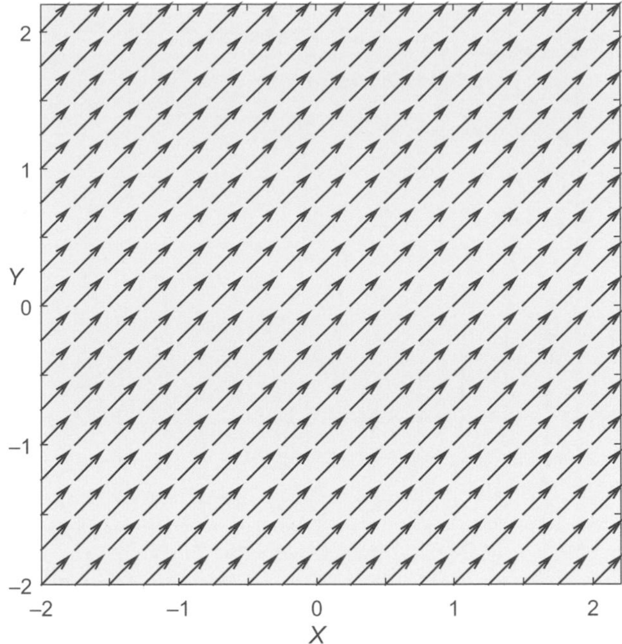

Figure 8.39 Displacement field with no *relative* displacements and no strain. This is a *pure translation*.

8.5 KINEMATICS OF FAULTING AND ASSOCIATED DEFORMATION

From kinematic relationships one can extract important insights about the relative displacement of neighboring particles and the displacement gradient at any point in the deforming rock mass. These insights are particularly helpful for building correct intuition about deformation near faults, so we develop the relevant concepts from kinematics here.

8.5.1 Relative Displacements and the Displacement Gradient

Recall from Section 4.9.2 that small strains are defined in terms of partial derivatives of the displacement components. These relationships are based on a referential description of motion in which the final position, $\mathbf{x}$, of any particle is a function of the initial (reference) position, $\mathbf{X}$, and the displacement vector, $\mathbf{u}$, such that $\mathbf{x} = \mathbf{X} + \mathbf{u}$. The partial derivatives of displacement components are taken with respect to the initial coordinates. In other words, we compare two configurations of particles, the initial state and final state, and we define the strain tensor components in terms of the reference state and the displacement field.

From the kinematic equations developed in Section 4.9.2, we understand that there is no strain associated with a *uniform* displacement field, because all partial derivatives of the displacement components are zero. For example, the displacement vector has the same magnitude and direction at every point in the two-dimensional displacement field in Figure 8.39. Such a uniform displacement field is a translation, one of two forms of rigid body motion. It should be clear from this example that *motion does not necessarily imply strain*.

One might conclude from the previous paragraph that relative displacements always are associated with strains. However, Figure 8.40 demonstrates this conjecture is *false*. The displacement vectors in this figure vary in both magnitude and direction, so relative displacements do occur between neighboring particles. This displacement field is a pure rotation, the second form of rigid body motion. At no point in this body would material lines stretch, nor would the angle between any two initially orthogonal

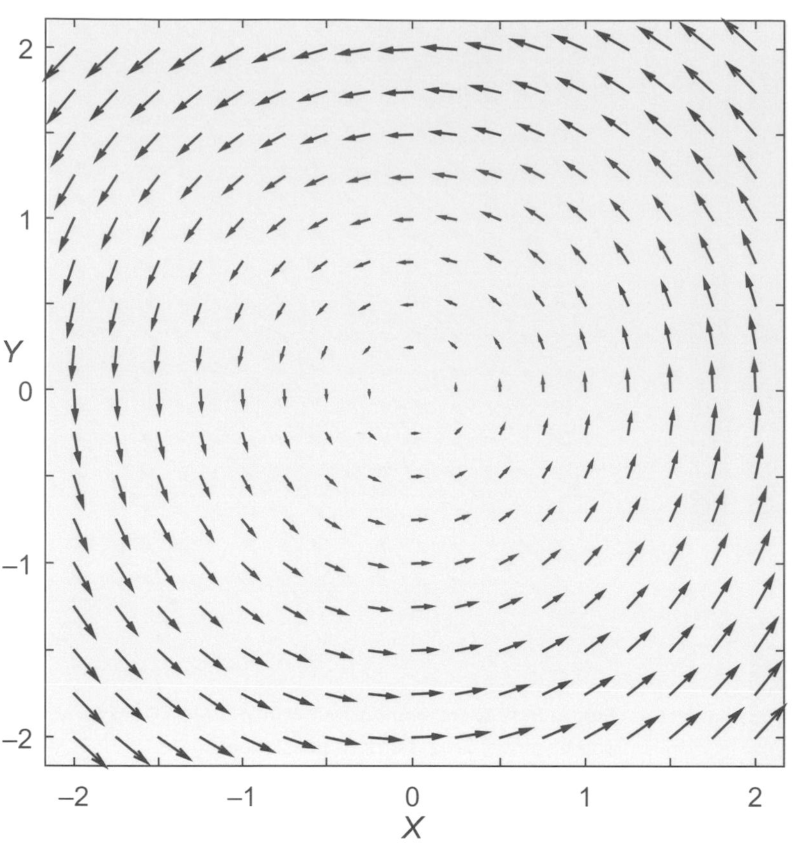

Figure 8.40 Displacement field with non-zero relative displacements (exaggerated for the sake of illustration), but no strain. This is a *pure rotation.*

material lines change, so the normal strains and shear strains are zero everywhere. This example shows that small strains do not arise from a rotation, even though neighboring particles displace relative to one another.

We have learned that small strain components are associated with certain kinds of relative displacements, but not with all kinds. Sorting out what kinds of relative displacements are associated with strains, and what kinds are not, is a fundamental problem of kinematics. We explore this problem in the next section in two dimensions and later generalize the results to three dimensions.

8.5.2 Small Rotations

Recall from Section 4.9.3 that the small strain matrix is equal to the matrix composed of one-half the *sum* of the displacement gradient matrix and its transpose:

$$[\varepsilon] = \tfrac{1}{2}\{[G] + [G']\} \tag{8.20}$$

Here we introduce the matrix that is equal to one-half the *difference* of the displacement gradient matrix and its transpose:

$$[\Omega] = \tfrac{1}{2}\{[G] - [G']\} \tag{8.21}$$

We show below that this tensor quantifies the small rotations at any point in the deforming material continuum.

The displacement gradient matrix is the sum of (8.20) and (8.21):

$$[G] = [\varepsilon] + [\Omega] \tag{8.22}$$

Writing out the elements of these matrices in terms of the partial derivatives of the displacement components we have:

$$\begin{bmatrix} \frac{\partial u_x}{\partial X} & \frac{\partial u_x}{\partial Y} \\ \frac{\partial u_y}{\partial X} & \frac{\partial u_y}{\partial Y} \end{bmatrix} = \begin{bmatrix} \frac{\partial u_x}{\partial X} & \frac{1}{2}\left(\frac{\partial u_x}{\partial Y} + \frac{\partial u_y}{\partial X}\right) \\ \frac{1}{2}\left(\frac{\partial u_y}{\partial X} + \frac{\partial u_x}{\partial Y}\right) & \frac{\partial u_y}{\partial Y} \end{bmatrix} + \begin{bmatrix} 0 & \frac{1}{2}\left(\frac{\partial u_x}{\partial Y} - \frac{\partial u_y}{\partial X}\right) \\ \frac{1}{2}\left(\frac{\partial u_y}{\partial X} - \frac{\partial u_x}{\partial Y}\right) & 0 \end{bmatrix} \tag{8.23}$$

Considering the elements on the secondary diagonals of the two matrices on the right side, we note that $[\varepsilon]$ is *symmetric* and $[\Omega]$ is the *skew-symmetric*. Skew-symmetric means the lower left element and the upper right element are equal in magnitude, but opposite in sign.

To make the connection between the partial derivatives of displacement in $[\Omega]$ and the small rotations, consider a particle at point P and a neighboring particle, arbitrarily located at point Q (**Figure 8.41**). The infinitesimal vector $d\mathbf{X}$ extends from P to Q in the initial state. Because we only are interested in the relative

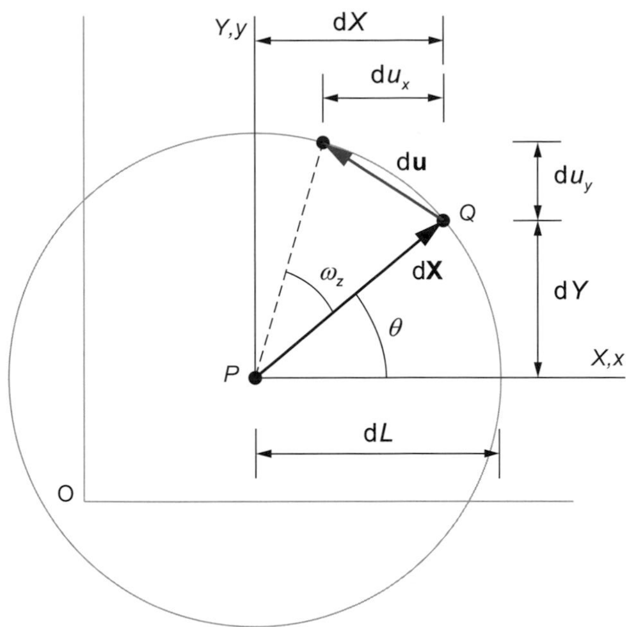

Figure 8.41 Geometry of the pure rotation through angle ω_z related to the relative displacement, **du**, of the particle at point Q, relative to the particle at P. The vector **dX** extends from P to Q, and all particles at a radial distance dL from P displace along the gray circle centered at P. Lengths of differential quantities are exaggerated for the sake of illustration.

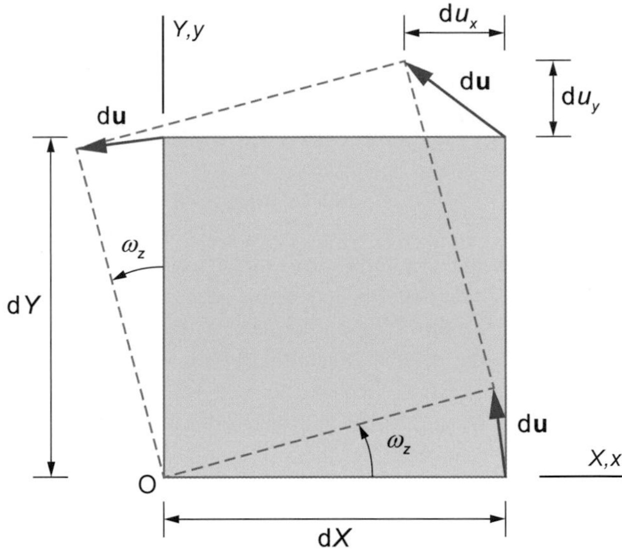

Figure 8.42 Geometry of the pure rotation of a square element through angle ω_z related to the relative displacement, Δ**u**, relative to the particle at the origin (exaggerated for the sake of illustration).

displacement, **du**, of the particle at Q with respect to that at P, we translate the origin of coordinates to P. The particle at Q has the coordinates (dX, dY) relative to the new origin, and those are the components of the vector **dX**. Because we are interested only in a *pure rotation*, the particle at Q must move relative to that at P along a *circular* path. Thus, the relative displacement, **du**, is such that the particle at Q is displaced to a location on a circle that passes through Q and has a center at P. The infinitesimal radial line from P to Q has a length dL, which is the radius of the circle.

The vector **dX** in Figure 8.41 makes an angle θ with the positive X-axis, and the vector **dX** + **du** makes an angle $\theta + \omega_z$ with the positive X-axis. Using these angles, the components of the relative displacement are:

$$\begin{aligned} du_x &= dL\cos(\theta + \omega_z) - dL\cos\theta \\ du_y &= dL\sin(\theta + \omega_z) - dL\sin\theta \end{aligned} \quad (8.24)$$

The angle ω_Z is the small rotation angle of the particle at Q relative to that at P. Using trigonometric identities for the sine and cosine of angle sums, the components become:

$$\begin{aligned} du_x &= dL(\cos\theta\cos\omega_z - \sin\theta\sin\omega_z - \cos\theta) \\ du_y &= dL(\sin\theta\cos\omega_z + \cos\theta\sin\omega_z - \sin\theta) \end{aligned} \quad (8.25)$$

From the trigonometry of Figure 8.41 note that $\cos\theta = dX/dL$ and $\sin\theta = dY/dL$. Substituting for the sine and cosine of θ, the components of relative displacements are:

$$\begin{aligned} du_x &= (\cos\omega_z - 1)dX - \sin\omega_z dY \\ du_y &= \sin\omega_z dX + (\cos\omega_z - 1)dY \end{aligned} \quad (8.26)$$

If the rotation angle is small, we can use the series expansions for sine and cosine of ω_z, and take only the first term as a good approximation:

$$\begin{aligned} \sin\omega_z &= \omega_z - \frac{1}{3!}\omega_z^3 + \frac{1}{5!}\omega_z^5 + \cdots \approx \omega_z \text{ for } \omega_z \ll 1 \text{ radian} \\ \cos\omega_z &= 1 - \frac{1}{2!}\omega_z^2 + \frac{1}{4!}\omega_z^4 + \cdots \approx 1 \text{ for } \omega_z \ll 1 \text{ radian} \end{aligned} \quad (8.27)$$

Using these approximations in (8.26), the components of relative displacement in matrix form are:

$$\begin{bmatrix} du_x \\ du_y \end{bmatrix} = \begin{bmatrix} 0 & -\omega_z \\ \omega_z & 0 \end{bmatrix} \begin{bmatrix} dX \\ dY \end{bmatrix} \quad (8.28)$$

The square matrix on the right side of (8.28) is skew-symmetric and the elements on the main diagonal are zero, so it has the same form as $[\Omega]$ on the right side of (8.23). Therefore, the partial derivatives of displacement in $[\Omega]$ can be associated with the small rotation angle about the Z-axis such that:

$$\omega_z = \tfrac{1}{2}\left(\frac{\partial u_y}{\partial X} - \frac{\partial u_x}{\partial Y}\right) \quad (8.29)$$

We illustrate the rotation of a small square element through angle ω_z in Figure 8.42. Note that the relative displacement, **du**, is different at the point (dX, dY) than at the point (dX, 0), because the distance from the origin is different. To be consistent with the notation for the strain components, we name the elements of the rotation matrix as follows:

$$\Omega_{xy} = -\omega_z \text{ and } \Omega_{yx} = \omega_z \tag{8.30}$$

The two non-zero elements of the rotation matrix are equal in magnitude and opposite in sign.

In summary, the displacement gradient matrix [G] may contain both small strains and small rotations. Put another way, if a displacement field is not a pure translation, it contains relative displacements, and these may give rise to: (1) small strains that are a measure of distortion; or (2) small rotations that are a measure of a rigid body motion; or (3) small strains and small rotations. In general, we expect the elements of [G], [ε], and [Ω] to vary from point to point in a brittle deforming rock mass.

No new concepts are required to write down the kinematic equations that relate rotation components to partial derivatives of displacement components in three dimensions. In general, the displacement vector is a function of three spatial coordinates, $\mathbf{u} = \mathbf{u}(X, Y, Z)$, as are each of the components of this vector. The small rotations are:

$$\Omega_{xy} = \frac{1}{2}\left(\frac{\partial u_x}{\partial Y} - \frac{\partial u_y}{\partial X}\right), \Omega_{yz} = \frac{1}{2}\left(\frac{\partial u_y}{\partial Z} - \frac{\partial u_z}{\partial Y}\right), \Omega_{zx} = \frac{1}{2}\left(\frac{\partial u_z}{\partial X} - \frac{\partial u_x}{\partial Z}\right)$$
$$\Omega_{yx} = \frac{1}{2}\left(\frac{\partial u_y}{\partial X} - \frac{\partial u_x}{\partial Y}\right), \Omega_{zy} = \frac{1}{2}\left(\frac{\partial u_z}{\partial Y} - \frac{\partial u_y}{\partial Z}\right), \Omega_{xz} = \frac{1}{2}\left(\frac{\partial u_x}{\partial Z} - \frac{\partial u_z}{\partial X}\right) \tag{8.31}$$

Rotations follow a right-hand rule: with the thumb of the right hand pointing in a positive coordinate direction the fingers curl in the direction of turning for a positive rotation about that coordinate axis.

Applications of the kinematic equations for small strains and small rotations require *small* partial derivatives of displacement, a condition generally met for rock that deforms in a brittle elastic manner. We emphasize that the *displacements* are not required to be small, but rather it is the *partial derivatives of displacement* that are required to be small relative to one. In some contexts the small strains and rotations are referred to as *infinitesimal* strains and rotations. This implies to some that these kinematic quantities are so small that they are inconsequential and can be ignored. Yet the majority of the deformation outside the immediate fault zone for most major earthquakes is quite adequately described using small strains and rotations. Clearly this deformation is not inconsequential, and indeed an understanding of small strains and rotations is a pre-requisite for investigating the development of many geologic structures.

Much of the deformation of rock in the brittle parts of Earth's crust is within the limits of small strain and small rotation kinematics, and that deformation is described adequately by the theory of elastic deformation. Structures such as fractures and faults may be treated as surfaces of displacement discontinuity, but the surrounding rock mass deforms such that the displacements are continuous and small strains and rotations are good approximations for the actual strains and rotations. In Chapter 5 we derived the kinematic equations without approximations because that is needed to investigate ductile deformation of rock. Under ductile conditions the partial derivatives of displacement may be large enough that using small strain and rotation kinematics would introduce significant errors. With equations for the actual strains and rotations one can quantify what is meant by "small partial derivatives of displacement" and evaluate the error.

8.5.3 Small Strains and Small Rotations During Faulting

To explore the relationships among the displacements, strains, and rotations during faulting we return to the canonical model of a two-dimensional fault (**Figure 8.43**) in the (x, y)-plane with a trace length $2a$. The displacement components in this plane are u_x and u_y. To conform to the notation used to describe that model fault, we use the coordinates (x, y) rather than (X, Y) used in the previous section, with the understanding that the strains and rotations are small. Slip on the model fault is driven by the shear stress drop, $\Delta\sigma_{\text{II}} = \sigma^{\text{r}}_{yx} - \sigma^{\text{f}}_{yx}$, the difference between the remote shear stress and the local shear stress acting on the fault. Slip on the model fault is resisted by the stiffness of the surrounding solid, where G is the shear modulus and ν is Poisson's ratio. The material surrounding the model fault is homogeneous and isotropic with respect to these elastic properties. The displacement vector field associated with slip on the model fault is illustrated in **Figure 8.34**, and it is for this displacement field that we calculate the small strain and small rotation.

To provide an instructive example we focus on a single point at the middle of one surface of the model fault, where $x = 0$ and $y = 0^+$ (**Figure 8.43b**). At that point we have two orthogonal

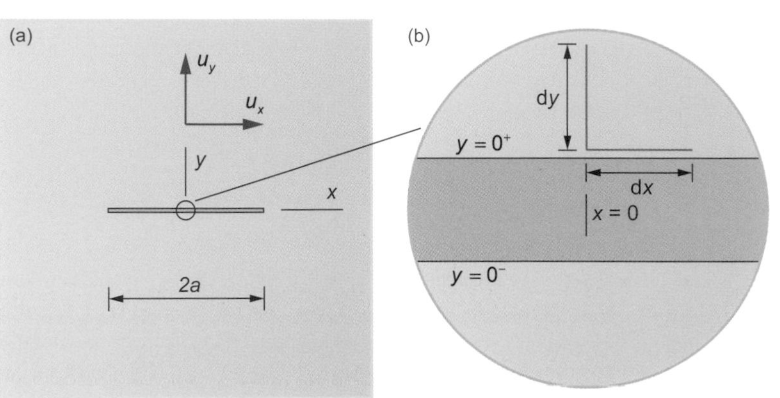

Figure 8.43 (a) Canonical fault of length $2a$ in the (x, y)-plane. The displacement components in this plane are u_x and u_y. (b) At the fault middle two material lines of length dx and dy are used to illustrate the small strain and rotation.

material lines of lengths dx and dy that are parallel to the x- and y-axes respectively. The small strains and rotation are given by partial derivatives of the displacement components as written on the right side of (8.23). The displacement components along the model fault surface are given in (8.16). Taking the partial derivatives of these components with respect to x and evaluating them at $x = 0$:

$$\frac{\partial u_x}{\partial x} = 0, \ \frac{\partial u_y}{\partial x} = \Delta\sigma_{II}\left(\frac{1-2\nu}{2G}\right), \quad \text{at } (x=0, y=0^+) \tag{8.32}$$

The displacement component along the positive y-axis, $x=0$, $y \geq 0^+$, are found from (8.15):

$$u_x = \Delta\sigma_{II}\left(\frac{1-\nu}{G}\right)\left[\left(a^2+y^2\right)^{1/2} - y\right] + \Delta\sigma_{II}\left(\frac{1}{2G}\right)\left[y^2\left(a^2+y^2\right)^{-1/2} - y\right], \quad u_y = 0$$

Taking the partial derivatives of these components with respect to y and evaluating at $y = 0^+$:

$$\frac{\partial u_x}{\partial y} = -\Delta\sigma_{II}\left(\frac{1-\nu}{G}\right) - \Delta\sigma_{II}\left(\frac{1}{2G}\right) = -\Delta\sigma_{II}\left(\frac{3-2\nu}{2G}\right), \ \frac{\partial u_y}{\partial y} = 0, \quad \text{at } (x=0, y=0^+) \tag{8.33}$$

From the first of (8.32) and the second of (8.33) we see that the small normal strains in the x- and y-directions both are zero:

$$\varepsilon_{xx} = \frac{\partial u_x}{\partial x} = 0 \text{ and } \varepsilon_{yy} = \frac{\partial u_y}{\partial y} = 0 \text{ at} (x=0, y=0^+) \tag{8.34}$$

When the model fault slips, material line elements at this point that are oriented parallel and perpendicular to the fault do not stretch. The small shear strains are found from the second of (8.32) and the first of (8.33):

$$\varepsilon_{xy} = \frac{1}{2}\left(\frac{\partial u_x}{\partial y} + \frac{\partial u_y}{\partial x}\right) = -\frac{\Delta\sigma_{II}}{2G} \text{ at } (x=0, y=0^+) \tag{8.35}$$

Fault slip induces negative shear strains at this point that are proportional to the stress drop and inversely proportional to twice the elastic shear modulus.

The small rotation is found using the second of (8.32) and the first of (8.33):

$$\Omega_{yx} = \frac{1}{2}\left(\frac{\partial u_y}{\partial x} - \frac{\partial u_x}{\partial y}\right) = \frac{\Delta\sigma_{II}}{G}(1-\nu), \quad \text{at } (x=0, y=0^+) \tag{8.36}$$

Fault slip induces a positive rotation at this point that is proportional to the driving stress and inversely proportional to the elastic shear modulus. The rotation also is proportional to one minus Poisson's ratio.

The small shear strain and rotation can be visualized by recalling that these quantities are related to the change in orientation of material lines that initially are orthogonal (**Figure 8.43b**). In **Figure 8.44** the shear strain and rotation are represented by somewhat exaggerated partial derivatives of displacement. The initial 90° angle between the two material lines increases, resulting in a negative shear strain in **Figure 8.44a**. The initial 90° angle between the two material lines does not change as both rotate in a counter-clockwise sense in **Figure 8.44b**.

The sum of small shear strain and small rotation depicted in **Figure 8.44c** suggests that the fault surfaces (coincident with the material line along the X-axis in that figure) rotate counter-clockwise when the fault slips in a right-lateral sense. According to the second of (8.32) the displacement component u_y acting perpendicular to the fault surfaces is linearly distributed along the fault. This confirms that the fault surfaces remain planar as they rotate. With Poisson's ratio set to zero, $u_y(x = a, y = 0^{\pm}) = \Delta\sigma_{II}a/2G$ at the fault tip. The ratio $u_y/a = \Delta\sigma_{II}/2G$ is a measure of the rotation of the entire fault (**Figure 8.35**). In this example we have shown that the local small strain and rotation at the fault middle is consistent with the overall rotation of the fault surfaces.

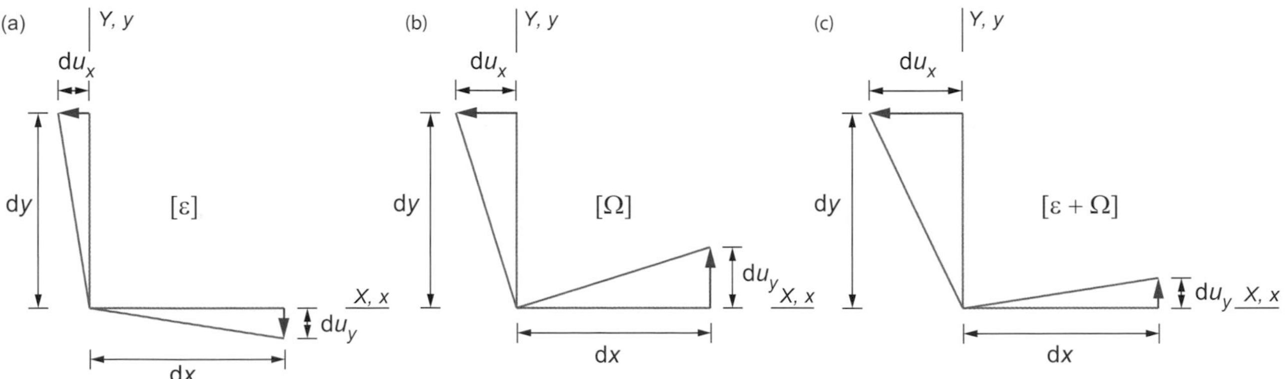

Figure 8.44 Relative displacements of material lines at the point ($x = 0$, $y = 0^+$) for the canonical fault model. Partial derivatives of displacement are exaggerated, so the small strain and rotation can be visualized. (a) Shear strain. (b) Rotation. (c) Shear strain plus rotation.

8.6 FOSSIL EARTHQUAKES

Mapping of newly ruptured fault traces after a major earthquake, and geophysical data taken before, during, and after the earthquake, document the association of fault slip and the generation of seismic waves that represent the earthquake. Although seismology is the primary discipline responsible for the monitoring and interpretation of modern earthquakes, structural geologists have a role to play by studying exhumed faults that may have generated earthquakes in the past (Box 8.2). We begin this section by reviewing seismic moment and earthquake magnitude, because these seismological quantities relate to geological quantities measured in outcrop, or on maps of faults. Then, we return to the faults in the Lake Edison Granodiorite, because some of them contain pseudotachylyte, which is a fingerprint of an ancient earthquake.

Box 8.2 Earthquake Geology

Earthquake geology is the study of faults, fault rocks, and associated structures using geological data to relate slip events to earthquakes. It includes the geological study of active faults with known slip events related to modern earthquakes (Yeats et al., 1997; see Further Reading), as well as the geological study of exhumed faults, where offsets of older structures document slip events that might have been associated with ancient earthquakes (Sibson, 1989; Cowan, 1999; Rowe and Griffith, 2015; see Further Reading). For inactive faults a central question is: what are the *geological fingerprints* of an earthquake? In other words, can geological data elucidate any aspects of unrecorded earthquake dynamics on faults that are inactive today? This is a challenging and important question for the structural geologist who wants to contribute to the understanding of tectonic history, or the mitigation of earthquake hazards. Addressing this question also provides a productive link between structural geology and earthquake seismology.

Table 8.2 provides calculated estimates for a structural geologist to relate representative slip on a fault, Δu, rupture length, $2R$, rupture area, πR^2, and moment magnitude, M_w. The calculations are based on relationships like (8.18), (8.37), and (8.38), and the shear stress drop on all of the model faults during the supposed earthquake is assumed to be 3 MPa. For the sake of constructing a simple relationship, faults up to 10 km in rupture length have a circular tipline with radius R. The two largest faults have a rectangular tipline with horizontal width, W, and down-dip height, H, so they are confined to a 15 km thick brittle lithosphere. For this table, the representative slip is associated with one earthquake, although the measured slip on actual faults may have accumulated during many events. These are rough estimates of moment magnitude, but they provide a useful bridge between the field data collected by structural geologists and the earthquake physics of the seismologists.

The most widely cited geological evidence for slip rates associated with earthquake generation is pseudotachylyte (Rowe and Griffith, 2015), a glassy material found in some faults (Section 8.6.3). Slip rates in the range of meters per second are associated with earthquakes (Section 8.6.2), and these rates can lead to frictional heating sufficient to melt at least those minerals with the lowest melting temperatures, thereby forming the pseudotachylyte (Figure 8.3c).

Table 8.2 Representative slip, rupture geometry, and moment magnitude

Representative slip: Δu	Rupture length: $2R$	Rupture area: πR^2	Moment magnitude: M_w
1 mm	30 m	10^3 m^2	1
4 mm	100 m	10^4 m^2	2
1 cm	300 m	10^5 m^2	3
4 cm	1 km	1 km^2	4
10 cm	3 km	10 km^2	5
40 cm	10 km	10^2 km^2	6
Δu	W & H	WH	M_w
1 m	60 & 15 km	10^3 km^2	7
4 m	650 & 15 km	10^4 km^2	8

From Sibson (1989)

Typical pseudotachylyte has sharp contacts with the rock hosting the fault, internal flow structures including foliation and spherulites (Figure 8.47), and relic clasts of host rock minerals with the highest melting temperatures such as quartz and feldspar. Injection veins, filled with pseudotachylyte, taper away from the fault (Figure 8.49). The matrix surrounding the spherulites and clasts appears to be glassy because its grain size is less than the resolution of the hand lens or optical microscope used to examine the pseudotachylyte. Because tachylyte is a volcanic glass of basaltic composition, the name *pseudo*tachylyte indicates this material is similar to a volcanic glass in appearance, but it is a quenched melt due to shear heating, not a quenched volcanic lava.

8.6.1 Earthquake Moment and Magnitude

The seismic moment is the product of the elastic shear modulus, G, for the rock surrounding the fault, the surface area of the fault, A, and the average tangential displacement discontinuity, Δu, across a fault during an earthquake (Kostrov and Das, 1988; see Further Reading):

$$M_0 = GA\Delta u \tag{8.37}$$

Recall that the shear modulus is a measure of elastic stiffness: the ratio of shear stress to twice the small shear strain (Section 4.10), so it has the same units as stress. Working out the SI units for the terms on the right side of (8.37), we have G [=] Pa = N m^{-2}, A [=] m^2, and Δu [=] m, so the units of seismic moment are those of force times distance, M_0 [=] N m. The seismic moment can be estimated from information on seismograms following an earthquake, and often is reported with units dyne cm. Values range from about 10^{12} dyne cm = 10^5 N m for *micro*-earthquakes, to about 10^{30} dyne cm = 10^{23} N m for *great* earthquakes.

The moment magnitude of an earthquake is an empirical quantity, based on the seismic moment. It is defined using the units dyne cm for M_0 in the following formula (Scholz, 1990; see Further Reading):

$$M_w = \frac{2}{3}\log_{10}(M_0) - 10.7 \tag{8.38}$$

The moment magnitude is a dimensionless and unit-less number. Noting that the units of M_0 are the units of work (and energy), the subscript "w" on M_w indicates this is a magnitude based on the work done (and energy released) during slip on the fault. The constants in (8.38) were designed by seismologists to supersede, but be consistent with, the earlier Richter magnitude scale for earthquake size. Taking the seismic moments mentioned in the preceding paragraph, the moment magnitudes calculated using (8.38) would range from about –2.7 for micro-earthquakes to 9.3 for great earthquakes. Most earthquake magnitudes reported today use the moment magnitude scale.

8.6.2 Slip Rate and Rupture Velocity

Two different velocities are key to understanding the physical relationships between faulting and earthquakes. One of these is a measure of the tangential velocity of particles on one side of the fault relative to particles on the other side, as the fault is slipping. In (8.17) we defined the tangential displacement discontinuity, or slip, for the canonical fault (**Figure 8.30**) as the difference between the tangential component of displacement on one side and that on the other side: $\Delta u_x(x) = u_x(x, y = 0^+) - u_x(x, y = 0^-)$. This slip varies with position, x, along the model fault from zero at the tips to a maximum value at the center (**Figure 8.35**). For this quasi-static problem, time was not a relevant variable: we only considered the relative position of particles before and after the slip event. Here, however, we consider the rate of tangential displacement, that is the velocity, v_x, during the slip event, and define the slip velocity as:

$$\Delta v_x(x, t) = v_x(x, y = 0^+, t) - v_x(x, y = 0^-, t) \tag{8.39}$$

At any one location along the fault the slip rate varies with time during a slip event, from zero through some range of values and back to zero. Also, the slip rate varies spatially along the fault, from zero at the tips through some range of values between the tips. For this discussion we ignore these variations in time and space and only consider a representative value, Δv.

The second velocity used to understand earthquakes and faulting is the rupture tip velocity. The conceptual model includes a pre-existing fault that is weak compared to the surrounding rock, and also is well oriented with respect to the tectonic stress state. A slip event nucleates at a point on this fault, and a shear rupture propagates along the fault. We use the word "rupture" to distinguish this phenomenon from the fault, which is the geological structure that develops as a result of many such ruptures. The half-length of the rupture increases as a function of time, from zero just before nucleation to a final value, a, when the rupture stops propagating. The rupture tip velocity, $v_t(x, t)$, correspondingly varies from zero, through some range of values, and back to zero as the rupture stops. For this discussion we ignore these variations in time and space, and only consider a representative value of the rupture tip velocity, v_t.

Representative values of rupture tip velocity and the corresponding slip velocity are plotted in **Figure 8.45** to characterize faulting in terms of these dynamic quantities. The graph, which spans eight orders of magnitude in rupture velocity and fourteen orders of magnitude in slip velocity, is populated with events taken from the geophysical literature. The most obvious message from this plot is that earthquakes occupy a limited region within these immense ranges. Representative earthquake rupture velocities vary from about 10^2 to 5×10^3 m s^{-1}, and earthquake slip velocities vary from about 10^{-4} to 10^1 m s^{-1}. Slower slip events occupy a middle range on this plot, and creep events occupy the lower left corner. A categorization based on seismicity broadly divides the plot into *aseismic*, *intermediate*, and *seismic* slip rates.

For structural geologists it is important to understand that all of the rates shown on **Figure 8.45** participate in the development of the faults on which they occur. While the earthquake seismologists may focus their attention on the upper right corner of **Figure 8.45**, to understand faults and faulting the structural geologist should consider the full ranges (and perhaps beyond…) of rupture velocity and slip velocity.

8.6.3 Pseudotachylyte Production During Frictional Slip

We now return to the question asked at the beginning of Section 8.6: are any of the observable characteristics of faults or fault rocks the fingerprints of an earthquake? Several different indicators are reviewed in Rowe and Griffith (2015), listed in Further Reading. For example, they describe chemical reaction products found in fault zones that indicate *frictional heating* during fast slip. These reaction products are associated with earthquakes in the upper right corner of **Figure 8.45**. Here we describe the occurrence and structural setting of pseudotachylyte in certain faults of the Lake Edison Granodiorite as an example of frictional heating and earthquake fingerprints.

Two small faults at the Waterfall site (**Figure 8.46**) offset a steeply dipping aplite dike 21 cm in the plane of the outcrop (see **Figure 8.4** for the outcrop location, WF). Quartz mylonite and younger epidote cataclasite are arranged in alternating zones along these faults. The pseudotachylyte usually is not visible at the outcrop scale, but it has been identified in thin sections from samples taken in those zones with little or no epidote cataclasite. The pseudotachylyte usually is found in tabular bands that are less than 300 μm thick, sub-parallel to the fault, and typically located between the quartz mylonite and the host granodiorite. The constituents of the pseudotachylyte are so small that a scanning electron microscope (SEM) is required to study them.

The pseudotachylyte found in the two small faults at the Waterfall site (**Figure 8.46**) has many of the features described in the literature of fault rocks, including sharp contacts with the host rock, flow structures, injection veins, clasts of host rock minerals with high melting temperatures such as quartz and feldspar, and spherulites made up of a central core composed of quartz and feldspar (**Figure 8.47**). The matrix surrounding the spherulites has a grain size less than the 200 nm resolution of the SEM, so would be described as *glassy* at that scale and at the

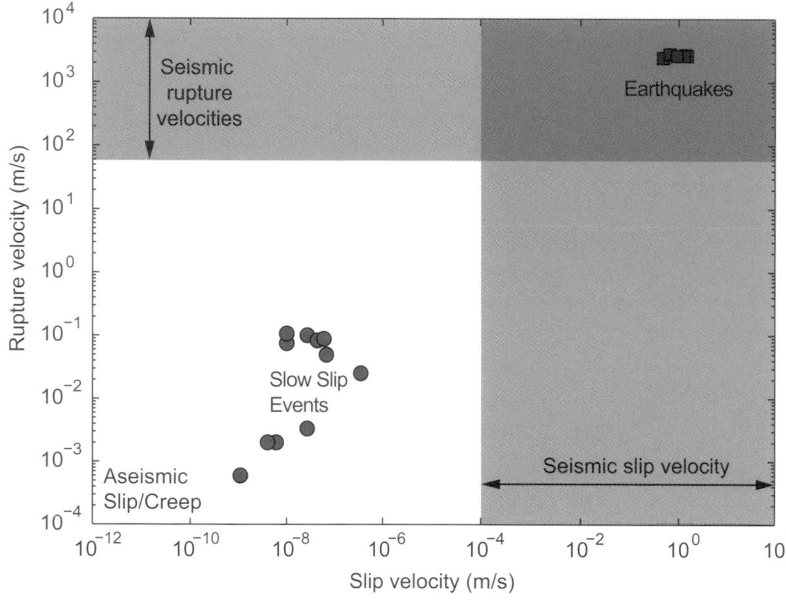

Figure 8.45 Velocity of the shear rupture tip plotted versus slip velocity across the rupture categorized as slow (aseismic), intermediate, and fast (seismic) slip events. Modified from Rowe and Griffith (2015), Figure 2.

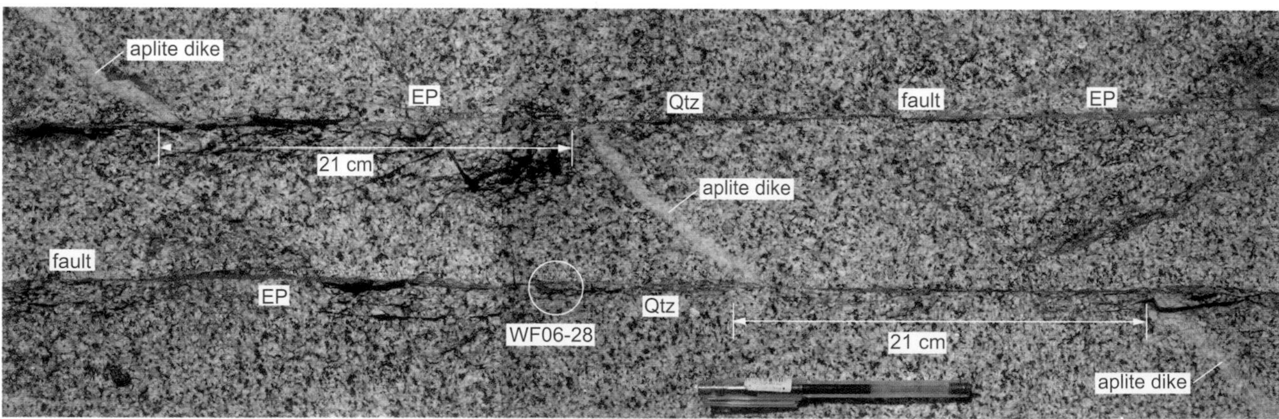

Figure 8.46 Two small faults east of the Pacific Coast Trail at the Waterfall site, WF, Lake Edison Granodiorite, Sierra Nevada, CA (see **Figure 8.4** for location WF). Pseudotachylyte sample was taken at WF06–28 (see **Figure 8.47**). The aplite dike has 21 cm strike separation on each small fault. EP and Qtz refer to alternating zones of predominantly epidote cataclasite and quartz mylonite. Pen is 13 cm long. Modified from Griffith et al. (2008), Figure 2a. Waterfall site UTM: 11 S 334022 m E, 4134542 m N.

resolution of an optical microscope or in a hand specimen. Volcanic glass of basaltic composition is called tachylyte, so the name *pseudo*tachylyte indicates this material is similar to a volcanic glass in appearance and composition, but it is a fault rock, not a quenched volcanic lava.

Four of the stages in the development of the faults at the Waterfall site (**Figure 8.46**) are illustrated schematically in **Figure 8.48** and are associated with estimates of the temperature range during their formation. Recall that at an earlier stage, not illustrated here, joints opened in the granodiorite and were filled with a suite of hydrothermal minerals (**Figure 8.6**). It is noteworthy that the faults are only about 0.5 cm thick and offset the aplite dike only 21 cm, yet the fault rock is composed of three dramatically different and successively younger products of shearing: mylonite, cataclasite, and pseudotachylyte (**Figure 8.48a,b,c**). Thus, these fault rocks bear witness to three different episodes of faulting that apparently cross the entire spectrum of rupture velocity and slip rate on **Figure 8.45**.

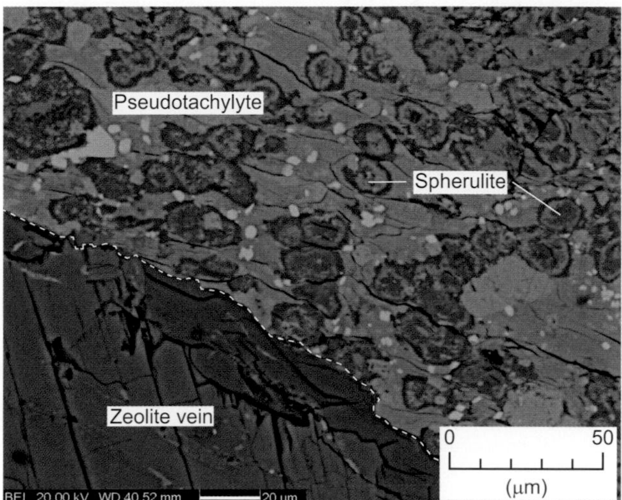

Figure 8.47 BSE-SEM image of fault rock from the sample WF06–28 (see **Figure 8.46** for sample location) on a small fault at the Waterfall site, Lake Edison Granodiorite, Sierra Nevada, CA (see **Figure 8.4** for WF outcrop location). Pseudotachylyte with spherulites containing a central core typically of quartz or plagioclase. Zeolite-filled vein cuts pseudotachylyte along the white dashed line and post-dates fault slip. Modified from Griffith et al. (2008), Figure 6d.

In the conceptual model (**Figure 8.48**) the quartz mylonite is associated with temperatures greater than 400 °C, whereas the epidote cataclasite is associated with temperatures in the range 200 to 300 °C. These episodes are interpreted as ductile creep, followed by grain-scale brittle fracture. During the inferred seismic slip, the *local* temperature of frictional melting was likely in the range 1000 to 1200 °C. Estimates of the slip required to raise the local temperature into this range are between 1 and 7 cm. Thus, the majority of the 21 cm of offset on the aplite dike (**Figure 8.46**) was accomplished before the seismic event that produced the pseudotachylyte. During exhumation and cooling below 200 °C (**Figure 8.48d**) fractures opened along the faults, crosscutting all of the previously formed fault rocks, and zeolites were deposited in these fractures (**Figure 8.47**).

At a few exceptional outcrops of faults in the Lake Edison Granodiorite (e.g. **Figure 8.49**), enough pseudotachylyte was produced to form tabular zones and injection veins that are visible to the naked eye. In this outcrop the zone is about 3 mm thick and the injection vein is about 13 mm long. Most occurrences of pseudotachylyte on these faults are only identifiable in thin section or SEM images, because the tabular zones are much thinner and the injection veins are much shorter. This suggests that frictional melting and seismic activity may have been more common on these faults than would be appreciated from the outcrop evidence alone.

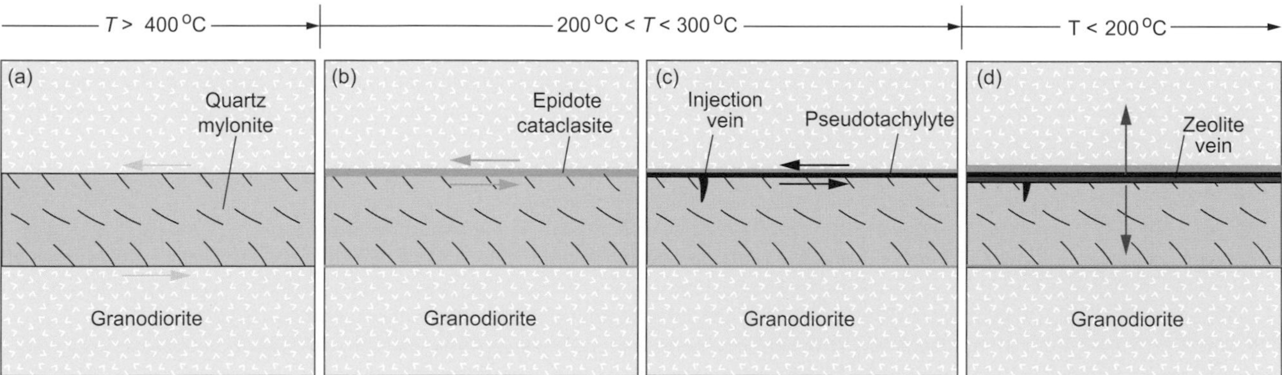

Figure 8.48 Conceptual model for evolution of faults in the Lake Edison Granodiorite, Sierra Nevada, CA (**Figure 8.4**). (a) Shearing of pre-existing joint forms quartz mylonite. (b) Shearing at interface between granodiorite and mylonite forms epidote cataclasite. (c) Seismic slip generates pseudotachylyte; injection veins crosscut cataclasite and mylonite. (d) Zeolite filled veins open after shearing has stopped. Modified from Griffith et al. (2008), Figure 7.

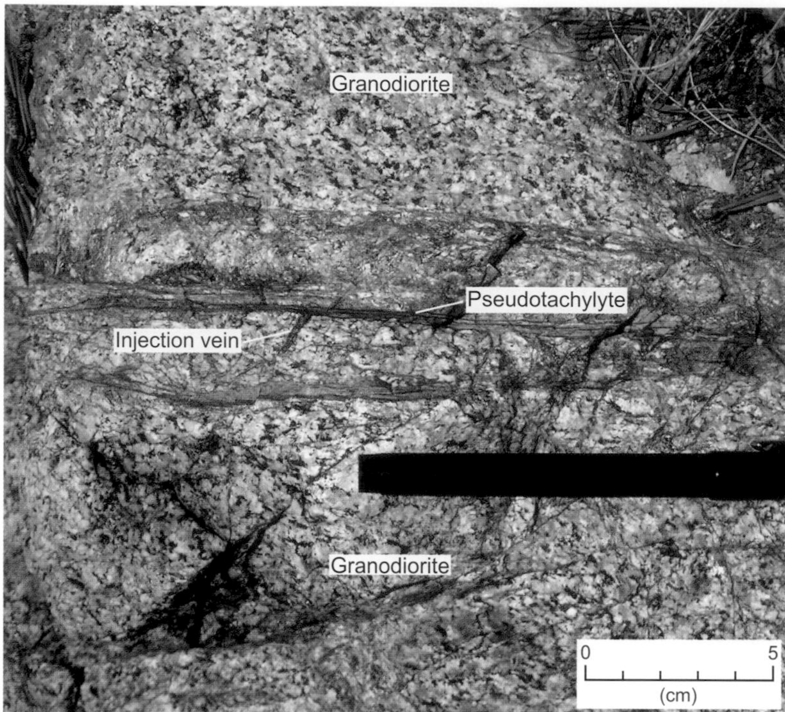

Figure 8.49 Pseudotachylyte injection vein on fault in the Lake Edison Granodiorite, Sierra Nevada, CA. See **Figure 8.4** for location on regional map near Ape Man (AM). Pen is 13.5 cm long. See Kirkpatrick et al. (2009) for description and interpretation. Approximate UTM: 11 S 335256.32 m E, 4132425.15 m N.

Recapitulation

Faults are distinguished from fractures because the relative motion of the rock mass on either side of a fault is dominantly *shearing*, whereas the relative motion for a fracture is dominantly *opening*. To begin to grasp the variety of mechanisms that can be involved in the development of faults we described faults at the outcrop scale in granitic rock from the Sierra Nevada, and in sandstone from the San Rafael Desert. The faults in granite began as opening fractures, were sealed with hydrothermal minerals, and later were sheared in a left-lateral sense to form small faults that grew in length by linking with opening fractures across left steps to other small faults. The faults in sandstone began as shearing deformation bands that formed side-by-side with other bands in narrow zones. Ultimately, slip surfaces developed on the margins of these zones, some with meters of offset.

Stepping up to kilometer-scale faults, four different sets of normal faults from the Chimney Rock area roughly fall into two *conjugate* sets. Using stratigraphic analysis, the sequence of thrust faulting in the Elk Hills oil field is deduced and these faults appear in 3D seismic imaging to be concave upward. Two of the faults appear to form a conjugate set, but one of these is distinctly younger than the other. The strike-slip (wrench) faults in the Imperial Valley of southern California are discontinuous and arranged in *echelon* patterns. For these right-lateral faults, right steps are bounded by normal faults, which play a similar role to the opening fractures at the meter scale for the Sierran faults.

Although acknowledging the complexities of fault geometry, the different mechanisms that are activated during faulting, and the complex kinematics of fault growth, we focus on the key geometric and kinematic features in order to identify a canonical model for faulting. This model adopts a *linear elastic* constitutive law and uniform tractions on a single planar fault with uniform remote stresses. The Cauchy equations of motion are reduced to the equilibrium equations and a single compatibility equation with stress components as the dependent variables. The displacement field for the canonical fault includes an elliptical distribution of displacement discontinuity across the fault that roughly corresponds to the slip distribution for a significant earthquake on the Superstition Hills fault. The stress field includes an antisymmetric distribution of normal stress acting parallel to the model fault that correlates with the deformation at fault steps. In these and other ways, this idealized model provides insights about faulting and deformation.

From some basic kinematic relationships for faulting we extracted important insights about the relative displacement of neighboring particles and the displacement gradient at any point in a deforming rock mass. These insights are particularly helpful for building correct intuition about deformation near faults. Deformation includes *small strains* and *small rigid-body rotations*, both of which are included in the displacement gradient matrix. Small strain components are associated with certain kinds of relative displacements, but not those that give a rigid rotation. The deformation near the center of the canonical fault includes small strains and small rotations.

This chapter concludes by describing aspects of faulting related to earthquakes. We define the *seismic moment* and the *moment magnitude* that provide quantitative measures of the size of an earthquake. Two kinematic quantities relevant to faulting and earthquakes are the slip velocity and the rupture velocity. Values for these velocities span many orders of magnitude and only those at the upper ends signal an earthquake. Faults can slip at creeping rates and ruptures can propagate at very modest velocities. However, pseudotachylyte production apparently requires frictional slip rates that may be associated with earthquakes.

REVIEW QUESTIONS

The following questions are designed to highlight the expected *learning outcomes* for this chapter. Each question is taken directly from the material in the chapter and, for the most part, in the same sequence that it appears in the chapter. If an answer is not forthcoming, students are advised to read the relevant section of the chapter and discover the answer.

8.1. Draw a block diagram of a fault that offsets a *linear structure* and use this to illustrate and define the concept of a piercing point, the slip vector, strike slip, and dip slip.

8.2. Draw a block diagram of a fault that offsets a *planar structure* and use this to illustrate and define the horizontal separation, the vertical separation, and a few possible slip vectors. Explain why one uses the term separation in this context and the term slip in the context of Question 8.1.

8.3. E. M. Anderson suggested three classes of faults: *wrench*, *normal*, and *reverse*. Distinguish the members of this classification based on the direction and sense of slip. In doing so explain what is meant by hanging wall and foot wall.

8.4. Among other rock types, faults may host *cataclasite*, *mylonite*, or *pseudotachylyte* (see **Figure 8.3**). Using one sentence for each, describe these different fault rocks and what they imply about the conditions during fault slip.

8.5. The microscopic image (**Figure 8.11**) shows fault rock from a boundary fault in the Lake Edison Granodiorite, Sierra Nevada, CA. Interpret the sequence of events on this fault and document your interpretation with descriptions of the fault rock.

8.6. The outcrop image (**Figure 8.16**) shows a zone of deformation bands from a normal fault in the Entrada Sandstone, San Rafael Desert, UT. Interpret the sequence of events on this fault and document your interpretation with descriptions of the image and related figures.

8.7. Starting at the south edge of the map in **Figure 8.20**, follow the 1680 m structure contour to the north edge of the map. What happens to this contour at the black faults? What happens to this contour at the red faults? Explain your observations based on the orientations of these normal faults.

8.8. The reverse (thrust) faults and anticlinal folds revealed on the cross section A–A′ from the Elk Hills oil field (**Figure 8.23**) are intimately related in space and time. Use the change in thickness of the stratigraphic interval between the middle Miocene McDonald horizon and the lower Pliocene Base Reef Ridge horizon to infer that slip was accumulating on these faults as the sediments in this interval were being deposited. Explain your reasoning.

8.9. The Coyote Creek fault takes a two kilometer *left step* at the Ocotillo Badlands, which is topographically high compared to the surrounding desert and contains numerous folds (**Figure 8.24a**). The Brawley and Imperial fault segments of the San Andreas system are separated by five kilometer *right step* at Mesquite Lake, which is a topographic depression surrounded by normal faults (**Figure 8.24b**). Explain why these steps along otherwise similar right-lateral strike-slip faults are so different.

8.10. The following Google Earth image is taken from the file: Question 08 10 Parashant normal fault. kmz. Open the file; click on the Placemark for this question to open the popup box; copy and paste the questions into your answer document; and address the assigned questions. The Google Earth image data are from Landsat / Copernicus.

8.11. Describe the geometry, boundary conditions, and constitutive law for the *canonical model* for faults (**Figure 8.32**). What aspect of some fault geometries suggests that using small strain kinematics is likely to be satisfactory?

8.12. Show how the *tractions* on the two surfaces of the canonical fracture (**Figure 8.32**) are related to the remotely applied *stresses* and the stress state at a point within the fault using Cauchy's formula (8.6).

8.13. The governing equations for the canonical fault model are the two equilibrium equations (8.9) and one compatibility equation (8.10). The stress components for this two-dimensional problem in elasticity theory are given in equation (8.11) by second derivatives of the *Airy stress function*. Show that these stress components satisfy equilibrium and derive the biharmonic governing equation.

8.14. The *tangential displacements* for the canonical fault surfaces (**Figure 8.35**) are given in equation (8.16). Describe the physical quantities in this equation and give representative values for the elastic moduli and shear stress drop for the segments of the Superstition Hills fault (**Figure 8.37**). Using these values, what is the maximum tangential displacement for a fault that is 25 km long?

8.15. The displacement component u_x is directed parallel to canonical fault surface (**Figure 8.31**), and the distribution of this displacement along the fault surface is plotted in **Figure 8.35**. Recalling that the longitudinal strain acting parallel to the fault is $\varepsilon_{xx} = \partial u_x / \partial x$ use the plot to infer the strain magnitude where the displacement is a maximum. Use the plot to infer the strain magnitude where the displacement is a minimum. Explain what you have inferred.

8.16. The *normal* stress component acting parallel to the canonical fault model is discontinuous across the fault (see **Figure 8.38a**). This discontinuity plays a major role in the development of secondary structures near faults. For example, use this discontinuity to explain the development of *wing cracks* shown in **Figure 8.8**. Also use this discontinuity to explain the development of *normal faults* at the right step between the Imperial and Brawley faults in **Figure 8.24**.

8.17. Explain why the displacement field depicted in **Figure 8.40** is associated with no strain, although clearly the displacements are changing in both magnitude and direction from point to point.

8.18. **Figure 8.41** shows that the relative displacement of the particle at Q with respect to the particle at P is defined as $\Delta\mathbf{u} = \mathbf{u}(Q) - \mathbf{u}(P)$, and the analysis of this quantity leads to the matrix equation written [R] = [G][D]. Here the elements of [R] are the relative displacements, the elements of [D] provide the direction from P to Q, and [G] is the displacement gradient matrix. Show how [G] is composed of a *symmetric* matrix with elements that are the small strains, and a *skew-symmetric* matrix with elements that account for a pure rotation.

8.19. Two different velocities play important roles in the dynamics of faulting from slow slip events to earthquakes. Explain what the *slip velocity* is and over what range of values it varies. Also explain what the *rupture tip velocity* is and over what range of values it varies.

MATLAB EXERCISES FOR CHAPTER 8: FAULTS

Documentation of the geometric and kinematic features of faults occurs at outcrops, on maps of various scales, and in seismic reflection surveys. The first exercise clarifies the difference between fault slip and fault separation, and distinguishes the displacement discontinuity vector from the slip vector. The second exercise compares fault development in granitic and sedimentary rock, emphasizing the differences arising from variations in rock type. Using GPS data from Chimney Rock, UT, we employ MATLAB to construct topographic and structure contour maps, and analyze the orientations of fault surfaces and slickenlines. Seismic reflection data from the Wytch Farm oil field, England, document the three-dimensional fault geometry and fault tiplines. The graphics capabilities of MATLAB visualize the fault geometry and allow inferences about the flow of hydrocarbons. For the canonical model, the two fault surfaces may slip over one another in frictional contact, or they may bound a thin zone of distributed shearing. We use MATLAB to calculate the displacement gradient matrix and to analyze the small strains and rotations associated with faulting. The exercises help students appreciate seismic observations of faults, use these data to build a canonical model, and interrogate model results to gain intuition about the behavior of faults. www.cambridge/SGAQI

FURTHER READING

For citations in figure captions see the reference list at the end of the book.

Anderson, E. M., 1972. *The Dynamics of Faulting and Dyke Formation with Applications to Britain*. Hafner Publishing Company, New York. (Facsimile reprint of 1951 edition.)
In this reprint of Anderson's classic book, originally published in 1942, the author summarizes and expands upon his earlier publications, dating back to the 1905 paper on *The Dynamics of Faulting*, and including those papers on dikes and the state of stress in Earth's crust.

Barton, N., 2007. *Rock Quality, Seismic Velocity, Attenuation and Anisotropy*. Taylor & Francis, London.
This reference book for geologists, engineering geologists, rock engineers, and geophysicists takes an integrated and interdisciplinary approach to rock quality and the mechanical properties of rock as revealed by seismic surveys.

Cooke, M. L., and Madden, E. H., 2014. Is the Earth lazy? A review of work minimization in fault evolution. *Journal of Structural Geology* 66, 334–346.
This review paper uses the principle of work minimization to gain insights about how fault systems evolve in different tectonic settings.

Cowan, D. S., 1999. Do faults preserve a record of seismic slip? A field geologist's opinion. *Journal of Structural Geology* 21, 995–1001.
Utilizing a field geologist's perspective, Cowan reviews the literature on faulting and earthquakes, and concludes that only the presence of pseudotachylyte provides physical evidence for seismic slip on exhumed faults.

Crider, J. G., 2015. The initiation of brittle faults in crystalline rock. *Journal of Structural Geology* 77, 159–174.
This review article makes the case for faults in the brittle lithosphere initiating from pre-existing structures, and developing by the mechanical interaction and linkage of these structures.

Hubbert, M. K., 1972. *Structural Geology*. Hafner Publishing Company, New York.
Seven papers by the author form the core of this book, many of which take up key aspects of the mechanical process of faulting, which is put in a useful historical context in the introduction.

Kattenhorn, S. A., Krantz, B., Walker, E. L., and Blakeslee, M. W., 2016. Evolution of the Hat Creek fault system, northern California, in: Krantz, B., Ormand, C., and Freeman, B., *3-D Structural Interpretation: Earth, Mind, and Machine*. Memoir 111. American Association of Petroleum Geologists, pp. 121–154.
This article describes a geometrically complex and segmented normal fault that evolved as the horizontal principal stresses rotated clockwise about 45 degrees over the past million years.

King, G. C. P., Stein, R. S., and Rundle, J. B., 1988. The growth of geological structures by repeated earthquakes 1. Conceptual framework. *Journal of Geophysical Research* 93, 13,307–13,318.
Models used to explain leveling data associated with the seismic cycle (coseismic slip and postseismic relaxation) are considered over time to show that the flexural rigidity of the crust is the most important parameter for controlling the width of geologic structures related to dip-slip faulting.

Kostrov, B. V., and Das, S., 1988. *Principles of Earthquake Source Mechanics*. Cambridge University Press, Cambridge.
The authors aim to provide a deeper understanding of the mechanics of the earthquake generation process in order to supplement the lack of comprehensive data used in purely empirical studies, and thereby make the applications of seismology to earthquake hazard mitigation more reliable.

Mandl, G., 1988. *Mechanics of Tectonic Faulting: Models and Basic Concepts*. Elsevier, Amsterdam.
The central issues considered in this book are the mechanical development of faults and their relationships to the state of tectonic stress with applications to structural geology, rock mechanics, and petroleum geology.

Price, N. J., 1966. *Fault and Joint Development in Brittle and Semi-brittle Rock*. Pergamon Press, Oxford, England.
This classic book on brittle and semi-brittle rock deformation uses laboratory test results, elasticity theory, and fracture mechanics to interpret fractures, with special attention to faults and faulting.

Rowe, C. D., and Griffith, W. A., 2015. Do faults preserve a record of seismic slip: a second opinion. *Journal of Structural Geology* 78, 1–26.
This review paper builds upon Sibson (1989) and Cowan (1999) and covers the range of reaction products that indicate frictional heating during fast slip on faults during earthquakes, as well as those features formed near the tip of a shear rupture propagating at high velocity.

Scholz, C. H., 1990. *The Mechanics of Earthquakes and Faulting*. Cambridge University Press, Cambridge.
Taking the perspective of rock mechanics laboratory investigations of brittle failure, this book integrates concepts and data from structural geology and geophysics to investigate the mechanics of earthquakes and faulting.

Segall, P., 2010. *Earthquake and Volcano Deformation*. Princeton University Press, Princeton, NJ.
This advanced textbook goes beyond the elementary mechanics and structural geology covered here, with a focus on geophysical data related to active deformation associated with earthquakes and faulting.

Sheriff, R. E., and Geldart, L. P., 1995. *Exploration Seismology*. Cambridge University Press, Cambridge.
Designed as both a reference book for the practicing geophysicist and a textbook for students, this comprehensive volume introduces the history of seismic exploration, covers the theory of seismic waves, describes the characteristics of reflected and refracted waves, and discusses both data processing and interpretation using an abundance of figures and examples.

Sibson, R. H., 1989. Earthquake faulting as a structural process. *Journal of Structural Geology* 11, 1–14.
The underlying hypothesis of this review paper is that deformation in the upper third to half of the continental crust is largely due to seismic slip on faults, which therefore must play a central role in the development of geologic structures.

Snoke, A. W., Tullis, J., and Todd, V. R., 1998. *Fault-related Rocks: A Photographic Atlas*. Princeton University Press, Princeton, NJ.
This is a wonderful introduction to the many different visual aspects of fault rocks as seen in outcrop and thin section, and includes descriptions, locations, and regional or experimental information that put the images in a useful context.

Stein, R. S., King, G. C. P., and Rundle, J. B., 1988. The growth of geological structures by repeated earthquakes 2. Field examples of continental dip-slip faults. *Journal of Geophysical Research* 93, 13,319–13,331.
Using three examples of crustal scale dip-slip faults, the authors add multiple episodes of deformation associated with the seismic cycle (coseismic slip and postseismic relaxation) to conclude that Earth's crust must weaken over the time of active faulting.

Wibberley, C., Faulkner, D., Schlische, R., Jackson, C., Lunn, R., and Shipton, Z. (Eds.), 2010. Fault Zones (Special Issue). *Journal of Structural Geology* 32, 1553–1864.
This special issue of the journal includes 23 papers on the structure of fault zones, their mechanics, and fluid flow properties.

Yeats, R. S., Sieh, K. E., and Allen, C. R., 1997. *The Geology of Earthquakes*. Oxford University Press, Oxford.
The geological investigation of active faults associated with earthquake ruptures is documented with basic principles as well as abundant case histories in this amply illustrated book.

Chapter 9
Folds

Introduction

Chapter 9 reviews fold terminology, describes fold profiles at outcrop scales, uses differential geometry to quantify folds in three dimensions at crustal scales, and builds canonical models for folding due to bending and buckling. For two-dimensional profiles of folds in sedimentary, metamorphic, and igneous rocks, we introduce the concept of curvature to characterize fold shapes. Then we describe the crustal-scale fold at Sheep Mountain, WY, and use a modern technology called Airborne Laser Swath Mapping (ALSM) to map the fold geometry in three dimensions. We employ differential geometry to characterize a three-dimensional bedding surface in this fold in terms of the principal curvatures at every point. The fundamental shapes at a given point on such a surface can be planar, elliptic, parabolic, or hyperbolic. These distinct shapes are used to define a classification scheme for folded geological surfaces based on curvature.

Folds at the kilometer scale in sedimentary rock play an important role in the oil and gas industry as hydrocarbon traps. For example, a porous and permeable sandstone layer could be a *reservoir*, holding the water and hydrocarbons in its pores. If the sandstone were overlain by a much less permeable shale, the shale *cap rock*, would retard fluids from migrating upward and away from the reservoir. For reservoir and cap layers folded into a convex upward form, the less dense hydrocarbons would flow into the crest of the fold, while the more dense water would flow downward along the limbs of the fold. Structural geologists work with geophysicists, who gather and process seismic reflection images of sedimentary strata to identify such hydrocarbon traps as targets for exploratory drilling. Structural geologists also work with petroleum engineers to design effective strategies to produce hydrocarbons from folded reservoirs that commonly are faulted and fractured.

9.1 FOLD TERMINOLOGY

A fold is a surface in a body of rock that has undergone a change in curvature. Typical examples include the surfaces of sedimentary bedding, the surfaces of lava flows, the surfaces of compositional layering in metamorphic rocks, and metamorphic foliations. Usually these surfaces were approximately planar before folding, and after folding they are curved. In some cases a surface is folded again at a later time, and with a different curvature. By identifying the different curvatures, a structural geologist may be able to sort out the different generations of folding and relate them to different tectonic events in the geologic history of the rock mass.

A *layer* of rock bounded by two surfaces also may be folded, for example, a sedimentary bed bounded by two bedding planes, or a vein bounded by two fracture surfaces. Typically, the layer in its initial state was approximately tabular, and after folding the layer is curved. Such a curved rock layer also is called a fold. Usually adjacent layers have similar distributions of curvature, and this collection of layers also may be called a fold, or more descriptively, a multilayer fold. Because we quantify the shapes of folds using the concept of curvature, and because that mathematical concept applies to a surface, we usually focus attention on the curved bounding surfaces when describing the folded layer or folded multilayer.

Folds have been identified and described in sedimentary, metamorphic, and igneous rock. They have been observed in thin section under the microscope at scales of millimeters; they are common at the meter scale in outcrops; and they are found on many geologic maps extending for tens of kilometers. Perhaps it is not surprising then, that different mechanisms lead to folding: two of these are readily demonstrated using a thin layer of soft elastic material. In **Figure 9.1a** the two ends of a thin elastic layer are pressed against the table top by wooden blocks, while the center of the layer is pushed upward by another block. The wooden blocks apply forces *perpendicular* to the plane of the layer that cause it to fold by bending. The applied forces cause a change in curvature of the surfaces of the layer, and therefore produce a fold.

In **Figure 9.1b** the elastic layer is bent slightly and held in place by a wooden block at each end. This provides a perturbation of the original planar geometry of the layer, which is a prerequisite for buckling as we will see in Section 9.7 of this chapter. In **Figure 9.1c** forces are applied by the wooden blocks *parallel* to the initial plane of the layer that cause it to fold by buckling.

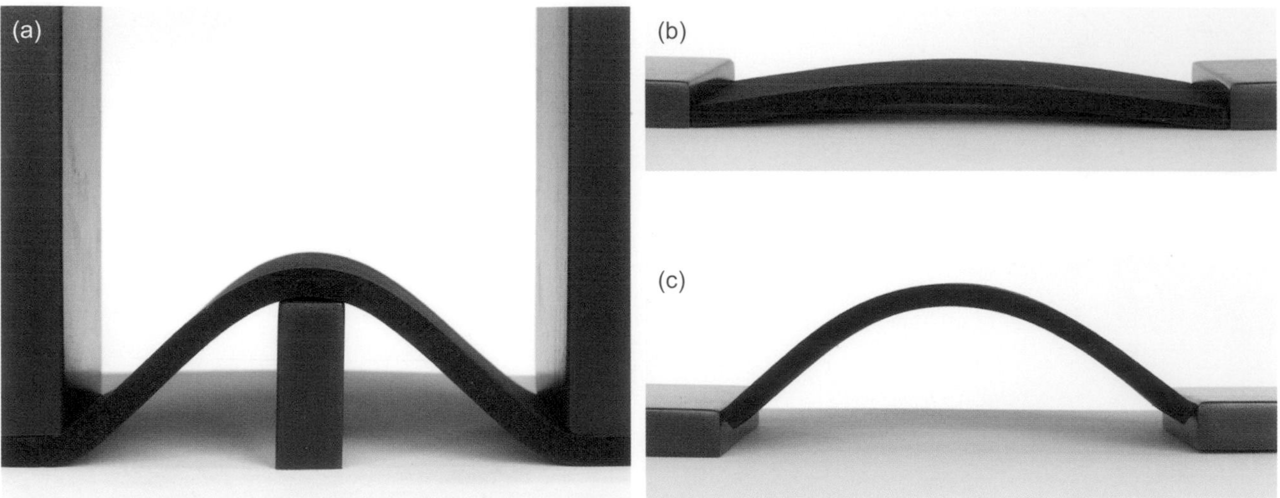

Figure 9.1 Two mechanisms for folding a thin elastic layer, in this case of foam neoprene rubber. (a) Forces applied perpendicular to the layer at both ends and the middle cause *bending*. (b) Layer is slightly bent out of its plane to facilitate folding. (c) Forces applied parallel to the layer shown in (b) cause *buckling*. Photography by Richard Stultz.

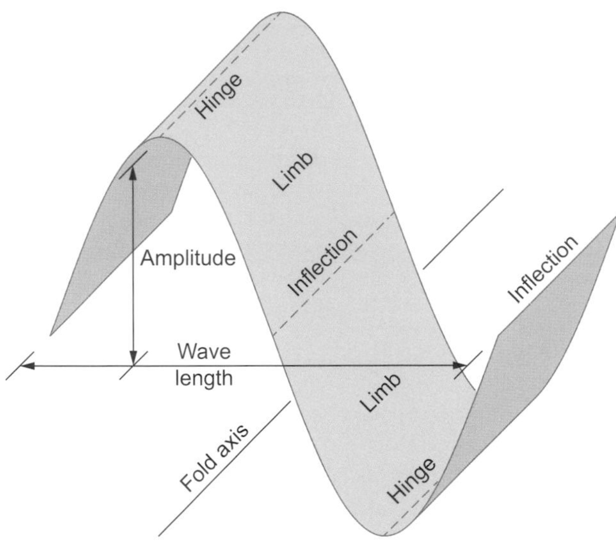

Figure 9.2 Qualitative nomenclature for a folded surface. See text for descriptions of the terms. This surface was generated by moving the fold axis perpendicular to itself following a sinusoidal curve. The folds so generated are cylindrical folds.

The applied forces cause a change in curvature of the surfaces of the layer and therefore produce a fold.

The experiments illustrated in **Figure 9.1** demonstrate that folding can occur when forces are applied either perpendicular or parallel to the plane of a thin elastic layer. The two mechanisms, bending and buckling, also apply to a stack of layers forming a multilayer fold, and to a single layer embedded in a material with different properties. For example, in Section 9.6 of this chapter we introduce a canonical model for bending of a stack of sedimentary strata over the top of an igneous intrusion called a laccolith. In Section 9.7 we also introduce a canonical model for buckling of a thin quartz vein embedded in a metamorphic schist. In that model the quartz vein and schist are treated as linear viscous materials, demonstrating that buckling is not just an elastic mechanism.

The basic terminology of folds is introduced using an idealized single folded surface (**Figure 9.2**). The portion of the folded surface with the greatest curvature is called the hinge, and the portions to either side with lesser curvature are called the limbs. The inflection is that portion of the surface where the limb of one fold passes into the limb of the adjacent fold and the curvature goes to zero. Thus, what we call a fold in this illustration is that portion of the surface from one inflection through the adjacent hinge to the next inflection. In this figure, the two folds differ in their curvature relative to up and down: the one on the left is *convex upward* and the one on the right is *concave upward*.

If a folded surface can be generated by a straight line (axis), that axis is called the fold axis, and the fold geometry is cylindrical. The cylindrical fold in **Figure 9.2** is generated by moving the fold axis along a perfect sinusoidal wave. For this special case the hinge lines are the dashed lines parallel to the fold axis where the surface is most curved. The inflection lines are the dashed lines parallel to the fold axis where the surface has zero curvature.

The two-dimensional trace of a cylindrical folded surface in a plane perpendicular to the fold axis is called the fold profile. In such a plane, the hinge line and inflection line appear as points, and the distance between two inflection points for the sinusoidal fold is one half the fold wavelength. The perpendicular distance from the plane containing the inflection lines to the plane containing the hinge lines is the fold amplitude.

The concept of a cylindrical fold is useful for making illustrative drawings like **Figure 9.2**, but folded surfaces in nature usually have more complicated geometries. Some folds may be adequately approximated over a limited distance along the fold axis by a cylindrical surface, but many are not, so we introduce methods later in this chapter to characterize the three-dimensional geometry of non-cylindrical folds using differential geometry.

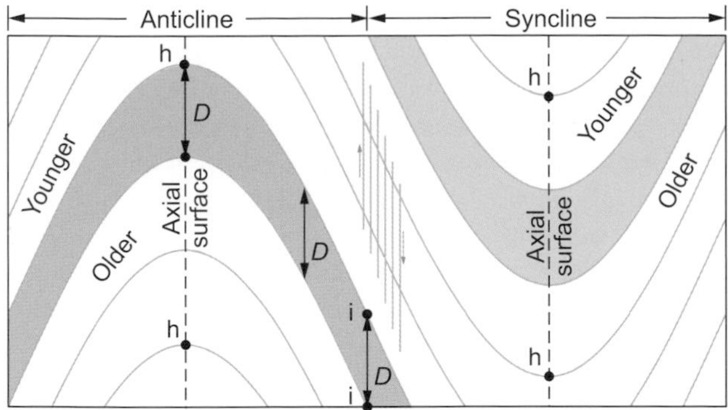

Figure 9.3 Profile of an idealized multilayer fold. The surfaces bounding all layers are the same sinusoidal curve, so these are *similar* folds. The hinge lines of a particular layer intersect the profile at points labeled, h; the inflection lines intersect at points labeled, i. The distances, *D*, are constant. The inset depicts shear planes parallel to the axial surfaces that represent a kinematic model for similar folds.

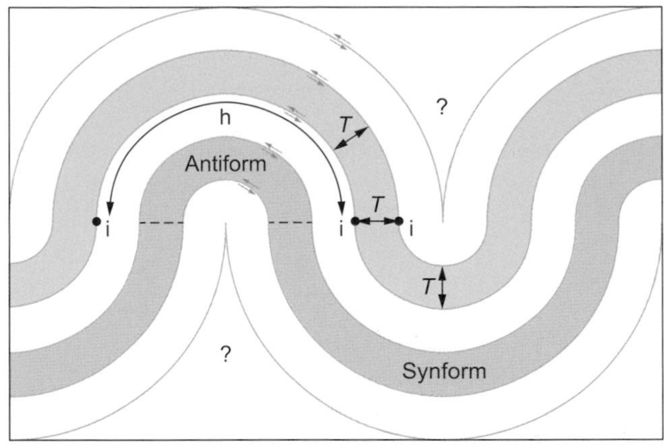

Figure 9.4 Profile of a multilayer fold. Relative ages of layers are not known, so concave downward layers are antiforms; concave upward layers are synforms. The surfaces bounding all layers are circular arcs, but they have different radii. These are called *parallel* folds. The inflection lines of a particular surface intersect the profile at points, i; the hinge encompasses the entire circular arc.

Some fold terminology refers to multilayer folds (**Figure 9.3**), rather than a single folded surface. A surface that contains successive hinge lines in a multilayer fold is an axial surface. If the rock layers get progressively older toward the axial surface, the fold is an anticline; if the layers get progressively younger, it is a syncline. **Figure 9.3** shows an anticline that is convex upward, and a syncline that is concave upward. In complexly folded terrain the normal stratigraphic succession may be overturned (imagine turning **Figure 9.3** upside down), permitting an anticline to be concave upward and a syncline to be convex upward. However, the names anticline and syncline still apply, because they depend upon the relative ages of strata in the multilayer fold. If the relative ages of layers in a multilayer fold are unknown, a convex upward fold is an antiform, and a concave upward fold is a synform.

Figure 9.3 is a *profile view* of multiple sinusoidal surfaces identical to the one shown in **Figure 9.2**. The hinge lines and inflection lines for these surfaces appear as points, h and i, respectively, in this view. In this case the axial surface is planar and intersects the profile plane in a straight line. Every layer of these folds are identical in shape and size. For any given layer the distance, *D*, measured parallel to the axial surface across the layer is *constant*. Folded layers with these geometric characteristics are similar folds. One could imagine similar folds forming by heterogeneous simple shearing with shear planes parallel to the axial surface (see inset **Figure 9.3**). The magnitude of shearing goes to zero at the axial surface and changes sign on the opposite limb. The layering moves *passively* with the shearing laminae. In other words, the mechanical properties of the layers have no effect on the folding. Because the shearing resembles laminar flow, this kinematic model is called flexural flow folding.

In the idealized profile of **Figure 9.4** every folded surface is a *circular arc*, but the radii of these circular arcs decrease as the arc length of the layers decrease. For this example, we suppose the relative ages of layers in the multilayer fold are unknown, so a convex upward fold is an antiform, and a concave upward fold is a synform. The inflection points for each surface mark the transition, for example, from an antiform to a synform. Along a folded surface between inflection points, the curvature is constant, so the hinge is not a point, but rather the entire arc of the folded surface excluding the inflection points.

For any given layer in **Figure 9.4** the thickness, *T*, measured perpendicular to the bounding surfaces of that layer, is *constant* throughout the fold. Folded layers with these geometric characteristics are parallel folds. One could imagine that as parallel folds form successive layers slip over one another along bedding contact (see pairs of arrows **Figure 9.4**). The slip magnitude goes to zero at the crest of the antiform, and changes sign on the opposing limb. This is the kinematic model for flexural slip folding.

The parallel fold geometry (**Figure 9.4**) cannot be continued indefinitely through a stack of layers. As the radius of successive arcs goes to zero, the length of the arc goes to zero, and the surface takes on a cusp-like geometry. This forces the next layer in succession to accommodate the folding with a very different geometry. Here we have placed a question mark in the regions that do not admit a continuation of parallel folding. A natural example of multilayer folding in the Castile Formation, described in Section 9.2.1, provides examples of how parallel fold geometries change near such a cusp.

Many fold styles described in the literature of structural geology are based on their shapes in profile. These include chevron, monocline, parasitic, box, ptygmatic, and sinusoidal to name a few (**Figure 9.5**). This abundant variety of fold shapes has led to many geometric classification and naming schemes based on certain geometric parameters, and the orientations of fold axes, limbs, or axial surfaces. We do not review these schemes here, because they do not lead to a quantification of the curvature of folded surface. However, in a later section we show how the concepts of differential geometry lead to such a quantitative classification for fold profiles in two dimensions, and for three-dimensional folded surfaces. Before introducing differential

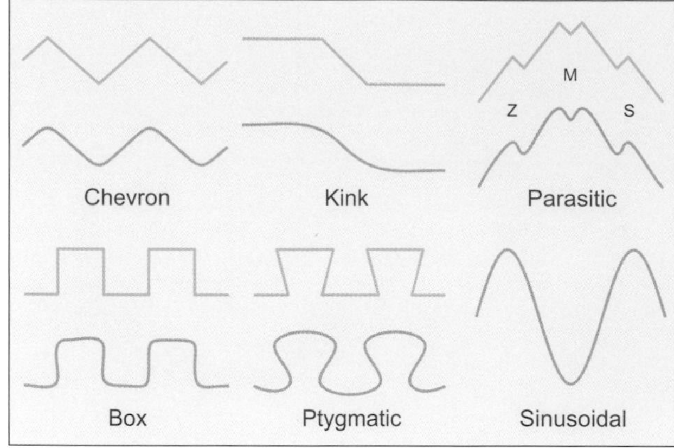

Figure 9.5 Some styles of folds as viewed in profile. For the first five styles a version with sharp hinges and another with rounded hinges are illustrated. For the parasitic folds the symmetry varies from Z to M to S across the larger fold.

geometry we describe folds in sedimentary, metamorphic, and igneous rock at scales from outcrop to crustal.

9.2 DESCRIPTIONS OF FOLDS AT THE OUTCROP SCALE

To introduce the diversity of geometric shapes and tectonic settings for folds at the outcrop scale, we describe examples from sedimentary, metamorphic, and igneous rocks. The most accessible folds, and the ones described in most detail, are those exposed in hand samples and outcrops. A typical outcrop might be in a road cut, or a natural cliff face, where cross sections of folded surfaces and layers are visible. If the exposure is approximately perpendicular to the axis of cylindrical folds, one can view the fold profile. However, if the exposure is oblique to the fold axis, the profile of the cylindrical fold is distorted.

9.2.1 Folds in Sedimentary Rock

The Upper Permian Castile Formation crops out in southeastern New Mexico and West Texas, where it is composed of calcite laminae, usually less than a few millimeters thick, alternating with anhydrite or gypsum laminae, usually less than a few centimeters thick. In some outcrops the Castile Formation is gently folded with wavelengths that vary from about 0.5 to 2.5 m. In the hinge regions of these folds the laminae are dramatically folded with wavelengths of a few centimeters or less (**Figure 9.6**). These multilayer folds are approximately cylindrical over lengths of several centimeters along the fold axis, and the plane of this photograph is roughly perpendicular to the fold axis.

The Castile folds (**Figure 9.6**) display an extraordinary diversity of shapes. For example, anhydrite layer A is nearly tabular, but shows subtle folding with different wavelengths along the top and bottom surfaces. In contrast, the entire layer B displays very regular folds with wavelengths of about 15 mm and amplitudes of about 5 mm. On the other hand, layer C is folded strongly on its bottom surface, but very little on its top surface. The package of layers labeled D has some of the distinctive features illustrated in **Figure 9.4** for parallel folds, where thin layers are pinched off in cusps and the multilayer fold abuts unfolded layers. Individual layers appear to have adopted the wavelength of the thickest layer at the middle of this package. Thinner layers within this package have broad horizontal sections that enable the accompanying antiforms or synforms to develop with a wavelength that matches the thickest layer. Inspection of the layers within the package labeled E suggests that fold wavelength increases with layer thickness. The thinnest layers have wavelengths of 3 to 4 mm, whereas the thickest layers have wavelengths of 10 to 15 mm. The diversity of fold sizes and shapes described here reflects the diversity of thicknesses and mechanical properties of the layers.

Figure 9.6 indicates that the thicker layers are less prone to folding, and if they do fold, they tend to influence the profile shapes of adjacent thinner layers. Also, certain calcite layers, perhaps the thicker ones, separated the rock mass into packages that folded more or less independently of adjacent packages. The fact that these features are repeated within this specimen, and also within specimens from different outcrops of the Castile Formation, indicates that they are systematic and indicative of the mechanical properties of a multilayer sequence. We consider a canonical model for buckling in a later section that addresses some of these observations.

9.2.2 Folds in Metamorphic Rock

In their review article on folds for the *Journal of Structural Geology*, Hudleston and Treagus (2010) illustrate some of the different geometries of folds in metamorphic rock with spectacular outcrop photographs. For example, the thinly bedded metasiltstones and slates that crop out near Boscastle, England (**Figure 9.7**) are folded together with sharply curved hinges and nearly planar limbs. The view is approximately parallel to the fold axis for these multilayer folds.

A prominent synform–antiform pair in the central left of **Figure 9.7** is identified by axial surfaces with traces marked by gently curved white lines that are inclined to the right about 15° from horizontal. The axial surface traces locally define a fold wavelength of about 30 cm and an amplitude of about 15 cm. Midway between the prominent synform and antiform, the otherwise nearly straight limbs are folded into two asymmetric folds that make a z-shaped pattern in profile. The upper limb of the antiform is folded into a pair of s-shaped asymmetric folds. Note that this designation refers to the *symmetry* of the letters z and s, and not to whether they are composed of straight or curved lines. This arrangement of asymmetric minor folds (e.g. synformal hinge – z – antiformal hinge – s) is very common in folds in many rock types.

9.2.3 Folds in Igneous Rock

One might suppose that folds in igneous rock are rare, if not absent, because igneous rock typically lacks the layering of sedimentary and metamorphic rock, which plays an important role in the development of folds. However, otherwise homogeneous bodies of igneous rock commonly have at least one set of fractures that can

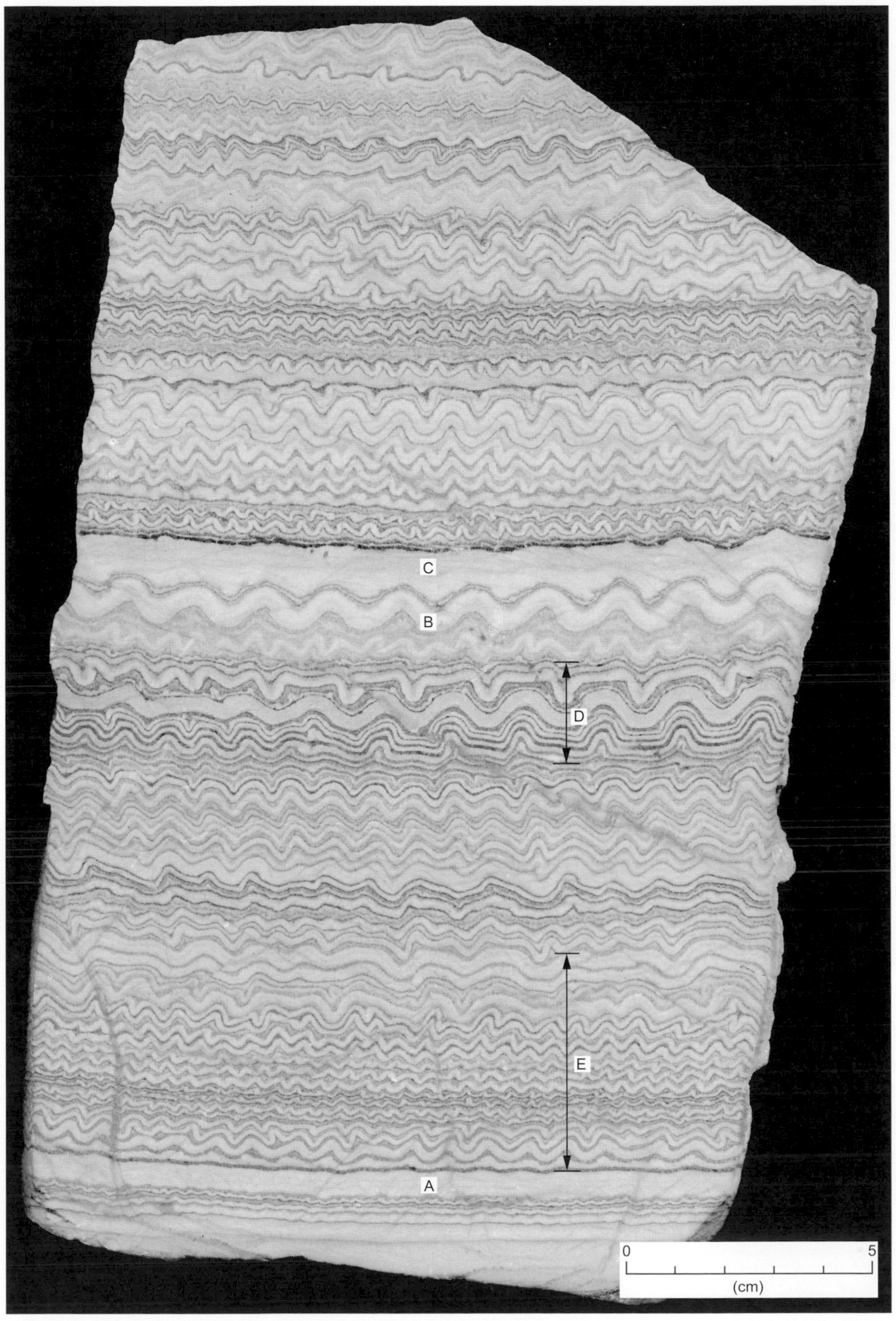

Figure 9.6 Hand specimen of the Castile Formation composed of anhydrite or gypsum (light) and calcite (dark) laminae with a variety of fold shapes. Specimen courtesy of A. Aydin. See Kirkland and Anderson (1970).

Figure 9.7 Multilayer folds in siltstone and slates near Boscastle, Cornwall, England. Slightly curved lines connect the hinge points, h, of surfaces in the prominent inclined synform and antiform. Smaller asymmetric folds with z-shapes and s-shapes are labeled. Modified from Hudleston and Treagus (2010), Figure 2d.

Figure 9.8 Photograph of a right-lateral kink band at the Trail Fork outcrop, TF, Lake Edison Granodiorite, Sierra Nevada, CA. See Figure 8.4 for geologic map of region. The sub-parallel fractures and small faults bend and change orientation by about 40° clockwise within the band. Scale bar is approximate in foreground. See Davies and Pollard (1986). UTM: 11 S 334866.00 m E, 4132905.00 m N.

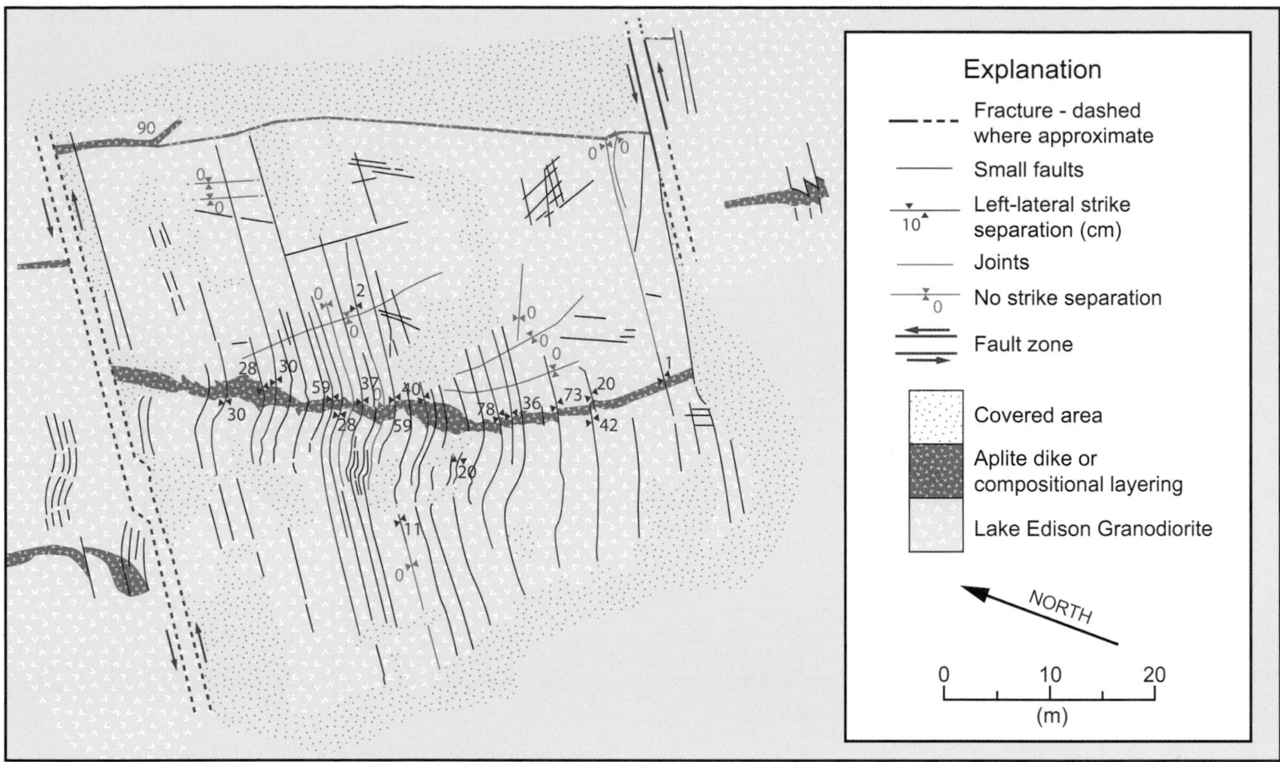

Figure 9.9 Structure map of the Trail Fork outcrop, Lake Edison Granodiorite, Sierra Nevada, CA. See Figure 8.4 for geologic map of region. Modified from Segall and Pollard (1983), Figure 9. UTM: 11 S 334849.00 m E, 4132896.00 m N.

provide a mechanical effect similar to bedding planes and foliations during folding of sedimentary and metamorphic rock. The example we use here is from the Lake Edison Granodiorite of the central Sierra Nevada, CA. The geology of this area was introduced in Section 8.2.1 to describe how a set of left-lateral faults developed from what originally were opening fractures bearing hydrothermal minerals. The mineralized fractures apparently were weaker than the surrounding granodiorite, and some slipped to form small faults with a left-lateral sense of relative motion.

Figure 9.8 shows the fractures and small faults cutting the Lake Edison Granodiorite at the Trail Fork outcrop (see location map, Figure 8.4). In the foreground the fractures and faults are sub-parallel, striking between 050 and 060. However, near the aplite dike that cuts across the outcrop, they bend clockwise about 40°, trend in this new direction for several meters, and then bend counter-clockwise back to their original orientation. A given fracture or small fault is folded into a style called a monocline or kink (Figure 9.5). Taken together, the set of surfaces are folded into a structure called a kink band.

The geological map of the Trail Fork kink band (Figure 9.9) reveals its structural setting and detailed geometry. The outcrop is bounded on the northwest and southeast by fault zones with 10 or more meters of left-lateral offset (Section 8.2.1). The aplite dike that crops out within the kink band is offset in a left-lateral sense along the small faults by as much as 0.8 m. Individual fractures and faults that cross the kink band are offset from 1.5 to 3.5 m in a right-lateral sense. This right-lateral kink band is more than 40 m long and strikes between 357° and 007°. Thus, the acute angle between the left-lateral small faults and the right-lateral kink band ranges from 43° to 63°.

Recall from Section 4.8.2 that the Coulomb criterion identifies two conjugate orientations of shear fractures, with opposite senses of slip, that are symmetrically inclined to the extreme principal stress axes. The Trail Fork map reveals a single set of left-lateral small faults and a conjugate right-lateral kink band. These two structures apparently accommodated the deformation in a manner somewhat analogous to the two sets of potential shear fractures from the Coulomb criterion.

9.3 QUANTIFYING FOLD PROFILES USING CURVATURE

In the introduction to this chapter we defined a fold as a surface in a body of rock that has undergone a *change in curvature*. To characterize folds, we need to understand how to describe, measure, and quantify the curvature of a surface. The mathematical tools for doing this come from differential geometry. In this section we introduce the elementary methods of differential geometry that are used to study and classify the geometry of folds in profile. Thus, we begin with a two-dimensional (plane) curve.

9.3.1 Curvature of a Plane Curve

A curve that lies in a plane is two-dimensional and is called a plane curve. For example, consider the curve in Figure 9.10a,

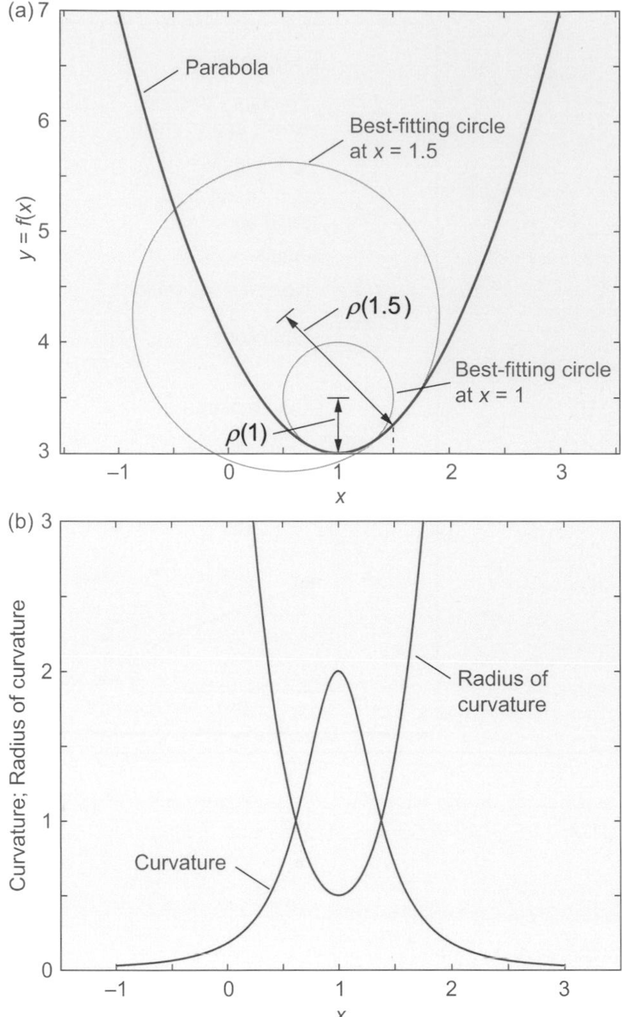

Figure 9.10 Measures of curvature for a plane curve. (a) Graph of the parabola (9.1) centered at x = 1. The radius of the best-fitting circle to the curve at x = 1 is $\rho = 0.5$, and at x = 1.5 it is $\rho = 1.4$. (b) Variations of curvature and radius of curvature with x for the parabola.

which is constructed using the mathematical equation for a parabola:

$$y = f(x) = x^2 - 2x + 4 \tag{9.1}$$

This equation yields a graph that is concave upward like a synform (**Figure 9.4**), but (9.1) cannot represent a fold in nature, because it is not *dimensionally homogeneous* (Section 3.3.1). If x on the right side is considered a length (e.g. distance along the x-axis), then the three terms would have dimensions L^2, L^1, and L^0. We do not prescribe dimensions or units in (9.1) because the quantities are thought of as *dimensionless* numbers. This equation was chosen because it is easy to manipulate mathematically, and therefore it facilitates a discussion of the concept of curvature. After this preliminary discussion, we employ differential geometry to introduce an equation with consistent dimensions to describe folds in nature, which could have similar shapes to the parabola.

In order to describe the shape of the parabola in (**Figure 9.10a**) consider finding the best-fitting circle at a particular point on the curve. The circle passes through that point and locally matches the shape of the curve. For example, at the apex of the parabola, $x = 1$, a circle with a radius $\rho(1) = 0.5$ matches the shape of the curve better than any other circle. The radius of the best-fitting circle is called the radius of curvature of the parabola at that point. A circle that best fits the parabola at $x = 1.5$ has a radius $\rho(1.5) = 1.4$, so the radius of curvature increases as the parabola becomes less curved. This relationship motivates the definition of the curvature of the plane curve as the *reciprocal* of the radius of curvature: $\kappa(x) = 1/\rho(x)$. Thus, at $x = 1$ the curvature of the parabola is $\kappa(1) = 2$, and at $x = 1.5$ the curvature is $\kappa(1.5) = 0.71$.

This discussion of curvature and radius of curvature begs the questions: how are these quantities calculated, given a *mathematical* equation for a curve? The curvature of the function $y = f(x)$ in (9.1) is found using the following equation, introduced in most introductory calculus courses (e.g. see Varberg et al., 2006, in Further Reading):

$$\kappa(x) = \left|\frac{d^2y}{dx^2}\right| \Big/ \left[1 + \left(\frac{dy}{dx}\right)^2\right]^{3/2}, \quad \text{plane curve} \tag{9.2}$$

In the special case where the first derivative (slope of the graph in **Figure 9.10a**) is zero, that is at $x = 1$, the curvature equals the absolute value of the second derivative, $\kappa(1) = |d^2y/dx^2|$. The absolute value of the second derivative, sometimes is referred to as the "curvature" of a plane curve, but the denominator in (9.2) is a correction factor that should be considered where the first derivative is not zero.

For the parabola (9.1), the first and second derivatives of the function with respect to x are $dy/dx = 2x - 2$ and $d^2y/dx^2 = 2$, so the curvature is written $\kappa(x) = 2/(4x^2 - 8x + 5)^{3/2}$. Note that this equation is not dimensionally homogeneous, if one were to scale x as a length. The curvature, plotted in **Figure 9.10b**, attains a maximum value at $x = 1$ where $\kappa(1) = 2$, and decreases symmetrically toward zero for lesser or greater values of x. Again, we prescribe no units on this figure because the quantities are dimensionless numbers. The radius of curvature is at a minimum for $x = 1$ and plots off of **Figure 9.10b** for $x < 0.25$ and $x > 1.75$. We find, for example, that $\rho(2) = 5.59$ and $\rho(3) = 35.05$. At $x = 3$ the parabola is difficult to distinguish by eye from a straight line because its curvature, $\kappa(3) = 0.0285$, is nearly zero.

9.3.2 Parametric Description of a Plane Curve

As a first step toward generalizing the concepts just developed for plane curves, consider a continuous set of points that are arranged so that *locally*, in the neighborhood of any given point, they approximate a line, rather than a plane or a volume. With the concept of a curve as a continuous set of points, it is natural to quantify the curve using a set of position vectors directed at those points. One such vector is shown in **Figure 9.11**. Progress along the curve is tracked using a single dimensionless parameter, t, and the set of vectors is defined as a vector-valued function, $\mathbf{c}(t)$.

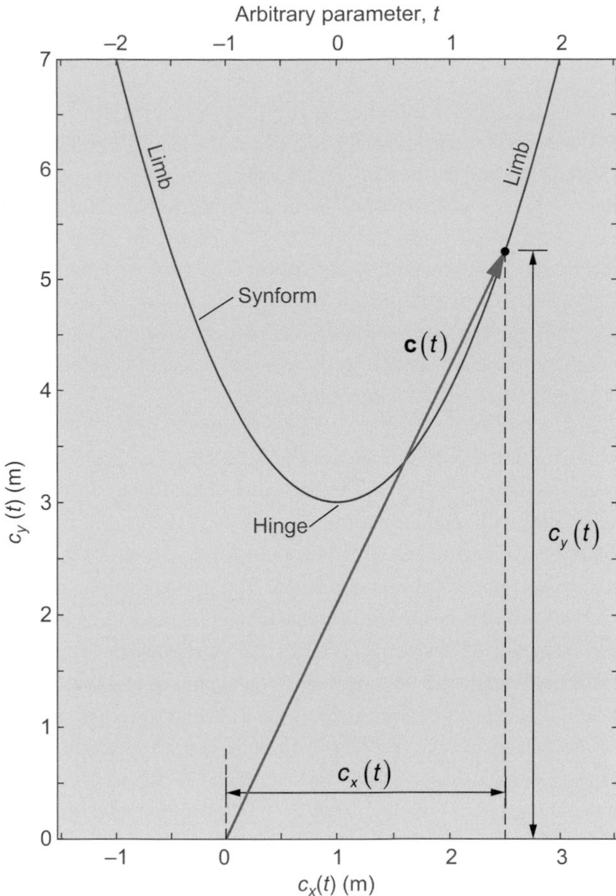

Figure 9.11 The plane curve (blue) representing the profile of a synform is constructed using position vectors defined by the vector-valued function **c**(*t*) and the arbitrary parameter, *t*, using (9.3). The vector components are scaled in meters. The green arrow is one such position vector.

The letter **c** reminds us this function represents a *curve*. The position vectors are scaled as lengths with appropriate dimensions and units: $\mathbf{c}\,\{=\}\,\mathrm{L}$ and $\mathbf{c}\,[=]\,\mathrm{m}$.

This generalization introduces the arbitrary parameter, t, to replace x in (9.1). For example, we could write $x = t + 1$, which centers the parabola on $t = 0$, and substitute for x in (9.1) to find $y(t) = t^2 + 3$. Then, the vector-valued function for the plane parabolic curve is:

$$\mathbf{c}(t) = c_x(t)\hat{\mathbf{i}} + c_y(t)\hat{\mathbf{j}} = (t+1)\hat{\mathbf{i}} + (t^2+3)\hat{\mathbf{j}} \tag{9.3}$$

Recall from Section 2.2.1 that $\hat{\mathbf{i}}$ and $\hat{\mathbf{j}}$ are the unit base vectors for the x and y axes. The components of the vector $\mathbf{c}(t)$ are $c_x(t)$ and $c_y(t)$, and both components are functions of the parameter t. The components carry the dimensions and units so, for example, $c_x\,\{=\}\,\mathrm{L}$ and $c_x\,[=]\,\mathrm{m}$, and the axes in **Figure 9.11** are scaled accordingly. As the parameter varies continuously over $-2 \le t \le +2$, the vector-valued function (9.3) defines the set of vectors with tails fixed at the origin of the (x, y)-plane, heads that trace out the curve, and lengths that are scaled in meters. These are the position vectors for all the points on the geometric model of a synform.

We introduce several different vector-valued functions in this chapter. Each function represents a set of vectors, but when referring to the function itself we may use the singular rather than the plural form of the word *vector*. For example, in the preceding paragraph we said: "The components of the vector $\mathbf{c}(t)$ are $c_x(t)$ and $c_y(t)$, ..." It should be understood that $\mathbf{c}(t)$ stands for an infinite number of vectors, not a single vector.

Typically, a fold profile would be described using a discrete set of points that are digitized from a photograph of an outcrop such as **Figure 9.7**, or from a map such as **Figure 9.9**, or taken from a data set gathered using surveying techniques such as GPS or LIDAR. In those cases the calculations we discuss here would be made using numerical methods and a computational engine such as MATLAB.

The dimensionless parameter t is called *arbitrary* because we could choose other ways to define it as a function of x, and preserve the geometry of the curve. For example, we could have taken $x = t$, and substituted in (9.1) to find $y(t) = t^2 - 2t + 4$. Then, the vector-valued function for the plane curve would be $\mathbf{c}(t) = t\hat{\mathbf{i}} + (t^2 - 2t + 4)\hat{\mathbf{j}}$, instead of the function given in (9.3), and the parabola would be centered on $t = 1$. The *shape* of the curve is not affected by this choice, so the distributions of curvature and radius of curvature are the same as those plotted in **Figure 9.10b**.

In addition to the position vector that traces out the curve, another vector provides important constraints on the geometry of the curve and is directly related to data gathered by structural geologists in the field. This is the unit tangent vector, $\hat{\mathbf{t}}(t)$, which we introduced in Section 2.5.1. The unit tangent vector is parallel to the best-fitting straight line at any point along the curve. This is consistent with the intuitive notion that the curve approximates a straight line in the neighborhood of a given point. The first derivative of the position vector, $\mathbf{c}(t)$, with respect to t yields a vector-valued function that defines tangent vectors for all points on a curve. To make these unit vectors, the first derivative is divided by its magnitude (Lipschultz, 1969, p. 62):

$$\hat{\mathbf{t}}(t) = \frac{d\mathbf{c}}{dt} \bigg/ \left|\frac{d\mathbf{c}}{dt}\right| \tag{9.4}$$

The unit tangent vectors are dimensionless.

For the synformal curve illustrated in **Figure 9.11**, and defined in (9.3), the first derivative of $\mathbf{c}(t)$ is $d\mathbf{c}/dt = 1\hat{\mathbf{i}} + 2t\hat{\mathbf{j}}$, and the magnitude of this vector is $|d\mathbf{c}/dt| = (1 + 4t^2)^{1/2}$, so the vector function for the tangent vectors is:

$$\hat{\mathbf{t}}(t) = \left(1\hat{\mathbf{i}} + 2t\hat{\mathbf{j}}\right) \Big/ \left(1 + 4t^2\right)^{1/2} \tag{9.5}$$

These unit tangent vectors lie in the (x, y)-plane, because they have no z-component. **Figure 9.12** illustrates a few examples of these vectors along with the parabolic curve. At the hinge of the fold, where the curvature is greatest, the curve diverges abruptly from the direction of the unit tangent vector. In the limbs of the fold, where the curvature is much less, the entire unit tangent vector appears to overlie the curve.

The curvature vector is the first derivative of the unit tangent vector, $\hat{\mathbf{t}}(t)$, normalized by the magnitude of the tangent vector:

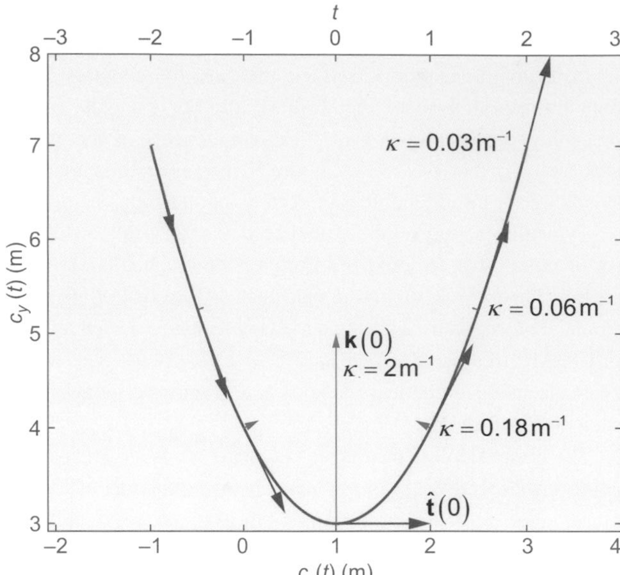

Figure 9.12 The plane curve (blue) representing the profile of a synform is accompanied by a few unit tangent vectors (red), $\hat{\mathbf{t}}(t)$, calculated using (9.5), and curvature vectors (green), $\mathbf{k}(t)$, calculated using (9.7). Scalar curvatures, $\kappa(t)$, are listed near the respective curvature vector.

$$\mathbf{k}(t) = \frac{\mathrm{d}\hat{\mathbf{t}}}{\mathrm{d}t} \bigg/ \left|\frac{\mathrm{d}\mathbf{c}}{\mathrm{d}t}\right| \tag{9.6}$$

The only dimensional quantity in this definition of the curvature vector is **c** in the denominator, which has dimensions of length, so the curvature vector has dimensions of reciprocal length and units of reciprocal meters: $\mathbf{k}\,\{=\}\,\mathrm{L}^{-1}$ and $\mathbf{k}\,[=]\,\mathrm{m}^{-1}$. The curvature vector is not a unit vector, and the magnitude of the curvature vector is the scalar curvature, $\kappa(t) = |\mathbf{k}(t)|$, which also carries the dimensions of reciprocal length. We use the adjective *scalar* to distinguish this quantity from the curvature *vector*.

As an example we calculate the curvature vector for the synformal curve illustrated in **Figure 9.11**. The first derivative of the unit tangent vector (9.5) is $\mathrm{d}\hat{\mathbf{t}}/\mathrm{d}t = \left(-4t\hat{\mathbf{i}} + 2\hat{\mathbf{j}}\right)/(1+4t^2)^{3/2}$ and, as we found above, $|\mathrm{d}\mathbf{c}/\mathrm{d}t| = (1 + 4t^2)^{1/2}$. Thus, using (9.6), the vector function for the curvature is:

$$\mathbf{k}(t) = \left(-4t\hat{\mathbf{i}} + 2\hat{\mathbf{j}}\right) \Big/ \left(1 + 4t^2\right)^2 \tag{9.7}$$

Like the unit tangent vectors for the synformal curve, the curvature vectors lie in the (x, y)-plane, because they have no z-component. **Figure 9.11** shows a few examples of these curvature vectors. Note the significant change in magnitude of these vectors from the hinge to the limbs of the fold. Taking the magnitude of the vector function for the curvature (9.7), gives the scalar curvature, $\kappa(t) = 2/(1 + 4t^2)^{3/2}$, as a function of position along the curve. The scalar curvature is greatest at $t = 0$, $c_x = 1$ m, $c_y = 3$ m where $\kappa(0) = 2\,\mathrm{m}^{-1}$. This defines the *hinge* of the synformal fold (**Figure 9.11**).

9.4 DESCRIPTIONS OF FOLDS IN THREE DIMENSIONS AT THE CRUSTAL SCALE

Folds occur across a broad range of observational scales from microscopic to crustal. Hand-sample and outcrop photographs (**Figure 9.6** and **Figure 9.7**) were appropriate for characterizing the centimeter to meter scale folds described in Section 9.2, but here we consider a fold at the ten-kilometer scale, so different technology and procedures are required to document the three-dimensional form of this fold. We begin this section by reviewing the geology and form of Sheep Mountain anticline, and then describe how a new structure contour map was constructed using Airborne Laser Swath Mapping (ALSM).

9.4.1 Sheep Mountain Anticline

The Sheep Mountain anticline (**Figure 9.13**) is located in north central Wyoming and trends from southeast to northwest along the eastern margin of the Bighorn Basin. The geological map reveals sedimentary rocks exposed in the anticline that range in age from Mississippian to Cretaceous. The Mississippian Madison Limestone is the oldest exposed unit and it crops out along the axial surface of the fold. The flanks of Sheep Mountain are held up by resistant units of the Pennsylvanian Tensleep Sandstone and the Permian Phosphoria Formation. Younger Triassic through Cretaceous formations also are part of the anticline. They crop out in the surrounding plain and provide good surfaces for mapping. A smaller anticline, called the Thumb, branches south from the main fold. The hinge of the Sheep Mountain anticline narrows as it plunges to the northwest.

About 3 km of sedimentary rocks were deposited in the Big Horn Basin during Paleozoic and Mesozoic times, and about 1 km of that stratigraphic section is exposed on Sheep Mountain and in the adjacent area (**Figure 9.13**). The Laramide orogeny, beginning in the late Cretaceous, is associated with a pronounced northeast–southwest horizontal compression that apparently led to thrust faulting and folding throughout this region. The strikes of thrust faults and the trends of fold axes are interpreted as being approximately perpendicular to the direction of the greatest horizontal tectonic compression.

Sheep Mountain anticline is distinctly asymmetric in cross section (**Figure 9.14**). This cross section was constructed using data from Airborne Laser Swath Mapping (**Box 9.1**). The northeastern limb is steeply inclined with dips typically greater than 60°, and locally approaching 90°. The southwestern limb typically dips less than 40°. The asymmetry of Sheep Mountain anticline suggests that the dip of an underlying thrust fault would be to the southwest. In that context, the NE limb is the forelimb of the fold, and the SW limb is the backlimb of the fold. However, a thrust fault is not exposed at the surface, and seismic data are not of sufficient quality to reveal the details of its location or geometry in the subsurface.

9.4.2 Structure Contour Mapping

Fourteen erosion resistant, continuous, and easily identified units from the top of the Madison Limestone to the Cretaceous Frontier Formation were chosen for mapping. These are identified as color

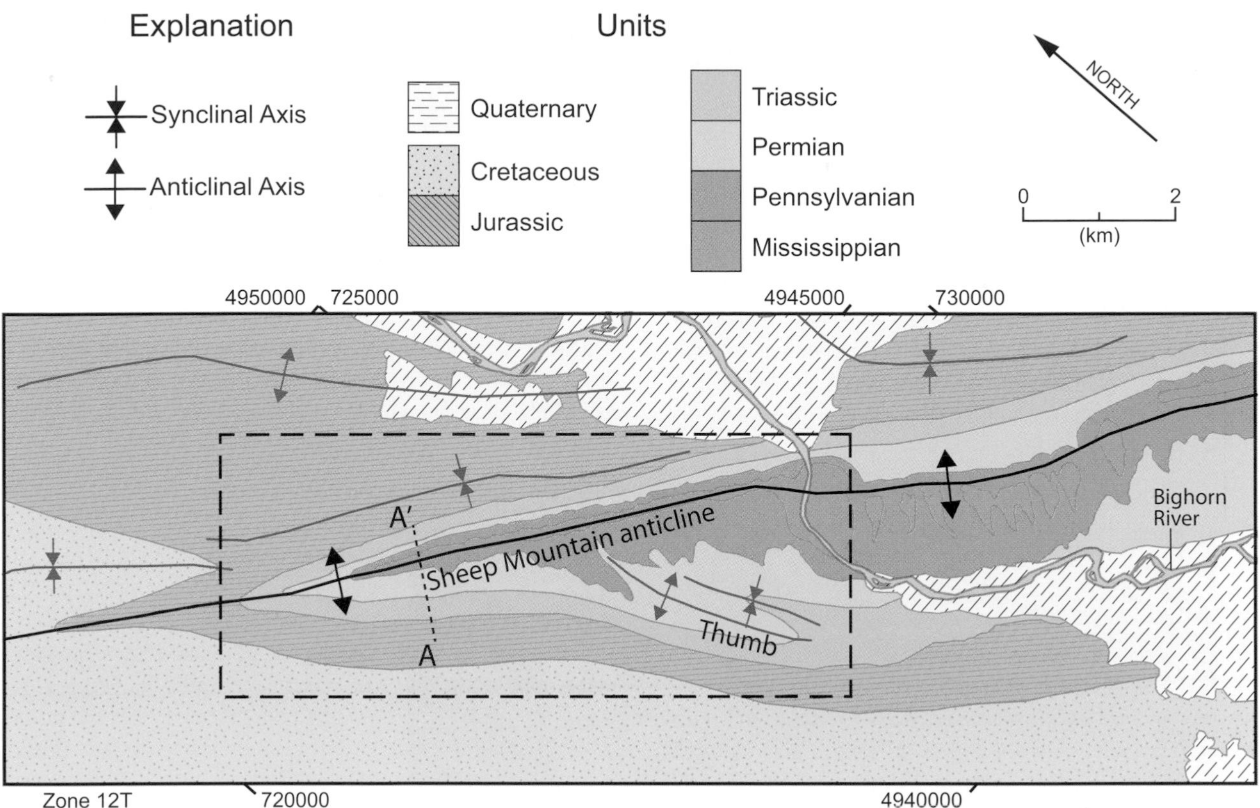

Figure 9.13 Geologic map of northwestern portion of Sheep Mountain anticline and adjacent structures. A–A′ is the location of the cross section shown in Figure 9.14. Dashed rectangle outlines the area of the interpolated fold surface shown in Figure 9.17. Modified from Lovely et al. (2010), Figure 1. Approximate map center UTM: 12 T 725829.78 m E, 4943672.34 m N.

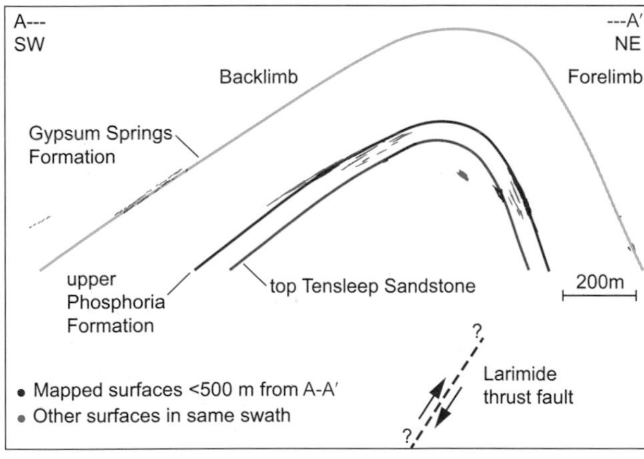

Figure 9.14 Cross section A–A′ (Figure 9.13) of three surfaces in the Sheep Mountain anticline: top Tensleep – Pennsylvanian; upper Phosphoria – Permian; Gypsum Springs – Jurassic. Colors are coded to the stratigraphic column (Figure 9.15). ALSM data shown as gray and black dots. View is down the plunge of the fold axis (326, 6). Modified from Lovely et al. (2010), Figure 8.

coded contacts on the stratigraphic column (Figure 9.15). If a single contact were exposed over the entire fold, the ALSM data could be used directly to define the three-dimensional geometry of that folded surface. However, as observed on the geologic map (Figure 9.13), and as shown for one swath of the ALSM data in cross section (Figure 9.14), exposures of particular contacts are not continuous across the fold. In fact, on any given cross section only fragments of each mapped unit are exposed. This makes the task of constructing a structure contour map of a representative folded surface very challenging.

The traditional method of constructing a structure contour map involves drafting by hand a series of parallel cross sections perpendicular to the fold axis (such as Figure 9.14) using data from a topographic map and a geologic map. Then, a particular stratigraphic contact is chosen and a three-dimensional surface is drafted as a structure contour map based on the cross sections. The method is necessarily subjective and interpretive where the data is sparse.

An alternative method for constructing the three-dimensional folded surface is to use the dense and precise topographic data from an ALSM survey, outcrop scale geologic mapping on an ALSM image, and a numerical interpolation method that is reproducible. The base of the Jurassic Sundance Formation (thick black horizon on Figure 9.15) was selected to describe the Sheep Mountain anticline, and data were projected from lower and higher in the stratigraphic section to that level using bedding-perpendicular thicknesses of the formations obtained from the ALSM data. The selected contact is about mid-way between the bottom and the top of the stratigraphic section (Figure 9.15), so

Box 9.1 Airborne Laser Swath Mapping (ALSM)

Several technologies employ a laser light source to direct a beam toward a target (e.g. a road cut, outcrop, or engineered structure) and measure the two-way travel time, t, of the light from the source to the target and back to the source. Given the velocity of the beam, v, the distance, d, from the source to the target is calculated as $d = vt/2$. The development of the laser in the middle of the twentieth century made these technologies possible. The general acronym LIDAR, standing for Light Imaging, Detection, And Ranging, represents all of these technologies. "Ranging" in this context refers to measuring the distance (range) from the source to the target.

LIDAR instruments can acquire many individual distances to different points on the target surface, which together form a point cloud of data. These data enable the construction of a three-dimensional representation of that surface by interpolating the point cloud onto a regular planar grid to form a digital surface model. For example, if the target is Earth's surface, the interpolated point cloud is a Digital Elevation Model (DEM), a digital version of the classic analogue topographic map. Instead of the contours of equal elevation found on a topographic map, the DEM associates a digital elevation with every grid point.

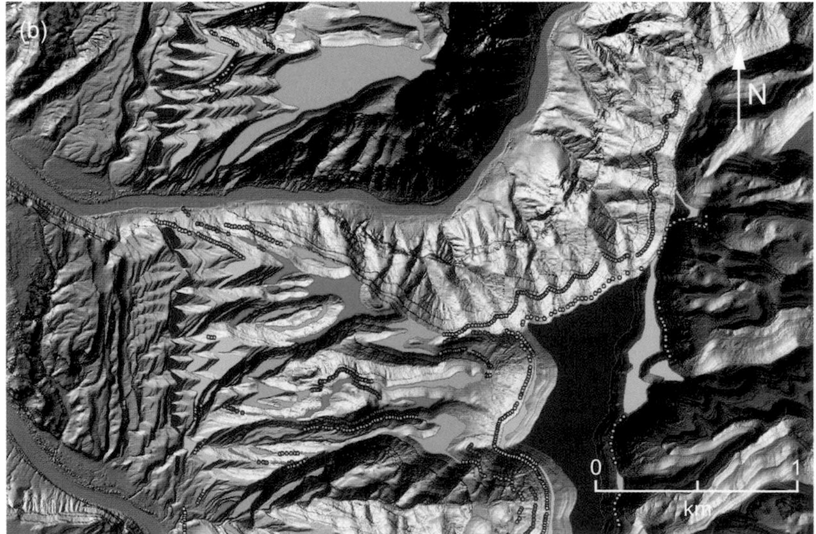

(c)

Unit	Thickness
McKim limestone	3–6 m
RB6	7–11 m
Baby oil sand	1–4 m
RB6	5–12 m
Goodrich oil sand	2–7 m
RB5	7–13 m
Shafer limestone	1–6 m
RB4	17–23 m
Third oil sand	1–3 m
RB3	11–15 m
Mendenhall oil sand	6–11 m
RB2	15–17 m
Unnamed limestone	2–8 m
RB1	5–15 m
Little Loop oil sand	6–8 m

Rico Formation

Figure B9.1 Geologic mapping at Raplee Ridge. (a) Hill shaded relief map using the digital elevation model prepared from the ALSM survey. (b) Geologic map overlain on (a) using five beds from the Rico Formation. Colored polygons are exposed pavements; trails of dots are narrow ledges. (c) Stratigraphic section for the Rico Formation with colored dots indicating the tops of beds mapped on (b).

LIDAR instruments find applications in many different disciplines that focus on processes and structures at Earth's surface, including structural geology, geomorphology, archaeology, forestry, and civil engineering. For terrestrial (land-based) applications, the instrument typically sits on a tripod or on a moving truck. For airborne applications, an airplane or helicopter carries the instrument, which scans Earth's surface back and forth, perpendicular to the flight path, gathering a *swath* of distance measurements. Using GPS to determine the precise location of the aircraft as each signal beams toward the surface, the instrument calculates the position of the point on the surface. This use of LIDAR technology is Airborne Laser Swath Mapping (ALSM). The US National Science Foundation organized and funded the National Center for Airborne Laser Mapping (NCALM) in 2003 to coordinate the collection and processing of ALSM data for scientific research (Slatton et al., 2007; see Further Reading).

The Digital Elevation Model (DEM) of Earth's surface obtained from ALSM surveys provide exceptional bases for geologic mapping. For example, ALSM technology employed by NCALM gathered a point cloud data set for Raplee Ridge, UT, enabling the precise mapping of this multilayer fold (Hilley et al., 2010). Using three or more points per square meter, the elevation data were interpolated to a one-meter grid and a hill-shaded relief map was prepared from this DEM. **Figure B9.1a** is a portion of the hill-shaded map where the San Juan River cuts the ridge in a deep gorge.

Five beds of limestone and sandstone within the Rico Formation highlight the fold geometry. These beds crop out across much of Raplee Ridge, where they form prominent ledges and expansive pavements that facilitate mapping in this rugged terrain. Colored markers on the stratigraphic column in Figure B9.1c identify the tops of these resistant beds (McKim limestone, Goodrich oil sand, Shafer limestone, Mendenhall oil sand, and Unnamed limestone). A geologic map of the five bed tops was prepared in the field on printouts of the hill-shaded map. Digitized polygons, filled with the designated color from the stratigraphic column, map the large bedding pavements on **Figure B9.1b**. Digitized trails of appropriately colored points map the narrow ledges. This project demonstrates the utility of ALSM technology for quantitative mapping of geologic structures.

the projected distances range from 416 m below to 578 m above the base Sundance. The projection used to constrain the structure contour map assumes the Sheep Mountain anticline is adequately represented as a parallel fold, in which beds retain a constant thickness from the hinge to the limbs (**Figure 9.4**).

The fourteen colored horizons identified on the stratigraphic column (**Figure 9.15**) were mapped in the field onto hill-shaded images of the ALSM data (e.g. **Figure 9.16**). This particular image covers just a portion of the backlimb of Sheep Mountain anticline, but provides a good example of the mapping method. Strike and dip measurements of bedding surfaces were recorded at each red dot on this image, and a total of 282 such orientations were taken over the mapped area (dashed rectangle on **Figure 9.13**) using a hand-held compass and inclinometer. Polygons defining the edge of each colored surface on the hill-shaded images were digitized, and 514,342 Digital Elevation Model (DEM) points were extracted from the entire ALSM data set to define the locations and elevations of points on the mapped outcrops. Strike and dip calculations, based on the classic three point problem, reviewed in Section 2.4.2, were performed using the DEM points on each outcrop. Because the DEM is on a 1 m by 1 m grid, most of the mapped outcrops have many points, so the mean gradient was used to define the orientation of bedding. These values were supplemented with direct measurements of bedding orientation for the smaller outcrops.

The structure contour map (**Figure 9.17**) of the base of the Sundance formation shows that this portion of Sheep Mountain anticline is *not* a cylindrical fold (**Figure 9.2**). For example, the hinge of the fold plunges to the northwest in the northwestern half of this map, becomes horizontal near the middle of the map, and plunges gently to the southeast over the other half. In contrast, the hinge of a cylindrical fold plunges only in one direction at a constant angle. In addition, the hinge of the fold narrows to the northwest and broadens to the southeast. For a cylindrical fold the width of the hinge does not vary with position along the fold axis: every profile of the cylindrical fold is identical. These and other interesting aspects of the fold geometry would not be captured by treating data from Sheep Mountain in the context of cylindrical folding.

The complexity of the structure contour map for the base of the Sundance formation (**Figure 9.17**) begs the question: is it possible to describe this fold quantitatively in three dimensions? The answer is yes, but the description requires additional concepts from differential geometry. We introduce those concepts in Section 9.5 and then use them to quantitatively describe the anticline at Sheep Mountain.

9.5 QUANTIFYING FOLDS IN THREE DIMENSIONS USING CURVATURE

In Section 9.4 we introduced the Sheep Mountain anticline and described how to map this fold at the kilometer scale. From the geologic and topographic maps we construct a three-dimensional surface that represents the base of the Jurassic Sundance Formation (**Figure 9.18**), and use that to investigate the geometry of the fold. This rendering of the surface is color-coded to elevation, and shows the steeply dipping northeastern limb, the less steeply dipping southwestern limb, and the hinge that narrows and plunges to the northwest. The folded surface is not cylindrical

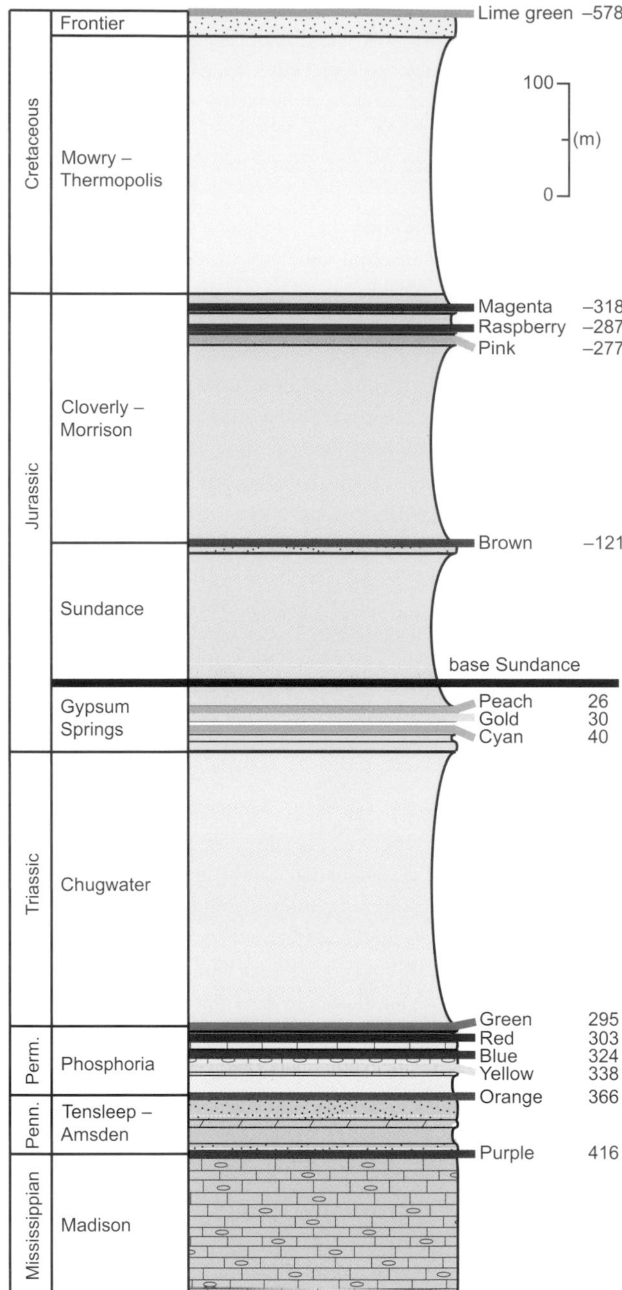

Figure 9.15 Stratigraphic section for Sheep Mountain anticline. Fourteen colored horizons were mapped onto ALSM images (e.g. **Figure 9.16**). Numbers on right are projected distances in meters to the base Sundance, which is the level used for the structure contour map (**Figure 9.17**). Modified from Lovely et al. (2010), Figure 3.

and has numerous interesting bumps and hollows. The challenge, taken up in this section, is to describe this complicated three-dimensional surfaces quantitatively.

The surface illustrated in **Figure 9.18** contains two sets of curves that form a warped 50 m grid. These curves collectively mimic the surface itself, so we can use their shape to quantify the shape of the surface at any point. Thus, the curvature of a surface is a concept that builds upon the curvature of curved lines, so we begin by defining three-dimensional curves.

9.5.1 Curvature of a Three-Dimensional Curve

Most curves lying in a three-dimensional folded surface (**Figure 9.18**) are three dimensional. Therefore, we take the two-dimensional concepts for curvature of a plane curve, developed in Section 9.3.2, and generalize them to three dimensions. The only change is that the vector-valued function describing the curve must include a z-component, so this function is:

$$\mathbf{c}(t) = c_x(t)\hat{\mathbf{i}} + c_y(t)\hat{\mathbf{j}} + c_z(t)\hat{\mathbf{k}} \tag{9.8}$$

This is the parametric representation of a three-dimensional curve. The components carry the dimensions and units of length so, for example, $c_z\,\{=\}\,\mathrm{L}$ and $c_z\,[=]\,\mathrm{m}$. The set of unit tangent vectors to this three-dimensional curve is given by (9.4), and the set of curvature vectors is given by (9.6).

The scalar curvature for a two- or three-dimensional curve may be calculated directly from derivatives of (9.8) as (Lipschultz, 1969, p. 64):

$$\kappa(t) = \left|\frac{\mathrm{d}\mathbf{c}}{\mathrm{d}t} \times \frac{\mathrm{d}^2\mathbf{c}}{\mathrm{d}t^2}\right| \bigg/ \left|\frac{\mathrm{d}\mathbf{c}}{\mathrm{d}t}\right|^3 \tag{9.9}$$

This is a general equation for the scalar curvature of a vector-valued function of the arbitrary parameter t. The power of (9.9) lies in the fact that it may be applied to any three-dimensional curve described by (9.8), providing one can take the first and second derivatives of its components. Because t is dimensionless and the position vector $\mathbf{c}$ has dimensions of length, the dimensions of the scalar curvature are reciprocal length: $\kappa\,\{=\}\,\mathrm{L}^{-1}$ and $\kappa\,[=]\,\mathrm{m}^{-1}$.

9.5.2 Parametric Description of a Three-Dimensional Surface

As a first step toward describing a three-dimensional surface think about a continuous set of points that, in the neighborhood of any given point on the surface, approximate a *plane*. This would be the tangent plane to the surface. To specify the position on any plane we need two variables, for example the x and y coordinates in the (x, y)-plane. Thus, we quantify a surface using a set of position vectors that are continuous vector functions of two scalar variables. We choose u and v as the two *dimensionless* parameters for the surface and define the three-dimensional vector-valued function for a surface as:

$$\mathbf{s}(u, v) = s_x(u, v)\hat{\mathbf{i}} + s_y(u, v)\hat{\mathbf{j}} + s_z(u, v)\hat{\mathbf{k}} \tag{9.10}$$

This is a parametric representation of a surface.

Each component of the vector function $\mathbf{s}(u, v)$ in (9.10) may be a function of the two parameters, and the letter $\mathbf{s}$ reminds us this represents a *surface*. The components carry the dimensions and units of length so, for example, $s_y\,\{=\}\,\mathrm{L}$ and $s_y\,[=]\,\mathrm{m}$. For a given (u, v) pair, the position vector is directed to one point on the surface. As the two parameters vary continuously, the heads of

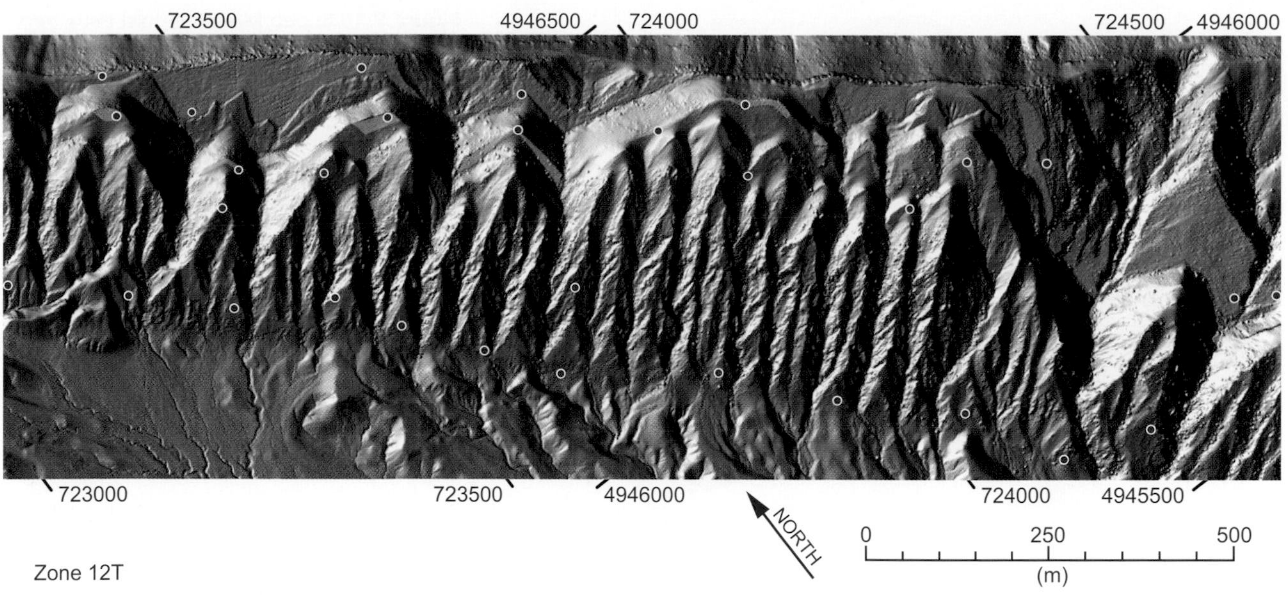

Figure 9.16 Hill shaded image from ALSM data covering a portion of the southwestern limb of Sheep Mountain anticline. For location see **Figure 9.13**. Colored polygons are mapped bedding surfaces with color corresponding to that on the stratigraphic section (**Figure 9.15**). The units mapped here are from the Permian Phosphoria Formation (Orange, Yellow, Blue, Red, Green). Small red dots are locations of strike and dip measurements. Modified from Lovely et al. (2010), Figure 4b.

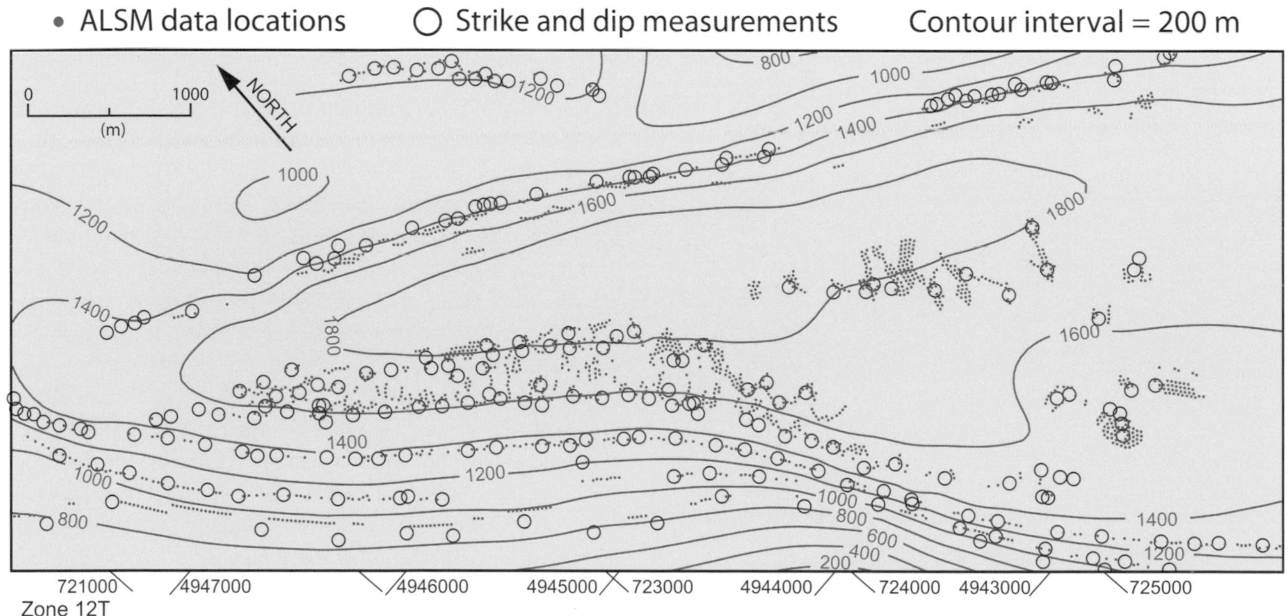

Figure 9.17 Structure contour map of the base of the Sundance formation (**Figure 9.15**). For location see **Figure 9.13**. Contour interval = 40 m. Green dots are ALSM data locations. Blue circles are locations of strike and dip measurements. Modified from Lovely et al. (2010), Figure 11b.

successive position vectors sweep out the points on the curved surface in three-dimensional space.

Recall that the folded surface representing the base of the Sundance Formation at Sheep Mountain (**Figure 9.18**) contains two sets of curves. Each curve is generated by holding one parameter constant in (9.10), and varying the other parameter. For example, setting $v = v_0$ and varying u continuously, the heads of successive position vectors sweep out the points on the curve, $\mathbf{c}(u, v_0)$, which lies *in* the surface $\mathbf{s}(u, v)$. This is called a u-parameter curve because u is the variable parameter.

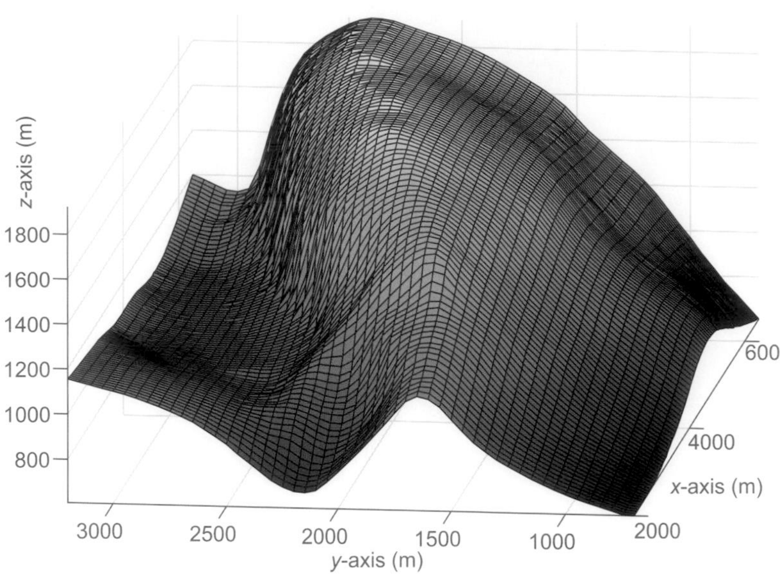

Figure 9.18 Sheep Mountain fold visualized using the surface at the base of the Sundance Formation (Figure 9.15) here viewed toward the southeast. Color correlates with elevation, and the two sets of curves form a wire-frame mesh on the surface. Modified from Lovely et al. (2010), Figure 11c.

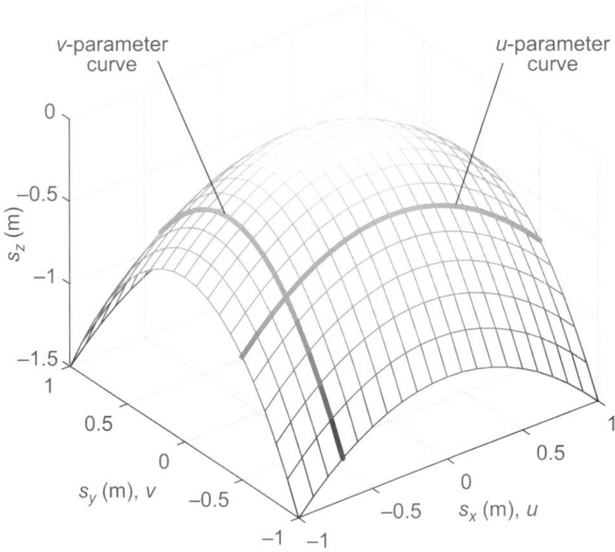

Figure 9.19 The elliptic paraboloid (idealized antiform) defined in (9.11) is represented here by a graph using the components of **s**(u,v). A selection of u- and v-parameter curves lie in that surface.

Alternatively, setting $u = u_0$, and varying v continuously, the position vectors sweep out the points on a v-parameter curve, $\mathbf{c}(u_0, v)$, which also lies *in* the surface. For a range of different values of u_0 and v_0, two families of curves are generated, all lying in the surface.

An example three-dimensional surface containing u- and v-parameter curves is illustrated in **Figure 9.19** and is defined using the following vector-valued function for an elliptic paraboloid:

$$\mathbf{s}(u, v) = u\hat{\mathbf{i}} + v\hat{\mathbf{j}} - \tfrac{1}{2}\left(u^2 + 2v^2\right)\hat{\mathbf{k}} \tag{9.11}$$

To think of the elliptic paraboloid as a fold we imagine that the surface was planar at some earlier time, and then was folded into this shape, which looks like an idealized antiform. The three vector components are scaled in meters. The family of u-parameter curves all lie in planes perpendicular to the v-axis. The family of v-parameter curves all lie in planes perpendicular to the u-axis. As the spacing between adjacent curves in either family goes to zero, those curves completely cover the surface. However, we need both families of curves to characterize the surface.

The three-dimensional surface defined in (9.11) is special in the sense that s_x is not a function of v and s_y is not a function of u, and in fact $s_x = u$ m and $s_y = v$ m. Thus, these two components of the vector function for the surface, $\mathbf{s}(u, v)$, take on the same values as the respective parameters, and the axes are labeled accordingly in **Figure 9.19**. Each point on the plane $s_z = -1.5$ m, is associated with a (u, v)-pair, so this is referred to as the parameter plane. We could have designated any plane perpendicular to the s_z-axis as the parameter plane for the purpose of constructing the graph.

For the curved surface defined in (9.11) and illustrated in **Figure 9.19** we take the first partial derivative of $\mathbf{s}(u, v)$ with respect to u while holding v constant, to define the set of tangent vectors to the u-parameter curves. The first partial derivative of $\mathbf{s}(u, v)$ with respect to v, taken while holding u constant, defines the set of tangent vectors to the v-parameter curves. The two tangent vector functions are:

$$\begin{aligned} \frac{\partial \mathbf{s}}{\partial u} &= \frac{\partial s_x}{\partial u}\hat{\mathbf{i}} + \frac{\partial s_y}{\partial u}\hat{\mathbf{j}} + \frac{\partial s_z}{\partial u}\hat{\mathbf{k}} \\ \frac{\partial \mathbf{s}}{\partial v} &= \frac{\partial s_x}{\partial v}\hat{\mathbf{i}} + \frac{\partial s_y}{\partial v}\hat{\mathbf{j}} + \frac{\partial s_z}{\partial v}\hat{\mathbf{k}} \end{aligned} \tag{9.12}$$

Because the u- and v-parameter curves lie in the three-dimensional surface, $\mathbf{s}(u, v)$, these vectors are tangent to that surface. The tangent vectors defined in (9.12) are not necessarily unit vectors and they, like the components, have dimensions of length.

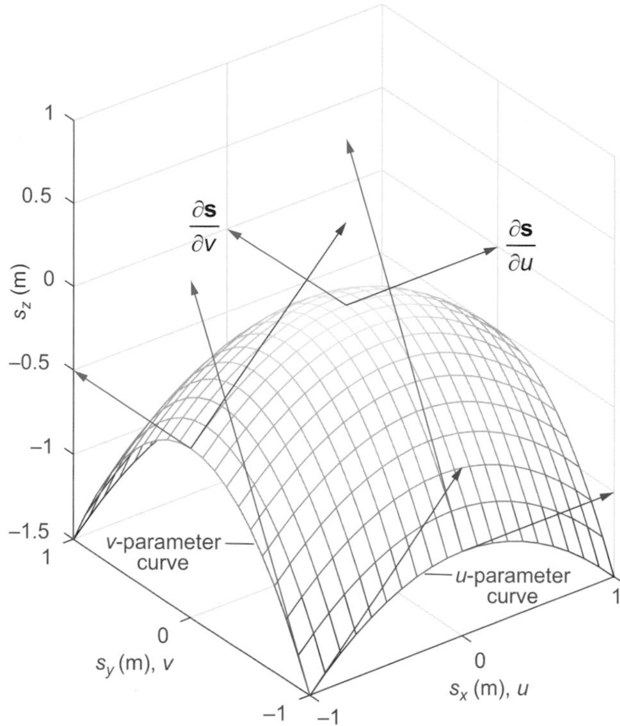

Figure 9.20 The elliptic paraboloid (idealized antiform) defined in (9.11) is represented here using the components of **s**(*u*,*v*) scaled in meters. Selected tangent vectors to the *u*-parameter curves (blue) and *v*-parameter curves (green) are defined using (9.12).

The vector function (9.11) has partial derivatives $\partial \mathbf{s}/\partial u = 1\hat{\mathbf{i}} - u\hat{\mathbf{k}}$ and $\partial \mathbf{s}/\partial v = 1\hat{\mathbf{j}} - 2v\hat{\mathbf{k}}$, which are the respective tangent vectors to the surface in the direction of the u- and v-parameter curves. A few examples of these tangent vectors are illustrated in **Figure 9.20**. Note that $\partial s/\partial u$ is independent of $\hat{\mathbf{j}}$ so the blue vectors lie in a plane perpendicular to the axis labeled s_y(m), v. Also, $\partial s/\partial v$ is independent of $\hat{\mathbf{i}}$ and therefore the green vectors lie in a plane perpendicular to the axis labeled s_x(m), u. Because the two tangent vectors at a particular point both are tangent to the surface, they may be used to define the unit normal vector to the surface at that point (Section 2.2.3), which is used when gathering orientation data in the field on folded surfaces.

The tangent vectors defined in (9.12) are directed parallel to the local u- and v-parameter curves (**Figure 9.20**), and the curvature of the surface in these directions is related to the curvature of these curves. However, to fully characterize the shape of a surface, we need to know the curvature of the surface in *all* directions at a given point. To specify a particular direction we use vectors that are tangent to the surface. In general, the tangent vector to an *arbitrarily directed* curve in a surface is defined as:

$$\mathbf{T} = \frac{\partial \mathbf{s}}{\partial u}\frac{\mathrm{du}}{\mathrm{dt}} + \frac{\partial \mathbf{s}}{\partial v}\frac{\mathrm{dv}}{\mathrm{dt}} \tag{9.13}$$

The two partial derivatives in (9.13) are the tangent vectors to the u- and v-parameter curves (**Figure 9.21**). These partial derivatives are scaled by the ordinary derivatives of the parameters u and v with respect to the parameter, t, along a line that is parallel to the projection of $\mathbf{T}$ onto the (u, v)-plane. On this plane, $\Delta u/\Delta t = \cos\theta$ and $\Delta v/\Delta t = \sin\theta$ are direction cosines that determine the orientation of the arbitrary curve. As the angle θ varies from 0 to $\pi/2$, the tangent vector varies smoothly and continuously from $\partial \mathbf{s}/\partial u$ to $\partial \mathbf{s}/\partial v$. Thus (9.13) describes all possible tangent vectors to the surface at a given point, some of which are shown in **Figure 9.21**.

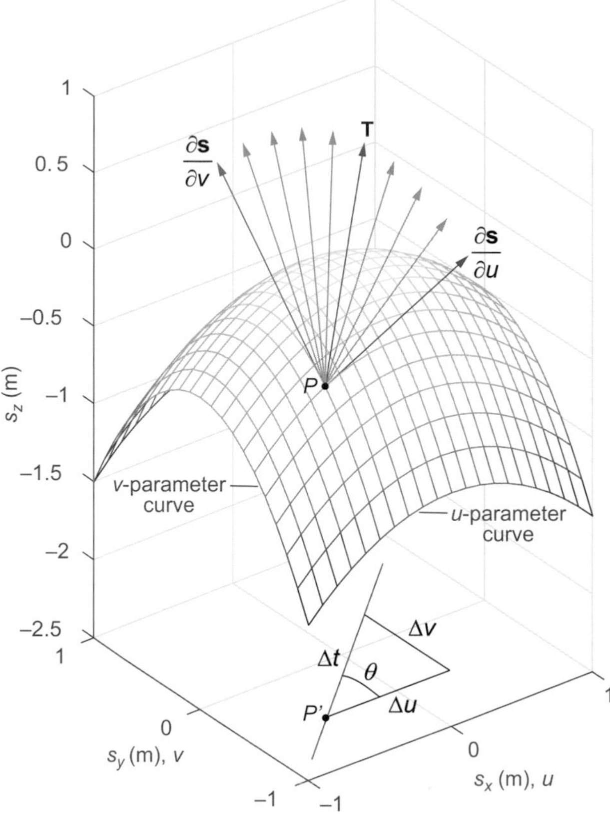

Figure 9.21 The antiformal surface defined in (9.11) with tangent vectors (9.12) at the point *P* for the *u*-parameter curve (blue) and the *v*-parameter curve (green). The tangent vector **T** (red) in an arbitrary direction is defined in (9.13). The point *P* projects onto the (*u*, *v*)-plane as *P*′ and the orientation of the arbitrary direction is determined by the angle θ in this plane.

9.5.3 Fundamental Forms of a Surface

To characterize the geometry of a folded surface, we focus attention on the neighborhood of a point and ask two questions:

(1) How is distance along the surface measured?
(2) What is the local shape of the surface?

The answer to each question provides an intrinsic property of the surface, and these are called the first and second fundamental forms of the surface. In this section we derive those forms and show how they are used to quantify and classify folded surfaces.

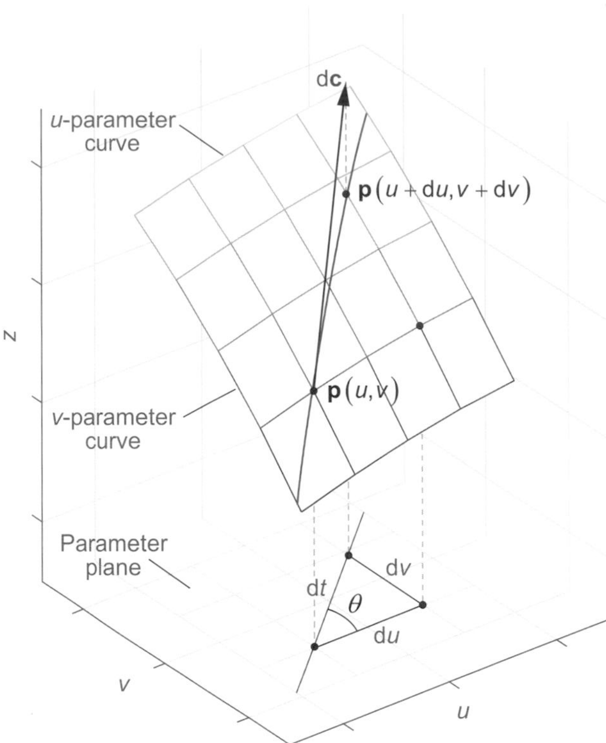

Figure 9.22 Neighborhood of the point **p**(u, v) on the surface **s** (u, v) as defined in (9.11), with a nearby point **p**(u + du, v + dv) on the highlighted curve (red). Differential quantities du, dv, and dt related to this curve are indicated on the parameter plane. The differential tangent vector in the direction of the curve is **dc**.

We derive the first and second fundamental forms using the antiformal surface illustrated in **Figure 9.21**. However, that figure portrays a *macroscopic* view of the surface **s**(u, v), as defined in (9.11) and here we focus on the *neighborhood* of a single point, **p**(u, v), using the differential quantities du and dv (**Figure 9.22**). A patch of the surface is shown in this figure with an exaggerated curvature and size, so the key geometric quantities can be distinguished. A few u- and v-parameter curves are shown, along with a highlighted curve (red) on the surface in an arbitrary direction that goes through the point **p**(u, v). A second point, **p**(u + du, v + dv), is a differential distance from **p**(u, v) along the highlighted curve. These two points project onto the parameter plane where they are separated by the distance dt. The tangent vector at the point **p**(u, v) in the direction of the highlighted curve is defined using (9.13) and is made a differential quantity by multiplying by dt:

$$\mathbf{dc} = \mathbf{T}\mathrm{d}t = \frac{\partial \mathbf{s}}{\partial u}\mathrm{du} + \frac{\partial \mathbf{s}}{\partial v}\mathrm{dv} \qquad (9.14)$$

This is the differential tangent vector, and it is parallel to **T** (**Figure 9.21**). The orientation of the highlighted curve is determined by $\theta = \tan^{-1}(\mathrm{d}v/\mathrm{d}u)$ on the (u, v)-plane.

The differential tangent vector in **Figure 9.22** is not directed exactly from **p**(u, v) to **p**(u + du, v + dv) because the highlighted curve is not a straight line. However, as the distance between the two points goes to zero, the differential tangent vector becomes parallel to the best-fitting line to the curve at **p**(u, v). In this limit the *magnitude* of the differential tangent vector, |**dc**|, is equal to the differential arc length of the highlighted curve. Because this curve lies in the surface **s**(u, v), |**dc**| measures arc length along the surface in the direction of the curve. As the direction of the curve changes with angle θ, the differential arc length may vary, and this variation is captured by the magnitude of the vector function defined in (9.14). This answers the first question we posed at the beginning of this section about how distance along the surface is measured.

The first fundamental form is defined as the scalar product of the differential tangent vector with itself:

$$\mathrm{I} = \mathbf{dc} \cdot \mathbf{dc} = E(\mathrm{d}u)^2 + 2F(\mathrm{d}u\mathrm{d}v) + G(\mathrm{d}v)^2 \qquad (9.15)$$

The scalar product of a vector with itself is equal to the squared magnitude of that vector, so the first fundamental form is equivalent to $|\mathbf{dc}|^2$: the differential arc length is the positive square root of I. The differential arc lengths along the surface in all possible directions at a given point are provided as du and dv vary in (9.15). The first fundamental form (9.15) is a differential quantity, and therefore is not calculated directly. Instead, it is used in *integration* formulae to calculate the finite *arc length* between two points on a surface, and also to calculate the finite *area* of a patch of a surface. Both of these calculations have important applications in structural geology, but are not pursued here (see Pollard and Fletcher, 2005, in Further Reading).

The coefficients E, F, and G of the first fundamental form in (9.15) are not differential quantities, and they are evaluated by using (9.14) to calculate the scalar products:

$$E = \frac{\partial \mathbf{s}}{\partial u} \cdot \frac{\partial \mathbf{s}}{\partial u}, \quad F = \frac{\partial \mathbf{s}}{\partial u} \cdot \frac{\partial \mathbf{s}}{\partial v}, \quad G = \frac{\partial \mathbf{s}}{\partial v} \cdot \frac{\partial \mathbf{s}}{\partial v} \qquad (9.16)$$

The two partial derivatives used in (9.16) are the tangent vectors to the u- and v-parameter curves that we defined in (9.12) and illustrated in **Figure 9.20**. This should make it clear that one needs to consider two families of curves to characterize the geometry of a surface. The coefficients are scalar quantities with dimensions of length squared, for example: $F\{=\}\mathrm{L}^2$ and $F[=]\mathrm{m}^2$. The parametric representation of the surface, **s**(u, v), must yield first partial derivatives to calculate these coefficients.

To quantify the *shape* of a folded surface we use the second fundamental form, which depends upon local changes in the orientation of the surface as determined by its unit normal vector. Recall from Section 2.2.3 that the vector product of two non-parallel vectors defines the normal vector to the plane in which the two vectors lie. Here we use the tangent vectors, $\partial\mathbf{s}/\partial u$ and $\partial\mathbf{s}/\partial v$, to the u- and v-parameter curves (9.12) and divide their vector product by the magnitude of that product to find the vector function for the unit normal vector to the surface:

$$\hat{\mathbf{N}} = \left(\frac{\partial \mathbf{s}}{\partial u} \times \frac{\partial \mathbf{s}}{\partial v}\right) \Big/ \left|\frac{\partial \mathbf{s}}{\partial u} \times \frac{\partial \mathbf{s}}{\partial v}\right| \qquad (9.17)$$

This is the general equation for the set of unit normals to the surface **s**(u, v) as defined in (9.10). For the antiformal surface

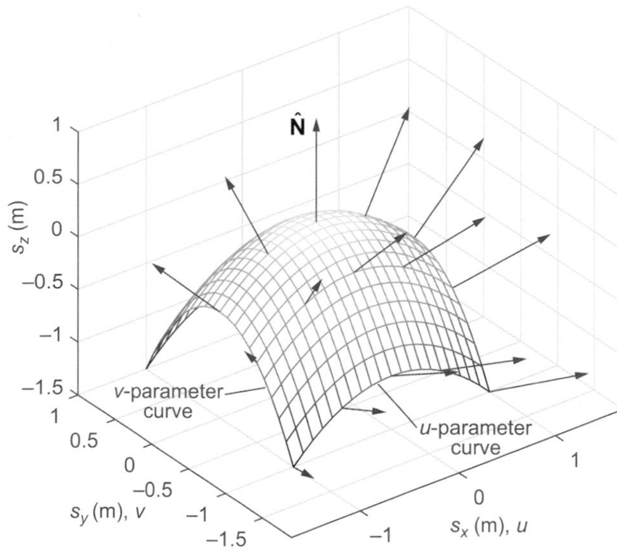

Figure 9.23 Unit normal vectors (blue) to the antiformal surface defined in (9.11) are plotted using (9.17) at selected intersections of *u*- and *v*-parameter curves.

illustrated in Figure 9.19 and defined in (9.11), the vector function for the unit normal is $\hat{\mathbf{N}} = \left(u\hat{\mathbf{i}} + 2v\hat{\mathbf{j}} + \hat{\mathbf{k}}\right) / \left|u\hat{\mathbf{i}} + 2v\hat{\mathbf{j}} + \hat{\mathbf{k}}\right|$. A selection of these normal vectors at the intersections of *u*- and *v*-parameter curves on the antiformal surface are illustrated in Figure 9.23.

The local change in direction of the unit normal vector with distance along the surface is a measure of the change in orientation of the surface itself. This can be appreciated on Figure 9.23, for example, by noting how the unit normal vectors change direction along a *u*- or *v*-parameter curve, and how the surface changes orientation accordingly. The local change in $\hat{\mathbf{N}}$ at any point on the surface is quantified using the differential normal vector:

$$\mathrm{d}\mathbf{N} = \frac{\partial \hat{\mathbf{N}}}{\partial u}\mathrm{d}u + \frac{\partial \hat{\mathbf{N}}}{\partial v}\mathrm{d}v \tag{9.18}$$

Because the vectors $\hat{\mathbf{N}}$ have constant (unit) magnitudes, each vector $\mathrm{d}\mathbf{N}$ is perpendicular to the associated unit normal vector. Thus, at any point on the surface the differential normal vector (9.18) and the differential tangent vector (9.14) are parallel to the tangent plane to the surface, but they are not necessarily parallel to each other.

To capture the component of $\mathrm{d}\mathbf{N}$ in the direction $\mathrm{d}\mathbf{c}$, the second fundamental form is defined using the scalar product of these two vector functions:

$$\mathrm{II} = -\,\mathrm{d}\mathbf{N} \cdot \mathrm{d}\mathbf{c} = L(\mathrm{d}u)^2 + 2M(\mathrm{d}u\mathrm{d}v) + N(\mathrm{d}v)^2 \tag{9.19}$$

The second fundamental form is a differential quantity, and therefore is not calculated directly. However, the coefficients *L*, *M*, and *N* in (9.19) are finite quantities that are used to characterize the shape of surfaces. These coefficients are found by taking the scalar product of (9.18) and (9.14):

$$L = \hat{\mathbf{N}} \cdot \frac{\partial^2 \mathbf{s}}{\partial u^2}, \quad M = \hat{\mathbf{N}} \cdot \frac{\partial^2 \mathbf{s}}{\partial u \partial v}, \quad N = \hat{\mathbf{N}} \cdot \frac{\partial^2 \mathbf{s}}{\partial v^2} \tag{9.20}$$

The coefficients are scalars with dimensions of length, for example: $M\{=\}\mathrm{L}$ and $M[=]\mathrm{m}$.

In the neighborhood of every point on a folded surface the shape can be categorized using the coefficients of the second fundamental form (9.20), organized as follows:

$$\begin{aligned} &L = M = N = 0, \text{planar point} \\ &LN - M^2 = \begin{cases} > 0, \text{elliptic point} \\ = 0, \text{parabolic point} \\ < 0, \text{hyperbolic point} \end{cases} \end{aligned} \tag{9.21}$$

This categorization includes the planar point, for which all the coefficients are zero. The shape near an elliptic point looks like a *basin* (Figure 9.24a); that near a parabolic point looks like a cylindrical *synform* (Figure 9.24b); and that near a hyperbolic point looks like a *saddle* (Figure 9.24c).

The four shapes defined in (9.21) cover *all* the possible second-order shapes for a curved surface, and therefore form the basis for categorizing such surfaces, including folds. For a mathematician that is the end of the story, but for a geologist it is important to acknowledge the orientation of the surface with respect to Earth's surface and Earth's gravitational field. This is accomplished using the gravitational acceleration vector to define *down* and choosing a *vertical* coordinate axis that is parallel to this vector. Throughout this text we have chosen the *z*-axis to be vertical and positive upward. With this added sense of up and down, the shape near an elliptic point (Figure 9.24a) can be distinguished as a basin or a dome, and the shape near a parabolic point (Figure 9.24b) can be distinguished as a cylindrical synform or antiform.

In all three of the renderings of the second-order shapes (Figure 9.24), some of the *u*- and *v*-parameter curves are drawn. These curves help to distinguish the different shapes. For example, both the *u*- and *v*-parameter curves for the elliptic point are concave upward in Figure 9.24a. For the parabolic point in Figure 9.24b the *u*-parameter curves are concave upward and the *v*-parameter curves are straight lines. For the hyperbolic point in Figure 9.24c the *u*-parameter curves are concave upward and the *v*-parameter curves are convex upward. These illustrations demonstrate that where a *u*- and *v*-parameter curve cross, the curvature in different orientations can be different. In the next section we look more closely at how the curvature of a surface varies with orientation at a given point and describe a classification scheme for folds based on curvature.

9.5.4 Geological Curvature Classification for Folds

The curvature of the base of the Jurassic Sundance formation (Figure 9.18), varies systematically from location to location on the fold. For example, perpendicular to the hinge of the anticline the curvature is greater in the foreground of this image than in the background. At a given location on the fold the curvature also varies systematically with orientation. For example, where the hinge intersects the edge of the image in the foreground, the curvature is much greater perpendicular to the hinge than parallel to the hinge. To quantify such variations in curvature we introduce the *shape operator*, which is a function of the coefficients of the first and second fundamental forms. Using the shape operator

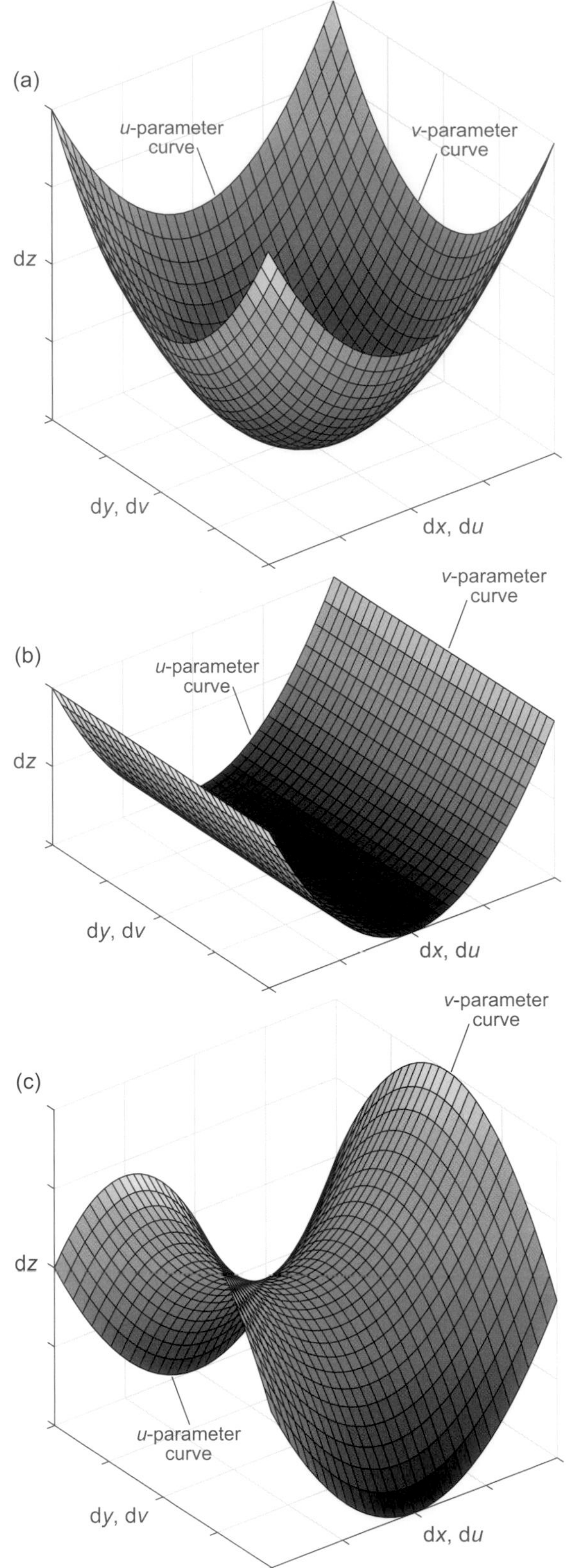

Figure 9.24 Surface shapes near a point, drawn with *u*- and *v*-parameter curves and colors scaled to elevation. Axes are differential quantities because this represents the shape in the neighborhood of a point, and the curvatures are greatly exaggerated. (a) Elliptic point. (b) Parabolic point. (c) Hyperbolic point. Planar point not shown.

we show how to calculate *principal values* and *orientations* of curvature, and we develop a geological classification for folds based on curvature.

The coefficients of the first and second fundamental forms are defined in (9.16) and (9.20) in terms of partial derivatives of the *u*- and *v*-parameter curves on the surface. The shape operator is defined by arranging these coefficients into two symmetric 2×2 matrices, and multiplying them as follows:

$$[\mathrm{O}] = \begin{bmatrix} E & F \\ F & G \end{bmatrix}^{-1} \begin{bmatrix} L & M \\ M & N \end{bmatrix} \tag{9.22}$$

The shape operator, [O], is a 2×2 matrix. Recall from the discussion in Section 2.6.5 that two-dimensional symmetric matrices, such as the stress or strain tensors, have principal values (greatest and least magnitudes) and principal axes. In the case of curvature, the principal values and principal axes are given by the eigenvalues and eigenvectors of the shape operator (9.22):

$$[\mathrm{D}] = \begin{bmatrix} \kappa_2 & 0 \\ 0 & \kappa_1 \end{bmatrix}, \; [\mathrm{V}] = \begin{bmatrix} m_x & l_x \\ m_y & l_y \end{bmatrix} \tag{9.23}$$

The elements on the principal diagonal of the matrix [D] are the principal values, that is the two extreme curvature magnitudes, κ_2 and κ_1. The elements in the columns of the full matrix [V] are the components of two unit vectors, $\hat{\mathbf{m}}$ and $\hat{\mathbf{l}}$, that are parallel to the principal axes of curvature *on* the parameter plane.

To provide an example, we return to the elliptic paraboloid defined in (9.11) and illustrated in **Figure 9.19**. The coefficients of the first and second fundamental forms for this antiformal surface are:

$$E = 1 + u^2, F = 2uv, G = 1 + 4v^2$$
$$L = -1/\left(u^2 + 4v^2 + 1\right)^{1/2}, M = 0, N = -2/\left(u^2 + 4v^2 + 1\right)^{1/2}$$

At the point $(u, v) = (0, 0)$ we have $E = 1$, $F = 0$, $G = 1$, $L = -1$, $M = 0$, and $N = -2$, so the shape operator is:

$$[\mathrm{O}] = \begin{bmatrix} 1 & 0 \\ 0 & 1 \end{bmatrix}^{-1} \begin{bmatrix} -1 & 0 \\ 0 & -2 \end{bmatrix} = \begin{bmatrix} -1 & 0 \\ 0 & -2 \end{bmatrix}$$

The eigenvalues and eigenvectors are:

$$[\mathrm{D}] = \begin{bmatrix} -2 & 0 \\ 0 & -1 \end{bmatrix}, \; [\mathrm{V}] = \begin{bmatrix} 0 & 1 \\ 1 & 0 \end{bmatrix}$$

The principal values at this point are $\kappa_1 = -1\ \mathrm{m}^{-1}$ and $\kappa_2 = -2\ \mathrm{m}^{-1}$. The principal axes are aligned, respectively, with the *u*- and *v*-axes on the parameter plane.

The variations of the principal curvatures in magnitude and orientation over the antiformal surface are illustrated in **Figure 9.25**. The magnitudes are indicated by color and they are negative everywhere because the unit normal vectors are directed upward (**Figure 9.23**) and the curvature vectors are directed downward. The magnitudes are most negative near the crest of the fold, and they become less negative on the limbs. The directions are indicated by the short tic marks at the intersections of the u- and v-parameter curves. The greatest principal curvature, κ_1, is parallel to the *u*-axis at the crest of the antiform,

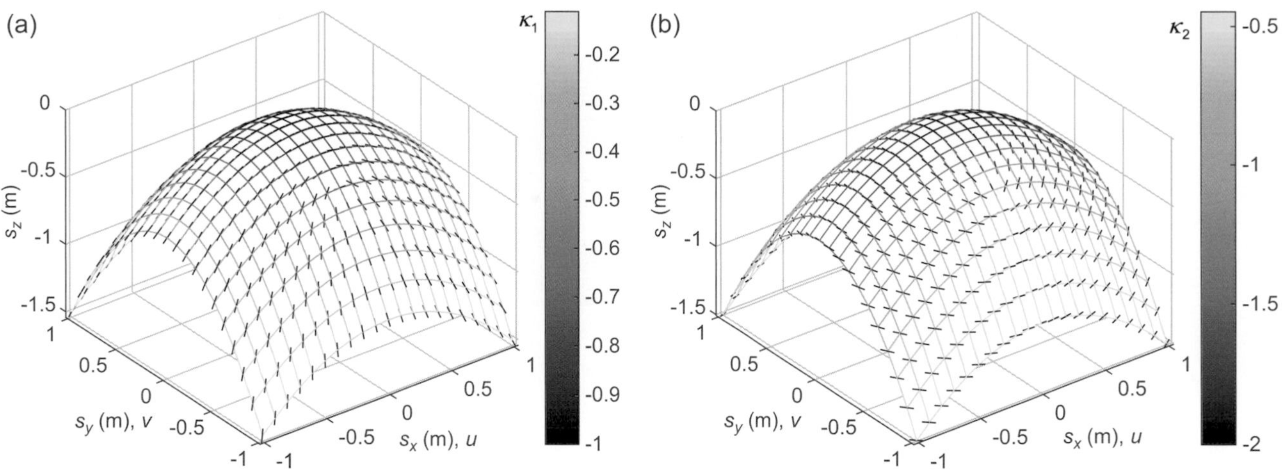

Figure 9.25 The elliptic paraboloid defined in (9.11) is represented here by a graph using the components of **s**(*u*,*v*). A selection of *u*- and *v*-parameter curves lie in that surface, which looks like an idealized antiform. (a) Tic marks are aligned with the orientation of the greatest curvature and colors correspond to values of κ_1. (b) Tic marks are aligned with the orientation of the least curvature and colors correspond to values of κ_2.

but rotates to become parallel to the *v*-axis on the limbs. The least principal curvature, κ_2, is perpendicular to the greatest principal curvature at all points on the surface.

The last step in sorting out how to describe three-dimensional folded surfaces is to define Gaussian curvature and the mean normal curvature using the principal normal curvatures:

$$\kappa_g = \kappa_1 \kappa_2, \quad \kappa_m = \tfrac{1}{2}(\kappa_1 + \kappa_2) \tag{9.24}$$

The Gaussian curvature, κ_g, is the product of the principal curvatures, and as the name implies, the mean normal curvature, κ_m, is the average value of the principal curvatures. These two measures of curvature are used to classify the different shapes of three-dimensional folded geologic surfaces by noting whether they are negative, zero, or positive (Figure 9.26). Negative Gaussian curvature is associated with hyperbolic shapes (saddles); zero Gaussian curvature is associated with parabolic and planar shapes (synforms, planes, and antiforms); and positive Gaussian curvature is associated with elliptic shapes (basins and domes). Distinguishing between negative and positive mean normal curvature separates basins from domes, synforms from antiforms, and synformal from antiformal saddles.

We return now to the folded surface that represents the base of the Sundance Formation at Sheep Mountain, WY (Figure 9.18). The partial derivatives necessary to calculate the coefficients of the first and second fundamental forms from (9.16) and (9.20) were calculated numerically using the elevation data for this surface. Then the shape operator (9.22) was calculated and the eigenvalue problem solved to determine the principal normal curvatures. Finally, the Gaussian and mean normal curvatures were calculated (9.24) and the appropriate color was assigned to that grid square using the classification table in Figure 9.26.

All of the colors assigned using the classification table (Figure 9.26) were plotted on the three-dimensional rendering of the surface to create Figure 9.27. If Sheep Mountain anticline were a cylindrical fold, all of the grid squares in Figure 9.27 would be filled with the same color, the light blue representing antiforms from Figure 9.26. Each of the different colors is associated with a unique shape, so this plot demonstrates how the folded surface changes shape from one grid square to another. Along the hinge of the anticline the shapes are predominantly domes and antiformal saddles. Along the hinge of the syncline to the northeast of Sheep Mountain the shapes are predominantly basins and synformal saddles.

Inspection of the folded surface in Figure 9.27 reveals that the only colors present are from the corners of the classification table (Figure 9.26), representing synformal and antiformal saddles, basins, and domes. The shapes along the central row or central column of the table are absent. The reason for their absence is that either the Gaussian or mean normal curvatures, or both curvatures, must be identically zero. Because the rendering of the surface involved actual geologic and topographic data, and not exact mathematical functions, it is not surprising that there were no zeros. On the other hand, some of the calculated curvatures were quite small numbers that could not be defended as different from zero, given the imprecision of the data. Taking this imprecision into account, and setting small curvatures to zero, produces a colorized rendering of the surface that includes the shapes along the central row and column of the table.

In this section we have introduced the concepts from differential geometry that are necessary for the quantitative characterization of the geometry of folded surfaces in three dimensions. The key concept is that curves lying in the surface provide a measure of the curvature, which can be used to calculate the normal curvature of the surface in all directions at any point. The variation of normal curvature with orientation ranges from the maximum to the minimum principal curvature. These principal values and the principal orientations are found by solving an eigenvalue

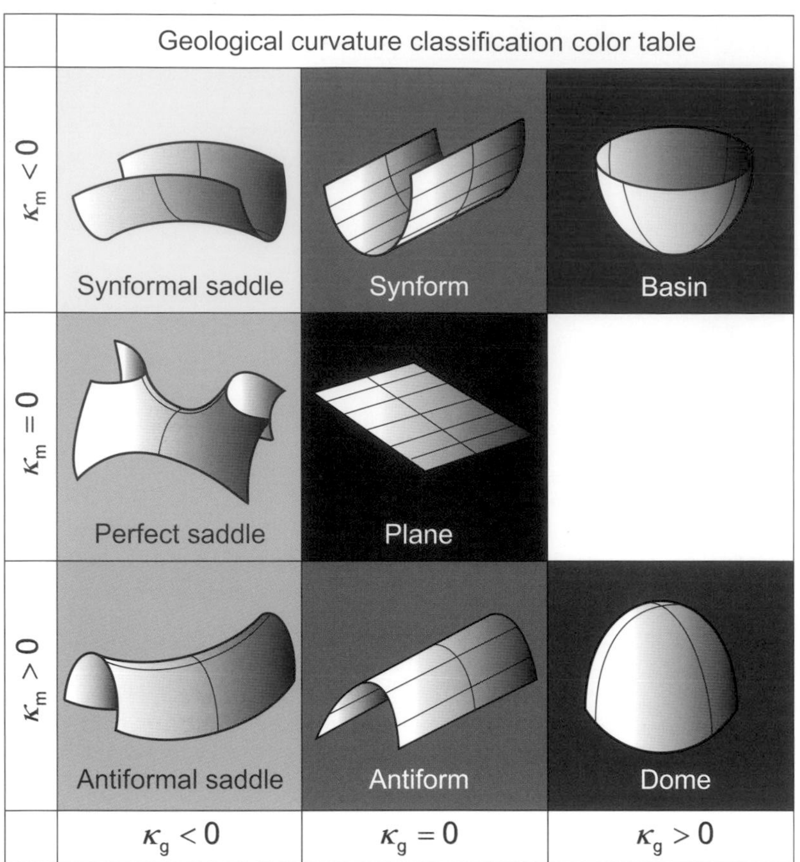

Figure 9.26 The signs of the Gaussian curvature, κ_g, and mean normal curvature, κ_m, are used to classify folded geological surfaces. Refer to **Figure 9.27** for an application of this color table to the fold at Sheep Mountain. Modified from Mynatt et al. (2007), Figure 2.

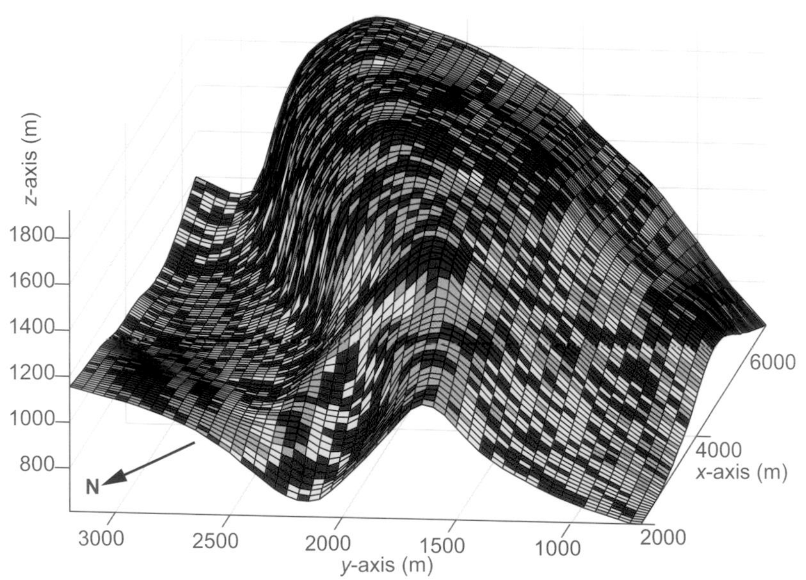

Figure 9.27 Sheep Mountain fold visualized using the surface at the base of the Sundance formation and viewed toward the southeast. See **Figure 9.15** for the stratigraphic section and **Figure 9.13** for location. Color in each grid square depends upon the signs of the Gaussian and mean normal curvatures computed using (9.24) and taken from the color table in **Figure 9.26**. Modified from Mynatt et al. (2007), Figure 5.

problem that uses the coefficients of the first and second fundamental forms for the surface, organized in two symmetric matrices. From the principal curvatures, the Gaussian and mean normal curvature are calculated, and they are used to classify all possible second-order fold shapes.

9.6 A CANONICAL MODEL FOR BENDING

In this and the next section we focus on two mechanisms, bending and buckling, that account for many folds observed in nature. Recall from the discussion of **Figure 9.1** that bending results from application of forces perpendicular to the plane of a layer, whereas buckling results from application of forces parallel to that plane. For each of these mechanisms we develop a canonical model that introduces the basic physical concepts and constitutive laws used to derive the governing equations. In both cases we describe a geologic structure that conforms, in some measure, to the conditions invoked for the canonical model, and we use that model to deduce insights about the formation of the structure.

9.6.1 Model Geometry

In the introduction to this chapter we described how you could *bend* a layer by applying forces *perpendicular* to its plane. The horizontal sedimentary strata in the Henry Mountains (**Figure 9.28**) apparently were pushed upward by vertical forces due to the magma pressure in an underlying intrusion called a laccolith. We use the laccolith to motivate an analysis of the physics of bending of thin layers.

The sedimentary domes of the Henry Mountains were interpreted by Gilbert as underlain by large laccoliths (**Figure 9.28**), although no lower contacts of the intrusion and no feeder dikes are exposed. At Mount Holmes (**Figure 9.29**) prominent outcrops of Navajo Sandstone dip away from the summit on the flanks of the mountain and are nearly horizontal at the top, where the reddish and buff-colored sedimentary rocks are cut by vertical dikes (gray). No upper contact of the inferred laccolith is exposed, but the Navajo Sandstone has been bent upward approximately 1.2 km relative to its elevation beyond the edge of the dome. The oldest tilted sedimentary unit exposed on Mount Holmes is the Triassic Chinle Formation, so the upper contact of the laccolith is inferred to be below the Chinle Formation.

The dome at Mount Holmes is roughly circular in map view, but we begin this investigation by considering the somewhat simpler, two-dimensional problem of a layer that bends into an anticline. This is generalized later to the three-dimensional problem of a circular layer that bends into a dome similar to that at Mount Holmes. To set up the two-dimensional model geometry (**Figure 9.30a**) we place the origin of the (x, z)-coordinate system at the center of a layer of length $2a$ and thickness h. The x-axis lies along the middle surface of the layer and the z-axis is vertical. Every cross section of the layer that is perpendicular to the y-axis is identical, and the layer is very long in the y-direction compared to $2a$, so we refer to the model fold is *cylindrical*. The model is based on postulates that depend upon the layer being *thin* relative

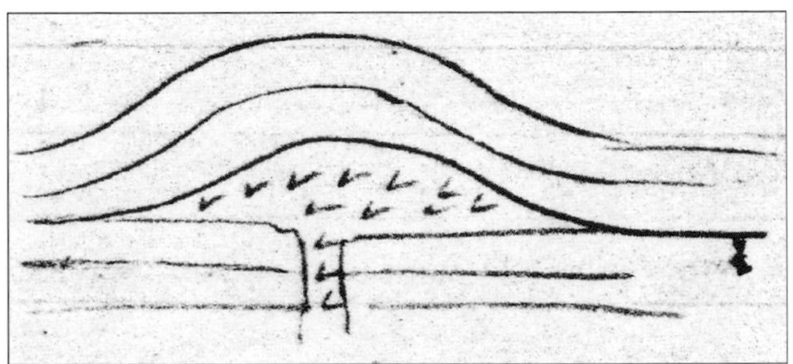

Figure 9.28 Interpretive sketch from field Note Book No. 3, p. 51, drawn on Sunday, Aug. 22, 1875, by G. K. Gilbert of the Powell Survey in the southern Henry Mountains. According to Gilbert's concept, the magma (checked) rose through the underlying sedimentary layers in a dike, insinuated between two layers as a sill, and then pushed up the overlying strata to form a laccolith. See Gilbert (1877).

Figure 9.29 Modern photograph looking south at Mount Holmes in the southern Henry Mountains from near Gilbert's campsite on August 22, 1875 where he conceived the concept of the laccolith (**Figure 9.28**). UTM: 12 S 531634.71 m E, 4187348.71 m N. See Jackson and Pollard (1988).

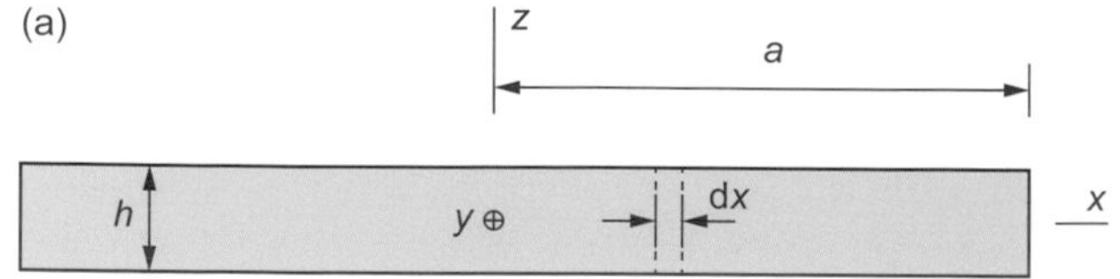

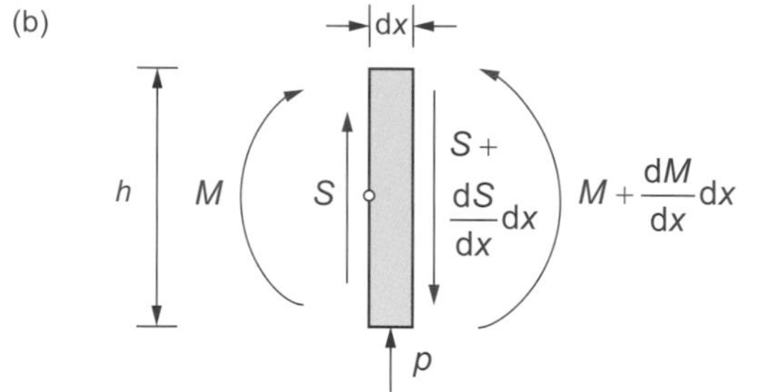

Figure 9.30 (a) A layer of length 2*a* and thickness *h*. (b) An element of width dx is isolated from this layer and shear forces, *S*, and bending moments, *M*, are assigned that vary from the left to the right cross section. A pressure, *p*, acts on the base of the layer.

to its length, so we consider layers with a ratio of thickness to length, $h/a < 1/10$. In the engineering literature this is called thin-plate theory, and we adopt the basic concepts presented in a classic textbook on the subject by Timoshenko and Woinowsky-Krieger (1959), which is listed in Further Reading.

For the canonical problems introduced in earlier chapters we began by invoking Cauchy's laws of motion to develop the governing equations (Section 7.2). Here we take a different approach, starting with an analysis of the forces and moments acting on cross sections of the bending layer. For static equilibrium, the sum of these forces in each coordinate direction must be zero, and the sum of the moments about each coordinate axis must be zero. At the end of Section 9.6 we circle back and show that the distribution of stress in the bending layer satisfies Cauchy's laws. This approach is taken because it introduces the classical methods used to study bending, and because it clarifies important relationships among the forces, moments, and stresses.

9.6.2 Equilibrium of Forces and Moments

A uniform pressure, *p* (force per unit area), acts in the positive *z*-direction on the base of the layer. For a horizontal sedimentary layer this pressure could be due, in part, to the weight of the overlying layers, and in part to the magma pressure in an underlying sill or laccolith. In this case, if p_r is the lithostatic pressure of the rock and p_m is the magma pressure, the pressure driving upward bending of the layer is $p = p_m - p_r \geq 0$. We also suppose that no shear tractions act on the top and bottom surfaces of the layer.

As the layer bends, shear forces, *S*, and bending moments, *M*, are induced on cross sections of the layer (**Figure 9.30b**). These are forces and moments *per unit length* perpendicular to the (*x*, *z*)-plane for this two-dimensional analysis. To analyze these forces and moments we isolate a very narrow element of the layer (**Figure 9.30b**) that has a differential width, d*x*, but includes the full height, *h*, of the layer. Anticipating that these forces and moments vary along the length of the layer, we assign the shear force *S* to the left side of the element, and $S + (\mathrm{d}S/\mathrm{d}x)\mathrm{d}x$ to the right side. Similarly, we assign the bending moment *M* to the left side and $M + (\mathrm{d}M/\mathrm{d}x)\mathrm{d}x$ to the right side.

The deformation of the layer is constrained by the static equilibrium of forces and moments according to Newton's laws of motion. To establish equilibrium, we sum the forces acting in the *z*-direction and set that sum to zero:

$$\sum F_z = 0 = S + p\mathrm{d}x - S - (\mathrm{d}S/\mathrm{d}x)\mathrm{d}x$$

The consequence of this balance of forces is that the variation in shear forces with position along the layer is equal to the applied uniform pressure acting at the base of the layer:

$$\frac{\mathrm{d}S}{\mathrm{d}x} = p \tag{9.25}$$

The moments are summed about a point where the middle surface of the layer intersects the left cross section (small circle on **Figure 9.30b**). Moments due to applied forces are the product of the force and the distance in *x* from the force to the designated point. Setting the sum of moments to zero we have:

$$\sum M = 0 = M + S\mathrm{d}x - M - (\mathrm{d}M/\mathrm{d}x)\mathrm{d}x + (\mathrm{d}S/\mathrm{d}x)\mathrm{d}x\mathrm{d}x - p\mathrm{d}x\mathrm{d}x/2$$

The last two terms are ignored because the products d*x*d*x* make them insignificant relative to the other terms. The consequence of this balance of moments is that the spatial variation in moments across the element is equal to the shear force:

$$\frac{\mathrm{d}M}{\mathrm{d}x} = S, \text{ so } \frac{\mathrm{d}^2M}{\mathrm{d}x^2} = p \tag{9.26}$$

The relationship between the second derivative of the moment and the pressure is found by substituting the first of (9.26) into (9.25). These relationships among the applied pressure, the shear force, and the bending moment guarantee that the layer remains in *equilibrium*.

9.6.3 Kinematics of Pure Bending

The kinematics of the bending layer are developed using the concept of *pure* bending. A rectangular (undeformed) element of the layer is isolated as we did to consider the static equilibrium (**Figure 9.30b**). This element has a height, *h*, equal to the entire height of the layer, but a width that is the differential quantity, d*x* (**Figure 9.31a**). For the bending layer we derive expressions for the strain of two material lines that parallel the *x*-axis. The *reference line* is at mid-height in the layer and the other line is at an arbitrary height, *z*, above or below the reference line. Both of these material lines have an initial length d*x* and both are straight. For a material line below the reference line, *z* is negative. Pure bending means that these straight lines change shape to become circular arcs.

For pure bending, lines that parallel the *x*-axis do not distort due to shearing along the cross sections of the layer. This appears

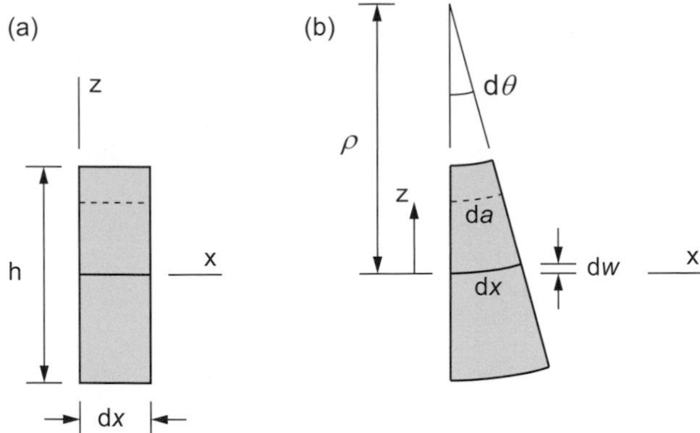

Figure 9.31 (a) An undeformed element with height h and width dx isolated from the thin layer (**Figure 9.30**a). (b) The element deformed by pure bending. The radius of curvature of the circular arc dx along the middle surface of the layer is ρ. See text for discussion of pure bending.

to contradict what we learned about the statics of bending (**Figure 9.30**), where we considered shear forces, S, to act on layer cross sections. These shear forces correspond to shear stresses, σ_{xz}, acting on the cross sections, which cause shear strains, ε_{xz}, that do distort the material lines. However, as we show below, the bending moments, M, correspond to normal stresses, σ_{xx}, which cause normal strains, ε_{xx}. These normal strains are the dominant strains within a bending layer that is *thin* compared to its length. Invoking pure bending means that the contribution of the shear strains to layer distortion are not included.

Recall from Section 4.9.2 that the small normal strain for a material line initially along the x-axis is defined as the fractional change in length. For the undeformed element shown in **Figure 9.31a** the original length of all material lines parallel to the x-axis is the differential quantity, dx. After pure bending the reference line retains its original length (no normal strain), but the line at height z has a new length da, so the small normal strain is $\varepsilon_{xx} = (\mathrm{d}a - \mathrm{d}x)/\mathrm{d}x$. Because both of these material lines are circular arcs, their lengths can be related to the radius of curvature, ρ, and the differential angle, dθ, between the projected sides of the deformed element. The lengths are $\mathrm{d}x = \rho\mathrm{d}\theta$ and $\mathrm{d}a = (\rho - z)\mathrm{d}\theta$. Using these expressions for the arc lengths, the normal strain for the line at z is:

$$\varepsilon_{xx} = [(\rho - z)\mathrm{d}\theta - \rho\mathrm{d}\theta]/\rho\mathrm{d}\theta = -z/\rho = -z\kappa \quad (9.27)$$

In the last step we used the fact that the radius of curvature is the inverse of the scalar curvature, κ, for a plane curve (Section 9.3.1). The normal strain parallel to the x-axis is proportional to the distance from the middle surface of the layer, z, and to the curvature of that surface. For a positive curvature (concave upward), the normal strain is negative (shortening) for positive z, and it is positive (lengthening) for negative z.

The curvature for a plane curve described by $y = f(x)$ is given in (9.2). For the bending layer we call the displacement of the middle surface of the layer in the z-direction w, and write $w = f(x)$. Invoking the restriction that the slope of the layer is small compared to one, the curvature is approximated using only the numerator in (9.2). For slopes less than 15° this introduces errors of less than 10% in calculations of curvature. For the element in **Figure 9.31b** the differential displacement of the middle surface is dw and the approximation for the curvature of that surface is $\kappa \approx \mathrm{d}^2w/\mathrm{d}x^2$. Using (9.27) the normal strain at a distance z from the middle surface is:

$$\varepsilon_{xx} = -z\frac{\mathrm{d}^2w}{\mathrm{d}x^2} \quad (9.28)$$

For concave upward bending (positive second derivative) material lines shorten above the middle surface and lengthen below the middle surface in proportion to their distance from that surface (**Figure 9.31**). The middle surface is called the neutral surface because it does not change length, so the normal strain is zero.

9.6.4 Constitutive Law and Flexural Rigidity

The normal strain (9.28) is related to the normal stress using a linear elastic constitutive law to approximate the material behavior of sedimentary strata in the upper lithosphere (Section 4.4). Also, for thin layers in the (x, y)-plane, changes in the normal stress σ_{zz} are negligible compared to changes in the normal stresses σ_{xx} and σ_{yy} acting parallel to the layer. Recalling Hooke's law for linear elasticity from Section 4.10, the normal strains are $\varepsilon_{xx} = (\sigma_{xx} - \nu\sigma_{yy})/E$ and $\varepsilon_{yy} = (\sigma_{yy} - \nu\sigma_{xx})/E$. Here E is Young's modulus and ν is Poisson's ratio. Because the layer is very long in the y-direction we invoke the condition of plane strain such that $\varepsilon_{yy} = 0$, so $\sigma_{yy} = \nu\sigma_{xx}$ and $\varepsilon_{xx} = \sigma_{xx}(1 - \nu^2)/E$. Thus, the normal stress acting parallel to the x-axis is related to the corresponding normal strain (9.28) as:

$$\sigma_{xx} = \left(\frac{E}{1 - \nu^2}\right)\varepsilon_{xx} = B\varepsilon_{xx} = -Bz\frac{\mathrm{d}^2w}{\mathrm{d}x^2} \quad (9.29)$$

Here we define a new elastic modulus, B, for bending as the combination of elastic constants in parentheses. In the last step of (9.29) we use (9.28) to substitute for the strain. The normal stress varies linearly across the layer, and it is proportional to the elastic modulus, B, and the approximation for the curvature (Section 9.3.1).

The normal stress (9.29) gives rise to the bending moment, M, because the stress is compressive (pushes against the element side) above the neutral surface and tensile (pulls on the element side) below the neutral surface (**Figure 9.32**). The bending moment is the integral over the cross section of the force per unit length, $-\sigma_{xx}\mathrm{d}z$, times the lever arm for this force, z:

$$M = -\int_{-h/2}^{+h/2} \sigma_{xx} z\,\mathrm{d}z = B\frac{\mathrm{d}^2w}{\mathrm{d}x^2}\int_{-h/2}^{+h/2} z^2\mathrm{d}z$$

In the second step we use (9.29) to substitute for the stress. The integral evaluates as:

$$\int_{-h/2}^{+h/2} z^2\mathrm{d}z = \frac{1}{3}z^3\Big|_{-h/2}^{+h/2} = h^3/12$$

This integral depends upon the cross-sectional shape of the layer: the value found here applies to a *rectangular* cross section. Completing the evaluation, we find:

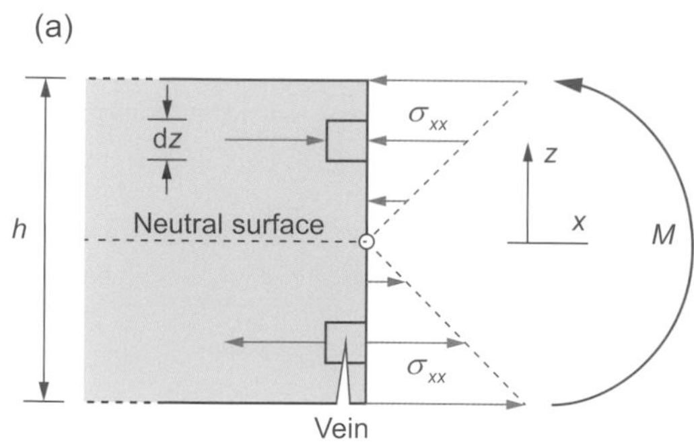

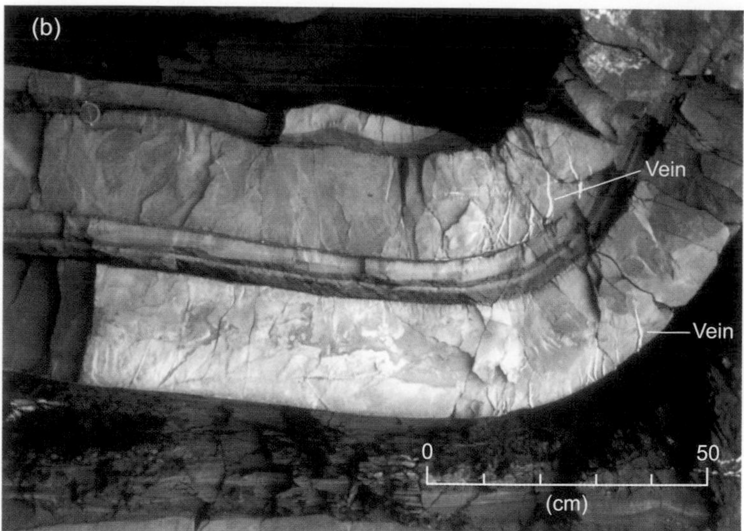

Figure 9.32 (a) Cross section of a bending layer with a linear distribution of normal stress, σ_{xx}, that goes to zero at the neutral surface (9.32). The resultant of this distribution of normal stress is the bending moment M. (b) Two folded limestone beds with calcite-filled veins concentrated in the outer arc of their hinges and penetrating upward toward the respective neutral surfaces.

$$M = \frac{Bh^3}{12}\frac{d^2w}{dx^2} = R\frac{d^2w}{dx^2} \tag{9.30}$$

The bending moment is proportional to the elastic modulus, the curvature of the bent layer, and the thickness cubed. The relationship to the thickness has profound implications, because small differences in thickness produce large differences in moment. In the second step we introduce the flexural rigidity of the layer:

$$R = Bh^3/12 \tag{9.31}$$

The name of this quantity implies that it is a measure of the resistance to flexure (bending): the layer resists bending in proportion to the elastic modulus and the thickness cubed.

Combining (9.29) and (9.30) we find the relationship between the normal stress on cross sections of the layer and the bending moment acting on that cross section:

$$\sigma_{xx} = -12Mz/h^3 \tag{9.32}$$

The normal stress varies linearly with distance, z, from the neutral surface (**Figure 9.32a**), and it is proportional to the bending moment, and is inversely proportional to the thickness cubed of the layer. A positive bending moment is the resultant of normal stresses that are compressive above and tensile below the neutral surface.

Stresses induced by bending may perturb the stress state enough to initiate secondary structures, such as opening fractures. In **Figure 9.32b** two limestone beds have calcite-filled veins that initiated on the lower bedding surfaces and propagated upward, but terminated short of the middle of these beds. Both the initiation and termination locations are consistent with the state of stress illustrated in **Figure 9.32a**, which shows that the tensile stress due to bending is greatest at the lower surface and decreases to zero at the middle of the layer. Also, the veins are concentrated in the hinge region of these folds, where the curvature of the beds in greatest, and they are nearly absent from the limbs where the curvature is least. This is consistent with the relationship in (9.29) that shows the stress is proportional to the curvature.

9.6.5 Governing Equation and its Solution

Returning to the bending of strata over a laccolith, and substituting the equation for the bending moment (9.30) into the equation relating the moment to the applied pressure (9.26), we find the governing equation for two-dimensional bending of a thin elastic layer subject to a uniform pressure acting perpendicular to the layer:

$$\frac{d^4w}{dx^4} = \frac{12p}{Bh^3} = \frac{p}{R} \tag{9.33}$$

Solving this fourth order ordinary differential equation requires four integrations with respect to x, and use of appropriate symmetry and boundary conditions to evaluate the constants of integration. The solution is the displacement distribution $w = f(x)$ for the middle surface of the anticlinal layer.

The first integration of (9.33) yields $d^3w/dx^3 = (p/R)(x + C_1)$, but recall from (9.26) that $S = dM/dx$, and from (9.30) we have $dM/dx = R(d^3w/dx^3)$, so $S = p(x + C_1)$. The boundary condition used to determine the constant of integration, C_1, follows from the symmetry of the layer and the applied loading: the magnitude of the shear force must be equal at equal distances from the layer center. Therefore, the shear force must go to zero at the center: at $x = 0$, $S = 0$, so $C_1 = 0$.

The second integration of (9.33) yields $d^2w/dx^2 = (p/R)(x^2/2 + C_2)$ and the third integration yields $dw/dx = (p/R)(x^3/6 + C_2x + C_3)$. Again, we use a boundary condition based on symmetry: the slope of the layer must be zero at the center: at $x = 0$, $dw/dx = 0$, so $C_3 = 0$. Also, at both ends the layer must return to its original horizontal orientation, so we invoke the following displacement boundary condition: at $x = \pm a$, $dw/dx = 0$. This requires $0 = (\pm a^3/6 \pm C_2 a)$ so $C_2 = -a^2/6$.

The fourth integration of (9.33) yields $w = (p/R)(x^4/24 - a^2x^2/12 + C_4)$. To determine the last constant of integration we invoke a displacement boundary condition: the vertical displacement of the layer goes to zero at both ends: at $x = \pm a$, $w = 0$. This requires $0 = (a^4/24 - a^4/12 + C_4)$ so $C_4 = a^4/24$. Thus, the solution to (9.33) for the specified symmetry and displacement boundary conditions is:

$$w = (p/R)\left(\tfrac{1}{24}x^4 - \tfrac{1}{12}a^2x^2 + \tfrac{1}{24}a^4\right)$$

Substituting for the flexural rigidity from (9.31) and rearranging this solution, we write the vertical displacement of the middle surface of the anticlinal layer as:

$$w = (p/2Bh^3)(x^4 - 2a^2x^2 + a^4) \quad (9.34)$$

The displacement is proportional to the uniform pressure, p, and the fourth power of the length, a^4. It is inversely proportional to the elastic modulus, B, and to the third power of the thickness, h^3. The displacement is much more sensitive to the layer geometry (i.e. length and thickness), than to the applied pressure or the elastic stiffness. Because the constant C_1 is zero, the shear force is proportional to the driving pressure and linearly distributed along the layer:

$$S = px \quad (9.35)$$

The bending moment is found from (9.30) and the result of the second integration in the previous paragraph:

$$M = R\frac{d^2w}{dx^2} = p\left(\tfrac{1}{2}x^2 - \tfrac{1}{6}a^2\right) \quad (9.36)$$

The bending moment also is proportional to the driving pressure, and it is distributed along the layer as the square of the distance.

Normalizing the displacement and distance in (9.34) by the half-length, a, we find the dimensionless displacement for the layer with an anticlinal shape, which is plotted versus dimensionless position in **Figure 9.33**:

$$\frac{2wBh^3}{pa^4} = \left[\left(\frac{x}{a}\right)^4 - 2\left(\frac{x}{a}\right)^2 + 1\right]$$

For comparison, the dimensionless displacement for a layer with a circular shape and a radius a is taken from Pollard and Johnson (1973) (see Further Reading):

$$\frac{2wBh^3}{pa^4} = \frac{3}{8}\left[\left(\frac{r}{a}\right)^4 - 2\left(\frac{r}{a}\right)^2 + 1\right]$$

Here r is the radial coordinate, which is analogous to x for the anticlinal shape. Changing the layer shape from anticlinal (infinitely long) to circular, while keeping the total length in x and r the same, makes no change in the form of the bending (terms in square brackets). In both cases the layer bends concave upward over the ends (edges) of the layer, and concave downward over the center. The greatest magnitude of displacement, at $x = 0 = r$, is less by a factor 3/8 for the circular shape. These results demonstrate the qualitative and quantitative value of two-dimensional analyses.

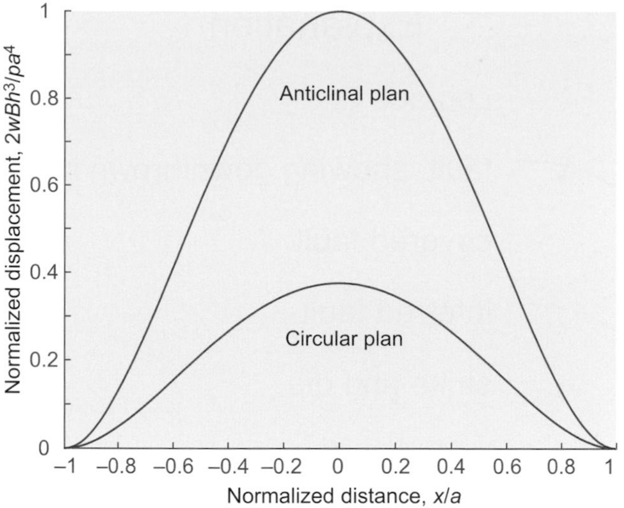

Figure 9.33 Normalized displacement for middle surface of bending layer plotted versus normalized distance. Anticlinal and circular refer to plan shape of layers, which are subject to uniform pressure, and boundary conditions are defined in the text. Displacement is greatly exaggerated relative to distance to clarify the form of bending.

9.6.6 Application to the Henry Mountain Laccoliths

To apply the model results just obtained, we compare the profiles of deformed layers illustrated in **Figure 9.33** to the cross-sectional form of strata in the dome at Mount Holmes in the southern Henry Mountains. The stratigraphic section with formation names for this region is provided in **Figure 9.34**. The cross section is constructed from the geological map of Mount Holmes (**Figure 9.35**), which reveals Mesozoic sedimentary strata cropping out around the mountain. For the most part older formations are exposed near the mountain center and younger formations are exposed on its flanks. Exceptions to this stratigraphic pattern are due largely to the complex interplay of very irregular topography with the more regular form of the strata.

A traverse from the mountain center to its flanks (A–A′ on **Figure 9.35**) crosses the Wingate Sandstone, the Kayenta Formation, the Navajo Sandstone (N), the Carmel Formation, and the Entrada Sandstone (E). This pattern, from older strata near the mountain center to younger strata on the flanks, supports the interpretation as an eroded sedimentary dome. That interpretation is reinforced by the orientation of the strata, which dip away from the center of the mountain in a radial pattern. Furthermore, the dip magnitudes near the traverse increase systematically from 5° near the mountain center, to 23° on the flanks, and then decrease to about 5° at radial distances of 4 to 5 km from the mountain center.

The cross section (**Figure 9.36**), constructed along the traverse A–A′ from the geologic map (**Figure 9.35**), extends from

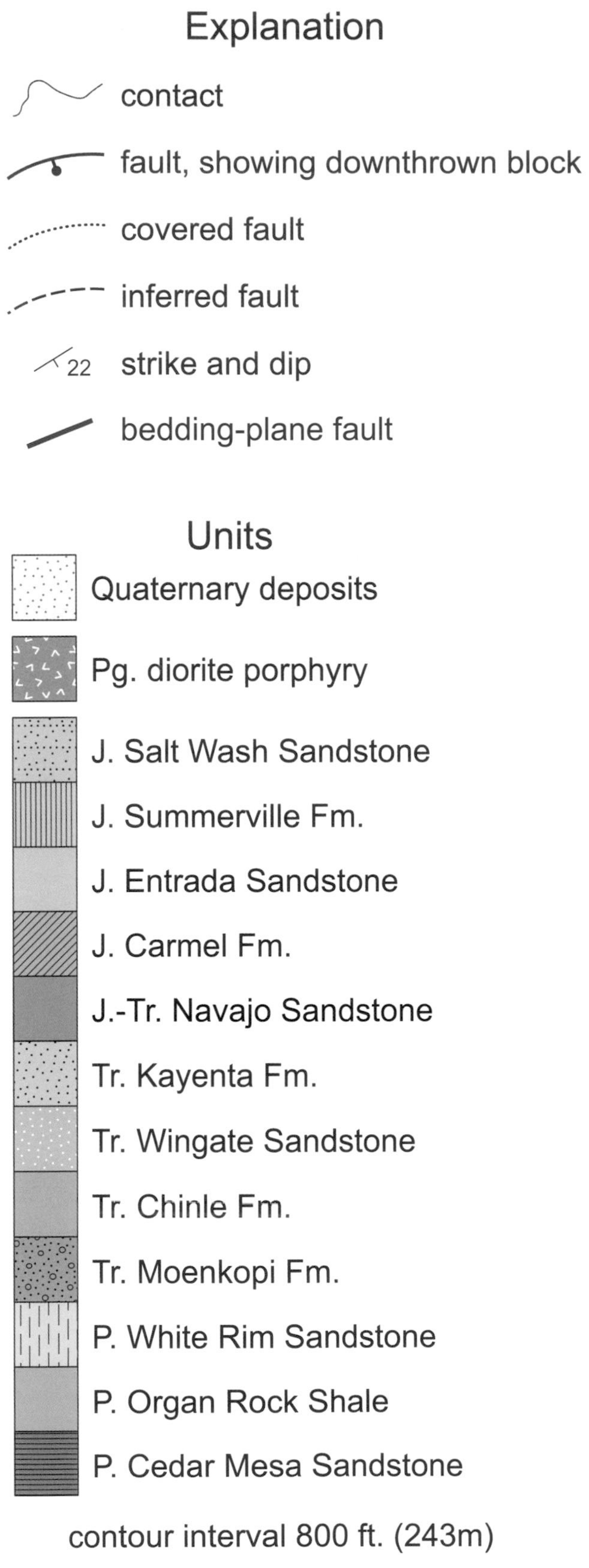

Explanation

contact

fault, showing downthrown block

covered fault

inferred fault

22 strike and dip

bedding-plane fault

Units

Quaternary deposits

Pg. diorite porphyry

J. Salt Wash Sandstone

J. Summerville Fm.

J. Entrada Sandstone

J. Carmel Fm.

J.-Tr. Navajo Sandstone

Tr. Kayenta Fm.

Tr. Wingate Sandstone

Tr. Chinle Fm.

Tr. Moenkopi Fm.

P. White Rim Sandstone

P. Organ Rock Shale

P. Cedar Mesa Sandstone

contour interval 800 ft. (243m)

Figure 9.34 Map explanation and stratigraphic section for geologic map of Mount Holmes (**Figure 9.35**). Modified from Jackson and Pollard (1988), Figure 6.

the mountain center to the distal limb and illustrates the domal structure of the mountain. The bending of individual formations is qualitatively similar to the bending of the circular elastic layer in **Figure 9.33**. In the upper hinge the strata are convex upward; in the central limb they are nearly straight; in the lower hinge they are concave upward; and in the peripheral limb they are nearly straight. The greatest dips occur at inflection points in the central limb of the dome. One obvious difference is the fact that the formations are not horizontal at the most distal locations from the mountain center. This could be due to smaller intrusions underlying the peripheral limb that are similar in size and form to the sills at Buckhorn Ridge, and at #1 and #2 on the geologic map (**Figure 9.35**).

The oldest rocks exposed at Mount Holmes are from the Triassic Chinle formation, but the contact with the top of the central intrusion is not exposed (**Figure 9.34**). All of the igneous rock exposed in dikes and sills is diorite porphyry of Tertiary age. At nearby Mount Ellsworth and Mount Hillers in the southern Henry Mountains the top of the central intrusion is exposed in the Permian Organ Rock Shale and Cedar Mesa Sandstone. Assuming the central intrusion at Mount Holmes was injected at this stratigraphic level, the thickness of the overburden would have included about 2.7 km of Mesozoic strata and another 1.0 km of Tertiary strata during the early to mid-Tertiary time, when the laccoliths formed. Thus, the total thickness of strata was about 4 km, and the total length subject to bending was about 10 to 12 km. This is *not* a thin layer! However, outcrops within the central limb of Mount Holmes (**Figure 9.36**) reveal bedding-plane fault zones at some formation contacts. These fault zones are less than 1 m thick and are made up of a network of deformation bands. The lack of markers precludes measuring slip on these faults, but slickenlines trend roughly radial to the mountain center. This suggests that the faults are related to the growth of the domes, in which case they would have subdivided the overlying strata into many layers that were thin compared to their length.

To help understand how the thick stack of sedimentary strata at Mount Holmes (**Figure 9.34**) could have bent into the dome mapped in **Figure 9.35**, we calculate the flexural rigidity of a stack of layers capable of sliding over one another during bending. The bedding-plane faults on the cross section (**Figure 9.36**) demonstrate that the strata did slide, and no doubt friction offered some resistance to sliding. However, to simplify the analysis we suppose the frictional resistance was negligible, and consider n layers, of thickness h_i and elastic modulus B_i. In this case the effective flexural rigidity, R_e, of the stack of layers is equal to the sum of all the individual flexural rigidities:

$$R_e = \sum_{i=1}^{n} \left(B_i h_i^3/12\right) \quad \textbf{(9.37)}$$

In general, the effective flexural rigidity is less than the flexural rigidity of the overburden treated as a single layer with elastic modulus B and thickness H: $R = BH^3/12$.

For the sake of a simple example, suppose the elastic moduli all are the same, so $B_i = B$, and both B and 12 can be taken out of the summation in (9.37). Then, the effective flexural rigidity

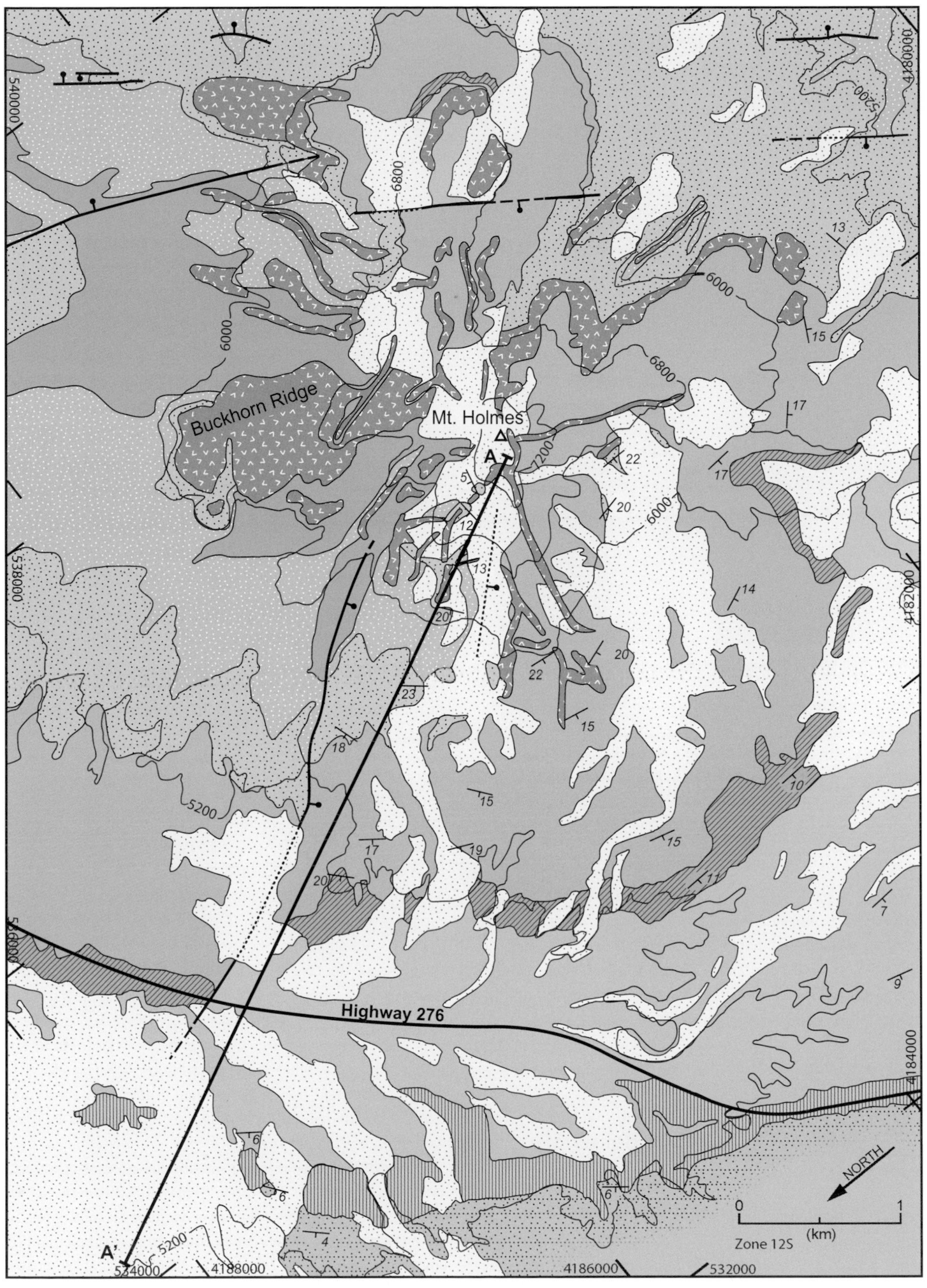

Figure 9.35 Geological map of Mount Holmes, southern Henry Mountains, UT. Cross section A–A′ is shown in Figure 9.36. Key is provided in Figure 9.34. Approximate map center UTM: 12 S 537021.08 m E, 4183470.76 m N. Modified from Jackson and Pollard (1988), Figure 7.

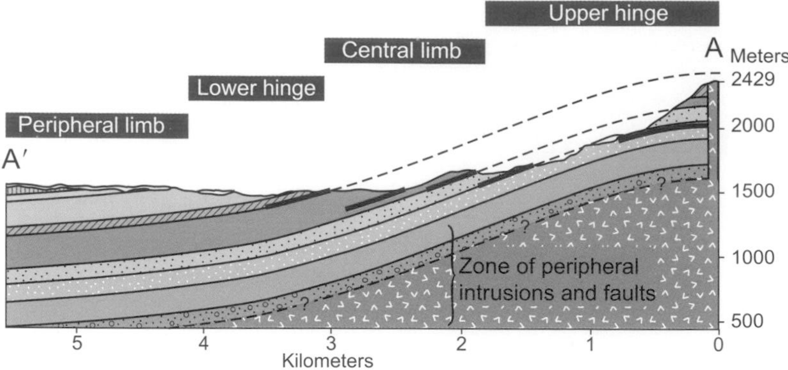

Figure 9.36 Interpretative cross section radial to the dome at Mount Holmes in the southern Henry Mountains, UT, based on the geological map (**Figure 9.35**). Short heavy red lines are bedding-plane fault zones. Stratigraphic labels (**Figure 9.34**) range from the Triassic Moenkopi Formation (TRm) to the Jurassic Salt Wash Sandstone of the Morrison Formation (Jms). Modified from Jackson and Pollard (1988), Figure 10a.

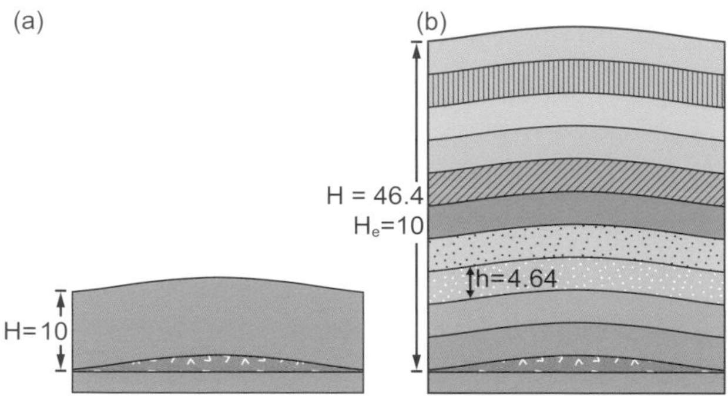

Figure 9.37 Cross sections of two model laccoliths displaying the effective thickness concept. (a) Actual thickness and effective thickness are equal: $H = 10$ units. (b) Actual thickness of each layer is $h = 4.64$ units; actual thickness of stack of ten layers is $H = 46.4$ units; effective thickness of stack of ten layers is $H_e = 10$ units, exactly the same as the effective thickness for the single layer in (a). Modified from Pollard and Johnson (1973), Figure 5.

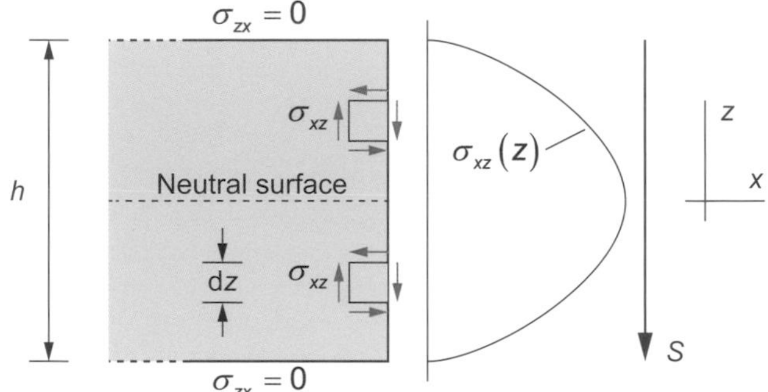

Figure 9.38 On an arbitrary cross section of a bending layer a distribution of shear stress, σ_{xz}, occurs that goes to zero at the top and bottom surface. The resultant of this distribution of shear stress is the shear force, S, on the cross section.

reduces to $R_e = (B/12)H_e^3$, where H_e is the effective thickness of the stack of layers. For this example, suppose the layers are equal in thickness, so the actual thickness of the stack is $H = nh$, and the effective thickness is $H_e = n^{1/3}h$. In **Figure 9.37** we show a single layer of thickness $H = 10$ units that has the same flexural rigidity as the stack of 10 freely sliding layers, each of thickness $h = 4.64$ units. During doming in the Henry Mountains, we suggest that the effective thickness was much less than 4 km, due to bedding-plane faulting.

9.6.7 Stress State Induced by Bending: Cauchy's Laws

Recall that the bending moment, M, is related to the resultant of normal stresses, σ_{xx}, acting on cross sections of a layer (**Figure 9.32**). We showed in (9.32) that the normal stresses are distributed linearly from zero at the neutral surface to greatest magnitudes at the top and bottom surfaces of the layer. Similarly, the shear force, S, is related to the resultant of shear stresses, σ_{xz}, acting on cross sections. Here we show how the shear stresses are distributed, and see whether the stresses in the layer satisfy Cauchy's first law.

We set up the bending layer model at the beginning of Section 9.6.2 by specifying that the top and bottom surfaces of the layer carry no shear stress, $\sigma_{zx} = 0$ on $z = \pm h/2$ (**Figure 9.38**). From Cauchy's second law (Section 3.7.1), we have $\sigma_{xz} = \sigma_{zx}$, so the shear stress on cross sections also must be zero where they intersect the top and bottom surfaces of the layer: $\sigma_{xz} = 0$ at $z = \pm h/2$. By symmetry, we would expect the shear stress on cross sections to attain the greatest magnitude at the neutral surface, where $z = 0$, but what is the distribution of shear stress on the cross section, and how does it relate to the shear force, S? Note that the positive shear force drawn in **Figure 9.38** is associated with a negative shear stress.

To derive the shear stress distribution on cross sections consider an element of the layer (**Figure 9.39**) with a variable bending moment. This requires a variable normal stress acting on the two cross sections. Then, consider a partial element (darker color) that includes the bottom surface of the layer, where $\sigma_{zx} = 0$, and an upper edge at some arbitrary level z, where $\sigma_{zx} \neq 0$. The force per unit length acting in the x-direction due to this shear stress is $\sigma_{zx}dx$. The force per unit length due to the normal stress acting on the left cross section is found using (9.32):

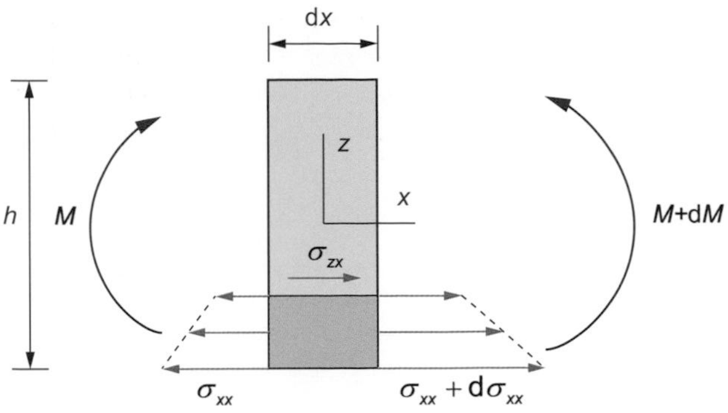

Figure 9.39 Moments and normal stresses on cross sections of a bending layer used to calculate the distribution of shear stress.

$\sigma_{xx}\mathrm{d}z = -12Mz\mathrm{d}z/h^3$. This must be integrated over the height of the partial element to find the net force in the x-direction:

$$\int_{-h/2}^{z} \sigma_{xx}\mathrm{d}z = -\frac{12M}{h^3}\int_{-h/2}^{z} z\mathrm{d}z$$

The net force per unit length acting on the right cross section of the partial element is:

$$\int_{-h/2}^{z} (\sigma_{xx} + \mathrm{d}\sigma_{xx})\mathrm{d}z = -\frac{12(M + \mathrm{d}M)}{h^3}\int_{-h/2}^{z} z\mathrm{d}z$$

These are all the forces acting on the partial element in the x-direction.

Setting the sum of the forces acting on the partial element in x to zero, we have:

$$\sigma_{zx}\mathrm{d}x - \frac{12\mathrm{d}M}{h^3}\int_{-h/2}^{z} z\mathrm{d}z = 0$$

Rearranging this equation and recalling from (9.26) that $\mathrm{d}M/\mathrm{d}x = S$, the shear stress is:

$$\sigma_{zx} = \frac{12S}{h^3}\int_{-h/2}^{z} z\mathrm{d}z = \frac{6S}{h^3}\left[z^2 - (h/2)^2\right] \quad \textbf{(9.38)}$$

The shear stress is proportional to the shear force per unit length and has a parabolic distribution over the cross section of the layer, as illustrated in **Figure 9.38**. The shear stress goes to zero at the top and bottom surface, where $z = \pm h/2$, and it attains a maximum value at the neutral surface where $\sigma_{zx} = -3S/2h$. A positive shear force is associated with a negative shear stress.

We now address the question raised at the beginning of Section 9.6: does the deformation of the bending layer satisfy Cauchy's laws of motion? Because the deformation is quasi-static, the equations of motion reduce to the equilibrium equations (Section 7.2.3). For the two-dimensional problem illustrated in **Figure 9.31**, equilibrium is satisfied if:

$$\frac{\partial \sigma_{xx}}{\partial x} + \frac{\partial \sigma_{zx}}{\partial z} = 0 \quad \textbf{(9.39)}$$

Substituting for the shear force in (9.38) using (9.35), the shear stress in the bending layer is:

$$\sigma_{zx} = \frac{6px}{h^3}\left[z^2 - \tfrac{1}{4}h^2\right]$$

Substituting for the bending moment in (9.32) using (9.36), the normal stress is:

$$\sigma_{xx} = -\frac{6pz}{h^3}\left(x^2 - \tfrac{1}{3}a^2\right)$$

Note that both the shear and the normal stress are proportional to the driving pressure and inversely proportional to the thickness cubed. Also, both stress components are functions of the two coordinates, x and z. Substituting the equations for the stress components into (9.39), we find that equilibrium is satisfied. The deformation leading to the displacement profile (9.34) that we compared to the bending over the laccolith in the Henry Mountains (**Figure 9.36**) is consistent with Cauchy's laws of motion.

9.7 A CANONICAL MODEL FOR BUCKLING

The review paper by Hudleston and Treagus (2010), listed in Further Reading, on mechanical information gleaned from folds, includes a beautiful photograph of a train of folds interpreted as resulting from buckling (**Figure 9.40**). This outcrop is from Cap de Creus in the Catalonia Autonomous Region of northeastern Spain, where the metamorphosed sedimentary rocks from the Late Proterozoic to Early Paleozoic are biotite schists, cut by quartz-filled veins. When the veins opened, they were tabular fractures filled with a hydrothermal fluid from which the quartz precipitated. The prominent foliation of the biotite schist at Cap de Creus is folded with wavelengths of about 1 to 10 m and meter-scale amplitudes. In contrast, the folded quartz vein in **Figure 9.40** has wavelengths of 1 to 2 cm and amplitudes of 0.5 to 1 cm. The two surfaces bounding this folded layer are the two surfaces of the tabular fracture that opened to form the vein.

The folded quartz vein in **Figure 9.40** is an example of a single layer fold that apparently was unencumbered by the folding of any nearby veins. The canonical model of folding elucidates the buckling of a single layer in homogeneous surroundings. Because the vein folded without fracturing we suppose it, and the surrounding schist, were ductile rather than brittle, and we choose a linear viscous constitutive law to model their deformation. The regularity of the waves suggests that some physical characteristics of the vein and schist promoted centimeter wavelengths, rather than wavelengths measured in millimeters or decimeters. In this section we develop a mechanical model for buckling that offers an explanation for this wavelength selection, and identifies the physical quantities related to the dominant wavelength.

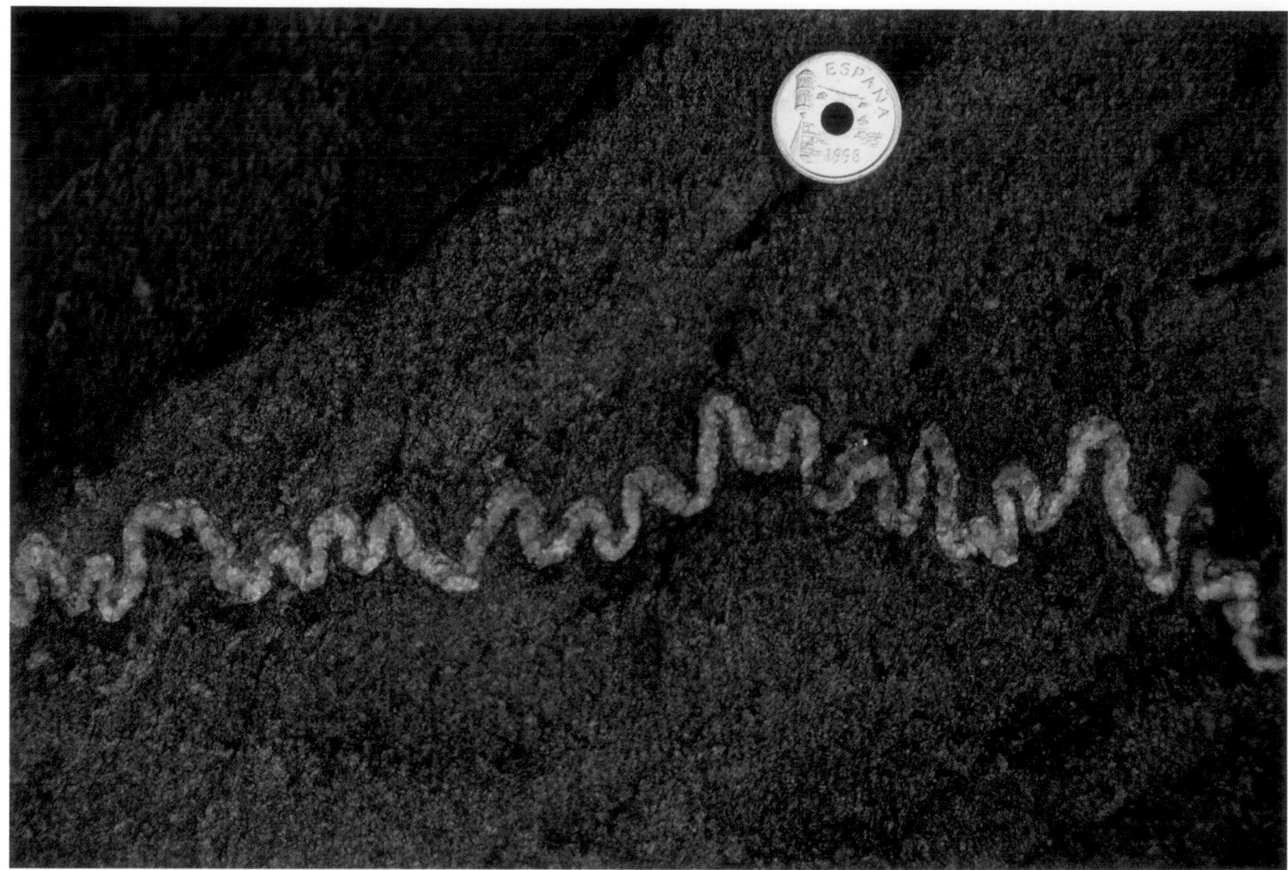

Figure 9.40 Single layer folded quartz vein in a biotite schist from Cap de Creus, Spain; coin diameter 2 cm. From Hudleston and Treagus (2010), Figure 1b.

The current wavelength, λ, of individual folds is the straight-line measurement from one hinge, past the adjacent hinge, to the next hinge (**Figure 9.41b**). Because the entire train of folds in **Figure 9.40** is about 18 cm long, with 17 to 19 waves (depending on how you count them), the average current wavelength is about 1 cm. However, the wavelength in the current state is not the wavelength as the folding began (**Figure 9.41a**). To estimate the initial wavelength, λ_o, we take a string and overlay the folded vein, carefully following the entire train of folds, and then measure the length of the straightened string as about 45 cm. This arc length, divided by the number of waves gives an average initial wavelength of about 2.5 cm, assuming the arc length has not changed due to deformation during the folding process.

The first information that we glean from **Figure 9.40** is an estimate of the stretch, S_{xx}, parallel to this fold train. Recall from the discussion of finite strain and stretch in Section 5.5.3 that the stretch is the final length of a material line divided by the initial length. If the final length is less than the initial length, the material line has shortened and the stretch is between zero and one. We take $n\lambda$ as the final length of the fold train, and $n\lambda_o$ as an estimate of the initial length, so the estimate of stretch parallel to the fold train is $S_{XX} = n\lambda/n\lambda_o = 0.4$. This is the bulk shortening of the rock mass in that direction. If the region around the fold train deformed by pure shearing (Section 5.5.3), where the dilation

(a)

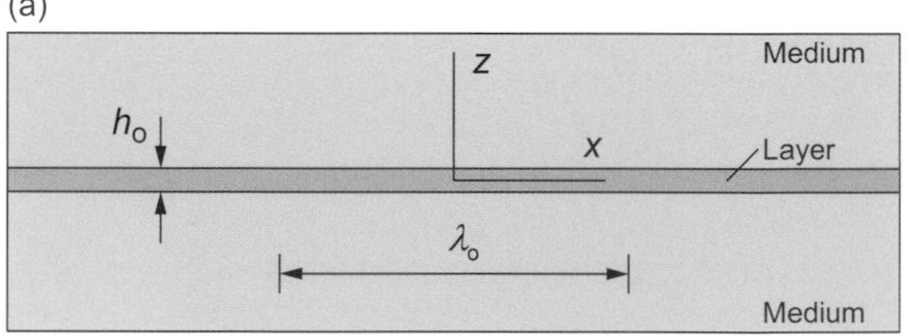

(b)

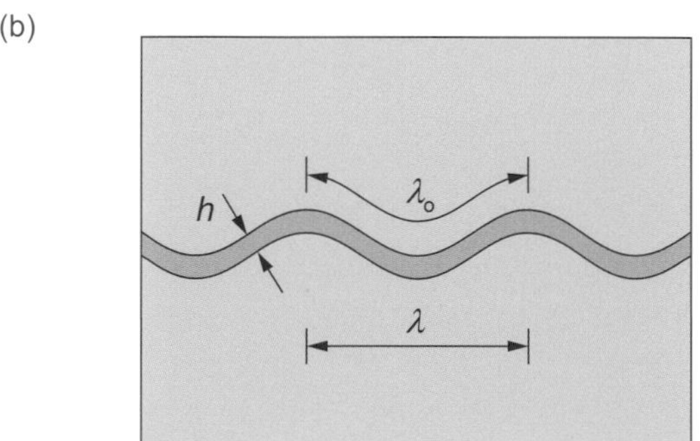

Figure 9.41 (a) Initial state: planar layer of thickness h_o in uniform medium. (b) Current state: layer of thickness h buckled into sinusoidal waves of wavelength λ. The arc length of one wave is an estimate for the initial wavelength, λ_o.

equals zero, the stretch perpendicular to the fold train is the reciprocal of that parallel to the train: $S_{ZZ} = 1/S_{XX} = 2.5$. This is the bulk lengthening. Recall that pure shear demands no area change in the (x, z)-plane, so $S_{XX}S_{ZZ} = 1$. We impose this condition on the canonical model.

Choice of the word *bulk* to modify shortening in the previous paragraph reflects the fact that we assumed the arc length of the folded vein did not change during folding. Rather, the length of the entire train of folds shortened, along with the surrounding rock mass. S_{XX} represents an average value over the *bulk* of the rock mass depicted in **Figure 9.40**. The quartz vein buckled during shortening, so we expect the deformation to be quite heterogeneous within the vein and in the immediate surroundings. Meanwhile, the biotite schist developed a foliation perpendicular to the bulk shortening. At perpendicular distances of several wavelengths from the fold train we expect the deformation of the schist to be more or less homogeneous, with the stretch state estimated in the previous paragraph.

To glean more information from folds we need to investigate a mechanical model for buckling. We take that up in the next three sections, utilizing the problem of a thin viscous layer in the less viscous surroundings. We begin be reviewing the basic concepts of viscous flow, introduced in Chapter 6, which are directly applicable to the buckling problem.

9.7.1 Two-Dimensional Viscous Flow

In Chapter 6 we laid out the governing equations for deformation of a viscous material, starting with Cauchy's laws of motion (Section 6.8). We recall one of the equations from the first law here:

$$\rho\frac{Dv_x}{Dt} = \frac{\partial\sigma_{xx}}{\partial x} + \frac{\partial\sigma_{yx}}{\partial y} + \frac{\partial\sigma_{zx}}{\partial z} + \rho g_x^* \tag{9.40}$$

The other two equations from the first law are found by cycling the subscripts, $x \to y$, $y \to z$, $z \to x$. Three of the dependent variables in these equations are the velocity components (v_x, v_y, v_z), one of which appears in each equation. The buckling problem illustrated in **Figure 9.41** is *two dimensional*, and is set up in the (x, z)-plane. Thus, the out-of-plane velocity, v_y, is zero and the two in-plane velocities are functions of x and z only:

$$v_x = v_x(x, z),\ v_y = 0,\ v_z = v_z(x, z) \tag{9.41}$$

This constraint on the velocity components has important implications for the other physical quantities relevant to the deformation, namely the rates of deformation and the stresses.

Recall that the kinematic equations (Section 6.6.3) relate derivatives of the velocity components to the rate of deformation. For two-dimensional flow constrained by (9.41) only four rates of deformation may be non-zero, and because the rate of deformation matrix is symmetric, $D_{zx} = D_{xz}$, the three independent rates are:

$$D_{xx} = \frac{\partial v_x}{\partial x}, D_{xz} = \frac{1}{2}\left(\frac{\partial v_x}{\partial z} + \frac{\partial v_z}{\partial x}\right), D_{zz} = \frac{\partial v_z}{\partial z} \tag{9.42}$$

We postulate that the model materials are incompressible: no change in volume at any point. As we showed in Section 6.7.3, this requires the three normal rates of deformation to sum to zero, $D_{xx} + D_{yy} + D_{zz} = 0$. However, $D_{yy} = 0$, so the two normal rates of deformation in the (x, z)-plane just compensate for one another: $D_{zz} = -D_{xx}$. Thus, for two-dimensional flow of an incompressible material only two independent rates of deformation arise, D_{xx} and D_{xz}.

The dependent variables in Cauchy's first law, for example (9.40), also include the stress components, and these are related to the rates of deformation using constitutive equations. For the canonical model of buckling we choose the simplest constitutive law that might describe the flow of rock as a fold develops. This is the constitutive law for a *linear* and *isotropic* viscous liquid. In Section 6.7.3 we showed that incompressibility leads to a simplification of the constitutive equations such that only five of the stress components may be non-zero, and the two non-zero shear stresses are equal, $\sigma_{xz} = \sigma_{zx}$, because of Cauchy's second law (Section 3.7.1) Thus, the four independent stress components are:

$$\sigma_{xx} = -\bar{p} + 2\eta D_{xx}, \sigma_{xz} = 2\eta D_{xz}, \sigma_{yy} = -\bar{p}, \sigma_{zz} = -\bar{p} - 2\eta D_{xx} \tag{9.43}$$

In the third equation we used the fact that $D_{yy} = 0$ for two-dimensional flow. In the last equation, we replaced D_{zz} with $-D_{xx}$ based on incompressibility. These constitutive equations relate the stress components to the mean normal pressure, $\bar{p}$, the Newtonian viscosity, η, and the two independent rates of deformation, D_{xx} and D_{xz}.

For the flowing incompressible liquid (see Section 6.7.3), the mean normal pressure, $\bar{p}$, defined using mechanical principles, is equal to the pressure, p, defined using thermodynamic principles. The constraints that two-dimensional flow of an incompressible viscous liquid place on the stress components do not determine this pressure. Thus, in the next section, we define the pressure independently. Then, given the viscosity from laboratory tests, and prescribing the two rates of deformation, the four stress components are determined using (9.43).

Cauchy's first law (Section 6.8) places constraints on the way the stress components are distributed in the deforming body. Referring to (9.40), for example, we only consider cases where the material time derivative of velocity on the left side is negligible compared to spatial gradients in the stresses on the right side. This would be the case for slow viscous flow at low Reynold's number, where the flow is laminar, not *turbulent*. We also neglect the body force term, for example on the right side of (9.40), compared to the stress gradients. This is appropriate for small folds, such as the centimeter-scale quartz vein in **Figure 9.40**. After eliminating these terms, the equilibrium equations remain, which reduce for two-dimensional flow in the (x, z)-plane to:

$$\frac{\partial\sigma_{xx}}{\partial x} + \frac{\partial\sigma_{zx}}{\partial z} = 0, \quad \frac{\partial\sigma_{xz}}{\partial x} + \frac{\partial\sigma_{zz}}{\partial z} = 0 \tag{9.44}$$

Note that these equilibrium equations are the same as the ones we used to check the stress distribution in the bending elastic layer over the laccolith in Section 9.6.7. This highlights one of the interesting commonalities between elastic deformation and slow viscous flow: the stress components in both cases must satisfy these equilibrium equations.

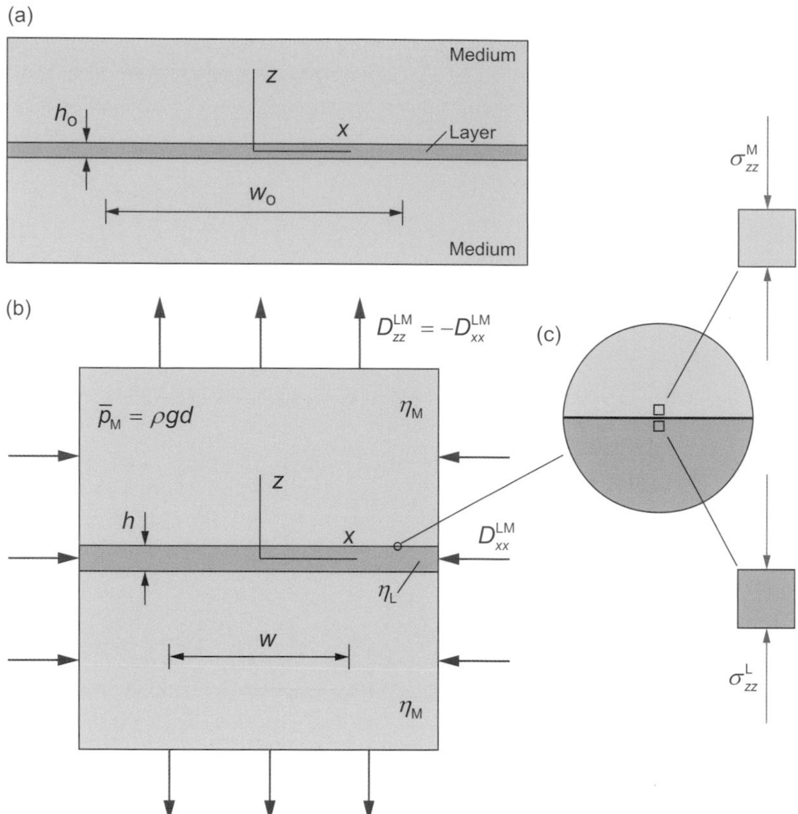

Figure 9.42 Base state deformation for single layer of viscosity η_L in an infinite medium with viscosity η_M subject to layer-parallel shortening rate D_{xx}^{LM} and lithostatic pressure $\bar{p}_M$ in the medium. (a) Initial configuration: representative width w_o and thickness h_o. (b) Current configuration: width w and thickness h with no buckling. (c) Zoom in showing normal stresses that are equal across the interface.

In summary, a complete description of two-dimensional flow in the (x, z)-plane of an incompressible, linear, and isotropic viscous material involves eight independent physical quantities: two velocities (v_x, v_z), two rates of deformation (D_{xx}, D_{xz}), three stresses $(\sigma_{xx}, \sigma_{xz}, \sigma_{zz})$, and the mean normal pressure, $\bar{p}$. In the next section we specialize the two-dimensional problem to that of a layer embedded in a medium with different viscosity, subject to bulk shortening parallel to the layer (**Figure 9.41**). The objective is to evaluate all of the physical quantities and to determine if the layer buckles.

9.7.2 The Base State for Buckling

For the canonical model of buckling we consider a single planar layer with viscosity η_L in an infinite medium with viscosity η_M (**Figure 9.42**). L, M, and LM are used as subscripts or superscripts to indicate the physical quantity is associated with the *layer*, the *medium*, and *both* the layer and the medium, respectively. We prescribe a *uniform* shortening rate (negative) parallel to the layer, and we prescribe *symmetric* deformation, so the shearing rate is zero in both the layer and the medium. Thus, the rates of deformation throughout the layer and medium for this base state are:

$$D_{xx}^{LM} = \text{constant} < 0, D_{xz}^{LM} = 0, D_{zz}^{LM} = -D_{xx}^{LM} \qquad (9.45)$$

The *uniform* lengthening rate perpendicular to the layer follows from the stipulation that the material is incompressible. For this problem we prescribe only one independent rate of deformation, D_{xx}^{LM}.

In order to have enough independent variables to solve this problem we also prescribe the pressure in the medium, based on Section 6.7.4:

$$\bar{p}_M = \rho g d \qquad (9.46)$$

Here ρ is average mass density of the overlying rock, g is acceleration of gravity, and d is depth. We refer to this quantity as the lithostatic pressure.

For an interfacial boundary condition we specify that the layer and the medium remain in *physical contact* along both interfaces during the deformation (**Figure 9.42b**). This means that the traction exerted on the medium by the layer must be equal, and oppositely directed, to the traction exerted on the layer by the medium. In the second of (9.45) we set the shearing rate of deformation to zero, so only the normal components of the tractions on the interface are non-zero. Recall from the discussion of Cauchy's formula in Section 2.6.5 that this constraint on the normal tractions requires the normal stress, σ_{zz}^L at the interface in the layer to be equal to the normal stress, σ_{zz}^M, on the other side of the interface in the medium (**Figure 9.42c**). This normal stress is homogeneous throughout the layer and medium:

$$\sigma_{zz}^L = \sigma_{zz}^M = \sigma_{zz}^{LM} \qquad (9.47)$$

The normal stresses acting parallel to the layer are not constrained by the condition on the tractions at the interface, and we show below that they are different in the layer and medium.

Using the last of (9.43) written first for the layer and then for the medium, and equating these normal stresses because of (9.47), we have $-\bar{p}_L - 2\eta_L D_{xx}^{LM} = -\bar{p}_M - 2\eta_M D_{xx}^{LM}$. Rearranging, and substituting for the pressure in the medium using (9.46), the pressure in the layer is:

$$\bar{p}_L = \rho g d - 2D_{xx}^{LM}(\eta_L - \eta_M) \tag{9.48}$$

We will learn later that buckling requires the viscosity of the layer to be greater than that of the medium, so the expression in parentheses is positive. For shortening parallel to the layer the rate of deformation is negative, so the pressure in the layer exceeds the pressure in the medium, $\bar{p}_L > \bar{p}_M$. As the viscosity of the layer approaches that of the medium, $\eta_L \to \eta_M$, the pressure in the layer approaches the lithostatic pressure, $\bar{p}_L \to \bar{p}_M = \rho g d$.

The normal stresses parallel to the layer in the y-direction, in the layer and in the medium, are the negative of the respective pressures from (9.48) and (9.46):

$$\sigma_{yy}^L = -\rho g d + 2D_{xx}^{LM}(\eta_L - \eta_M), \sigma_{yy}^M = -\rho g d \tag{9.49}$$

The normal stress perpendicular to the layer is found by writing the last of (9.43) either for the layer or the medium, and substituting from (9.48) or (9.46) for the appropriate pressure:

$$\sigma_{zz}^{LM} = -\rho g d - 2\eta_M D_{xx}^{LM} \tag{9.50}$$

For shortening parallel to the layer the rate of deformation is negative, so the second term makes this normal stress less compressive than that due to the lithostatic pressure. This is consistent with lengthening perpendicular to the layer.

The normal stresses parallel to the layer in the x-direction, both in the layer and in the medium, are found from the first of (9.43) with the appropriate substitutions for the pressures from (9.48) and (9.46):

$$\sigma_{xx}^L = -\rho g d + 2D_{xx}^{LM}(2\eta_L - \eta_M), \sigma_{xx}^M = -\rho g d + 2D_{xx}^{LM}\eta_M \tag{9.51}$$

The rate of deformation is negative, and the expression in parentheses is positive, so both second terms are negative and both normal stresses are more compressive than that due to the lithostatic pressure. This is consistent with shortening parallel to the layer. The normal stress in the layer is more compressive than that in the medium, which is consistent with the layer being more viscous. As the viscosity of the layer approaches that of the medium, $\eta_L \to \eta_M$, the layer parallel normal stress in the layer approaches that in the medium: $\sigma_{xx}^L \to \sigma_{xx}^M$.

The stress components must satisfy the equilibrium equations (9.44). Because the shear stresses are zero throughout the layer and medium the equilibrium equations reduce to $\partial\sigma_{xx}/\partial x = 0$ and $\partial\sigma_{zz}/\partial z = 0$. Because the normal stresses in (9.50) and (9.51) are uniform (not varying with x or z), equilibrium is satisfied.

The velocity components follow from the kinematic equations (9.42):

$$v_x = \int D_{xx}^{LM} dx = D_{xx}^{LM} x + C_1, \quad v_z = \int D_{zz}^{LM} dz = D_{zz}^{LM} z + C_2$$

Taking the origin of coordinates as the reference, $v_x = 0$ at $x = 0$ and $v_z = 0$ at $z = 0$, so both constants of integration are zero: $C_1 = 0$, $C_2 = 0$. Also, from the incompressibility condition, $D_{zz}^{LM} = -D_{xx}^{LM}$, so the velocity components are:

$$v_x = D_{xx}^{LM} x, \quad v_z = -D_{xx}^{LM} z \tag{9.52}$$

Each component is proportional to the prescribed rate of shortening in the x-direction, and varies linearly with the respective coordinate. This velocity field is illustrated in **Figure 9.43**. Note that the layer and the medium deform together with the same rates of deformation. The velocities go to zero at the origin and increase with distance from the origin. Because the rate of deformation is negative (shortening), v_x is directed toward the z-axis and v_z is directed away from the x-axis.

As the deformation proceeds, both the layer and the medium shorten parallel to the x-axis and thicken parallel to the z-axis (**Figure 9.43**). Nothing in the analysis leads to any buckling of the layer. This outcome is acceptable for the idealized layer, which is perfectly planar, and homogeneous with respect to viscosity. Neither the geometry, nor the properties, provide a perturbation that might lead to buckling, but the prescribed shortening parallel to the layer is a pre-requisite for buckling. Note that the velocity vectors for the base state *increase* in magnitude with distance from the layer. On the other hand our expectation for buckling is that the velocity vectors should *decrease* in magnitude with distance from the layer. In the next section we introduce geometric irregularities that could lead to buckling during this shortening and show that the perturbed velocity vectors do decrease with distance.

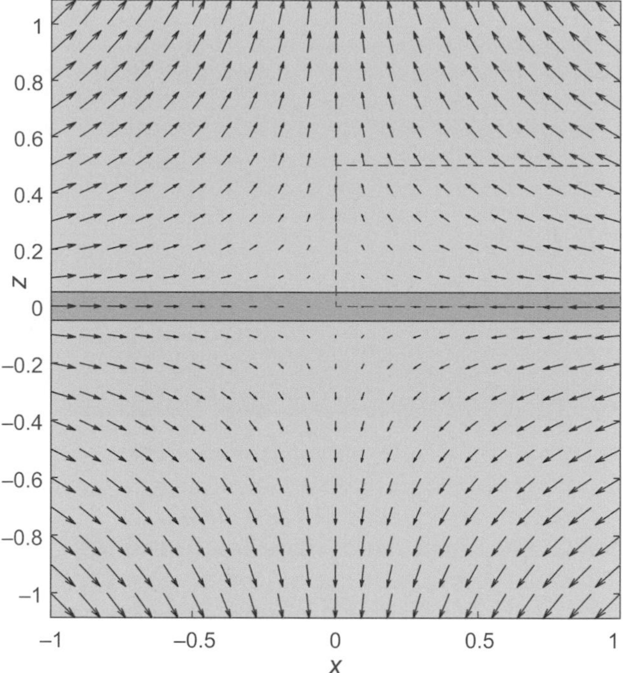

Figure 9.43 The velocity field for the base state rates of deformation (9.52) are associated with a shortening and thickening of the layer. Material flows toward the z-axis and away from the x-axis to maintain constant volume (incompressible) deformation. Dashed box refers to **Figure 9.45**.

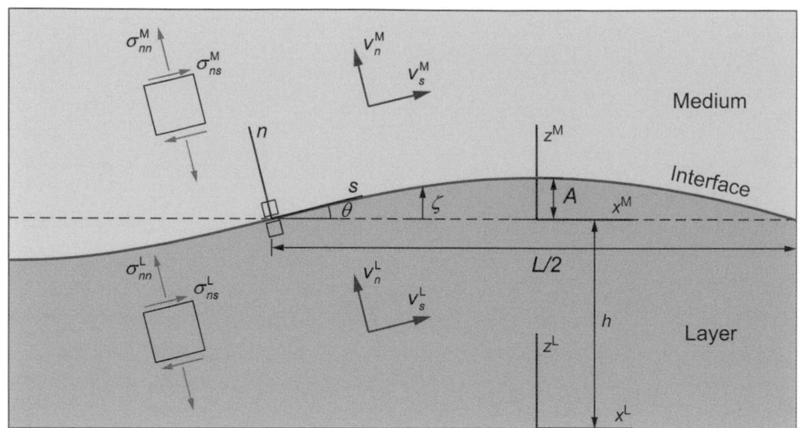

Figure 9.44 Coordinates for the layer (x^L, z^L) and for the medium (x^M, z^M). Cosinusoidal form of the bonded interface has amplitude *A* and wavelength *L*. Coordinates (*n*, *s*) on the interface make angle θ with Cartesian coordinates. Matching velocity components and stress components from (9.56) and (9.57) are shown near the point on the interface where they apply.

9.7.3 The Perturbed State for Buckling

A perturbed state that gives rise to buckling must be added to the base state derived in the previous section to solve the canonical model for this folding mechanism. For the perturbed state the applied rates of deformation are the same as for pure shearing (**Figure 9.42**), and the layer and medium have the same linear viscous and incompressible properties as those of the base state. The difference is that the interfaces between the layer and the medium are wavy instead of planar. Recall from **Figure 9.1b** that the layer was not perfectly planar, before initiating buckling. We show in this section that such a geometric perturbation may give rise to buckling.

The velocity and stress components in the layer and medium are the sum of the respective components for the base and perturbed states. For example, in the layer:

$$v^L_x = \bar{v}^L_x + \tilde{v}^L_x, v^L_z = \bar{v}^L_z + \tilde{v}^L_z$$
$$\sigma^L_{xx} = \bar{\sigma}^L_{xx} + \tilde{\sigma}^L_{xx}, \sigma^L_{xz} = \tilde{\sigma}^L_{xz}, \sigma^L_{zz} = \bar{\sigma}^L_{zz} + \tilde{\sigma}^L_{zz} \quad (9.53)$$

The overbar ($\bar{\sigma}$) indicates the base state and the tilde ($\tilde{\sigma}$) indicates the perturbed state. For the medium a similar set of equations is found by replacing the superscript L with M in (9.53). Note that the shear stress in the base state is zero throughout, because the shearing rate of deformation, D^{LM}_{xz}, is zero throughout (9.45). We have derived the base state velocity and stress components in the previous section, so we focus here on the perturbed state.

To quantify the shape of the two interfaces between the layer and the medium, we choose a cosinusoidal form:

$$\zeta = A \cos \lambda x \quad (9.54)$$

This form is identical for the interfaces at the top and bottom of the layer. We take advantage of this symmetry, and place the origin of coordinates (x^L, z^L) at the middle of the layer in its base state (rectangular) configuration (**Figure 9.44**). The origin of coordinates (x^M, z^M) for the medium is at the top of the layer in its base state configuration, a distance h above the origin for the layer (**Figure 9.44**). A point along the upper interface shares the same x-coordinate, $x^L = x^M = x$, but the z-coordinate is $z^L = h + \zeta$ when referred to the layer, and $z^M = \zeta$ when referred to the medium.

The layer thickness is $2h$ and the amplitude of the wave is A (**Figure 9.44**). The wave length is the value of x where $\lambda x = 2\pi$, which is $L = 2\pi/\lambda$. The slope of the interface is:

$$\theta \cong \tan \theta = d\zeta/dx = -A\lambda \sin \lambda x \quad (9.55)$$

In the first step we approximate the tangent with the first term in the series expansion: $\tan\theta = \theta + \theta^3/3 + \ldots$. Minimum and maximum values of the slope are found where $\lambda x = \pi/2$ and $3\pi/2$, which gives $\mp A\lambda$. In what follows we only consider wavy interfaces with slopes that are small enough to make the approximation in (9.55) acceptable (i.e., slopes less than about 15°). The objective of this section is to consider different wave lengths, determine which one grows most rapidly in terms of its maximum slope, and note what physical quantities this depends upon.

Given the shape of the interface (9.54), we establish matching boundary conditions on the velocity vectors acting at neighboring points across the interface using a local coordinate system (n, s) where n is perpendicular and s is parallel to the interface (**Figure 9.44**). The s-axis makes an angle θ with the x-axis, and this angle is the slope of the interface (9.55). For the interface we postulate a no separation boundary condition and a no slip boundary condition. The no separation condition means that the normal velocity components for neighboring points across the interface must be equal in magnitude and direction. The no slip condition means that the tangential velocity components at these points must be equal. Thus, both velocity components are *continuous* across the interface. These conditions are written for the upper interface as:

$$v^L_n(x, h+\zeta) = v^M_n(x, \zeta), v^L_s(x, h+\zeta) = v^M_s(x, \zeta) \quad (9.56)$$

Similar conditions apply to the lower interface. The respective components in the Cartesian coordinate system also must be equal, so (9.56) implies: $v^L_x(x, h+\zeta) = v^M_x(x, \zeta)$ and $v^L_z(x, h+\zeta) = v^M_z(x, \zeta)$.

By matching the velocity components (9.56) we establish what is referred to as a bonded interface. For the bonded interface the traction vectors acting at neighboring points across the interface must be equal in magnitude and oppositely directed. In other words, the traction that the layer imposes on the medium must be equal and opposite to the traction that the medium imposes on the layer. Equal and opposite tractions means the respective normal stresses acting *perpendicular* to the interface must match, and the respective shear stresses acting *tangential* to the interface must match (**Figure 9.44**):

$$\sigma_{nn}^{\mathrm{L}}(x,h+\zeta)=\sigma_{nn}^{\mathrm{M}}(x,\zeta),\sigma_{ns}^{\mathrm{L}}(x,h+\zeta)=\sigma_{ns}^{\mathrm{M}}(x,\zeta) \tag{9.57}$$

The normal and shear stress components can be written in terms of those referenced to the Cartesian coordinate system using the slope angle θ (**Figure 9.44**). These expressions are simplified for small slope angles to write approximate matching conditions for the base and perturbed state stress components. The derivation is lengthy and little new physical insight is provided, so we omit the complete derivation.

The governing equations for the perturbed flow include the incompressible condition for two-dimensional flow in the (x, z)-plane (Section 6.7.3):

$$\frac{\partial \tilde{v}_x}{\partial x}+\frac{\partial \tilde{v}_z}{\partial z}=0 \tag{9.58}$$

One of the classic methods for solving two-dimensional problems in fluid dynamics exploits the stream function, $\psi(x, z)$, which is related to the velocity components as:

$$\tilde{v}_x=+\frac{\partial \psi}{\partial z},\tilde{v}_z=-\frac{\partial \psi}{\partial x} \tag{9.59}$$

Substitution of (9.59) into (9.58) shows that this relationship between the velocity components and the stream function guarantees incompressible flow.

The stress components for the perturbed flow are written in terms of the stream function by starting with the constitutive equations (9.43), substituting for the rates of deformation using the kinematic equations (9.42), and then substituting for the velocity components using (9.59):

$$\begin{aligned}\tilde{\sigma}_{xx}&=-\bar{p}+2\eta\frac{\partial^2\psi}{\partial x\partial z},\tilde{\sigma}_{xz}=2\eta\left(\frac{\partial^2\psi}{\partial z^2}-\frac{\partial^2\psi}{\partial x^2}\right),\tilde{\sigma}_{zz}\\&=-\bar{p}-2\eta\frac{\partial^2\psi}{\partial x\partial z}\end{aligned} \tag{9.60}$$

These stress components are substituted into the two equilibrium equations (9.44) and the pressure is eliminated between them to find the biharmonic equation for the stream function:

$$\frac{\partial^4\psi}{\partial x^4}+2\frac{\partial^4\psi}{\partial x^2\partial z^2}+\frac{\partial^4\psi}{\partial z^2}=0 \tag{9.61}$$

This equation governs two-dimensional flow of an incompressible viscous fluid. Recall that the biharmonic equation for the Airy stress function was derived in Section 7.2.4 to solve the two-dimensional quasi-static problem of linear elasticity with traction (stress) boundary conditions.

The biharmonic equation for the stream function (9.61) is solved, subject to the boundary conditions on the interfaces and in the remote field, yielding $\psi(x, z)$ for the perturbed flow. To solve the problem of the wavy viscous layer in viscous surroundings, we determine the stream function in two different regions: one represents the layer and the other represents the medium above the layer. Then, symmetry determines the stream function in the medium below the layer.

The stream function that satisfies (9.61) in both regions is:

$$\begin{aligned}\psi(x,z)=(1/\lambda)\{[a+b(\lambda z-1)]e^{\lambda z}-\\ [c+d(\lambda z+1)]e^{-\lambda z}\}\sin\lambda x\end{aligned} \tag{9.62}$$

The four constants of integration (a, b, c, and d) must be determined by the boundary conditions. In general, the perturbed velocity components are found using (9.59):

$$\begin{aligned}\tilde{v}_x&=\left[(a+b\lambda z)e^{\lambda z}+(c+d\lambda z)e^{-\lambda z}\right]\sin\lambda x\\ \tilde{v}_z&=-\left\{[a+b(\lambda z-1)]e^{\lambda z}-[c+d(\lambda z+1)]e^{-\lambda z}\right\}\cos\lambda x\end{aligned} \tag{9.63}$$

Note that the velocity components vary with the sine and cosine of λx in keeping with the sinusoidal waviness of the interface (**Figure 9.44**). Both components of velocity vary with z as increasing and decreasing exponential terms.

To capture the behavior of the layer, we insist that symmetric points on opposite sides of the x^{L}-axis (**Figure 9.44**) must have the same velocity vectors: $v_x^{\mathrm{L}}(x^{\mathrm{L}},-z^{\mathrm{L}})=v_x^{\mathrm{L}}(x^{\mathrm{L}},z^{\mathrm{L}})$ and $v_z^{\mathrm{L}}(x^{\mathrm{L}},-z^{\mathrm{L}})=v_z^{\mathrm{L}}(x^{\mathrm{L}},z^{\mathrm{L}})$. This condition is satisfied if $c = -a$ and $d = b$ in (9.63), so the velocity field in the layer is:

$$\begin{aligned}\tilde{v}_x^{\mathrm{L}}&=\left[(a+b\lambda z^{\mathrm{L}})e^{\lambda z^{\mathrm{L}}}+(-a+b\lambda z^{\mathrm{L}})e^{-\lambda z^{\mathrm{L}}}\right]\sin\lambda x^{\mathrm{L}}\\ \tilde{v}_z^{\mathrm{L}}&=-\left\{\left[a+b(\lambda z^{\mathrm{L}}-1)\right]e^{\lambda z^{\mathrm{L}}}-\left[-a+b(\lambda z^{\mathrm{L}}+1)\right]e^{-\lambda z^{\mathrm{L}}}\right\}\cos\lambda x^{\mathrm{L}}\end{aligned} \tag{9.64}$$

This is different from the base state velocity field in the layer (**Figure 9.43**), where the x-components on opposite sides of the x^{L}-axis are equal, but the z-components are equal and oppositely directed.

The medium above the layer is an infinite half-space with a lower boundary defined by the wavy interface at $z^{\mathrm{M}} = \zeta$. Both velocity components in the medium must go to zero with distance from the interface: $v_x^{\mathrm{M}}\to 0$ as $z^{\mathrm{M}}\to+\infty$ and $v_z^{\mathrm{M}}\to 0$ as $z^{\mathrm{M}}\to+\infty$. This condition is satisfied if the positive exponential terms are removed in (9.63). In other words, $a = 0 = b$, so the velocity field in the medium is:

$$\begin{aligned}\tilde{v}_x^{\mathrm{M}}&=\left[(c+d\lambda z^{\mathrm{M}})e^{-\lambda z^{\mathrm{M}}}\right]\sin\lambda x^{\mathrm{M}}\\ \tilde{v}_z^{\mathrm{M}}&=\left\{\left[c+d(\lambda z^{\mathrm{M}}+1)\right]e^{-\lambda z^{\mathrm{M}}}\right\}\cos\lambda x^{\mathrm{M}}\end{aligned} \tag{9.65}$$

Again, this is different from the base state velocity field in the medium (**Figure 9.43**), where both components increase without bounds with distance from the interface. Recall that at the beginning of Section 9.7 we suggested that at perpendicular distances of several wavelengths from the fold train we expect the deformation to be more or less homogeneous. Here we see quantitatively how the perturbed velocity decreases with distance, z, from the buckling layer. The decrease scales with $e^{-\lambda z^{\mathrm{M}}}$.

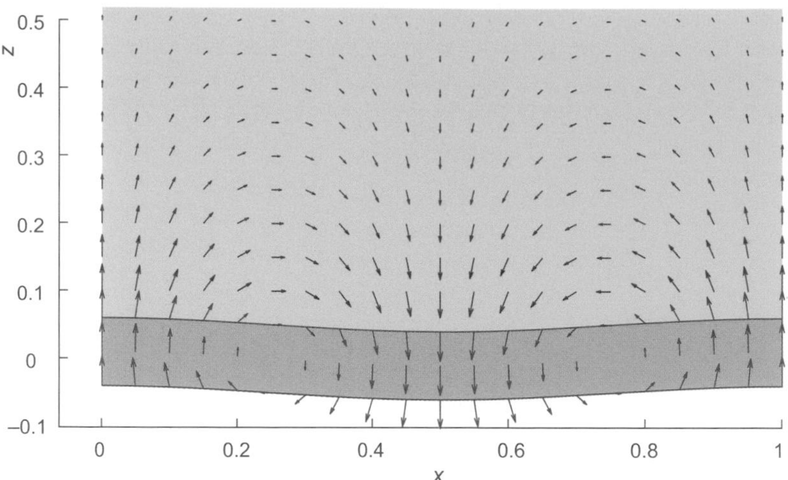

Figure 9.45 Velocity fields for layer (blue arrows) and medium above the layer (purple arrows). Perturbed wavelength $L = 1$; amplitude $A = L/100$; layer thickness $H = L/10$; viscosity ratio $R = 0.01$; rate of deformation $D_{xx} = -1$.

The stress components in the layer and in the medium can be written using the constitutive equations (9.60) with the stream function (9.62). Similar symmetry conditions to those leading to (9.64) reduce the stress components *in the layer* to functions of the two constants, a and b. Also, requiring the perturbed stress components *in the medium* to go to zero far from the interface reduces them to functions of the two constants, c and d. These four constants are determined using the matching conditions on velocity components at the interface (9.56), and the matching conditions on the normal and shear stress components at the interface (9.57). The algebra does not provide further conceptual understanding, so the details are omitted.

Using the following shorthand for the sum and difference of the common exponential terms reduces the size of the equations for the constants: $F = e^{\lambda h} + e^{-\lambda h}$, $G = e^{\lambda h} - e^{-\lambda h}$, $I = e^{2\lambda h} + e^{-2\lambda h}$, and $J = e^{2\lambda h} - e^{-2\lambda h}$. We also introduce the ratio of viscosity in the medium to that in the layer as $R = \eta_M/\eta_L$. Finally, the common denominator in equations for a and b is: $X = -4\lambda h(1 - R^2) + J(1 + R^2) + 2IR$. With these definitions, the constants for the layer are:

$$
\begin{aligned}
a &\simeq -2D_{xx}^{\mathrm{LM}}A(1-R)[\lambda h(F+RG)-(G+RF)]/X \\
b &\simeq 2D_{xx}^{\mathrm{LM}}A(1-R)(G+RF)/X
\end{aligned}
\tag{9.66}
$$

The displacement components in the layer (**Figure 9.45**) are found using these constants in (9.64). The constants for the medium are:

$$
\begin{aligned}
c &\simeq aG + b\lambda hF \\
d &\simeq -(a+b\lambda h)(F+G)+bG
\end{aligned}
\tag{9.67}
$$

Note that evaluation of c and d require substitution of a and b from (9.66). Using those values in (9.64) provides the velocity field in the medium (**Figure 9.45**).

The velocity vector fields for the layer and medium (**Figure 9.45**) clearly match across the interface, so the boundary conditions of *no slip* and *no separation* are met. However, the base state velocity field (**Figure 9.43**) and the perturbed velocity field (**Figure 9.45**) are quite different. The sinusoidal wave prescribed for the layer appears to be amplified by velocity vectors directed upward over the antiformal hinge and downward over the synformal hinge. Over the limbs of the nascent fold the velocity vectors turn to form a circulating pattern with upwelling over the antiform and downwelling over the synform. Note in particular that the perturbed velocity components decrease in magnitude with distance from the layer in the z-direction. This is consistent with the conditions imposed to derive (9.65).

To be a viable mechanism for folding, the waviness, defined as the geometric perturbation of the straight interfaces, must grow with time. In other words, the maximum slope of the limbs, λA must increase with time. The derivative of the product λA is:

$$
\frac{\mathrm{d}(\lambda A)}{\mathrm{d}t} = \lambda\frac{\mathrm{d}A}{\mathrm{d}t} + A\frac{\mathrm{d}\lambda}{\mathrm{d}t}
\tag{9.68}
$$

The last term is $-D_{xx}^{\mathrm{LM}}\lambda A$. The time derivative of the maximum slope is:

$$
\begin{aligned}
\frac{\mathrm{d}(\lambda A)}{\mathrm{d}t} &= q\left|D_{xx}^{\mathrm{LM}}\right|(\lambda A), \text{ where} \\
q &= 2 - \frac{2(1-R)}{(1-R^2) - \left[(1-R^2)J + 2RI\right]/4\lambda h}
\end{aligned}
\tag{9.69}
$$

The factor q is plotted in **Figure 9.46** for different values of the inverse viscosity ratio $1/R = \eta_L/\eta_M$. In that figure q is plotted versus $2\lambda h = 4\pi H/L$, where H/L is the ratio of layer thickness to fold wavelength.

In **Figure 9.46** note that as the viscosity of the layer increases relative to that of the medium, the amplification factor, q, for the limb slope increases. The maximum value for each curve occurs where the thickness to wavelength ratio provides the greatest amplification. This ratio, or rather its inverse, is referred to as the dominant wavelength, because this wavelength is the one that grows the most rapidly. If the curve has a well-defined and sharp peak, one would expect the dominant wavelength to be prominent in a train of folds, such as that shown in **Figure 9.40**. This behavior is favored for greater ratios of layer to medium

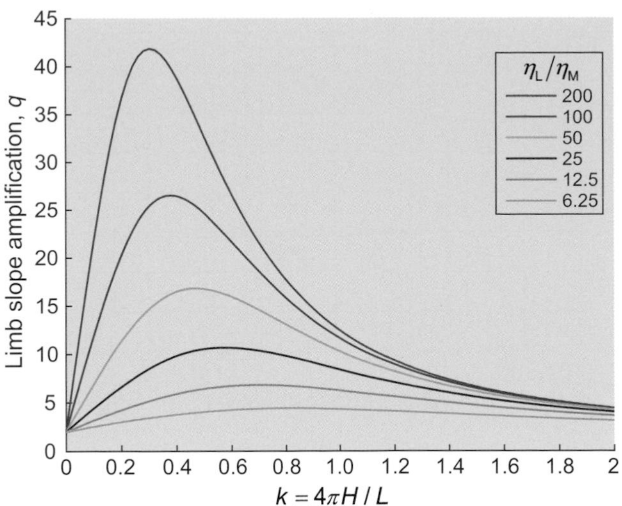

Figure 9.46 Plot of the limb slope amplification factor, q, versus $k = 2\lambda h = 4\pi H/L$ for six different values of the inverse viscosity ratio.

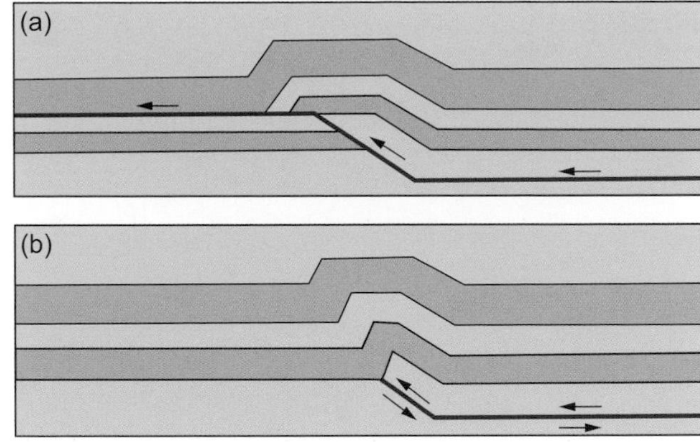

Figure 9.47 Kinematic models for fault-cored anticlines. (a) Fault-bend fold forms as the overlying block moves to the left on a through-going fault (red line) with an inclined step. (b) Fault-tip fold forms as slip accumulates on a terminating fault (red line) with an inclined tip. Modified from Johnson (2018).

viscosity. Note also that the layer thickness to wavelength ratio, k, at the dominant wavelength, decreases as the ratio of layer viscosity to medium viscosity increases. Put another way, for layers with the same thickness, the dominant wavelength increases as the ratio of layer viscosity to medium viscosity increases.

9.8 FAULT-CORED ANTICLINES

Relationships between thrust faults and anticlinal folds have captured the attention of structural geologists for decades. One motivation for this is the presence of hydrocarbon traps in anticlinal folds associated with thrust faults. For example, more than 2,000 wells and very detailed seismic surveys reveal this association of faults and folds in the Elk Hills oil field (Section 8.3.2). The cross section (**Figure 8.23**) shows a major thrust fault underlying each anticlinal crest, a geometry that is a characteristic of this field, which yielded more than 1.2 billion barrels of oil by 2010. That geometry, repeated in many oil fields around the world, provides the title for this section, fault-cored anticline. It also gives rise to interpretations, even where the underlying thrust fault is not exposed. For example, the cross section of Sheep Mountain anticline (**Figure 9.14**) includes a speculative underlying thrust fault.

Another motivation for the attention given to fault-cored anticlines by geologists and geophysicists is the association of blind thrust faults with seismic hazards (Johnson, 2018; see Further Reading). The name "blind" implies the fault is not exposed at Earth's surface, but the overlying anticline may be exposed, or its geometry may be established from shallow wells and seismic surveys. Thus, models that link the mechanical behavior of the anticline to slip on the fault enable one to make inferences about the underlying fault, and possible seismic hazards, from data on the anticline.

Much of the geological literature on fault-cored anticlines employs kinematic models to investigate the geometry of these structures (Brandes and Tanner, 2014). Two well-studied examples are fault-bend folds (**Figure 9.47a**) and fault-tip folds (**Figure 9.47b**). The model fault-bend folds in this illustration form over a horizontal fault with an inclined ramp. Here, the overlying strata move to the left without deforming over the horizontal parts of the fault, but deform into kink folds (**Figure 9.5**) over the ramp. The model fault-tip folds in this illustration form over a horizontal fault with an inclined tip. Deformation associated with the reduction in slip at or near the fault tip deforms the overlying strata into kink folds. The fault tip in this case is stationary, but this class of kinematic models also includes fault tips that propagate and form fault-propagation folds.

Fletcher and Pollard (1999) review the basic elements of kinematic models, which typically begin with geometric idealizations, such as the horizontal strata and the horizontal fault with a ramp that was the initial stage for the fault-bend fold in **Figure 9.47a**. Then, one assumes an evolutionary concept that prescribes the temporal sequence of events during folding. This may include, for example, constraints on fold limb lengths, and distributions of slip on the fault. Finally, purely geometric and trigonometric relationships lead to a detailed kinematics and a final geometry. Kinematic models are appealing when they lead to approximations of the observed geometry, or permit restoration of the hypothesized initial geometry, so they find many applications in both academia and industry (Brandes and Tanner, 2014). However, demonstrating that the kinematic models adhere to Newton's laws is problematic, because the kinematic quantities are not linked to the causative stresses (Sections 3.5 and 3.6).

To highlight some of the insights gained by investigating a mechanical model for fault-cored anticlines, we use results from Johnson (2018). The model includes slip on a reverse fault and layers that can accommodate flexural slip folding on frictional bedding contacts. Thus, the model is well suited to evaluate the contributions to fold growth due to fault slip, relative to fold growth due to buckling associated with bedding-contact slip. Equations of motion that follow from Cauchy's first and second

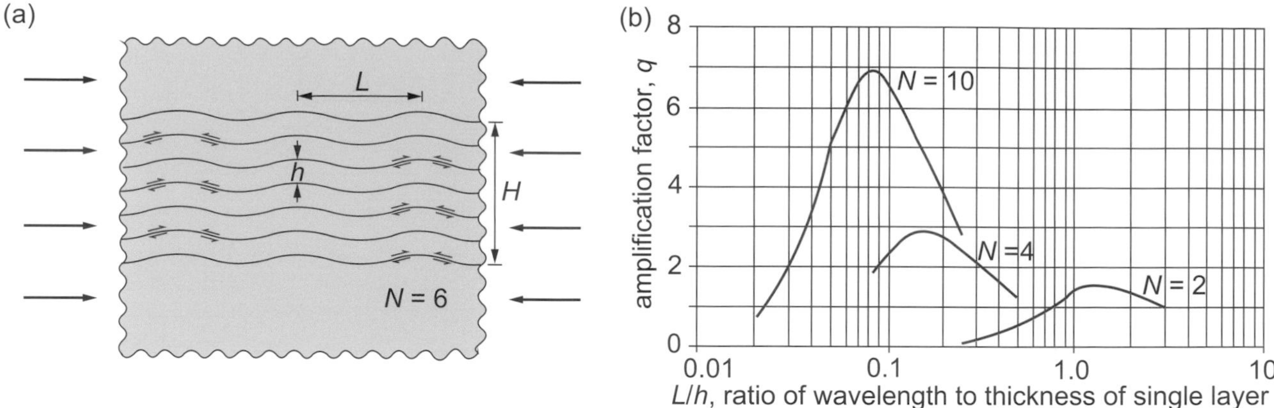

Figure 9.48 Mechanical behavior of multilayer viscous folding. (a) Six layers each of thickness h with fold wavelength L making up multilayer of total thickness H. (b) Amplification factor, q, plotted versus wavelength to thickness ratio. N is number of layers. Modified from Johnson (2018).

laws (Sections 3.5–3.7) govern the behavior of the model. The material properties of the model are viscoelastic, including both linear viscous (Section 6.7) and linear elastic (Section 4.10) behaviors, and the ratio of Newtonian viscosity, η, to elastic shear modulus, G, quantifies the relative importance of these behaviors. Stresses induced by slip on the fault and folding of the multilayer decay exponentially in time, with a characteristic relaxation time given by the ratio, η/G. For a given slip rate on the fault, the behavior of the multilayer is mostly elastic for long relaxation times, and mostly viscous for short relaxation times. Thus, the model admits a useful comparison of these behaviors.

Recall from Section 9.7.3 that an initial sinusoidal perturbation of a single viscous layer grows in amplitude under horizontal shortening according to the amplification factor for the limb slope (Figure 9.46). In that case, the dominant wavelength is more prominent in a train of folds for greater ratios of layer to medium viscosities. A similar phenomenon exists for a multilayer fold with a given total thickness, H (Figure 9.48a). The amplification factor increases as the number of layers, N, increases, and the thickness, h, of individual layers decreases (Figure 9.48b). In addition, the dominant wavelength decreases as the number of layers increases. These results do not include the action of an underlying reverse fault, but suggest that buckling could enhance the growth of a fault-cored anticline.

Physical experiments using rubber strips, and theoretical investigations of multilayers with elastic or viscous constitutive properties, show that bedding contact friction, μ, effects the form of the folds subject to layer parallel shortening (Johnson, 2018). An initial sinusoidal perturbation of the layers grows into a train of rounded, parallel folds, if the contact friction is near zero ($\mu \sim 0$). If the contact friction is moderate ($\mu = 0.6$), the sinusoidal perturbations grow into more localized and kinked fold forms in conjugate sets. These models, however, do not include the effects of slip on an underlying fault. In the case of fault-cored anticlines, slip on the fault induces some layer parallel shortening, but the distribution of stress in the overlying layers is complicated, so mechanical modeling that includes a fault is required to understand how bedding contact friction and fold form are related.

The mechanical model employed by Johnson (2018) uses the Boundary Element Method (BEM) to solve for the displacement field throughout the multilayer and the slip distributions on the layer contacts. The details of BEM are beyond the scope of this textbook, so we refer interested readers to the textbooks by Crouch and Starfield (1983) and Segall (2010), see Further Reading. None-the-less, the concepts and results illustrated by the BEM models are central to the objectives of this chapter, so we review some of those concepts and results here.

Figure 9.49 evaluates how friction between layers influences the form of fault-cored anticlines for a horizontal fault with an upturned tip and no fault propagation. For this example, the viscoelastic layers have a high ratio of viscosity to elastic shear modulus, so the behavior is dominantly elastic. For freely slipping bedding contacts (Figure 9.49a), a rounded, parallel fold develops over the upturned portion of the fault. The bedding slip is right lateral for the fold limb over the fault tip, and left lateral for the fold limb over the bend in the fault. In contrast, for moderate friction on bedding contacts (Figure 9.49b), a pair of conjugate kink folds develop, separated by nearly horizontal layers. The left-facing kink is over the fault tip and develops first; the right-facing kink is over the fault bend and develops later. The left-facing kink dips more steeply than the right-facing kink. Despite the more complicated stress distribution associated with faulting, the relationship between bedding contact friction and fold form is similar to that found for layer-parallel shortening without a fault.

The form of the fault-cored anticline in the models by Johnson (2018) also depends on the elastic stiffness and the viscosity. For example, with little or no viscous relaxation, a stiffer elastic multilayer ($G = 10$ GPa) deforms by pervasive flexural slip into rounded, parallel folds similar to those in Figure 9.49a. For a given slip on the fault, the induced stresses are greater, because the elastic modulus is greater, and these stresses overcome the frictional resistance to bedding contact slip throughout the fold. In contrast, a softer multilayer ($G = 1$ GPa) deforms into localized conjugate kink bands with a relatively flat crest, because the

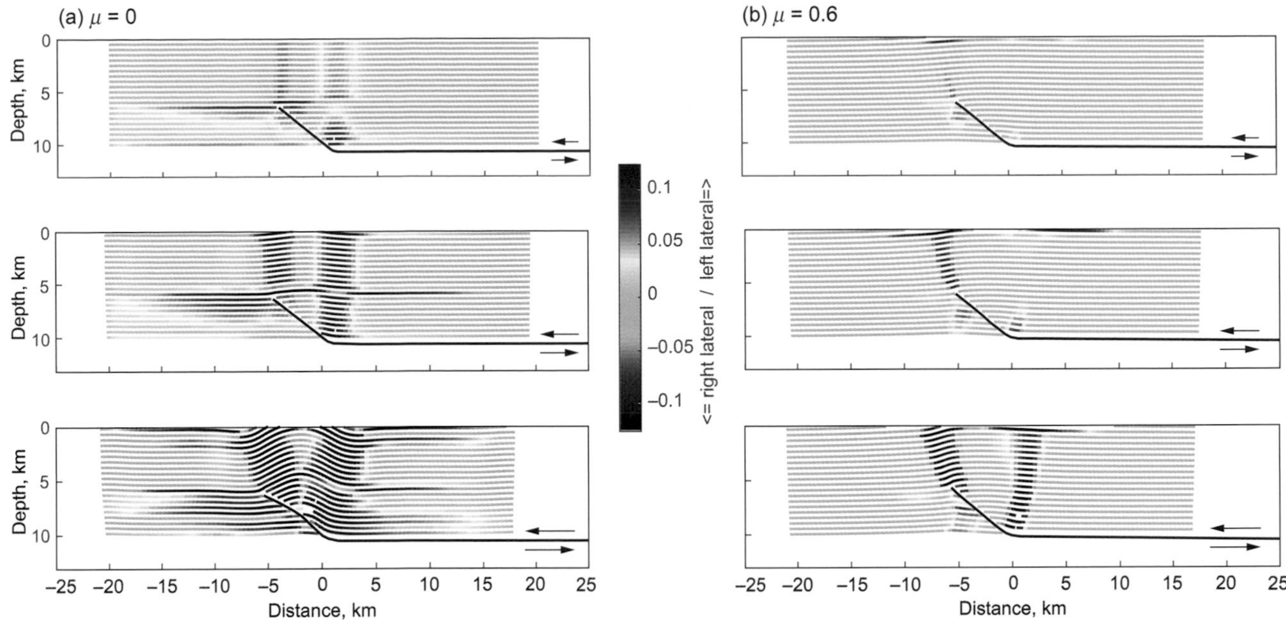

Figure 9.49 Effects of friction at layer contacts on the form of fault-cored folds. Slip on fault increases from top to bottom. (a) Layer contacts are freely slipping, μ = 0. (b) Layer contacts are frictional, μ = 0.6. Modified from Johnson (2018).

stresses induced by a given slip on the fault are an order of magnitude less, and only locally overcome the frictional resistance to bedding contact slip.

The BEM models of fault-cored anticlines demonstrate the advantages of a mechanical analysis (**Figure 9.49**) compared to a purely kinematic analysis (**Figure 9.47**). The explicit linkage of stress to strain and stress to rate of deformation through constitutive laws enables one to assess the geometry of the folds in terms of material properties such as friction, elastic modulus, and viscosity. Furthermore, the mechanical models are consistent with the fundamental physics incorporated in Cauchy's laws of motion.

Recapitulation

We began Chapter 9 by defining a fold as a surface in a body of rock that has undergone a change in *curvature*. The folded surface could be the top of a sandstone bed, foliation in a metamorphic schist, or a fracture surface in a massive igneous rock. The simplest idealization of a fold is one that is sinusoidal in cross section with a *hinge* where the curvature is greatest, two adjoining *limbs* where the curvature is reduced, and two *inflections* where the surface has zero curvature. A straight line, called the *fold axis*, can generate a cylindrical fold in a thought experiment in which this axis moves from one limb through the hinge to the next limb while remaining parallel to itself. The two-dimensional trace of the folded surface in a plane perpendicular to the fold axis is called the fold *profile*. In such a plane the distance between two inflections for the sinusoidal fold is one half the fold *wavelength*. The perpendicular distance from the line containing the inflections to the line containing the hinges is the fold *amplitude*.

Multilayer folds may be idealized as *similar*, if each surface has an identical shape, and the perpendicular distance between surfaces is less in the limbs than at the hinge. Or, the multilayer may be idealized as *parallel*, if surfaces maintain a constant perpendicular distance from one another. An imaginary surface that passes through the hinges on successive surfaces is called the *axial surface*. If the rock layers get progressively older toward the axial surface, the fold is called an *anticline*; if the layers get progressively younger, it is a *syncline*. If the relative ages of layers in a multilayer fold are not known, a convex upward fold is called an *antiform*, and a concave upward fold is called a *synform*.

Folds span length scales from microscopic to crustal, that is from fractions of a millimeter to tens of kilometers. Examples are given of outcrop scale maps prepared with tape and compass, and of fold profiles captured in photographs. We turn to *differential geometry* to find mathematical tools and concepts for describing the geometry of these fold profiles in terms of the *curvature* of a plane curve. Mapping and describing crustal scale folds requires different tools, and an example is given of mapping on images taken by Airborne Laser Swath Mapping (ALSM). Folds also may be characterized using the Global Positioning System (GPS) to quantify their geometry in three dimensions. Again, curvature is the key geometric property, so we extend what was learned for plane curves to a three-dimensional curve using a *parametric representation*. Those concepts enable us to define a fold that is composed of two sets of curves, the *u*- and *v*-parameter curves for that surface. The tangent vectors to the *u*- and *v*-parameter curves define the *tangent plane* to the surface, which is used by structural geologists to measure the strike and dip. The tangent vectors also are used to define the *unit normal vector* to the surface, which quantifies its orientation in three-dimensional space.

To characterize the geometry of a folded surface, we focus attention on the neighborhood of a point and answer two questions: (1) How is an infinitesimal distance along the surface measured? (2) What is the local shape of the surface? The answers to these questions give us the *first* and *second fundamental forms* of the surface. The coefficients of the second fundamental form are used to categorize the local shape of any folded surface as *planar*, *elliptic*, *parabolic*, or *hyperbolic*. The coefficients of the first and second fundamental forms also are used to define a symmetric 2 by 2 matrix called the *shape operator*. Solving the eigenvalue problem for the shape operator defines the two principal curvature magnitudes and directions for any point on the folded surface. The product of the principal magnitudes is called the *Gaussian* curvature. Using the Gaussian curvature and mean curvature, a table of fold shapes is produced that serves to classify every point on a fold in three dimensions in terms of eight fundamental shapes.

Given the different geologic and tectonic settings where folds are found, and the broad range of length scales, it is not surprising that different mechanisms exist for the development of folds. We develop a canonical model for folding based on the mechanism called *bending* and consider this for a thin *elastic* layer. Unlike every other canonical model reviewed in this textbook, for the bending model we do not start with *Cauchy's equations of motion*. Instead, we set up the model geometry and invoke equilibrium of forces and moments on a small element. The kinematics of pure bending and the elastic constitutive law are combined with the equilibrium equations to find the *governing equation for bending*. This equation is solved for bending driven be pressure applied across the layer and the resulting shape of the bent layer is compared favorably to the shape of strata over a laccolith in the Henry Mountains of southeastern Utah. For the laccolith the driving pressure is supplied by magma that intrudes between two sedimentary layers, spreads laterally, and then bends the overlying layers upward. After deriving the stress distributions in the bent layer, we circle back and show that these distributions do indeed satisfy Cauchy's equations as reduced to the case of quasi-static equilibrium.

The second canonical model considers the folding mechanism called *buckling*, and this is applied to a single layer in a surrounding medium, both of which deform by *viscous* flow. The more viscous layer and the less viscous surroundings are shortened parallel to the plane of the layer and lengthened perpendicular to the layer with rates of deformation appropriate for incompressible flow in two dimensions. *Cauchy's first law of motion* is reduced to the equilibrium equations for *slow viscous flow*. In order to solve for the stress components a lithostatic pressure is chosen for the medium. The pressure in the layer and the stress components in the layer and the medium are dependent on the respective viscosities and the rates of deformation. For a perfectly planar layer with symmetric shortening and lengthening, no buckling occurs. This *base state* is added to a *perturbed state* in which the layer is prescribed to have a sinusoidal waviness with low amplitude. The *stream function* is introduced and the *biharmonic* governing equation is solved with matching velocity and stress conditions along the interfaces between the layer and the medium. The sinusoidal wave increases in amplitude with time as a function of the ratio of viscosities and the ratio of layer thickness to wavelength, so this mechanism is capable of producing buckle folds.

Motivations for studying *fault-cored anticlines* include the presence of hydrocarbon traps in these structures, and their association with seismic hazards. A mechanical model for fault-cored anticlines includes slip on a reverse fault and viscoelastic layers that can accommodate flexural slip folding on frictional bedding contacts. This model provides an evaluation of the contributions to fold growth due to fault slip, relative to fold growth due to buckling associated with bedding-contact slip. The layers grow into rounded, parallel folds, if the contact friction is near zero, and into more localized and kinked fold forms in conjugate sets, if the contact friction is moderate. For example, with little or no viscous relaxation, a stiffer elastic multilayer deforms by pervasive flexural slip into rounded, parallel folds. In contrast, a softer multilayer deforms into localized conjugate kink bands with a relatively flat crest. The explicit linkage of stress to strain and stress to rate of deformation through constitutive laws enables one to assess the geometry of the folds in terms of material properties such as friction, elastic modulus, and viscosity.

REVIEW QUESTIONS

The following questions are designed to highlight the expected *learning outcomes* for this chapter. Each question is taken directly from the material in the chapter and, for the most part, in the same sequence that it appears in the chapter. If an answer is not forthcoming, students are advised to read the relevant section of the chapter and discover the answer.

9.1. We define a fold as a surface in a body of rock that has undergone a change in curvature. Describe five surfaces, at least one each in sedimentary, metamorphic, and igneous rocks that respond to tectonic events by folding.

9.2. Two common folding mechanisms are *bending* and *buckling*. Illustrate each using a sketch of the folded layer and describe how the loading distinguishes the two mechanisms.

9.3. Use a three-dimensional sketch to illustrate the geometry of a sinusoidal cylindrical fold and distinguish the *hinge*, *limb*, and *inflection*. Also, point out the *amplitude* and *wavelength*, and draw a line that is parallel to the *fold axis*.

9.4. Draw a multilayer with parasitic folds in profile including Z, M, and S shapes. Identify the axial surfaces and distinguish *antiforms* from *synforms*.

9.5. Use a plane curve to represent the profile of the fold and describe the *radius of curvature* and the *curvature* at a few points along the profile. What is the mathematical relationship between these two geometric quantities?

9.6. Use the vector-valued function $\mathbf{c}(t) = t\hat{\mathbf{i}} + (t^2 - 2t + 4)\hat{\mathbf{j}}$ for a *plane curve* representing a fold and derive the vector-valued functions for the *unit tangent vector* and the *curvature vector*. Assuming the curve was originally a straight line, use the equation for the curvature vector to find the scalar curvature and identify the position of the fold hinge. What are the scalar curvature and radius of curvature at the hinge?

9.7. An idealized antiform is defined using the three-dimensional vector-valued function $\mathbf{s}(u, v) = u\hat{\mathbf{i}} + v\hat{\mathbf{j}} - \frac{1}{2}(u^2 + 2v^2)\hat{\mathbf{k}}$. Use this function to describe the u-parameter curves and the v-parameter curves that completely cover, and thus define, the three-dimensional antiformal surface.

9.8. Using the same antiformal surface defined in Question 9.7, describe how one calculates the *tangent vectors* to the u-parameter curves and the v-parameter curves, and explain how these are used to define the *unit normal vector* to the surface. Then explain how this unit normal vector is related to the strike and dip of the folded surface, which would be measured at an outcrop.

9.9. The following Google Earth image is taken from the file: Question 09 09 Sheep Mountain anticline. kmz. Open the file; click on Edit in the Menu Bar; select Properties to open the Edit Placemark dialogue box; and address the questions in this box. The Google Earth image data are from Landsat / Copernicus.

9.10. The *first fundamental form* for a three-dimensional folded surface is defined as: $\mathrm{I} = d\mathbf{c} \cdot d\mathbf{c} = E(du)^2 + 2F(dudv) + G(dv)^2$. Describe and sketch the differential tangent vector, $d\mathbf{c}$, and explain how this is a measure of arc length on the surface.

9.11. The second fundamental form for a three-dimensional folded surface is defined as: $\mathrm{II} = -d\mathbf{N} \cdot d\mathbf{c} = L(du)^2 + 2M(dudv) + N(dv)^2$. Describe the differential normal vector, $d\mathbf{N}$, and explain with words and a sketch how the second fundamental form captures the component of $d\mathbf{N}$ in the direction $d\mathbf{c}$.

9.12. In the neighborhood of every point on a three-dimensional folded surface the shape can be categorized using the coefficients of the second fundamental form: L, M, and N. Use these coefficients to define the four possible shapes and give a geologic name for each.

9.13. Why do structural geologists add three more shapes to the mathematical construction of four possible shapes defined in Question 9.12? How is this done?

9.14. At every location on a typical three-dimensional fold in nature the curvature varies with orientation. To quantify this variation one uses the coefficients of the first and second fundamental forms to define the shape operator. Explain how the shape operator is calculated, and how it is used to characterize the variation in curvature with orientation.

9.15. The Gaussian curvature and the mean normal curvature are used to construct a color table (**Figure 9.26**) that classifies folded geological surfaces. Describe these two curvatures, and explain how they are used in the color table.

9.16. Use the geometry depicted in **Figure 9.31** to describe how the normal strain in the plane of a bending layer is related to the curvature and the distance from the neutral plane. Include Hooke's law to show how the normal stress is related to these quantities.

9.17. The governing equation for the bending layer over a laccolith is:

$$\frac{d^4w}{dx^4} = \frac{12p}{Bh^3} = \frac{p}{R}$$

Identify the SI units of all the physical quantities and show that the units are consistent from term to term. Integrate four times and use boundary conditions for the laccolith to solve for the vertical displacement. Describe the sensitivity of the displacement to the physical parameters in your solution. How would these sensitivities guide your field work on laccoliths?

MATLAB EXERCISES FOR CHAPTER 9: FOLDS

Chapter 9 characterizes folds as geological surfaces that have undergone a change in curvature. Documentation of the geometric and kinematic features of folds takes place at outcrops, on maps of various scales, and in seismic reflection surveys. The first exercise reviews the terminology for folds. Then, MATLAB illustrates how vector-valued functions quantify fold profiles in two dimensions using plane curves and the scalar curvature. We extend these concepts to three-dimensional curves and quantify the base of the Sundance formation at Sheep Mountain anticline. Two exercises focus on the second fundamental form for such a surface, which quantifies the local shape as planar, elliptic, parabolic, or hyperbolic. We explore how the unit normal vector to a surface, and the curvature vectors to curves lying in that surface, lead to the principal values and principal orientations of normal curvature. The Gaussian curvature and mean normal curvature form the basis for a geological curvature classification table. Application of this table to the Sundance formation demonstrates that different patches are basins and domes, as well as synformal and antiformal saddles. These exercises help students understand how differential geometry provides the mathematical tools for characterizing and classifying folds based on the principal normal curvatures. www.cambridge/SGAQI

FURTHER READING

For citations in figure captions see the reference list at the end of the book.

Crouch, S. L., and Starfield, A. M., 1983. *Boundary Element Methods in Solid Mechanics: With Applications in Rock Mechanics and Geological Engineering*. George Allen & Unwin, London.
This textbook introduces the boundary element method for solving problems in solid mechanics, and emphasizes the physical aspects of the methods rather than the mathematical difficulties.

Hilley, G., Mynatt, I., and Pollard, D., 2010. Structural geometry of Raplee Ridge monocline and thrust fault imaged using inverse Boundary Element Modeling and ALSM data. *Journal of Structural Geology* 32, 45–58.
This paper describes the application of ALSM data to mapping the fold at Raplee Ridge, UT, and uses continuum mechanics models to constrain the geometry of an underlying fault.

Hudleston, P. J., and Treagus, S. H., 2010. Information from folds: a review. *Journal of Structural Geology* 32, 2042–2071.
The authors review the information one can interpret from observations of folds, predominantly at the outcrop scale in metamorphic rocks, including the rheological properties and bulk strain, and provide an extensive reference list on folding.

Jackson, M. D., and Pollard, D. D., 1988. The laccolith-stock controversy: new results from the southern Henry Mountains, Utah. *Geological Society of America Bulletin* 100, 117–139.
New geologic mapping and geophysical data shed light on the shapes of the central intrusions in the Henry Mountains, suggesting that they are laccoliths, as conceived by Gilbert rather than stocks.

Johnson, A. M., and Fletcher, R. C., 1994. *Folding of Viscous Layers: Mechanical Analysis and Interpretation of Structures in Deformed Rock*. Columbia University Press, New York.

The principles of fluid mechanics are used to analyze the buckling of layered viscous materials under various geometric configurations and boundary conditions with applications to natural folds in stratified rock.

Johnson, K. M., 2018. Growth of fault-cored anticlines by flexural slip folding: analysis by boundary element modeling. *Journal of Geophysical Research: Solid Earth* 123, 2426–2447.
This paper reviews the published kinematic and mechanical models for fault-cored anticlines and presents new results using boundary element methods for viscoelastic layers in frictional contact with an embedded fault.

Pollard, D. D., and Johnson, A. M., 1973. Mechanics of growth of some laccolithic intrusions in the Henry Mountains, Utah, II. *Tectonophysics* 18, 311–354.
A stack of thin elastic plates provides a model for bending the sedimentary strata over a laccolith and for estimating the effective thickness of the overburden.

Ragan, D. M., 2009. *Structural Geology: An Introduction to Geometrical Techniques*, 4th edition. Cambridge University Press, Cambridge.
The geometry and geometric analysis of folds are described and analyzed in this introductory textbook of techniques for the structural geologist.

Ramberg, H., 1967. *Gravity, Deformation and the Earth's Crust*. Academic Press, London.
Combining theoretical analysis of deformation with laboratory centrifuged models, this text considers tectonic deformation at the scale of Earth's crust where gravitational forces are important.

Ramsay, J. G., 1967. *Folding and Fracturing of Rocks*. McGraw-Hill, New York.
The geometry of folds and superimposed folds, and the kinematics of folding are described and analyzed in this textbook with particular attention to the state of strain within and outside of folded layers.

Savage, H. M., Shackleton, J. R., Cooke, M. L., and Riedel, J. J., 2010. Insights into fold growth using fold-related joint patterns and mechanical stratigraphy. *Journal of Structural Geology* 32, 1466–1475.
This paper uses mechanical stratigraphy and joint pattern data to show how jointing evolves as the Sheep Mountain anticline grew laterally along its axis.

Segall, P., 2010. *Earthquake and Volcano Deformation*. Princeton University Press, Princeton, NJ.
This advanced textbook goes beyond the elementary mechanics and structural geology covered here, with a focus on geophysical data related to active deformation associated with volcanic and earthquake hazards.

Slatton, K. C., Carter, W. E., Shrestha, R. L., and Dietrich, W., 2007. Airborne laser swath mapping: achieving the resolution and accuracy required for geosurficial research. *Geophysical Research Letters* 34, L23S1023.
This article reviews the history and technical challenges of ALSM, and the rationale for setting up a National Center devoted to the acquisition and archiving of ALSM data.

Timoshenko, S., and Woinowsky-Krieger, S., 1959. *Theory of Plates and Shells*. McGraw-Hill Book Company, New York.
This classic textbook on bending of thin elastic plates and shells is written for engineers. It contains basic derivations of the governing equations, and hundreds of practical solutions for a wide variety of geometries, loading conditions, and boundary conditions.

Chapter 10
Fabrics

Introduction

Chapter 10 describes rock fabrics at outcrop and thin section scales in sedimentary, metamorphic, and igneous rock, introduces the kinematics of ductile deformation, builds a canonical model for a viscous shear zone, and relates fabric development to plastic deformation at fault steps. We revisit the spherical particles called ooliths that were introduced in Section 2.1 and use their alignment in the South Mountain–Blue Ridge Uplift to quantify the deformation that produced a planar fabric called cleavage. Next, we describe outcrops from the Sierra Nevada where shear zones initiated along aplite dikes that localized the strain, and where a fabric developed between the overlapping tips of echelon fault segments. To quantify the kinematics relevant to fabric development, we introduce the concepts of progressive deformation, and the rate of deformation and spin in a shear zone. Then, we develop the canonical model for a viscous shear zone from the equations of motion using the constitutive law for a linear viscous material. Finally, we investigate fabric development at a fault step in granitic rock, and show how a model that includes elastic deformation of the surrounding rock, frictional slip on the fault segments, and plastic deformation of rock within the step matches geometric observations from the outcrop.

10.1 DESCRIPTIONS OF ROCK FABRICS FROM OUTCROP TO THIN SECTION

The fabric of a rock is the configuration of all the geometric elements making up that rock. Because rock is composed of minerals, those geometric elements include the crystal structure of individual minerals, and the shapes of the mineral grains. Some rocks include aggregates of mineral grains that have distinct geometries, for example the ooliths from South Mountain fold (Section 2.1), or the cobbles in a metamorphic conglomerate (**Figure 10.1**). Thus, characterization of fabrics depends upon the scale of observation: some fabric elements are visible and measurable at the outcrop scale, but others require an optical microscope or an electron microscope.

At the broadest level of descriptive classification, structural geologists distinguish planar fabrics, called foliations, and linear fabrics, called lineations. For example, mineral grains with platy shapes, such as mica, define a foliation if they meet two criteria. First, most of the grains have a similar orientation. Second, the grains have a uniform distribution throughout the rock mass. In this case, the mica grains define a foliation by their *preferred* orientation and their *penetrative* distribution. The deformed cobbles (**Figure 10.1**) are distributed throughout the outcrop, so their distribution is penetrative, and their long axes have a preferred orientation, so they define a lineation.

At the outcrop scale the geometric elements making up the rock fabric might include the axial lengths, A and C, of the deformed cobbles of a metamorphosed conglomerate (**Figure 10.1**). Here we approximate the shapes of these cobbles as ellipsoids with orthogonal axes $A \geq B \geq C$. We don't see the third axial length, B, in this photograph, but if that length approximates the *shorter* axial length in the plane of the outcrop, the cobble shape would resemble a fat pencil, with one long axis, A, and two approximately equal shorter axes, $A >> B \approx C$. In that case the alignment of the pencil-like objects defines a linear fabric, and the orientation of the lineation (the orientation of the A-axes) would be recorded as a trend and plunge at this outcrop (Section 2.3.2).

If the cobbles in **Figure 10.1** were stretched perpendicular to the plane of the outcrop to lengths comparable to the *longer* axial lengths in the outcrop plane, their shapes would resemble pancakes. In this case the ellipsoid that approximated a cobble would have two long axes and one orthogonal shorter axis, $A \approx B >> C$. The alignment of the pancake-like objects would define a planar fabric, and the orientation of those planes would be recorded as a strike and dip at this outcrop (Section 2.3.2). Settling the question about the geometry of this fabric (linear or planar) would require additional outcrops that revealed the third dimension of the metamorphic conglomerate. If this were a microscopic fabric, additional thin sections cut orthogonal to the initial section would be required.

The stretch of the metamorphic conglomerate (**Figure 10.1**) in the orientation of the long axis of the ellipsoid is the ratio of the current to the original length of a material line in that orientation. In

Figure 10.1 Outcrop of stretched cobbles in a metamorphosed conglomerate from the Blackcap Mountain quadrangle, Sierra Nevada, CA. Cobble shapes are approximated as elliptical (inset) with long axis, A, and short axis, C. Photograph by S. Martel.

this case we do not know the original length, but the principal stretches can be estimated if the cobbles were approximately equidimensional in their undeformed state; there was no stretching perpendicular to the outcrop plane; and there was little volume change during deformation. This is a special case of deformation called pure shear strain (Section 5.5.1), with no stretch perpendicular to the plane that contains the extreme stretches, and the extreme stretches are reciprocals of each other. For the sake of this example, we use these assumptions to characterize the deformation.

The range of ratios of long axis, A, to short axis, C, in the outcrop plane (**Figure 10.1**) for 10 cobbles is $4.3 \leq A/C \leq 10.7$. The principal stretches of a particular cobble are $S_1 = A/R$ and $S_3 = C/R$, where R is the *spherical* cobble radius in the undeformed state, so $S_1/S_3 = A/C$. With no stretching perpendicular to the outcrop plane, $S_2 = 1$, and to preserve volume $S_3 = 1/S_1$, so $S_1 = (A/C)^{1/2}$. Thus, the range of greatest principal stretch is $2.1 \leq S_1 \leq 3.3$, and the range of least principal stretch is $0.3 \leq S_3 \leq 0.5$. The assumptions leading to these results need to be tested, but the estimates of principal stretches are a good starting point for relating the observed fabric to the deformation. This makes the point that fabric characterization is *not* an end in itself, but rather it is a first step toward quantifying the deformation.

For an example of fabric at the thin section scale consider **Figure 10.2**. This thin section is taken from the center of a right step between two echelon fault segments in granodiorite from the Sierra Nevada that we describe in Section 10.2.3. The elements making up the planar fabric include individual mineral grains and aggregates of quartz, feldspar, and mafic minerals aligned in tabular zones that are approximately parallel to the scale bar. We analyze the deformation using a numerical model in Section 10.5. At the sample location, the model gives a greatest principal stretch, $S_1 = 4.7$, oriented parallel to the fabric plane, and a least principal stretch, $S_3 = 0.21$, oriented perpendicular to the fabric plane. Again, one of the objectives of fabric analysis that we address in this

Figure 10.2 Thin section photograph from sample SG10–03 of sheared Lake Edison Granodiorite, Sierra Nevada, CA. Aggregates of deformed quartz, feldspar, and mafic minerals define a planar fabric wrapping around porphyroclasts. Box outlines area of EBSD analysis (**Figure 10.4**). Modified from Nevitt et al. (2017b), Figure 6a.

chapter is to associate the configuration of fabric elements with quantitative measures of deformation, such as the principal stretches.

At the scale of individual mineral grains, the crystal structure provides important fabric elements. Minerals are composed of regular arrangements of atoms or groups of atoms that form a repetitive pattern in three-dimensional space (see Klein and Philpotts, 2013, in Further Reading). Based on the symmetries of these repetitive patterns, six different crystal systems are distinguished. For example, quartz is a member of the *hexagonal* crystal system. The hexagonal geometry is displayed beautifully in the shape of quartz crystals (**Figure 10.3a**) that were free to grow without the impediments of surrounding crystals. This hexagonal geometry persists down to the scale of the smallest repetitive grouping of atoms, the so-called *unit cell* (**Figure 10.3b**). A set of crystallographic axes are defined with the c-axis parallel to the longest dimension of the unit cell. Three other axes, called a_1, a_2, and a_3, are arranged in a plane perpendicular to the c-axis, and are oriented at 120° from one another. The unit cell has dimensions $a_1 = a_2 = a_3 = 4.91 \times 10^{-10}$ m and $c = 5.41 \times 10^{-10}$ m.

When a quartz grain is forced to grow in competition with surrounding minerals, for example in a cooling magma, its exterior shape usually will not reflect the hexagonal geometry seen in **Figure 10.3a**, but the internal structure of quartz grains will maintain this geometry. Using a petrographic or electron microscope, the orientation of the crystallographic axes can be measured for individual grains, and if many grains in the same rock have similar orientations, the rock is said to have a Crystallographic Preferred Orientation (CPO; see Blenkinsop, 2000, in Further Reading). This phenomenon also is called a Lattice Preferred Orientation (LPO; see Passchier and Trouw, 1996, also in Further Reading). Although the determination of a CPO is quite tedious using a universal stage on a petrographic microscope, modern electron microscope techniques and analysis software make this determination much easier. This technology has brought new attention to the crystallographic axes as an important fabric element for the structural geologist. Quantitative relationships between a CPO and the local deformation (e.g. principal stretches) is the subject of current research, but the basic relationships between the symmetry elements of any physical property of a crystal and the symmetry elements of the crystal structure are well established (see Nye, 1985, in Further Reading).

As an example of fabric analysis using the orientations of c-axes of quartz grains, we reconsider the sample SG10–03 (**Figure 10.2**) from the Lake Edison Granodiorite. This sample was sheared between two fault segments and transformed into a mylonite, a strongly sheared rock with a planar fabric. The outcrop and the geologic context are described in more detail in Section 10.2.3, so we just highlight the fabric analysis here. The laboratory technique, called Electron Backscatter Diffraction (EBSD), produces a map (**Figure 10.4a**) of a plane through the

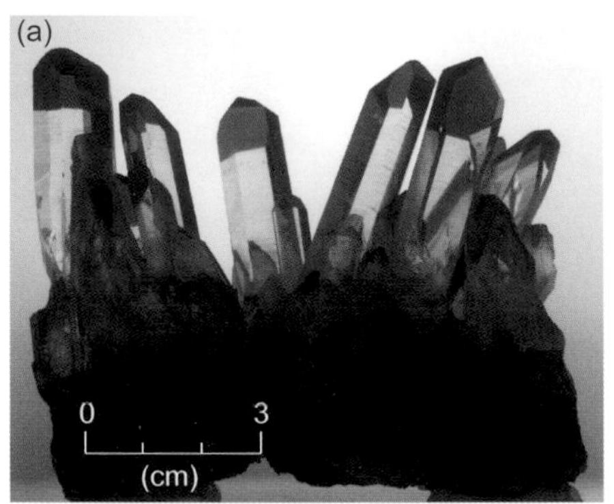

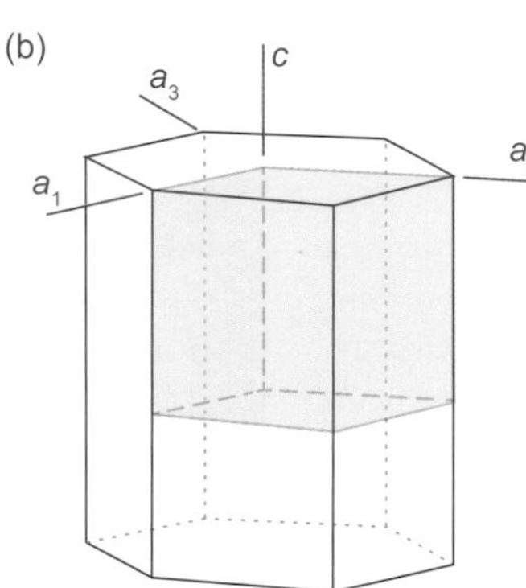

Figure 10.3 (a) Quartz crystals from Montgomery County, Arkansas, some showing almost ideal shapes as hexagonal prisms terminating in six-sided pyramids. Specimen from the Branner Library collection, Stanford University. (b) Crystallographic axes for quartz and the unit cell (tan). The c-axis is perpendicular to the plane containing the a-axes; successive a-axes subtend an angle of 120°.

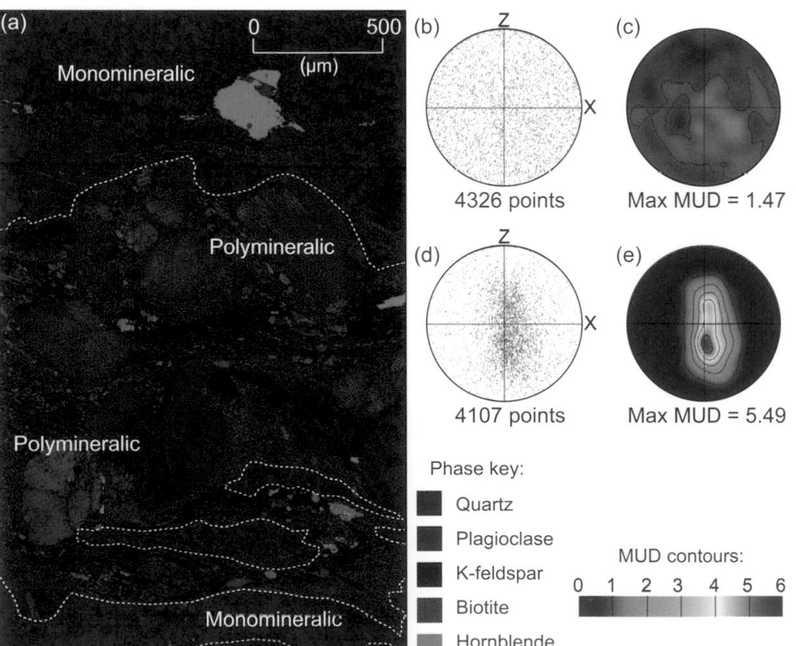

Figure 10.4 Electron backscatter diffraction analysis for deformed Lake Edison Granodiorite from sample SG10–03 (**Figure 10.19**). (a) Map of five phases indicated by color in Phase key. (b) c-axis pole figure for quartz from polymineralic aggregates. (c) Contoured pole figure for reduced data from (b). (d) c-axis pole figure for quartz from monomineralic lenses. (e) Contoured pole figure for reduced data from (d). See text for data reduction method and description of multiples of uniform density (MUD). Modified from Nevitt, et al. (2017b), Figure 8.

sample of sheared granodiorite with color coding corresponding to five different mineral phases: quartz, plagioclase, K-feldspar, biotite, and hornblende. These *polymineralic* aggregates are surrounded by lenses that are almost exclusively composed of quartz, and therefore are referred to as *monomineralic*.

Electron backscatter diffraction was used to measure the *c*-axis orientations of quartz grains at 4,326 sample points within the polymineralic aggregates, and at 4,107 sample points within the monomineralic lenses of the mylonite. These orientations are plotted as points on the lower hemisphere of stereographic projections (**Figure 10.4b, d**). Because the distance between sample points was about 1 μm, and the average quartz grain size is 11 μm in the aggregates and 18 μm in the lenses, individual grains are represented by many neighboring points with very similar *c*-axis orientations. The many orientations from a grain were reduced to a single average orientation representing that grain. Then these average orientations for all the grains were plotted on stereographic projections and contoured (**Figure 10.4c, e**).

The preferred orientation for quartz, if any, is quantified using Multiples of a Uniform Density, called MUD. This density is defined by counting the number of plotted points within a small circle on the stereographic projections of the reduced data. This operation is repeated on a grid of such points over the entire projection, and the densities are normalized by dividing by the density for a uniform distribution. The resulting MUDs are contoured and the maximum value is recorded. MUD = 1 means no preferred orientation; MUD > 1 means some preferred orientation. For quartz grains in the polymineralic portion of the mylonite (**Figure 10.4c**) the contours appear to wander aimlessly, and the maximum MUD is 1.47. In contrast, for quartz grains in the monomineralic lenses (**Figure 10.4e**) the contours form an elongate bull's eye pattern and the maximum MUD is 5.49. The contrast in maximum MUDs, and in the two contour patterns, suggests that fabric development in this mylonite was concentrated in the monomineralic lenses.

10.2 DESCRIPTIONS OF ROCK FABRICS IN SEDIMENTARY AND IGNEOUS ROCK

We begin with a description of fabrics in the Precambrian and Paleozoic sedimentary rocks that are associated with folding during the formation of the Appalachian mountain belt of the Eastern United States. This provides a classic example of using strain markers to quantify the deformation and relate that to a planar fabric in these rocks. Then we turn attention to shear zones that developed in the Cretaceous Mono Creek Granite from the Sierra Nevada of California and investigate how the deformation was localized in a narrow tabular zone. Older dikes are crosscut by the shear zones and provide measurements of the shear offset. These data, along with measurements of the zone thickness, lead to estimates of the shear strain variation along the zones. Finally, we describe planar fabrics that developed at a step between echelon fault segments in the Cretaceous Lake Edison Granodiorite from the Sierra Nevada. This locality exposes the entire structural context of fabric development in a single outcrop, so gradients in deformation can be documented. These data enable one to associate the spatially varying magnitude of strain with different degrees of fabric development.

10.2.1 Ooliths and Cleavage: South Mountain–Blue Ridge Uplift

In an early example of microtectonics, Ernst Cloos used ooliths in limestone from the South Mountain–Blue Ridge uplift of the central Appalachian mountain belt to investigate rock deformation during folding and mountain building (**Figures 2.1** and **2.2**; see Cloos, 1971, in Further Reading). The objective was to relate outcrop and microscopic observations of deformed objects with known undeformed shapes to the regional tectonics, particularly the folding of the Paleozoic sedimentary rocks. The original shape of an oolith is known to be roughly spherical, so measurements of their current shapes, which are approximately ellipsoidal, provides quantitative data to estimate the principal stretches. The long axes of the ooliths defines a linear fabric, called a lineation, in these rocks.

Some of the oolitic limestone beds from the central Appalachians also have a weak planar fabric, called cleavage, which is defined by the tendency for the rock to fracture (or *cleave*) along a set of closely spaced planes. The shales that are interlayered with these limestone beds often have a well-developed cleavage that usually is oblique to the bedding. For example, the bedding in **Figure 10.5** dips steeply to the left and the cleavage dips steeply to the right. The shale bed in the center of this

Figure 10.5 Interlayered sandstone and shale dip steeply to the left in this coastal exposure. Finer-grained units display a well-developed cleavage that dips to the right, whereas coarser-grained units display only poorly developed cleavage.

photograph literally is falling apart along cleavage planes, due to wave action along this coastal exposure in the western UK.

Anisotropy with respect to fracture, which is a characteristic of rock with cleavage, usually is caused by the alignment of platy minerals, such as clay or mica. The obliquity of cleavage to bedding (**Figure 10.5**) supports the interpretation that cleavage results from tectonic deformation, rather than sedimentary processes like compaction, which would align platy minerals parallel to bedding. The occurrence of cleavage does not mean that the rock fractured to form the cleavage in a brittle process. Instead, the cleavage formed without loss of cohesion across the cleavage planes, presumably by a ductile process that re-oriented the mineral grains as these rocks were folded. This begs the question: what are the magnitudes and orientations of the principal stretches associated with cleavage production? For the Paleozoic rocks of the South Mountain–Blue Ridge uplift, these magnitudes and orientations can be estimated from the shapes of the deformed ooliths. A portion of the geological map of this region with the structural data is shown in **Figure 10.6** to illustrate the consistency of the orientations of the linear and planar fabric elements along the Blue Ridge uplift.

Where undeformed, the ooliths and mud pellets are approximately spherical in three dimensions, and approximately circular in cross section. Thin sections from outcrops in the South Mountain–Blue Ridge uplift, however, display elliptical ooliths and elliptical mud pellets in cross section (**Figure 10.7**). These elliptical shapes are interpreted as a record of deformation and their aligned geometries provide a measurable rock fabric. Although these objects are small, they are not infinitesimal, so one must assume that the deformation was *homogeneous*, at least at the scale of an oolith or pellet, to apply the kinematic concepts described here.

The geometric relationships among the sedimentary bedding, cleavage, and ooliths that Cloos found along the South Mountain–Blue Ridge uplift are illustrated in **Figure 10.8**. Cleavage planes usually are oblique to bedding, and the line formed by the intersection of a cleavage plane and a bedding surface is approximately parallel to the axis of folding. Recall that the fold axis is parallel to the straight line used to generate a cylindrical fold (Section 9.1). The ooliths are approximately ellipsoidal in shape and the axes of the best-fitting ellipsoid for a given oolith are assigned the letters $A \geq B \geq C$. The A and B axes typically lie in cleavage planes with the B-axes approximately parallel to the axis of folding, and the long axes, A, defining a lineation on the cleavage planes. The extreme axes of the ellipsoids, A and C, are found in planes that are approximately perpendicular to the fold axis.

To analyze the kinematics of the deformed ooliths, we assume that there was *no change in volume* during their deformation. The volume of an ellipsoidal oolith in the deformed (final) state is $V_f = \frac{4}{3}\pi ABC$, and the volume in the undeformed (initial) state

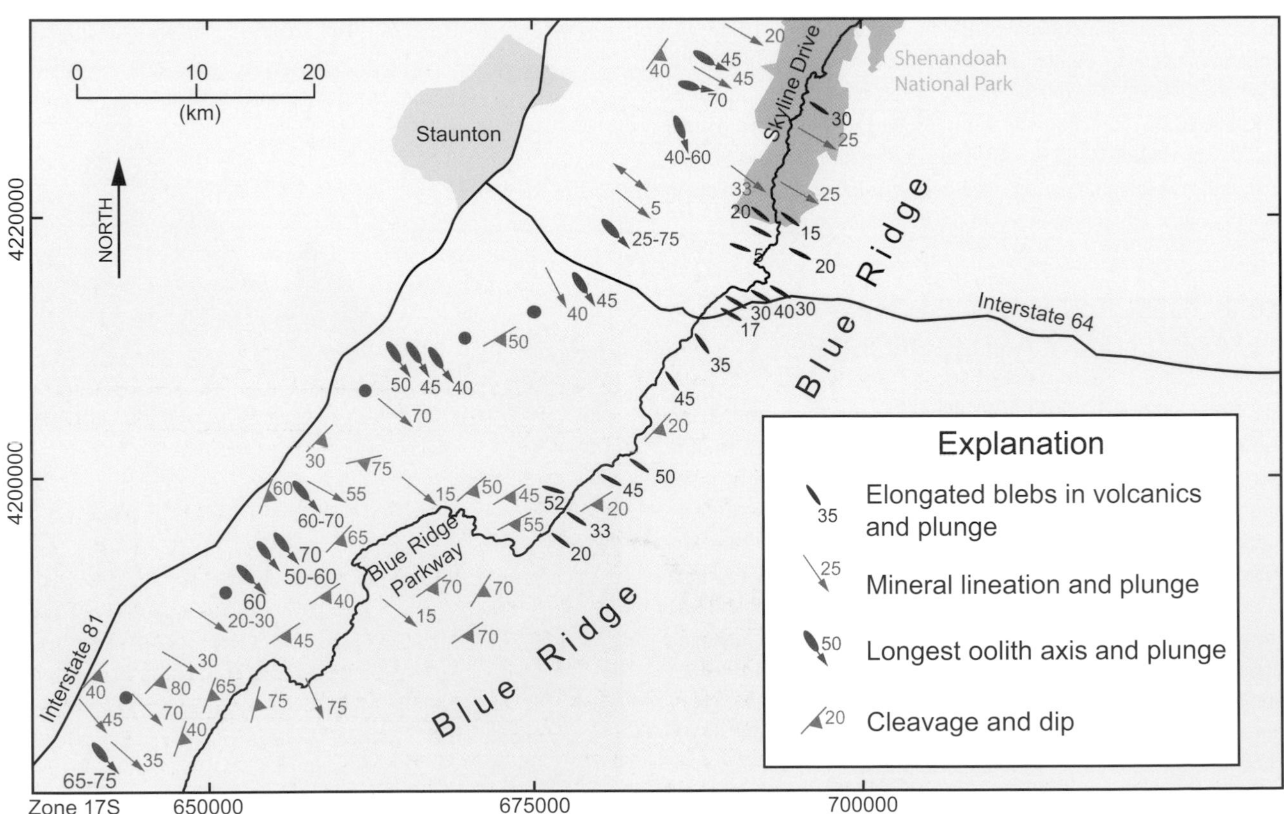

Figure 10.6 Map of linear fabric elements (elongate blebs and minerals, and longest axis of ooliths) and planar fabric elements (cleavage) along the Blue Ridge uplift in Virginia. Approximate map center UTM: 17 S 684684.77 m E, 4204489.15 m N. Modified from Cloos (1971), Plate 1.

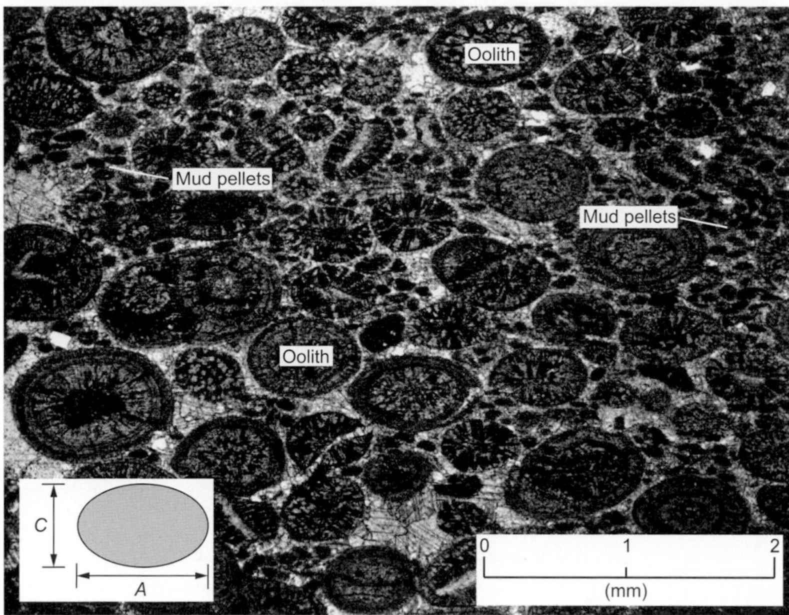

Figure 10.7 Thin section image of limestone containing ooliths and mud pellets sampled from near Lexington, Virginia. At this outcrop 100 deformed ooliths have an average axial ratio $A/C = 1.5$, and 50 deformed mud pellets have an average axial ratio $A/C = 2.3$. Modified from Cloos (1971), Plate 37.

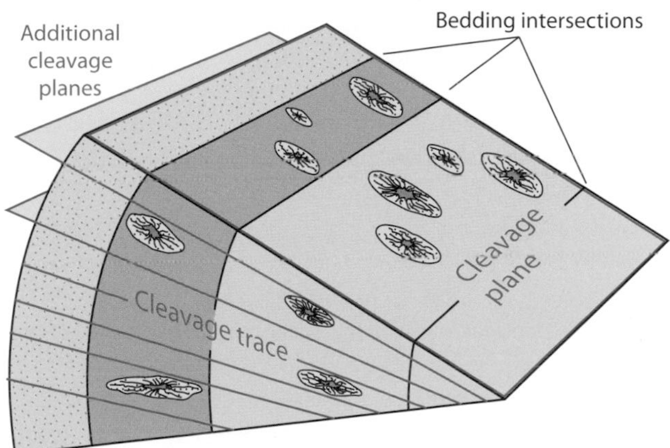

Figure 10.8 Schematic geometric relationships for bedding, cleavage, and ooliths of the South Mountain–Blue Ridge uplift. Modified from Cloos (1947), Figure 15.

is $V_i = \frac{4}{3}\pi R^3$, where the radial length is R. If volume is maintained throughout the deformation, $V_f = V_i$ and $ABC = R^3$. Normalizing all lengths by the initial radius yields $(A/R)(B/R)(C/R) = 1$. In other words, the product of the ratios of final to initial lengths of the axes must equal unity. The principal stretches are the ratios of final to initial lengths of the axes, so the constraint of no volume change means that $S_1S_2S_3 = 1$. A solution for the principal stretches that preserves volume is:

$$S_1 = A/R,\ S_2 = B/R = 1,\ S_3 = C/R = 1/S_1$$

This is a special case of deformation called pure shear strain (Section 5.5.2) with no stretch perpendicular to the plane that contains the extreme stretches, $S_2 = 1$, and the extreme stretches are reciprocals of each other. The assumptions of no volume change and pure shear deformation should be tested, but we accept them provisionally for the sake of this example.

To analyze one oolith set its center at the origin of coordinates and align S_2 with the Z-axis. An arbitrary position vector $\mathbf{X} = \langle X, Y\rangle$ overlays a material line in the initial state (**Figure 10.9**) that is a radius of a unit circle (the model undeformed oolith). The corresponding position vector $\mathbf{x} = \langle x, y\rangle$ overlays the same material line in the final state, where the oolith perimeter is elliptical. Here the deformation gradient tensor, introduced in Section 5.5.1, is taken as pure shear with the following matrix elements:

$$[\mathrm{F}] = \begin{bmatrix} 1.25 & 0 \\ 0 & 0.8 \end{bmatrix}$$

The diagonal elements are the stretches along the X and Y axes and these are the principal stretches. The ratio of long to short axis, A/C, for the model oolith is about 1.5, the average ratio for the ooliths in **Figure 10.7**.

The extreme axial ratio for ooliths in the South Mountain–Blue Ridge uplift covers the range $1.11 \leq A/C \leq 5.67$. Cleavage is prominent in the shales of the Blue Ridge, and it is recognizable in the limestones, if the ooliths have axial ratios $A/C \geq 2$. The orientations of the oolith axes are measurable if $A/C \geq 1.2$, and otherwise they appear to be random. The relationship between oolith shapes and the development of cleavage is illustrated in **Figure 10.10**. From left to right in this figure the greatest principal stretch varies as $S_1 = A/R = 1, 1.5, 2$ and the least principal stretch varies as $S_3 = C/R = 1, 0.67, 0.5$. With no deformation perpendicular to the A/C-plane, $S_2 = 1$. At the final stage the block has doubled in length and halved in height, and the cleavage is intense in the shale and recognizable in the oolitic limestone.

Cleavage development in the rocks of the South Mountain–Blue Ridge uplift is intimately associated with folding. However, strata are not continuously exposed through an entire fold, so fold shapes and relationships to cleavage and deformation were reconstructed from isolated outcrops. For example, within two miles of Shepherdstown, West Virginia, outcrops were found to reconstruct the fold profile in **Figure 10.11**. Eleven samples from oolitic limestones in these exposures were sectioned and examined microscopically to provide data for the average *A*/*C* ratios given in this figure. These ratios range from 1.55 to 2.91, but do not appear to vary systematically across the fold profile. The *A*- and *C*-axes of the ooliths lie in planes perpendicular to the fold axes (**Figure 10.8**). The cleavage planes are systematically oriented across the fold profile, forming a cleavage fan. On the western limb of the reconstructed fold the cleavage dips to the east, and on the eastern limb it dips to the west.

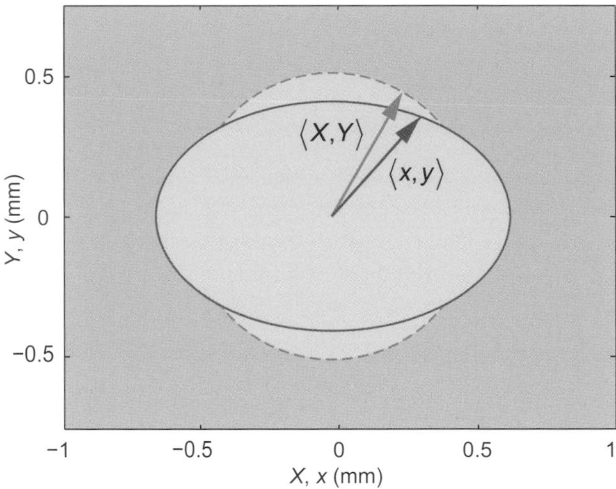

Figure 10.9 Homogeneous pure shear on the millimeter scale approximates deformation of ooliths in **Figure 10.7** (see also Section 5.5.2). The circle in the initial state deforms to the ellipse in the final state. Example material line vector has components $\langle X, Y \rangle$ in initial state, and $\langle x, y \rangle$ in final state.

10.2.2 The White Bark Shear Zone, Sierra Nevada

The White Bark shear zone provides an exceptional example of fabric development in igneous rock and its intimate relationship to faulting. Recall that in the upper brittle parts of the lithosphere frictional sliding and cataclasis are prominent deformation mechanisms, and they are sensitive to confining pressure, or more specifically to the compressive stress acting across the fault (Section 5.4.1). The shear stress required for sliding is approximately proportional to this compressive stress. However, at greater depths ductile mechanisms prevail, such as intracrystalline plasticity due to dislocation motion, and these mechanisms are sensitive to temperature (Section 5.4.2). Although exceptions exist, as depth increases in Earth's lithosphere, localized deformation by faulting gives way to distributed deformation in shear zones.

The White Bark shear zone, shown on the structure map in **Figure 10.12**, crops out in the Mono Creek Granite, with an age of about 86 Ma near the contact with the Lake Edison Granodiorite, which has an age of about 88 Ma. Structural relationships along the contact show that the granite intrudes the granodiorite, thereby confirming that the Mono Creek Granite is the younger pluton. The shear zones near the contact are offset by small faults of the same set that we described from the nearby Lake Edison pluton (Section 8.2.1). These fractures initiated as joints about 85 Ma and were reactivated by shearing about 83 Ma to become small faults (Section 10.2.3). The mylonitic shear zones developed after pluton emplacement, but before this faulting, when the temperature of the granite probably was in the range of 400 to 500 °C.

Two sets of dikes cut the Mono Creek Granite (**Figure 10.12**) and members of one of these sets served as the locus of shearing for the mylonitic shear zones. Dikes of the older set strike from 140 to 180, approximately parallel to the contact with the Lake Edison pluton, and include both fine-grained members (aplite) and very coarse-grained members (pegmatite). The younger dikes are mostly aplites and they vary in strike from 000 to 130. Among this set are nearly vertical dikes that strike from about 070 to 110,

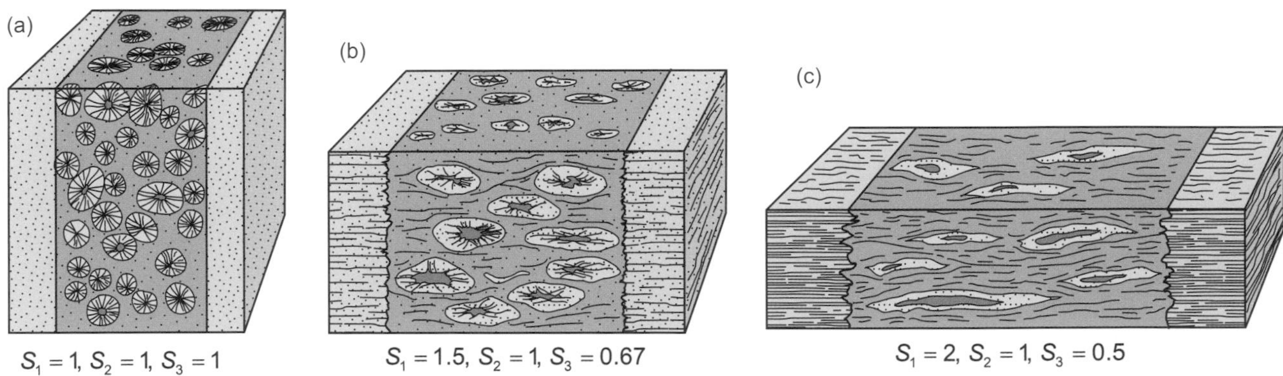

Figure 10.10 Schematic illustration of oolith deformation and cleavage development in relationship between oolith shapes and the development of cleavage in shale and limestone of the South Mountain–Blue Ridge uplift. Block doubles in horizontal length and halves in height from (a) to (c). Modified from Cloos (1947), Figure 21.

display mylonitic textures, and offset the older dikes in a left-lateral sense. In the vicinity of the White Bark outcrop the sheared dikes are 3 to 10 cm thick, can exceed 200 m in length, and offset the older dikes by as much as 5 m.

The South Shear Zone (**Figure 10.12**) is exposed over 235 m horizontally and about 90 m vertically. At its western tip (location A, **Figure 10.12**) an opening fracture filled with quartz extends from the shear zone toward the southwest along an azimuth of 245. Its aperture tapers from 6 cm at the shear zone boundary to a hairline fracture over a distance of about 4 m. The location and orientation of this fracture relative to the shear zone is similar to that of the wing cracks associated with the tips of sheared joints in the Lake Edison Granodiorite (Section 7.1.4), although it is much thicker and longer. The presence of this opening fracture indicates that brittle deformation of the Mono Creek Granite accompanied the ductile deformation of the shear zone. We take the 6 cm opening at the base of this wing crack, where it truncates against the shear zone, as a measure of the offset across the shear zone at this location. We also take the orientation of the wing crack, 245, as parallel to the direction of maximum compression in the horizontal plane as the shear zone developed.

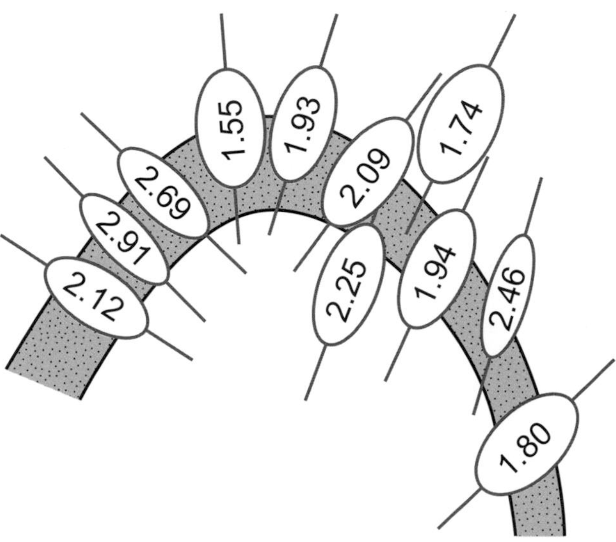

Figure 10.11 Fold profile normal to fold axis reconstructed from isolated fold fragments within 2 miles of Shepherdstown, West Virginia (**Figure 2.1**). Cross sections of ellipses with *A/C* ratios given are averages from cut surface and thin section measurements of ooliths. Extensions of *A*-axes indicate orientation of cleavage, which forms a cleavage fan on this fold profile. Modified from Cloos (1971), Figure 28.

Toward the center of the South Shear Zone outcrop (location B, **Figure 10.12**) the shear zone offsets an older dike by 432.8 ± 10.8 cm. This dike provides an *apparent* offset, but solution of a three-point problem (Section 2.4.2) using two differently oriented dikes at a nearby location on the North Shear Zone demonstrates that the *true* offset vector rakes 15° toward the west. Based on this shallow rake we take the apparent offsets of single older dikes as good estimates of the true offset on these shear zones. Near the eastern extent of the shear zone (location C, **Figure 10.12**) the offset decreases to 25 ± 25 cm. Thus, over the 200 meter length of the shear zone the offset increases from less than a decimeter to more than four meters, and then decreases to less than a few decimeters. The distribution of offset along the South Shear Zone (**Figure 10.13**) supports this general trend of values increasing toward the middle of the outcrop, along with an increase in the width of the zone. Irregularities in the offset distribution are attributed to the segmented nature of the shear zone.

The hand sample taken from location B (**Figure 10.12**) displays the entire 6 cm wide South Shear Zone at that point (**Figure 10.14**). The granite is composed of large crystals of alkali feldspar, up to 2 cm across, set in a groundmass of plagioclase, quartz, and alkali feldspar with typical grain sizes from 1 to 5 mm. From either border of the shear zone the granite grades into a mylonite with reduced

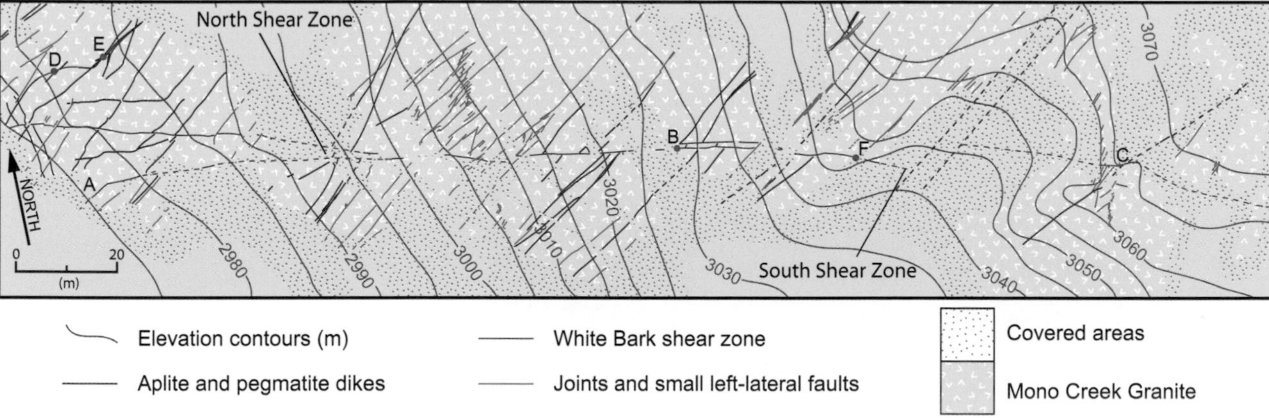

Figure 10.12 White Bark (WB) outcrop map with North, South, and Curved Shear Zones (red lines; black dashed lines where covered). Covered area shaded. Aplite and pegmatite dikes (blue); Joints and small left-lateral faults (green). Dotted lines are elevation contours (meters). Letters A–F referred to in text for location. Approximate UTM: 11 S 335578.63 m E, 4135519.79 m N. Modified from Christiansen and Pollard (1997), Figure 2.

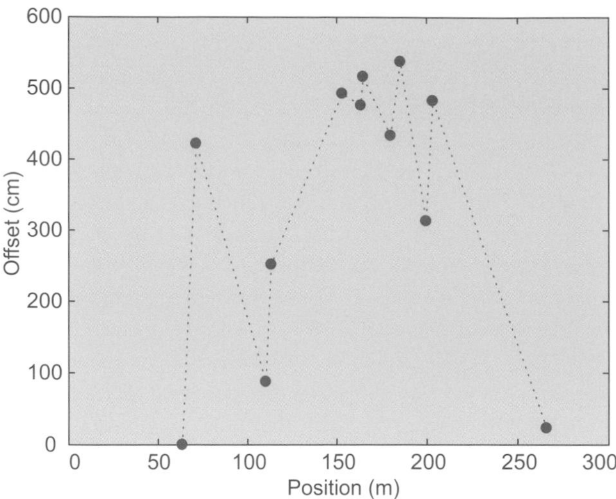

Figure 10.13 Offset of aplite dikes by South Shear Zone at White Bark outcrop plotted versus position along shear zone. Modified from Christiansen and Pollard (1997), Figure 7a.

grain sizes, aligned biotites, quartz and plagioclase stretched out into narrow stringers, and some large blocky alkali feldspars. Relatively large grains in a fine-grained matrix are called porphyroclasts. Toward the center of the shear zone the grain size is reduced to less than a few hundred microns and the structure consists of millimeter thick alternating gray and black layers.

Some of the shear zones in the White Bark area have a sharper, less gradational border than that shown in **Figure 10.14**, but evidence for discrete slip as might be expected on a fault is lacking. Recall that we introduced the displacement discontinuity vector to quantify the magnitude and direction of slip on a model fault (Section 8.4.4). Neighboring particles on the two opposed surfaces of the fault displace in opposite directions, thereby creating a *discontinuity* in the displacement field. In contrast, for the shear zone in **Figure 10.14** the displacements are interpreted as *continuous* across the zone.

As illustrated in **Figure 10.15**, the offset, o, is defined as the tangential separation of older planar markers that are truncated at the two edges of the shear zone. For the White Bark shear zones the planar markers are the surfaces of older aplite or pegmatite dikes. Because the shear zones initiated along younger aplite

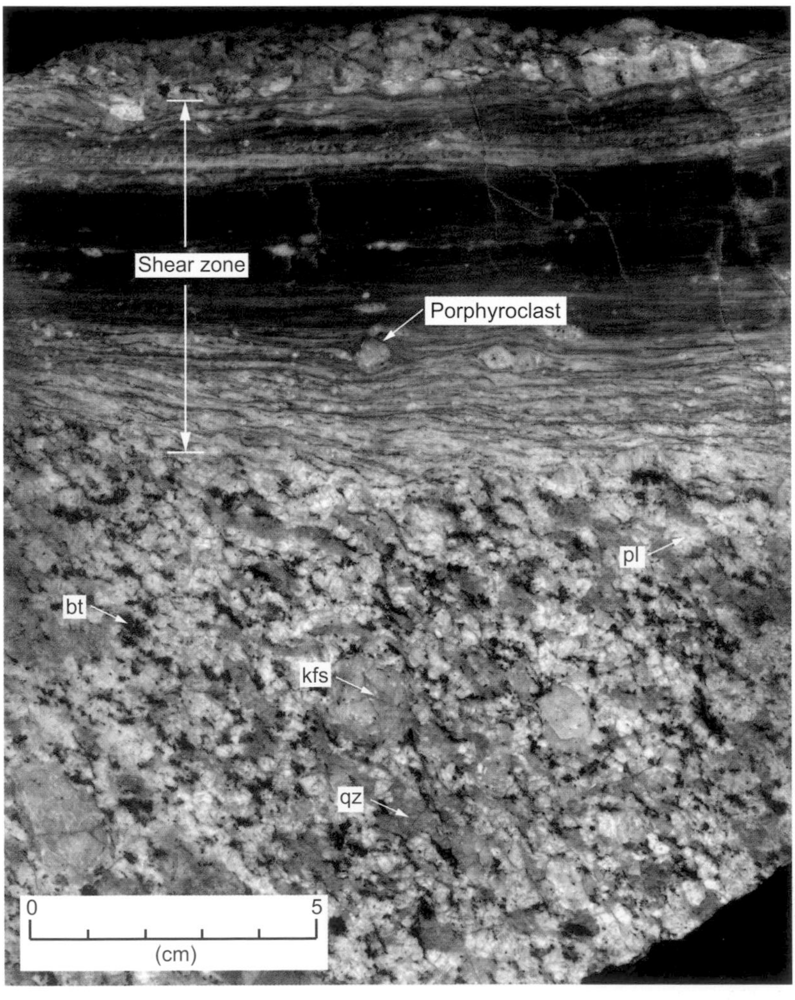

Figure 10.14 Saw-cut face of the hand sample of Mono Creek Granite with a left-lateral mylonitic shear zone (location B, Figure 10.12). The zone is about 6 cm wide, and from bottom to top of the sample undeformed granite grades to mylonite with relic feldspar clasts and then to more homogeneous mylonite and back through this sequence to granite. Prominent minerals are quartz (qz, gray); alkali-feldspar (kfs, pinkish); plagioclase (pl, white); biotite (bt, black).

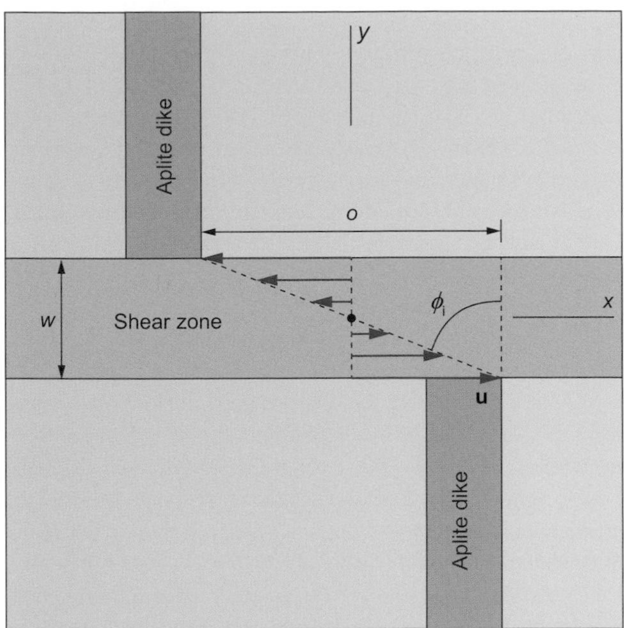

Figure 10.15 Schematic illustration of a shear zone that initiated along younger aplite dike and offsets an older aplite dike. In this two-dimensional diagram the displacement vectors, **u**, are in the (*x*, *y*)-plane of view. Shear zone width, *w*, and offset, *o*. Compare to **Figure 8.30**.

dikes that cut and separated the older dikes, sheared remnants of the older dikes are not found crossing the shear zones. Because the older dikes are truncated at the edges of the shear zones (**Figure 10.15**), they do not provide markers that might reveal displacement discontinuities within the zone.

The average shear strain within the shear zone is calculated from measurements of the offset and width (**Figure 10.15**):

$$\tan \phi_i \approx o/w \qquad (10.1)$$

Using the center of the South Shear Zone (**Figure 10.14**) as the reference frame, the displacement vector, **u**, is directed to the left with a magnitude of about 215 cm at the top of the hand sample; decreases to zero at the center of the shear zone; is directed to the right below that point; and increases in magnitude to about 215 cm at the bottom of the zone. The total offset of the older aplite dike is about 430 cm, so the average shear strain is $\phi_1 \approx 430/6 \approx 72$ at this location. Average shear strains have been estimated elsewhere along the shear zones using outcrop measurements of zone width and offset. For the Central Shear Zone (CSZ), the North Shear Zone (NSZ), and the South Shear Zone (SSZ) mapped in **Figure 10.12** shear strains range from 0.1 to 100.

Additional evidence that the White Bark shear zones initiated along favorably oriented aplite dikes is found at locations D and E on **Figure 10.12**. There, an aplite dike curves continuously so the strike varies from 040 to 100 and back to 040. Where the strike ranges from about 040 to 075 older structures are not offset and sample E, collected where the strike is 057, is representative of the un-sheared dike. In thin section (**Figure 10.16a**) the different mineral grains are roughly equal in size with an average

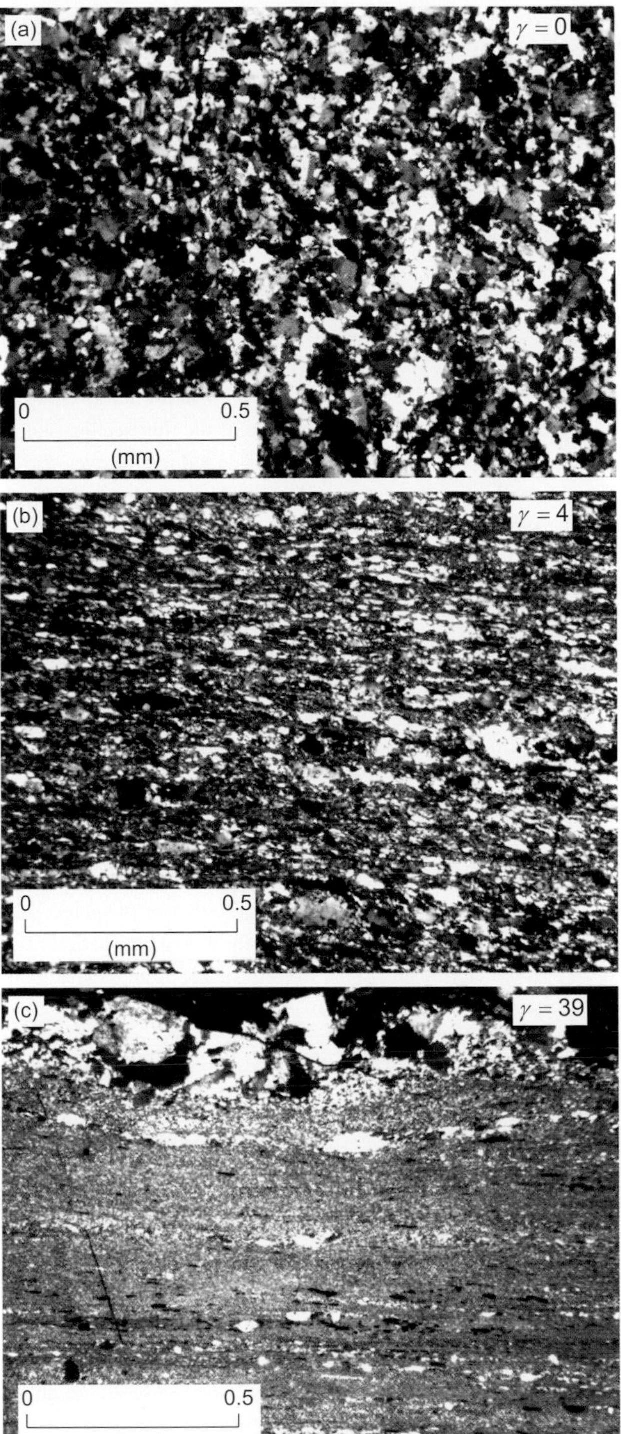

Figure 10.16 Comparisons of fabrics from White Bark shear zones with measured offsets and widths providing average shear strains, γ. Fields of view are 1.1 mm by 1.5 mm. (a) Sample D: no discernable fabric, $\gamma \approx 0$. (b) Sample E: grain size reduction and proto-mylonitic fabric, $\gamma \approx 4$. (c) Sample F: further grain size reduction and well-developed mylonitic fabric, $\gamma \approx 39$. Mono Creek Granite at top of photograph. Modified from Christiansen and Pollard (1997), Figure 4.

grain size of about 230 μm and no fabric is discernable. At the location of sample D the strike of the dike is about 085, its width is 1.4 cm and the estimated offset is 6 cm, so the average shear strain is about 4. The outcrop displays a discernable, but weak mylonitic fabric roughly parallel to the contact of the dike with the host granite. The average grain size observed in thin section (**Figure 10.16b**) is reduced to only 73 μm, and the foliation is defined by aligned aggregates of quartz grains and the alignment of elongate feldspar grains.

Sample F comes from the South Shear Zone on **Figure 10.12** where the strike of the zone is about 110. The zone width is 12.5 cm and the offset is about 480 cm, giving an average shear strain of about 39. The shear zone has a sharp contact with the host granite seen along the upper edge of the thin section photograph (**Figure 10.16c**). The grain size is reduced further within the zone at this greater strain, and this reduction imparts the black color seen in outcrop and hand samples (**Figure 10.14**). Although this is the second greatest offset measured at the White Bark outcrop, it also is the greatest shear zone width, so the average shear strain is about the median value of those measured. This serves to emphasize the fact that although shear strain is proportional to offset, it is inversely proportional to width (10.1).

The White Bark outcrop provides evidence that the mylonitic shear zones initiated within steeply dipping aplite dikes that strike in the range 090 ± 020. Dikes outside this range are not sheared, and curved dikes have planar fabrics where their strikes are within this range, but not where strikes are outside this range. Although the shearing incorporated granite in some locations, the geometry of a given shear zone is largely that of the antecedent aplite dike in terms of thickness, length, and segmentation. The shear strain, determined by the ratio of offset of an older dike to zone thickness, correlates systematically with the development of the planar fabric. These observations lead to the conclusion that the aplite dikes were weaker than the surrounding Mono Creek Granite. Those dikes most favorably oriented with respect to the remote principal stresses deformed by shearing in preference to deformation of the granite.

10.2.3 Mylonite Foliation in Fault Steps, Sierra Nevada

Many geometric configurations of fabric at the outcrop scale and microstructures in thin section are interpreted in terms of the orientation of planes of shearing and the sense of shearing. Monographs on deformation microstructures and microtectonics (see Blenkinsop, 2000; Passchier and Trouw, 1996, in Further Reading) review these interpretive techniques. This approach does not necessarily place the deformation in a context that provides an independent measure of the magnitude of strain. In addition, samples or photographs may document the deformation on a profile *across* a shear zone, but gradients in strain *along* a shear zone and the structures at the termination of a shear zone rarely are described. The Bear Creek fault steps expose the entire structural context of fabric development in a single outcrop, so the extent of the shear zone is known, and gradients in deformation have been documented. These data enable one to associate the spatially varying magnitude of strain with different degrees of development of the mylonitic foliation.

At the base of Seven Gables, along the south bank of the East Fork of Bear Creek, an outcrop of Lake Edison Granodiorite exposes a left-lateral fault composed of about 15 segments, ranging in length from 1 to 4 m (refer to **Figure 8.4** for location). Here we focus on one of the right steps between two of the fault segments, which provides a remarkable opportunity to investigate the localized fabric in the granodiorite that apparently was induced by faulting. Recall from Section 8.4.5 that stresses are concentrated near the tips of a fault, and we would expect the stresses to be doubly concentrated within the step between two tips. At this locality that stress concentration led to the development of a fabric that is localized within the step. This fault is a member of the set of faults that we described from the Lake Edison pluton in Section 8.2.1, so the reader should review the geologic map, structural setting, and detailed description found there before reading this section.

To assess the microstructural changes in Lake Edison Granodiorite when the local fabric developed at the fault step, we compare samples from within the step to samples not associated with a step. The hand sample shown in **Figure 10.17** contains a fault (Section 8.2.1) and an aplite dike that has a left-lateral separation of about 9 cm. The mineralogical makeup of the granodiorite varies throughout the pluton, but usually contains 30–50% plagioclase (white), 15–20% K-feldspar (pinkish), 20–30% quartz (gray), and 15–25% biotite and hornblende (black). In this hand sample some elongate biotite grains are oriented sub-perpendicular to the fault, and this alignment helps to define the weak regional fabric of the Lake Edison pluton.

We use the thin section (**Figure 10.18**) of the Lake Edison Granodiorite, not associated with a fault tip or step, as a basis for comparison to thin sections from a right step where dramatic changes occurred due to overprinting of the weak regional fabric by a very strong local fabric. Because those changes involved, among other things, a reduction in grain size, we note that the average grain size of quartz is about 0.2 mm for specimens only effected by the regional fabric. Feldspars typically are much larger, ranging up to about 5 mm. The unsheared granodiorite has the classic granular texture of an igneous rock with roughly equal grain sizes, no glassy material, interlocking grains, and little evidence of deformation.

A photograph of the right step (**Figure 10.19a**) reveals the ends of two fault segments and an aplite dike that cuts the granodiorite exactly at the step. The orientations and dimensions of key structural features at this step are provided on the outcrop map (**Figure 10.19b**). Outside the step, the strike of the two fault segments is 259 and the dip is 76. Slickenlines on these fault segments plunge about 10°, so the relative motion approximates strike slip. The total separation of the dike is made up of about 12 cm of slip on each fault segment and about 30 cm of shearing within the step. Thus, localized slip on the fault segments very near their tips accounts for 40% of the relative motion of the granodiorite from one side of the segmented fault to the other, whereas distributed shear strain within the step accounts for 60%. This partitioning of the relative motion highlights the

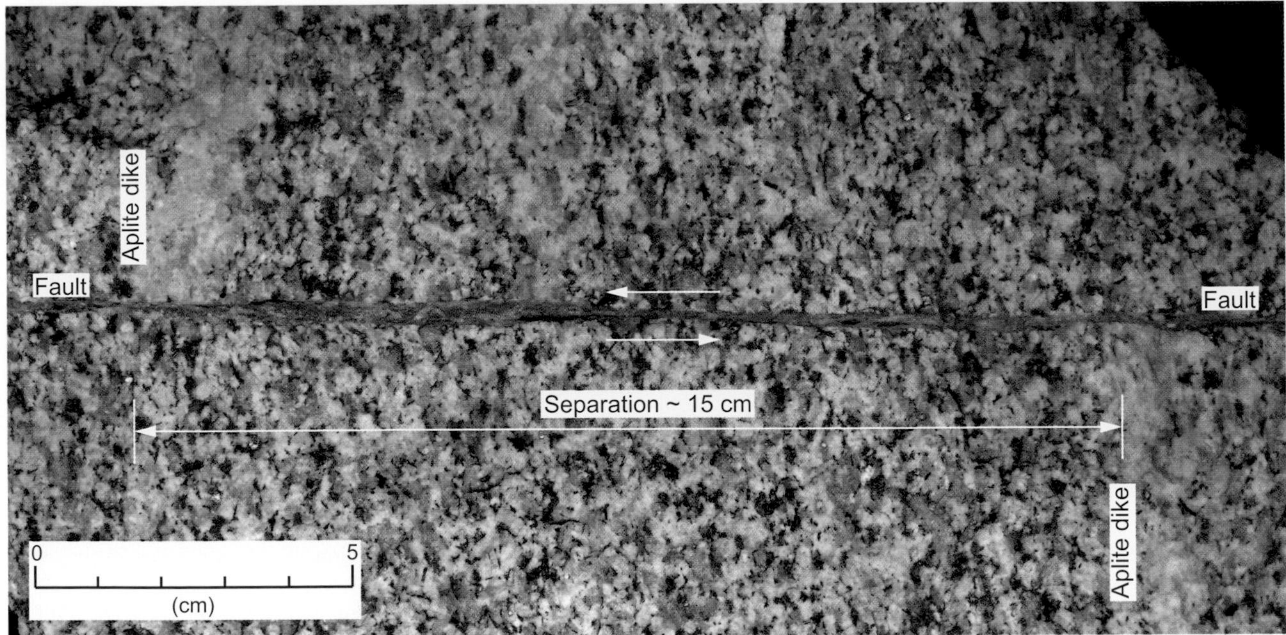

Figure 10.17 Lake Edison Granodiorite. (a) Hand sample containing a fault and an older aplite dike with about 15 cm of left-lateral separation. Elongate biotite grains (black) define a weak foliation sub-perpendicular to the fault.

Figure 10.18 Thin section photograph of the Lake Edison Granodiorite with only the weak regional fabric. Compare to thin section photographs in Figure 10.20 that show local mylonite fabric. Minerals include plagioclase (pl), K-feldspar (kfs), quartz (qz), biotite (bt), and hornblende (hb). Modified from Nevitt et al. (2017b), Figure 5a.

contributions of both brittle deformation (fault slip) and ductile deformation (mylonitization) during faulting, and focuses attention on the importance of the strong fabric in the fault step.

Sample SG10–03 was taken from the center of the right step (Figure 10.19a), where the local mylonitic fabric is most highly developed. Compare the photographs of thin sections from this sample (Figure 10.20) to those from the granodiorite with only the regional fabric (Figure 10.18), while noting the scale differences. The most obvious change accompanying the development of the mylonitic fabric in sample SG10–03 is the reduced grain size of certain minerals. For example, the average grain size of quartz in the mylonite is 0.01 to 0.02 mm, an order of magnitude less than that in the granodiorite, but very little evidence exists for fracturing of quartz grains, and they often have rounded and interlocking boundaries (Figure 10.20a). In contrast, the much larger feldspar grains have ragged boundaries and internal fractures, and fragments of former single grains are strung out along the foliation (Figure 10.20b). The contrast between the quartz and feldspar grain in terms of size reduction and shape changes suggests that the feldspars were stronger and more brittle during the development of the mylonite. Studies of rock fabrics using thin sections like these are the primary focus of structural petrology and microtectonics (Box 10.1).

Using the thin section photographs (Figure 10.20) we introduce a few terms from the literature of microtectonics (see Passchier and Trouw, 1996, in Further Reading). This rock displays a fabric called an *S-C* mylonite (Figure 10.20a). The letter *S* stands for the French word *schistosité*, and the foliation is primarily defined in this thin section by elongate quartz grains with a preferred orientation that is oblique to the *C*-planes. The letter *C* stands for the French word *cisaillement*, which refers to the shear planes, so this foliation is parallel to what is inferred to be the planes of greatest shearing. The *C* foliation in this thin section is composed of lenses of fine-grained feldspar and biotite.

Relatively large grains in a fine-grained matrix are called porphyroclasts (see Blenkinsop, 2000, in Further Reading). For example, the large plagioclase grains in Figure 10.20c are porphyroclasts that are surrounded by anastomosing lenses of much finer grained quartz, feldspar, and biotite. In

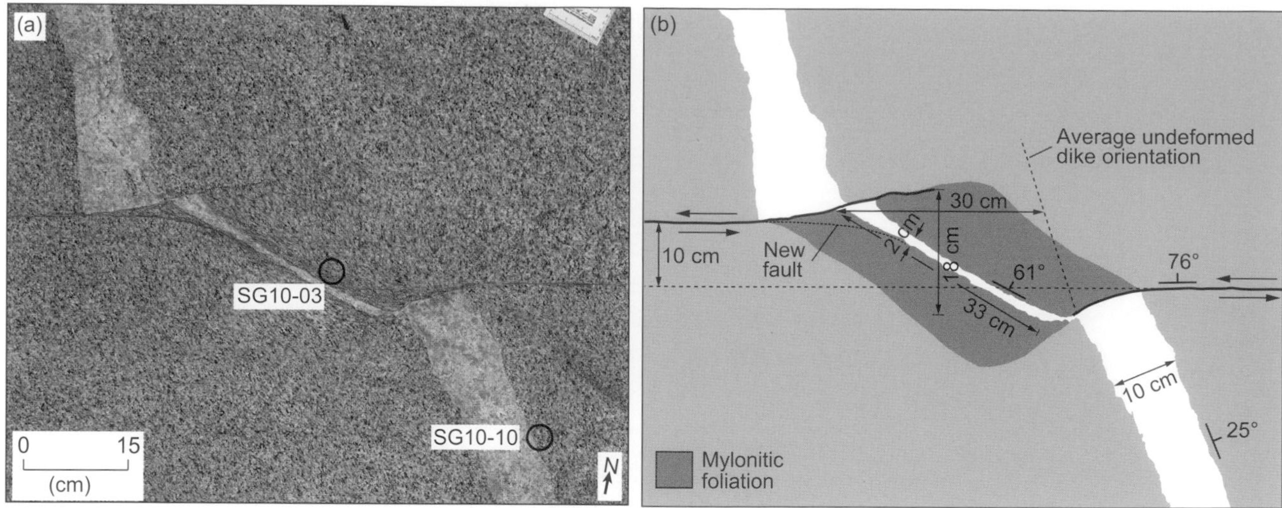

Figure 10.19 (a) At a right step between two left-lateral fault segments an aplite dike is offset along the faults and stretched, thinned, and rotated in the step. Mylonite foliation is localized in the step. Seven Gables (SG) outcrop UTM: 11 S 336592.23 m E, 4132431.64 m N. See **Figure 8.4** for geologic map of region. (b) Outcrop map with key structural data. Modified from Nevitt et al. (2017b), Figure 2.

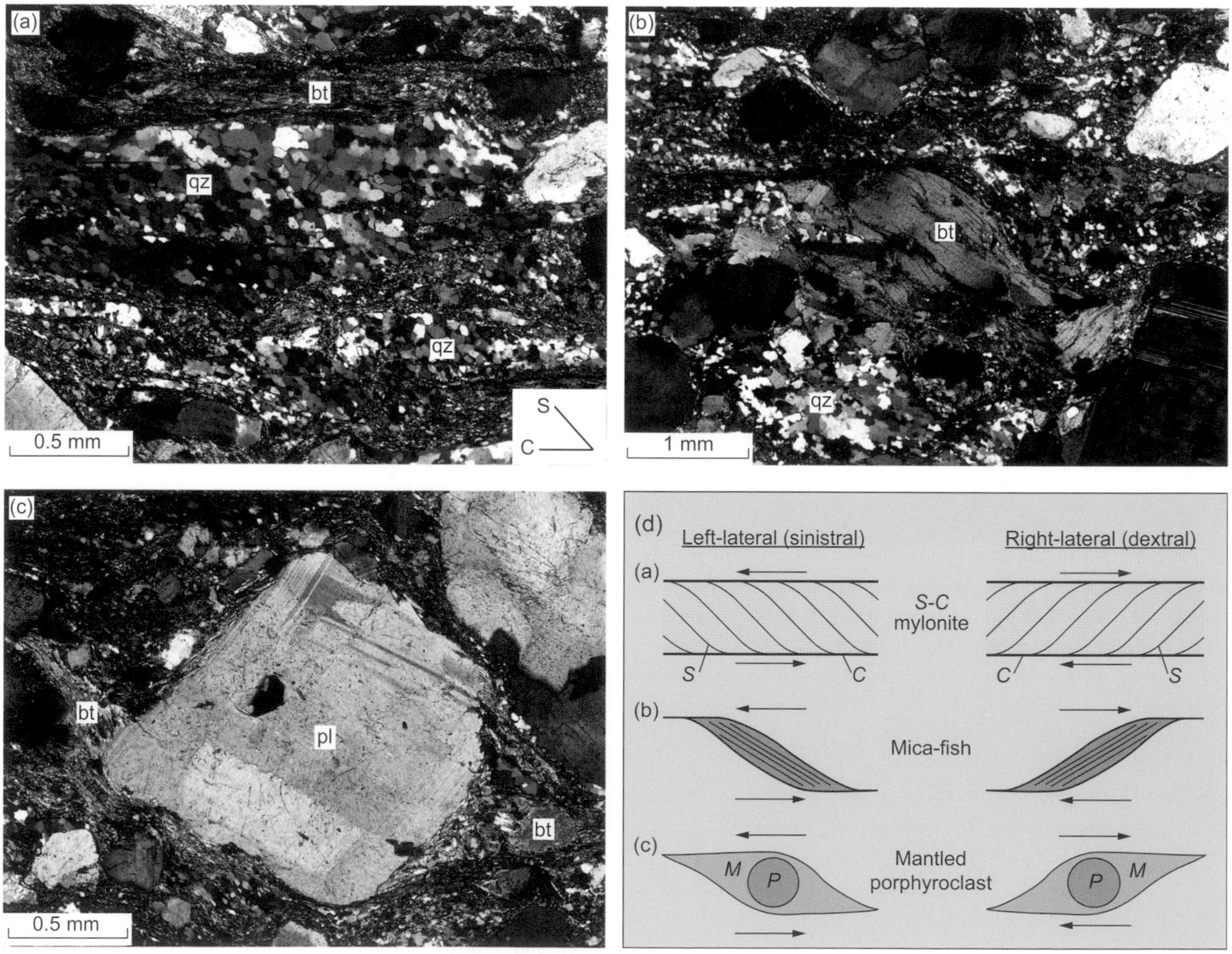

Figure 10.20 Thin sections of Lake Edison Granodiorite from sample SG10–03 taken from the center of the right step shown in **Figure 10.19**. Compare to thin section in **Figure 10.18** that has only the weak regional fabric. (a) *S-C* mylonite. (b) Mica-fish. (c) Plagioclase porphyroclast with mantle of biotite. (d) Schematic illustration of shear sense indicators for both left- and right-lateral shearing of *S-C* mylonite, mica-fish, and mantled porphyroclast. Modified from Nevitt et al. (2017b), Figure 6.

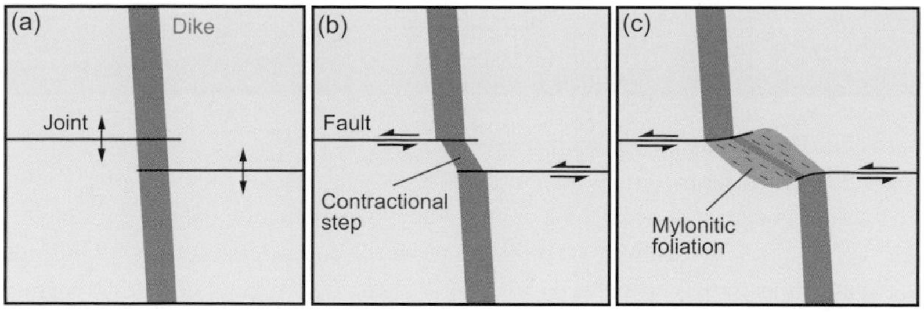

Figure 10.21 Conceptual model of three stages in the deformation of granodiorite and dike near a contractional fault step. Gray mylonitic foliation. (a) Dike formation followed by jointing. (b) Slip on the joints to form small faults and offset the dike. (c) Continued fault slip with ductile shearing within the fault step accompanied by mylonitic fabric development. Modified from Nevitt et al. (2017b), Figure 4.

Figure 10.20b a large biotite grain has a characteristic geometry that is called mica-fish. These grains are oblique to the *C*-planes and have tails of finer grained biotite that extend along the shear planes. In **Figure 10.20c** a plagioclase porphyroclast has a mantle that tapers laterally away from the clast and is composed mostly of much smaller biotite grains.

The obliquity of the *S* foliation relative to the *C* foliation can be used to infer the sense of shearing (**Figure 10.20d**) where other independent measures, such as offset older structures, are lacking. A counter-clockwise orientation of *S* relative to *C* suggests a left-lateral (also called sinistral) sense of shearing. For the right step depicted in **Figure 10.19** we have offset markers that demonstrate the fault is left lateral, confirming the sense of shearing inferred from the *S-C* mylonite. The geometry of mica-fish provides a means to interpret the sense of shearing, and that also is consistent with left-lateral (*sinistral*) shearing. Some mantled porphyroclasts also may be used as shear sense indicators. Of course these indicators may display a right-lateral (also called *dextral*) sense of shearing. Monographs on microtectonics describe many other shear sense indicators (see Borradaile et al., 1982, in Further Reading).

The total separation across the segmented fault (**Figure 10.19**) may be restored by moving the northern piece of the dike 42 cm to the east along the fault. This provides an estimate of the original length of the dike between the two segments. This length was increased by a factor of 3.3 during the deformation. Using the fact that the dike dips 25° outside the step and 61° inside, and the apparent width of the dike measured on the outcrop, we estimate that the width decreased from 4.2 cm outside the step to 1.7 cm inside the step. The dike inside the step is rotated about 44° counter-clockwise in the plane of the outcrop. The mylonitic foliation associated with faulting changes strike from about 305 near the lateral edges of the step to become parallel to the dike, striking about 290 near the center of the step.

Recall from the laboratory results described in Section 5.4 that the yield point separating elastic from inelastic deformation in conventional triaxial compression tests of Solnhofen limestone occurred at axial strains of about –0.01 to –0.02. Tests of granitic rock find similar limits on strain separating elastic from inelastic deformation. Using the change in width of the dike in **Figure 10.19**, the contractional strain is estimated as $\varepsilon = (1.7\text{cm} - 4.2\text{cm})/4.2\text{cm} = -0.6$, roughly 30 times the strain at yielding during laboratory tests, so we infer that the Lake Edison Granodiorite was deformed well beyond its elastic limit.

The lack of fractures within the right step (**Figure 10.19**) suggests that the inelastic deformation occurred while the rock was ductile. Recall from the laboratory tests described in Chapter 5 that this means the slope of the differential stress plotted versus the axial strain was positive. In contrast, inelastic deformation that occurs while the rock was brittle exhibits a negative slope on such a plot. In what follows we associate mylonitic fabric observed in the granodiorite with ductile deformation. The strong mylonitic foliation within the step (**Figure 10.20**) thus correlates with a concentration of inelastic ductile deformation within the step.

A conceptual model for deformation near the right step shown in **Figure 10.19** begins with the opening and filling of the fracture that becomes the aplite dike, followed by the opening of two echelon joints that are sealed with hydrothermal minerals (**Figure 10.21a**). Later slip in a left-lateral sense converts the opening joints into small faults (**Figure 10.21b**). The piece of the dike within the right step is stretched and rotated in response to this faulting, and this deformation is accompanied by a decrease in the amount of overlap of the echelon fault segments. In the final stage (**Figure 10.21c**) slip continues to accumulate on the fault segments and the dike between them continues to stretch, thin, and rotate counter-clockwise. The tips of both faults also rotate away from each other in response to deformation within the step, where the granodiorite develops a strong foliation that is sub-parallel to the rotated dike.

The Bear Creek fault step (**Figure 10.19**) provides a well-documented example of fabric development in granodiorite, where the pre-existing fabric is known from the surrounding outcrops, and the cause of the deformation is clear from geometric relationships at this outcrop. Furthermore, the sense of shearing and the magnitude of the shear strain can be estimated from the offset dike. Although the geometry of the fault step is somewhat more complicated than the geometry of the shear zones at the White Bark outcrop (**Figure 10.12**), the fault step offers another opportunity to quantify the relationships between deformation and fabric. At the end of this chapter we return to this fault step and describe a modeling investigation that tests different constitutive laws for the mechanical behavior of the granodiorite and the aplite dike.

Box 10.1 Structural Petrology and Microtectonics

Rock fabrics commonly are described from thin section images observed using a petrographic microscopic. For example, **Figure B10.1** is a thin section image at a scale of a few hundred microns of the aplite dike in sample SG10–03 (**Figure 10.19a**), located within the right step between two left-lateral faults. The deformation there produced a strong *S-C* mylonitic fabric consistent with left-lateral shearing (**Figure 10.20d**). The microscopic study of rock fabrics blossomed in the mid-twentieth century when petrologists turned their attention to the characterization of grain shapes and orientations, and to preferred orientations of crystallographic axes in deformed rock. Structural petrology (den Tex, 1972) emerged from this examination of rock fabrics using optical microscopes, and now straddles the disciplines of petrology and structural geology. Bruno Sander, who was the Professor of Mineralogy and Petrography at the University of Innsbruck from 1922 to 1955, receives credit for much of the development of structural petrology.

Sander adopted the familiar coordinate system from crystallography with three mutually perpendicular axes (a, b, c) and applied it to rock fabrics at all scales of observation, not just under the microscope. For example, in outcrop or crustal scale folds with a prominent cleavage (Section 10.2.1), the cleavage plane is associated with the ab-plane, the line parallel to the intersection of bedding and cleavage is the b-axis, and the normal to cleavage is the c-axis. Typically, the b-axis is associated with the fold axis of cylindrical folds (Section 9.1). The fabric axes provide a useful reference system for recording and plotting orientation data obtained from the microscope and from field measurements. Structural geologists have exploited this coordinate system and developed visualization tools such as the stereographic projection (Section 2.3.3) to analyze orientation data. For example, the normals to cleavage (c-axes) for a cleavage fan (e.g. **Figure 10.11**) may plot along a great circle (plane) on the stereographic projection, in which case the pole (normal) to this great circle would be interpreted as the fold axis.

Some geologists have associated fabric axes with kinematic quantities (movement directions, principal strain axes), or dynamic quantities (principal stress axes), without using the physical constraints imposed by Cauchy's laws of motion (Sections 3.6 and 3.7), or the relationships defined in constitutive laws (Sections 4.10, 5.7, and 6.7). This is analogous to supposing that one could directly associate the crystallographic axes of a mineral grain, undergoing crystal plastic deformation, with the average strain tensor or applied stress tensor, without identifying the deformation mechanism (e.g. dislocation motion), and without employing the physics of dislocations (Section 5.6). One of the authors' goals for this textbook is to use those physical constraints to show how the applied stresses cause the motions and strains that produce the observed geometries of structures. Fabric data can help to set up such a mechanical analysis by characterizing the current geometry of a structure, but geometric data are not sufficient to address the kinematics and dynamics of the deformation that produced the structure.

The textbook *Structural Analysis of Metamorphic Tectonites* (Turner and Weiss, 1963; see Further Reading) describes and elaborates Sander's methods for the structural geologist, and applies them to metamorphic rocks at scales from mineral grains to mountain belts. Following Sander, the focus is on domains of deformed rock that are statistically homogeneous

Figure B10.1 Thin section image of a *S-C* mylonite from the Lake Edison Granodiorite, Sierra Nevada. Fine-grained feldspar (pl + kfs) defines the *C*-plane and shape-preferred orientation of quartz (qz) defines the *S*-plane. Shear sense is top to the left. Modified from Nevitt et al. (2017b).

with respect to their fabric. The basic assumption of this method of structural analysis is that the symmetry of the fabric within such a domain relates to the symmetry of the movements and strains during the deformation, and to the symmetry of the stresses that caused the deformation. Thus, after characterizing the fabric, the symmetry assumption leads to inferences about the symmetry of the movements, strains, and stresses. While attractive in its simplicity, this methodology ignores the physical constraints on deformation imposed by the conservation laws of mass, momentum, and energy (Sections 3.5–3.8). These laws, when reduced to the appropriate equations of motion (Sections 4.11 and 6.8), enable one to shrink Sander's homogeneous domains to points in the material continuum and to use continuum mechanics to analyze heterogeneous deformation at all scales. The equations of motion constrain how the applied stresses cause the motions and strains that lead to the observed geometries and symmetries of structures.

Today, fabrics identified under the petrographic microscope are called deformation microstructures (Blenkinsop, 2000; see Further Reading), and the discipline devoted to their description and interpretation is microtectonics (Passchier and Trouw, 1996; see Further Reading). These two referenced textbooks describe deformation microstructures found in deformed rocks and classify them into three broad categories: (1) microfractures; (2) microstructures indicative of material removal and deposition (e.g. microstylolites and microveins); and (3) microstructures indicative of crystal lattice distortions (e.g. undulatory extinction and twinning). These three descriptive categories align with the three grain-scale deformation mechanisms introduced in Section 5.6: (1) cataclasis; (2) pressure solution, diffusive mass transport, and precipitation; and (3) intracrystalline plasticity. By identifying the deformation mechanisms and understanding their relationships to physical variables such as confining pressure, temperature, pore fluid pressure, and strain rate (Section 5.4), one can focus on appropriate constitutive laws and select equations of motion to undertake a mechanical analysis (e.g. Mancktelow, 2011; Jiang, 2014; see Further Reading).

10.3 KINEMATICS OF DUCTILE DEFORMATION

Rock fabrics in outcrop (e.g. **Figure 10.1** and **Figure 10.14**), and in thin section (e.g. **Figure 10.7** and **Figure 10.20**), are evocative of the kinematics of deformation. One can make qualitative deductions about the stretching of cobbles, or shearing of quartz grains, directly from the geometry of the deformed rocks and minerals. Indeed, a voluminous literature exists on microtectonics and deformation microstructures related to the kinematics of ductile deformation, which relies primarily on geometric relationships (see e.g. Cloos, 1946; Borradaile et al., 1982; Paterson et al., 1989; Trouw et al., 2009, in Further Reading). This literature usually is not explicit about a constitutive law relating strain to stress, or rate of deformation to stress, and rarely considers the spatial and temporal distribution of deformation as constrained by the equations of motion. As a result, these kinematic descriptions commonly lack physical constraints and often are not testable on physical grounds. None-the-less, kinematics do play an important role in any investigation of the mechanics of deformation, so we describe some basic principles and concepts in this section before introducing constitutive laws and equations of motion in Section 10.4.

10.3.1 Progressive Deformation in a Shear Zone

In this section and Section 10.3.2, we consider two different formulations of shear zone kinematics. In this section, we use a *referential* description of motion, comparing an initial and final configuration. In the next section, we use a *spatial* description of motion, comparing the kinematics at different positions at a given time. In this section, the kinematic quantity of interest is the displacement, but it is the velocity in the next section. We do not specify a constitutive law in either section, so our focus is strictly on the kinematics. However, the formulation in Section 10.3.2 provides a straightforward path for introducing a constitutive law, which we take up in Section 10.4.

We introduced the deformation gradient tensor in Section 5.5.1 by focusing on a *straight* material line of infinitesimal length at position **X** in the initial state, and using the vector d**X** to represent that line (**Figure 10.22**). After stretching, rotation, and translation that material line is at position **x** in the final state, and is represented by the vector d**x**. The material line in the final state depends on its configuration in the initial state according to a referential description of motion. The material line is taken as *infinitesimal* because, if it were of finite length, it could be curved in the final state, and then would be represented poorly by the vector d**x**.

As described in Section 5.5.1, the total differentials for dx and dy relate the components of d**x** (**Figure 10.22**) to the components of d**X**. These linear relationships are written in matrix form as:

$$\begin{bmatrix} dx \\ dy \end{bmatrix} = \begin{bmatrix} F_{xx} & F_{xy} \\ F_{yx} & F_{yy} \end{bmatrix} \begin{bmatrix} dX \\ dY \end{bmatrix}, \quad 0 < F_{xx}, F_{yy} < +\infty, \quad -\infty < F_{xy}, F_{yx} < +\infty \tag{10.2}$$

The square matrix contains the components of the deformation gradient tensor, **F**, which are dimensionless numbers composed of partial derivatives such as $F_{xy} = \partial x/\partial Y$ and $F_{yy} = \partial y/\partial Y$. The elements of [F] on the primary diagonal must be positive numbers, and those on the secondary diagonal can be negative, zero, or positive. In general, the elements of [F] are functions of the spatial coordinates: for example, $F_{xy} = F_{xy}(X, Y, Z)$. These relationships describe the heterogeneous deformation at any point in the material continuum because the elements of [F] vary from point to point.

Straight material lines of finite length in the initial state are straight in the final state only if the deformation does not vary on the length scale of that line. Thus, for homogeneous deformation, the elements of the deformation gradient matrix are *uniform* in space over a designated volume. Employing this assumption, we rewrite the matrix equation (10.2) as:

$$\begin{bmatrix} x \\ y \end{bmatrix} = \begin{bmatrix} I & J \\ K & L \end{bmatrix} \begin{bmatrix} X \\ Y \end{bmatrix}, \quad 0 < I, L < +\infty, \ -\infty < J, K < +\infty \tag{10.3}$$

The elements of this square matrix are dimensionless numbers with the same restrictions on their ranges, but they are not functions of the spatial coordinates within the designated volume. For homogeneous deformation the infinitesimal vector components, $\langle dx, dy \rangle$ and $\langle dX, dY \rangle$, on the left and right sides of (10.2), are replaced in (10.3) by the components of position vectors in the final and initial states, $\langle x, y \rangle$ and $\langle X, Y \rangle$.

Instead of focusing only on an initial and final state, we consider successive applications of the same or perhaps different deformation gradient tensors. Each application produces additional deformation and a new geometry, eventually leading to the final state. The geometry at each stage in this progression is like a frame of a movie that depicts the evolution of the structure. This is progressive deformation. This method still employs the referential description of motion at each stage, and the kinematic quantity is displacement, not velocity, so time is not explicitly considered.

As an example of homogeneous progressive deformation consider a model shear zone with simple shear strain in the (X, Y)-plane. We introduced particular examples of simple shear in Section 5.5.1 and illustrated these deformations in **Figures 5.18d, e,** and **f.** In the initial (undeformed) state a square of side length 2 m (**Figure 10.23a**) and a circle of radius 1 m

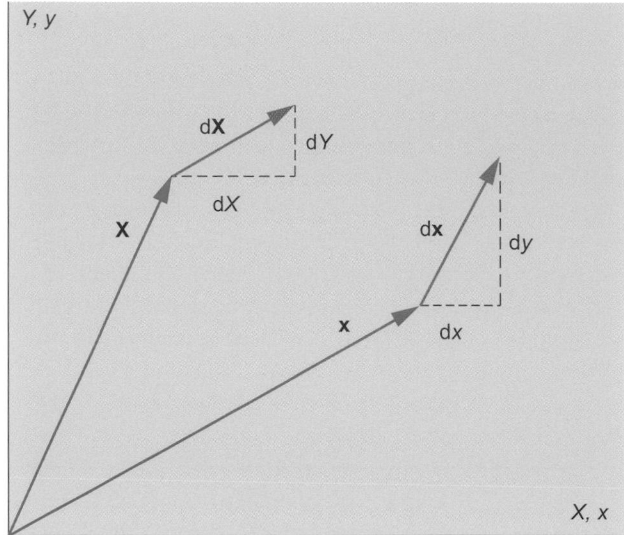

Figure 10.22 Vectors used to define deformation in the infinitesimal neighborhood of a particle at position **X** in the initial state and at position **x** in the final state. The vector d**X** represents the infinitesimal material line at **X**, and the vector d**x** represents the same material line after stretching, rotation, and translation to **x**. Lengths of the differential vectors are greatly exaggerated relative to the position vectors.

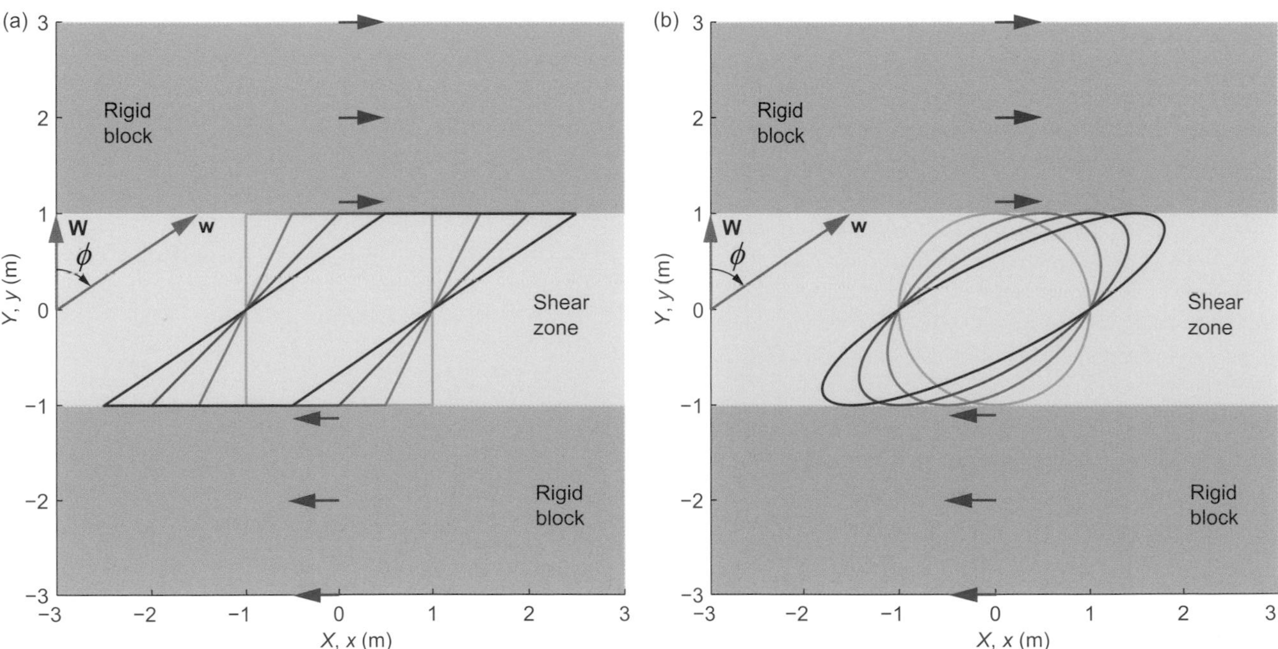

Figure 10.23 Illustrations of the deformation of a square (a) and a circle (b) using a homogeneous progressive deformation across a 2 m wide model shear zone. Displacement vectors (blue) for the bounding rigid blocks record the last increment. The vectors **W** and **w** represent a material line in the initial and final states. The angle of shearing is ϕ.

(**Figure 10.23b**) are drawn on the model shear zone, which is subject to the following arbitrarily defined deformation gradient tensors:

$$[F] = \begin{bmatrix} 1 & 0.5 \\ 0 & 1 \end{bmatrix}\begin{bmatrix} 1 & 0.5 \\ 0 & 1 \end{bmatrix}\begin{bmatrix} 1 & 0.5 \\ 0 & 1 \end{bmatrix} = \begin{bmatrix} 1 & 1.5 \\ 0 & 1 \end{bmatrix}$$

Three successive identical deformations occur, with sufficiently large shearing so each increment is distinctly illustrated. Here the incremental displacement vectors (blue) for the bounding blocks have a magnitude of 0.5 m. One could make the increments arbitrarily small, and arbitrarily numerous, without introducing new concepts. For this example, the progressive deformation leads to a shear strain, $\gamma = 1.5$, given by the element in the first row and second column of the final deformation gradient matrix.

Consider the effects of homogeneous simple shear strain on the material line represented by the vector $\mathbf{W} = \langle W_x, W_y \rangle = \langle 0, 1 \text{ m} \rangle$. The deformed material line is represented by the vector, $\mathbf{w} = \langle w_x, w_y \rangle$, but here the two vectors are not required to be infinitesimal, as in (10.2), because the deformation is homogeneous throughout the model shear zone:

$$\begin{bmatrix} w_x \\ w_y \end{bmatrix} = \begin{bmatrix} I & J \\ K & L \end{bmatrix}\begin{bmatrix} W_X \\ W_Y \end{bmatrix} = \begin{bmatrix} 1 & 1.5 \\ 0 & 1 \end{bmatrix}\begin{bmatrix} 0 \text{ m} \\ 1 \text{ m} \end{bmatrix} = \begin{bmatrix} 1.5 \text{ m} \\ 1 \text{ m} \end{bmatrix}$$

The angle of shear strain in the final stage is $\phi = \tan^{-1}(w_x/w_y) = 56.3°$ and the total offset across the shear zone, which extends from $y = -1$ m to $y = +1$ m is $2w_x = 3$ m. For homogeneous simple shear, the angle of shearing is related to the shear strain as $\tan \phi = \gamma$.

10.3.2 Rate of Deformation and Spin in a Shear Zone

In this section, we reconsider shear zones, but use a spatial description of motion. Again, we do not specify a constitutive law, so our focus is strictly on the kinematics. Time was not explicit in the previous section, because we only compared the configuration of particles in two states, and the kinematic quantity of interest was the displacement. However, in this section time is explicit, because the kinematic quantity of interest is the velocity. We use the concepts developed in this section to consider the complete mechanical analysis of a shear zone, including a constitutive law, in Section 10.4.

Recall from the discussion of the kinematics of flow in Section 6.6.1 that a spatial description of motion utilizes the velocity vector, **v**, defined at an arbitrary position in a flowing fluid designated by the position vector, **x** (**Figure 10.24**). The velocity vector is tangent to the particle path through that position. We focus on a specified time during the flow, called the *current time*, and ignore the initial configuration of particles. At an arbitrary neighboring position, **x** + d**x**, the velocity vector is **v** + d**v**, which is tangent to the particle path passing through that position. Using the total differentials for $\text{d}v_x$ and $\text{d}v_y$, the components of the relative velocity, d**v**, depend linearly on the components of the relative position, d**x**. In matrix form these relationships are:

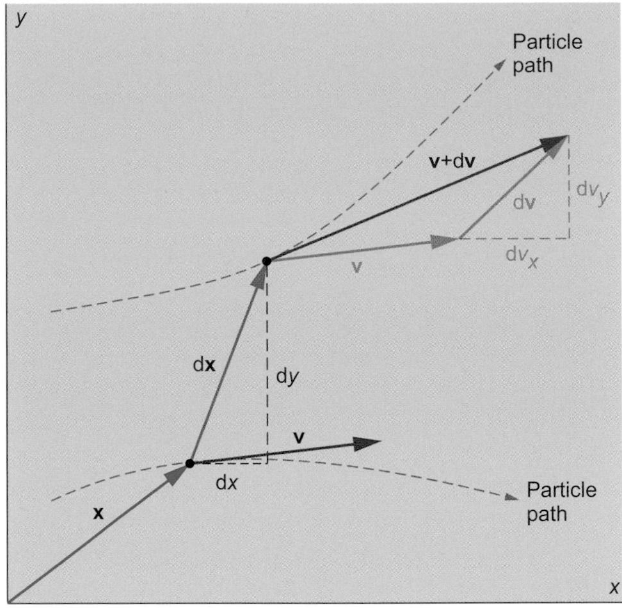

Figure 10.24 In this spatial description of motion the velocity, **v**, at **x** is compared to the velocity, **v** + d**v**, at the neighboring position **x** + d**x**. Brown dashed curves are particle paths. Lengths of the differential vectors, d**x** and d**v**, are greatly exaggerated relative to finite vectors, **x** and **v**.

$$\begin{bmatrix} \text{d}v_x \\ \text{d}v_y \end{bmatrix} = \begin{bmatrix} L_{xx} & L_{xy} \\ L_{yx} & L_{yy} \end{bmatrix}\begin{bmatrix} \text{d}x \\ \text{d}y \end{bmatrix}, \quad -\infty < L_{xx}, L_{xy}, L_{yx}, L_{yy} < +\infty \tag{10.4}$$

Each element of the square matrix in (10.4) is a partial derivative of a velocity component taken with respect to a coordinate in the current state, for example, $L_{xy} = \partial v_x/\partial y$ and $L_{yy} = \partial v_y/\partial y$.

The elements of the square matrix [L] in (10.4) are the components of the spatial gradient of velocity tensor, **L**, introduced in Section 6.6.2, and these components have dimensions of inverse time, T^{-1}. This tensor is not necessarily symmetric, so L_{xy} is not necessarily equal to L_{yx}. In general, the elements of [L] are functions of the spatial coordinates and time, for example, $L_{xy} = L_{xy}(x, y, z, t)$, and the sign or magnitude of these elements are not restricted.

The spatial gradient of velocity tensor, [L], plays a parallel role to the deformation gradient tensor, [F], described in Section 10.3.1, but notable differences exist. [F] compares initial and final states using a referential description of motion, whereas [L] is focused on the current state using a spatial description of motion. The components of [F] are partial derivatives of *final* position with respect to *initial* position (e.g. $F_{xy} = \partial x/\partial Y$), whereas components of [L] are partial derivatives of *velocity* with respect to *current* position (e.g. $L_{xy} = \text{d}v_x/\text{d}y$). Partial derivatives of the velocity components in [L] are components of the velocity gradient as defined in Section 2.6.1.

We use the spatial gradient of velocity (10.4) to characterize the kinematics of the idealized shear zone illustrated in **Figure 10.25**. The flow regime is postulated to be *steady*, so it does not change with time and what is shown here applies for all

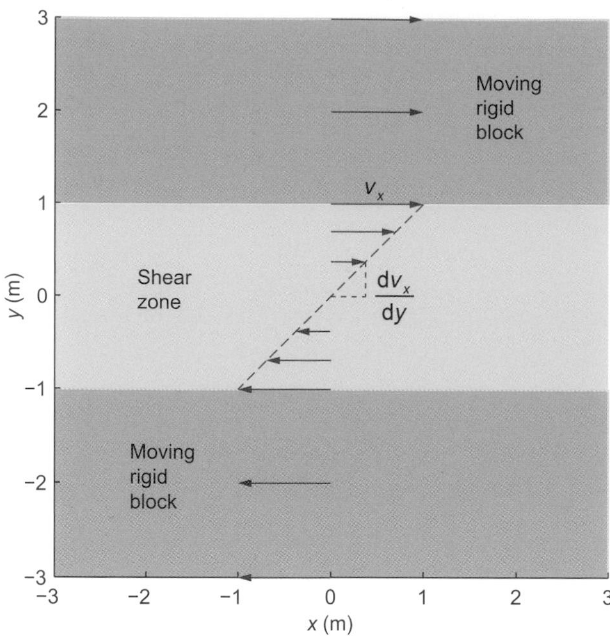

Figure 10.25 Model shear zone using a spatial description of motion with one non-zero velocity component, v_x, which is only a function of y. The flow is steady in time and the gradient in velocity is uniform across the zone. Compare to the model shear zone using a referential description of motion in Figure 10.23.

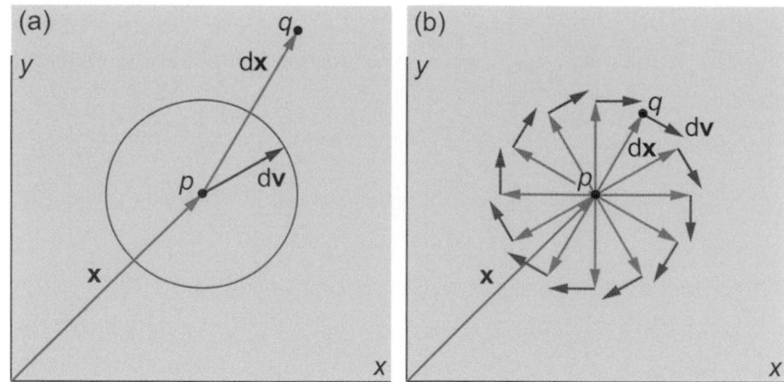

Figure 10.26 (a) Neighborhood of the particle at any position **x** in the model shear zone (Figure 10.25). A neighboring particle at **x** + **dx** is moving with the relative velocity **dv**. All neighboring particles and their velocities are represented by the purple circle. (b) Same neighborhood with relative velocity vectors (purple) illustrating the spin (vorticity) in the shear zone. The lengths of the differential vectors, **dx** and **dv**, are greatly exaggerated relative to the finite vector, **x**.

times. For the model shear zone two of the velocity components are zero, $v_y = 0 = v_z$, and the velocity component in the x-direction varies only in the y-direction: $v_x = v_x(y)$. Thus, the spatial gradient of velocity matrix (10.4) has just one non-zero element, $L_{xy} = dv_x/dy$, and this is postulated to be *uniform* everywhere in the shear zone. Note that we use the total derivative here, instead of the partial derivative, because v_x is *only* a function of y: it is not a function of the other two coordinates and time.

Recall from Section 6.6.2 that the square matrix [L] in (10.4) can be decomposed into a symmetric matrix [D] and a skew-symmetric matrix [W], such that [L] = [D] + [W]. To analyze the kinematics of the shear zone we begin by considering the matrix [D]. The elements of this matrix are the components of the rate of deformation tensor. Given the restrictions on the velocity components, the rate of deformation everywhere in the shear zone is:

$$[D] = \begin{bmatrix} D_{xx} & D_{xy} \\ D_{yx} & D_{yy} \end{bmatrix} = \begin{bmatrix} 0 & \frac{1}{2}(dv_x/dy) \\ \frac{1}{2}(dv_x/dy) & 0 \end{bmatrix} = \begin{bmatrix} 0 & 0.5 \\ 0.5 & 0 \end{bmatrix} \tag{10.5}$$

This *symmetric* matrix has only two non-zero elements, and both are equal to one-half the uniform and constant gradient of velocity across the shear zone.

For the sake of this example we choose a unit gradient in velocity and display the relevant vectors in Figure 10.26a. The velocity vectors are not shown, and usually would not be differential (small) quantities. Because the flow regime is *steady* (does not change with time), we do not need to specify a particular time. Because the velocity gradient is *uniform* from one side of the shear zone to the other, this figure applies to every position within the shear zone. The shape swept out by the relative velocity vectors is *circular*. This means the relative velocity **dv** has the same *magnitude* for all particles surrounding the particle at position **x**. The *direction* of **dv** varies systematically with the position of the point at **x** + **dx**, but is not necessarily parallel to the material line represented by **dx**. Recall from the more general example described in Section 6.6.2, that the shape swept out by the relative velocity vectors was elliptical (Figure 6.22), so the vectors varied in magnitude with orientation of the material lines. The uniform magnitude illustrated in Figure 10.26a is associated with the special flow regime where $D_{xx} = 0 = D_{yy}$, and $D_{xy} = D_{yx}$.

For the model shear zone the skew-symmetric matrix, [W], has only two non-zero elements: they are of opposite sign and equal in magnitude to one-half the gradient of velocity across the shear zone:

$$[W] = \begin{bmatrix} 0 & W_{xy} \\ W_{yx} & 0 \end{bmatrix} = \begin{bmatrix} 0 & -\frac{1}{2}(dv_x/dy) \\ \frac{1}{2}(dv_x/dy) & 0 \end{bmatrix} = \begin{bmatrix} 0 & -0.5 \\ 0.5 & 0 \end{bmatrix}$$

The elements of this matrix are the components of the rate of spin tensor, which also is called the vorticity tensor and was introduced in Section 6.2.2. The velocity is illustrated by plotting the components of the relative velocity vector, **dv**, with their tails at the neighboring particles. In other words, at the distal end of the material line **dx**. A sampling of the relevant vectors are illustrated in Figure 10.26b. Each relative velocity vector, **dv**, is perpendicular to the associated vector, **dx**, representing the material line, so these material lines do not stretch. This part of the motion is a uniform clockwise *rotation* of all the neighboring points with no stretching of the infinitesimal material lines between **x** and the neighboring points. In other words, [W] is a pure rotation.

The shear strain in the model shear zone described in Section 10.3.1, and the rates of deformation and spin in the model shear zone described in this section are defined without invoking a constitutive law for the material. Therefore, these kinematic quantities may be useful for any material that flows. Much of the structural geology literature about shear zones begins and ends with these kinematics: simple shear is invoked as a leading assumption, and measurements of geometric relationships observed in the field (e.g. offset markers or spatial variations in the orientation of foliation) are used to constrain the sense of shearing, and perhaps the magnitude of the shear strain. Assuming a velocity gradient across the shear zone, the components of a rate of deformation and spin may be calculated. In the next section we put these kinematics concepts into a *complete* description of the mechanics of a shear zone.

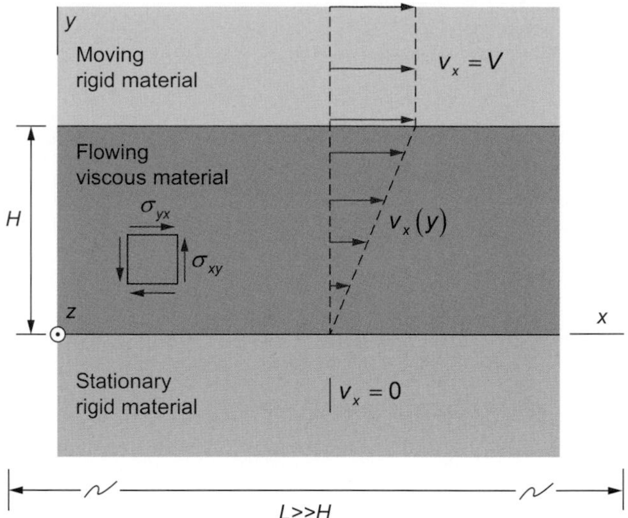

Figure 10.27 The model viscous shear zone. Two rigid bodies bound a thin zone of constant thickness *H*, filled with a flowing viscous material. The upper rigid body moves with velocity *V* relative to the stationary lower body, and the length, *L*, is much greater than the thickness.

10.4 A CANONICAL MODEL FOR A VISCOUS SHEAR ZONE

We are motivated to expand upon the analyses presented in Sections 10.3.1 and 10.3.2 because kinematics alone cannot address important questions about the process of shearing. For example, what were the stresses that *caused* the shearing? What were the material properties of the rock that *resisted* the shearing? How *long* did it take for the offset to develop across the shear zone? To address these questions we introduce a canonical model by defining the geometry of the model shear zone and then we invoke the equations of motion, choose a constitutive law, prescribe boundary conditions, and *solve* for the relevant kinematic quantity, which is the velocity.

Shear zones are tabular structures within which the shearing deformation is intense compared to the deformation in the surrounding rock mass (**Figure 10.14**). In other words, the deformation is localized within the shear zone. In outcrop, shear zones typically are thin compared to their trace length (**Figure 10.12**), and they have roughly parallel sides. Inspection at the outcrop scale usually does not serve to identify the constitutive equations that would best approximate the behavior of rock within the shear zone. Here we employ Newtonian viscosity as an example. The viscous material within the model shear zone is constrained between two rigid bodies, just as in Newton's hypothesis (Section 6.1). The rigid bodies play the role of the relatively undeformed rock on either side of the shear zone. The two shear zone boundaries are parallel and planar surfaces, and the zone has a uniform thickness equal to *H*.

10.4.1 The Equations of Motion

Recall from Section 6.8 that Cauchy's laws of motion reduce to the Navier–Stokes equations, which describe the motion of a linear, isotropic, and incompressible viscous liquid with uniform mass density. To simplify these equations of motion, we postulate that the shearing is entirely in the (x, y)-plane with no variation in the z-direction (**Figure 10.27**). This reduces the problem to two-dimensional flow, so the velocity component in the z-direction is zero, and the other two velocity components do not vary with z. For low Reynolds number (Section 6.8.2) the flow is laminar and parallel to the boundaries, so the velocity component in the y-direction is zero. Thus, the only non-zero velocity component is in the x-direction and it only varies in the y-direction.

From Section 6.8.1 the one Navier–Stokes equation for low Reynolds number flow that involves the velocity component in the x-direction is:

$$0 = -\frac{\partial \bar{p}}{\partial x} + \eta\left(\frac{\partial^2 v_x}{\partial x^2} + \frac{\partial^2 v_x}{\partial y^2} + \frac{\partial^2 v_x}{\partial z^2}\right) + \rho g_x^* \tag{10.6}$$

For two dimensions flow, v_x does not vary in the z-direction, so the third term in parentheses is zero. From the continuity equation for incompressible flow (Section 6.7.3), we have $\partial v_x/\partial x = 0$, so the x component of velocity does not vary in the x-direction and the first term in parentheses is zero. Summarizing the constraints on the velocity components, we have:

$$v_x = v_x(y),\ v_y = 0,\ v_z = 0 \tag{10.7}$$

We postulate that gravitational body forces may be ignored at the small scale of the shear zone, so the last term on the right side of (10.6) is zero. Finally we assert that no gradient in the pressure exists along the shear zone, so the first term on the right side of (10.6) is zero. Using these constraints the Navier–Stokes equation for slow viscous flow reduces to:

$$\eta\left(\frac{\mathrm{d}^2 v_x}{\mathrm{d}y^2}\right) = 0 \tag{10.8}$$

Note that the partial derivative has been rewritten as a total derivative, because the dependent variable, v_x, is only a function of one independent variable, y.

Integrating (10.8) twice with respect to y introduces two constants:

$$\eta v_x = \mathrm{C}_1 y + \mathrm{C}_2 \tag{10.9}$$

We invoke a no slip boundary condition on the velocity of the viscous material in contact with the rigid bodies (**Figure 10.27**):

$$\text{at } y = 0,\ v_x = 0 \text{ and at } y = H,\ v_x = V \tag{10.10}$$

In other words, the viscous material sticks to the two parallel boundaries of the rigid bodies. This is one of the classic boundary conditions in fluid dynamics. The first boundary condition requires $\mathrm{C}_2 = 0$, and the second boundary condition leads to $\mathrm{C}_1 = \eta V/H$. In this way the no slip boundary conditions provide the constraints necessary to determine the constants of integration.

Substituting for the constants of integration in (10.9), the velocity distribution across the viscous shear zone is:

$$v_x(y) = \frac{Vy}{H} \tag{10.11}$$

The velocity distribution is linear in y, proportional to the velocity, V, of one side of the shear zone relative to the other side, and inversely proportional to the thickness, H, of the shear zone (**Figure 10.27**). Interestingly, the viscosity drops out of this relationship, which contains only the kinematic and geometric quantities.

One could have simply postulated that the velocity varies linearly across the shear zone, as we did in Section 10.3.2 that only dealt with geometry and kinematics. However, here we have shown how this velocity distribution may be derived from the most general equations of motion for a linear, isotropic, and incompressible viscous liquid with uniform mass density. Not only does the velocity distribution depend upon these assumptions about the material properties, it also depends upon ruling out a pressure variation along the zone, and any effects due to gravitational body forces. By understanding all of these material and loading constraints, one gains important insights about the physics of the model shear zone. One also acquires a clearly defined path for relaxing some of these constraints in order to build a model that would represent natural shear zones more accurately.

Taking the derivative of velocity with respect to y in (10.11), we find that the velocity gradient is uniform across the shear zone, just as illustrated in **Figure 10.27**:

$$\frac{\mathrm{d}v_x}{\mathrm{d}y} = \frac{V}{H} \tag{10.12}$$

Recalling the rate of deformation tensor from Section 6.6.3, we find that the conditions imposed on the velocity by (10.7) leaves only two non-zero rate of deformation components:

$$D_{xy} = \frac{1}{2}\frac{\mathrm{d}v_x}{\mathrm{d}y} = D_{yx}$$

Substituting for the velocity gradient in (10.12) we have:

$$D_{xy} = \frac{1}{2}\frac{V}{H} \tag{10.13}$$

Again, this relationship is independent of the viscosity. Having imposed the relative velocity, V, of the rigid bodies through the boundary conditions (10.10), the rate of deformation is proportional to that velocity and inversely proportional to the thickness of the shear zone.

10.4.2 The Constitutive Equation

The constitutive equations for slow viscous flow (Section 6.7.3) provide the following relationship between the one non-zero shear stress, σ_{xy}, and the associated rate of deformation:

$$\sigma_{xy} = 2\eta D_{xy} \tag{10.14}$$

Substituting for the rate of deformation from (10.13), we have:

$$\sigma_{xy} = \frac{\eta V}{H}$$

The shear stress is uniform across the shear zone, proportional to the viscosity and the relative velocity of the rigid bodies, and inversely proportional to the thickness of the shear zone.

Finally, solving for the relative velocity of the rigid bodies bordering the shear zone:

$$V = \frac{\sigma_{xy} H}{\eta} \tag{10.15}$$

The velocity of one side of the shear zone relative to the other is proportional to the applied shear stress and the thickness of the shear zone, and inversely proportional to the viscosity. This relationship shows how the geometry, H, and the kinematics, V, are related to the shear stress, σ_{xy}, driving the relative motion, and the material viscosity, η, resisting the relative motion.

In summary, we deduced all of the mechanical features of the idealized viscous shear zone using a constitutive law for a linear and isotropic viscous material, the equation of continuity for an incompressible material with uniform density, and the equations of motion for slow viscous flow with negligible effects of gravity and no pressure variation along the zone. We idealized the shear zone as a parallel-sided conduit with rigid bounding bodies that move relative to one another with a constant velocity parallel to the model shear zone. Each of these constraints can be relaxed to model a shear zone with more complex geometry, material properties, and boundary conditions, but this example provides the basis for building those models, and a special case that can be used to test those models. In the next section we consider a shear zone with a less regular geometry, and compare the deformation associated with two different elastic–plastic constitutive laws.

10.5 RELATING FABRIC TO PLASTIC DEFORMATION AT FAULT STEPS

In most cases where a region of plastic deformation is surrounded by a region of elastic deformation, solutions to the governing equations do not exist in elementary forms. Recall that we did describe a solution in Section 5.8 for a plastically deforming model salt layer flowing into a diapir and loaded by rigid blocks representing the sedimentary layers above and below the salt. For that special case we were able to write down elementary algebraic equations that show how the stress and velocity are distributed throughout the salt.

No such solutions exist for the plastically deformed granodiorite caught between two left-lateral fault segments in **Figure 10.28**. Instead, numerical methods implemented on computers are required to construct a model and solve for the relevant mechanical variables, which are displayed graphically. Here we provide an example using the Finite Element Method (FEM) to model localized plastic deformation associated with slip on echelon fault segments, and relate the mylonitic foliation in right steps to the orientation and magnitude of the principal stretches, and to the constitutive properties of the granodiorite.

Following intrusion of the Lake Edison Granodiorite at about 88 Ma, a single set of steeply dipping fractures formed with strikes varying from about 050° to 070° (**Figure 10.29a**). These fractures are segmented and the segments are arranged in echelon patterns. Many of the fractures crosscut, but do not offset older aplite dikes, demonstrating that they are opening-mode fractures called joints. The joints were sealed by hydrothermal minerals, including quartz, epidote, and chlorite, so some would call them veins. Later, some of the sealed joints were sheared in a left-lateral sense (**Figure 10.29b**), deforming the hydrothermal minerals and offsetting the older dikes up to about 2 m. We call these structures faults, although they also could be called sheared

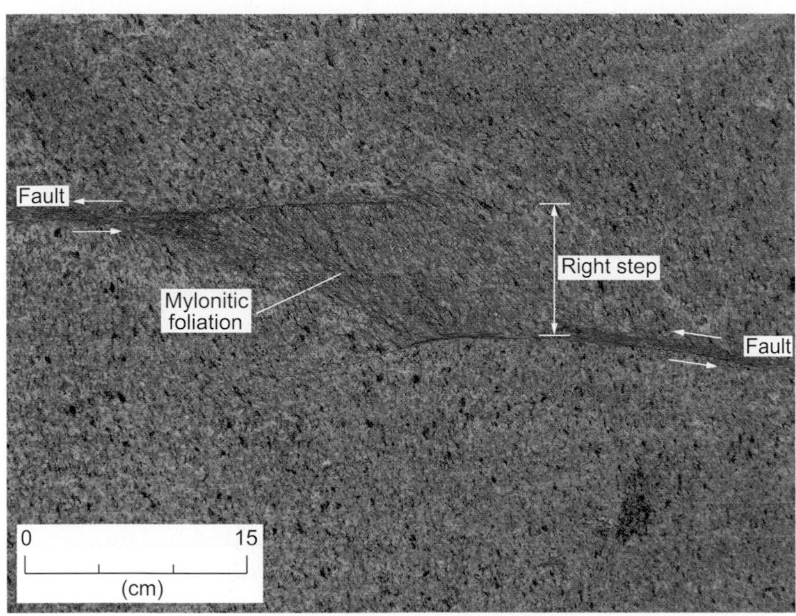

Figure 10.28 Example of mylonitic foliation within a right step between two left-lateral fault segments in the Lake Edison Granodiorite, Sierra Nevada, CA.

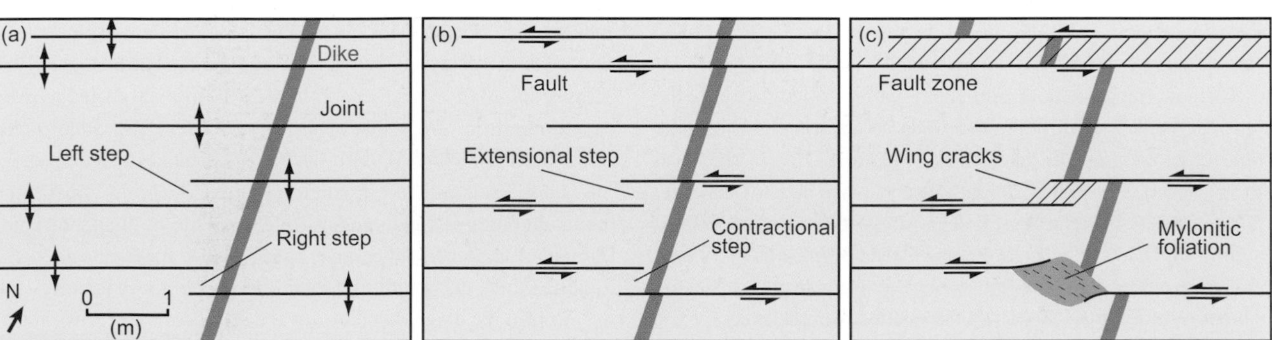

Figure 10.29 Schematic development of left-lateral faults in the Lake Edison Granodiorite, Sierra Nevada, CA. (a) One set of joints form as opening cracks. (b) Some joints are sheared to become left-lateral small faults. (c) Secondary brittle structures develop at left steps (wing cracks) and secondary ductile structures develop at right steps (mylonitic foliation). Two adjacent small faults evolve into a fault zone.

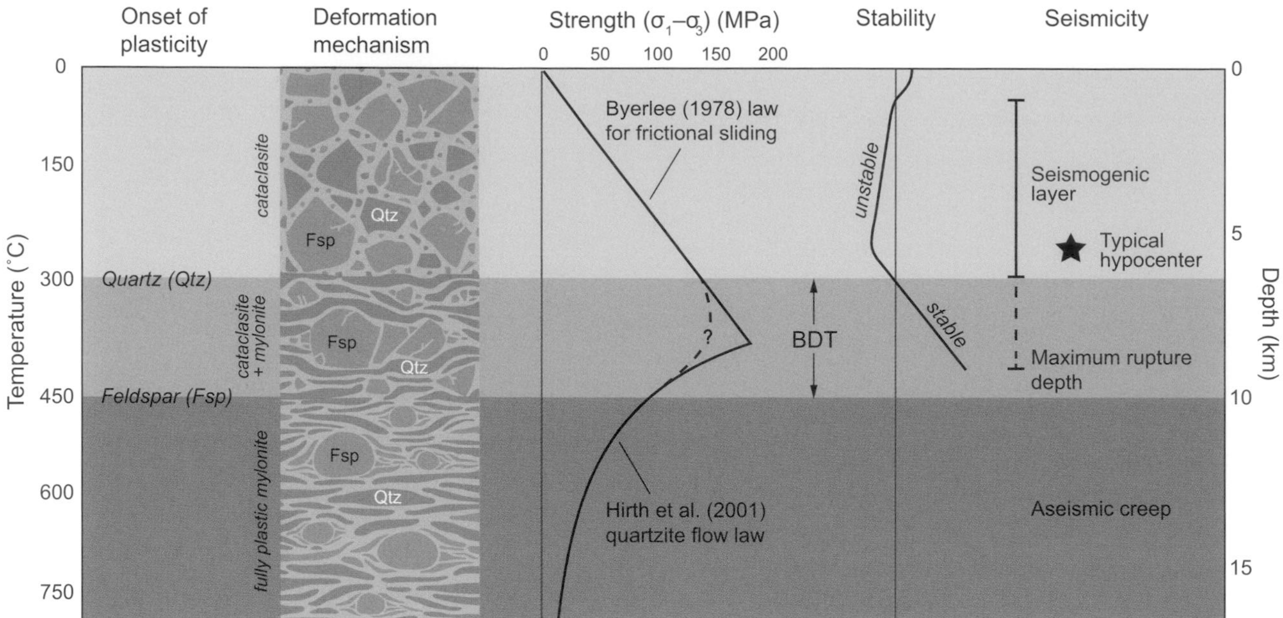

Figure 10.30 Temperature and depth of the brittle–ductile transition (BDT) identified with onset of plasticity in quartz and feldspar. The transitions from strength dominated by friction to that dominated by flow, from unstable to stable frictional sliding, and from seismogenic faulting to aseismic creep on faults also are indicated. Modified from Nevitt et al. (2017b), Figure 1. See also Scholz (1988), Byerlee (1978), and Hirth et al. (2001).

joints. The important points here are that these structures formed as opening fractures, were sealed, and later were sheared.

The inherited echelon geometry of the small faults provides many examples of both right steps and left steps between adjacent fault segments (**Figure 10.29c**). These also are referred to as contractional steps and extensional steps, respectively, based on the kinematics of left-lateral slip that pushed rock into the contractional step, and pulled rock out of the extensional step. We expect contraction to be associated with an increase in the local confining pressure, and extension to be associated with a decrease. In this regard recall that in laboratory tests, increases in confining pressure are associated with a transition from brittle to ductile deformation (Section 5.4.1). This is consistent with the fact that many contractional steps host a mylonitic foliation (**Figure 10.28**), diagnostic of ductile deformation, whereas most extensional steps host oblique opening-mode fractures called wing cracks (Section 8.2.1), diagnostic of brittle deformation.

The occurrence of both brittle and ductile structures along the same segmented small fault, and along adjacent faults within the same outcrop, indicate that different perturbations in the local stress state due to the different sense of step led to markedly different styles of deformation. This suggests that at the time of faulting this part of the Lake Edison Granodiorite was in a range of depths associated with the *transition* from brittle to ductile behavior. In **Figure 10.30** this transition is bracketed by the onset of plasticity in quartz at about 300 °C and the onset of plasticity in feldspar at about 450 °C. From Earth's surface to the top of the brittle–ductile transition (BDT), the temperature ranges from 0 to 300 °C, and brittle deformation is characterized there by opening and shearing cracks to form a cataclasite (Section 8.1). Frictional sliding on faults is governed by Byerlee's law, and can be unstable, leading to earthquakes with hypocenter depths extending partly into the BDT region (**Figure 10.30**). Where Byerlee's law applies, the strength increases linearly with depth as the confining pressure increases.

From the bottom of the BDT downwards (**Figure 10.30**), at temperatures exceeding 450 °C, ductile deformation is characterized by crystal plastic mechanisms and the formation of mylonite. This deformation is represented by a quartz flow law for which the strength *decreases* non-linearly with depth as the temperature increases. Frictional sliding on faults at these depths is stable, limiting typical earthquake hypocenters to depths less than 10 km, and promoting aseismic creep on deeper faults.

Within the BDT the behavior of quartz can be profoundly different from that of feldspar. For example, the thin section photograph (**Figure 10.31**) shows plagioclase porphyroclasts up to about 1.5 mm in size that have irregular borders and offset fragments. Many of these porphyroclasts have been reduced by fracturing from sizes that range up 5 mm in the undeformed granodiorite (**Figure 10.18**). On the other hand the average size of quartz grains in the mylonite is about 0.02 mm: smaller by about an order of magnitude relative to quartz in the undeformed granodiorite. Furthermore, no evidence exists for fracturing in the quartz, which apparently deformed by plastic mechanisms.

To provide an explanation for the localization of mylonite in right steps (**Figure 10.19**, **Figure 10.28**) we refer to the rock mechanics testing described in Section 5.4 where changing experimental conditions were shown to lead to ductile deformation. Perhaps the most obvious such change is an increase in temperature: differential compressive strengths decrease with increased temperature, and ductile deformation proceeds without fracture at elevated temperature. The temperature of the Lake Edison Granodiorite during faulting has been estimated to be between 300 and 500 °C, but no local heat sources (on a scale of several meters) have been identified that could raise the temperature in contractional steps, but not in the adjacent granodiorite. Therefore, we rule out variable temperature as the primary cause of localized ductile deformation in right steps.

Rock mechanics test data reviewed in Section 5.4 also show brittle fracture giving way to ductile deformation as pore fluid pressure is decreased. Again, no local sources (on a scale of several meters) of varying pore pressure have been identified that would lower pore pressure in contractional steps. In fact, one would expect the contraction of pores and microcracks to raise the fluid pressure in right steps, producing the opposite behavior from what is inferred from these outcrops. Rock mechanics test data also show that rocks change from brittle to ductile behavior as the strain rate is decreased. However, the strain rate experienced by Lake Edison Granodiorite at contractional and extensional steps along the same fault are unlikely to be very different, but the extensional steps display brittle secondary deformation, so we rule out differing strain rates as a primary cause of the different behaviors.

Figure 10.31 Thin section photograph of Lake Edison Granodiorite from sample SG10–03 from within a right step between two echelon fault segments (**Figure 10.19**). A plagioclase feldspar grain (pl) is fractured and offset indicating brittle deformation. Ductile deformation of quartz grains (qz) is indicated by reduced grain size. Modified from Nevitt et al. (2017b), Figure 6d.

For most rock types, including granitic rock, increasing the confining pressure in conventional triaxial tests increases the differential strength and the ductility (Section 5.4.1). The Lake Edison Granodiorite within fault steps (**Figure 10.28**) was not subject to a stress state as simple as the axisymmetric state of laboratory tests, but we consider the variation of mean normal stress, $\sigma_m = (\sigma_1 + \sigma_2 + \sigma_3)/3$, within the fault steps as a proxy for variation in confining pressure, P_c. Recall from Section 4.7 that two of the principal stresses in conventional triaxial tests are related to the confining pressure as $\sigma_1 = \sigma_2 = -P_c$. Also, the differential stress and confining pressure are related to the third principal stress as $\sigma_3 = \Delta\sigma - P_c$. From these relationships we understand that the mean normal stress in conventional triaxial tests is related to confining pressure as $\sigma_m = -P_c + \Delta\sigma/3$. Therefore, increasing confining pressure, and the associated increase in ductility, should correspond to the mean normal stress becoming more compressive. This provides a possible mechanism for the mylonitic fabrics observed at right steps (**Figure 10.28**) along faults in the Lake Edison Granodiorite.

To evaluate this mechanism we consider the variation of mean normal stress near a pair of left-lateral model faults arranged in the echelon pattern illustrated in **Figure 10.32**. The stress is not plotted in a small circular region around each fault tip to avoid extreme values associated with that tip and the likely plastic deformation there. Here we consider only elastic deformation in order to understand the effects of fault geometry on the distribution of mean normal stress. Each fault segment has a unit half-length, $a = 1$, and a prescribed uniform slip, $\Delta u_x = 0.001a$. Young's modulus is 75 GPa and Poisson's ratio is 0.27, values chosen to match those used in the elastic–plastic model described later in this section. The displayed mean normal stress is that due only to slip on the fault segments: about 250 MPa of all-around

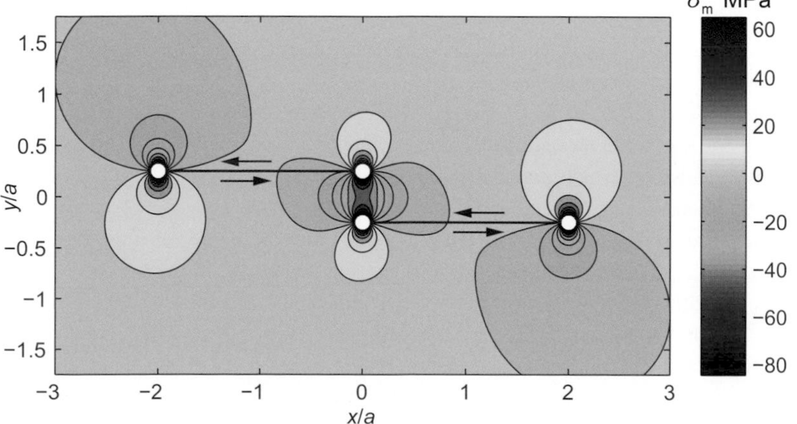

Figure 10.32 Mean normal stress change for two echelon left-lateral model faults in an elastic material. The stress state is that due to slip on the faults only. See text for model parameters and interpretation.

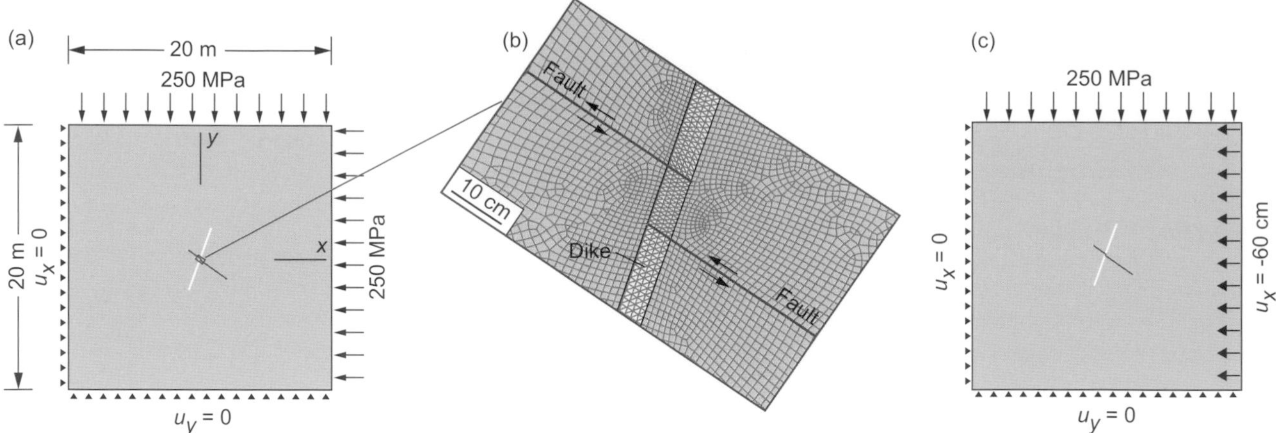

Figure 10.33 Finite Element Method (FEM) model geometry and boundary conditions for two echelon fault segments (red) and an aplite dike (white). (a) Entire model with all-around 250 MPa confining pressure. (b) Detail of two fault segments and dike with FEM mesh. (c) Entire model with –60 cm displacement boundary condition on right side. Modified from Nevitt et al. (2017b).

compression would have to be added to account for the lithostatic stress from the weight of the overlying rock.

For the model fault segments (**Figure 10.32**), the compressive stress lobes overlap in the right step and create an even greater compressive stress concentration than at the distal tips. To the extent that this mean normal stress is a proxy for elevated confining pressure in laboratory tests, the ductility of the rock would be enhanced there, and indeed this is the region where mylonitic fabric is observed in the Lake Edison Granodiorite (**Figure 10.28**). Thus, the distribution of mean normal stress from this elastic model provides a rationale for the occurrence of ductile deformation at contractional steps along segmented faults.

To evaluate the contribution of plasticity to fabric development at right steps, we focus on the example described in Section 10.2.3, and illustrated in **Figure 10.19** because an older aplite dike crosses the step and provides additional geometric data relevant to the faulting. In particular, the initial geometry of the fault segments and the dike can be inferred accurately from the outcrop observations. The conceptual model (**Figure 10.21**) suggests that three mechanisms contributed to the offset and deformation of the aplite dike that crosses the right step: (1) elastic deformation of the granodiorite and dike; (2) frictional sliding on the two fault segments; and (3) plastic deformation within the right step. We employ FEM to solve the governing equations and evaluate these mechanisms. The details of FEM are beyond the scope of this textbook, but the concepts and results illustrated by this solution are central to the objectives of this chapter. Therefore, we refer interested readers to the works listed in Further Reading, which includes textbooks devoted to FEM.

The two-dimensional plane strain model geometry includes the two fault segments and the dike at the decimeter scale in a configuration that matches the inferred initial state before fault slip (**Figure 10.33a**). Plane strain is appropriate for the nearly vertical fault segments, but does not accurate capture the geometry of the dipping dike, so the change in dip of the dike is not modeled here. These structures are set in a much larger body, 20 m on a side. The orientations of the remote boundaries (**Figure 10.33b, c**) correspond to the directions of the remote principal stresses. These directions are inferred from the orientations of wing cracks that extend from near the tips of the small faults (Section 8.2.1).

The model granodiorite material properties are: Young's modulus 74.8 GPa, Poisson's ratio 0.2735, and yield strength 377 MPa. The model aplite dike material properties are: Young's modulus 60 GPa, Poisson's ratio 0.2, and yield strength 377 MPa. Two steps of loading on the boundary are depicted in **Figure 10.33b** and **c**, with an initial lithostatic pressure of 250 MPa, followed by a lateral displacement of the right side, $u_x = -60$ cm. The pressure simulates a depth of about 10 km and the lateral displacement has been adjusted to produce a representative slip on the model faults.

Recall from the discussion of uniaxial loading of an elastic–plastic material in Section 5.1 that linear and recoverable elastic deformation gives way to non-linear and irrecoverable deformation at a particular normal stress, the magnitude of which is equal to a material property called the yield strength, σ_{ys}. For example, at the limit of purely elastic deformation under uniaxial compression the axial normal stress is the least principal stress, so $|\sigma_a| = |\sigma_3| = \sigma_{ys}$. For more general states of stress irrecoverable deformation occurs when the stress state satisfies a yield criterion, which may depend on all three principal stresses.

Several different yield criteria have been proposed for rock and for engineering materials and these depend upon the stress invariants. The stress tensor has three invariants, which are groupings of the stress components, and these grouping do not change value as the coordinate system changes. Because the yield strength is a property of the material itself, the yield criterion should not depend upon the arbitrary choice of a coordinate system, and therefore should depend only on the principal stresses or the stress invariants. The first invariant of stress plays a prominent role in yield criteria and it is defined as:

$$I_1 = (\sigma_1 + \sigma_2 + \sigma_3) = 3\sigma_m \quad (10.16)$$

One third the sum of principal stresses is the mean normal stress, σ_m, so I_1 is a measure of the *typical* normal stress.

The deviatoric stress also plays an important role in yield criteria. The deviatoric stress is defined by subtracting the mean normal stress from each of the normal components of the stress tensor. When considering only the principal stresses, all the shear stresses are zero, so the principal components of the deviatoric stress are:

$$\begin{bmatrix} \sigma'_1 & 0 & 0 \\ 0 & \sigma'_2 & 0 \\ 0 & 0 & \sigma'_3 \end{bmatrix} = \begin{bmatrix} \sigma_1 - \sigma_m & 0 & 0 \\ 0 & \sigma_2 - \sigma_m & 0 \\ 0 & 0 & \sigma_3 - \sigma_m \end{bmatrix}$$

The deviatoric stress also has three invariants, and we use the second invariant in the yield criteria considered here. The second invariant of deviatoric stress is:

$$J_2 = \tfrac{1}{2}\left(\sigma'^2_1 + \sigma'^2_2 + \sigma'^2_3\right)$$

Because both the first stress invariant (10.16) and this second deviatoric stress invariant are used in the same yield criterion, we rewrite J_2 in terms of the principal stresses rather than the deviatoric principal stresses:

$$J_2 = \tfrac{1}{6}\left[(\sigma_1 - \sigma_2)^2 + (\sigma_2 - \sigma_3)^2 + (\sigma_3 - \sigma_1)^2\right] \quad (10.17)$$

Each principal stress difference in this equation is equal to twice the maximum shear stress in the plane containing those normal stresses, so (10.17) is a measure of these extreme values of the shear stress.

The von Mises yield criterion specifies that plastic yielding occurs where and when the second invariant of the deviatoric stress, J_2, reaches a critical value. A practical critical value to employ is the yield strength, σ_{ys}, measured in a uniaxial compression test in the laboratory. For that test the only non-zero principal stress is σ_3, so (10.17) reduces to $J_2 = \sigma_3^2/3$. Substituting the yield strength for σ_3 and taking the square root of both sides of this equation, the von Mises yield criterion is written:

$$\sqrt{J_2} = \sigma_{ys}/\sqrt{3} \quad (10.18)$$

Using (10.17) this criterion is illustrated in **Figure 10.34** as a cylinder in principal stress space with a radius that is proportional to the yield strength in uniaxial compression. To represent typical conditions in Earth's crust, the yield surface is plotted in the octant of principal stress space where all three principal stresses are compressive (negative). For a material adhering to von Mises yield criterion any stress state that plots inside the cylinder is attainable with purely elastic deformation, and stress states outside the cylinder are not attainable. For stress states on the cylinder the material is yielding plastically.

The Drucker–Prager yield criterion appeals to the fact that rock in conventional triaxial tests gains strength with increasing confining pressure. This criterion uses the first stress invariant (10.16) as a proxy for confining pressure and is defined with the following linear relationship:

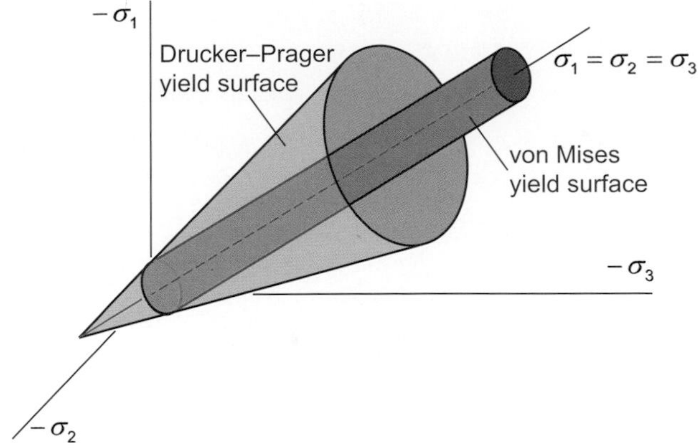

Figure 10.34 Von Mises (cylinder) and Drucker–Prager (cone) yield surfaces plotted in principal stress space. The axis of the cylindrical and conical yield surfaces is the isotropic line where all three principal stresses are equal. Modified from Nevitt et al. (2017b).

$$\sqrt{J_2} = A + BI_1 \quad (10.19)$$

Here A and B are constants that depend on the properties of the material. Using (10.16) and (10.17) to introduce the principal stresses, this criterion is illustrated in **Figure 10.34** as a cone in principal stress space. In what follows we compare the deformation at a right step in a model fault using the von Mises yield criterion to that using the Drucker–Prager criterion in order to evaluate which of these criteria might represent the behavior of the Lake Edison Granodiorite.

Recall that the slip on the model fault segments (**Figure 10.33**) is resisted by friction, which we quantify using the Coulomb criterion. This was introduced in Section 4.8.2 in the context of shear fracturing, and was written there as $|\sigma_s| = S_o - \mu_i\sigma_n$. In this equation σ_s and σ_n are the shear and normal stresses acting on the prospective fracture plane, and S_o and μ_i are the inherent shear strength and coefficient of internal friction. However, for the frictional slip considered here, the faults are designed into the FEM mesh: no shear fracturing occurs, only slip on the two fault surfaces. Therefore, we write the Coulomb criterion as:

$$|\sigma_s| = S_f - \mu_c\sigma_n \quad (10.20)$$

Here S_f and μ_c are the frictional strength and coefficient of friction. The changes in the names of these material constants recognizes the importance of the different physical situation: slip on pre-existing faults, rather than shear fracture of intact rock. We take the frictional strength as zero, and the static coefficient of friction as $\mu_c = 0.4$.

The FEM model results employing the von Mises yield criterion (10.18) are illustrated in **Figure 10.35** using contours of the principal stretches and white tic marks for the principal stretch

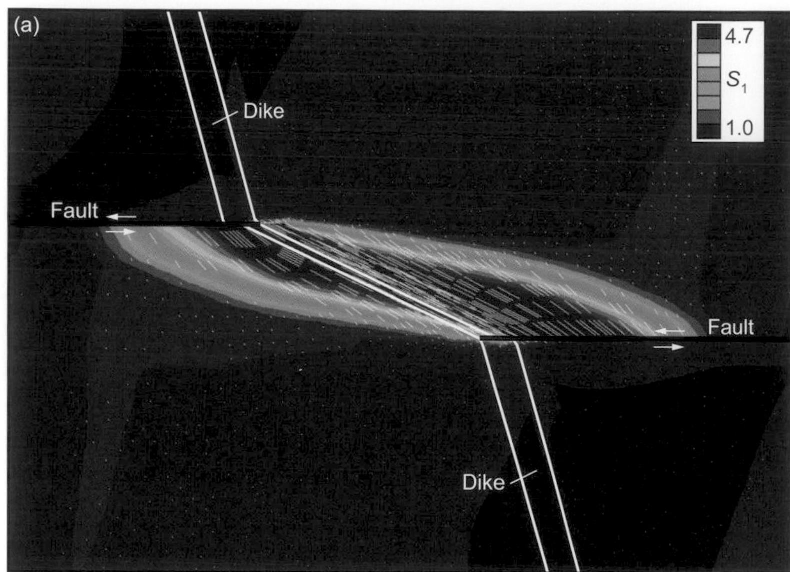

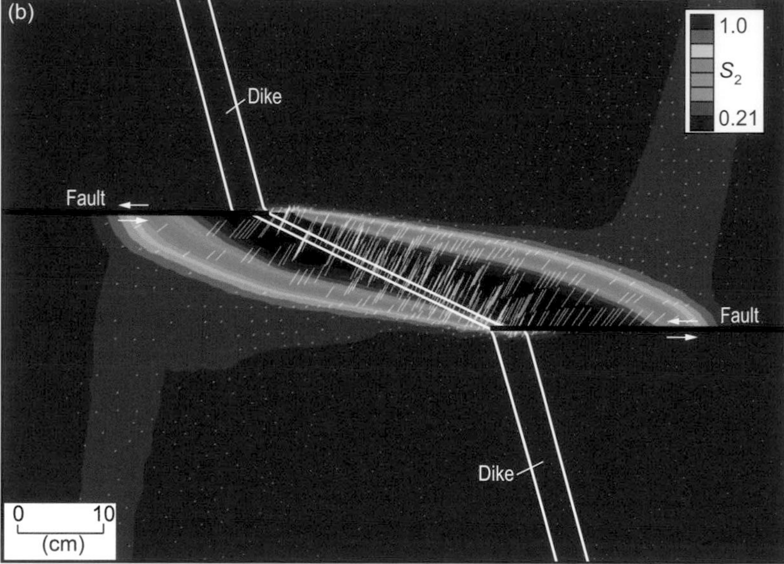

Figure 10.35 Contour maps of principal stretches near a model right (contractional) step between two left-lateral fault segments with a dike that crosses the faults through the step. Model parameters described in text. White tic marks align with orientations of the respective principal stretches. (a) Maximum principal stretch. Compare alignment of S1 with foliation in **Figure 10.19a**. (b) Minimum principal stretch. Compare model dike geometry to dike geometry in **Figure 10.19a**. See Nevitt et al. (2017b).

axes. Notice that the maximum stretch ranges from 1 to 4.7 and the minimum stretch ranges from 0.21 to 1. The most obvious correlation between these model results and the field observations is the close match between the region of large inelastic stretching within the model step, and the region of mylonitic foliation within the step between the two fault segments (**Figure 10.19a**). Perhaps of more importance, the measured geometries of the dike and step also match those of the model quite well. The dike strike is 290° and the model dike strike is 286°; the dike lengthening is 3.3 and the model dike lengthening is 3.3; the dike thinning is 0.4 and the model dike thinning is 0.3; the step widening is 1.8 and the model step widening is 1.7; the dike offset within the step is 30 cm and the model dike offset is 29 cm.

To evaluate the sensitivity of the FEM model results (**Figure 10.35**) to the choice of material properties, we compare models that are similar in every respect except for the yield criterion. To compare the von Mises yield criterion and the Drucker–Prager yield criterion we plot contours of the maximum principal logarithmic strain for the right step in **Figure 10.36a** and **b**. For the von Mises material the contours for this measure of the strain mimic the distribution of the mylonitic foliation and its orientation (**Figure 10.19**). However, the strain distribution for the Drucker–Prager material correlates poorly with the fabric distribution because it is concentrated outside the step instead of inside the step. Furthermore, the dike is not stretched, thinned, and rotated as it was in nature, and the fault tips bend towards each other rather than away from each other. Finally, note that the plastic deformation within the step accounts for considerable left-lateral offset of the two fault tips for the von Mises material, and very little offset for the Drucker–Prager material.

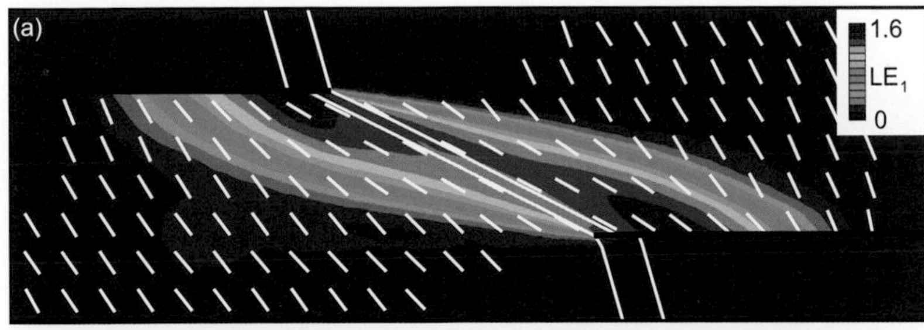

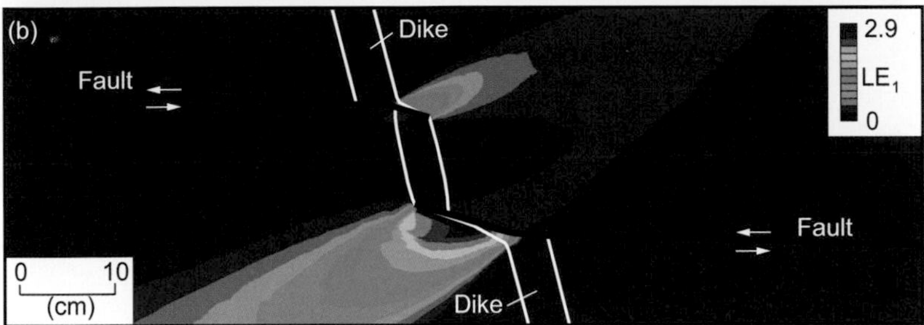

Figure 10.36 Comparison of maximum principal logarithmic strains for right step between left-lateral fault segments. FEM results using model illustrated in **Figure 10.33**. (a) von Mises yield criterion. (b) Drucker–Prager yield criterion. Modified from Nevitt et al. (2017b), Figure 12.

While the comparison of the FEM model results (**Figure 10.36a, b**) to the outcrop data (**Figure 10.19**) provides a strong case for ruling *out* the Drucker–Prager yield criterion (10.19), it should not be used to conclude that the von Mises yield criterion (10.18) is necessarily the *best* choice. Other yield criteria might perform as well for the conditions imposed, and perhaps some would perform even better under other conditions. This motivates additional modeling that seeks to rule out other yield criteria. For an insightful discussion of the evaluation of numerical model results see Oreskes et al. (1994) in Further Reading. We do conclude that the von Mises yield criterion is *consistent* with the deformation observed in this outcrop under the imposed conditions of the FEM models.

An important concept that the model results emphasize is the fact that 99% of the region considered in the model (**Figure 10.33**) experiences only elastic deformation. The plastic deformation is confined to the right step between the fault segments with an area of less than 0.1 m^2 compared to the entire model space with an area of 400 m^2. In other words, the plastic deformation is completely *contained* in a surrounding model rock mass in which the stress does not exceed the yield strength, and in which the strain does not exceed the limit imposed by the small strain kinematic equations.

If small increments of the 60 cm of displacement are imposed on the right boundary of the model (**Figure 10.33**), elastic deformation prevails during much of the loading history, even within the fault step. As the loading became sufficient for the Coulomb criterion (10.20) to be satisfied on the faults, they would begin to slip, and stress would be concentrated near the fault tips. Eventually this stress concentration would satisfy the von Mises criterion (10.18) and plastic yielding would begin in the fault step. It should be clear from this example that elastic deformation is the precursory behavior to ductile deformation, and that it acts to constrain the ductile deformation. Therefore, elasticity should not be ignored when attempting to understand how rock deforms to large stretches that induce visible fabrics such as the mylonitic foliation in this example.

Recapitulation

We began by describing rock fabric as the configuration of all the geometric elements making up that rock, including the crystal structure of individual mineral grains and the shapes of those grains, as well as much larger aggregates of grains such as cobbles in a metamorphic conglomerate. Thus, rock fabrics can be observed over length scales ranging from microns to meters. Because much of the study of rock fabrics focuses on the grain scale or smaller, optical and electron microscopes are a primary tool and these investigations are referred to as *microtectonics.*

Fabrics can be described in two classes. The alignment of geometric elements with pencil-like shapes defines a *linear fabric*, and the orientation of the lineation is recorded as a *trend* and *plunge* at the outcrop scale. The alignment of geometric elements with pancake-like shapes defines a *planar fabric*, and the orientation of the planes is recorded as a *strike* and *dip* at the outcrop scale. Where these elements are known to have been spherical before deformation (e.g. ooliths), and now are ellipsoidal, their axial lengths can be used to calculate the *principal stretches*, $S_1 \geq S_2 \geq S_3$. These are the greatest, intermediate, and least ratios of the current to original lengths of material lines. Those lines are in three orthogonal orientations that coincide with the axes of the ellipsoidal objects.

In many cases fabrics are made up of objects of unknown initial shapes. Although the orientation of the fabric commonly can be measured, quantifying the associated principal stretches is a challenge. Nearby fabric elements that do yield the principal stretches can provide guidance. For example, shale beds with a prominent planar fabric called *cleavage* are interbedded with the oolitic limestones of the Blue Ridge uplift in the central Appalachian mountain belt. Based on the assumptions of *pure shear* deformation and *no volume change* the ooliths yield principal stretches of about $1 \leq S_1 \leq 2$, $S_2 = 1$, and $0.5 \leq S_3 \leq 1$. The cleavage is recognizable in the limestones, if the ooliths have principal stretch ratios $S_1/S_3 \geq 2$. The long and intermediate axes of the deformed ooliths typically lie in cleavage planes, with the intermediate axis approximately parallel to the axis of folding and the long axis defining a *lineation* on the cleavage planes. The long and short axes of the ooliths are found in planes that are approximately perpendicular to fold axes. Because the cleavage is more developed in the shales we conclude that the estimates of the principal stretches from the ooliths provide lower bounds on the deformation in the shales.

Fabrics may be localized in *shear zones* with offset markers that provide data to estimate the amount of shearing. For example, in the Mono Creek Granite of the Sierra Nevada of California, older aplite dikes are cut by the White Bark shear zones and are offset from less than a centimeter to almost five meters. Ratios of dike offset to shear zone width give shear strains ranging from near zero to more than 100. These shear zones are characterized by reduced grain sizes, aligned biotites, and quartz and plagioclase stretched out into narrow stringers. Weak *mylonitic fabrics* are associate with average shear strains of about 4, whereas well-developed fabrics are associate with shear strains of about 40 or greater.

In other cases fabrics may be localized in structural settings where models can be used to estimate the magnitude and distribution of the deformation. For example, in the Lake Edison Granodiorite, also from the Sierra Nevada, right steps between segments of left-lateral faults are associated with local mylonitic fabrics. The average grain size of quartz in these steps is about 0.01 mm, an order of magnitude less than the size of grains distant from the steps, but with little evidence for fracturing. In contrast, feldspar grains are fractured, but remain as mm-sized *porphyroclasts*, and biotites have a characteristic shape called *mica-fish*. The arrangement of elongate quartz grains oblique to the prominent foliation, composed of lenses of fine-grained feldspar and biotite, identifies this as an *S-C mylonite*. The geometry of all these fabric elements in the Lake Edison Granodiorite is consistent with a left-lateral (sinistral) sense of shearing. Offset aplite dikes can be used to constrain models of this deformation and help to identify the constitutive properties of the granodiorite.

Two quite different formulations are used to describe shear zone kinematics. They differ in the type of reference frame used, and in the primary kinematic variables used. They are similar in that both are independent of the constitutive law of the sheared rock. One formulation uses a *referential description of motion*, comparing an initial and final configuration, and using the displacement as the primary kinematic variable. The *deformation gradient tensor* is used to quantify how an infinitesimal material line is stretched, rotated, and translated during the deformation. If the deformation is *homogeneous*, material lines of any length may be deformed through successive stages, like frames of a movie, in what is called *progressive deformation*. Although "progressive" seems to imply time, time is not explicit in this formulation, so the speed of the movie projector is completely arbitrary. For the model shear zone we use progressive *simple shear* to show, for example, how initially circular objects become elliptical objects with axes that rotate in the direction of shearing. Because the model shear zone is defined without invoking a constitutive law for the material, this model is applicable to any material that deforms according to the prescribed kinematics.

The other formulation uses a *spatial description of motion*, comparing the kinematics at different positions at a given time, and using the velocity as the primary kinematic variable. The *spatial gradient of velocity tensor* is used to quantify how the velocity changes in different directions near the position of interest. This tensor may be decomposed into a symmetric part, called the *rate of deformation* (or stretching), and a skew-symmetric part, called the *rate of spin* (or vorticity). For the model shear zone the gradient in velocity is uniform across the zone. Time enters this formulation through the velocity, which is arbitrarily set by choosing the relative velocity of the two rigid bodies bordering the shear zone. Because the rates of deformation and spin in the model shear zone are defined without invoking a constitutive law for the material, this model is applicable to any material that flows according to the prescribed kinematics.

The answers to key questions about shear zones require an approach that goes beyond kinematics. For example, what were the stresses that *caused* the shearing? What were the material properties of the rock that *resisted* the shearing? How *long* was required for the offset to develop across the shear zone and the fabric to form? To address these questions one can define the geometry of the model shear zone, invoke the equations of motion, choose a constitutive law, define the boundary conditions, and *solve* for the kinematic quantities (e.g. displacement or velocity). In other words, the kinematic quantities are calculated, not prescribed.

The canonical model for a shear zone treats a linearly viscous, isotropic, and incompressible liquid with uniform density deforming at low Reynolds number, so the flow is laminar. The constitutive law constrains the rate of deformation to be proportional to the shear stress, and the Newtonian viscosity, η, is the proportionality constant: $\sigma_{xy} = 2\eta D_{xy}$. The equations of motion constrain the relative velocity of the edges of the shear zone to be $V = \sigma_{xy} H/\eta$, where H is the thickness of the shear zone. The postulates used to solve this problem can be relaxed to model a shear zone with more complex geometry, material properties, and boundary conditions, but this example provides the methodology for building those models, and a special case that can be used to test those models.

The right steps between echelon fault segments in the Lake Edison Granodiorite contain mylonitic fabric that apparently developed because of left-lateral slip on the faults. Some of these steps also are crossed by an aplite dike that provides geometric data to constrain the deformation. A numerical model employing the finite element method provides a tool for solving the relevant equations of motion for the orientations and magnitudes of the principal stretches, and finding a constitutive law that is consistent with the location and distribution of the fabric and the deformed geometry of the dike. Using a *von Mises plastic yielding criterion*, the maximum stretch is as high as 4.7 in the center of the right step, and the minimum stretch is as low as 0.21. The region of inelastic stretching in the model step corresponds to the region of mylonitic foliation between the two fault segments, and the deformed geometry of the model dike and fault segments match those documented at the outcrop.

Of the central topics in the five chapters on geologic structures (fractures, faults, folds, fabrics, and intrusions), we suggest that fabrics have the greatest potential for significant advancement using the methodology developed in this textbook.

REVIEW QUESTIONS

The following questions are designed to highlight the expected *learning outcomes* for this chapter. Each question is taken directly from the material in the chapter and, for the most part, in the same sequence that it appears in the chapter. If an answer is not forthcoming, students are advised to read the relevant section of the chapter and discover the answer.

10.1. Fabric is the configuration of all geometric elements making up a rock. Name and describe fabrics at the *grain* scale, the *hand specimen* scale, and the *outcrop* scale.

10.2. Describe the measurements made at an outcrop to record the orientation of a *linear fabric* and the orientation of a *planar fabric*. What property of the planar fabric is used to record its orientation using a unit vector?

10.3. Define the kinematic variable called the *stretch*. What is the range of stretch associated with lengthening of a material line? What is the range of stretch associated with shortening of a material line?

10.4. Describe what is meant by a *crystallographic preferred orientation* or CPO. Explain how a CPO is visualized using a *c*-axis pole figure on a stereographic projection, including a description of the MUD.

10.5. What is rock *cleavage*? Use the Paleozoic rocks of the South Mountain–Blue Ridge uplift (**Figure 10.6**) to quantify the ranges of *principal stretches* associated with the development of cleavage. What assumptions have you used to address this question?

10.6. Explain how the *average shear strain* is estimated for shear zones at the White Bark outcrop (**Figure 10.12**). Include a discussion of the assumptions you have used.

10.7. Observations at the White Bark outcrop (**Figure 10.12**) led to the interpretation that the shear zones initiated along favorably oriented *aplite dikes*. Describe the key observations and discuss the implications for the relative yield strengths of the Mono Creek Granite and the aplite dikes.

10.8. The right step captured in the image and map of **Figure 10.19** separates two left-lateral faults in the Lake Edison Granodiorite. Use the image and map to characterize the deformation, distinguishing *brittle* and *ductile* deformation, as well as *localized* and *distributed* deformation. Point to specific features that support your characterizations.

10.9. Fabric elements from the right step between faults in the Lake Edison Granodiorite **Figure 10.19** include *S-C mylonite*, *mica-fish*, and *mantled porphyroclasts*. Describe what these look like in thin section, and explain how these are used to interpret the sense of shearing.

10.10. In **Figure 10.22** a *referential* description of motion is used to define the *deformation gradient tensor*. Describe how this is done; why it is important to consider the material lines to be infinitesimal; and why their infinitesimal length does not restrict the deformation to be small.

10.11. Compare and contrast the deformation gradient tensor in (10.2) with that in (10.3). In doing so describe what is meant by *heterogeneous* and *homogeneous* deformation.

10.12. Use the depiction of a shear zone in **Figure 10.23** to describe *homogeneous progressive deformation*, and explain how this depends on a referential description of motion, but does not depend upon time.

10.13. **Figure 10.24** illustrates a *spatial* description of motion and is used to define the *spatial gradient of velocity tensor*. Describe how this is done; why it is important to consider the distance between neighboring positions to be infinitesimal; and why this does not restrict the velocities to be small.

10.14. Compare and contrast the *deformation gradient tensor* and the *spatial gradient of velocity tensor*. Pay particular attention to the description of motion used for each tensor.

10.15. Use the depiction of a shear zone in **Figure 10.25** to describe *steady flow* with a *uniform velocity gradient*, and explain how this depends on a *spatial* description of motion, and how it depends upon time.

10.16. Using the shear zone model depicted in **Figure 10.25** explain how the spatial gradient of velocity, [L] in (10.4) is decomposed into a symmetric matrix [D], called *the rate of deformation*, and a skew-symmetric matrix [W], called the *rate of spin*. Describe the kinematics captured by [D] and [W].

10.17. The Navier–Stokes equations describe the flow of a linear, isotropic, and incompressible viscous liquid with uniform mass density. Reduce these equations to consider a parallel-sided shear zone with low Reynolds number flow (**Figure 10.27**), finding:

$$\eta\left(\frac{\mathrm{d}^2 v_x}{\mathrm{d}y^2}\right) = 0$$

Show that the constraints embodied in the equation of motion lead to the linear velocity distribution, $v_x = Vy/H$, in the shear zone.

10.18. Use the linear velocity distribution, $v_x = Vy/H$, and introduce the relevant component of the rate of deformation, $D_{xy} = \frac{1}{2}(V/H)$. Then, use the constitutive equations for a linear viscous material to find $V = \sigma_{xy}H/\eta$. Use this relationship to describe how the relative velocity of the rock mass on either side of the shear zone depends upon loading, geometry, and viscosity.

MATLAB EXERCISES FOR CHAPTER 10: FABRICS

Rock fabric is the configuration of all the geometric elements making up that rock. In the first exercise we distinguish planar from linear fabrics, and foliations from lineations. Then, MATLAB is employed to study the deformation of spherulites using the deformation gradient tensor, and to show that material lines in different orientations have different stretches, even though the deformation is homogeneous. Students use MATLAB to visualize progressive simple shear using a circle in the initial state that transforms into an ellipse in the deformed state. Then we take the kinematic model for a shear zone to the next level using the spatial gradient of velocity tensor. In a companion exercise, students evaluate the rate of deformation and rate of spin tensors using MATLAB. Questions remain, however: what were the stresses that caused the shearing; what were the material properties that resisted the shearing; and how long did it take for the offset to develop? Students answer these questions by investigating the canonical model for a shear zone using a linear viscous constitutive law. The exercises for this chapter help students understand how continuum mechanics provides the mathematical tools and physical concepts for quantifying ductile deformation. www.cambridge/SGAQI

FURTHER READING

For citations in figure captions see the reference list at the end of the book.

Blenkinsop, T., 2000. *Deformation Microstructures and Mechanisms in Minerals and Rocks*. Kluwer Academic Publishers, Dordrecht.
Microscopic structures resulting from deformation of rocks and minerals are described and illustrated with abundant thin section photographs in both black and white and color. The deformation mechanisms leading to the formation of rock fabrics are reviewed and used to interpret the fabrics.

Borradaile, G. J., Bayly, M. B., and Powell, C. M., 1982. *Atlas of Deformational and Metamorphic Rock Fabrics*. Springer Verlag, Berlin.
This introduction to many different visual aspects of rock fabrics as seen in outcrop, hand sample, and thin section includes descriptions, locations, and regional or experimental information that put the images in a useful context.

Cai, W., and Nix, W. D., 2016. *Imperfections in Crystalline Solids*. Cambridge University Press, Cambridge.
The principles of mechanics and thermodynamics are used in this textbook to introduce students to the behavior of defects in crystalline solids, which provide important mechanisms for plastic deformation and fabric development.

Cloos, E., 1946. *Lineation – A Critical Review and Annotated Bibliography*. Memoir 18. Geological Society of America, Ann Arbor, MI.
This monograph describes many different kinds of lineations observed in rock, including those attributed to shearing, rotation, intersection of planar structures, and mineral growth.

1971. *Microtectonics along the Western Edge of the Blue Ridge, Maryland and Virginia*. The Johns Hopkins Press, Baltimore, MD.
This classic study of deformed ooliths in sedimentary rocks of the Appalachian mountain belt set the stage for the development of microtectonics as an integral part of structural geology.

den Tex, E., 1972. Chapter XIII, in: Hills, E. S., *Elements of Structural Geology*. New York, John Wiley & Sons, Inc.
Chapter XIII, written by E. den Tex, of this textbook by Hills reviews the geometrical properties of rock fabrics and the relationships between fabrics and kinematics as developed by B. Sander.

Hobbs, B. E., and Ord, A., 2015. *Structural Geology: The Mechanics of Deforming Metamorphic Rocks*. Elsevier, Amsterdam.
This monograph covers the physical and chemical processes leading to the structures that develop in metamorphic rocks including deformation, mineral reactions, fluid flow, heat transport, and microstructural adjustments.

Jiang, D., 2007. Numerical modeling of the motion of deformable ellipsoidal objects in slow viscous flows. *Journal of Structural Geology* 29, 435–452.
This paper provides the equations and modeling tools to investigate the deformation of an ellipsoidal object during laminar viscous flow for applications to investigations of shape fabrics and preferred orientation fabrics.

2014. Structural geology meets micromechanics: a self-consistent model for the multiscale deformation and fabric development in Earth's ductile lithosphere. *Journal of Structural Geology* 68, 247–272.
The concept of embedding inhomogeneities within much larger inhomogeneities facilitates the analysis of multiscale deformation with application to fabric development in a shear zone.

Klein, C., and Philpotts, A. R., 2013, *Earth Materials, Introduction to Mineralogy and Petrology*. Cambridge University Press.
A colorful and well-organized textbook on mineralogy and petrology, including basic crystallography.

Mancktelow, N. S., 2011. Deformation of an elliptical inclusion in two-dimensional incompressible power-law viscous flow. *Journal of Structural Geology* 33, 1378–1393.
This paper considers the deformation of a linear viscous elliptical inclusion in a linear viscous matrix, and extends that using the finite element method to a power-law viscous inclusion and matrix.

Mura, T., 1987. *Micromechanics of Defects in Solids*, 2nd edition. Martinus Nijhoff, Dordrecht, The Netherlands.
This textbook covers the materials science field in which microstructures are analyzed using the continuum theory of elasticity, but applications are made, with certain restrictions, to the mechanical behavior of plastic and composite materials.

Nye, J. F., 1985. *Physical Properties of Crystals*. Oxford University Press, Oxford.
The classic treatment of physical properties of crystals using tensor notation and thermodynamic relations.

Oreskes, N., Shrader-Frechette, K., and Belitz, K., 1994. Verification, validation, and confirmation of numerical models in the earth sciences. *Science* 263, 641–646.
This insightful paper explains why verification and validation of numerical models of natural systems is impossible, so the primary value of models is heuristic.

Passchier, C. W., and Trouw, R. A. J., 1996. *Microtectonics*. Springer-Verlag, Berlin.

This is a comprehensive summary of microscopic fabrics of rock with abundant illustrations and thin section photographs in black and white of different fabrics and their kinematic interpretation.

Paterson, S. R., Vernon, R. H., and Tobisch, O. T., 1989. A review of criteria for the identification of magmatic and tectonic foliations in granitoids. *Journal of Structural Geology* 11, 349–363.

This review paper provides criteria for distinguishing the origins of foliations in granitic rocks including formation by flow of a viscous magma, by high-temperature solid-state deformation, and by moderate-to-low temperature solid-state deformation.

Qu, J., and Cherkaoui, M., 2006. *Fundamentals of Micromechanics of Solids*. John Wiley & Sons, Hoboken, NJ.

This textbook for students of materials science and the engineering mechanics of materials covers the fundamental concepts of micromechanics with thorough mathematical derivations and many exercises that introduce both linear and non-linear behaviors.

Trouw, R. A., Passchier, C. W., and Wiersma, D. J., 2009. *Atlas of Mylonites and Related Microstructures*. Springer, Heidelberg.

This is a wonderfully illustrated companion to Passchier and Trouw (1996), with abundant color images of microstructures at the millimeter scale of cataclasites, mylonites, pseudotachylytes, and other highly deformed rocks.

Turner, F. J., and Weiss, L. E., 1963. *Structural Analysis of Metamorphic Tectonites*. McGraw-Hill, New York.

This textbook reviews Sander's method of structural analysis and emphasizes symmetry principles and homogeneous deformation, rather than the fundamental conservation laws of physics and heterogeneous deformation.

Chapter 11
Intrusions

Introduction

Chapter 11 begins by defining intrusions, then describes the different characteristic forms of intrusions, and ends by deriving the solution for the rise of a spherical body of viscous liquid in a more dense viscous liquid. This classic solution from fluid dynamics is the canonical model for the intrusion of salt in sedimentary basins. In general, an intrusion is a body of rock that, in a former more mobile state, was injected into and deformed the surrounding host rock. Intrusions take the form of dikes, sills, laccoliths, stocks, plutons, and diapirs. The intruded material could be magma, rising due to buoyancy from deep in Earth's asthenosphere (Section 1.1.2), or magma injected laterally from a shallow pressurized chamber in Earth's lithosphere (Section 6.8.2). The intruded material could be salt, moving upward in a diapir due to buoyancy (Section 5.8), or a mobilized slurry of sand injected into the surrounding sedimentary rock (Section 11.1). The intruded material also could be molten rock, formed due to frictional heating on a fault during an earthquake (Section 8.6.3). The diversity of intrusions makes them an interesting and challenging topic for structural geologists.

The different forms of intrusions, and the different properties of the intruding materials, motivate the discussion of problems in mechanics, including conductive cooling, flow of non-linear viscous liquids, and thermal stresses. Some of the problems in mechanics that are relevant to intrusions were introduced in earlier chapters. For example, the flow of a linear viscous (Newtonian) liquid in a parallel-sided conduit, described in Section 6.8, is the canonical model for magma flow in a sill. The opening fracture, described in Section 7.2, is the canonical model for elastic deformation of the host rock accompanying the opening of a dike. The bending elastic plate, introduced as a model for folding in Section 9.6, is the canonical model for bending strata over a laccolith. The variety of problems that are relevant to intrusions encourages structural geologists to broaden their understanding of mechanics.

The mobile state of the intruding material may involve brittle, ductile, and/or viscous deformation, so intrusions may encompass the full spectrum of deformation mechanisms described in Chapters 4 through 6. Fractures, faults, folds, or fabrics may initiate and grow in the host rock as it deforms in response to the injection of the mobile material, so intrusions may be associated with any one, or all, of the basic geologic structures considered in Chapters 7 through 10. Fractures, faults, folds, and fabrics also may be found within the intruded rock mass. Thus, the documentation and study of intrusions may require all of the tools in the structural geologist's toolbox. For these reasons, we place the chapter on intrusions at the end of this textbook.

11.1 DIKES

Dikes are discordant intrusions. In other words, they crosscut the sedimentary strata or metamorphic foliation of the older rock into which the younger material is injected. For example, a vertical dike filled with igneous rock of early Miocene age crosscuts horizontal Cretaceous shale near the town of Walsenburg in southeastern Colorado (**Figure 11.1**). Tabular intrusions in a rock mass without bedding or foliation, such as in a stock or pluton of relatively uniform igneous rock, also are called dikes. The book by E. M. Anderson (1972) is a classic monograph on the formation of dikes, and the review paper by Rivalta et al. (2015) summarizes modern research on mechanical models of dike propagation. Both are listed in Further Reading at the end of the chapter.

The thickness of most dikes is much less than their dimensions measured in the plane of the intrusion; both dimensions are important to understand dike behavior. For example, the outcrop length of the Walsen dike (**Figure 11.1**) is 11.5 km and typical thicknesses along the eastern half of the outcrop range from 5 to 7 m, so the ratio of thickness to length is roughly 1:2000. Over the western half of the outcrop this dike tapers gradually to less than 1 m. Because dikes and sills typically resemble thin, planar to gently curved sheets, the more general name sheet intrusion is applied to both. However, sills do not crosscut sedimentary strata or metamorphic foliation, so they are distinguished as concordant intrusions (see Section 11.2).

Figure 11.1 The vertical Walsen dike cuts and bakes the nearly horizontal Pierre Shale near Walsenburg, CO. Petrologic studies have identified three distinct magmas that were injected into this dike, so it is referred to as a composite dike. UTM: 13 S 518561.00 m E, 4165430.00 m N. See Pollard and Muller (1976).

Dike contacts are the interfaces between the intrusive material and the host rock. For the Walsen dike the contacts are interfaces between igneous rock and baked shale (Figure 11.1). Prior to the intrusion, these contacts may have been the two sides of a pre-existing fracture, such as a joint in the shale, or they may have been bonded together as continuous sedimentary layers until fractured by the intrusive event. In either case the dike *opened* as material was injected into the fracture and an opening displacement discontinuity developed, so dikes are classified as opening fractures (Section 7.3.2). Recall that opening fractures are called mode I, and that they are distinguished from mode II and mode III fractures, which have shearing and tearing displacement discontinuities.

In a material that is isotropic with respect to fracture toughness, an opening fracture propagates in a plane that is perpendicular to the greatest principal stress, σ_1 (Section 7.4.3). At depth in the Earth all three principal stresses may be compressive, in which case mode I fractures propagate perpendicular to the *least compressive* stress, also σ_1, and are pushed open by an internal fluid pressure that exceeds this stress in magnitude. To the extent that this relationship describes the propagation of sheet intrusions, we expect dikes to be good predictors of the orientation of the least compressive stress. For example, the strike of the Walsen dike (Figure 11.1) is 068, so we infer that the azimuth of the least compressive stress at the time of intrusion was 158.

Dikes may be injected by magma, salt, or a sedimentary slurry, but they share the geometric attributes of being discordant sheet intrusions. Where magma is the injected fluid, the host rock may be locally altered by heat flowing from the dike in a process called contact metamorphism. In Section 11.1.5 we consider a model for conductive heat flow from a dike and show that the temperature just outside the contact quickly rises to about midway between the temperature of the magma and the temperature of the host rock. Thus, for magma at 1000 °C injected into a cold host rock at 0 °C, heating would raise the local temperature along the contact to about 500 °C. Typical consequences can be seen within a few meters of the Walsen dike (Figure 11.1) where the Pierre Shale has been baked and hardened, so it is more resistant to erosion than the unbaked shale farther from the contact.

On the other hand, where salt or a sedimentary slurry is the injected fluid, the temperature of the fluid and host rock may be nearly identical, so little or no contact metamorphism occurs. For example, in Figure 11.2b a steeply dipping sandstone dike about 13 cm thick crosscut the gently dipping Mowry Formation of Cretaceous age with no visible signs of contact metamorphism. This and other clastic dikes crop out near Sheep Mountain anticline in north central Wyoming. The dikes range in outcrop length from several tens of meters to more than 1 km, and they have maximum thicknesses from about a decimeter to less than 3 m. Reported thickness to length ratios for six of the dikes range from 1:34 to 1:140. These ratios are greater than that for the Walsen dike (Figure 11.1), but are not untypical for dikes in general.

Geologic evidence at Sheep Mountain anticline provide insight into the process by which the clastic dikes were intruded. The fold formed during the Laramide orogeny that lasted from the latest Cretaceous to the early Miocene. The current orientations of the clastic dikes (Figure 11.2) do not appear to be systematic, and many are inclined, but upon restoring the sedimentary strata to horizontal by unfolding, the dikes become approximately vertical, and their orientations define two sets: one striking northwest and the other striking northeast. This suggests that vertical dikes formed prior to folding, or in the earliest stages of folding, and then *rotated* with the sedimentary strata as the fold developed.

The sand grains and chert pebbles of the clastic dikes are similar in composition and geometry to those in the overlying Peay Sandstone (Figure 11.2a), and four of the dikes can be traced upward into that unit, so the Peay is identified as the source of the clastic material. Apparently, the dikes propagated *downward* into the Upper Mowry Formation, but what mechanism led

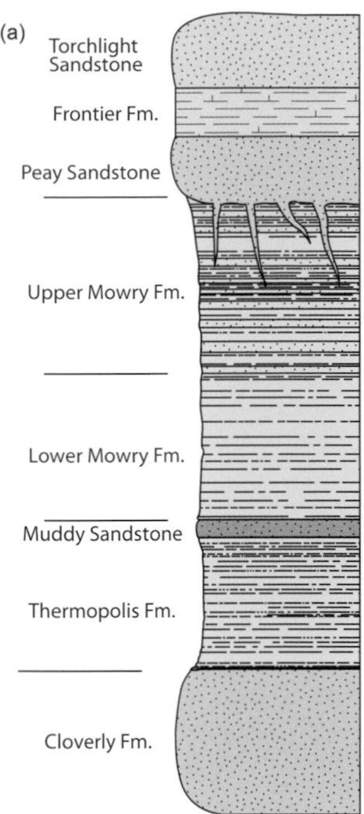

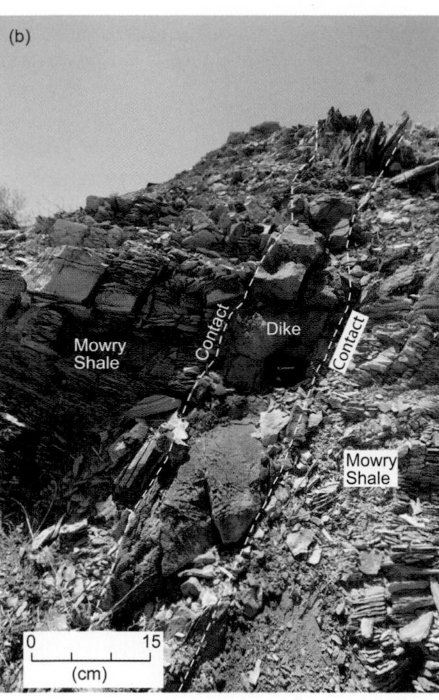

Figure 11.2 (a) Stratigraphic section in region with clastic dikes near Sheep Mountain, WY. (b) Clastic dike cuts the Upper Mowry Formation. Dashed white lines mark the contacts. Structural and petrographic studies conclude that the clastic material filling this dike was injected downward from the overlying Peay Sandstone at the base of the Frontier Formation. Approximate UTM: 12 T 723996.00 m E, 4943711.00 m N. Modified from Beyer and Griffith (2016), Figure 1.

to the fluidization of the sandstone, so it could flow into the dikes?

To form a slurry capable of flowing into the dikes, the pores of the Peay Sandstone must have been saturated with groundwater, and the water pressure must have exceeded the normal compressive stresses holding the sand grains together. In other words, the effective stress must have been positive. Recall from Section 4.7.3, that the effective stress is defined by adding the pore fluid pressure, P_p, to all normal stress components and leaving the shear stress components unchanged. For example, typical normal and shear stress components are $\sigma^e_{xx} = \sigma_{xx} + P_p$ and $\sigma^e_{xy} = \sigma_{xy}$. At the time the dikes formed, the normal stresses in the Peay Sandstone were negative (compressive), due to the weight of the overlying strata, and pore pressure always is positive. The addition of sufficient pore pressure can make the effective normal stress positive.

From Section 6.7.4 the gradient in normal compressive stress with depth is about –25 MPa/km, but the gradient in pore pressure is only +10 MPa/km. Therefore, to achieve a positive effective stress, the pore pressure must be elevated well above hydrostatic. This is a common occurrence where sandstone is over- and underlain by relatively impermeable shale, which prevents the escape of the water as the sandstone is compacted due to the overburden weight. The Frontier Formation at Sheep Mountain is underlain by the 200 m thick Mowry Formation and it is overlain by the 700 m thick Cody Shale, so the conditions were appropriate for generating excess pore pressure. As the pore pressure increased, the stress state in the sandstone was driven toward a positive effective stress. One can think of the fluid pressure pushing the rock grains apart, or the effective tension pulling them apart, but in either case this facilitates the fluidization of the sandstone.

11.1.1 Echelon Dikes

The two contacts of a dike come together at the tipline, the curved line that marks the most distal extent of the opening fracture that contains the intrusive material. The simplest idealization of a dike prescribes the tipline as circular, so the three-dimensional geometry of a dike could be described as a circular disk, or *penny-shaped*. If a penny-shaped dike were exposed, its tipline would intersect Earth's surface at two points, each called a termination of the dike in outcrop. While the penny-shaped dike is useful for models and scaling, most dikes in nature have more complicated tiplines. In this section we describe a dike with many terminations that provide important information about the three-dimensional shape of the tipline.

The Northeastern dike (**Figure 11.3**), near the spectacular volcanic neck called Ship Rock, serves to illustrate an important geometric feature of dikes: in outcrop they commonly consist of *discontinuous segments*, each with two terminations. For this igneous dike the segment length was measured as the horizontal distance from one termination to the other, and thickness was measured at about one meter intervals along the segments as the perpendicular distance from one contact to the other. The 35 discrete segments of the Northeastern dike range from 8 to 395 m in length and from 0.6 to 4.6 m in average thickness. The horizontal distance from the nearest termination in the foreground of **Figure 11.3** to the most distant in the background is 2,901 m, and the average of 2,726 thickness measurements is 2.3 m. Thus, the ratio of average thickness to total outcrop length is about 1:1,260.

When the Northeastern dike is viewed toward the northeast, as in **Figure 11.3**, each more distant segment steps to the left, but is nearly parallel to the adjacent segments. A line through the

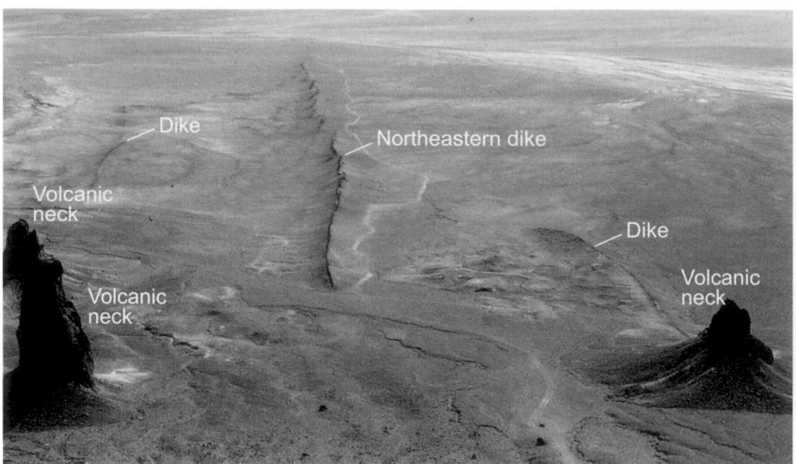

Figure 11.3 Oblique photograph toward the northeast of three volcanic necks, each about 20 to 30 m in diameter, and three dikes intruded into Mancos Shale near Ship Rock, NM. The Northeastern dike is composed of 35 discrete segments and is 2.9 km in outcrop length. UTM 12 S 693312.58 m E, 4062782.83 m N. See Delaney and Pollard (1981).

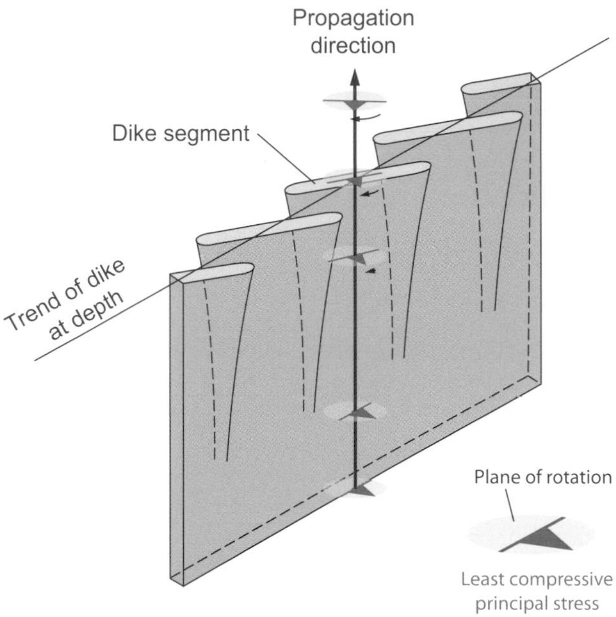

Figure 11.4 Schematic illustration of vertical dike with echelon segments that twist in strike upward and overlap adjacent segments. Tiplines of adjacent echelon segments diverge from the breakdown zone. Modified from Delaney and Pollard (1981), Figure 29.

middle of adjacent segments is approximately parallel to the overall trend of the dike, which is 056. This geometry of dike segments is referred to as an echelon pattern, and it is similar to the echelon arrangement of joint segments near the distal margin, or fringe, of joints as described in Section 7.1.3. Recall that in Section 7.4.3 we explained how fractures, such as joints, propagate into an echelon arrangement of segments when subject to a dominant opening with minor shearing parallel to the tipline. In other words, this geometry is induced by the propagation of a mixed mode I–III fracture.

Given the mixed mode I–III interpretation of the echelon dike segments, we suppose that at some depth, below the current outcrop, the Northeastern dike is vertical and continuous (**Figure 11.4**). At that depth the dike was perpendicular to the least compressive principal stress, σ_1, as the dike opened. Thus, the least compressive stress was horizontal and directed along an azimuth of about 146. At shallower depths, however, the least compressive stress was in a different orientation, rotated clockwise in the horizontal plane to between 150 and 155. This induced a mode III stress intensity at the dike tip, which caused it to break down into echelon segments (**Figure 7.36**). As the dike segments propagated upward they twisted clockwise about a vertical axis to remain perpendicular to the least compressive stress.

The detailed geometry of the echelon dike segments shown on the geological map in **Figure 11.5** depends upon the distance between adjacent segments, measured perpendicular to their strike. We refer to this distance as the step because the segments in map view look like the treads on a stair viewed in cross section. The steps for the mapped segments vary from 5.7 m (16–17), to 11.0 m (17–18/19), to 16.6 m (21–22). The overlap of the segment terminations vary from 5.4 m to 16.3 m to 32.9 m. Similar relations are found for the other echelon segments of this dike, with overlap systematically increasing with step. An analysis based on solutions to the linear elastic boundary-value problem for two echelon pressurized cracks shows that the opening mode stress intensity increases as the crack tips approach one another, and then decreases precipitously as the tips overlap. In other words, the cracks enhance each other's propagation when underlapped, but compete with one another when overlapped. When the stress intensity falls below the fracture toughness (Section 7.4.2), crack propagation ceases. The competition is less severe for greater steps, so the amount of overlap is greater. This effect can be seen by comparing the overlapped tips of segments 21 and 22 with the overlapped tips of segments 16 and 17.

The shapes of the echelon segment tips also vary with the size of the step (**Figure 11.5**). For small steps (e.g. segments 16–17) the tips taper abruptly to a rounded termination. For greater steps (e.g. segments 21–22) the tips taper gradually to a sharper termination. The tips also have a distinctive asymmetry. The outer contacts are nearly straight and parallel to the length of the

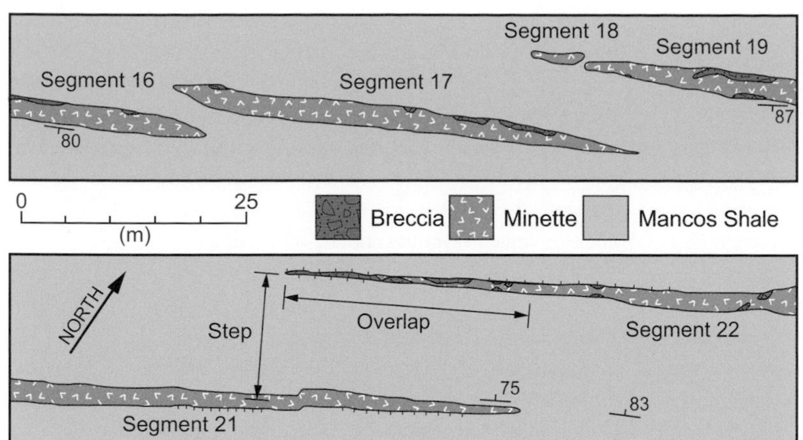

Figure 11.5 Geologic maps of selected echelon segments of the Northeastern dike at Ship Rock, NM, displaying characteristic geometries for the segment tips. (a) Segment 17 middle UTM: 12 S 694901.33 m E, 4063856.43 m N. (b) Step and overlap are defined using segments 21 and 22. UTM: 12 S 695225.71 m E, 4064093.10 m N. Maps modified from Delaney and Pollard (1981), Plate 1.

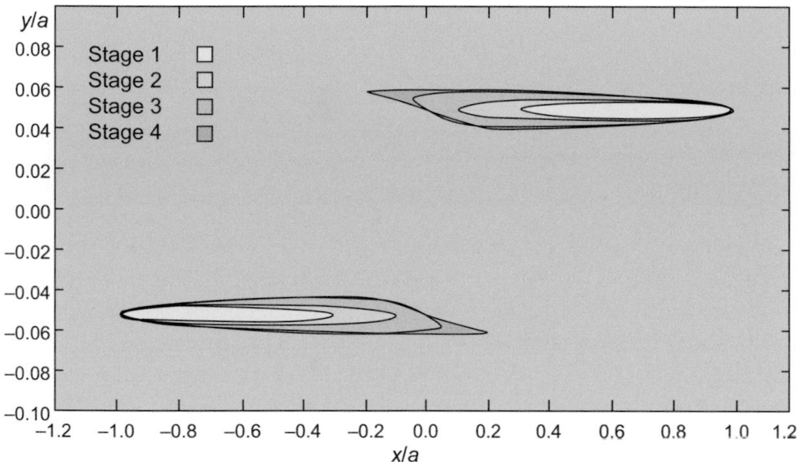

Figure 11.6 Geometry of pairs of pressurized echelon fractures in a linear elastic material. Note exaggeration of ordinate scale to clearly reveal segment shapes. Compare asymmetric overlapped crack tips to dike segment tips in Figure 11.5. Modified from Delaney and Pollard (1981), Figure 23B.

segment, whereas the inner contacts are slightly bent and trend slightly oblique to the length of the segment. The tapering in segment thickness is almost entirely due to the obliquity of the inner contacts.

The shapes of dike segments described in the preceding paragraph can be reproduced using the elastic solution for two pressurized cracks (**Figure 11.6**). Each pair of cracks in this figure has the same step and fixed distal tips. Successive pairs of cracks display different geometries as the proximal tips vary in position from underlapped to overlapped. The two most underlapped cracks are nearly elliptical in shape with symmetric tips. The two cracks with the greatest overlap have asymmetric tips with nearly straight outer contacts and oblique and slightly curved inner contacts. These characteristics are similar to those described for the mapped segments (**Figure 11.5**), suggesting that elastic deformation of the Mancos Shale was responsible for these aspects of the geometry of the echelon dike segments.

11.1.2 Volcanic Necks Grow from Dikes

The view of the Northeastern dike in **Figure 11.3** includes three volcanic necks. These are roughly cylindrical structures with vertical axes and diameters that range from 20 to 30 m. Two of the necks are associated with dikes that are approximately parallel in strike to the Northeastern dike. Outcrops along the Northeastern dike hold clues that inform us about the growth of the volcanic necks. Even though along most of the dike segments the two contacts are nearly parallel (e.g. see **Figure 11.5**), in a few locations where breccia is present, the contact bulges slightly outward causing a minor thickening of the dike. At the location along dike segment 19, pictured in **Figure 11.7a**, the northwest contact bulges sharply outward, but the southeast contact remains almost straight. We call this protuberance a bud and suggest that it represents an early stage in the formation of a volcanic neck.

The thickness of dike segment 19 is about 3.1 m where the dike is in direct contact with the Mancos Shale, but on the geological map (**Figure 11.7b**) it increases to 5.3 m at the bud. Because the conceptual model for dikes is that they are opening fractures, we envision reversing this process by removing the intrusive material in segment 19, and closing the two contacts. This leaves a rectangular hole at the bud that is about 5 m long and 2 m wide. What happened to the Mancos Shale and siltstone that formerly occupied this space?

Two mechanisms for removal of the host rock from the bud were identified at the outcrop, and they were documented using the technology called Structure from Motion (**Box 11.1**). The first mechanism relates to the abundance of breccia near the bud. Some of this breccia is composed of a complex mixture of shale, siltstone, and basalt, labeled Heterobreccia (**Figure 11.7b**). This breccia may have formed due to heating the groundwater in the Mancos Shale by the magma, and the accompanying change in effective stress (Section 4.7.3). Because the normal stresses next to the dike would be negative (compressive) and the pore fluid pressure is positive, as the pore pressure increases due to heating, the stress state is driven toward an effective tension, which facilitates brecciation. The breccia apparently behaved like a fluid that was able to mix with the magma and flow out of the bud.

The second mechanism for removing host rock from the bud is related to the two sets of vertical joints that occur in the baked sedimentary rock near the contact (**Figure 11.8a**). One set is parallel to the contact and the other set is perpendicular to the contact. These two joint sets, along with the bedding planes, divide the host rock into rectangular prisms that were plucked from the contact by the flowing magma in a process called stoping. An example of a stoped block is shown in **Figure 11.8b**. Note that this block is completely surrounded by igneous rock, but still is oriented with nearly horizontal bedding and vertical joints that strike approximately as they did before this block separated from the contact of the dike.

With removal of host rock by brecciation and stoping, the dike thickness increased from about 3 m to 5 m at the bud (**Figure 11.7**). This would have had a significant effect on the local flow rate of the magma. Recall from Section 6.8.3 that the volumetric flow rate, Q, of a viscous fluid in a parallel-sided conduit is proportional to the cube of the thickness T:

$$Q = -\frac{2LT^3}{3\eta}\frac{\partial p}{\partial x}$$

The flow rate is proportional to the conduit length, L, and pressure gradient, $\partial p/\partial x$, in the direction of flow, and inversely

Box 11.1 Structure from Motion (SfM)

This technology uses two-dimensional data from multiple photographs of the same scene (e.g. an outcrop, landscape, or engineered structure) from different camera locations to compute the three-dimensional structure of the scene. Computer algorithms that process the photographic data accomplish what the human brain does when it takes visual data and recognizes the three-dimensional structure of the objects viewed, as the viewer moves among them. Thus, Structure from Motion (SfM) is a photogrammetric technique that has much in common with a person's visual perception, and with other older photogrammetric techniques such as stereo vision using overlapping pairs of aerial photographs. SfM has grown rapidly since the 1980s and is an attractive alternative under some circumstances to other ranging technologies, such as ALSM (see **Box 9.1**), because it is less expensive, faster, and applicable across many length scales (Carrivick, et al., 2016).

A consumer-grade digital camera takes photographs suitable for SfM, although cameras with better optics lead to better resolution of three-dimensional structures. The digital camera can be mounted on a tethered helium balloon (**Figure B11.1a**), or on an unmanned aerial vehicle (UAV). Computer algorithms identify matching features in overlapping digital images and then calculate the position and orientation of the camera relative to that feature. An internal or external GPS locates the camera position for each image, so the position of a feature on the ground can be located relative to the camera. Repeating this calculation for many different features produces a coarse, three-dimensional point cloud for the scene.

Multi View Stereo (MVS) computational methods (Carrivick et al., 2016; see Further Reading) enhance the resolution of the three-dimensional model. If the scene is Earth's surface, the products are a high-resolution Digital Elevation Model (DEM) and a stitched-together composite and orthorectified photograph of the scene (**Figure B11.1b**). Orthorectification involves the geometric correction of the image for distortions due to the camera and topography, so the entire two-dimensional image has a uniform scale. The DEM associates a digital elevation with each grid point, forming a high-resolution digital version of a classic analogue topographic map. By differencing the DEMs prepared from photographs taken at different times, a geologist can quantify very subtle changes in topography that reveal physical processes in action on Earth's surface.

Applications of SfM in the geosciences are abundant in geomorphology, fluvial hydrology, sedimentology, active tectonics, and structural geology. For example, SfM generated DEMs with 2 to 3 centimeter grids enable the detection of very subtle features in the topography due to *historic* earthquake ruptures (Johnson et al., 2014). In addition, this technology facilitates the rapid mapping and analysis of topographic distortions that accompany *current* earthquake ruptures. In an application related to volcanology and structural geology, SfM provided the 4.7 mm/pixel DEM and orthorectified photograph (**Figure B11.1b**) to map fractures associated with the emplacement of igneous dikes at Ship Rock, NM (Townsend et al., 2015). A 12-megapixel camera with built-in GPS hung from the helium balloon (**Figure B11.1a**) and took an image every 5 seconds as the operator walked along the dike outcrop, while keeping the balloon height at about 13 meters. Surveying the locations of colored discs, each 20 cm in diameter, provided ground control points for stitching and orthorectifying the 138 overlapping images.

Figure B11.1 Mapping the northeastern dike at Ship Rock, NM, using Structure from Motion (SfM). (a) Tethered helium balloon taking photographs as operator walks along the dike. (b) Image of dike and shale host rock with two surveyed colored disks used for orthorectification.

proportional to the viscosity, η. The cubic relationship for thickness means that changes in this variable will have much more significant effects on flow rate than comparable changes in the other variables. As the bud formed the local flow rate would have increased by almost a factor of five: $(5/3)^3 \approx 4.6$. The increased flow rate would have retarded the cooling of the magma due to conductive heat loss (Section 11.1.5), and this would have retarded the increase in viscosity due to a drop in temperature (Section 6.4.2). These effects suggest that magma flow through a growing bud might continue, even as flow in the adjacent dike diminished due to heat loss and increased viscosity. If the dike breeched Earth's surface, so there was a place to disgorge the breccia and stoped blocks, this process might have resulted in the formation of a volcanic neck (**Figure 11.3**).

11.1.3 Giant Dikes

While many dikes are several kilometers in outcrop length (**Figure 11.3**), some are much longer. For example, the Mackenzie dikes (**Figure 11.9**) crop out from the Arctic coast of northern Canada to the shore of Hudson Bay and into Ontario Province, distances greater than 2,100 km (see Ernst et al., 2001, in Further Reading). These dikes are grouped in what is referred to as a dike swarm, based on similarities in geochemistry, petrology, and radiometric age (1,272–1,275 ma). Members of this swarm average about 30 m in thickness and form a fan-like pattern that radiates from a focal point north of the Arctic coastline. The concentric circles at 500 km intervals on the map are approximately perpendicular to the traces of individual dikes. The astonishing length of these dikes raises important questions. Where did the magma come from? Was there a magma source region underlying the entire dike swarm? Or, was the source limited to a much smaller region, directly beneath the focus of the swarm? We address these questions by investigating the direction of magma flow in the dikes.

The igneous rocks of the Mackenzie dikes lack a distinct petrographic fabric (e.g. preferred orientation of mineral grains), but they do have a magnetic fabric. The magnetic susceptibility of these rocks is anisotropic, so it can be represented by a triaxial ellipsoid with least, intermediate, and greatest principal axes. In some rocks the magnetic fabric has been shown to correlate with the petrographic fabric, so where the petrographic fabric is not distinct the Anisotropy of Magnetic Susceptibility (AMS) can provide a good proxy. For the Mackenzie dikes the magnetic

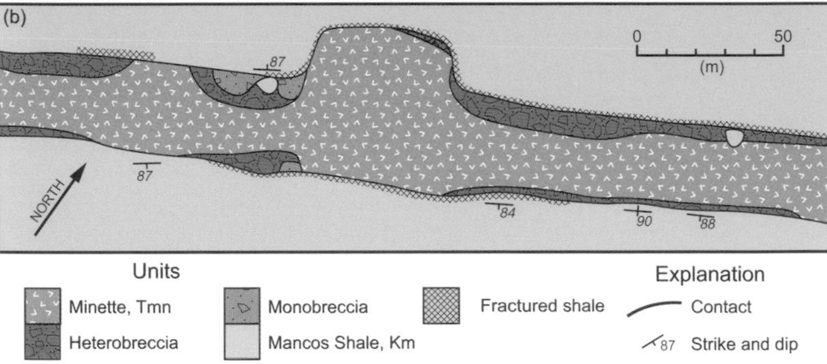

Figure 11.7 (a) View to northeast along segment 19 of the Northeastern dike (see Figure 11.3). The northwest (left) contact bulges outward to form a bud, but the southeast contact (right) is nearly straight. From Townsend et al. (2015). (b) Map of bud along segment 19 viewed in (a). UTM: 12 S 694970.74 m E, 4063907.42 m N. Modified from Delaney and Pollard (1981), Figure 13.

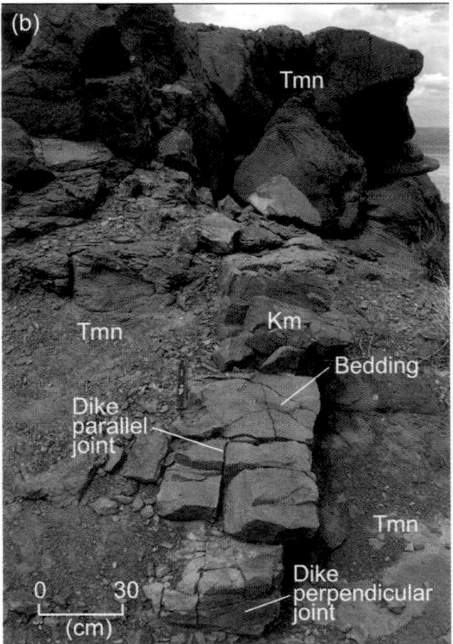

Figure 11.8 (a) Vertical aerial photograph of Northeastern dike segment 6: dike contact (white); dike-parallel joints (blue); dike-perpendicular joints (red); Mancos Shale (Km); Tertiary basalt (Tmn). (b) Stoped blocks of Mancos siltstone entrained in basalt and shaped as rectangular prisms bounded by the two joint sets mapped in (a) and bedding planes. UTM: 12 S 694514.43 m E, 4063551.60 m N. Modified from Townsend et al. (2015).

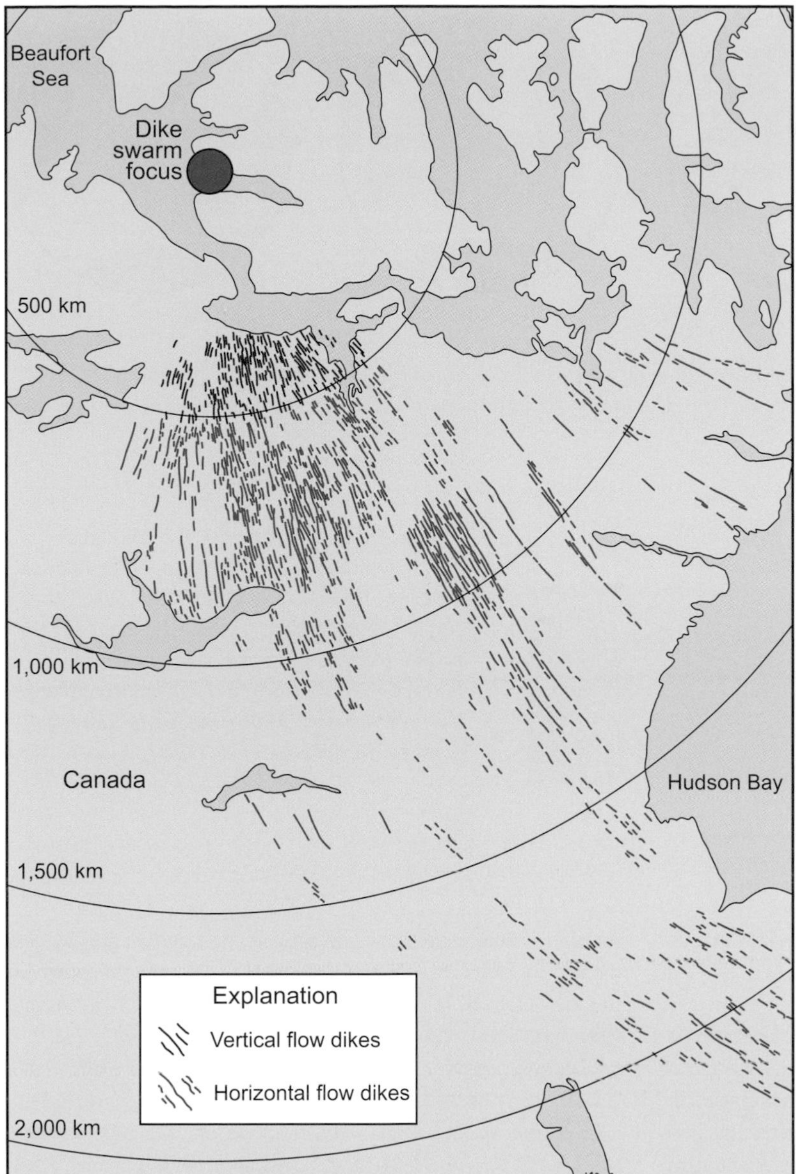

Figure 11.9 Structure map of north central Canada showing regional distribution of the Mackenzie dike swarm. The outcrop traces of the dikes converge on a central focus (red circle) with a location uncertainty of less than 50 km. Approximate map center UTM: 14 V 394430.20 m E, 6436595.56 m N. Modified from Ernst and Baragar (1992), Figure 1; Townsend et al. (2017)

fabric probably is caused by the subtle alignment of non-equant ferromagnetic minerals by the flowing magma. The AMS of these rocks is arranged with the least axis perpendicular to the dike contacts, and the greatest axis is interpreted as aligned in the flow direction of the magma. Samples from 44 different dikes located between 400 and 2,100 km from the focus of the swarm (**Figure 11.9**) were analyzed and the average anisotropy was 3%. As interpreted, these data reveal a systematic change from steeply inclined flow to approximately horizontal flow at a radial distance of about 500 to 600 km from the focus of the swarm.

A subset of data from the AMS study is shown in **Figure 11.10**, where the orientations of the greatest principal axis are plotted on stereographic projections and their densities are contoured. Because dikes in the swarm have different orientations, the greatest principal axes have been rotated to a common north–south vertical plane to facilitate comparisons at given radial distances. At a radial distance of 400 km the axes are clustered around a single mean orientation, so the contours enclose a well-defined point maximum that plunges about 70°. A similar steeply inclined point maximum occurs for the axes at 500 km distance. At 600 km and 1000 km there also is a strong point maximum, but the plunge is approximately zero. Taking these maxima as the flow directions for the magma, we conclude that flow changed from steeply inclined to horizontal between 500 km and 600 km from the swarm focus.

The AMS data (**Figure 11.10**) do not support the existence of a source region for the magma that underlies the entire Mackenzie swarm. Instead, these data are consistent with a more localized source region in the vicinity of the swarm focus. The source is interpreted to be a mantle plume centered under the focus, with an outer boundary of melt generation that coincides approximately with the transition from steeply inclined to horizontal

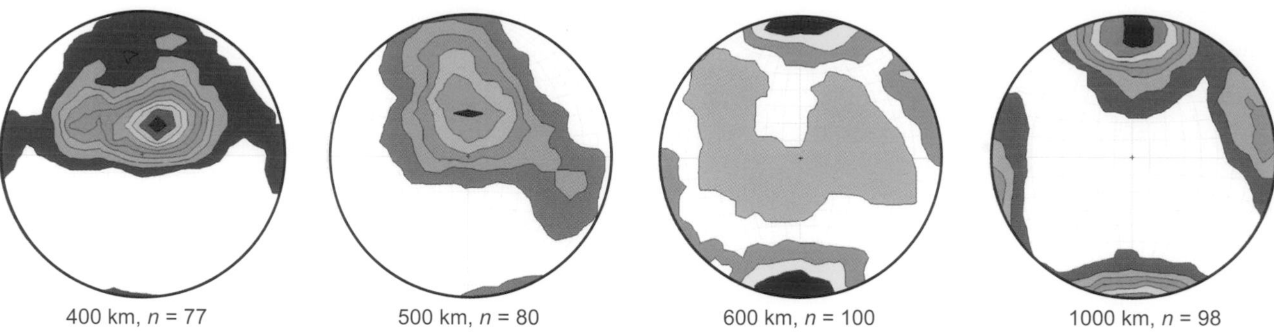

Figure 11.10 Stereographic projections of greatest principal axes for the anisotropy of magnetic susceptibility (AMS) of the Mackenzie dikes. Samples are grouped by distance from the dike swarm focus (Figure 11.9). Number of samples in each group is n. Modified from Ernst and Baragar (1992), Figure 2.

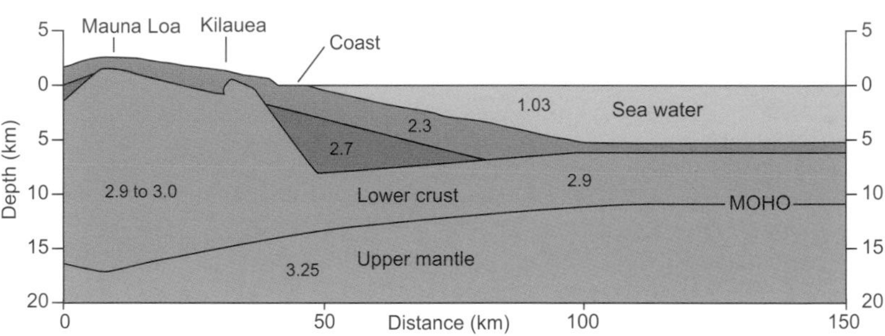

Figure 11.11 Distribution of mass density (multiply numbers by 10^3 kg m^{-3}) for Kilauea Volcano, Hawaii. This distribution was obtained using data from a seismic refraction profile to determine the velocity structure, and then constructing a distribution of density that is consistent with a measured gravity profile. From Zucca et al. (1982), Figure 10.

flow. Thus, the horizontal dimensions and shape of the mantle plume corresponds approximately to the boundary circle labeled 500 km on Figure 11.9.

The AMS data provide support for a local source beneath the dike swarm focus, but why did the dikes extend such great distances from this source, and not simply propagate upward to Earth's surface? We address this question in the next section, and it is relevant to dikes in volcanic rifts as well as giant dike swarms.

11.1.4 Limits for Dike Heights

Given field data on the outcrop length of a dike, we might suppose that its height is approximately equal to its length. In other words, the dike could be roughly an equidimensional sheet intrusion. While this is plausible for the dikes at Ship Rock that are a few kilometers in outcrop length (Figure 11.3), it is implausible for dikes of the Mackenzie swarm that are up to 2000 km in length (Figure 11.9). Equidimensional shapes also are implausible for dikes that built the east rift zone of Kilauea Volcano, Hawaii. This rift zone extends eastward about 100 km from the summit of the volcano, and some dikes must have reached the distal parts of the rift to build it up with new lava flows. However, geophysical data associated with recent injections of magma into this rift zone are consistent with dike heights of a few to several kilometers. This suggests that the dikes propagate laterally at shallow depths for distances that are many times their heights. Here we investigate the role that gravity, and the density structure of Earth's lithosphere, can play in constraining the height of a dike (see Townsend et al., 2017, in Further Reading).

Typical densities of rocks making up the lithosphere are $2 \times 10^3 \leq \rho \leq 3 \times 10^3$ kg m^{-3} and the acceleration of gravity near Earth's surface is $g^* = 9.8$ m s^{-2}, so unit weights, $\gamma = \rho g^*$, are $2 \times 10^4 \leq \gamma \leq 3 \times 10^4$ N m^{-3}. Because we use spatial gradients in stress here, it is convenient to convert N m^{-3} to MPa m^{-1}. The conversion is accomplished by noting that 1 N m^{-3} = 1 Pa m^{-1} = 1×10^{-6} MPa m^{-1}. Thus, the range of unit weights for common rocks converts to stress gradients, g, in the range $2 \times 10^{-2} \leq g \leq 3 \times 10^{-2}$ MPa m^{-1}. Taking the middle of this range as an example, we would expect the normal stress to become more compressive with depth at a spatial rate of about 0.025 MPa m^{-1} or 25 MPa km^{-1}. This is consistent with Anderson's standard state of stress in Earth's lithosphere as described in Section 6.7.4.

For Kilauea Volcano on the island of Hawaii the distribution of density, as determined from geophysical data, is illustrated in Figure 11.11. The central core of the volcano, from near sea level to depths of about 15 km is composed of rock that matches the lower oceanic crust in density at 2.9×10^3 kg m^{-3}. This is overlain by less dense rock with an average density of 2.3×10^3 kg m^{-3}. For comparison, the density of basaltic magma is 2.7×10^3 kg m^{-3}. The important point is that the magma is less dense than rock in the deep parts of the volcano, and it is more dense than rock in the shallow parts. This density stratification provides a mechanism for constraining the heights of dikes: they become trapped near the Level of Neutral Buoyancy (LNB), where the magma density matches the host rock density.

To quantify the mechanism that constrains the height of a dike, we consider just two layers of different density with a dike centered at their interface and use an elastic solution that includes gradients in magma pressure and in rock stress due to density. For reference, recall the canonical model for opening fractures, introduced in Chapter 7, that considers two boundary conditions, a uniform pressure exerted by a fluid within the fracture, and a uniform remote stress acting normal to the fracture plane (Section 7.2.5). For application to dikes, the remote stress is the least compressive principal stress, σ_1, and the fluid is magma. If the magma pressure exceeds the magnitude of the remote least compressive stress, the model dike opens, and under these uniform loading conditions the shape of the displaced dike contact is *elliptical*. With the addition of gradients in pressure and stress, we anticipate that the shape of the dike contact will depart from elliptical, and the amount of opening will depend on the gradients.

The mechanism that constrains the height of a dike depends upon the stress concentration at the dike tips. Recall that this stress concentration is measured by the mode I stress intensity, K_I (Section 7.3.3), and the fracture propagation occurs when the stress intensity equals the mode I fracture toughness, $K_I = K_{IC}$ (Section 7.4.2). Therefore, one possible constraint on dike height is that the rock toughness prevents further growth. However, if the magma is taking advantage of pre-existing fractures, or the rock is very weak in tension, the constraint reduces to $K_I = 0$. If the stress intensity falls below zero, the dike tip will close and its height reduces. If the stress intensity exceeds zero, the pre-existing fracture will open, magma will flow into the opening, and the dike height will increase. This constraint gives the *greatest* height the dike can attain under a given loading condition.

The boundary conditions used to study the shape and height of dikes subject to uniform and linearly varying normal tractions are illustrated in **Figure 11.12a**. This schematic figure shows five contributions to the distribution of normal tractions acting on the *right* contact of a model dike. Normal tractions equal in magnitude, but oppositely directed, also act on the *left* contact (not shown). On the right contact the coordinate n is directed to the left, away from the elastic solid and into the dike. The normal component, t_n, of the traction vector, is positive when directed in the positive n direction. Starting on the left, the first two distributions of the contributions to the total normal traction are uniform, and equal in magnitude to the magma pressure and the remote least compressive stress acting at the dike center, $x = 0$. Symbolically we write $t_n = -p_m + p_r$, and these tractions are applied over the entire height of the model dike, where $+c \geq x \geq -c$. These are the traction boundary conditions applied to the

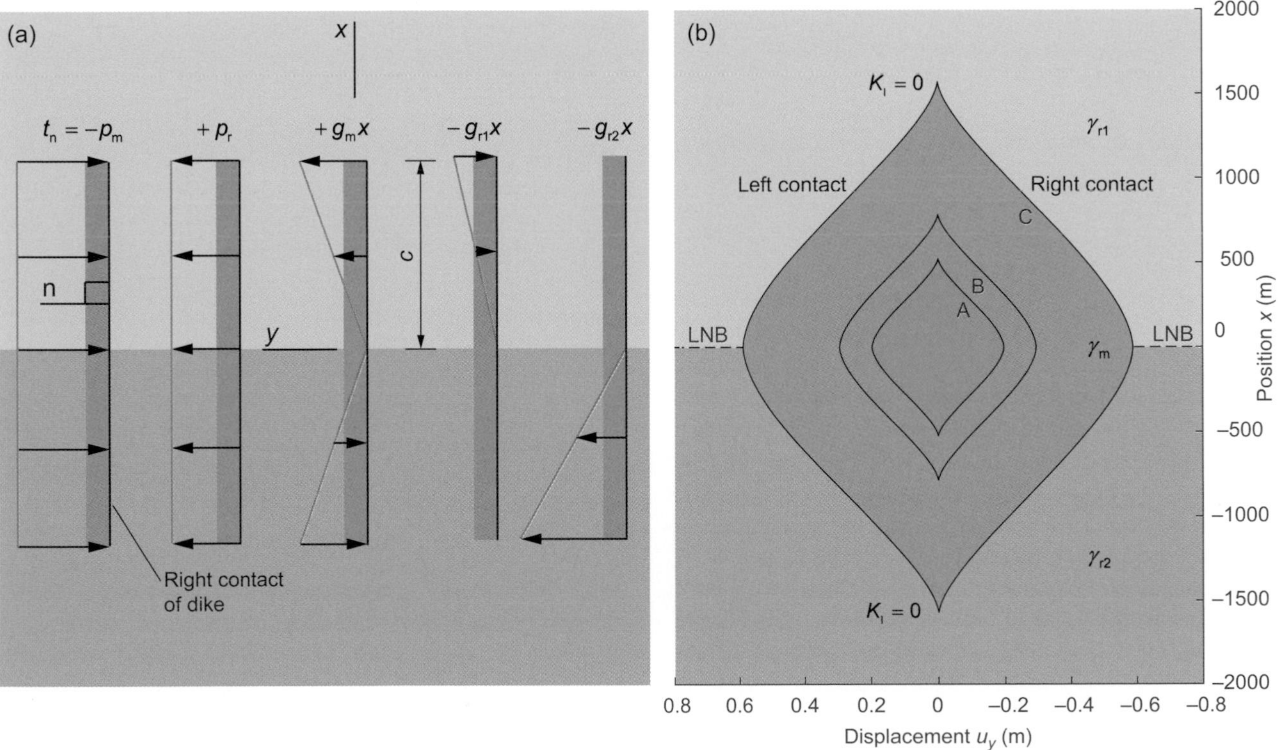

Figure 11.12 Elastic boundary-value problem for model dike at level of neutral buoyancy. (a) Normal tractions acting on right contact of dike. The light orange color represents the magma on the left side of the right contact of the dike. (b) Opening shape of dike for three different boundary conditions, which produce zero stress intensity at dike tips. The horizontal scale of the plot is exaggerated relative to the vertical scale. See text for details. Modified from Pollard and Townsend (2018).

canonical fracture model in Chapter 7. The negative traction due to the magma pressure *pushes* to the right, against the contact, and tends to open the dike; the positive traction due to the remote compression *pulls* to the left on the contact and tends to close the dike. In **Figure 11.12a** the remotely applied compression is converted to the equivalent traction pulling to the left on the dike contact.

The traction $t_n = +g_m x$ (**Figure 11.12a**) accounts for the linear gradient in magma pressure due to the magma density. This traction is positive (pulls left on the contact) above the dike center, because the pressure decreases upward; it is negative (pushes right on the contact) below the center, because pressure increases downward. The traction $t_n = -g_{r1}x$ accounts for the linear gradient in remote stress due to rock density in the upper layer, so it only acts where $+c \geq x \geq 0$, and it pushes right on the contact because the compressive stress decreases upward. The traction $t_n = -g_{r2}x$ accounts for the linear gradient in stress due to rock density in the lower layer, so it only acts where $0 \geq x \geq -c$, and it pulls left on the contact because the compressive stress increases downward.

The five different traction distributions illustrated in **Figure 11.12a** are the loading used to analyze the limits on dike height, which we assess using the stress concentrations at the dike tips. The mode I stress intensity at the upper and lower tips of the model dike are:

$$K_I(+c) = \sqrt{\pi c}\left[p_m - p_r - \tfrac{1}{2}g_m c - g_{r1}c\left(-\tfrac{1}{4} - \tfrac{1}{\pi}\right) - g_{r2}c\left(-\tfrac{1}{4} + \tfrac{1}{\pi}\right)\right]$$
$$K_I(-c) = \sqrt{\pi c}\left[p_m - p_r + \tfrac{1}{2}g_m c - g_{r1}c\left(+\tfrac{1}{4} - \tfrac{1}{\pi}\right) - g_{r2}c\left(+\tfrac{1}{4} + \tfrac{1}{\pi}\right)\right] \quad \textbf{(11.1)}$$

If we require these two stress intensities to be equal, $K_I(+c) = K_I(-c)$, then the gradients must relate to one another as:

$$g_m = \tfrac{1}{2}(g_{r1} + g_{r2}) \quad \textbf{(11.2)}$$

In other words, for the dike to attain its greatest height, the pressure gradient in the magma must be the average of the stress gradients in the two rock layers. This means that the magma density is equal to the average density of the rock layers. For this reason we label the interface between the two layers in **Figure 11.12b** the Level of Neutral Buoyancy (LNB). We return to this concept at the end of this section, but note here that it is the center of the dike that resides at the LNB: the upper tip extends above, and the lower tip extends below the LNB.

If the stress intensities at both tips are zero, $K_I(+c) = 0 = K_I(-c)$, the maximum dike height, $2c_c$, is found using (11.2) in (11.1):

$$2c_c = 2\pi(p_m - p_r)/(g_{r2} - g_{r1}) \quad \textbf{(11.3)}$$

The maximum height is proportional to the difference between the magma pressure and the remote compressive stress at the LNB, and it is inversely proportional to the difference in stress gradients due to density of the two layers.

Figure 11.12b shows the cross section shapes of three different vertical dikes at their maximum heights. For these dikes (A, B, C) the stress gradients for the layers are $g_{r1} = 0.024$, 0.025, 0.026 MPa m^{-1} and $g_{r2} = 0.030$, 0.029, 0.028 MPa m^{-1}, and the pressure gradient for the magma is their average: $g_m = 0.027$ MPa m^{-1}. The difference in magnitudes between the magma pressure and least compressive stress, $p_e = p_m - p_r$, is referred to as the *excess pressure*, which is taken as $p_e = 1$ MPa for all three models. **Figure 11.12b** shows that the opening profile of a dike and its maximum height are sensitive to relatively modest changes in the stress gradients and rock densities.

Because the magma has a greater density than the upper layer, the respective gradients are $g_m > g_{r1}$ (**Figure 11.12a**), so the magma pressure decreases toward the upper tip *more* rapidly than the magnitude of the compressive stress. At the maximum height the upper dike tip takes on a cusp-like form and the stress intensity drops to zero (**Figure 11.12b**). Because the magma has a lesser density than the lower layer, their respective gradients are $g_m < g_{r2}$, and the magma pressure increases toward the lower tip *less* rapidly than the magnitude of the compressive stress. At the maximum height the lower tip takes on a cusp-like form and the stress intensity drops to zero. Note that the maximum height increases from dike A to C as the difference between rock densities, and the difference in respective gradients, $g_{r2} - g_{r1}$, decreases (**Figure 11.12b**). As this difference goes to zero, the maximum height becomes unbounded. Dike heights in a host rock with no density stratification depend upon fracture toughness to limit vertical propagation. For the two-layer model, dike heights less than the maximum height given by (11.3) are associated with non-zero values of fracture toughness.

Recall from Section 1.1.2 that the downward directed gravitational force acting on a body is the product of its mass and the acceleration of gravity, $\mathbf{F}_{grav} = m\mathbf{g}$. The upward directed buoyant force acting on this body is the product of the displaced mass and the acceleration of gravity, $\mathbf{F}_{buoy} = -m_d\mathbf{g}$. Thus, the net vertical force is proportional to the difference in mass between the body and its surroundings. The intuitive demonstration of this principle considers a small blob of fluid of constant density in a fluid that is continuously stratified with respect to density. The small blob rises or falls until its density matches that of the surrounding fluid, so $m - m_d = 0$. At this level the blob attains neutral buoyancy.

For the model dikes illustrated in **Figure 11.12b** the mass of magma exactly equals the mass of the displaced host rock, because the opening displacement discontinuity is symmetric about the dike center, which is located at the interface, and the magma density is the average densities of the two rock layers. This justifies calling the interface the level of neutral buoyancy (LNB), but it should be understood that the stability of the dike at that level depends upon the horizontal forces (normal tractions acting on the vertical dike contact) and the stress concentrations they induce at the dike tips. These stress concentrations are characterized by the stress intensity associated with each tip, which depends on the dike height and all the tractions applied to the dike contact (**Figure 11.12a**). The condition that the dike has a fixed height (no closure and no propagation) with its center at the interface, requires the stress intensities to be zero.

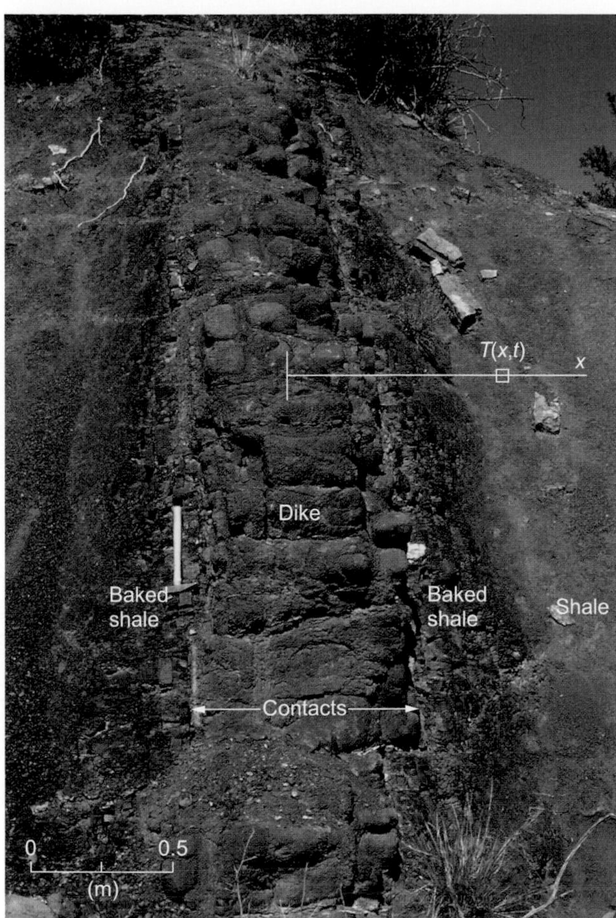

Figure 11.13 Vertical igneous dike cutting shale near Walsenburg, CO. Shale within ~0.5 m of the dike contact is darker in color, harder, and more resistant to erosion than shale farther from the dike, all indications of baking by heat flow from the dike. Fractures cutting the dike rock perpendicular to the contact may be related to thermal stresses.

11.1.5 Cooling Dikes

Igneous intrusions cool when they come into contact with colder rock. Heat flows from the magma into the surrounding rock. The cooling of an intrusion can generate thermal stresses sufficiently large to fracture the igneous rock (Section 11.4). In addition, warming the host rock can lead to metamorphism (Section 11.1.2) and the development of valuable ore deposits. These processes are important for structural geologists to understand, and the flow of heat is a central issue. Outcrops of dikes, such as that shown in **Figure 11.13**, illustrate the baking of the host rock and the fracturing of the igneous rock, and observations such as this provide the motivation for the heat flow problem considered here (see Carslaw and Jaeger, 1986, in Further Reading).

We consider heat flow and temperature changes where a single igneous dike intrudes colder host rock (**Figure 11.13**). We postulate that heat is transferred exclusively by conduction (energy transmission by molecular motion), but recognize that in nature flow of groundwater in the host rock also can transport heat by convection, so this model represents one end member in a spectrum of potentially relevant behaviors. Field observations (e.g. **Figure 11.3**, **Figure 11.5**, **Figure 11.9**) consistently show that dike lengths greatly exceed dike thicknesses, and that the contacts are approximately parallel. These observations suggest that nearly all the conductive flow of heat from a dike would be in a direction perpendicular to the dike plane, so we postulate that the conductive heat flow is one-dimensional and investigate the distribution of temperature, $T(x, t)$, as a function of distance from the dike center and time.

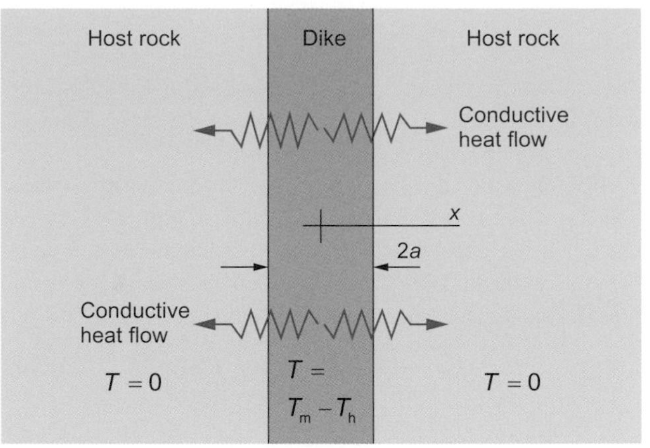

Figure 11.14 Conceptual model for a dike of constant thickness 2*a* filled rapidly with magma at an initial temperature $T_m - T_h$ in colder host rock at zero initial temperature. Heat flows by conduction only in the positive and negative *x*-directions.

The only relevant length scale in this problem (**Figure 11.14**) is the dike thickness, $2a$. Acquiring data on dike thickness can be problematic because dikes depicted on most geologic maps reveal the outcrop length, sometimes the orientation, but rarely the thickness. Assuming we have measured the dike thickness, we also need to establish an appropriate reference frame. We choose the origin of the reference frame as halfway between the dike contacts, with the x-axis perpendicular to the dike contacts. We postulate that the model dike (**Figure 11.14**) intruded *rapidly*, so insignificant heat was lost to the host rock until after dike propagation ceased and magma stopped flowing. Later in this section we clarify what intruded "rapidly" means. We start the clock at the moment magma stops flowing, $t = 0$, and consider the initial temperature of the magma to be uniform. Because heat flow by conduction depends only on temperature differences, the host rock temperature, T_h, is subtracted from the uniform temperature of the magma, T_m, to prescribe the initial condition on temperature within the dike as a function of position and time:

$$\text{for } |x| < a \text{ at } t = 0,\ T = T_m - T_h \tag{11.4}$$

Because the dike intruded rapidly, insignificant heat was conducted into the host rock before the dike stopped, so the model host rock has a uniform, but cooler initial temperature, T_h. This temperature is subtracted from itself to prescribe the initial condition outside the model dike:

$$\text{for } |x| > a \text{ at } t = 0,\ T = 0 \tag{11.5}$$

These conditions would have to be modified if significant heat were lost to the host rock as the dike intruded, or if magma continued to flow into the dike after $t = 0$.

We suppose that a finite amount of thermal energy is contained per unit height in the cross section of **Figure 11.14** and this ultimately is transferred out of the dike into the host, which is of infinite extent. Thus, the remote boundary condition for all time, including and subsequent to the initial time, is:

$$\text{for } x = \pm\infty \text{ and } t \geq 0,\ T = 0 \tag{11.6}$$

Accordingly, the relative temperature in the host rock very far from the dike should not change as the dike cools.

The governing equation for the temperature as a function of distance and time, $T = T(x, t)$, was derived in Section 3.8 using Fourier's law for heat conduction and the principle of conservation of energy to find:

$$\rho c \frac{\partial T}{\partial t} = k \frac{\partial^2 T}{\partial x^2}$$

Here ρ [=] kg m^{-3} is mass density, c [=] J kg^{-1} K^{-1} is specific heat capacity, and k [=] J m^{-1} s^{-1} K^{-1} is thermal conductivity. We postulate that all three of these physical quantities are uniform in space, including both the dike and host rock, and they are constant in time. These constants are combined into a single material property, $\kappa = k/\rho c$, so the one-dimensional equation for heat conduction is:

$$\frac{\partial T}{\partial t} = \kappa \frac{\partial^2 T}{\partial x^2} \tag{11.7}$$

Thermal diffusivity, κ [=] m^2 s^{-1}, measures how quickly heat is conducted from a hot region to a cold region in order to reach thermal equilibrium, and it too is uniform in space and constant in time. The governing equation (11.7) requires the rate of increase of temperature with respect to time, at any point in the model space, to be proportional to the second partial derivative of temperature with respect to position. This second derivative is a measure of the *curvature* of the graph of temperature versus position. If temperature varies linearly near a particular position, the curvature is zero, and the temperature there will not vary with time. In this case, the flow of heat is just sufficient to maintain the temperature distribution. On the other hand, if the temperature varies non-linearly near a particular position, the curvature is not zero, and the temperature there will vary with time. The rate of temperature variation is proportional to the thermal diffusivity.

Given the symmetry about the plane $x = 0$ for the problem defined in **Figure 11.14**, we would expect the solution for the temperature to be an *even* function of x. This means that $T(x, t) = T(-x, t)$. This insight is corroborated qualitatively for the dike in **Figure 11.13**, which has baked zones that are roughly equal in thickness on either side of the dike. Because the solution for T is an even function of x, the temperature at the midline of the model dike, $T(x = 0, t)$, must be either a local maximum or minimum for all times $t \geq 0$. Because the initial temperature of the dike is greater than that of the surroundings, $T_m - T_h > 0$, the temperature at the midline is a maximum. At the maximum, where $x = 0$, the graph of temperature versus position must have zero slope: $\partial T/\partial x = 0$. With no spatial gradient in temperature, no heat can flow across the midline of the dike. These conjectures about the temperature distribution in space and time demonstrate that one can gain insight from the governing equation (11.7) before solving a particular problem.

Carslaw and Jaeger (1986), in Further Reading, give the solution that satisfies the governing equation (11.7), and provides a distribution of temperature in space and time that satisfies the initial conditions, (11.4) and (11.5), and the remote boundary condition (11.6):

$$T(x,t) = \tfrac{1}{2}(T_m - T_h)\left[erf\left(\frac{a-x}{2\sqrt{\kappa t}}\right) + erf\left(\frac{a+x}{2\sqrt{\kappa t}}\right)\right], \quad -\infty < x < +\infty, t \geq 0 \tag{11.8}$$

The function *erf*() is called the error function: it is tabulated in standard reference books and built into MATLAB. The error function has a sigmoidal shape somewhat akin to the arctangent function. Equation (11.8) is *normalized* by dividing both sides by the initial temperature difference between the dike and the surroundings:

$$\frac{T(x,t)}{T_m - T_h} = \frac{1}{2}\left[erf\left(\frac{a-x}{2\sqrt{\kappa t}}\right) + erf\left(\frac{a+x}{2\sqrt{\kappa t}}\right)\right], \quad -\infty < x < +\infty, t \geq 0 \tag{11.9}$$

The error function has the following properties, which enable us to confirm that (11.9) satisfies the initial and boundary conditions for the problem defined here:

$$erf(0) = 0, \quad erf(\infty) = 1, \quad erf(-x) = -erf(x)$$

Inside the dike, as t approaches zero, the arguments of both error functions in (11.9) approach positive infinity, because both numerators are positive for $|x| < a$. Thus, the sum of the error functions approaches 2 and the normalized temperature approaches 1. Outside the dike, where $|x| > a$, the numerators of the two error functions in (11.9) are of opposite sign, and the arguments approach $-\infty$ and $+\infty$ as t approaches zero. Thus, the sum of the error functions approaches 0 as does the normalized temperature. Graphically, the initial conditions appear as a step function: for $\kappa t/a^2 = 0$, the temperature inside the dike is $T_m - T_h$, and outside the dike it is zero (**Figure 11.15**). The remote boundary condition is confirmed by noting that as $x \to \pm\infty$, the sum of the two error functions in (11.9) approaches 0, as does the normalized temperature.

Figure 11.15 is a plot of the normalized temperature versus normalized position, x/a for the dike heat flow problem. The variable parameter is dimensionless time, defined as $\kappa t/a^2$. This plot shows something that might be unexpected: the maximum temperature in the host is one half the temperature difference, $(T_m - T_h)/2$. One might expect the temperature just outside the contact to rise initially to values close to the magma temperature. However, at the contact, where $x = \pm a$, the argument of one of the error functions is zero, while the argument of the other approaches positive infinity as t approaches 0. Thus, the two error

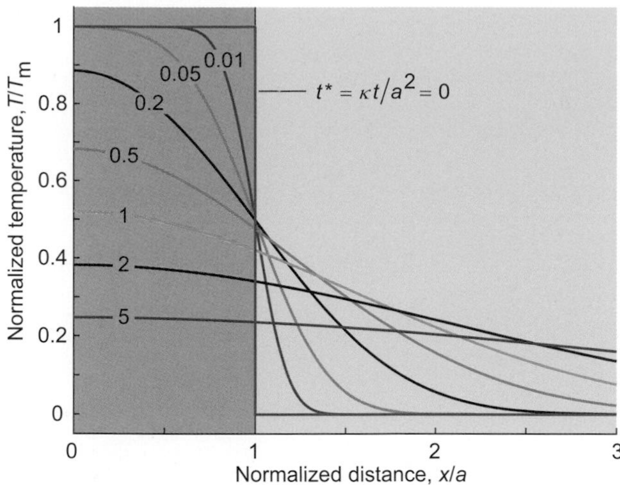

Figure 11.15 Conductive cooling of the model dike from an initial normalized temperature of 1, and heating of the host rock from an initial normalized temperature of 0. On the abscissa the mid-line of the dike is at 0 and the contact is at the normalized distance of 1. Dimensionless time, t*, ranges from 0 (initial time) to 5. See Carslaw and Jaeger (1986).

functions have values of 0 and 1, respectively, and this yields a normalized temperature of ½. For all subsequent times the temperature at the contact is less than ½. Also note where the temperature profile for a given time is most highly curved: this is where the temperature change between time steps is greatest. In addition, Figure 11.15 shows that where the temperature profile is *convex*, the temperature decreases with time, and where it is *concave*, the temperature increases with time. Each of these effects is consistent with the governing equation (11.7).

In order to build intuition for dike cooling we evaluate the dimensionless time used to construct Figure 11.15. Thermal diffusivity does not vary much for common rock types, so we select a representative value of 10^{-6} m^2/s, so time is calculated as $t = t^* a^2 \times 10^6$ s m^{-2}. Noticeable conductive cooling of the dike extends almost to the dike center, $x/a = 0$, after a dimensionless time of $t^* = 0.05$. At that same dimensionless time conductive heating of the host rock has penetrated almost to a normalized distance from the dike center of $x/a = 2$. This would take about half a day for a dike that is 2 m thick ($a = 1$ m), and 58 days for a dike that is 20 m thick ($a = 10$ m). Magma temperature at the dike center would drop to about half its initial temperature, $T/T_m = 0.5$, after a dimensionless time of $t^* = 1$. This would take about 12 days for the 2 m thick dike and about 1,157 days (~3 years) for the 20 m thick dike. We conclude that cooling durations are sensitive to the dike thickness.

We now return to what "rapid" intrusion of a dike means in terms of heat loss. The temperature distributions in Figure 11.15 depend upon the postulate that the model dike intruded so rapidly that the magma and host rock temperatures at the initial time, $\kappa t/a^2 = 0$, remain uniform. We evaluate this postulate by comparing the rate of heat carried along the dike by advection to the rate of heat transported into the host rock by conduction (Figure 11.16). The section of dike in Figure 11.16 has a height h, a thickness

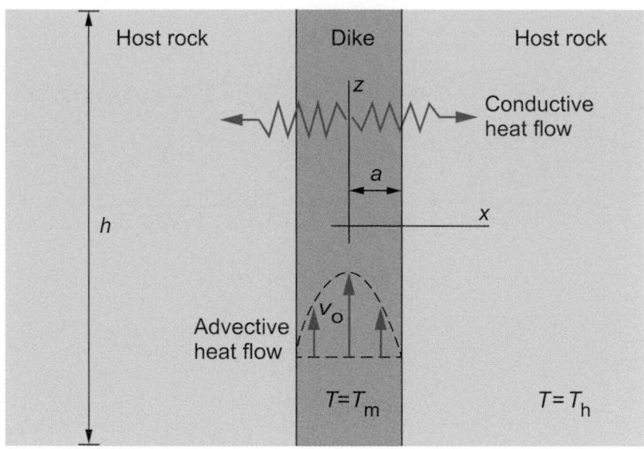

Figure 11.16 Geometry used to compare the advective heat transport due to vertical magma flow within a dike to horizontal conductive heat transport into the host rock. The characteristic velocity is v_o.

$2a$, and a depth w in and out of the page. The rate of heat advection in the vertical direction is estimated as the product of the horizontal cross-sectional area of the dike, $2aw$, a characteristic velocity of the magma, v_o, and the excess heat content of the magma, $\rho c(T_m - T_h)$:

$$q_y \approx 2awv_o\rho c(T_m - T_h) \tag{11.10}$$

The conductive transport in the horizontal direction is estimated by referring to Fourier's law (Section 3.8), written for one-dimensional heat flow in the x-direction as:

$$q_x = -\rho c\kappa \frac{\partial T}{\partial x} \tag{11.11}$$

Here q_x is the component in the x-direction of the heat flux vector per unit time and per unit area, and $\rho c\kappa = k$ is the thermal conductivity of the magma. The partial derivative on the right-hand side is estimated as $(T_m - T_h)/a$. The rate of heat loss by conduction in the horizontal direction is estimated as the product of the right-hand side of (11.11) and the surface area of the two contacts, $2wh$:

$$q_x \approx 2wh\rho c\kappa(T_m - T_h)/a \tag{11.12}$$

We use (11.10) and (11.12) to establish the conditions for which the heat transport by advection exceeds the heat loss by conduction (Figure 11.16). For $q_y > q_x$, we have:

$$2awv_o\rho c(T_m - T_h) > 2wh\rho c\kappa(T_m - T_h)/a$$

This condition would ensure a relatively uniform temperature in both the dike and the host rock before the magma stops flowing and heat transfer by conduction begins. Solving this inequality for the velocity, we have:

$$v_o > h\kappa/a^2 \tag{11.13}$$

For a dike that is 10 km high and 2 m thick, with a thermal diffusivity of 10^{-6} m^2 s^{-1}, we find the magma velocity must exceed about 10^{-2} m s^{-1}. Under these conditions the magma would experience no significant heat loss and the uniform initial

Figure 11.17 Oblique aerial photograph of sills filled with igneous rock interleaved with sedimentary strata on the San Rafael Swell in southeastern Utah.

temperature postulate used to compute the temperature profiles in **Figure 11.15** would be honored approximately.

Note that the limiting velocity defined in (11.13) is one to two orders of magnitude less than velocities reported for magma in propagating dikes on Kilauea Volcano, so one would expect the temperature distributions of **Figure 11.15** to apply there. However, these calculations ignore the fact that magma viscosity increases very rapidly with decreases in temperature (Section 6.4.2). The immediate 50% decrease in temperature at the contact means the magma there would behave essentially as a solid material. This layer of immobilized magma would increase in thickness as heat is lost to the host rock and ultimately would choke the flow in the interior of the dike.

11.2 SILLS

Sills are concordant intrusions because their contacts conform to the sedimentary strata or the metamorphic foliation. For example, the prominent sills pictured in **Figure 11.17**, from the San Rafael Swell in southern Utah, may include minor steps across sedimentary bedding, but they remain essentially concordant over distances of hundreds of meters to kilometers. Because they conform to the bedding and are more resistant to erosion, the sills of the San Rafael Swell cap the mesas in this region. The thickness of these sills, measured as the perpendicular (vertical) distance between the contacts, is small compared to their horizontal dimensions, measured in the plane of the intrusion. For example, the maximum thickness to length ratio of the sill in the foreground of **Figure 11.17** is less than 1:25. Because their shapes resemble nearly planar sheets, sills like dikes are called sheet intrusions. Sills may by filled with magma, salt, or a sedimentary slurry, but they share the geometric attributes of being concordant sheet intrusions. Sills also play an important role in some volcanic eruptions (see Segall, 2010, in Further Reading)

The two contacts of a sill join at a tipline that marks the distal extent of the intrusion. Prior to sill intrusion in a sedimentary sequence, these contacts were bedding interfaces. Sills *open* as the intruding material is injected between the beds and an opening displacement discontinuity develops, so sills are classified as opening fractures (Section 7.3.2). If the fracture toughness of the bedding interface is negligible, sill propagation is similar to dike propagation into pre-existing fractures. In other words, the stress intensity at the sill tip during propagation would be nearly zero.

Before a sill is formed in horizontal sedimentary rock (**Figure 11.17**), the stress acting across the bedding interface well below the ground surface would be that due to the weight of the overlying strata. The magma pressure must exceed the magnitude of this compressive stress for the sill to open. In a material that is isotropic with respect to fracture toughness, an opening mode fracture propagates in a plane that is perpendicular to the least compressive stress, σ_1 (Section 7.4.3). To the extent that this applies to horizontal sills, we would interpret σ_1 to be vertical at the time of intrusion. This interpretation does not take into account anisotropy with respect to strength or stiffness due to the stratified nature of the host rock, which may play a role in sill formation.

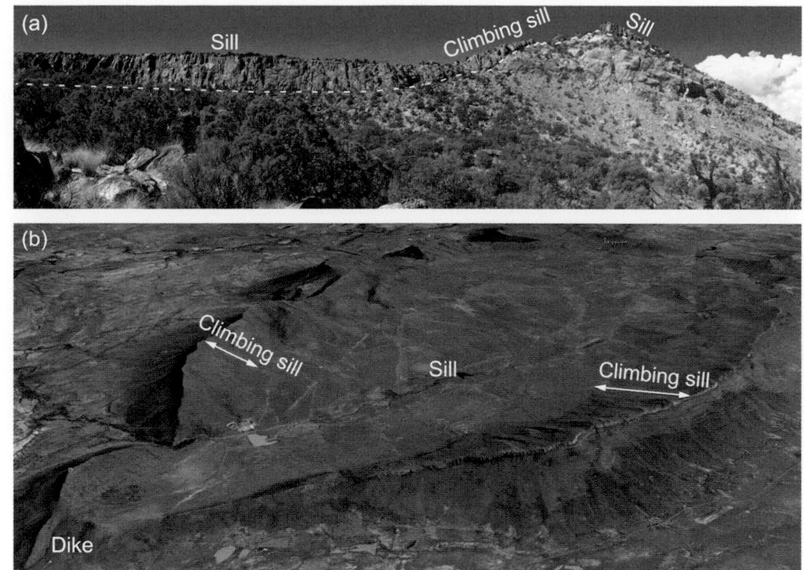

Figure 11.18 Climbing sills. (a) Buckhorn Ridge sill, Mount Holmes, Henry Mountains, UT. The concordant sill is about 20 m thick in the left center of this photograph. It cuts obliquely across the sedimentary layering to the right of center, a dip of about 45°, and becomes concordant near the ridge summit. UTM: 12 S 537994.42 m E, 4184504.78 m N. From Johnson and Pollard (1973). (b) Oblique Google Earth image of Golden Valley sill near Queenstown, South Africa. Approximate center UTM: 35 J 430359.41 m E, 6467878.49 m S. Note the flat inner portion where the sheet intrusion is concordant, and the distal portion that cuts upward across the sedimentary strata as an inclined dike.

11.2.1 Climbing Sills

Many sills are concordant over their entire length, but some are concordant centrally and cut upward across the sedimentary layering near their periphery. In other words, they climb up through the stratigraphic section as they propagate laterally. The Buckhorn Ridge sill on the east slope of Mount Holmes in the Henry Mountains (**Figure 11.18a**) is a well-exposed example of a climbing sill. The central concordant part of the sill is about 1 km in length with a maximum thickness of about 40 m. Near its distal termination the sill turns abruptly upward and cuts obliquely across the sedimentary strata as a dike with a dip of about 45°. At the summit of the ridge the dike turns into a bedding plane and continues as a sill to its termination. The three-dimensional geometry of another climbing sill (**Figure 11.18b**) is exposed in the Karoo Basin near Queenstown, South Africa. The Golden Valley sill is shaped like a *saucer*, with a relatively flat inner base, and a distal inclined portion that approximates the shape of a conical frustum.

The tendency for sills to become discordant near their periphery and climb to a higher stratigraphic position is indicative of mechanical interaction between the opening sill and Earth's traction-free surface. To quantify this concept we calculate the stress intensities at the tips of a model sill at depth d in an elastic half space (**Figure 11.19**, upper inset). Recall from Section 7.3.3 that the stress intensity is a measure of the stress concentration very near the tip of a fracture. The sill is opened by internal fluid pressure, p, and this opening induces a stress concentration at the tips. If the ratio of depth to sill half-length, d/c, is infinite, the mode I stress intensity is $K_I = p(\pi c)^{1/2} = K^\infty$ and the mode II stress intensity is zero. We use K^∞ to normalize the stress intensities in **Figure 11.19**. Normalized stress intensity factors greater than unity for K_I and different from zero for K_{II} reflect the degree of mechanical interaction between the sill and Earth's surface.

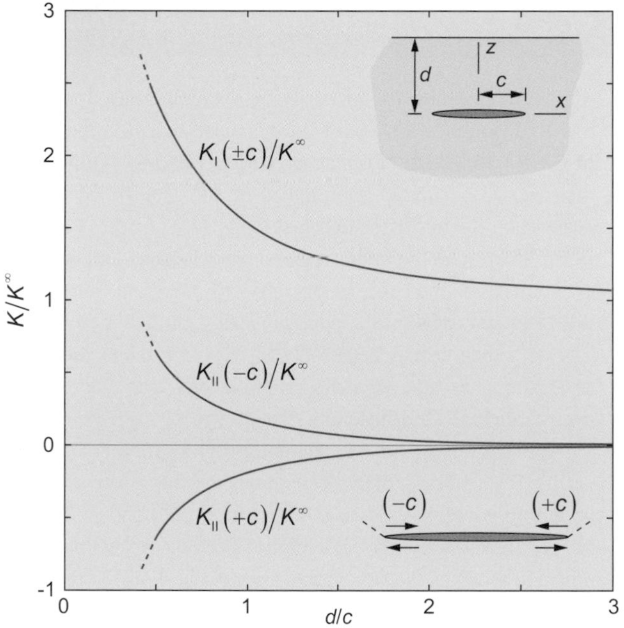

Figure 11.19 Mode I and mode II stress intensity plotted versus the ratio of depth to center and half-length of the model sill. Non-zero mode II stress intensity promotes upward climb of the sill. Modified from Pollard and Holzhausen (1979), Figure 12.

The graph in **Figure 11.19** shows that as d/c decreases to values less than 3, the normalized mode I intensity increases to values greater than 1. This change is induced by the mechanical interaction of the traction-free surface with the opening sill. Because K^∞ is proportional to the square root of the sill half length, we expect the stress intensity to increase as the sill propagates in an infinite body, but the normalized stress intensity is maintained at a value of one. Here we show that the mode I intensity increases more rapidly due to the mechanical interaction, reaching a normalized value of about 1.5 when $d/c = 1$, and about 2.4 when $d/c = 0.5$.

As d/c decreases to values less than 3, the results plotted in **Figure 11.19** show that the magnitudes of the normalized mode II intensities diverge from zero. This indicates a tendency for the sill to propagate out of the bedding plane. Note that K_{II} is positive at the left tip and negative at the right tip. Positive values are associated with clockwise rotation of the propagation direction and negative values are associated with counter-clockwise rotation (see dashed lines in **Figure 11.19**, lower inset). Thus, both tips of the sill have a tendency to cut upward towards the traction-free surface: the mechanical interaction with the traction-free surface provides an explanation for climbing sills.

11.2.2 Growth of a Sill

Modern intrusive events provide insight about the depth and size of the magma body, and the rate of influx of magma, and they serve to illustrate the utility of an important technology (**Box 11.2**) called Synthetic Aperture Radar Interferometry (InSAR). Sierra Negra Volcano is one of the most active volcanoes making up the Galápogos Islands (**Figure 11.20**) with 11 historic eruptions. It has the largest summit caldera (~60 km^2) of the six active volcanoes on the islands. The satellite radar data recorded about 1.6 m of uplift of the caldera floor from 1992 to 1997, presumably caused by the intrusion of magma into the volcano at shallow depths. In this section we describe InSAR data from a second uplift event between September 1998 and March 1999.

One of the interferograms for the caldera of Sierra Negra Volcano over the seven months of the inflation from 1998 to 1999 is shown in **Figure 11.21a**. The pattern of fringes has a remarkable radial symmetry that closely matches the shape and size of the caldera and suggests an intimate relationship between intrusions of this kind and the structural development of the caldera. The symmetry suggests that the intrusion is a sill with a circular tipline. The margin of the caldera is composed of an array of normal faults. As the caldera developed, these faults

Box 11.2 Synthetic Aperture Radar Interferometry (InSAR)

Synthetic Aperture Radar (SAR) is a satellite-based technology that provides images of Earth's surface with a resolution of tens of meters and Digital Elevation Models (DEM) with meter-scale accuracy (Bürgmann et al., 2000). Synthetic Aperture Radar Interferometry (InSAR) uses two such radar images of the same region, taken before and after an event that changed the topography, to calculate patterns of displacement of Earth's surface with centimeter accuracy. The deformation may be associated with tectonic events, such as earthquakes and volcanic eruptions, or with geomorphic events, such as landslides and glacial flow.

Applications of InSAR technology grew rapidly from the early 1990s across many disciplines in the earth sciences, including geophysics, geodesy, structural geology, geomorphology, volcanology, and glaciology. Because a satellite transmits and receives the signal, this technology does not require a ground-based operator like GPS (**Box 2.1**) and SfM (**Box 11.1**). This is particularly important for remote or dangerous locations, where logistics could be challenging and safety of the operators could be imperiled. InSAR has opened up many regions of Earth's surface to deformation monitoring, hazard assessment, and modeling of tectonic and geomorphic processes.

The SAR signal travels from the satellite to the ground surface, and then reflects back to the satellite, which records the travel time and signal amplitude from multiple reflectors. Knowledge of the travel time and speed of the signal provide the *range*, or distance from the satellite to each reflective site on the surface. A typical survey images a *swath* of Earth's surface that is about 100 km wide and resolves the range for points (pixels) that are 20 to 100 m wide (Bürgmann et al., 2000). Subsequent images may collect new images each time the satellite passes over that region. If the reflectivity of the surface has not changed significantly, the difference in ranges for comparable pixels and satellite positions allow the calculation of the displacement component directed along the line of sight from the satellite to the reflector. The map of these differences is a radar interferogram. Note that the data contained on the interferogram are not the displacement vector at each pixel, but one component of that vector. A typical interferogram (**Figure B11.2a**) shows a pattern of *fringes*, which are colored bands representing contours of the displacement component with each cycle of colors representing a given change, usually about 5 to 10 cm.

The May 13, 1995 Kozani-Grevena earthquake provides a good example of how surface displacement data from an InSAR interferogram (**Figure B11.2a**), integrated with aftershock locations (**Figure B11.2c**), reveal the subsurface structure in a region where sparse surface faulting hindered the assessment of seismic hazards (Resor et al., 2005). The interferogram uses images from before and after the earthquake, acquired on November 16, 1993 and November 9, 1995. The displacement field in the satellite's line of sight (**Figure B11.2b**) shows a broad region of increased range (positive, warm colors) with a maximum subsidence of about 70 cm, and a less well defined region of decreased range (negative, cool colors) with a maximum uplift of about 10 cm. The region of subsidence extends from a few kilometers south of the surface rupture to more than 20 km north of the rupture. This pattern is broadly consistent with local west striking and north dipping normal faults, and north–south extension of the region, but does not reveal the details of the fault geometry.

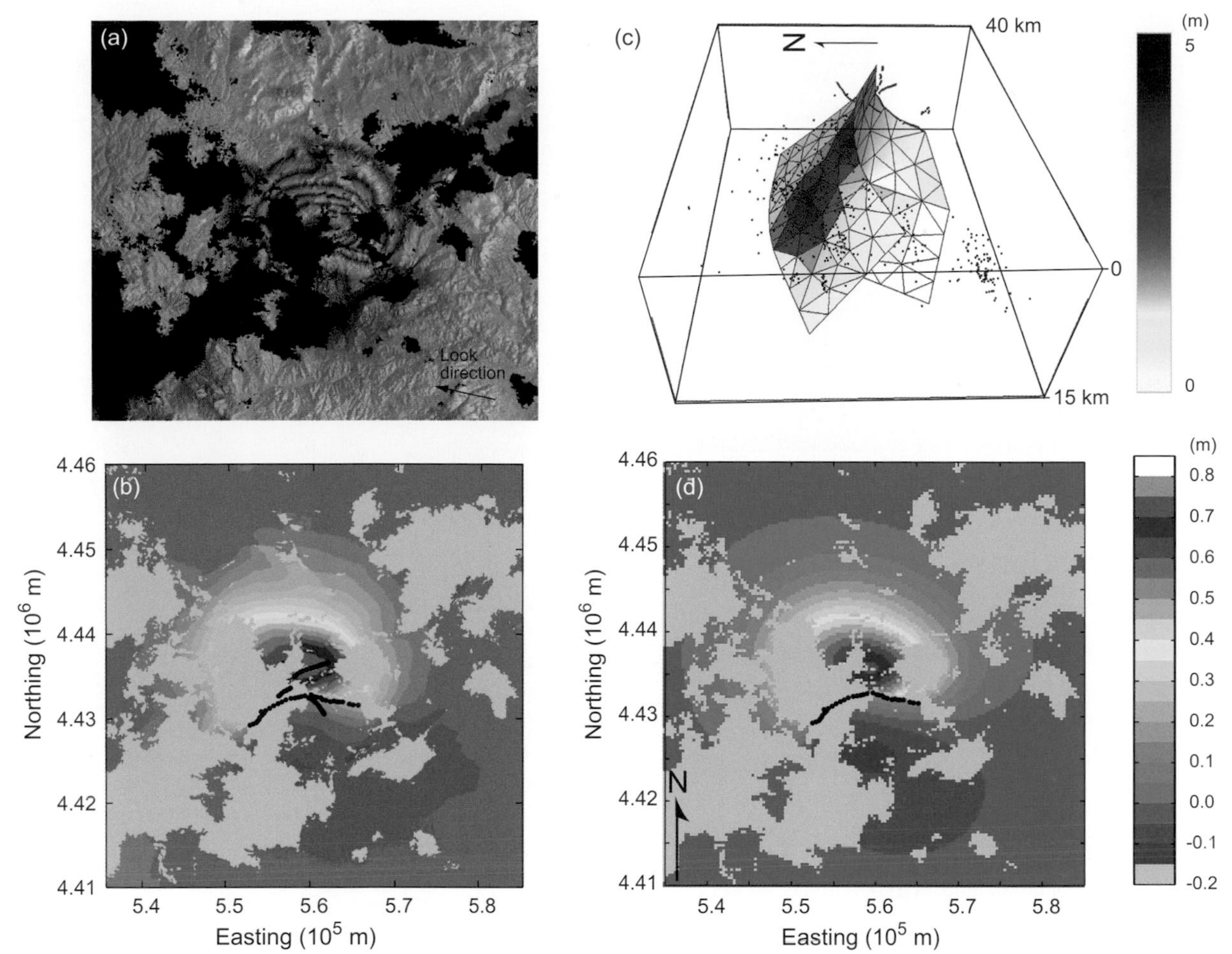

Figure B11.2 InSAR data and elastic model for the 1995 Kozani-Grevena Earthquake, Greece. (a) Interferogram with each fringe equivalent to 10 cm of displacement in the satellite's line of sight. (b) Color contours of surface displacement from the InSAR data. Color bar on the right gives displacement magnitude with increasing range (subsidence) positive. Black lines mark surface ruptures. (c) Subsurface fault model interpreted from aftershock locations (black dots) and surface ruptures. Color bar (right) gives slip on fault segments. (d) Surface displacements from three faults in an elastic half-space model. Compare to displacements from the InSAR data in (b). From Resor et al. (2005).

The fault model (**Figure B11.2c**) for the Kozani-Grevena earthquake as interpreted from the aftershock locations includes three faults (Resor et al., 2005). The most significant of these, fault 1, strikes to the west, dips about 43° to the north, and extends from about 6 km to 15 km depth, so it did not breech the surface. Fault 2 has a curving strike, dips to the northwest, and appears to connect to the curving surface rupture. Fault 3 strikes to the east, dips about 55° to the south, and extends from about 4 km to 10 km depth, so it did not breech the surface. Triangular elements approximate the geometry of the three faults and the InSAR data constrain the slip distributions on these faults. For the best fitting model, Fault 1 has a maximum 4.7 m of normal slip near its upper tip. The distribution of surface displacements due to slip on the three model faults (**Figure B11.2d**) is similar to that of the InSAR data (**Figure B11.2b**). Using aftershock locations to infer fault geometry, and using InSAR data to constrain models of surface displacements due to slip on the model faults, provides insights that may inform future hazard assessments in this earthquake prone region.

have dropped the floor of the caldera down about 110 m relative to the surrounding summit of the volcano. Some of these faults were active with as much as 1.5 m of slip in the period 1997 to 1998.

The model intrusion is a horizontal sill in a linear elastic material with a pressure boundary condition on the upper and lower contacts. Although this model is three dimensional and includes a traction-free surface above the sill, it is similar in all

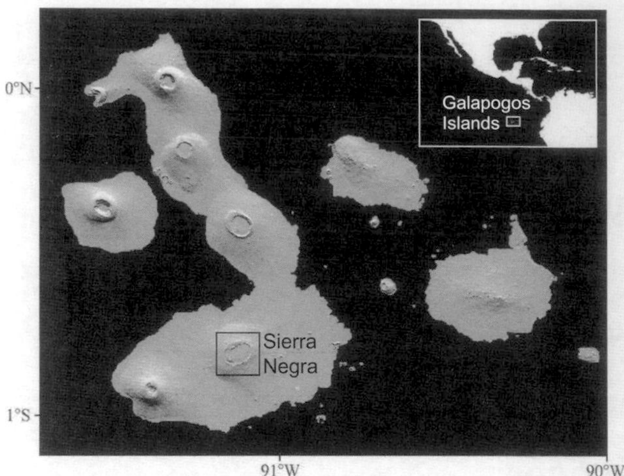

Figure 11.20 Shaded relief topographic map of Galápogos Islands (see inset for location). The caldera of Sierra Negra Volcano is within the black box, which outlines the area of the interferograms used in Figure 11.21. UTM: 15 M 707312.78 m E, 9909927.34 m S. From Yun et al. (2006), Figure 1.

other respects to the canonical model for two-dimensional opening fractures described in Section 7.2. The driving pressure is the difference between the magma pressure, P_M, and the lithostatic pressure, P_L, due to the weight of the overlying rock, so $\Delta\sigma_I = P_M - P_L$. The opening displacement discontinuity, Δu_y, for the canonical opening fracture is proportional to the driving stress, $\Delta\sigma_I$, to the fracture half-length, a, and to $(1 - \nu)/G$, where ν is Poisson's ratio and G is the elastic shear modulus. The greatest opening is $\max(\Delta u_y) = 2\Delta\sigma_I a(1 - \nu)/G$. For the model sill, the relevant length, a, is the radius. We would expect the greatest vertical displacement of the upper sill contact to be *approximately* proportional to $\Delta\sigma_I a(1 - \nu)/G$. The displacement for the model sill would be somewhat less than that for the canonical opening fracture because the sill tipline is roughly circular. This reduction in displacement due to the finite area of the sill would be somewhat offset by an increase in displacement arising from the interaction of the sill with the nearby traction-free surface.

An inversion procedure was used to find the model sill that best fit the InSAR data (Figure 11.21). The synthetic InSAR pattern generated for this model matches the data quite well, with residual differences between data and model being no greater than 2.5 cm. The opening distribution for the best-fit model sill has a maximum of 50 cm and the driving pressure is 4.5 MPa. Over the seven-month period analyzed using the InSAR data, the greatest upward displacement of the caldera floor was 30 cm, and this deformation was attributed to the intrusion of a sill at about 1.9 km depth. The maximum uplift is a little more than half the maximum opening of the model sill. The horizontal dimensions of the sill are about 6.5 km (east–west) by 5.5 km (north–south). The thickness to length ratio of the sill is about 1:12,000 and the inferred total influx of magma was 6.7 million cubic meters at an average rate of about 1.1 million cubic meters per month.

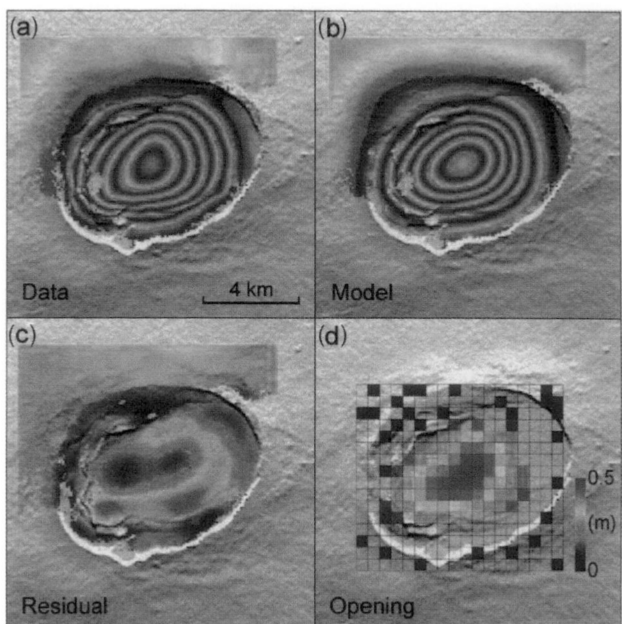

Figure 11.21 InSAR data and synthetic InSAR from a sill model for the uplift event in Sierra Negra caldera from 09/26/1998 to 03/20/1999. See Figure 11.20 for location. (a) Observed interferogram: one color cycle represents a displacement of 5 cm. (b) Simulated interferogram for the best-fit sill model. (c) Residual differences between data and model do not exceed 2.5 cm. (d) Opening displacements of pressurized model sill. Isolated dark blue segments can be ignored because their contribution to opening is not significant. From Yun et al. (2006), Figure 6.

11.2.3 Flow of Power-Law Magma in a Sill

In Section 6.7.3 we introduced the constitutive equations for an incompressible and isotropic linear viscous fluid. The nine relationships between stress components and rate of deformation components take the following forms:

$$\sigma_{xx} = -\bar{p} + 2\eta D_{xx}, \sigma_{xz} = 2\eta D_{xz} \tag{11.14}$$

Here $\bar{p}$ is the mean normal pressure and η is the Newtonian viscosity. Pressure is inherently positive, so we take the negative of the pressure to correspond to a compressive (negative) stress. Thus, the mean normal pressure is related to the negative of the mean normal stress as $\bar{p} = -\sigma_m = -\frac{1}{3}(\sigma_1 + \sigma_2 + \sigma_3)$, which is an invariant with respect to orientation of the coordinate system. The constitutive relationships in (11.14) are linear in that each stress component is proportional to the corresponding rate of deformation component, and the proportionality constant is 2η. The normal stress components reduce to the negative of the mean pressure when the corresponding rate of deformation goes to zero. The shear stress components go to zero when the corresponding rate of deformation goes to zero.

Magma may have constitutive properties that are more complex than the linear viscous fluid. To build upon that simpler model we introduce a power-law fluid, with a constitutive law that relates the stress, raised to a power n, to the rate of deformation. We begin this discussion by introducing the concepts of *deviatoric* stress and *deviatoric* rate of deformation, which are used in the constitutive equations for the power-law fluid. Then, we use the power law for steady laminar flow of magma in a sill and compare that to flow of a linear viscous (Newtonian) fluid.

By appealing to intuition, we suppose that a change in volume of a deforming material is associated with a change in pressure, whereas a change in shape is associated with that part of the stress tensor excluding the pressure. To account for the pressure we need to compose a stress tensor made up only of the pressure. This is done by multiplying the negative of the mean normal pressure by the identity matrix [I], which was introduced in Section 2.6.2:

$$-\bar{p}[\mathrm{I}] = -\bar{p}\begin{bmatrix} 1 & 0 & 0 \\ 0 & 1 & 0 \\ 0 & 0 & 1 \end{bmatrix} = \begin{bmatrix} -\bar{p} & 0 & 0 \\ 0 & -\bar{p} & 0 \\ 0 & 0 & -\bar{p} \end{bmatrix}$$

This isotropic compressive stress is subtracted from the stress, $[\sigma]$, to define the deviatoric stress, $[\sigma']$:

$$[\sigma'] = [\sigma] - (-\bar{p}[\mathrm{I}]) = \begin{bmatrix} \sigma_{xx} + \bar{p} & \sigma_{xy} & \sigma_{xz} \\ \sigma_{yx} & \sigma_{yy} + \bar{p} & \sigma_{yz} \\ \sigma_{zx} & \sigma_{zy} & \sigma_{zz} + \bar{p} \end{bmatrix} = \begin{bmatrix} \sigma'_{xx} & \sigma'_{xy} & \sigma'_{xz} \\ \sigma'_{yx} & \sigma'_{yy} & \sigma'_{yz} \\ \sigma'_{zx} & \sigma'_{zy} & \sigma'_{zz} \end{bmatrix} \quad (11.15)$$

The deviatoric stress has nine components, six of which are independent because the conjugate shear stresses are equal: $\sigma'_{xy} = \sigma'_{yx}, \sigma'_{yz} = \sigma'_{zy}, \sigma'_{xz} = \sigma'_{zx}$. This follows from Cauchy's second law, which was introduced in Section 3.7.1. Thus, the deviatoric stress components compose a *symmetric* matrix. Any state of stress may be decomposed into a deviatoric and an isotropic part: $[\sigma] = [\sigma'] - \bar{p}[\mathrm{I}]$. The change in volume is associated with the isotropic part, and the change in shape is associated with the deviatoric part.

Similarly, the rate of deformation tensor can be split into an isotropic part and a deviatoric part. To derive these quantities, the mean normal rate of deformation is defined as:

$$\bar{D} = \tfrac{1}{3}(D_{xx} + D_{yy} + D_{zz}) = \tfrac{1}{3}(D_1 + D_2 + D_3) \quad (11.16)$$

The deviatoric rate of deformation, $[D']$, is found by subtracting the mean value from each normal component, while leaving the shearing components unchanged:

$$[D'] = [D] - \bar{D}[\mathrm{I}] = \begin{bmatrix} D_{xx} - \bar{D} & D_{xy} & D_{xz} \\ D_{yx} & D_{yy} - \bar{D} & D_{yz} \\ D_{zx} & D_{zy} & D_{zz} - \bar{D} \end{bmatrix} = \begin{bmatrix} D'_{xx} & D'_{xy} & D'_{xz} \\ D'_{yx} & D'_{yy} & D'_{yz} \\ D'_{zx} & D'_{zy} & D'_{zz} \end{bmatrix} \quad (11.17)$$

The deviatoric rate of deformation tensor is *symmetric* because, for example, $D'_{xy} = D'_{yx}$.

Using the deviatoric stress and deviatoric rate of deformation, the two example constitutive equations for the linear isotropic viscous liquid given in (11.14) reduce to:

$$\sigma'_{xx} = 2\eta D'_{xx} \text{ and } \sigma'_{xy} = 2\eta D'_{xy} \quad (11.18)$$

A total of three deviatoric normal stress equations and six deviatoric shear stress equations exists. Each deviatoric stress component is proportional to the respective deviatoric rate of deformation component and the proportionality constant is the same for all components. Unlike (11.14), the mean normal pressure plays no role in these relationships.

For the power-law fluid we raise each stress component to the power n in (11.18) and the other constitutive equations. Again, for example, just writing one normal and one shear equation, we have:

$$\left(\sigma'_{xx}\right)^n = AD'_{xx} \text{ and } \left(\sigma'_{xy}\right)^n = AD'_{xy} \quad (11.19)$$

For $n = 1$ we take $A = 2\eta$ and recover (11.18), the deviatoric form of the constitutive equations for the linear viscous fluid. For $n \neq 1$ the deviatoric rate of deformation is related *non-linearly* to the deviatoric stress, so this defines a power-law fluid. Recall that in Section 5.6.4 we described laboratory experiments on olivine that deformed by dislocation creep and gave stress exponents of about $n = 3$. For such non-linear relationships the proportionality constant A must take on units that are appropriate for the value of n. For example, the units of each quantity in the second of (11.19) are:

$$\left(\sigma'_{xy}\right)^n [=] \mathrm{Pa}^n, A [=] \mathrm{Pa}^n \mathrm{s}, D'_{xy} [=] \mathrm{s}^{-1} \quad (11.20)$$

Note that the constant A has units that include Pa raised to the power n to compensate for the fact that the stress term on the left side of (11.19) is raised to the power n. These units result in dimensionally consistent equations of the form (11.19) for any power n. The constant A has units of viscosity, that is Pa s, only when $n = 1$.

To understand the effect of the exponent n on flow behavior described by a power-law relationship like (11.19) we plot deviatoric stress versus deviatoric rate of deformation in **Figure 11.22** for $A = 1$ (Pa)^{n}s and values of n ranging from 1 to 4. For $n = 1$ the graph is a straight line with a slope of 1 Pa s, but for $n = 2$, 3, and 4 the graphs are curves with initial slopes that are greater than that for the linear relationship. For deviatoric stress less than 1 Pa, a given increment of stress produces successively *lesser* increments of the deviatoric rate of deformation as n increases from 2 to 3 to 4. All the curves pass through the point ($\sigma' = 1$ Pa, $D' = 1$ s^{-1}). For deviatoric stress greater than 1 Pa the curves for $n = 2$, 3, and 4 have successively lesser slopes. In other words, a given increment of the deviatoric stress produces successively *greater* increments in the deviatoric rate of deformation as n increases from 2 to 3 to 4. These relationships suggest that distributions of velocity, for example for flow in a sill (**Figure 11.17**), could be quite different for a power-law magma, than for the Newtonian magma.

In Section 6.8.3 we found that steady laminar flow in the x-direction of a linear viscous liquid in a model sill of thickness $2h$, has a *parabolic* velocity distribution with a maximum velocity, v_0, at mid-height ($z = 0$):

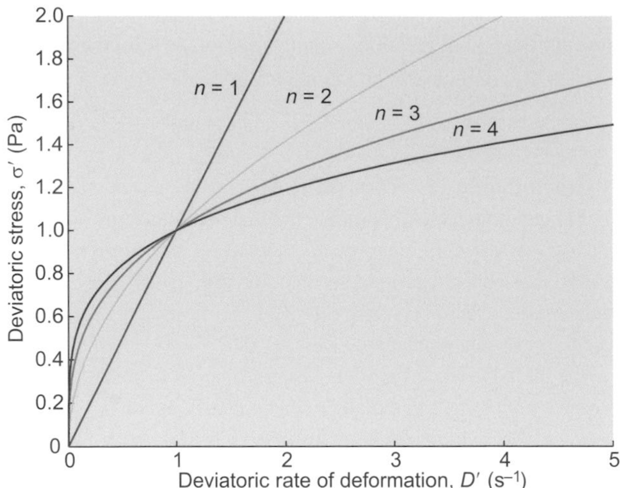

Figure 11.22 Deviatoric stress plotted versus deviatoric rate of deformation for the power-law liquid (11.19) with $A = 1$ (Pa)^{n}s and the exponent n ranging from 1 to 4.

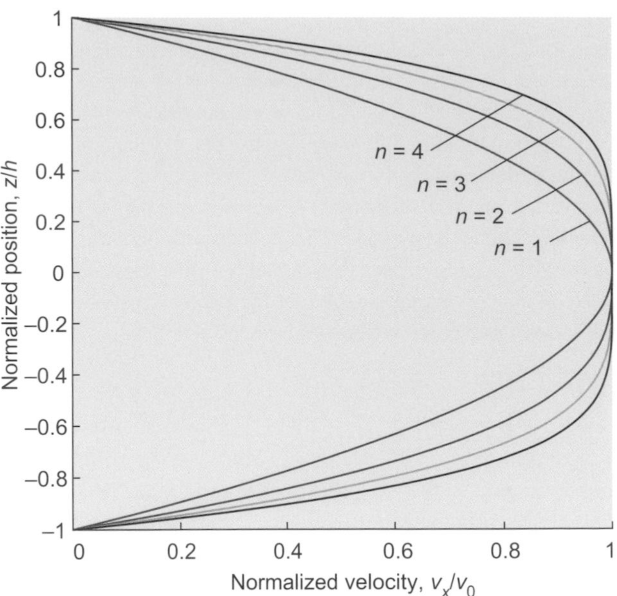

Figure 11.23 Velocity normalized by central velocity (11.22) plotted versus position normalized by sill half-thickness for a power-law liquid with stress exponent n ranging from 1 to 4.

$$v_x = \frac{1}{2\eta}\frac{\partial \bar{p}}{\partial x}\left(z^2 - h^2\right), \quad v_0 = -\frac{h^2}{2\eta}\frac{\partial \bar{p}}{\partial x} \tag{11.21}$$

The velocity is directly proportional to the pressure gradient in x, $\partial\bar{p}/\partial x$, and inversely proportional to the Newtonian viscosity, η. The velocity is positive if the pressure gradient is negative, so the pressure decreases in the positive x-direction, which is the direction of flow. The maximum velocity is proportional to the square of the sill half-thickness, and to the pressure gradient along the length of the model sill. The maximum velocity is inversely proportional to the Newtonian viscosity. In terms of cause and effect, we say that the pressure gradient *drives* the flow, and the viscosity *resists* the flow.

The velocity distribution and the maximum velocity for flow of a power-law fluid in a parallel-sided conduit (the model sill) are:

$$v_x = \frac{2}{A(n+1)}\left(\frac{\partial p}{\partial x}\right)^n\left(z^{(n+1)} - h^{(n+1)}\right), \quad v_0 = -\frac{2h^{(n+1)}}{A(n+1)}\left(\frac{\partial p}{\partial x}\right)^n \tag{11.22}$$

The maximum velocity is directly proportional to the sill half-thickness raised to the power $n+1$, and to the pressure gradient along the length of the sill raised to the power n. The maximum velocity is inversely proportional to the power-law constant A. For $n = 1$ (11.22) reduces to (11.21). Velocity profiles for power-law liquids flowing in the model sill and normalized by the maximum velocity, v_0, are shown in **Figure 11.23**.

The distribution of velocity across the model sill (**Figure 11.23**) is parabolic for the Newtonian liquid ($n = 1$), but flattens near the sill center as n increases. For $n = 4$ the central 40% of the sill has a nearly uniform velocity. In other words, the flow approaches what is called plug flow, where the central portion is flowing as a nearly rigid plug. Within the central plug the liquid is shearing very little, because the deviatoric shear stress is low, so the deviatoric rate of shearing is *very low* (**Figure 11.22**). In contrast, near the sill margin drag against the boundary creates sufficiently great deviatoric shear stress that the deviatoric rate of shearing is *very high*. This example demonstrates that power-law liquids can, if n is great enough, have distinctly different velocity distributions than Newtonian liquids for flow in the model sill.

11.3 LACCOLITHS

In Section 9.6.1 we introduced the sketch from Gilbert's field notebook that captures the conceptual laccolith model that he formulated in the Henry Mountains of southeastern Utah in August of 1875. His concept was that the diorite porphyry magma flowed upward through the horizontal sedimentary strata in a dike, turned and intruded between two layers, spread laterally for some distance as a thin sill, and then bent the overlying strata upward to form a much thicker laccolith.

Gilbert's conceptual model was modified and formalized in a drawing (**Figure 11.24a**) that was published in 1877 in his report to the US Geological Survey. In this drawing the laccolith is a concordant intrusion that has attained a greater thickness-to-length ratio than nearby sills by *bending* the overlying strata upward, while maintaining a relatively *flat* lower contact. In this drawing the bending is concentrated as a monoclinal flexure over the periphery of the laccolith. The greater expanse of strata on top of the laccolith is arched, but not as acutely bent as in his field sketch. Thinner sills lie under the laccolith, and over the periphery, but they do not attain the lateral dimensions of the laccolith.

The intrusion at Black Mesa in the Henry Mountains (**Figure 11.24b**) is consistent with Gilbert's conceptual model (**Figure 11.24a**) in that a nearly flat lower contact may be inferred, the thickness to length ratio greatly exceeds that of nearby sills, the distal margin of the intrusion is very steep, and the overlying

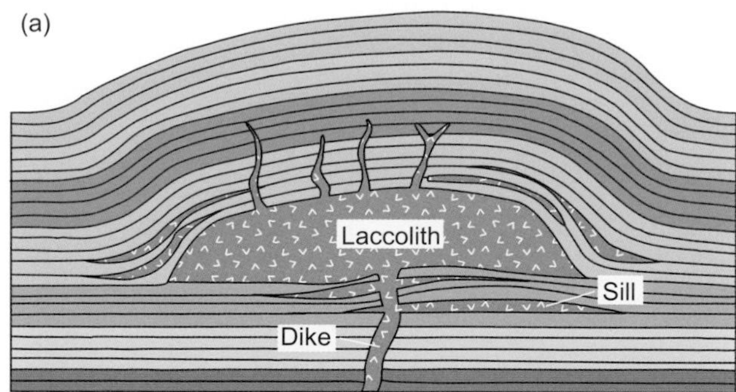

Figure 11.24 (a) Idealized laccolith with narrow feeder dike, nearly flat concordant lower contact, and strata bent over the arched and concordant upper contact. Modified from Gilbert (1877). Sills are upturned over the flanks of the laccolith and dikes emerge from its upper contact. (b) Oblique aerial photograph of Black Mesa laccolith composed of diorite porphyry on the eastern flank of Mount Hillers in the southern Henry Mountains. UTM: 12 S 535302.33 m E, 4196896.58 m N. See Saint-Blanquat et al. (2006) for detailed maps, cross sections, photographs and fabrics.

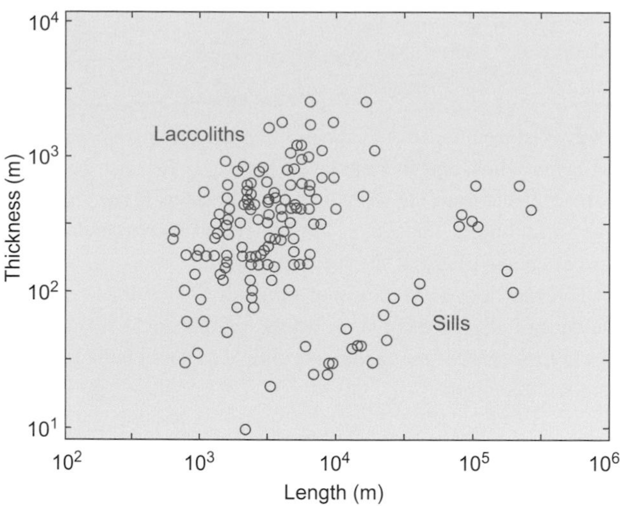

Figure 11.25 Plot of intrusion thickness versus length compiled from the literature for laccoliths and large mafic sills. Modified from Bunger and Cruden (2011), Figure 2.

sedimentary strata (now largely eroded) arched gently over the central portion of the upper contact. In this section we examine the transition from a thin sill to a thicker laccolith when bending takes over from elastic compression of the overlying strata. This occurs when the sill starts to interact mechanically with Earth's surface, which carries a traction-free boundary condition. We also present geologic data from the Henry Mountains that support Gilbert's interpretation of the development of laccoliths.

11.3.1 The Sill–Laccolith Transition

Gilbert's concept (**Figure 11.24**) suggests that laccoliths begin as sills that spread laterally as thin sheet intrusions, but eventually they increase significantly in thickness by bending the overlying strata. The thickness to length ratios for large mafic sills reported in **Figure 11.25** range from about 1:2000 to 1:200. In contrast, this ratio for laccoliths is distinctly greater, ranging from about 1:45 to 1:2. Points on this graph representing laccoliths form a cluster that does not overlap with the cluster of points representing large mafic sills. Gilbert's concept and these data bring up the question: what are the physical conditions that lead to the sill–laccolith transition? We use the example of the Black Mesa laccolith (**Figure 11.24b**) to address this question.

We begin by referring to the displacement field for the canonical model of an opening fracture in an infinite elastic body, described in Section 7.2.5. Taking the opening fracture as a model for a horizontal sill that is deeply buried relative to its length, we note that the top contact is displaced *upward* exactly as much as the bottom contact is displaced *downward* (**Figure 7.19**). In other words, the mechanism for opening the sill is the elastic displacement of the host rock, and this displacement is symmetric about the horizontal plane of the sill. The rock below and above the sill is compressed elastically as the sill opens. Because the Black Mesa laccolith is roughly circular in plan shape, we consider an opening fracture with a circular tipline. The vertical displacements of the model sill contacts are:

$$u_z = \pm \frac{4\Delta p(1 - \nu^2)}{\pi E}\left[a^2 - r^2\right]^{1/2} \qquad (11.23)$$

Here (r, z) are the horizontal and vertical cylindrical coordinates centered at the middle of the sill. The driving pressure is the difference between the magma pressure, P_M, and the lithostatic pressure, P_L, due to the weight of the overlying rock, so $\Delta p = P_M - P_L$. The sill radius is a, and (E, ν) are Young's modulus and Poisson's ratio, respectively. The plus and minus signs refer to the top and bottom contacts. The distribution of displacements from the terms in square brackets is elliptical. For driving pressures of a few MPa and Young's modulus of a few tens of GPa, the maximum thickness to length ratio, u_z/a, for the model sill would be of order 1:1000. This is a very eccentric ellipse, so the top and bottom contacts of the model sill would be almost parallel, except near the tips.

Next, we consider the displacement for the canonical model of a bending layer over a laccolith, described in Section 9.6.5. Again, taking the layer to be circular in shape to approximate the Black Mesa laccolith, the vertical displacement of the middle surface of the layer is:

$$u_z = \frac{3\Delta p(1-\nu^2)}{16Eh^3}\left[a^2 - r^2\right]^2 \tag{11.24}$$

Here (r, z) are the horizontal and vertical cylindrical coordinates centered at the middle of the layer, Δp is the driving pressure, a is the layer radius, and h is the layer thickness. The distribution of displacements from the terms in square brackets is a *bell-shaped* curve that bulges upward in the center and tapers gradually to zero at the periphery of the model laccolith.

To characterize the transition from sill to laccolith we equate the upward displacement at the center of the model sill (11.23) to that at the center of the model laccolith (11.24). In both cases we take $r = 0$ and find:

$$\frac{4\Delta p(1-\nu^2)a}{\pi E} = \frac{3\Delta p(1-\nu^2)a^4}{16Eh^3}$$

Recall from the application of the bending layer model to laccoliths in the Henry Mountains that the sedimentary strata over the laccoliths apparently behaved like a *stack* of layers, and that slip between some layers greatly reduced the resistance to bending. To account for this behavior we write the layer thickness as h_e, meaning the effective thickness. This is the thickness of a single layer that has the same resistance to bending as the entire stack of layers. Then, we write the layer radius as a_t, meaning the sill–laccolith transition radius, and use the above equation to solve for this radius:

$$a_t = 4h_e/(3\pi)^{1/3} \approx 2h_e \tag{11.25}$$

When sills are sufficiently deep to not interact mechanically with Earth's traction-free surface, they gain thickness by *deforming* the host rock both below and above the plane of the sill, as in (11.23). However, if sills spread far enough laterally relative to their depth, they gain sufficient leverage to *bend* the overlying strata upward. In that case (11.25) marks the transition from a sill to a laccolith, which occurs when the radius of the circular intrusion is about twice the effective thickness of the overlying strata.

We illustrate the sill–laccolith transition in **Figure 11.26** using Black Mesa intrusion (**Figure 11.24b**), which today is exposed at Earth's surface with a radius of about 850 m and a thickness of about 200 m. Suppose the total overburden thickness at the time of intrusion was 2 to 3 km, but the effective thickness, h_e, was only 300 m. We take the excess magma pressure, Δp, in the dike feeding the intrusion at this depth as 5 MPa. The dike turned at the base of the Morrison Formation and a sill spread laterally, displacing the underlying strata downward and the overlying strata upward. The sill thickness increased linearly (11.23) in proportion to the sill radius, maintaining an upward displacement to radius ratio of about 1:50 as the host rock deformed elastically.

From (11.25) the sill–laccolith transition depicted in **Figure 11.26** occurred at a radius of about 600 m and the laccolith thickness now started to increase non-linearly (11.24) as the intrusion continued to grow laterally. Also, because the intrusion now was mechanically interacting with Earth's traction-free surface, a surficial dome started to form. As the laccolith radius approached 850 m and the thickness approached 50 m, inelastic deformation in the overlying strata included jointing and bedding-plane faulting. This deformation *localized* into a nearly vertical peripheral fault zone that propagated toward the surface, and the laccolith began to increase in thickness without further growth in radius. Black Mesa laccolith ceased growing when the thickness reached about 200 m (**Figure 11.24b**).

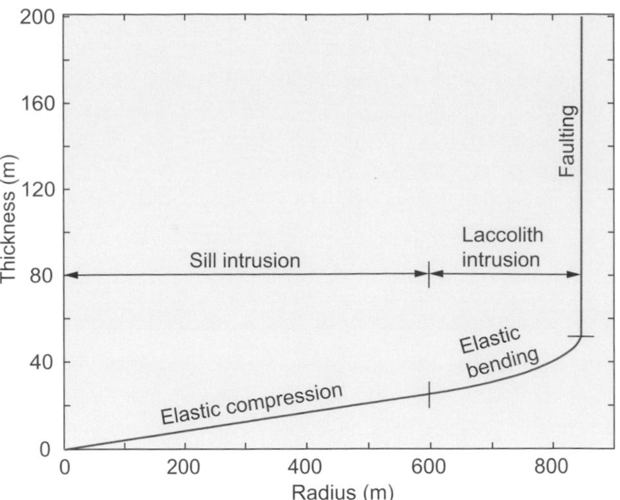

Figure 11.26 Plot of vertical thickness versus horizontal radius for Black Mesa intrusion. Three distinct stages are sill intrusion, laccolith intrusion, and peripheral faulting. Modified from Pollard and Johnson (1973).

The mafic sills plotted in **Figure 11.25** range in length up to 270 km. Some of these sills apparently spread laterally for very great distances without passing through the transition from sill to laccolith illustrated in **Figure 11.26**, where bending of the overlying strata becomes significant. The tendency to bend the overlying strata apparently was mitigated by the greater density of the magma, which promoted lateral propagation, and enabled these mafic sills to maintain lesser ratios of thickness to length as they spread horizontally to great distances. In contrast, the sills of the Henry Mountains spread laterally only from several hundred meters to a few kilometers before the transition to laccolithic bending of the overlying strata began (**Figure 11.26**).

In summary, the transition from growth of a sill to the early stages in the development of a laccolith depends upon the mechanical interaction between the intrusion and Earth's traction-free surface. The key concept is that the sill behaves as if it were in an infinite elastic body, so to increase in thickness the sill has to compress the overlying *and* underlying sedimentary strata. In contrast, the laccolith increases in thickness by bending the overlying strata upward, and this process depends upon the flexural rigidity of those strata. Bending is enhanced if the overlying strata can slip over one another on bedding-plane faults, thus reducing the effective thickness of the overburden. In the next section we document the bedding-plane faults and the form of bending at Mount Ellsworth in the southern Henry Mountains.

11.3.2 Growth of Laccoliths

In Section 9.6.6 we document the form of bending of the sedimentary strata over the laccolith at Mount Holmes, in the southern

Henry Mountains, UT, where about 4 km of overlying strata are bent into a dome over a radial distance of 5 to 6 km (**Figure 9.35**). Within this thick stack of sedimentary rocks bedding-plane faults subdivided the overburden and apparently enabled it to bend, with an effective thickness that was less than a kilometer, through a maximum vertical displacement of about 1.2 km (**Figure 9.36**). Sedimentary units in the limb of the dome dip as steeply as 23° in radial directions. Here we document the next stage in the growth of laccoliths using the nearby intrusions at Mount Ellsworth.

A brief overview of the stratigraphy at Mount Ellsworth is needed to set the stage. On the geological map (**Figure 11.27**) the sedimentary strata dip radially away from the mountain summit. As a result, individual formations crop out in a circular pattern around the mountain. **Figure 9.34** provides a stratigraphic column with formation names for the geologic map. The different map units crop out systematically from older near the center to younger on the flanks, making this a classic example of an eroded sedimentary dome. A large mass of diorite porphyry, cropping out near the summit is interpreted as the upper extent of the laccolith underlying the dome. Remnants of Permian White Rim Sandstone and Organ Rock Shale are the oldest sedimentary rocks in contact with the top of the laccolith. This suggests that the unexposed *base* of the laccolith lies near the top of the Permian Cedar Mesa Sandstone. The radius of the sedimentary dome at Mount Ellsworth is about 6 to 7 km, and the maximum vertical displacement is about 1.8 km. On the flanks of Mount Ellsworth, and within the central limb of the dome, the sedimentary units dip between 50° and 56° in radial directions. Minor intrusions, including radial dikes and concordant sills, crop out around the dome above the upper contact of the central intrusion.

The cross section (**Figure 11.28**) extends from B to B′ toward the northwest from the center of the dome at Mount Ellsworth and reveals the shape of the dome in profile. Strata in the peripheral limb dip gently to the northwest, are curved concave upward in the lower hinge, dip steeply to the northwest in the central limb, and bend concave downward in the upper hinge. Sills interleave with the sedimentary strata just above the upper contact of the central intrusion, and bedding-plane fault zones are exposed in the central limb and lower hinge. The greater upward displacement of the overlying sedimentary units and the more steeply dipping strata in the central limb indicate that the intrusive process reached a more advanced stage at Mount Ellsworth than at Mount Holmes (**Figure 9.36**).

Figure 11.29 depicts three stages in the growth of laccoliths in the southern Henry Mountains using vertical profiles that are *schematic*, but are based on the geologic maps (e.g. **Figure 11.27**) and cross sections (e.g. **Figure 11.28**) for Mount Holmes, Mount Ellsworth, and Mount Hillers. In stage 1 sills insinuated along bedding planes in the horizontal Paleozoic and Mesozoic sedimentary rocks. These intrusions of diorite porphyry were fed by vertical dikes from a much deeper magma reservoir. Many of the sills were tongue-shaped with their long axes radial to the intrusive center (**Figure 11.29a**), and they did not attain a sufficient horizontal dimension to mechanically interact with Earth's surface. Thus they maintained a small thickness to length ratio, characteristic of opening cracks in an infinite elastic solid.

Stage 2 depicted in **Figure 11.29b** began when one sill did spread laterally far enough to interact mechanically with Earth's surface and thereafter gained thickness by bending the overlying strata to form a laccolith. The upward flexure of the overburden created a dome, and this doming was facilitated by the development of bedding-plane faults that reduced the effective thickness and the flexural rigidity of the overlying strata. In stage 3 (**Figure 11.29c**) the laccolith grew in volume to perhaps 30 km^3, the lower and upper hinges of the dome increased in curvature, while the limbs of the dome steepened. In this more advanced stage vertical dikes extended from the upper contact of the growing intrusion and crosscut the older sills which were bent along with the strata over the dome.

11.4 STOCKS

Stocks are igneous intrusions with roughly circular to elliptical shapes in map view, and steeply dipping to vertical contacts that suggest a cylindrical geometry in three dimensions. The cylindrical geometry is reminiscent of the volcanic necks shown in **Figure 11.3**, but stocks generally have much greater diameters. Typical dimensions in map view range from a few kilometers to a several tens of kilometers. The height of stocks may not be well constrained by geological data because exposures often are limited to a narrow range of elevations. Usually the steeply dipping contacts are discordant with the host sedimentary or metamorphic rock, which suggests intrusive mechanisms such as stoping (Section 11.1.2) or melting.

11.4.1 The Laccolith–Stock Controversy

The conceptual model for laccoliths (**Figure 11.24a**) proposed by Gilbert in 1877 involves a process that begins with a vertical dike feeding a horizontal and concordant sill that grows into a laccolith by bending the overlying strata upward into a dome. The lack of exposure of concordant lower contacts and feeder dikes for the central intrusions under the large domes in the Henry Mountains (Section 11.3.2) motivated the interpretation by Hunt in 1953 that each is an upright cylindrical stock (**Figure 11.30**). Under this interpretation the dikes, sills, and smaller laccoliths, exposed at the current level of erosion, propagated laterally into the surrounding sedimentary rock from the central stock. The stocks presumably extend downward to a deep magma reservoir and serve as the feeding conduit for all the shallow intrusions.

A set of orthogonal cross sections, collectively called a fence diagram (**Figure 11.30**), shows the central intrusion as a stock at Mount Ellsworth. Similar fence diagrams for the intrusions at Mount Holmes and Mount Hillers interpret the central bodies of diorite porphyry as stocks. In contrast to Gilbert's conceptual model (**Figure 11.24**), neither a horizontal lower contact nor a feeder dike are included. Furthermore, the surrounding sedimentary units truncate against the stock, so they are discordant intrusions with respect to these strata. The smaller concordant intrusions on the flanks of Mount Ellsworth are sills. According to this interpretation these concordant intrusions radiate from the

Figure 11.27 Geologic map of Mount Ellsworth, southern Henry Mountains, UT. Cross section B–B′ is shown in Figure 11.28. Key for map units is provided in Figure 9.34. Mountain peak at UTM: 12 S 533571.49 m E, 4177703.70 m N. Modified from Jackson and Pollard (1988).

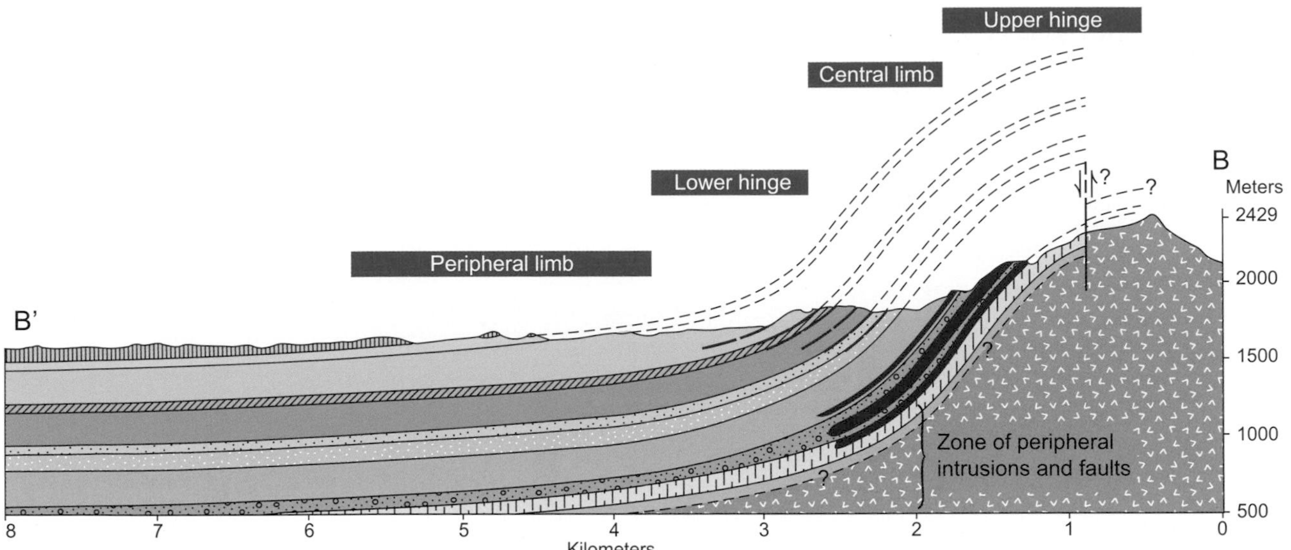

Figure 11.28 Interpretive cross section radial to the dome at Mount Ellsworth in the southern Henry Mountains, UT, based on the geological map (Figure 11.27). Thin red lines are bedding-plane fault zones; thick red lines are sills. Stratigraphic labels range from the Permian Organ Rock Shale (Pcor) to the Jurassic Salt Wash Sandstone of the Morrison Formation (Jms). Key is provided in Figure 9.34. Modified from Jackson and Pollard (1988).

central stock with tongue-like shapes. The laccolith–stock controversy arose from the different interpretations of the subsurface structure of the Henry Mountains intrusions, and the different mechanisms responsible for their formation.

Only the crests and flanks of the central intrusions are exposed for the three southern Henry Mountains, so a concordant lower contact, which would be diagnostic evidence for a laccolithic origin is not available. On the other hand, the cross sections for Mount Holmes and Mount Ellsworth (Figure 9.36 and Figure 11.28), prepared from the structural data on the geologic maps (Figure 9.35 and Figure 11.27) show the overlying strata arched into doubly hinged flexures that are concave upward at the lower hinge and concave downward at the upper hinge. Furthermore, where the sedimentary strata are exposed in contact with the central intrusion they are predominantly concordant. These relationships *are* consistent with Gilbert's conceptual model of a laccolith (Figure 11.24); they *are not* consistent with Hunt's conceptual model of a stock (Figure 11.30).

11.4.2 Cooling and Fracturing of a Stock

The Diamond Joe stock crops out over about 50 km^2 in western central Arizona (Figure 11.31) and is surrounded by Precambrian metamorphic rocks. The geological map shows the igneous rock concentrically zoned from a marginal granodiorite (Lgd), to an intermediate quartz monzonite (Lpqm), to a central quartz monzonite porphyry (Lqmp). With an age of about 72 m.y., this stock is associated with the Laramide orogeny that produced many similar intrusions in the southwestern United States. The internal contacts dip steeply outward and for the most part are gradational. The different lithologies of this intrusion are interpreted to have formed by differentiation in place of a single intrusive unit.

Fractures at more than 300 localities (Figure 11.31) are characterized by fillings of hydrothermal minerals (dominantly quartz, muscovite, microcline, and pyrite) and alteration of minerals in the adjacent igneous rock. Fractures vary in aperture from several microns to one meter, and in trace length from centimeters to at least 30 m. Lineaments on aerial photographs suggest that some fractures may be kilometers in length. The majority of the fractures are interpreted as *opening*, and because they are filled with hydrothermal minerals, we refer to them as veins. The veins form a radial pattern focused on the center of the stock. Northeast of the Gunsight fault the pattern is not perfectly radial, but it is symmetric about the Silvertrails vein. With distance from the center of the stock, fractures to the northwest of this vein diverge to a more northerly strike, whereas those to the southwest diverge to a more southerly strike.

The pattern of fractures at the Diamond Joe stock (Figure 11.31) is analyzed by supposing that fractures formed after the magma solidified and the stress state reached the tensile strength of the rock. If the stock cooled by horizontal heat flow from the center to the margin, fractures would form first near the margin and the stress state there would be influenced by: (1) thermal stresses due to cooling of the solidified outer shell of the stock constrained by its surroundings; (2) pressure from still molten magma in the core of the stock; and (3) regional tectonic stresses acting on the stock. The problem considered here addresses the stress state in a vertical cylindrical body cooling by horizontal conduction, as a possible model for the generation of the radial veins of the Diamond Joe stock.

We consider the cooling of a long cylindrical intrusions (Figure 11.32). By "long" we mean the height of the cylinder is much greater than its radius, $h >> b$, so the heat flow is in the radial direction only. Furthermore we suppose the stock was emplaced on a short time scale compared to that of cooling, so

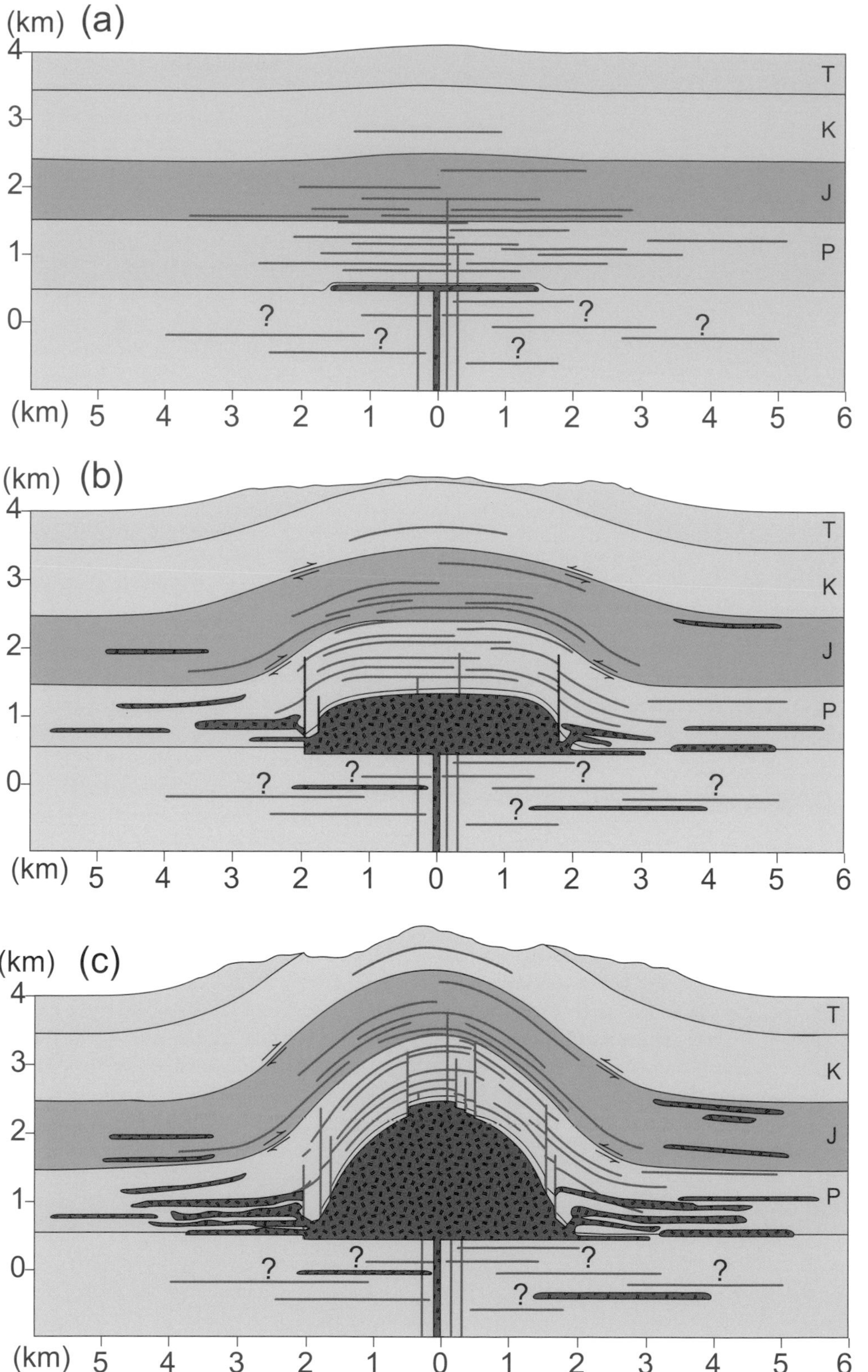

Figure 11.29 Vertical cross sections illustrating stages in the growth of sills and laccoliths in the Henry Mountains, UT. (a) Stage 1: radial dikes feed sills that spread laterally along the dikes with a tongue-like shape and longest dimensions radial to the intrusive center. (b) Stage 2: one sill passes through the sill–laccolith transition and begins to grow in thickness by bending the overlying strata. (c) Stage 3: the central laccolith thickens and the dome of sedimentary rocks grows in amplitude. Modified from Jackson and Pollard (1988).

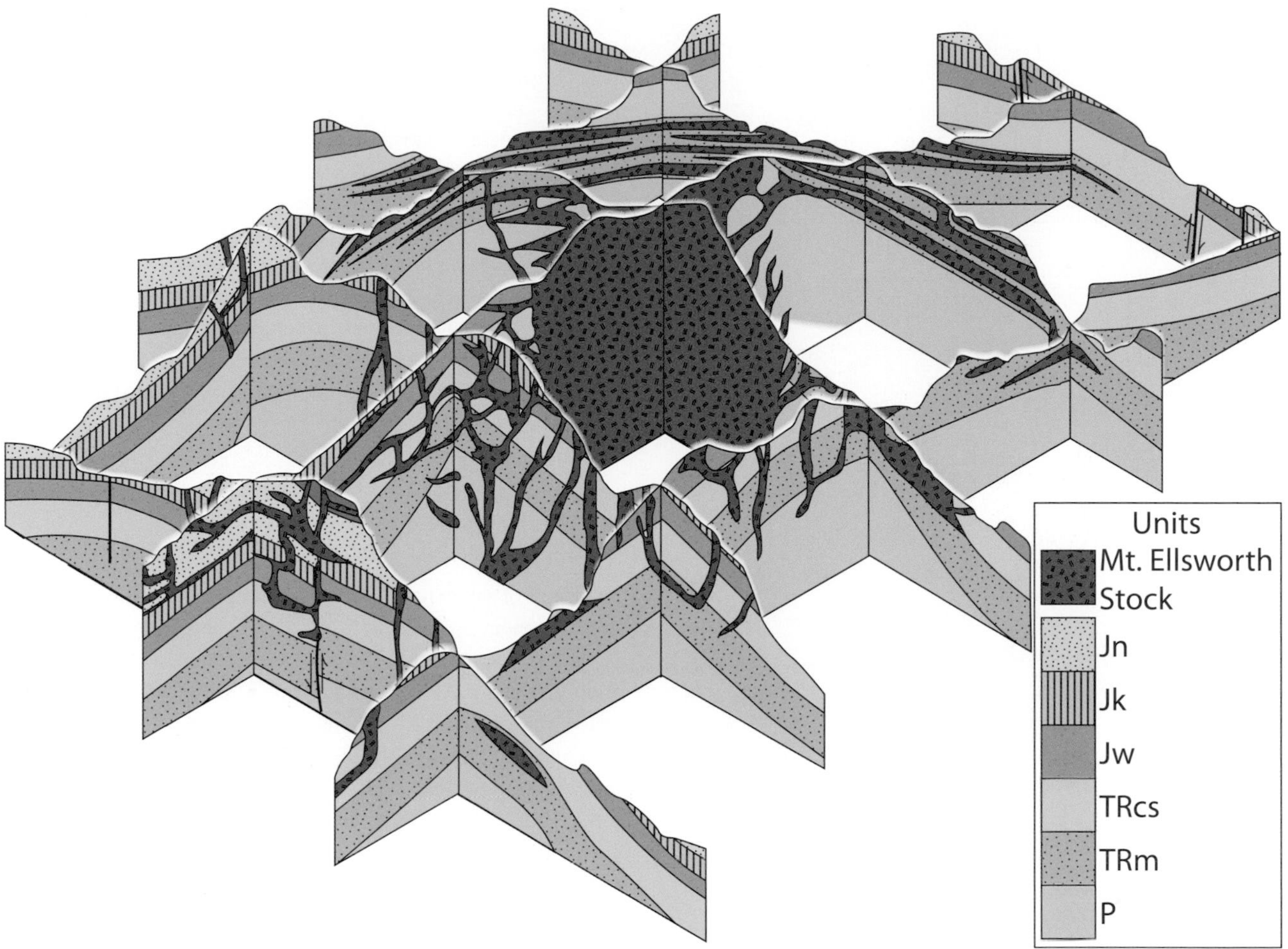

Figure 11.30 Interpretive fence diagram of Mount Ellsworth illustrating concept of a central stock with surrounding dikes and tongue-shaped sills. Modified from Hunt (1953), Plate 16.

both the stock and the surrounding rock have an initial uniform temperature. In this axisymmetric case, the temperatures and stresses depend only on the radial distance from the axis of the cylinder. Although the governing equation for thermal stresses are in general partial differential equations, for these axisymmetric cases the governing equations are ordinary differential equations, depending only on the radial distance r.

For the case of no strain in the axial direction the stress components are (Gerla, 1988):

$$\sigma_{rr}(r) = \frac{\alpha E}{1-\nu}\left(\frac{1}{b^2}\int_0^b Tr\,\mathrm{d}r - \frac{1}{r^2}\int_0^r Tr\,\mathrm{d}r\right)$$

$$\sigma_{\theta\theta}(r) = \frac{\alpha E}{1-\nu}\left[\frac{1}{b^2}\int_0^b Tr\,\mathrm{d}r + \frac{1}{r^2}\int_0^r Tr\,\mathrm{d}r - T(r)\right]$$

$$\sigma_{zz}(r) = \frac{\alpha E}{1-\nu}\left[\frac{2\nu}{b^2}\int_0^b Tr\,\mathrm{d}r - T(r)\right]$$

The mean temperature, $\bar{T}$, in a concentric sub-cylinder of radius $r \leq b$ and height h is the integrated temperature of the sub-cylinder divided by its volume:

$$\bar{T} = \frac{1}{\pi r^2 h}\int_0^r T2\pi rh\,\mathrm{d}r = \frac{2}{r^2}\int_0^r Tr\,\mathrm{d}r$$

This allows the equations for the stress components to be rewritten in terms of the mean temperature in the whole cylinder, $\bar{T}_c$, and that in a sub-cylinder, $\bar{T}_s$:

$$\sigma_{rr}(r) = \frac{\alpha E}{1-\nu}\left(\tfrac{1}{2}\bar{T}_c - \tfrac{1}{2}\bar{T}_s\right) \tag{11.26}$$

$$\sigma_{\theta\theta}(r) = \frac{\alpha E}{1-\nu}\left[\tfrac{1}{2}\bar{T}_c + \tfrac{1}{2}\bar{T}_s - T(r)\right] \tag{11.27}$$

$$\sigma_{zz}(r) = \frac{\alpha E}{1-\nu}\left[\nu\bar{T}_c - T(r)\right] \tag{11.28}$$

The radial stress at a distance $r \leq b$ from the center depends on the difference between the mean temperature of the whole cylinder of radius b and the mean temperature of the sub-cylinder. The circumferential stress at a distance $r \leq b$ depends on the sum of these two quantities less the temperature at that distance.

To evaluate the relative magnitudes of the thermal stresses consider the following differences:

$$\sigma_{\theta\theta}(r) - \sigma_{rr}(r) = \frac{\alpha E}{1-\nu}[\bar{T}_s - T(r)] \quad (11.29)$$

$$\sigma_{\theta\theta}(r) - \sigma_{zz}(r) = \frac{\alpha E}{1-\nu}\left[\left(\frac{1}{2}-\nu\right)\bar{T}_c + \frac{1}{2}\bar{T}_s\right] \quad (11.30)$$

These equations have direct implications for predicting fracture patterns in cylindrical stocks. Because the average temperature of a cooling sub-cylinder of the stock is greater than the temperature at its surface, (11.29) shows that $\sigma_{\theta\theta}(r) > \sigma_{rr}(r)$. Because $\nu \leq ½$, (11.30) shows that $\sigma_{\theta\theta}(r) > \sigma_{zz}(r)$. Thus, the circumferential stress is greater than the radial stress; it is greater than the axial stress, and it is positive (tensile). If the ambient state due to tectonic and gravitational stresses were an isotropic compression characterized by Anderson's standard state we would expect the introduction of thermal stresses, (11.26) to (11.28), to drive the circumferential stress to the tensile strength of the rock and induce radial fractures, akin to those at Diamond Joe stock (**Figure 11.31**). The fracture pattern at the Diamond Joe stock is not perfectly radial, however, and this may be due to the fact that the stock is not perfectly cylindrical, or to the addition of tectonic stresses.

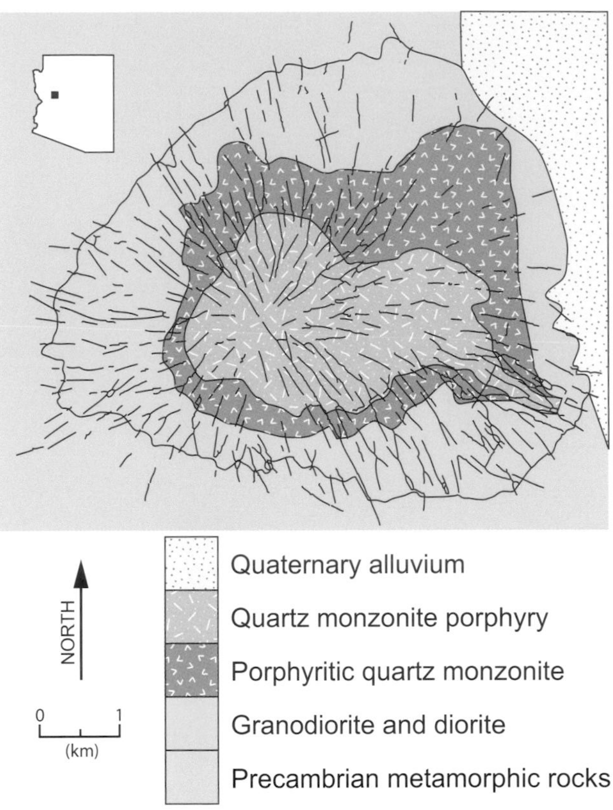

Figure 11.31 Geological and structural map of the Diamond Joe stock, western Arizona. Lithologic symbols include: Lqmp = quartz monzonite porphyry; Lpqm = Porphyritic quartz monzonite; Lgd = granodiorite. Structure map showing the strike of the dominant and/or most altered fracture set at each sample locality. Mountain peak UTM: 12 S 247896.70 m E, 3858397.58 m N. Modified from Gerla (1988), Figure 2.

11.5 SALT DIAPIRS

Intrusions of salt may be concordant or discordant with respect to the surrounding sedimentary strata, and some even extrude onto Earth's surface to form salt glaciers. Examples of salt glaciers from the Zagros Mountains are shown in **Figure 6.7**. As illustrated in **Figure 11.33** salt intrusions initiate as low amplitude anticlines or pillows on the top of the source salt layer. These are concordant intrusions with the overlying strata, but they may evolve into diapirs that intrude and crosscut strata, thereby becoming discordant intrusions. Diapirs range in shape from elongate walls of salt to roughly spherical bulbs rising above a narrow stem: some may disconnect entirely from the stem, rising as detached diapirs. The mechanics of salt intrusion and the role of tectonic extension and shortening are reviewed by Hudec and Jackson (2007), listed in Further Reading.

11.5.1 Buoyancy and Rheology in Diapiric Intrusion

Diapiric salt intrusions owe their existence, at least in part, to the unusual physical properties of salt. Salt flows readily at much lower temperatures and differential stresses than other sedimentary rocks, and it maintains a relatively low density during burial, while other sedimentary rocks increase in density. If the overlying sedimentary rocks attain a greater density, salt becomes buoyant, and

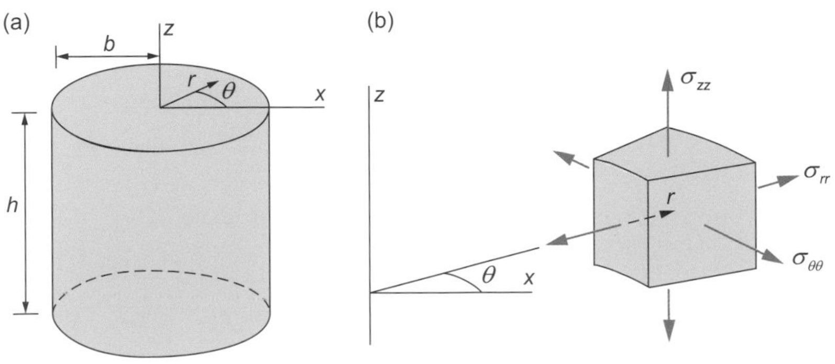

Figure 11.32 (a) Cylindrical geometry used for the thermal stress problem. (b) Normal stress components in cylindrical coordinates.

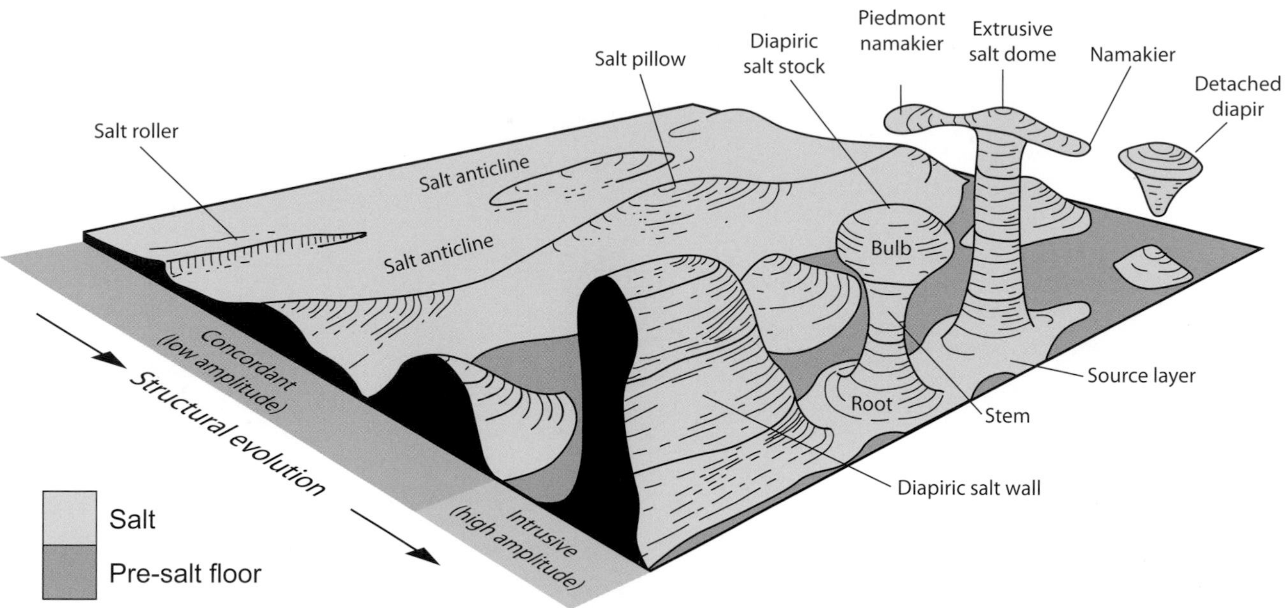

Figure 11.33 Summary of salt structures. Structure contours in arbitrary units. Diameters of salt diapirs typically range from 5 km to several tens of kilometers. Modified from Jackson and Talbot (1986), Figure 1.

this provides the driving force for salt diapirs to rise toward Earth's surface. Recall from Section 1.1.2 that the gravitational body force, $\mathbf{F}_{\text{grav}}$, acting on a detached salt diapir (**Figure 11.33**) is:

$$\mathbf{F}_{\text{grav}} = m\mathbf{g} \quad (11.31)$$

Here m is the total mass of salt in the diapir and the gravitational force vector is directed downward, in the same direction as the gravitational acceleration, $\mathbf{g}$. The buoyant body force is equal to the weight of the surrounding rock displaced by the salt, such that:

$$\mathbf{F}_{\text{buoy}} = -m_{\text{d}}\mathbf{g} \quad (11.32)$$

Here m_{d} is the displaced mass, and the buoyant force vector is directed upward: it buoys up the less dense body.

Taking the z-coordinate as positive upward, the gravitational acceleration is $\mathbf{g} = -g^*\hat{\mathbf{k}}$, and a representative value for Earth's crust is $g^* = 9.8\ \text{m s}^{-2}$. Considering a spherical diapir, and using the average density, ρ_{i}, of salt inside the diapir, the mass is $m = 4\pi a^3\rho_{\text{i}}/3$. Using the average density, ρ_{o}, of the rock outside the diapir, the displaced mass is $m_{\text{d}} = 4\pi a^3\rho_{\text{o}}/3$. Adding the gravitational (11.31) and buoyant (11.32) forces, and substituting for the respective masses and gravitational acceleration, the net body force acting on the spherical salt diapir is:

$$\mathbf{F}_{\text{buoy}} + \mathbf{F}_{\text{grav}} = (m - m_{\text{d}})\mathbf{g} = \tfrac{4}{3}\pi a^3(\rho_{\text{o}} - \rho_{\text{i}})g^*\hat{\mathbf{k}} \quad (11.33)$$

If the inside (salt) density is less than the outside (host rock) density, the body force is positive, the diapir is buoyant, and it would tend to rise.

The mass density of salt at Earth's surface is about 2,200 kg m^{-3} (**Figure 11.34**). The volumetric expansion of salt with increased temperature slightly exceeds the volumetric contraction

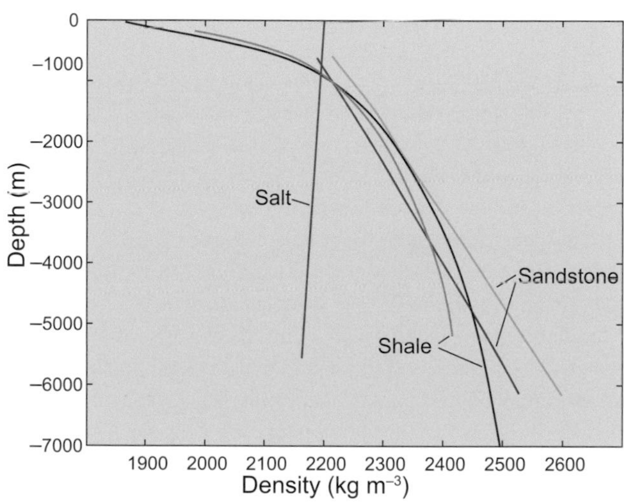

Figure 11.34 Plot of mass density versus depth for salt, brine-filled sandstone, and brine-filled shale compiled by Jackson and Talbot (1986), Figure 6.

with increased pressure, so the mass density decreases slightly with burial depth. On the other hand, clastic sediments typical of rapidly filling basins, such as the United States Gulf Coast, undergo significant density increases with burial. Mud and sand saturated with brine are less dense than salt at very shallow depths, but between 500 m and 1000 m depth the processes of compaction, dehydration, and cementation increase the density of these clastic sediments to that of salt as they transform into shale and sandstone. At this depth the salt is neutrally buoyant with respect to the surrounding sediment. At depths of about 5 km the range of estimated densities for shale and sandstone is between

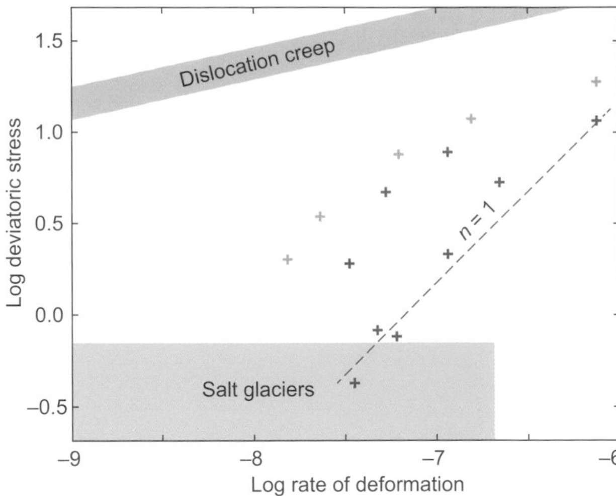

Figure 11.35 Log stress plotted versus log strain rate for laboratory data on wet synthetic salt. Experimental conditions (temperature, grain size, confining pressure, pore fluid pressure): yellow, $T = 20\,°C$, $d = 0.2\,mm$, $P = 21\,MPa$, $P_f = 3\,MPa$; red, $T = 70\,°C$, $d = 0.2\,mm$, $P = 21\,MPa$, $P_f = 3\,MPa$; blue, $T = 20\,°C$, $d = 0.08\,mm$, $P = 5.1\,MPa$, $P_f = 0.1\,MPa$. Colored areas indicate expected behavior for salt deforming by dislocation creep and conditions for salt glaciers. Modified from Urai et al. (1986), Figure 4.

2,400 and 2,550 $kg\,m^{-3}$. Salt at this depth would be buoyant and in the following example we take a difference between the sediment and salt density representative of this depth: $\rho_o - \rho_i = 250\ kg\,m^{-3}$. As this difference increases the buoyant force increases and the salt would tend to rise faster.

The density inversion depicted in **Figure 11.34** would be insufficient to create diapirs if salt could not flow readily, or if the strength of the overlying sedimentary rock were too great. Mechanisms for salt deformation have been suggested by observations of salt glaciers (**Figure 6.7**), which exhibit rapid accelerations after rainfall. Laboratory data from experiments that include brine in the salt pore spaces are consistent with these observations. Wet samples exhibit a marked weakening at low strain rates, compared to dry samples that deform by dislocation creep (**Figure 11.35**). For wet samples the rate of deformation at a given deviatoric stress increases as temperature increases (yellow to red symbols). Also, the rate of deformation increases as grain size decreases (yellow to blue symbols). The slopes of lines fitting the experimental data are approximately consistent with a linear viscous constitutive law in which the rate of deformation is linearly related to the deviatoric stress, such that $n \approx 1$ in the power-law relationship. This is the behavior of a linear viscous fluid, which is used for the canonical model of a salt diapir described in Section 11.6.

11.5.2 Styles of Flow Within Diapirs

Abundant geological and geophysical data, gathered to provide engineering solutions for the mining of salt, the production of hydrocarbons in traps associated with salt, and the storage of

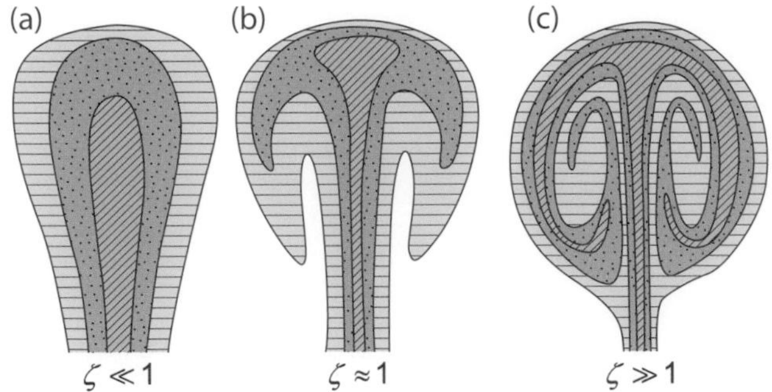

Figure 11.36 Schematic illustration of the geometry of three passive layers (gray, pink, and orange) caught up in the flow within a rising diapir. Here $\zeta = \eta_o/\eta_i$ is the ratio of viscosity of material *outside* the diapir to the viscosity of material *inside*. Modified from Talbot and Jackson (1987), Figure 14.

radioactive waste provide considerable evidence for the internal structure of salt diapirs. A noteworthy example is the Hänigsen salt diapir in northwestern Germany, which was introduced in Section 2.5.5. While considerably more complex, the Hänigsen salt diapir displays structures that are reminiscent of structures observed in laboratory experiments and numerical experiments to investigate the buoyant rise of one viscous material in another (**Figure 11.36**).

In **Figure 11.36** the layers within the model diapir identified with different colors have the same viscosity and the same density, so they are passive markers of the deformation. The ratio of the viscosity of the surrounding material to that of the intruding material is $\zeta = \eta_o/\eta_i$. For $\zeta << 1$ the model diapir rises with a more-or-less cylindrical shape, similar to a stock (Section 11.4). In this case the passive markers maintain their original stratigraphic relationships: younger over older at the top of the diapir, and older in the center with younger at the margin along the sides of the diapir. For $\zeta \approx 1$ the model diapir has a mushroom-shaped top and the younger passive markers overturn and form downward-facing pendants. For $\zeta >> 1$ the model diapir rises like a spherical bulb and circulation of the material within the bulb results in systematic, but very complex structural and stratigraphic relationships. In the following section we describe a model, based on fluid dynamics, with a velocity field that has a similar circulation pattern.

11.6 A CANONICAL MODEL FOR A RISING DIAPIR

Some salt diapirs and some igneous plutons are conceptualized (**Figure 11.33**) as discreet masses rising through the host rock. As pointed out in Section 11.5.1 these intrusions are subject to the upward force of buoyancy and the downward force of gravity. If the fluid within the diapir or pluton is less dense than the surrounding fluid, the buoyant force exceeds the gravitational force, thus impelling the body upward. Review papers listed in Further Reading on the emplacement of magmatic plutons include those by Glazner et al. (2004) and Menand et al. (2011).

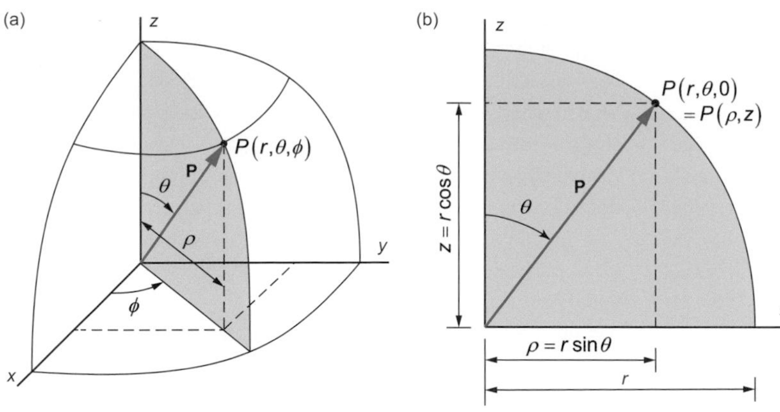

Figure 11.37 Spherical coordinates and meridian plane (taupe). (a) First octant with position vector **P** locating point *P* on a sphere of radius *r*. Coordinates are length, *r*, and angles, θ and ϕ. (b) Part of the meridian plane (ϕ = 0) with different position vector **P**. Length, ρ, and angle, θ, are used to define point *P* on a circle of radius *r* that is the trace of the sphere on the meridian plane. Modified from Happel and Brenner (1965).

Here we describe a canonical model for such a process that relates the density and viscosity contrasts between salt and host sediment in a sedimentary basin, or between magma and host rock, to the size of a spherical model diapir or pluton and its velocity relative to the host. We also look for insights about the deformation and velocity fields within the diapir or pluton and the surrounding host rock. Such relationships and insights are important for understanding how large bodies of salt or magma flow upward through Earth's lithosphere.

11.6.1 The Stokes Problem

The first solution to investigate the canonical model considered here was derived by G. G. Stokes and published in 1851. His famous solution from fluid dynamics, now called the Stokes Problem, considers the motion of a solid sphere in a linear viscous fluid. However, this ignores the possible deformation within the sphere and we know that the salt or magma in a diapir does deform. The Stokes problem was generalized to a sphere of viscous fluid surrounded by a fluid of a different viscosity by Hadamard and by Rybczynski in 1911. That solution is presented here as the canonical model for salt domes and plutons following the treatment by Happel and Brenner (1965), listed in Further Reading).

Before proceeding to this problem in fluid dynamics, we introduce spherical coordinates and the Stokes stream function, both of which play key roles in the solution method. Spherical coordinates are employed because both the governing equations and the boundary conditions are written most effectively for coordinates that conform to the geometry of the spherical interface between the two fluids. Three coordinates are used to locate any point in the spherical system, and these coordinates are the components of the position vector for that point (**Figure 11.37a**):

$$\mathbf{P} = \langle r, \theta, \phi \rangle \quad 0 \le r < \infty, 0 \le \theta \le \pi, 0 \le \phi < 2\pi \qquad (11.34)$$

The specified ranges restrict the values of the coordinates so a particular point is not represented by more than one set of coordinates. For example, the point $P(r, \theta, \pi)$ would be the same as the point $P(r, \theta, 3\pi)$, but this is precluded from consideration by the range restriction in (11.34). The z-axis is a special case along which the coordinate ϕ is not defined, but with $\theta = 0$ or $\theta = \pi$, points along this axis are given by $z = r$ or $z = -r$, respectively.

Many problems in spherical coordinates can be simplified from three dimensions to two dimensions because of symmetry. In doing so, one focuses on a so-called meridian plane, which is a half-plane with the straight edge defined by the z-axis. For example, in **Figure 11.37b** we illustrate part of the meridian plane $\phi = 0$ in the first quadrant of the (x, z)-plane. On that, or any other meridian plane the position vector is defined with two components:

$$\mathbf{P} = \langle \rho, z \rangle \text{ where } \rho = r \sin\theta,\ z = r\cos\theta \qquad (11.35)$$

Note that ρ is the length of the projection of **P** onto the (x, y)-plane. Given (11.35) and the trigonometric relations illustrated in **Figure 11.37a**, the three Cartesian coordinates are found from the three spherical coordinates of (11.34) as:

$$x = r\sin\theta\cos\phi,\ y = r\sin\theta\sin\phi,\ z = r\cos\theta \qquad (11.36)$$

Given the Cartesian coordinates, and using $\rho = (x^2 + y^2)^{1/2}$, the spherical coordinates are:

$$r = (\rho^2 + z^2)^{1/2},\ \theta = \tan^{-1}(\rho/z),\ \phi = \tan^{-1}(y/x) \qquad (11.37)$$

Spheres and some cones can be described simply using spherical coordinates. Surfaces defined by r = constant are spheres with a common center at the origin: traces of these spheres appear as quarter-circles on any meridian plane (**Figure 11.37b**). The surfaces defined by θ = constant are right-circular cones with a common apex at the origin, and traces of these cones appear as radial lines on any meridian plane. Each sphere and cone is a body of revolution because their traces on every meridian plane are identical. The respective sphere and cone can be generated by taking the circle and radial line shown in **Figure 11.37b** and revolving them about the z-axis. For the problem of the viscous sphere in a viscous surrounding we choose a particular sphere, $r = R$, as the boundary between the two fluids.

The problem Stokes solved was that of a solid sphere moving slowly relative to a viscous incompressible fluid. Following the concepts developed in Section 6.8.2, slowly implies that this is laminar flow, not turbulent flow. The motion can be thought about in two ways: (1) the fluid is motionless far from the sphere, while the sphere rises or settles; or (2) the sphere is motionless, while the surrounding fluid streams past it. For either case Stokes recognized that the symmetry of the sphere led to a significant simplification of the velocity field. In particular, this three-dimensional problem may be reduced to a two-dimensional problem, because the velocity field is identical on every meridian plane. This reduction to a two-dimensional problem may be invoked for all bodies of revolution if the velocity field meets two symmetry conditions (see Happel and Brenner, 1965, in Further Reading):

$$\frac{\partial \mathbf{v}}{\partial \phi} = 0, \text{and } v_\phi = 0 \qquad (11.38)$$

The first condition specifies that the velocity vector must be independent of the angle ϕ. The second condition specifies that the velocity component in the ϕ-direction must be zero. Here ϕ is the coordinate angle measured in the (x, y)-plane that defines the orientation of the meridian plane relative to the x-axis. The z-axis is the axis of revolution, and each meridian half-plane is defined by ϕ = constant (**Figure 11.37a**).

Velocity fields that meet the requirements in (11.38) are called axisymmetric velocity fields, and it is for these special three-dimensional velocity fields that Stokes was able to define a function, called a stream function, that facilitates the solution for the velocity field. To understand the Stokes' stream function consider the meridian plane $\phi = 0$ (**Figure 11.38**) and draw a curve, C_1, from any point on the z-axis with position vector $\mathbf{P}(r_1, 0)$, to an arbitrary point with position vector $\mathbf{P}(r, \theta)$. Next draw a curve, C_2, from a different point on the z-axis, $\mathbf{P}(r_2, 0)$, to that same point. Both curves lie in the meridian plane and intersect the z-axis such that $r_2 > r_1$, but the curves do not cross one another. Now rotate these two curves about the z-axis through an angle $\phi = 2\pi$ to form a body of revolution with lower surface and upper surface formed by the respective rotated curves. These are imaginary surfaces, constructed so we can consider the rate of fluid flowing through them for an axisymmetric velocity field.

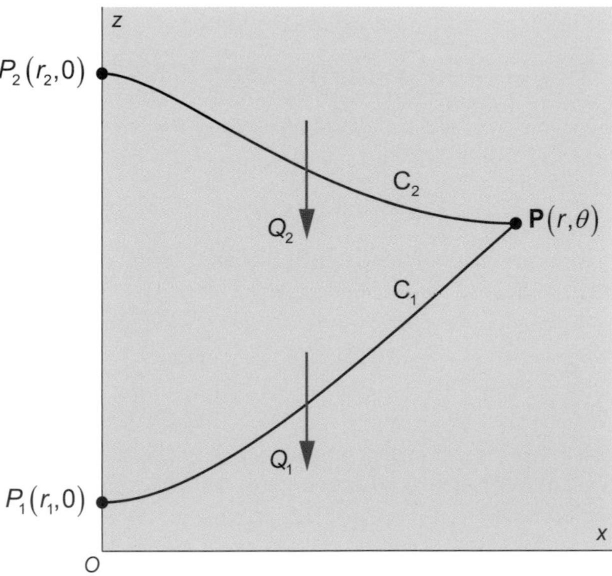

Figure 11.38 Meridinal plane $\phi = 0$ with arbitrary curves C_1 and C_2 from two points on the z-axis to a common point off that axis. Rotation of these curves about the z-axis form two surfaces of an imaginary body of revolution. Q_1 and Q_2 are volumetric flow rates through the lower and upper surfaces at a given time.

Suppose the volumetric rate of flow downward through the lower surface at some instant in time is Q_1, and that through the upper surface in that same instant is Q_2. Because the fluid is incompressible we have, from the continuity equation described in Section 6.7.3:

$$\frac{\partial v_x}{\partial x} + \frac{\partial v_y}{\partial y} + \frac{\partial v_z}{\partial z} = 0 \qquad (11.39)$$

Continuity for an incompressible fluid requires that the density in an infinitesimal element at any point does not change with time. The fluid in this element can deform, but stretching in one coordinate direction is exactly balanced by the necessary shortening in the other coordinate directions, so the density is unchanged. Here we extend that concept to the body of revolution: fluid leaving through the lower surface must be exactly compensated for by fluid entering through the upper surface. In terms of the instantaneous volumetric flow rates, $Q_1 = Q_2$.

If the location of $\mathbf{P}(r, \theta)$ is changed, the value of the flow rate across the newly defined lower and upper surfaces of the resulting body of revolution may change, but they must remain equal, so the unique flow rate, $Q(\mathbf{P}, t)$, is associated with each position vector, and therefore with each point in the meridian plane. The Stokes' stream function, $\psi(\mathbf{P}, t)$, is defined using this *scalar* quantity, which is a function of position in a meridian plane and time (see Happel and Brenner, 1965, in Further Reading):

$$\psi(\mathbf{P}, t) = \frac{Q(\mathbf{P}, t)}{2\pi} \qquad (11.40)$$

If the location of $\mathbf{P}(r, \theta)$ is on the z-axis, that is $\mathbf{P}(r, 0)$, the net flow through the body of revolution is zero and the Stokes' stream function is zero, $\psi(r, 0, t) = 0$.

Although Stokes' stream function is applicable to problems of axisymmetric flow in various coordinate systems of revolution, we focus here on spherical coordinates, where $\psi = \psi(r, \theta, t)$, and the two velocity components in any meridian plane are related to the stream function as (see Happel and Brenner, 1965, in Further Reading):

$$v_r = -\frac{1}{r^2 \sin\theta} \frac{\partial \psi}{\partial \theta}, \; v_\theta = -\frac{1}{r \sin\theta} \frac{\partial \psi}{\partial r} \qquad (11.41)$$

Recall from the symmetry conditions (11.38) that the third component of velocity, v_ϕ, is identically zero. Once the stream function is found, the velocity components are determined by differentiation.

11.6.2 Buoyant Viscous Sphere in Viscous Surroundings

Suppose a spherical body of fluid of radius a with uniform and constant density, ρ_i, and viscosity, η_i, resides in an infinite body of fluid with uniform and constant density, ρ_o, and viscosity η_o.

The subscripts i and o refer to *inside* and *outside* the spherical interface between the two fluids. Because the ratio of viscosities appears in the equations for the stream function and the velocity components, we define this viscosity ratio as:

$$\zeta = \eta_o / \eta_i \tag{11.42}$$

Laboratory and numerical modeling (e.g. **Figure 11.36**) indicate that the shape of salt diapirs and the structures within the diapir are dependent upon the ratio of viscosities.

We now consider the boundary conditions for the problem of the viscous sphere. Suppose the sphere rises in the positive z coordinate direction with a steady velocity V_s because $\rho_i < \rho_o$. The velocity field does not change with time, and for the arbitrary instant at which the field is visualized, let the center of the sphere be at the origin of the spherical coordinates. As mentioned above, the flow is considered laminar both inside and outside the sphere and the two fluids are incompressible. This problem is really two boundary-value problems: one for flow inside the sphere, and the other for flow outside, each with boundary conditions written in terms of the velocity components. For the fluid inside the sphere the boundary conditions are those specified on the spherical interface. For the surrounding fluid the boundary conditions must be specified on this interface and at an infinite distance. At the interface the simplest condition is that the two fluids stay in *mutual contact* without interpenetration, opening, or sliding. Preventing interpenetration of the two fluids seems an obvious constraint, but if one of the fluids were permeable with respect to the other, this would have to be questioned. Preventing the opening of a gap between the fluids would depend upon sufficient compressive stress at depth, but circumstances might prevail, such as during the formation of veins, where gaps open. Sliding across the interface between a salt diapir and the surrounding sediment would have to be considered, if field evidence such as slickenlines at the contact supports such a phenomenon.

To prevent interpenetration or opening at the interface, the normal components of velocity for neighboring particles on opposite sides of the interface must be equal. Because the interface is spherical the normal component is v_r, so we write the no interpenetration boundary condition as:

$$v_r(r=a^-, \theta) = v_r(r=a^+, \theta),\ 0 \le \theta < 2\pi \tag{11.43}$$

Here a^- and a^+ are infinitesimally less than and greater than a, respectively, so the velocity components in (11.43) are for neighboring particles just inside and just outside the interface. To prevent sliding at the interface, the tangential components of velocity for adjacent particles across the interface must be equal. For the spherical interface the tangential component is v_θ, so we write the no slip boundary condition as:

$$v_\theta(r=a^-, \theta) = v_\theta(r=a^+, \theta),\ 0 \le \theta < 2\pi \tag{11.44}$$

We specify a no flow boundary condition at an infinite distance as:

$$\text{as } r \to \infty,\ v_r \to 0, \text{ and } v_\theta \to 0,\ 0 \le \theta < 2\pi \tag{11.45}$$

The rising sphere locally perturbs the surrounding fluid into motion, but at a great distance relative to the sphere radius the fluid is at rest. Because of this remote boundary condition, the velocity field is the one observed by a *stationary viewer* a great distance from the sphere relative to its radius.

The Stokes' stream function for the fluid *inside* the sphere is:

$$\psi_i = \tfrac{1}{2} V_s r^2 \sin^2\theta \left[\tfrac{1}{2}\left(\frac{\zeta}{1+\zeta}\right)\left(\frac{r}{a}\right)^2 - \tfrac{1}{2}\left(\frac{\zeta}{1+\zeta}\right)\frac{r}{a} - 1 \right],\ r \le a \tag{11.46}$$

Employing (11.41), the radial and tangential velocity components *inside* the sphere are:

$$u_r = V_s \cos\theta \left[-\tfrac{1}{2}\left(\frac{\zeta}{1+\zeta}\right)\left(\frac{r}{a}\right)^2 + \tfrac{1}{2}\left(\frac{\zeta}{1+\zeta}\right) + 1 \right] \tag{11.47}$$

$$u_\theta = -V_s \sin\theta \left[-\left(\frac{\zeta}{1+\zeta}\right)\left(\frac{r}{a}\right)^2 + \tfrac{1}{2}\left(\frac{\zeta}{1+\zeta}\right) + 1 \right] \tag{11.48}$$

These velocity components are relative to the velocity at infinity, which is prescribed to be zero by the boundary condition (11.45). For the solid sphere, $\eta_i = \infty$, so $\zeta = 0$, and the components reduce to $u_r = V_s \cos\theta$, $u_\theta = -V_s \sin\theta$, which is a uniform velocity field in the positive z-direction. However, if the sphere is not solid, the velocity field within the sphere is non-uniform and this is illustrated in **Figure 11.39** for a viscosity ratio $\zeta = 10$ and unit rise velocity, $V_s = 1$. The velocity is greatest along the center line (z-axis), and it is symmetric about this axis. The velocity vectors are vertical along the center line, and along the horizontal line $z = 0$.

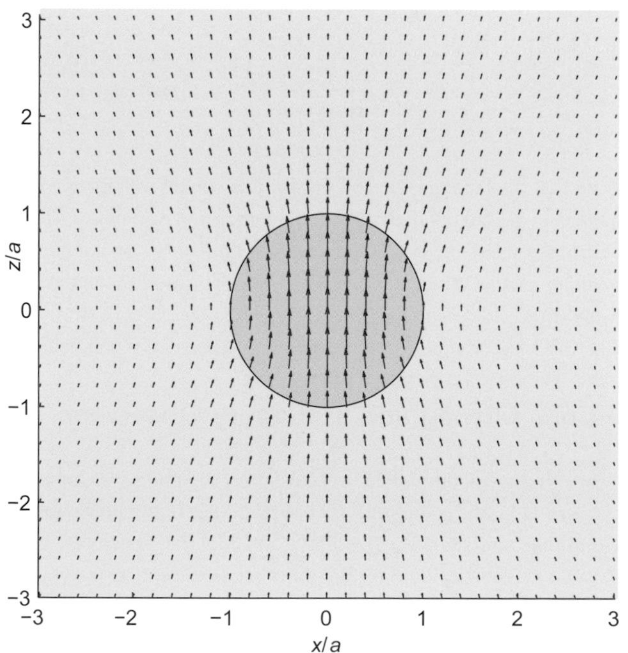

Figure 11.39 Velocity vector field inside and outside viscous sphere rising in viscous surroundings. Velocity is relative to stationary observer far from the sphere.

They are directed away from the center line in the upper half of the sphere, and toward the center line in the lower half.

The Stokes' stream function for the fluid *outside* the sphere is:

$$\psi_{\text{o}} = \frac{1}{2} V_{\text{s}} r^2 \sin^2\theta \left[\frac{1}{2} \left(\frac{1}{1+\zeta} \right) \left(\frac{a}{r} \right)^3 - \frac{3}{2} \left(\frac{1+\frac{2}{3}\zeta}{1+\zeta} \right) \frac{a}{r} \right] \tag{11.49}$$

Employing (11.41), the radial and tangential velocity components *outside* the sphere are:

$$u_r = V_{\text{s}} \cos\theta \left[-\frac{1}{2} \left(\frac{1}{1+\zeta} \right) \left(\frac{a}{r} \right)^3 + \frac{3}{2} \left(\frac{1+\frac{2}{3}\zeta}{1+\zeta} \right) \frac{a}{r} \right] \tag{11.50}$$

$$u_\theta = -V_{\text{s}} \sin\theta \left[\frac{1}{4} \left(\frac{1}{1+\zeta} \right) \left(\frac{a}{r} \right)^3 + \frac{3}{4} \left(\frac{1+\frac{2}{3}\zeta}{1+\zeta} \right) \frac{a}{r} \right] \tag{11.51}$$

Both velocity components go to zero as $r \to \infty$. This part of the field also is illustrated in **Figure 11.39** for a viscosity ratio $\zeta = 10$ and unit rise velocity, $V_{\text{s}} = 1$. This part of the field corresponds in many ways to that inside the sphere. For example, the velocity vectors are vertical along the center line, and along the horizontal line $z = 0$. They are directed away from the sphere above the line $z = 0$, and toward the sphere below this line.

To confirm the boundary conditions at the interface between the two fluids, (11.43) and (11.44), we reduce the velocity components to their respective values just inside and just outside the interface:

$$v_r(a^-, \theta) = V_{\text{s}} \cos\theta = v_r(a^+, \theta)$$

$$v_\theta(a^-, \theta) = -V_{\text{s}} \sin\theta \left(\frac{1+\frac{1}{2}\zeta}{1+\zeta} \right) = v_\theta(a^+, \theta) \tag{11.52}$$

Note that the fields inside and outside the sphere in **Figure 11.39** appear to be continuous with one another and smoothly varying across the interface, consistent with the boundary conditions. The velocity of the sphere relative to a distant stationary observer is found from (11.52) at the top of the sphere where $\theta = 0$, so $v_r = V_{\text{s}}$, $v_\theta = 0$. The vector plotted at that position represents a unit velocity.

11.6.3 Moving and Stationary Observers

While the velocity field relative to a stationary observer (**Figure 11.39**) is helpful to confirm the interface boundary conditions and to visualize flow outside the sphere, a more instructive velocity field for the fluid inside the sphere is obtained by letting the observer move with the same rise velocity as the sphere itself. We have already noted that for the solid sphere the inside field is uniform, so this observer would perceive that every particle of the solid is stationary. However, if the sphere is a viscous fluid, the inside velocity field is non uniform and the perceived field is quite interesting.

To calculate the velocity field relative to the moving observer, we subtract a uniform velocity field, V_{s}, in the positive z-direction from the velocity components in (11.47) and (11.48). The uniform field is obtained from the Stokes' stream function:

$$\psi_{\text{u}} = -\frac{1}{2} V_{\text{s}} r^2 \sin^2\theta \tag{11.53}$$

The corresponding velocity components are:

$$u_r = V_{\text{s}} \cos\theta, \; u_\theta = -V_{\text{s}} \sin\theta \tag{11.54}$$

The field inside the sphere is illustrated in **Figure 11.40**. Recall that this velocity field was illustrated in Section 1.1.2 to motivate thinking about the rise of diapirs. This field has two circulation loops that are symmetric about the z-axis, with fluid flowing upward along the z-axis and downward along the perimeter of the sphere. Note that this velocity field does not appear to be in continuity with the outside field illustrated in **Figure 11.39**, but that is because we have subtracted the rise velocity. This demonstrates how difficult it can be to interpret velocity fields unless the reference point is known and the effects of changing the position of that point are understood.

The circulation pattern seen in the velocity field shown in **Figure 11.40** provides an explanation for the pattern of passive layers that develops in model diapir experiments where the viscosity of the material outside the sphere is greater than that inside (**Figure 11.36c**). The particles rise up along the center line of the model diapir, turn to either side, and descend along the perimeter of the diapir, only to rise again along the center line. This velocity field also provides an explanation for the pattern of salt layers mapped in the Hänigsen salt diapir in northwestern Germany,

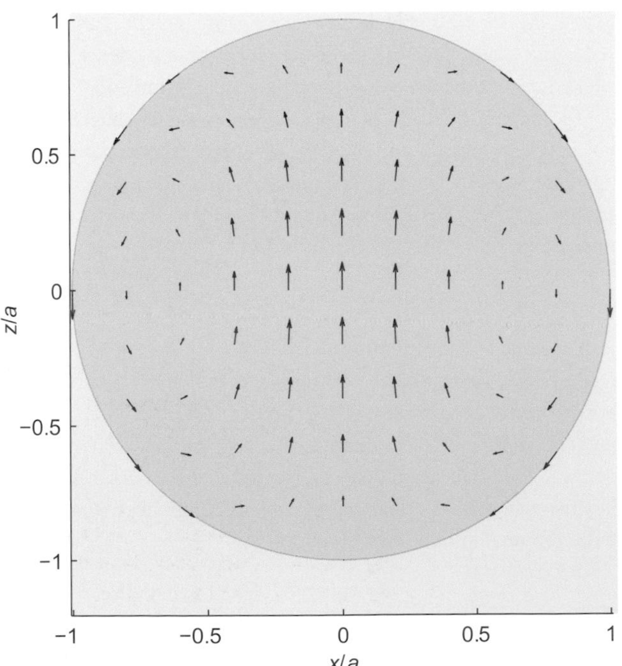

Figure 11.40 Velocity vector field inside the viscous sphere rising in a viscous surroundings. Velocity is relative to observer traveling at the rise velocity of the sphere, V_{s}.

which was introduced in Section 2.5.5. There, the overturning salt layers only developed on one side of the diapir, but the flow on that side clearly followed the pattern of rising, turning to the side, and descending along the perimeter.

11.6.4 Rise Velocity of the Viscous Sphere

Thus far we have not specified the magnitude of the rise velocity of the viscous sphere, but merely referred to it as V_s. The rise velocity is found by considering the forces acting on the sphere. The net body force due to buoyancy and gravity was given in (11.33) and this force acts parallel to the vertical z-axis. If the fluid inside the sphere is less dense than the surrounding fluid the net force is directed upward and the sphere rises. However that motion is resisted by the viscous drag of the sphere against the surrounding fluid. Because this is an axisymmetric flow, the drag also acts parallel to the axis of revolution, but downward in this case. The drag force exerted by the surrounding fluid on the sphere is found by integrating the traction vector over the surface of the sphere:

$$\mathbf{F}_{\text{drag}} = -6\pi\eta_o a V_s \left(\frac{1 + \frac{2}{3}\zeta}{1 + \zeta} \right) \hat{\mathbf{k}} \tag{11.55}$$

For equilibrium of forces the magnitudes of the downward drag force (11.55) and the upward net body force (11.33) must be equal:

$$6\pi\eta_o a V_s \left(\frac{1 + \frac{2}{3}\zeta}{1 + \zeta} \right) = \frac{4}{3}\pi a^3 (\rho_o - \rho_i) g^*$$

Solving for the rise velocity we have:

$$V_s = \frac{2a^2(\rho_o - \rho_i) g^* (1 + \zeta)}{9\eta_o \left(1 + \frac{2}{3}\zeta\right)} \tag{11.56}$$

This provides the scaling relations for the rise velocity, which is proportional to the square of the sphere radius, to the density contrast, and to the gravitational acceleration. It is inversely proportional to the viscosity of the outer fluid.

The viscosity ratio, ζ, of the viscosity outside the sphere to that inside the sphere enters the scaling relation (11.56) in a somewhat more complicated form. If the viscosity inside is much greater than that outside, $\zeta \to 0$, and the rise velocity derived by Stokes in 1851 for the solid sphere is obtained:

$$V_s = \frac{2a^2(\rho_o - \rho_i) g^*}{9\eta_o}, \zeta = 0 \tag{11.57}$$

Here the rise velocity is simply inversely proportional to the outside viscosity. If the viscosity inside is much less than that outside, $\zeta \to \infty$, and in the limit the rise velocity is:

$$V_s = \frac{a^2(\rho_o - \rho_i) g^*}{3\eta_o}, \zeta = \infty \tag{11.58}$$

This would be relevant to the rise of gas bubbles in a viscous magma, for example. The rise velocity still is inversely proportional to the viscosity outside the sphere. It is noteworthy that varying the viscosity ratio from zero to infinity modifies the rise velocity only by the factor 3/2. On the other hand, doubling the radius for either the solid sphere, or the gaseous sphere, increases the velocity by the factor 4. We understand from this result that the rise velocity is most sensitive to the size of the sphere.

Recapitulation

We began this chapter by describing an intrusion as a body of rock that, in a former more mobile state, was injected into and deformed the surrounding host rock. Examples were mentioned involving intrusion of magma, salt, a sedimentary slurry, and melt generated by frictional heating on a fault, illustrating the rich variety of intrusions in nature. We also pointed out that brittle, ductile, and viscous deformation can play roles in the formation of an intrusion. All of the deformation mechanisms we highlighted in Chapters 4 through 6, and all of the basic geologic structures considered in Chapters 7 through 10 come into play.

Dikes are magma-filled *opening fractures*, so they have very small thickness to length ratios and they share many geometric features with other opening fractures like veins and joints, including the tendency to break down into echelon segments. Dikes are an effective way for magma to open a conduit to Earth's surface because of the tensile stress concentration at the dike tip. However, they are not an effective conduit for magma flow, because their large surface area to volume ratio promotes rapid heat loss and solidification of the magma. An analysis of the cooling of a dike by conductive heat flow into cold host rock shows that a two meter thick dike would cool by 50%, well below the solidus temperature of the magma, in about 12 hours. However, the evolution of cylindrical volcanic necks from dikes by *brecciation* and *stoping* of the host rock ultimately creates a pipe-shaped conduit with much smaller surface area to volume ratio, and this promotes sustained magma flow to Earth's surface.

While dikes might be only hundreds of meters to a few kilometers long in outcrop, some giant dikes in the Mackenzie *dike swarm* extend more than 2000 km. Fabric analysis shows that magma flow was sub-horizontal over much of this length. To understand how dikes can propagate over these great distances without breaking through to Earth's surface we investigate a model for *stable* dike propagation that depends upon the magma having a density that is greater than the shallow host rock, and lesser than the deeper host rock. The stability of the dike is quantified by calculating the stress intensity at both the upper and lower tips. These stress intensities both drop to zero at the stable height. Under these conditions the gravitationally induced loads trap the dike near the level of *neutral buoyancy*.

Both dikes and sills are opening fractures with thickness to length ratios consistent with the general name *sheet intrusion*, but sills are *concordant* (they interleave with sedimentary strata), while dikes are *discordant* (they cut across sedimentary strata). Sills also have a tendency to climb to higher stratigraphic levels as they propagate laterally in horizontal sedimentary rocks. This climb is attributed to the mechanical interaction of the sill with Earth's traction-free surface. We describe a modern sill intrusion that was captured by InSAR data at Sierra Negra Volcano in the Galápogos Islands. Over a seven-month period 6.7 million cubic meters of lava was injected at a depth of 1.9 km and filled a sill with a horizontal dimension of about 6 km. To understand some of the effects of the constitutive properties of magma on flow in sheet intrusions, we compare the flow of magma with a Newtonian viscosity to flow of magma with a power-law viscosity. The *parabolic* velocity profile for a Newtonian fluid gives way to *plug flow* in which the shearing is concentrated near the contacts of the magma with the host rock as the exponent for the deviatoric stress in the power law changes from 1 to 4.

The conceptual model for a *laccolith* was conceived in the Henry Mountains, UT, in 1875 by Gilbert. He based this model on his observations of different magmatic intrusions that he interpreted as different temporal stages in the intrusive process. Thus, magma flowed upward in a dike, turned and insinuated between two sedimentary layers, spread laterally as a thin sill until mechanical interaction with Earth's traction-free surface began, and then bent the overlying strata upward to form a much thicker laccolith. The laccolith is largely a *concordant* intrusion, although some faulting above the periphery and over the top can produce minor discordances. The *transition* length at which a thin sill begins to thicken into a laccolith is about twice the *effective thickness* of overburden. The lack of exposure of concordant lower contacts and feeder dikes for the central intrusions under the large domes in the Henry Mountains motivated the interpretation by Hunt in 1953 that each is an upright cylindrical *stock* with discordant lateral contacts. However, structural mapping of the domes shows that the overlying strata are largely concordant and arch into doubly hinged flexures that are concave upward at the lower hinge and concave downward at the upper hinge, similar to Gilbert's concept.

As a stock cools, the contraction of the igneous rock may lead to tensile stresses that cause fractures to develop. We describe a model for the thermal stresses induced by cooling of the Diamond Joe stock in west central Arizona and compare the direction of the most tensile principal stress to the orientation of veins mapped in this stock. The model stock is an upright circular cylinder and the circumferential stress is the greatest principal stress. If this exceeds the tensile strength of the rock we would expect radial opening fractures to form. The majority of fractures in the Diamond Joe stock are veins, filled with hydrothermal minerals, and oriented in a radial pattern that is consistent with the thermal stress analysis.

Salt intrusions initiate as concordant ridges on the top of the source salt layer and may grow into discordant diapirs that cut upward across the overlying strata, thereby becoming discordant intrusions. These structures owe their existence to the lower density of salt relative to that of clastic sedimentary rocks at a few kilometers depth, and to the fact that salt can flow at modest temperatures and differential stresses. Mud and sand saturated with brine are less dense than salt at very shallow depths, but at about 1 km depth compaction, dehydration, and cementation increase their density to that of salt as they transform into sedimentary rock. At greater depths salt is buoyant relative to the overlying strata and will tend to rise. Laboratory models demonstrate that if the viscosity of the surrounding material is less than that of the diapir, it rises with a more-or-less cylindrical shape, similar to a stock. However, if the viscosity of the surrounding material is greater than that of the model diapir, it rises like a spherical bulb and the material within the bulb circulates to form systematic, but very convoluted structures.

The canonical model for the salt diapir or magmatic pluton is the fluid dynamics of a viscous sphere in surroundings with different viscosity and greater mass density. A special case of this problem was solved by Stokes in 1851, and later was generalized to the problem reviewed here. The slow viscous flow in both the sphere and surroundings is treated as laminar and incompressible, and the axisymmetric geometry leads to a solution using Stokes' stream function.

Two views of the resulting velocity field are instructive. For the first, the observer is *stationary* with respect to the rising sphere, which would happen from an observation point far from the sphere relative to its radius. In this case the velocity field inside and outside the sphere appears to be continuous and smoothly varying across the interface. The surrounding fluid above the center of the sphere moves up and away from it (to get out of its way), and that below the sphere moves up and toward it (to fill in what otherwise would be a gap). For the second view of the velocity field the observer moves upward with the *rise velocity* of the sphere. In this case the velocity field within the sphere has two circulation loops that are symmetric about the vertical axis of the sphere. Fluid flows upward along the axis and downward along the perimeter of the sphere, similar to the pattern seen in laboratory experiments and in some natural salt diapirs.

REVIEW QUESTIONS

The following questions are designed to highlight the expected *learning outcomes* for this chapter. Each question is taken directly from the material in the chapter and, for the most part, in the same sequence that it appears in the chapter. If an answer is not forthcoming, students are advised to read the relevant section of the chapter and discover the answer.

11.1. An intrusion is a body of rock that, in a former more *mobile* state, was injected into and deformed the surrounding rock. Describe three different types of mobile materials, only one of which involves melting of rock.

11.2. Dikes and sills both are *sheet intrusions*, but how are they distinguished? Explain what is meant by "sheet"; how this form of intrusion is related to fractures; and how they are related to the least compressive principal stress.

11.3. Vertical dikes can be trapped near the *level of neutral buoyancy* (LNB) in Earth's lithosphere. Describe what is meant by the LNB and use (11.3) to explain what physical quantities the maximum dike height depends upon.

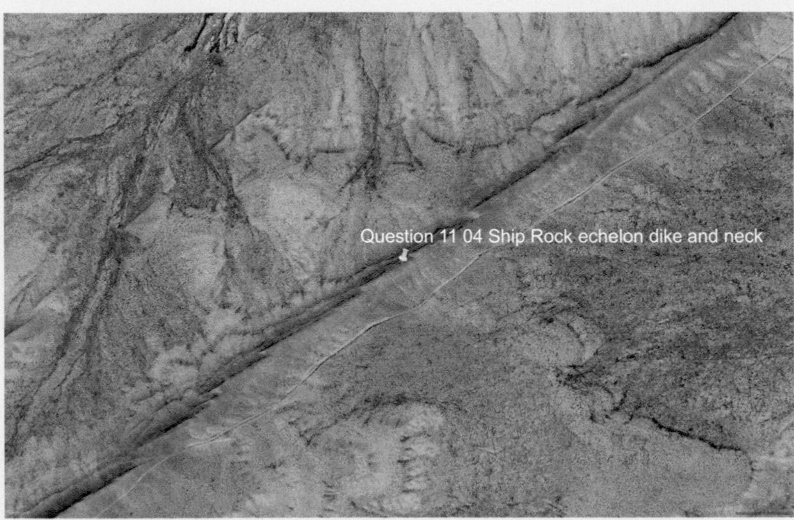

11.4. The following Google Earth image is taken from the file: Question 11 04 Ship Rock echelon dike and neck.kmz. Open the file; click on the Placemark for this question; copy and paste the questions into your answer document; and address the assigned questions. The Google Earth image data are from Landsat / Copernicus.

11.5. Heat *conduction* from a rapidly intruded dike is analyzed using the following relationship:

$$\rho c \frac{\partial T}{\partial t} = k \frac{\partial^2 T}{\partial x^2}$$

Name and give the SI units for each physical quantity in this relationship. Is this equation homogeneous with respect to units? What are the dependent and independent variables, and what assumptions are made about the material properties?

11.6. Describe what is plotted in **Figure 11.15**, including the quantities on the ordinate and on the abscissa. Describe the parameter used to determine the different colored curves. Explain how the plot is used to determine the temperature as a function of time at $x/a=2$.

11.7. Use **Figure 11.19** to explain the geometry of *climbing* sills illustrated in **Figure 11.18**.

11.8. Compare and contrast the *constitutive* equations in (11.14) and (11.18). Explain why the mean normal pressure plays no role in the second set of equations.

11.9. Use the *power-law* constitutive equations in (11.19) to explain the velocity profiles illustrated in **Figure 11.23** for magma flow in a sill.

11.10. The *transition* from growth of a sill to growth of a laccolith depends upon the resistance to deformation of the overlying strata. One way to address this resistance is to calculate the *effective thickness* of the overburden. Explain what the effective thickness is, and use this concept to discuss the stages in the development of the Black Mesa intrusion illustrated in **Figure 11.26**.

11.11. *Gravity* and *buoyancy* play pivotal roles in the growth of salt diapirs. The gravitational and buoyant forces acting on a spherical diapir are given in (11.33):

$$\mathbf{F}_{\text{buoy}} + \mathbf{F}_{\text{grav}} = (m - m_{\text{d}})\mathbf{g} = \tfrac{4}{3}\pi a^3(\rho_{\text{o}} - \rho_{\text{i}})g^*\mathbf{k}$$

Describe each of the physical quantities in this equation and provide the SI units. Use **Figure 11.34** to discuss the buoyancy of salt as a function of depth of burial.

11.12. The canonical model for a rising salt diapir considers a sphere of viscous material surrounded by a different viscous material. The *boundary conditions* at the surface of the sphere are written in terms of the radial and tangential velocity components as:

$$v_r(r=a^-,\theta) = v_r(r=a^+,\theta),\ 0 \le \theta < 2\pi$$

$$v_\theta(r=a^-,\theta) = v_\theta(r=a^+,\theta),\ 0 \le \theta < 2\pi$$

Describe what these boundary conditions mean in terms of the behavior of the two viscous materials at the surface of the sphere. In doing so, indicate what the consequences would be for the interface if the respective velocities were not equal and suggest geologic phenomena that might be relevant for such conditions.

11.13. Compare and contrast the velocity fields inside the viscous sphere as depicted in **Figure 11.39** and **Figure 11.40**. In particular, describe the location and velocity of the observer who would witness these fields.

MATLAB EXERCISES FOR CHAPTER 11: INTRUSIONS

An intrusion is a body of rock that, in a former more mobile state, was injected into and deformed the surrounding host rock. The first two exercises review the terminology of intrusions, including dikes, sills, volcanic necks, laccoliths, stocks, and diapirs. The role of mass density is explored by considering what determines the stable height of a dike. The conductive cooling of magma in sheet intrusions is quantified in an exercise using Fourier's law, along with MATLAB, to calculate and plot temperature histories. To appreciate the effects of non-linear viscous properties of magma, students compare the behaviors of Newtonian and power-law fluids flowing in a sill. The canonical problem of a sphere of viscous fluid surrounded by a fluid of different viscosity and density is the basis for an exercise that employs MATLAB to visualize the velocity fields inside and outside the sphere. The velocity field for an observer moving with the same velocity as the sphere provides a compelling explanation for the pattern of salt layers mapped in the Hänigsen salt diapir. These exercises help students understand that the study of intrusions includes both solid and fluid mechanics, as well as heat flow by advection and conduction. www.cambridge/SGAQI

FURTHER READING

For citations in figure captions see the reference list at the end of the book.

Anderson, E. M., 1972. *The Dynamics of Faulting and Dyke Formation with Applications to Britain*. Hafner Publishing Company, New York. (Facsimile reprint of 1951 edition.)
In this reprint of Anderson's classic book, originally published in 1942, the author summarizes and expands upon his earlier publications on dikes, dating back to his 1938 paper on the dynamics of sheet intrusions and the state of stress in Earth's crust.

Braunstein, J., and O'Brien, G. D., 1968. *Diapirism and Diapirs*. Memoir 8. American Association of Petroleum Geologists.
This edited volume has individual chapters that focus on many of the known examples of diapirs from around the world, as well as chapters on mechanical properties of salt and models of diapirism.

Brown, M., 2013. Granite: from genesis to emplacement. *Geological Society of America Bulletin* 125, 1079–1113.

Beginning with the historical literature, this review paper covers melting of Earth's crust, melt segregation and extraction, magma ascent, and emplacement of granite in the shallow crust, and concludes with challenges for future research.

Bürgmann, R., Rosen, P. A., and Fielding, E. J., 2000. Synthetic aperture radar interferometry to measure Earth's surface topography and its deformation. *Annual Review of Earth and Planetary Sciences* 28, 169–209.

This review paper provides an introductory summary of InSAR for earth scientists, illustrated applications to topographic studies as well as earthquake and volcanic deformation, and a helpful assessment of strengths and limitations.

Carrivick, J. L., Smith, M. W., and Quincey, D. J., 2016. *Structure from Motion in the Geosciences*. Wiley-Blackwell, Oxford.

Recognizing that structure from motion is a relatively new technology, this book is designed as a primer for geoscientists and environmental consultants looking for a comparison to other technologies, a detailed workflow, and both theoretical and practical information.

Carslaw, H. S., and Jaeger, J. C., 1986. *Conduction of Heat in Solids*, 2nd edition. Clarendon Press, Oxford.

This classic textbook on conduction of heat in solids includes examples of solved problems with relevance to the conductive cooling of magmatic intrusions.

Cashman, K. V., and Sparks, R. S. J., 2013. How volcanoes work: a 25 year perspective. *Geological Society of America Bulletin* 125, 664–690.

This review paper, motivated by the need to reduce vulnerability to volcanic eruptions, considers the physical processes involved in the intrusion of magma into the upper crust, the flow of magma from there to Earth's surface, and the types of eruptive activity.

Ernst, R. E., Grosfils, E. B., and Mege, D., 2001. Giant dike swarms: Earth, Venus, and Mars. *Annual Review of Earth and Planetary Sciences* 29, 489–534.

Dikes in giant radiating swarms, documented in this review paper, are up to 2000 km in trace length.

Fialko, Y. A., and Rubin, A. M., 1999. Thermal and mechanical aspects of magma emplacement in giant dike swarms. *Journal of Geophysical Research: Solid Earth* 104, 23,033–23,049.

This paper investigates the thermal and mechanical aspects of giant dike emplacement including the lateral propagation distance before thermal lock-up and thermal erosion of dike walls.

Glazner, A. F., Bartley, J. M., Coleman, D. S., Gray, W., and Taylor, R. Z., 2004. Are plutons assembled over millions of years by amalgamation from small magma chambers? *GSA Today* 14, 4–12.

This review paper considers geological and geophysical evidence relevant to the origin of plutons and suggests they may form incrementally.

Gudmundsson, A., 2011. *Rock Fractures in Geological Processes*. Cambridge University Press, Cambridge.

This textbook integrates abundant illustrations and photographs of fractures in the field with basic mechanical results to explain the initiation and propagation of dikes, sills, and other fractures in rock.

Happel, J., and Brenner, H., 1965. *Low Reynolds Number Hydrodynamics: With Special Applications to Particulate Media*. Prentice-Hall, Inc., Englewood Cliffs, NJ.

In this monograph for chemical engineers the authors focus on slow viscous flow of a fluid containing particles and develop solutions that have applications to magma dynamics.

Hudec, M. R., and Jackson, M. P., 2007. Terra infirma: understanding salt tectonics. *Earth Science Reviews* 82, 1–28.

This abundantly illustrated review paper covers the global distributions of salt basins, the mechanics of salt flow, the role of host rock strength in initiating diapirism, and the interplay of salt with regional tectonic extension or shortening.

Johnson, K., Nissen, E., Saripalli, S., Arrowsmith, J. R., McGarey, P., Scharer, K., Williams, P., and Blisniuk, K., 2014. Rapid mapping of ultrafine fault zone topography with structure from motion. *Geosphere* 10, 969–986.

This review paper covers the application of structure from motion to mapping fault zone topography for the purposes of active tectonics investigations, and includes valuable comparisons to airborne and terrestrial LiDAR.

Menand, T., de Saint-Blanquat, M., and Annen, C. (Eds.), 2011. Emplacement of magma pulses and growth of magma bodies. *Tectonophysics* 500, 1–112.

This special issue includes eight papers on various thermal, mechanical, and observational aspects of intrusion, with emphasis on multiple increments of magma injection in dikes and sills to build laccoliths and plutons.

Rivalta, E., Taisne, B., Bunger, A., and Katz, R., 2015. A review of mechanical models of dike propagation: schools of thought, results and future directions. *Tectonophysics* 638, 1–42.

The review paper highlights two schools of thought on vertical dike propagation: one is largely controlled by the fluid dynamics of magma flow; the other is largely controlled by the fracture mechanics of dike propagation.

Segall, P., 2010. *Earthquake and Volcano Deformation*. Princeton University Press, Princeton, NJ.

This advanced textbook goes beyond the elementary mechanics and structural geology covered here, with a focus on geophysical data related to active deformation associated with intrusion and eruption of magma.

Sigurdsson, H., Houghton, B., McNutt, S., Rymer, H., and Stix, J., 2015. *The Encyclopedia of Volcanoes*, 2nd edition. Elsevier, Amsterdam.

This mammoth volume is composed of 78 chapters written by experts in each of 9 different topical categories from magma origin and transport, to eruptions, to hazard mitigation, to economic benefits.

Townsend, M. R., Pollard, D. D., and Smith, R.P., 2017. Mechanical models for dikes: a third school of thought. *Techonophysics* 703–704, 98–118.

This review paper introduces a third school of thought for vertical dikes that emphasizes the horizontal flow of magma in dikes with heights constrained by gravity, induced stress gradients in the surrounding rock and pressure gradients in the magma.

White, S. M., Crisp, J. A., and Spera, F. J., 2006. Long-term volumetric eruption rates and magma budgets. *Geochemistry, Geophysics, Geosystems* 7, 1–20.

While focusing on a global compilation of 170 time-averaged eruption rates, this review of the literature finds that a ratio of 5:1 for intrusive to extrusive volumes may be common within many magmatic systems, although the range extends at least from 1:1 to 200:1 with considerable uncertainty.

REFERENCES

In this reference list we include publications from which we have used concepts, derivations, data, figures, maps, and cross sections for this textbook. The authors gratefully acknowledge the authors of these papers for their contributions.

Agterberg, F., 2014. *Geomathematics: Theoretical Foundations, Applications and Future Developments*. Springer, New York.

Allmendinger, R. W., Cardozo, N., and Fisher, D. M., 2012. *Structural Geology Algorithms: Vectors and Tensors*. Cambridge University Press, Cambridge.

Anderson, E. M., 1972. *The Dynamics of Faulting and Dyke Formation with Applications to Britain*. Hafner Publishing Company, New York. (Facsimile reprint of 1951 edition.)

Anderson, T. L., 1995. *Fracture Mechanics: Fundamentals and Applications*. CRC Press, Boca Raton, FL.

Atkinson, B. K. (Ed.), 1987. *Fracture Mechanics of Rock*. Academic Press, London.

Atkinson, B. K., and Meredith, P. G., 1987. Experimental fracture mechanics data for rocks and minerals, Chapter 11, in: Atkinson, B. K. (Ed.) *Fracture Mechanics of Rock*. Academic Press, London, pp. 477–525.

Aydin, A., 1977. *Faulting in Sandstone*. Stanford University Press, Stanford, California.

Bankwitz, P., 1965. Uber Klüfte I. Beobachtungen im Thüringischen Schiefergebirge. *Geologie* 14, 241–253.

Barber, J. R., 2010. *Elasticity*. Springer, New York.

Barton, N., 2007. *Rock Quality, Seismic Velocity, Attenuation and Anisotropy*. Taylor & Francis, London.

Bellahsen, N., Fiore, P., and Pollard, D. D., 2006. The role of fractures in the structural interpretation of Sheep Mountain Anticline, Wyoming. *Journal of Structural Geology* 28, 850–867.

Bergbauer, S., and Pollard, D. D., 2003. How to calculate normal curvatures of sampled geological surfaces. *Journal of Structural Geology* 25, 277–289.

Bergbauer, S., and Pollard, D. D., 2004. A new conceptual fold-fracture model including prefolding joints, based on the Emigrant Gap anticline, Wyoming. *Geological Society of America Bulletin* 116, 294–307.

Beyer, J., and Griffith, W. A., 2016. Influence of mechanical stratigraphy on clastic injectite growth at Sheep Mountain anticline, Wyoming: a case study of natural hydraulic fracture containment. *Geosphere* 12, 1633–1655.

Bieniawski, Z. T., 1984. *Rock Mechanics Design in Mining and Tunneling*. A. A. Balkema, Rotterdam.

Bird, R. B., Stewart, W. E., and Lightfoot, E. N., 2007. *Transport Phenomena*. John Wiley & Sons, Inc., New York.

Blenkinsop, T., 2000. *Deformation Microstructures and Mechanisms in Minerals and Rocks*. Kluwer Academic Publishers, Dordrecht.

Borradaile, G. J., Bayly, M. B., and Powell, C. M., 1982. *Atlas of Deformational and Metamorphic Rock Fabrics*. Springer Verlag, Berlin.

Brace, W. F., 1964. Brittle fracture of rock, in: Judd, W. R. (Ed.) *State of Stress in the Earth's Crust*. American Elsevier Publishing Company, Inc., pp. 111–178.

Brandes, C., and Tanner, D. C., 2014. Fault-related folding: a review of kinematic models and their application. *Earth-Science Reviews* 138, 352–370.

Braunstein, J., and O'Brien, G. D., 1968. *Diapirism and Diapirs*. Memoir 8. American Association of Petroleum Geologists.

Brown, M., 2013. Granite: from genesis to emplacement. *Geological Society of America Bulletin* 125, 1079–1113.

Bunger, A., and Cruden, A., 2011. Modeling the growth of laccoliths and large mafic sills: role of magma body forces. *Journal of Geophysical Research: Solid Earth* 116.

Bürgmann, R., Rosen, P. A., and Fielding, E. J., 2000. Synthetic aperture radar interferometry to measure Earth's surface topography and its deformation. *Annual Review of Earth and Planetary Sciences* 28, 169–209.

Byerlee, J., 1978. Friction of rocks. *Pageoph*. 116, 615–126.

Cai, W., and Nix, W. D., 2016. *Imperfections in Crystalline Solids*. Cambridge University Press, Cambridge.

Callister, W., and Rethwisch, D., 2014. *Materials Science and Engineering*. John Wiley & Sons, Inc., New York.

Carrivick, J. L., Smith, M. W., and Quincey, D. J., 2016. *Structure from Motion in the Geosciences*. Wiley-Blackwell, Oxford.

Carslaw, H. S., and Jaeger, J. C., 1986. *Conduction of Heat in Solids*, 2nd edition. Clarendon Press, Oxford.

Cashman, K. V., and Sparks, R. S. J., 2013. How volcanoes work: a 25 year perspective. *Geological Society of America Bulletin* 125, 664–690.

Chinnery, M. A., 1961. The deformation of the ground around surface faults. *Seismological Society of America Bulletin* 51, 355–372.

Christensen, R. M., 2013. *The Theory of Materials Failure*. Oxford University Press, Oxford.

Christiansen, P. P., and Pollard, D. D., 1997. Nucleation, growth and structural development of mylonitic shear zones in granitic rock. *Journal of Structural Geology* 19, 1159–1172.

Clark, S. P., (Ed.), 1966. *Handbook of Physical Constants*. Memoir 97. The Geological Society of America, Inc., New York.

Cloos, E., 1946. *Lineation – A Critical Review and Annotated Bibliography*. Memoir 18. Geological Society of America, Ann Arbor, MI.

Cloos, E., 1947. Oolite deformation in the South Mountain Fold, Maryland. *Geological Society of America Bulletin* 58, 843–918.

Cloos, E., 1971. *Microtectonics Along the Western Edge of the Blue Ridge, Maryland and Virginia*. The Johns Hopkins Press, Baltimore, Maryland.

Cooke, M. L., and Madden, E. H., 2014. Is the Earth lazy? A review of work minimization in fault evolution. *Journal of Structural Geology* 66, 334–346.

Cosgrove, J. W., and Hudson, J. A., 2016. *Structural Geology and Rock Engineering*. World Scientific Publishing Company, Inc, London.

Cowan, D. S., 1999. Do faults preserve a record of seismic slip? A field geologist's opinion. *Journal of Structural Geology* 21, 995–1001.

Crider, J. G., 2015. The initiation of brittle faults in crystalline rock. *Journal of Structural Geology* 77, 159–174.

Crouch, S. L., and Starfield, A. M., 1983. *Boundary Element Methods in Solid Mechanics: With Applications in Rock Mechanics and Geological Engineering*. George Allen & Unwin, London.

Cruikshank, K. M., and Aydin, A., 1995. Unweaving the joint in Entrada Sandstone, Arches National Park, Utah, U.S.A. *Journal of Structural Geology* 17, 409–421.

Davatzes, N., and Aydin, A., 2003. Overprinting faulting mechanisms in high porosity sandstones of SE Utah. *Journal of Structural Geology* 25, 1795–1813.

Davies, R. K., and Pollard, D. D., 1986. Relations between left-lateral strike-slip faults and right-lateral monoclinal kink bands in granodiorite, Mt. Abbot quadrangle, Sierra Nevada, California. *Pure and Applied Geophysics* 124, 177–201.

Davis, J. C., 2002. *Statistics and Data Analysis in Geology*. John Wiley & Sons, New York.

Davis, J. R., and Titus, S. J., 2017. Modern methods of analysis for three-dimensional orientational data. *Journal of Structural Geology* 96, 65–89.

de Saint-Blanquat, M., Habert, G., Horsman, E., Morgan, S. S., Tikoff, B., Launeau, P., and Gleizes, G., 2006. Mechanisms and duration of non-tectonically assisted magma emplacement in the upper crust: the Black Mesa pluton, Henry Mountains, Utah. *Techonophysics* 428, 1–31.

Decker, R. W., Wright, T. L., and Stauffer, P. H. (Eds.), 1987. *Volcanism in Hawaii*. US Geological Survey, Professional Paper 1350.

DeGraff, J. M., and Aydin, A., 1987. Surface morphology of columnar joints and its significance to mechanics and directions of joint growth. *Geological Society of America Bulletin* 99, 605–617.

Delaney, P. T., and Gartner, A. E., 1997. Physical processes of shallow mafic dike emplacement near the San Rafael Swell, Utah. *Geological Society of America Bulletin* 109, 1177–1192.

Delaney, P. T., and Pollard, D. D., 1981. Deformation of host rocks and flow of magma during growth of minette dikes and breccia-bearing intrusions near Ship Rock, New Mexico, in: Department of the Interior (Ed.) *Interior*. United States Government Printing Office, Washington, p. 61.

den Tex, E., 1972. Chapter XIII, in: Hills, E. S., *Elements of Structural Geology*. New York, John Wiley & Sons, Inc.

Di Toro, G., and Pennacchioni, G., 2005. Fault plane processes and mesoscopic structure of a strong-type seismogenic fault in tonalites (Adamello batholith, Southern Alps). *Techonophysics* 402, 55–80.

Duba, A. G., Durham, W. B., Handin, J. W., and Wand, H. F. (Eds.), 1990. *The Brittle-Ductile Transition in Rocks (The Heard Volume)*. Geophysical Monograph 56. American Geophysical Union, Washington, DC.

Engelder, T., and Peacock, D. C., 2001. Joint development normal to regional compression during flexural-flow folding: the Lilstock buttress anticline, Somerset, England. *Journal of Structural Geology* 23, 259–277.

Ernst, R. E., and Baragar, W. R. A., 1992. Evidence from magnetic fabric for the flow pattern of magma in the Mackenzie giant radiating dyke swarm. *Nature* 356, 511–513.

Ernst, R. E., Grosfils, E. B., and Mege, D., 2001. Giant dike swarms: Earth, Venus, and Mars. *Annual Review of Earth and Planetary Sciences* 29, 489–534.

Evans, B., and Kohlstedt, D. L., 1995. Rheology of rocks, in: Ahrens, T. J. (Ed.) *Rock Physics & Phase Relations: A Handbook of Physical Constants*. American Geophysical Union, Washington, DC, pp. 148–165.

Ferguson, J., 1994. *Introduction to Linear Algebra in Geology*. Chapman & Hall, London.

Fialko, Y. A., and Rubin, A. M., 1999. Thermal and mechanical aspects of magma emplacement in giant dike swarms. *Journal of Geophysical Research: Solid Earth* 104, 23,033–23,049.

Fiore, P. E., Pollard, D. D., Currin, W. R., and Miner, D. M., 2007. Mechanical and stratigraphic constraints on the evolution of faulting at Elk Hills, California. *AAPG bulletin* 91, 321–341.

Fisher, N. I., Lewis, T., and Embleton, B. J. J., 1987. *Statistical Analysis of Spherical Data*. Cambridge University Press, Cambridge.

Fletcher, R. C., and Pollard, D. D., 1999. Can we understand structural and tectonic processes and their products without appeal to a complete mechanics? *Journal of Structural Geology*, 21, 1071–1088.

Flugge, W., 1967. *Viscoelasticity*. Blaisdell Publishing Co., Waltham, MA.

Freed, A. M., and Bürgmann, R., 2004. Evidence of power-law flow in the Mojave desert mantle. *Nature* 430, 548–551.

Frocht, M. M., 1948. *Photoelasticity*. John Wiley & Sons, Inc., New York.

Fung, Y. C., 1969. *A First Course in Continuum Mechanics*. Prentice-Hall, Inc., Englewood Cliffs, NJ.

Gallagher, J. J., Friedmann, M., Handin, J., and Sowers, G. M., 1974. Experimental studies relating to microfracture in sandstone. *Techonophysics* 21, 203–247.

Gerla, P. J., 1988. Stress and fracture evolution in a cooling pluton: an example from the Diamond Joe stock, western Arizona, USA. *Journal of Volcanology and Geothermal Research* 34, 267–282.

Gilbert, G. K., 1877. Report on the geology of the Henry Mountains. US Government Printing Office, Washington.

Glazner, A. F., Bartley, J. M., Coleman, D. S., Gray, W., and Taylor, R. Z., 2004. Are plutons assembled over millions of years by amalgamation from small magma chambers? *GSA Today* 14, 4–12.

Gordon, J. E., 1976. *The New Science of Strong Materials, or Why You Don't Fall Through the Floor*, 2nd edition. Princeton University Press, Princeton, NJ.

Gratier, J-P., Dysthe, D. K., and Renard, F., 2013. The role of pressure solution creep in the ductility of the Earth's upper crust. *Advances in Geophysics*, 54, 47–179.

Griffith, W. A., Di Toro, G., Pennacchioni, G., and Pollard, D. D., 2008. Thin pseudotachylytes in faults of the Mt. Abbot quadrangle, Sierra Nevada: physical constraints for small seismic slip events. *Journal of Structural Geology* 30, 1086–1094.

Griggs, D. T., and Handin, J., 1960. *Rock Deformation*. Memoir 79. Geological Society of America, New York, p. 382.

Gudmundsson, A., 2011. *Rock Fractures in Geological Processes*. Cambridge University Press, Cambridge.

Hafner, W., 1951. Stress distributions and faulting. *Geological Society of America Bulletin* 62, 373–398.

Hambrey, M. J., and Lawson, W., 2000. Structural styles and deformation fields in glaciers: a review, in: Maltman, A., Hubbard, B., and Hambrey, M. (Eds.) *Deformation of Glacial Materials*. Geological Society, London, pp. 59–83.

Handin, J., Hager, R. V., Friedman, M., and Feather, J. N., 1963. Experimental deformation of sedimentary rocks under confining pressure: pore pressure tests. *American Association of Petroleum Geologists Bulletin* 47, 717–755.

Happel, J., and Brenner, H., 1965. *Low Reynolds Number Hydrodynamics: with Special Applications to Particulate Media*. Prentice-Hall, Inc., Englewood Cliffs, NJ.

Harris, R. A., and Segall, P., 1987. Detection of a locked zone at depth on the Parkfield, California, segment of the San Andreas fault. *Journal of Geophysical Research* 92, 7945–7962.

Heard, H. C., 1960. Transition from brittle fracture to ductile flow in Solenhofen limestone as a function of temperature, confining pressure, and interstitial fluid pressure, in: Griggs, D.T., and Handin, J. (Eds.) *Rock Deformation*. Geological Society of America, Washington, DC, pp. 193–226.

Hilley, G., Mynatt, I., and Pollard, D., 2010. Structural geometry of Raplee Ridge monocline and thrust fault imaged using inverse Boundary Element Modeling and ALSM data. *Journal of Structural Geology* 32, 45–58.

Hirth, G., and Kohlstedt, D., 2003. Rheology of the upper mantle and the mantle wedge: a view from the experimentalists, in: Eiler, J. (Ed.) *Inside the Subduction Factory*. American Geophysical Union, Washington, DC, pp. 83–105.

Hirth, G., Teyssier, C., and Dunlap, J. W., 2001. An evaluation of quartzite flow laws based on comparisons between experimentally and naturally deformed rocks. *International Journal of Earth Sciences* 90, 77–87.

Hobbs, B.E., and Ord, A., 2015. *Structural Geology: The Mechanics of Deforming Metamorphic Rocks*. Elsevier, Amsterdam.

Hoek, E., and Martin, C., 2014. Fracture initiation and propagation in intact rock–a review. *Journal of Rock Mechanics and Geotechnical Engineering* 6, 287–300.

Hofmann-Wellenhof, B., Lichtenegger, H., and Collins, J., 2001. *Global Positioning System: Theory and Practice*, 4th edition. Springer-Verlag, Wien, Germany.

Hoshino, K., Koide, H., Inami, K., Iwamura, S., and Mitsui, S., 1972. Mechanical properties of Japanese tertiary sedimentary rocks under high confining pressures, Report. Geological Survey of Japan.

Hubbert, M. K., 1951. Mechanical basis for certain familiar geologic structures. *Geological Society of America Bulletin* 62, 355–372.

Hubbert, M. K., 1972. *Structural Geology*. Hafner Publishing Company, New York.

Hudec, M. R., and Jackson, M. P., 2007. Terra infirma: understanding salt tectonics. *Earth Science Reviews* 82, 1–28.

Hudleston, P. J., and Treagus, S. H., 2010. Information from folds: a review. *Journal of Structural Geology* 32, 2042–2071.

Hunt, C. B., 1953. *Geology and Geography of the Henry Mountains Region, Utah*. US Government Printing Office, Washington.

Hytch, M. J., Putaux, J.-L., and Penisson, J.-M., 2003. Measurement of the displacement field of dislocations to 0.03 A by electron microscopy. *Nature* 423, 270–273.

Iwasaki, T., and Matsu'ura, M., 1982. Quasi-static crustal deformations due to a surface load. *Journal of Physics of the Earth* 30, 469–508.

Jackson, M. D., and Pollard, D. D., 1988. The laccolith-stock controversy: new results from the southern Henry Mountains, Utah. *Geological Society of America Bulletin* 100, 117–139.

Jackson, M. P., and Talbot, C. J., 1986. External shapes, strain rates, and dynamics of salt structures. *Geological Society of America Bulletin* 97, 305–323.

Jaeger, J. C., and Cook, N. G. W., 1969. *Fundamentals of Rock Mechanics*. Methuen & Co. Ltd., London.

Jaeger, J. C., and Hoskins, E. R., 1966. Stresses and failure in rings of rock loaded in diametral tension or compression. *British Journal of Applied Physics* 17, 685–692.

Jaeger, J. C., Cook, N. G. W., and Zimmerman, R. W., 2007. *Fundamentals of Rock Mechanics*, 4th edition. Blackwell Publishing, Oxford.

Jiang, D., 2007. Numerical modeling of the motion of deformable ellipsoidal objects in slow viscous flows. *Journal of Structural Geology*, 29, 435–452.

Jiang, D., 2014. Structural geology meets micromechanics: a self-consistent model for the multiscale deformation and fabric development in Earth's ductile lithosphere. *Journal of Structural Geology* 68, 247–272.

Johnson, A. M., and Fletcher, R. C., 1994. *Folding of Viscous Layers: Mechanical Analysis and Interpretation of Structures in Deformed Rock*. Columbia University Press, New York.

Johnson, A. M., and Pollard, D. D., 1973. Mechanics of growth of some laccolithic intrusions in the Henry Mtns., Utah: Part I, Field observations, Gilbert's model, physical properties and flow of magma. *Techonophysics* 18, 261–309.

Johnson, K. M., 2018. Growth of fault-cored anticlines by flexural slip folding. *Journal of Geophysical Research: Solid Earth*, 123, 2426–2447.

Johnson, K., Nissen, E., Saripalli, S., Arrowsmith, J. R., McGarey, P., Scharer, K., Williams, P., and Blisniuk, K., 2014. Rapid mapping of ultrafine fault zone topography with structure from motion. *Geosphere* 10, 969–986.

Jónsson, S., Zebker, H. A., Segall, P., and Amelung, F., 2002. Fault slip distribution of the 1999 M-w 7.1 Hector Mine, California, earthquake, estimated from satellite radar and GPS measurements. *Bulletin of the Seismological Society of America*, 92.

Kachanov, L. M., 1974. *Fundamentals of the Theory of Plasticity*. M. Konyaeva (trans.), MIR Publishers, Moscow.

Karato, S., 2008. *Deformation of Earth Materials: An Introduction to the Rheology of Solid Earth*. Cambridge University Press, Cambridge.

Kattenhorn, S. A., Krantz, B., Walker, E. L., and Blakeslee, M. W., 2016. Evolution of the Hat Creek fault system, northern California, in: Krantz, B., Ormand, C., and Freeman, B. (Eds.) *3-D Structural Interpretation: Earth, Mind, and Machine*. Memoir 111. American Association of Petroleum Geologists, pp. 121–154.

King, G. C. P., Stein, R. S., and Rundle, J. B., 1988. The growth of geological structures by repeated earthquakes 1. Conceptual framework. *Journal of Geophysical Research* 93, 13,307–13,318.

Kirkland, D. W., and Anderson, R. Y., 1970. Microfolding in the Castile and Todilto evaporites, Texas and New Mexico. *Geological Society of America Bulletin* 81, 3259–3282.

Kirkpatrick, J., Shipton, Z., and Persano, C., 2009. Pseudotachylytes: rarely generated, rarely preserved, or rarely reported? *Bulletin of the Seismological Society of America* 99, 382–388.

Klein, C., and Philpotts, A. R., 2013, *Earth Materials, Introduction to Mineralogy and Petrology*. Cambridge University Press, Cambridge.

Kohlstedt D. L., and Hansen, L. N., 2015. Constitutive equations, rheological behavior, and viscosity of rocks, in: *Treatise on Geophysics*, 2nd edition. Elsevier, pp. 441–472.

Kostrov, B. V., and Das, S., 1988. *Principles of Earthquake Source Mechanics*. Cambridge University Press, Cambridge.

Kundu, P. K., Cohen, I. M., and Dowling, D. R., 2015. *Fluid Mechanics*. Academic Press, London.

Kwasniewski, M., Li, X., and Takahashi, M., 2013. True triaxial testing of rocks, in: Kwasniewski, M. (Ed.) *Geomechanics Research Series*, Volume 4. CRC Press/Balkema, Leiden, The Netherlands.

Lai, W. M., Rubin, D. H., and Krempl, E., 2010. *Introduction to Continuum Mechanics*. Elsevier, New York.

Langlois, W. E., 1964. *Slow Viscous Flow*. The Macmillan Co., New York.

Lawn, B. R., 1993. *Fracture of Brittle Solids*, 2nd edition. Cambridge University Press, New York.

Lawn, B. R., and Wilshaw, T. R., 1975. *Fracture of Brittle Solids*. Cambridge University Press, Cambridge.

Lipman, P. W., and Mullineaux, D. R. (Eds.), 1981. *The 1980 Eruptions of Mount St. Helens*. US Geological Survey Professional Paper 1250, Washington.

Lipschultz, M. M., 1969. *Theory and Problems of Differential Geometry*. McGraw-Hill, New York.

Lockner, D. A., 1995. Rock failure, in: Ahrens, T. J. (Ed.) *Rock Physics & Phase Relations: A Handbook of Physical Constants*. American Geophysical Union, Washington, DC, pp. 127–147.

Lockner, D. A., and Beeler, N. M., 2003. Rock failure and earthquakes, Chapter 32, in: Lee, W. H. K., Kanamori, H., Jennings, P. D., and Kisslinger, C. (Eds.) *International Handbook of Earthquake and Engineering Seismology*. Academic Press, San Diego, CA, pp. 505–537.

Lovely, P., Zahasky, C., and Pollard, D. D., 2010. Fold geometry at Sheep Mountain anticline, Wyoming, constructed using airborne laser swath mapping data, outcrop-scale geologic mapping, and numerical interpolation. *Journal of Geophysical Research: Solid Earth* 115, 1–22.

Macdonald, G. A., 1955. Hawaiian volcanoes during 1952. *Geological Survey Bulletin* 1021-B, 1–108.

Macdonald, G. A., and Abbott, A. T., 1970. *Volcanoes in the Sea*. University of Hawaii Press, Honolulu.

Maerten, F., Resor, P., Pollard, D., and Maerten, L., 2005. Inverting for slip on three-dimensional fault surfaces using angular dislocations. *Bulletin of the Seismological Society of America* 95, 1654–1665.

Maerten, L., Pollard, D. D., and Maerten, F., 2001. Digital mapping of three-dimensional structures of the Chimney Rock fault system, central Utah. *Journal of Structural Geology* 23, 585–592.

Malvern, L. E., 1969. *Introduction to the Mechanics of a Continuous Medium*. Prentice-Hall, Inc., Englewood Cliffs, NJ.

Mancktelow, N. S., 2011. Deformation of an elliptical inclusion in two-dimensional incompressible power-law viscous flow. *Journal of Structural Geology* 33, 1378–1393.

Mandl, G., 1988. *Mechanics of Tectonic Faulting: Models and Basic Concepts*. Elsevier, Amsterdam.

Mandl, G., 2005. *Rock Joints – The Mechanical Genesis*. Springer, Berlin.

Mardia, K. V., and Jupp, P. E., 2000. *Directional Statistics*. John Wiley & Sons, New York.

Martel, S., 1990. Formation of compound strike-slip fault zones, Mount Abbot quadrangle, California. *Journal of Structural Geology* 12, 869–882.

Martel, S., Pollard, D. D., and Segall, P., 1988. Development of simple strike-slip fault zones in granitic rock, Mount Abbot quadrangle, Sierra Nevada, California. *Geological Society of America Bulletin* 99, 1451–1465.

McIntyre, D. B., and McKirdy, A., 2012. *James Hutton: The Founder of Modern Geology*. National Museums Scotland, Edinburgh, Scotland.

Mei, S., and Kohlstedt, D., 2000a. Influence of water on plastic deformation of olivine aggregates: 1. Diffusion creep regime. *Journal of Geophysical Research: Solid Earth* 105, 21,457–21,469.

Mei, S., and Kohlstedt, D., 2000b. Influence of water on plastic deformation of olivine aggregates: 2. Dislocation creep regime. *Journal of Geophysical Research: Solid Earth* 105 (B9), 21,471–21,481.

Menand, T., de Saint-Blanquat, M., and Annen, C. (Eds.), 2011. Emplacement of magma pulses and growth of magma bodies. *Tectonophysics* 500, 1–112.

Morgan, S., Stanik, A., Horsman, E., Tikoff, B., de Saint Blanquat, M., and Habert, G., 2008. Emplacement of multiple magma sheets and wall rock deformation: Trachyte Mesa intrusion, Henry Mountains, Utah. *Journal of Structural Geology* 30, 491–512.

Mukherjee, S., 2015. *Atlas of Structural Geology*. Elsevier, Amsterdam.

Mura, T., 1987. *Micromechanics of Defects in Solids*, 2nd edition. Martinus Nijhoff, Dordrecht, The Netherlands.

Murase, T., and McBirney, A. R., 1973. Properties of some common igneous rocks and their melts at high temperatures. *Geological Society of America Bulletin* 84, 3563–3592.

Muskhelishvili, N. I., 1975. *Some Basic Problems of the Mathematical Theory of Elasticity*. Noordhoff International Publishing, Leyden.

Myers, R., and Aydin, A., 2004. The evolution of faults formed by shearing across joint zones in sandstone. *Journal of Structural Geology* 26, 947–966.

Mynatt, I., Bergbauer, S., and Pollard, D. D., 2007. Using differential geometry to describe 3-D folds. *Journal of Structural Geology* 29, 1256–1266.

Nevitt, J. M., Pollard, D. D., and Warren, J. M., 2014. Evaluation of transtension and transpression within contractional fault steps: comparing kinematic and mechanical models to field data. *Journal of Structural Geology* 60, 55–69.

Nevitt, J. M., Warren, J. M., Kidder, S., and Pollard, D. D., 2017a. Comparison of thermal modeling, microstructural analysis, and Ti-in-quartz thermobarometry to constrain the thermal history of a cooling pluton during deformation in the Mount Abbot Quadrangle, CA. *Geochemistry, Geophysics, Geosystems* 18, 1270–1297.

Nevitt, J. M., Warren, J. M., and Pollard, D. D., 2017b. Testing constitutive equations for brittle-ductile deformation associated with faulting in granitic rock. *Journal of Geophysical Research: Solid Earth* 122, 6269–6293.

Newton, Isaac, Cohen, I. B., and Whitman, A., 1999. *The Principia, Mathematical Principles of Natural Philosophy, a New Translation by I. Bernard Cohen and Anne Whitman*. University of California Press, Berkeley, CA.

Nye, J. F., 1985. *Physical Properties of Crystals*. Oxford University Press, Oxford.

Olson, J. E., 2004. *Predicting Fracture Swarms – The Influence of Subcritical Crack Growth and The Crack-tip Process Zone on Joint Spacing in Rock*. Geological Society, London, Special Publications 231, pp. 73–88.

Olver, F. W. J., Olde Daalhuis, A. B., Lozier, D. W., Schneider, B. I., Boisvert, R. F., Clark, C. W., Miller, B. R., and Saunders, B. V. (Eds.) *Release 1.0.17 of 2017-12-22, [DLMF] NIST Digital Library of Mathematical Functions*. National Institute of Standards and Technology, US Department of Commerce.

Oreskes, N., Shrader-Frechette, K., and Belitz, K., 1994. Verification, validation, and confirmation of numerical models in the earth sciences. *Science* 263, 641–646.

Passchier, C. W., and Trouw, R. A. J., 1996. *Microtectonics*. Springer-Verlag, Berlin.

Paterson, M. S., and Wong, T.-f., 2005. *Experimental Rock Deformation: The Brittle Field*. Springer-Verlag, Berlin.

Paterson, S. R., Vernon, R. H., and Tobisch, O. T., 1989. A review of criteria for the identification of magmatic and tectonic foliations in granitoids. *Journal of Structural Geology* 11, 349–363.

Peirce, C. S., 2011. *Philosophical Writings of Peirce*, Buchler, J. (Ed.). Dover Publications, New York.

Petit, J.-P., Auzias, V., Rawnsley, K., and Rives, T., 2000. Development of joint sets in the vicinity of faults, in: Lehner, F.K.a.U., J.L. (Ed.) *Aspects of Tectonic Faulting*. Springer, Berlin, pp. 167–183.

Poirier, J.-P., 1985. *Creep of Crystals, High-temperature Deformation Processes in Metals, Ceramics and Minerals*. Cambridge University Press, London.

Pollard, D. D., and Aydin, A., 1984. Propagation and linkage of oceanic ridge segments. *Journal of Geophysical Research* 89, 10,017–10,028.

Pollard, D. D., and Aydin, A., 1988. Progress in understanding jointing over the past century. *Geological Society of America Bulletin* 100, 1181–1204.

Pollard, D. D., Delaney, P. T., Duffield, W. A., Endo, E. T., and Okamura, A. T., 1983. Surface deformation in volcanic rift zones. *Techonophysics* 94, 541–584.

Pollard, D. D., and Fletcher, R. C., 2005. *Fundamentals of Structural Geology*. Cambridge University Press, Cambridge.

Pollard, D. D., and Holzhausen, G., 1979. On the mechanical interaction between a fluid-filled fracture and the Earth's surface. *Techonophysics* 53, 27–57.

Pollard, D. D., and Johnson, A. M., 1973. Mechanics of growth of some laccolith intrusions in the Henry Mountains, Utah, II: bending and failure of overburden layers and sill formation. *Techonophysics* 18, 311–354.

Pollard, D. D., and Muller, O. H., 1976. The effect of gradients in regional stress and magma pressure on the form of sheet intrusions in cross section. *Journal of Geophysical Research* 81, 975–984.

Pollard, D. D., Muller, O. H., and Dockstader, D. R., 1975. The form and growth of fingered sheet intrusions. *Geological Society of America Bulletin* 86, 351–363.

Pollard, D. D., and Segall, P., 1987. Theoretical displacements and stresses near fractures in rock: with applications to faults, joints, veins, dikes, and solution surfaces, in: Atkinson, B. K. (Ed.) *Fracture Mechanics of Rock*. Academic Press Inc., London, pp. 277–349.

Pollard, D. D., Segall, P., and Delaney, P. T., 1982. Formation and interpretation of dilatant echelon cracks. *Geological Society of America Bulletin* 93, 1291–1303.

Pollard, D. D., and Townsend, M. R., 2018. Fluid-filled fractures in Earth's lithosphere: gravitational loading, interpenetration, and stable height of dikes and veins. *Journal of Structural Geology* 109, 38–54.

Press, F., Siever, R., Grotzinger, J., and Jordan, T., 2004. *Understanding Earth*. W.H. Freeman and Company, New York.

Price, N. J., 1966. *Fault and Joint Development in Brittle and Semi-brittle Rock*. Pergamon Press, Oxford, England.

Qu, J., and Cherkaoui, M., 2006. *Fundamentals of Micromechanics of Solids*. John Wiley & Sons, Hoboken, NJ.

Ragan, D. M., 2009. *Structural Geology: An Introduction to Geometrical Techniques*, 4th edition. Cambridge University Press, Cambridge.

Ramberg, H., 1967. *Gravity, Deformation and the Earth's Crust*. Academic Press, London.

Ramsay, J. G., 1967. *Folding and Fracturing of Rocks*. McGraw-Hill, New York.

Ramsay, J. G., and Lisle, R., 2000. *The Techniques of Modern Structural Geology. Volume 3: Applications of Continuum Mechanics in Structural Geology*. Academic Press, New York, pp. 701–1061.

Ranalli, G., 1987. *Rheology of the Earth: Deformation and Flow Processes in Geophysics and Geodynamics*. Allen & Unwin, London.

Rawnsley, K. D., Peacock, D. C. P., Rives, T., and Petit, J.-P., 1998. Joints in the Mesozoic sediments around the Bristol Channel Basin. *Journal of Structural Geology* 20, 1641–1661.

Resnick, R., and Halliday, D. 1977. *Physics*, Part 1, 3rd edition, John Wiley & Sons, New York.

Resor, P. G., 2008. Deformation associated with a continental normal fault system, western Grand Canyon, Arizona. *Geological Society of America Bulletin* 120, 414–430.

Resor, P. G., and Pollard, D. D., 2012. Reverse drag revisited: why footwall deformation may be the key to inferring listric fault geometry. *Journal of Structural Geology* 41, 98–109.

Resor, P. G., Pollard, D. D., Wright, T. J., and Beroza, G. C., 2005. Integrating high-precision aftershock locations and geodetic observations to model coseismic deformation associated with the 1995 Kozani-Grevena earthquake, Greece. *Journal of Geophysical Research* 110, doi:10.1029/2004JB003263.

Reynolds, O., 1883. An experimental investigation of the circumstances which determine whether the motion of water shall be direct or sinuous, and the laws of resistance in parallel channels. *Philosophical Transactions of the Royal Society* 174.

Rivalta, E., Taisne, B., Bunger, A., and Katz, R., 2015. A review of mechanical models of dike propagation: schools of thought, results and future directions. *Tectonophysics* 638, 1–42.

Rowe, C. D., and Griffith, W. A., 2015. Do faults preserve a record of seismic slip: a second opinion. *Journal of Structural Geology* 78, 1–26.

Rubin, A. M., 1995. Propagation of magma-filled cracks. *Annual Review of Earth and Planetary Sciences* 23(1), 287–336.

Rutter, E. H., 1974. The influence of temperature, strain rate and interstitial water in the experimental deformation of calcite rocks. *Techonophysics* 22, 311–334.

Rutter, E., 1986. On the nomenclature of mode of failure transitions in rocks. *Techonophysics* 122, 381–387.

Rymer, M. J., 1989. Surface rupture in a fault stepover on the Superstition Hills Fault, California, in: Schwartz, D. P., and Sibson, R. H. (Eds.) *Fault Segmentation and Controls of Rupture Initiation and Termination*. US Geological Survey Open-File Report 89–315, pp. 309–323.

Savage, H. M., Shackleton, J. R., Cooke, M. L., and Riedel, J. J., 2010. Insights into fold growth using fold-related joint patterns and mechanical stratigraphy. *Journal of Structural Geology* 32, 1466–1475.

Scholz, C., 1988. The brittle-plastic transition and the depth of seismic faulting. *Geologische Rundschau* 77, 319–328.

Scholz, C. H., 1990. *The Mechanics of Earthquakes and Faulting*. Cambridge University Press, Cambridge.

Segall, P., 2010. *Earthquake and Volcano Deformation*. Princeton University Press, Princeton, NJ.

Segall, P., and Pollard, D. D., 1980. Mechanics of discontinuous faults. *Journal of Geophysical Research* 85, 4337–4350.

Segall, P., and Pollard, D.D., 1983. Nucleation and growth of strike slip faults in granite. *Journal of Geophysical Research* 88, 555–568.

Sharp, R. V., Budding, K. E., Boatwright, J., Ader, M. J., Bonilla, M. G., Clark, M. M., Fumal, T. E., Harms, K. K., Lienkaemper, J. J., Morton, D. M., O'Neill, B. J., Ostergren, C. L., Ponti, D. J., Rymer, M. J., Saxton, J. L., and Sims, J. D., 1989. Surface faulting along the Superstition Hills fault zone and nearby faults associated with the earthquakes of 24 November 1987. *Bulletin of the Seismological Society of America* 79, 252–281.

Shaw, H. R., 1963. Obsidian-H_2O viscosities at 1000 and 2000 bars in the temperature range 700³ to 900³. *Journal of Geophysical Research* 68, 6337–6343.

Sheriff, R. E., and Geldart, L. P. 1995, *Exploration Seismology*. Cambridge University Press, Cambridge.

Shewmon, P. (Ed.), 2016. *Diffusion in Solids*. The Minerals, Metals & Materials Series, Springer, Switzerland.

Sibson, R. H., 1989. Earthquake faulting as a structural process. *Journal of Structural Geology* 11, 1–14.

Sigurdsson, H., Houghton, B., McNutt, S., Rymer, H., and Stix, J., 2015. *The Encyclopedia of Volcanoes*, 2nd edition. Elsevier, Amsterdam.

Slatton, K. C., Carter, W. E., Shrestha, R. L., and Dietrich, W., 2007. Airborne laser swath mapping: achieving the resolution and accuracy required for geosurficial research. *Geophysical Research Letters* 34, L23S1023.

Snoke, A. W., Tullis, J., and Todd, V. R., 1998. *Fault-related Rocks: A Photographic Atlas*. Princeton University Press, Princeton, NJ.

Stein, R. S., King, G. C. P., and Rundle, J. B., 1988. The growth of geological structures by repeated earthquakes 2. Field examples of continental dip-slip faults. *Journal of Geophysical Research* 93, 13,319–13,331.

Sternlof, K. R., Rudnicki, J.W., and Pollard, D.D., 2005. Anticrack inclusion model for compaction bands in sandstone. *Journal of Geophysical Research: Solid Earth* 110, 1–16.

Strang, G., 2016. *Introduction to Linear Algebra*. Wellesley-Cambridge Press, Wellesley, MA.

Tada, H., Paris, P. C., and Irwin, G. R., 2000. *The Stress Analysis of Cracks Handbook*, 3rd edition. The American Society of Mechanical Engineers Press, New York.

Talbot, C., and Jackson, M., 1987. Internal kinematics of salt diapirs. *AAPG Bulletin* 71, 1068–1093.

Terzaghi, K, 1943. *Theoretical Soil Mechanics*. John Wiley and Sons, New York.

Thomas, A. L., and Pollard, D. D., 1993. The geometry of echelon fractures in rock: implications from laboratory and numerical experiments. *Journal of Structural Geology* 15, 323–334.

Thompson, A., and Taylor, B. N., 2008, *Guide for the Use of the International System of Units (SI)*. National Institute of Standards and Technology, Special Publication 811, US Department of Commerce.

Tilling, R.I., Heliker, C., and Swanson D.A., 2010. *Eruptions of Hawaiian Volcanoes; Past, Present, and Future*. US Geological Survey General Information Product 117, online at http://pubs.usgs.gov/gip/117/

Timoshenko, S. P., and Goodier, J. N., 1970. *Theory of Elasticity*, 3rd edition. McGraw-Hill Book Company, New York.

Timoshenko, S., and Woinowsky-Krieger, S., 1959. *Theory of Plates and Shells*. McGraw-Hill Book Company, New York.

Townsend, M., Pollard, D. D., Johnson, K., and Culha, C., 2015. Jointing around magmatic dikes as a precursor to the development of volcanic plugs. *Bulletin of Volcanology* 77, 92.

Townsend, M. R., Pollard, D. D., and Smith, R. P., 2017. Mechanical models for dikes: a third school of thought. *Techonophysics* 703–704, 98–118.

Treiman, J. A., 2009. Surface faulting and deformation. *Assessment & Mitigation: Shlemon Specialty Conference*. Association of Environmental & Engineering Geologists, p. 37.

Treiman, J., Kendrick, K., Bryant, W., Rockwell, T., and McGill, S., 2002. Primary surface rupture associated with the *M*w 7.1 16 October 1999 Hector Mine earthquake, San Bernardinno County, California. *Bulletin of the Seismological Society of America* 92, 1171–1191.

Trouw, R. A., Passchier, C. W., and Wiersma, D. J., 2009. *Atlas of Mylonites and Related Microstructures*. Springer, Heidelberg.

Turcotte, D. L., and Schubert, G., 2002, *Geodynamics*, 2nd edition. Cambridge University Press, Cambridge.

Turner, F. J., and Weiss, L. E., 1963. *Structural Analysis of Metamorphic Tectonites*. McGraw-Hill, New York.

Urai, J. L., Spiers, C. J., Zwart, H. J., and Lister, G. S., 1986. Weakening of rock salt by water during long-term creep. *Nature* 324, 554–557.

Van Dyke, M., 1982, *An Album of Fluid Motion*. The Parabolic Press, Stanford, CA.

Varberg, D. E., Purcell, E. J., and Rigdon, S. E., 2006. *Calculus*, 9th edition. Pearson/Prentice Hall, New York.

Varberg, D. E., Purcell, E. J., and Rigdon, S. E., 2007. *Calculus with Differential Equations*. Pearson/Prentice Hall, New York.

Watts, A. B. 2001. *Isostasy and Flexure of the Lithosphere*. Cambridge University Press, Cambridge.

Wawersik, W. R., 1968. *Detailed Analysis of Rock Failure in Laboratory Compression Tests*. University of Minnesota, Minneapolis, MN.

Wawersik, W., and Fairhurst, C., 1970. A study of brittle rock fracture in laboratory compression experiments. *International Journal of Rock Mechanics and Mining Sciences & Geomechanics Abstracts*. Pergamon Press, pp. 561–575.

Weertman, J., and Weertman, J. R., 1964. *Elementary Dislocation Theory*. The Macmillan Company, New York.

White, S. M., Crisp, J. A., and Spera, F. J., 2006. Long-term volumetric eruption rates and magma budgets. *Geochemistry, Geophysics, Geosystems* 7, 1–20.

Wibberley, C., Faulkner, D., Schlische, R., Jackson, C., Lunn, R., and Shipton, Z. (Eds.), 2010. Fault Zones (Special Issue). *Journal of Structural Geology* 32, 1553–1864.

Wong, T.-f., and Baud, P., 2012. The brittle-ductile transition in porous rock: a review. *Journal of Structural Geology* 44, 25–53.

Woodworth, J. B., 1896. On the fracture system of joints, with remarks on certain great fractures. *Boston Society of Natural History, Proceedings* 27, 163–183.

Yamaguti, S., 1937. Deformation of the earth's crust in Idu Peninsula in connection with the destructive Idu earthquake of Nov. 26, 1930. *Bulletin of the Earthquake Research Institute of Tokyo University* 15, 899–934.

Yeats, R. S., Sieh, K. E., and Allen, C. R., 1997. *The Geology of Earthquakes*. Oxford University Press, Oxford.

Yun, S., Segall, P., and Zebker, H., 2006. Constraints on magma chamber geometry at Sierra Negra Volcano, Galápagos Islands, based on InSAR observations. *Journal of Volcanology and Geothermal Research* 150, 232–243.

Zoback, M. D., 2010. *Reservoir Geomechanics*. Cambridge University Press, Cambridge.

Zucca, J. J., Hill, D. P., and Kovach, R. L., 1982. Crustal structure of Mauna Loa Volcano, Hawaii, from seismic refraction and gravity-data. *Bulletin of the Seismological Society of America* 72, 1535–1550.

INDEX